Studies in Fuzziness and Soft Computing

Volume 319

Series editor

Janusz Kacprzyk, Polish Academy of Sciences, Warsaw, Poland
e-mail: kacprzyk@ibspan.waw.pl

About this Series

The series "Studies in Fuzziness and Soft Computing" contains publications on various topics in the area of soft computing, which include fuzzy sets, rough sets, neural networks, evolutionary computation, probabilistic and evidential reasoning, multi-valued logic, and related fields. The publications within "Studies in Fuzziness and Soft Computing" are primarily monographs and edited volumes. They cover significant recent developments in the field, both of a foundational and applicable character. An important feature of the series is its short publication time and world-wide distribution. This permits a rapid and broad dissemination of research results.

More information about this series at http://www.springer.com/series/2941

Quanmin Zhu · Ahmad Taher Azar
Editors

Complex System Modelling and Control Through Intelligent Soft Computations

Springer

Editors
Quanmin Zhu
Department of Engineering Design
and Mathematics
University of the West of England
Bristol
UK

Ahmad Taher Azar
Faculty of Computers and Information
Benha University
Benha
Egypt

ISSN 1434-9922 ISSN 1860-0808 (electronic)
Studies in Fuzziness and Soft Computing
ISBN 978-3-319-12882-5 ISBN 978-3-319-12883-2 (eBook)
DOI 10.1007/978-3-319-12883-2

Library of Congress Control Number: 2014957496

Springer Cham Heidelberg New York Dordrecht London

Printed on acid-free paper

Springer International Publishing AG Switzerland is part of Springer Science+Business Media (www.springer.com)

Preface

Soft computing-based inductive approaches are concerned with the use of theories of fuzzy logic, neural networks and evolutionary computing to solve real-world problems that cannot be satisfactorily solved using conventional crisp computing techniques. Representation and processing of human knowledge, qualitative and approximate reasoning, computational intelligence, computing with words, and biological models of problem solving and optimization form key characteristics of soft computing, and are directly related to intelligent systems and applications. In recent years there has been rapid growth in the development and implementation of soft computing techniques in a wide range of applications, particularly those related to natural and man-made science and engineering systems.

This book is intended to present important applications of soft computing as reported from both analytical and practical points of view. The material is organized into 29 chapters. In its chapters, the book gives a prime introduction to soft computing with its principal components of fuzzy logic, neural networks, genetic algorithms and genetic programming with a self-contained, simple, readable approach. The book also includes a few of representative papers to cover industrial and development effort in the applications of intelligent systems through soft computing, which is given to guide the interested readers on their ad hoc applications. Advanced topics and future challenges are addressed as well, with the researchers in the field in mind. The introductory material, application-oriented techniques, and case studies should be particularly useful to practicing professionals. In brief summary, this book provides a general foundation for soft computing-based inductive methodologies/algorithms as well as their applications, in terms of providing multidisciplinary solutions in complex system modelling and control.

As the editors, we hope that the chapters in this book will stimulate further research in Complex system modelling and utilize them in real-world applications. We hope that this book, covering so many different aspects, will be of value to all readers.

The editors would like to take this opportunity to thank all the authors for their contributions to this textbook. Without the hard work of our contributors, this book would not have been possible. The encouragement and patience of Series Editor,

Prof. Janusz Kacprzyk and Dr. Leontina Di Cecco is very much appreciated. Without their continuous help and assistance during the entire course of this project, the production of the book would have taken a great deal longer. Special thanks to Holger Schaepe for her great effort during the publication process.

Bristol, UK
Benha, Egypt

Quanmin Zhu
Ahmad Taher Azar

Contents

Design and Modeling of Anti Wind Up PID Controllers

Ahmad Taher Azar and Fernando E. Serrano

Abstract In this chapter several anti windup control strategies for SISO and MIMO systems are proposed to diminish or eliminate the unwanted effects produced by this phenomena, when it occurs in PI or PID controllers. Windup is a phenomena found in PI and PID controllers due to the increase in the integral action when the input of the system is saturated according to the actuator limits. As it is known, the actuators have physical limits, for this reason, the input of the controller must be saturated in order to avoid damages. When a PI or PID controller saturates, the integral part of the controller increases its magnitude producing performance deterioration or even instability. In this chapter several anti windup controllers are proposed to eliminate the effects yielded by this phenomena. The first part of the chapter is devoted to explain classical anti windup architectures implemented in SISO and MIMO systems. Then in the second part of the chapter, the development of an anti windup controller for SISO systems is shown based on the approximation of the saturation model. The derivation of PID SISO (single input single output) anti windup controllers for continuous and discrete time systems is implemented adding an anti windup compensator in the feedback loop, so the unwanted effects are eliminated and the system performance is improved. Some illustrative examples are shown to test and compare the performance of the proposed techniques. In the third part of this chapter, the derivation of a suitable anti windup PID control architecture is shown for MIMO (multiple input multiple output) continuous and discrete time systems. These strategies consist in finding the controller parameters by static output feedback (SOF) solving the necessary linear matrix inequalities (LMI's) by an appropriate anti windup control scheme. In order to obtain the control gains and parameters, the saturation is modeled with describing functions for the continuous time case and a suitable model to deal with this nonlinearity in

A.T. Azar (✉)
Faculty of Computers and Information, Benha University, Benha, Egypt
e-mail: ahmad_t_azar@ieee.org

F.E. Serrano
Department of Electrical Engineering, Florida International University,
10555 West Flagler St, Miami, FL 33174, USA
e-mail: fserr002@fiu.edu

Q. Zhu and A.T. Azar (eds.), *Complex System Modelling and Control Through Intelligent Soft Computations*, Studies in Fuzziness and Soft Computing 319,
DOI 10.1007/978-3-319-12883-2_1

the discrete time case. Finally a discussion and conclusions sections are shown in this chapter to analyze the advantages and other characteristics of the proposed control algorithms explained in this work.

1 Introduction

In this chapter several control architectures of anti windup controllers are shown for the stabilization of SISO and MIMO systems in their discrete and continuous forms. Windup is a phenomena found in different kind of systems, when a PI or PID controller is implemented, produced by the integral action of the controller. This phenomenon occurs when the input of the system saturates increasing the magnitude of the integrator producing unwanted effects on the system like high overshoot and long settling time. There are several techniques and architectures found in literature to deal with this problem, for the SISO and MIMO cases, usually by suppressing the integral action of the PI or PID controller with input saturation.

For the SISO continuous case, different anti windup controller architectures are found in literature such as the tracking anti windup, conditional integration and limited integrator (Bohn and Atherton 1995), these are some of the classical anti windup control architectures implemented to eliminate the unwanted effects of windup. These classical techniques usually consist in adding an extra feedback loop to the controller from the saturated output so the effects of windup can be cancelled by implementing these control models. The back—calculation techniques is a common anti windup control architecture that ensures the system stability when the input is saturated, improving the system performance by producing smaller overshoot and acceptable settling time (Tu and Ho 2011). One issue that makes it difficult to obtain a suitable anti windup control architecture is the nonlinearity introduced by the actuator saturation, one way to design an appropriate control system when this nonlinearity is found, is the introduction of a saturation model which includes all the properties of this nonlinearity (Saeki and Wada 1996). This consideration is very important in the design of anti windup controllers for SISO and MIMO systems in the continuous and discrete time cases respectively, allowing the development of appropriate controllers including a saturation model.

In the case of SISO discrete system, there are similar anti windup control techniques as the continuous counterpart that can be implemented when a discretized model of the system is available. One of the control architectures that is very popular in the control community is the back calculation model, where the saturated signal is feedback to the controller integrator in order to suppress the windup effects yielded by the integrator action (Wittenmark 1989). Apart from this anti windup control architecture for discrete time SISO systems, the anti windup controller design by the frequency response of the model is usually implemented where a discrete time controller is obtained by the design of a continuous time SISO controller and then this controller is transformed to discrete time by one of the several well known methods (Lambeck and Sawodny 2004).

In the case of MIMO continuous and discrete time systems several anti windup controllers are synthesized usually by static output feedback (SOF) and then the controller is found by the solution of the respective linear matrix inequalities (LMI's). The SOF control law can be found by solving the LMI's to ensure the stability of the system by traditional ways or by an H_∞ controller (Wu et al. 2005; Henrion et al. 1999), allowing a flexible anti windup controller design when the input of the system is saturated.

Based on the previous explanation of different kind of anti windup controller architectures, this chapter is divided in the following sections so the first part of the chapter is devoted to SISO continuous and discrete time systems and the second part of this chapter is devoted to MIMO continuous and discrete time systems. In Sect. 2, the explanation of popular anti windup control techniques is explained to introduce the proposed strategies shown in this article, where some continuous and discrete time classical anti windup techniques found in literature are explained. It is important to notice that in this chapter, the main objective is to design and obtain stable PID controllers for the SISO and MIMO case, so in the following sections this problem is considered for analysis. Based on the previous explanation, in Sect. 3 the design of an internal model anti windup controller for continuous time systems is explained, showing that is possible to obtain a desired anti windup PID controller with an internal model controller (IMC) characteristics. In Sect. 4 an internal model anti windup controller for discrete SISO system is shown where a similar technique like the continuous counterpart is developed to eliminate the unwanted effects produced by the system saturation by implementing a scalar sign function approach (Zhang et al. 2011); an illustrative example is shown to compare the performance of the system. In Sect. 5 the derivations of an anti windup PID controller are done by SOF applying LMI's that includes the saturation of the system. The SOF control law is obtained by the stability characteristics of the system and by a H_∞ design, so the controller and system performance can be compared by the solution of these control problems. In Sect. 6 an anti windup PID controller for MIMO discrete time systems is shown and similar to its continuous counterpart, a SOF controller is implemented and then solving the LMI's based on the system stability or H_∞ the respective PID gains are found when the input of the system is saturated; in this section an illustrative example is shown to compare the systems performance. Finally, in Sects. 7 and 8 the discussion and conclusions of this chapter are shown respectively so a complete analysis of all the proposed schemes is done and then the conclusions are analyzed at the end of this chapter.

2 Previous Work

As explained in the previous section, the windup phenomena is caused by the integral action of a PI or PID controllers when the input of the system is saturated, then the performance of the system is deteriorated by the increasing of the integral action of the controller, yielding many unwanted effects such as a higher overshoot, a long settling time and even instability. This phenomena is found in SISO and MIMO systems in the continuous and discrete time representations when the input is saturated due to the physical limits of different kind of actuators such as mechanical, hydraulic and electrical systems.

In the case of SISO continuous systems there are some classical architectures implemented to avoid this unwanted effect, some of them, are based on the back calculation of the integral action and other are based on the feedback of the saturated input to the PID controller. The tracking anti windup controller is one of the well known control strategies implemented to avoid the deterioration of the system when this phenomena is found; it consists of a feedback loop generated by the saturated and non saturated inputs and then this signal is used to reduce the integrator input (Bohn and Atherton 1995). In Fig. 1 the tracking anti windup controller is shown where as can be noticed the difference of the non saturated and saturated inputs are fedback to the integrator.

Another method is conditional integration, which consist in turning on and off the integrator according on higher values of the control and error inputs (Bohn and Atherton 1995). Another anti windup control architecture is the limited integrator, this technique consist in feed the integrator output through a dead zone with high gain, reducing the effects of windup when the input saturates (Bohn and Atherton 1995).

The anti windup control architectures for discrete time SISO system are similar to their continuous counterpart, for example in Chen et al. (2003) an anti windup cascade control technique is implemented to suppress the unwanted effects yielded by windup in digital control systems, proving that is an efficient control architecture when the input is saturated. In Lambeck and Sawodny (2004) an anti windup control architecture is derived when the input of the system is constrained, the development of this strategy is based on the frequency response characteristics of

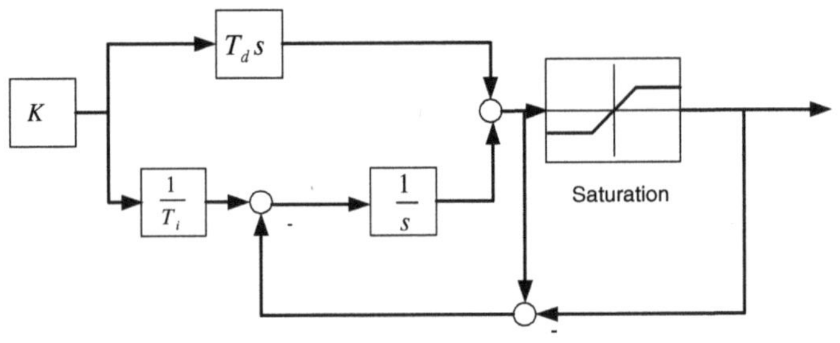

Fig. 1 Tracking anti windup controller

the closed loop system obtaining the digital controller by the conversion of an analog to digital controller. In (Wittenmark 1989) the development of different anti windup controllers are explained such as PID and cascade control for digital control system with constrained input, where the back calculation is implemented similar to the analog counterparts.

One of the common architectures for MIMO system with constrained inputs is the design of an output feedback control law that stabilizes the system while reducing the effects of windup, these control architectures can be applied in continuous and discrete MIMO systems such as explained in (Rehan et al. 2013) where an output feedback controller is implemented and the gains of the controller are found by solving the LMI's for continuous time systems. Another anti windup controller design technique is found in (Saeki and Wada 1996) where an output feedback controller is found by solving the LMI's for continuous MIMO systems with saturated inputs, the controller gains are found by solving the H_∞ optimal LMI's.

With this review about some commons anti windup architectures, in the following sections the development of this kind of novel configuration is shown, where in the first part of this chapter internal model anti windup architectures are developed for the SISO continuous and discrete cases, and the second part of the chapter, some anti windup techniques for MIMO continuous and discrete time systems are shown with illustrative examples to evince the performance of these control strategies.

3 Internal Model Anti Windup Control of Continuous SISO Systems

In this section an anti windup control architecture is developed by implementing an internal model controller (IMC). Internal model control is a technique that consists in designing an appropriate controller according to the internal stability of the system, therefore, as it is proved in this section, this control strategy is convenient for the design of an antiwindup control architecture, reducing the unwanted effects yielded by this phenomena and improving the system performance. The anti windup control strategy shown in this section is developed by feedback the saturated input to the internal model controller so the effects of windup are minimized. The IMC PID controller synthesis is done by the minimization of the H_∞ norm of the error signal as explained in (Morari and Zafiriou 1989; Lee et al. 1998; Tu and Ho 2011) when a unit step input is implemented as a reference signal (Cockbum and Bailey 1991; Doyle III 1999). With this control technique, the resulting PID controller has anti windup properties while maintaining its robustness, so this control strategy is ideal to avoid the unwanted effects yielded by windup. In this section the derivation of an IMC PID anti windup controller is shown step by step ensuring the internal stability of the system while reducing the unwanted effects yielded by the integral action of the PID controller.

3.1 IMC PID Anti Windup Controller for Continuous Time SISO Systems

The anti windup controller architecture implemented in this section is defined in Fig. 2 and it is based on the controller architecture explained in (Saeki and Wada 1996) where a compensator is added to the feedback loop from the saturation input of the system. In Fig. 2 the description of each block is the following; $G_p(s)$ is the plant transfer function that is represented by a first order plus time delay model (FOPTD), $G_c(s)$ is the internal model PID controller and $R(s)$ is the anti windup compensator filter.

In order to obtain a simplified model of the saturation nonlinearity, it is necessary to represent this model by the following equation (Saeki and Wada 1996):

$$\begin{aligned} U &= (\alpha + \beta \Delta_\phi)\tilde{U} \\ \|\Delta_\phi\| &< 1 \\ \alpha + \beta &= 1 \\ \alpha - \beta &= a \end{aligned} \tag{1}$$

Where the saturation nonlinearity is considered to be in the interval $[a, 1]$. The filter $R(s)$ is defined by a first order system as described below:

$$R(s) = \frac{1}{a_1 s + a_0} \tag{2}$$

Then the equivalent transfer function of the nonlinearity (1) and the filter (2), depicted in Fig. 2, is given as $G_{sat}(s)$ as shown in Fig. 3

$$G_{sat}(s) = \frac{\alpha + \beta \Delta_\phi}{1 - R(s)(\alpha + \beta \Delta_\phi)} \tag{3}$$

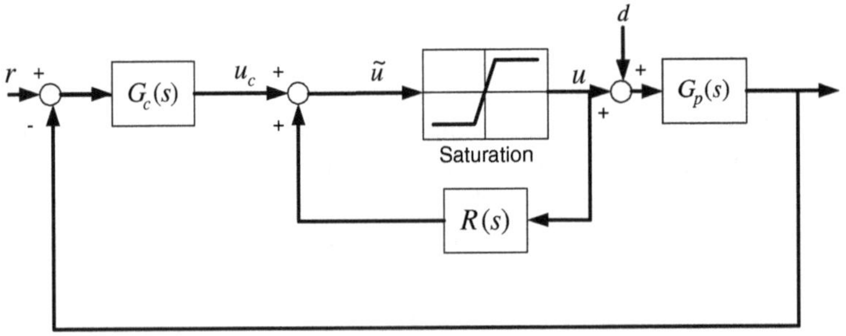

Fig. 2 Anti windup controller architecture

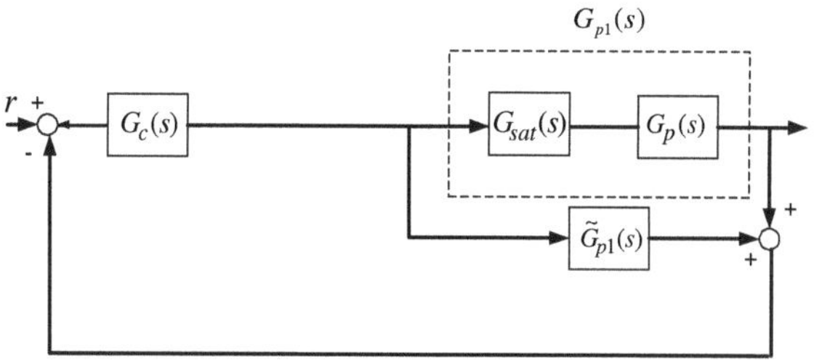

Fig. 3 Anti windup IMC PID architecture

Then the equivalent internal model anti windup control system is shown in Fig. 3. Where $G_{p1}(s)$ is the equivalent plant given by $G_{p1}(s) = G_{sat}(s)G_p(s)$

$G_p(s)$ is represented by a first order plus time delay function given by:

$$G_p(s) = \frac{ke^{-\theta s}}{\tau s + 1} \tag{4}$$

where k is the gain of the transfer function, θ is the time delay and τ is the time constant of the transfer function.

After finishing the explanation of the anti windup controller by implementing a model of the saturation nonlinearity, the IMC PID anti windup controller design can be derived using the equivalent transfer functions of the original system, considering the saturation effects on the model. To start this process it is necessary to obtain the equivalent transfer function of the anti windup controller, basically after obtaining this transfer function G_{p1}, the design of the IMC PID controller is straightforward because the equivalent transfer function is completely linear due to the implementation of an equivalent model of the saturation nonlinearity. Considering the equivalent transfer function G_{p1}

$$G_{p1}(s) = \frac{k(\alpha + \beta \Delta_\phi)(a_1 s + a_0)e^{-\theta s}}{(a_1 s + a_0 - (\alpha + \beta \Delta_\phi))(\tau s + 1)} \tag{5}$$

Then an IMC controller is obtained (Morari and Zafiriou 1989; Shamsuzzoha and Lee 2007) dividing first the transfer function G_{p1} into two parts as the process for designing a IMC controller with anti windup properties

$$G_{p1}(s) = p_{1m} p_{1A} \tag{6}$$

where p_{1a} contains all the RHP poles and zeros with time delay and the portion p_{1m} includes the rest of the transfer function. Now, define the IMC controller q_1 as shown in the following equation, considering a unit step input as the reference:

$$q_1 = p_{1m}^{-1} f \tag{7}$$

where f is a filter selected by the designer in the following form:

$$f = \frac{1}{(\lambda s + 1)^r} \tag{8}$$

for some positive constant r. The IMC PID anti windup controller $G_c(s)$ is given by the following formulae

$$G_c(s) = \frac{q_1}{1 - G_{p1} q_1} \tag{9}$$

where this controller is transformed into a PID form as shown in the rest of this section. The transfer function G_{p1} is divided in the following parts as explained in (6)

$$\begin{aligned} p_{1A}(s) &= e^{-\theta s} \\ p_{1m}(s) &= \frac{k(\alpha + \beta \Delta_\phi)(a_1 s + a_0)}{(a_1 s + a_0 - (\alpha + \beta \Delta_\phi))(\tau s + 1)} \end{aligned} \tag{10}$$

Based on these equations q_1 is given by:

$$q_1(s) = \frac{(a_1 s + a_0 - (\alpha + \beta \Delta_\phi))(\tau s + 1)}{k(\alpha + \beta \Delta_\phi)(a_1 s + a_0)(\lambda s + 1)^r} \tag{11}$$

Using these equations the controller $G_c(s)$ is given by:

$$G_c(s) = \frac{1}{p_{1m}((\lambda s + 1)^r - p_{1a})} \tag{12}$$

Substituting the functions p_{1m} and p_{1A} the following IMC anti windup controller is found:

$$G_c(s) = \frac{(a_1 s + a_0 - (\alpha + \beta \Delta_\phi))(\tau s + 1)}{k(\alpha + \beta \Delta_\phi)(a_1 s + a_0)((\lambda s + 1)^r - e^{-\theta s})} \tag{13}$$

For the PID anti windup controller synthesis it is necessary to consider a PID controller for $G_c(s)$ and then by Mclaurin series expansion the IMC anti windup controller parameters are found (Shamsuzzoha and Lee 2007). For this purposes, consider the following PID controller

$$G_c(s) = K_c\left(1 + \frac{1}{\tau_i s} + \tau_d s\right) \tag{14}$$

where K_c is the controller gain, τ_i and τ_d are the integral and derivative time constant that must be obtained in order to get the IMC anti windup controller time constants. The time constants of the IMC anti windup controller are found by the Mclaurin series expansion as shown in (Shamsuzzoha and Lee 2007). The IMC gain and constants are obtained as follow (Lee et al. 1998):

$$
\begin{aligned}
K_c &= \dot{f}(0) \\
\tau_i &= \frac{\dot{f}(0)}{f(0)} \\
\tau_d &= \frac{\ddot{f}(0)}{2\dot{f}(0)}
\end{aligned}
\tag{15}
$$

The IMC anti windup controller gains are given in detail in Appendix 1, so the reader can refer to this section for detailed information. The function *f* and its derivatives are defined in this section according to the formulas given in (Lee et al. 1998).

With the derivation and design of an IMC anti windup controller for SISO system, the internal stability of the system while suppressing the unwanted effects of windup is ensured with the addition of a feedback loop which includes the saturated input signal through a filter that improves the system performance when the input is saturated and windup occurs in the PID controller. As it is verified later this control strategy is efficient when saturation occurs in the model, as it is noticed, this strategy is based on the implementation of a saturation model that includes all the properties of this nonlinearity. In the following section an illustrative and comparative example is done in order to test the performance of the IMC anti windup controller, the conclusions of this section are shown in order to compare the system performance with anti windup compensation and no compensation.

3.2 Example 1

In this subsection an illustrative example of the internal model anti windup controller for SISO continuous time system is shown. Consider the following FOPTD system:

$$
G_p(s) = \frac{e^{-0.0000001s}}{0.001s + 1} \tag{16}
$$

and the following parameters for the anti windup filter and saturation model as shown in Table 1.

Now implementing the formulae found in Appendix 1, the following IMC parameters are found for the IMC PID controller with anti windup compensation and when there is no anti windup compensation. These parameters are shown in Table 2.

The system response of the IMC anti windup controller is depicted in Fig. 4.

It can be noticed that when the AWC is implemented the system response has almost no overshoot and a small settling time in comparison when no AWC is

Table 1 Filter and saturation parameters

Parameter	Value
a_0	5
a_1	0.1
α	1
β	0.06
Δ	100
λ	0.9
r	1

Table 2 Parameters with AW and no AW compensation

Parameter	Value with AW compensation	Value with no AW compensation
K_c	0.0635	0.0036
τ_i	0.9503	33,327.8
τ_d	1	0.0005

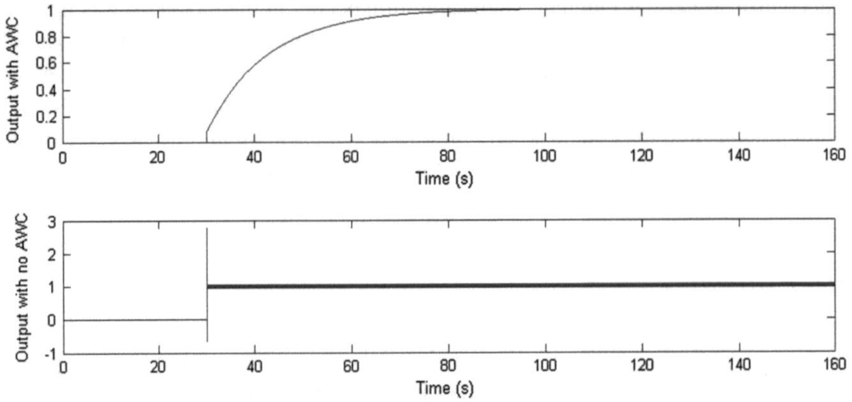

Fig. 4 System response with the IMC AWC (*upper*) and with no AWC (*lower*)

implemented where a high overshoot, a large settling time and higher oscillations are shown proving that the system has a better performance when the anti windup controller is implemented. These results are yielded due to the feedback compensation applied to the PID controller reducing the unwanted effects produced by windup, in comparison when there is not compensation where the system performance is deteriorated due to the increasing in the integrator output when the input of the system is saturated.

In Fig. 5 the input $\tilde{U}$ for the system with AWC is shown where the input is generated according to the reference signal. This signal is the non saturated signal generated by the IMC PID AWC, so the signal follows a designated trajectory according to the required control input necessary to control the system.

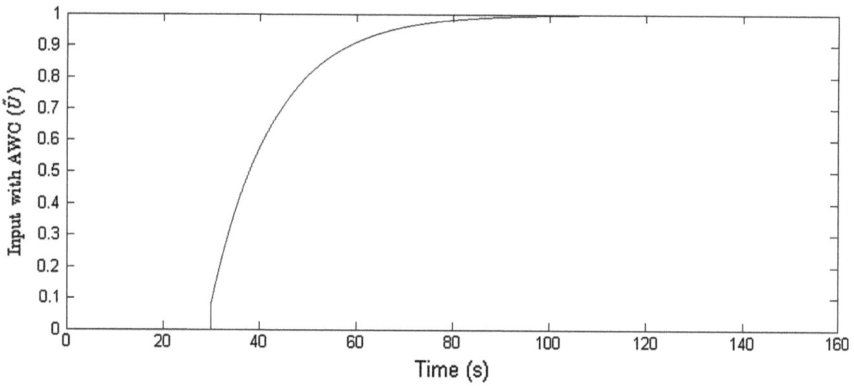

Fig. 5 Control input $\tilde{u}$ of the anti windup controller

In Fig. 6 the control input $\tilde{U}$ with no AWC is shown, where the nonsaturated signal applied to the system is depicted proving that this signal is more irregular than in the AWC version due to the increasing of the integral action producing an abrupt change in the input signal deteriorating the system response.

As it is corroborated in Figs. 7 and 8 these results are affected by the non saturated signals, especially when there is not AWC compensation due to the compensators improves the system performance considerably in comparison when there is no AWC compensation.

In Figs. 7 and 8 the respective control inputs with AWC and AWC compensation are shown, where as it is expected, the control input of the saturated system with no anti windup compensation is deteriorated due to the increasing of the integral action when the input of the system is saturated. This effect is improved by the IMC PID AWC compensation, because the extra feedback added to the model reduces the integral action when the system is saturated.

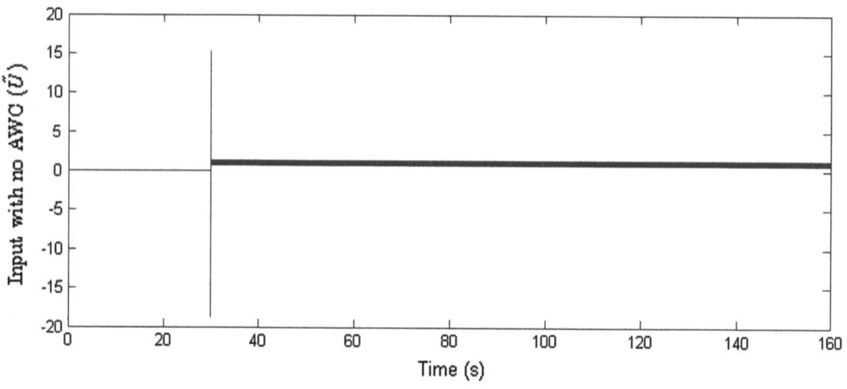

Fig. 6 Control input $\tilde{u}$ when there is no anti windup controller compensation

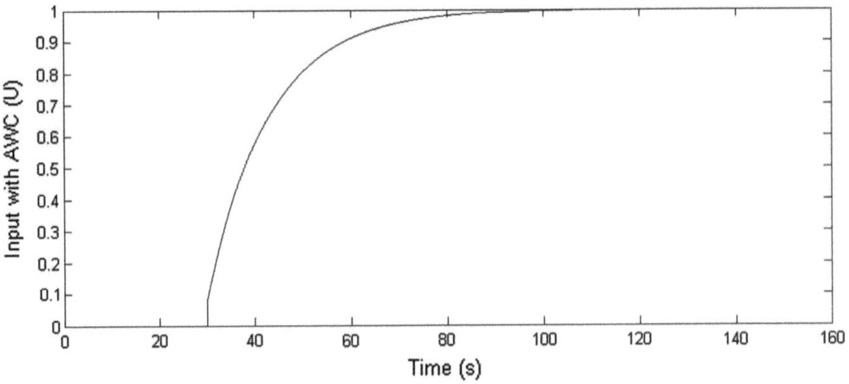

Fig. 7 Saturated input with AWC controller compensation

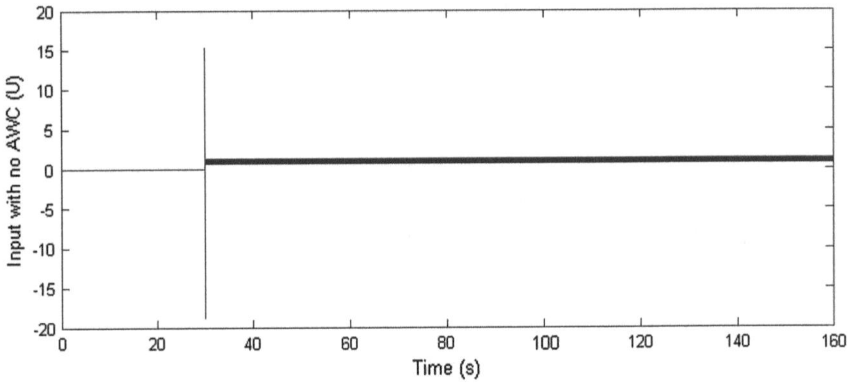

Fig. 8 Saturated input with no AWC controller compensation

These unwanted effects lead to the system performance deterioration, as explained before, Therefore a correction signal send to the internal model controller corrects and improves the system performance deterioration, yielding better system characteristics in comparison when there is no anti windup compensation.

Finally, as a conclusion of this section, it was proved that is possible to stabilize a saturated system by anti windup control compensation, when the system is a single input single output continuous time model, independently of the saturation and the unwanted effects yielded by the windup, generated by the increasing of the integral action. In the next section the discrete time counterpart of the IMC PID anti windup controller is derived, following the internal model control guidelines for the design of an appropriate anti windup controller for this kind of models.

4 Internal Model Anti Windup Control of Discrete Time SISO Systems

In this section the design of a discrete time anti windup controller for discrete time SISO system is explained. In this case, an internal model PID controller compensator is proposed to suppress the unwanted effects yielded by windup when the integrator output is increased due to the actuator saturation. The nonlinearities found in many control systems, specially saturation, deteriorates the system performance similar as it occurs in the SISO continuous time counterpart. As explained before, there are several anti windup control architectures for discrete time systems, some of them are derived from the system frequency response as shown in (Chen et al. 2003; Lambeck and Sawodny 2004) where the design of a frequency response method in a cascade configuration, eliminates the effects yielded by windup. As shown in (Wittenmark 1989) the incorporation of a back calculation compensator improves the system performance and reduces the unwanted effects yielded by saturation. This anti windup controller compensation is shown in (Baheti 1989) where a digital PID controller implementation is used to eliminate the unwanted effects of windup when the input of the system is saturated.

The anti windup controller strategy shown in this section is based on the theoretical background shown in (Morales et al. 2009) where a standard IMC anti windup compensator is implemented where the robustness of the control system is analyzed and the stabilization of the system is done by an internal model controller. The proposed strategy shown in this section is based on an IMC PID anti windup compensator, where due to integral characteristic of the PID controller it is necessary to cancel the windup effects yielded by saturation. The saturation nonlinearity model is obtained by a scalar function approach as explained in (Zhang et al. 2011), so the IMC PID controller can be derived in order to avoid the unwanted effects yielded by windup.

4.1 *IMC PID Anti Windup Controller for Discrete Time SISO Systems*

The anti windup internal model PID controller architecture is shown in Fig. 9.

Where $G_c(z)$ is the digital internal model controller, $R(z)$ is the anti windup compensator filter, $G_p(s)$ is the continuous time transfer function discretized by a sampler and $p_\gamma^*(z)$ is the equivalent discrete time transfer function implemented in the internal model PID anti windup controller design. In Fig. 10 the equivalent discrete time transfer function is shown, where this transfer function is obtained by the implementation of the scalar sign function approach.

The resulting transfer function $p_\gamma^*(z)$ is implemented to design the anti windup internal model controller with the robustness and internal stability requirements

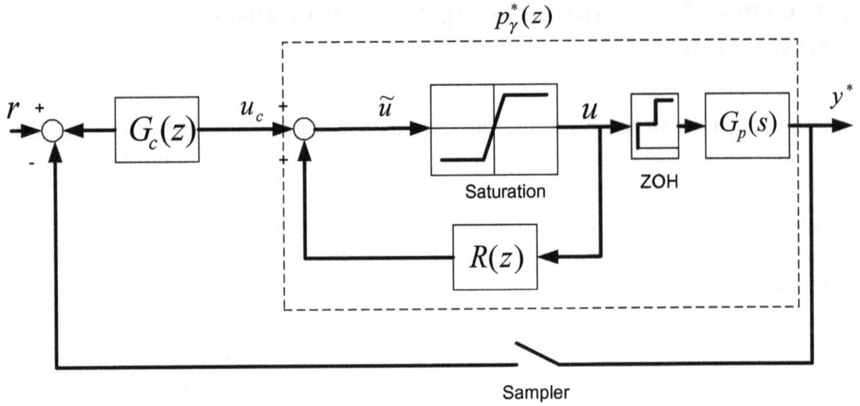

Fig. 9 Anti windup controller architecture

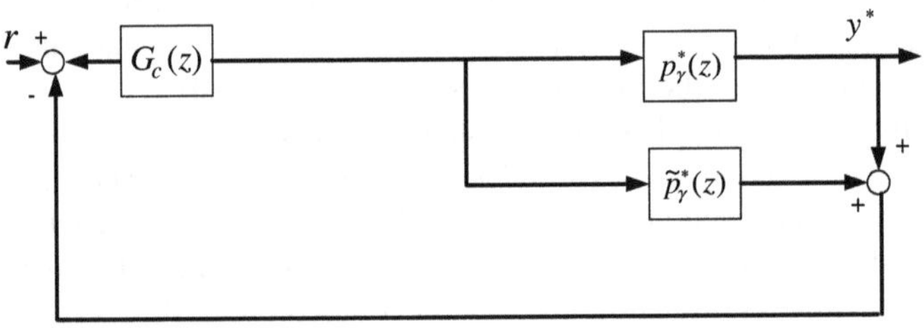

Fig. 10 Equivalent IMC controller architecture

including the anti windup compensator to eliminate the unwanted effects yielded by windup when the input signal is saturated.

Similar as the continuous time counterpart can be divided into two parts, $p_{\gamma A}^{*}(z)$ and $p_{\gamma M}^{*}(z)$ as shown in the following equation:

$$p_{\gamma}^{*}(z) = p_{\gamma A}^{*}(z) p_{\gamma M}^{*}(z) \tag{17}$$

where

$$p_{\gamma A}^{*}(z) = z^{-N} \prod_{j=1}^{h} \frac{(1 - (\zeta_j^H)^{-1})(z - \zeta_j)}{(1 - \zeta_j)(z - (\zeta_j^H)^{-1})} \tag{18}$$

ζ_j are the zeros of $p_{\gamma A}^{*}(z)$ outside the unit circle for $j = 1 \ldots h$. N is selected to make $p_{\gamma M}^{*}(z)$ semiproper and H denotes the complex conjugate (Morari and Zafiriou 1989). In order to design the internal model anti windup controller the following controller must be implemented:

$$G_c(z) = \frac{q(z)}{1 - p_\gamma^*(z)q(z)} \tag{19}$$

where

$$q(z) = p_{\gamma M}^{*-1}(z)f(z) \tag{20}$$

and the filter $f(z)$ is given by:

$$f(z) = \frac{(1-\alpha)z}{z-\alpha} \tag{21}$$

for a given value of α. Meanwhile, the anti windup compensator filter $R(z)$ is given by:

$$R(z) = \frac{1}{a_1 z + a_0} \tag{22}$$

where a_1, $a_0 > 0$. The saturation function is obtained by the scalar sign function (Zhang et al. 2011) taking into account the following *sign* function representation:

$$sign(z) = \begin{cases} 1 & \text{if Re(z)} > 0 \\ -1 & \text{if Re(z)} < 0 \end{cases} \tag{23}$$

so for $j = 1$ the following representation of the saturation model is implemented:

$$saturation(z) = U_{max} sign_1(z) \tag{24}$$

where U_{max} is the saturation limit and

$$sign_1(z) = z \tag{25}$$

In this section in order to design the anti windup control system for discrete time models the following first order plus time delay discrete time model is implemented:

$$G_p(z) = \frac{k}{\tau z + 1} z^{-N} \tag{26}$$

Where k is the system gain, τ is the time constant and $N > 0$ is an integer which indicates the number of time delays. In order to obtain the internal model controller it is necessary to get the equivalent transfer function $p_\gamma^*(z)$ taking in count the compensator and saturation in order to obtain this transfer function:

$$p_{\gamma}^{*}(z) = \frac{k(U_{max}a_1 z^{2-N} + U_{max}a_0 z^{1-N})}{((a_1 - U_{max})z + a_0)(\tau z + 1)} \tag{27}$$

With a sample time T. Where this transfer function is divided as explained in (17) and (18) as:

$$p_{\gamma M}^{*}(z) = \frac{k(U_{max}a_1 z^2 + U_{max}a_0 z)}{((a_1 - U_{max})z + a_0)(\tau z + 1)} \tag{28}$$

$$p_{\gamma A}^{*}(z) = z^{-N} \tag{29}$$

Then using (19) the following internal model controller is obtained:

$$G_c(z) = \frac{(1-\alpha)z((a_1 - U_{max})z + a_0)(\tau z + 1)}{(U_{max}a_1 z^2 + U_{max}a_0 z)(z - \alpha) - (U_{max}a_1 z^{3-N} + U_{max}a_0 z^{2-N})} \tag{30}$$

In order to obtain the internal model anti windup controller, it is necessary to define the following standard PID controller:

$$G_c(z) = K_c(1 + \frac{1}{\tau_i(z-1)} + \tau_d(z-1)) \tag{31}$$

Due to the integral term of $G_c(z)$ the controller gain and parameters using a similar procedure like the continuous time counterpart. Implementing the Taylor series expansion, similar as the previous section the following constant and time constants of the PID controller are found:

$$\begin{aligned} K_c &= f'(1) \\ \tau_i &= \frac{f'(1)}{f(1)} \\ \tau_d &= \frac{f''(1)}{2f'(1)} \end{aligned} \tag{32}$$

where f and its derivatives are defined in Appendix 2. This equations are valid for any sampling period T and the resulting equations are shown in Appendix 2. The proposed control strategy explained in this section meets the robustness and internal stability properties that make them suitable for the anti windup control of discrete time SISO systems. In the next subsection, an illustrative example is shown, to test the system performance by a numerical example.

4.2 Example 2

In this subsection the stabilization of a discrete time SISO system is done when saturation is found in the model. The system to be stabilized is the following:

$$G_p(z) = \frac{z^{-2}}{10z + 1} \tag{33}$$

with the following saturation, filter and anti windup compensator parameters (Table 3).

Using the formulae shown in Appendix 2and a sampling period of $T = 1$ s, the controller gain and parameters are found as shown in Table 4.

With this control systems parameters, the system output with AWC compensation and with no AWC compensation are shown in the figure below.

The system response shown in Fig. 11 corroborates that a small overshoot and small settling time is obtained when an internal model anti windup controller is implemented, in contrast when there is not anti windup compensation. These results are expected due to the anti windup compensator reduces the integral action when the system is saturated, so a smaller overshoot and smaller settling time is obtained when the internal model controller is implemented.

In Fig. 12 the non saturated input of the system with anti windup compensation is shown where this signal reaches the necessary output value to obtain the required value.

In Fig. 13 the non saturated input, when there is no anti windup compensation, is shown. As can be noticed, the required input signal is applied to the system until the required output value is obtained.

In Fig. 14 the saturated input value is depicted, where the limiter imposed by the saturation makes the system to reach the desired value and as it is compared with Fig. 15 the saturated signal is better when an anti windup controller is implemented.

Table 3 Filter, compensator and saturation parameters

Parameter	Value
α	0.2
a_0	4×10^{-7}
a_1	5×10^{-7}
U_{max}	10.9

Table 4 Parameters with AW and no AW compensation

Parameter	Value with AW compensation	Value with no AW compensation
K_c	3.4362×10^7	3.1062×10^8
τ_i	1.6869	2
τ_d	0.4265	1×10^{-6}

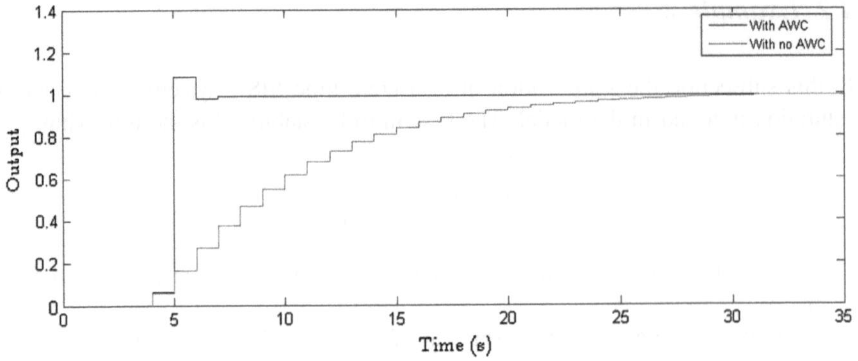

Fig. 11 System output with AWC and no AWC

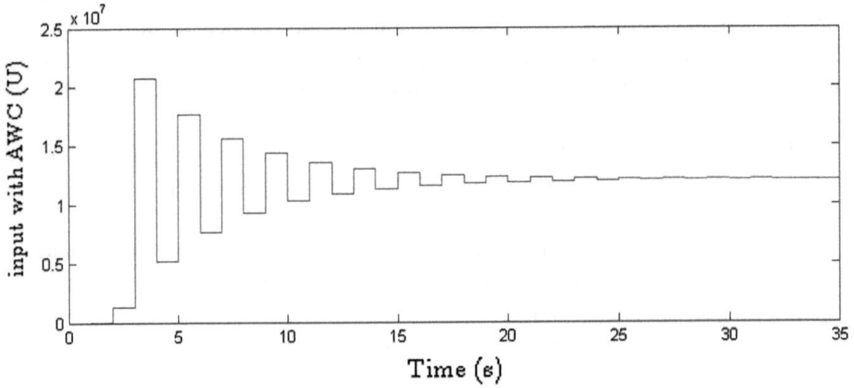

Fig. 12 Non saturated input with AWC

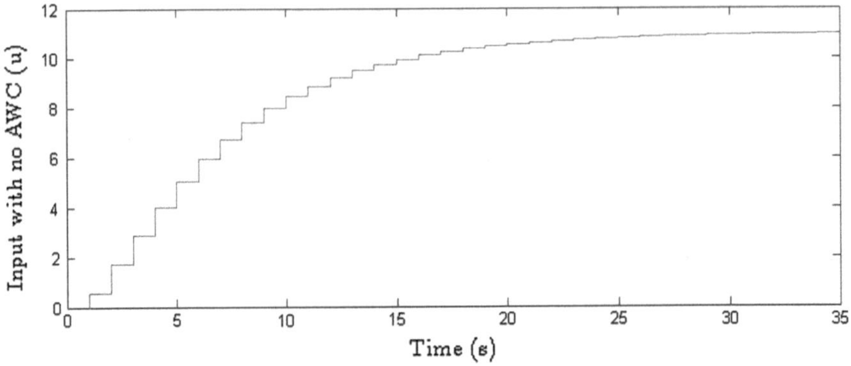

Fig. 13 Non saturated input with no AWC

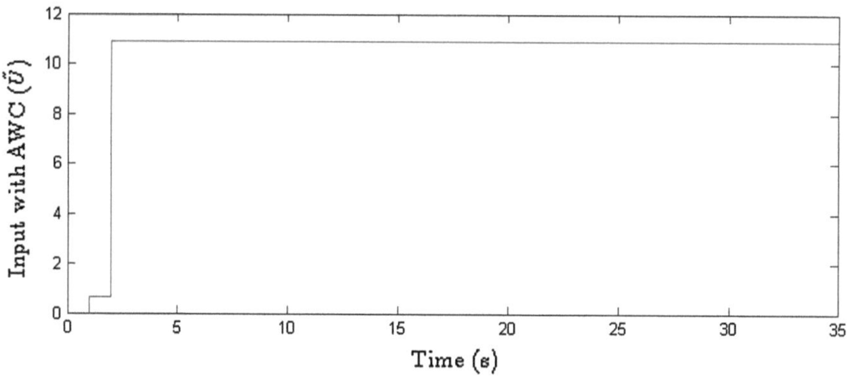

Fig. 14 Saturated input with AWC

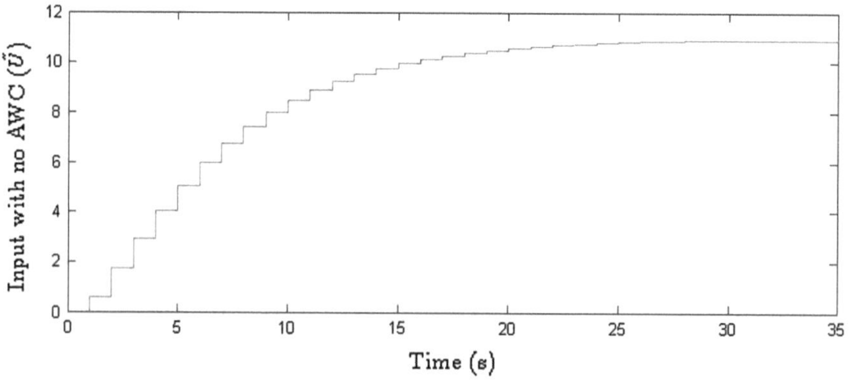

Fig. 15 Saturated input with AWC

In this section an internal model PID anti windup controller is designed in order to improve the system performance by reducing the windup effect. As it is noticed, the controller design is very similar to the continuous time counterpart taking in account the saturated signal and then this signal is sent through a feedback loop by a compensator. The main idea behind this controller is to apply the robust controller characteristics of internal model control in order to obtain the desired gain and time constants of the PID controller to make the system to follow a step reference signal. With the control strategies derived in Sects. 3 and 4 a complete design and analysis of anti windup controllers for continuous and discrete time SISO systems is deployed. Where it was proved that efficient anti windup control strategies can be derived implementing the internal model control strategy for any kind of SISO systems while ensuring internal stability and the improvement of the system output performance.

In the following sections, the design and analysis of anti windup techniques for discrete and continuous time MIMO systems is shown, where different approaches are implemented in order to improve the control system performance when saturation or constrained inputs are present in the system. Generally, the design of anti windup control strategies for MIMO systems are more difficult than the anti windup control of SISO system, for this reason, the solution of this problem is done by static output feedback control law design, where MIMO PID controllers are designed in the continuous and discrete time cases.

It will be proved that as similar to the SISO system cases, the modeling and design of effective anti windup control techniques is possible improving the system performance when some kind of compensation is added to the controller.

5 Anti Windup Control of Continuous MIMO Systems by Static Output Feedback (SOF)

In this subsection the design of an anti windup PID controller for continuous time MIMO system is derived based on static output feedback (SOF) controller. This work is based on the solution of the specified linear matrix inequalities (Cao et al. 2002; Wu et al. 2005; Rehan et al. 2013) where a static output feedback controller is defined in order to improve the anti windup characteristics of this MIMO controller (Neto and Kucera 1991; Henrion et al. 1999; Fujimori 2004; He and Wang 2006). A PID control law is obtained by solving the required LMI's in order to find the PID controller gains. The controller gains are found by two static output feedback solutions, by solving an standard LMI and a H_∞ problem. With these two control strategies it is possible to find appropriate controller gains for the PID anti windup controller taking in count the saturation nonlinearity.

In order to design the anti windup PID controller it is necessary to model the saturation nonlinearity by a describing function approach (Taylor and O'Donnell 1990) in order to deal with the nonlinearities added to the system by the actuators saturation.

The intention of this control approach is to design an efficient anti windup controller system for MIMO continuous time systems when the inputs are constrained or saturated. It is proved that solving the system constraints by LMI's in order to obtain a stable PID control law, the addition of anti windup compensation similar as the SISO time systems, improves the system performance and avoids the deterioration of the output signal. In the following subsections the design of an anti windup controller is explained in detail, and in order to test the system performance an illustrative example of the stabilization and control of a DC motor is evinced.

5.1 PID Anti Windup Controller Design for MIMO Continuous Time Systems

The PID anti windup controller design for MIMO continuous time systems, consist of a back calculation PID controller by a loop which includes the saturated and non saturated input signal of a linear time invariant MIMO system. In Fig. 16 the anti windup controller is shown where v is the back calculation signal that is implemented to avoid the windup effects which deteriorates the system performance. The MIMO system is represented by $G(s)$ and the anti windup PID controller and compensator is represented by $G_c(s)$.

The anti windup controller takes the non saturated and saturated difference signal v to suppress the increment of the integral action when the system saturates. As occurs in the SISO case, the windup phenomena yields unwanted effects that deteriorates the system performance; the settling time and overshoot generally are damage when the input signal is saturated, so the compensator corrects the effects of windup by the back calculation of the input signal added by an extra loop.

The approach explained in this section consist in obtaining a saturation representation by a describing function approach (Taylor and O'Donnell 1990), where this method simplifies the PID controller synthesis and provides an accurate representation of the equivalent control systems.

In Fig. 17 the saturation model is depicted in order to be represented by a describing function that helps to obtain an equivalent anti windup controller.

The saturation model shown in Fig. 17 depicts the parts in which this model is divided in order to obtain the Fourier series coefficients of the describing function. $n(e)$ is the saturation output, e is the saturation input, α is the input limit and M is the saturation output limit. The describing function is given by the following transfer function:

$$\phi(s) = \frac{a_1 + b_1 s}{E} \tag{34}$$

where a_1 and b_1 are the Fourier series coefficients when a sinusoidal input signal with amplitude E is implemented. The coefficients of the Fourier series implemented in this analysis are:

Fig. 16 Multivariable control system

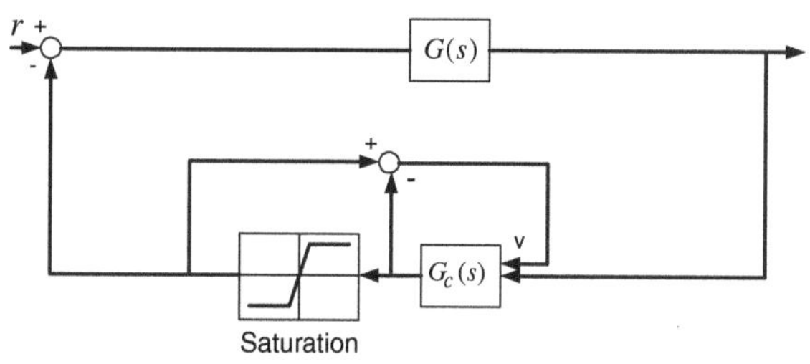

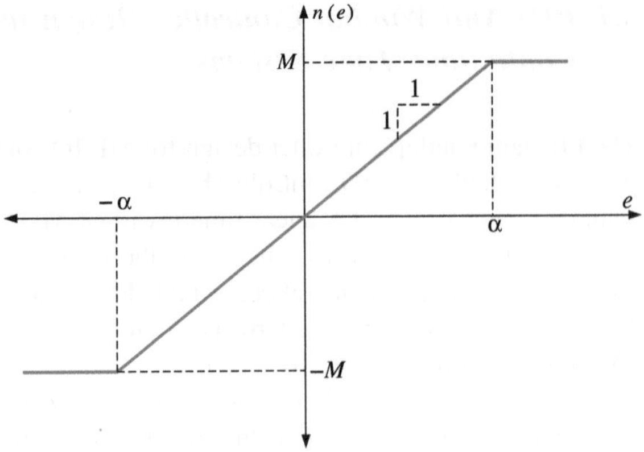

Fig. 17 Saturation model

$$
\begin{aligned}
a_1 &= \frac{1}{T}\int_{-T}^{T} \sin(\omega t)n(t)dt \\
b_1 &= \frac{1}{T}\int_{-T}^{T} \cos(\omega t)n(t)dt
\end{aligned}
\tag{35}
$$

where ω is the angular frequency of the input signal e and in order to obtain the Fourier series coefficients a sinusoidal input signal of amplitude E and period T must be assumed as the input of the saturation. Considering that e is 2π periodic or $T = 2\pi$ the following Fourier coefficients are obtained:

$$
\begin{aligned}
a_1 &= (\frac{-\alpha}{2E}\sqrt{1+(\frac{\alpha}{E})^2} + sin^{-1}(\frac{\alpha}{E}) + \frac{\pi}{4}) + \frac{M}{2\pi}(\frac{2\alpha}{E}) \\
b_1 &= \frac{-E}{4\pi}\left(2\left(1-\left(\frac{\alpha}{E}\right)^2\right) - \left(\frac{\alpha}{E}\right)^2 - 1\right) - \frac{M}{\pi}\left(\frac{\alpha}{E}\right)
\end{aligned}
\tag{36}
$$

Considering the following PID controller:

$$
y_c(s) = F_1 u_c(s) + F_2 \frac{u_c(s)}{S} + F_3 u_c(s)\left(\frac{s-1}{s}\right) \tag{37}
$$

where F_1, F_2 and F_3 are diagonal matrices of appropriate dimensions (He and Wang 2006) for the proportional, integral and derivative parts of the controller.

In order to obtain the anti wind up controller, consider the following linear time invariant system $G(c)$ given by:

$$\begin{aligned} \dot{x} &= Ax - B\phi(u) \\ y &= Cx \end{aligned} \tag{38}$$

where $x \in \Re^n, u \in \Re^m$ and A, B and C are matrices of appropriate dimensions. From (37) The controller and anti windup compensator can be represented in state space by (Cao et al. 2002):

$$\begin{aligned} \dot{x}_c &= k_1 y + \Delta(\phi(u) - u) \\ u &= x_c + k_2 y = x_c + k_2 Cx \end{aligned} \tag{39}$$

where $k_1 = F_2 - F_3$, $k_2 = F_1 + F_3$, and Δ is a positive definite diagonal matrix that is part of the anti windup controller and compensator. Usually a correction term $\Delta(\phi(u) - u)$ is needed in the controller to compensate the saturation effects.

With (38) and (39) a closed loop augmented control system is obtained as:

$$\begin{aligned} \dot{\overline{x}} &= \overline{A}\overline{x} + \overline{B}w \\ u &= F\overline{x} \end{aligned} \tag{40}$$

where:

$$\begin{aligned} \overline{x} &= \begin{bmatrix} x \\ x_c \end{bmatrix} \\ \overline{A} &= \begin{bmatrix} A & 0 \\ k_1 C & 0 \end{bmatrix} \\ \overline{B} &= \begin{bmatrix} -B & 0 \\ \Delta & -\Delta \end{bmatrix} \\ F &= \begin{bmatrix} k_2 C & I \end{bmatrix} \\ w &= \begin{bmatrix} \phi(u) & u \end{bmatrix}^T \end{aligned} \tag{41}$$

Using the saturation model ϕ, the obtained input vector w is:

$$w = \begin{bmatrix} F(\frac{a_1}{E}\overline{x} + \frac{b_1}{E}\dot{\overline{x}}) \\ F\overline{x} \end{bmatrix}$$

Making another change of variable with $z = \begin{bmatrix} \overline{x} & \dot{\overline{x}} \end{bmatrix}^T$ the following system is obtained:

$$\dot{z} = A'z + B'Mz \tag{42}$$

that yields the following system's equation:

$$\dot{z} = (A' + B'M)z \tag{43}$$

where:

$$A' = \begin{bmatrix} \overline{A} & 0 \\ 0 & 0 \end{bmatrix}$$
$$B' = \begin{bmatrix} \overline{B} \\ 0 \end{bmatrix} \quad (44)$$
$$M = \begin{bmatrix} Fa_1/E & Fb_1/E \\ F & 0 \end{bmatrix}$$

In order to obtain the linear matrix inequality to solve the gains of the PID anti windup compensator the following Lyapunov function must be considered:

$$V(z) = z^T P z \quad (45)$$

where P is a positive definite matrix, used in order to ensure the stability of the system. Deriving the Lyapunov function the following result is obtained:

$$\dot{V}(z) = z^T (A' + B'M)^T P z + z^T P (A' + B'M) z \quad (46)$$

where in linear matrix inequality form (45) is represented as:

$$(A' + B'M)^T P + P(A' + B'M) < 0 \quad (47)$$

So by solving the following LMI the controller parameters of the equivalent system are found (Fujimori 2004):

$$\begin{bmatrix} (A' + B'M)^T P + P(A' + B'M) & 0 \\ 0 & 0 \end{bmatrix} < 0 \quad (48)$$

For the H_∞ synthesis, a similar approach is implemented to find the controller gains, considering the following criteria:

$$\|T_{2\omega}(s)\|_\infty < \gamma \quad (49)$$

where $T_{2\omega}(s)$ is the closed loop transfer function of the model (Fujimori 2004; He and Wang 2006; Rehan et al. 2013) and $\gamma > 0$ is a positive constant that indicates the desired performance. Then the respective LMI is needed to find the gain F and the solutions of the anti windup PID controller.

$$\begin{bmatrix} PA_{cl} + P^T A_{cl} & 0 & C_{cl} \\ 0 & -\gamma I & 0 \\ C_{cl} & 0 & -\gamma I \end{bmatrix} < 0 \quad (50)$$

where:

$$A_{cl} = (A' + B'M)$$
$$C_{cl} = \begin{bmatrix} I & 0 \\ 0 & 0 \end{bmatrix} \tag{51}$$

where I is an identity matrix of an appropriate dimension. Solving the LMI shown in (48) and (50) for P and M the controller parameters can be extracted from M obtaining all the PID controller gain matrices found by these optimization techniques.

In this subsection is proved that an anti windup PID controller for MIMO continuous time system can be implemented by solving a LMI based optimization problem. In the following subsection, the control of a DC motor is done in order to show by an illustrative example the application of these control strategies, it is proved that finding the respective matrix M the rest of the controller variables can be obtained. The solution of these LMI can be obtained by several numerical methods found in literature, such as shown in (He and Wang 2006) for example.

5.2 *Example 3*

In this section the stabilization and control of a DC motor by an anti windup PID controller for MIMO systems is shown to illustrate the advantages of the proposed technique.

Consider the following DC motor transfer function (Cockbum and Bailey 1991):

$$\frac{\omega_L(s)}{v_a(s)} = \frac{k_m}{(J_mL + J_LL)s^2 + (J_mR + J_LB_L)s + k_m^2} \tag{52}$$

where ω_L is the angular velocity of the model, v_a is the applied armature voltage, J_L is the inertial load, J_m is the motor inertia, L is the inductance, R is the resistance, B_L is the viscous friction constant and k_m is the motor constant. Converting (52) to state space the following equation is obtained:

$$\begin{bmatrix} \dot{x}_1 \\ \dot{x}_2 \end{bmatrix} = \begin{bmatrix} \frac{J_mR+J_LB_L}{J_m+J_LL} & 1 \\ -\frac{K_m^2}{J_m+J_LL} & 0 \end{bmatrix} \begin{bmatrix} x_1 \\ x_2 \end{bmatrix} + \begin{bmatrix} 0 & 0 \\ 0 & \frac{K_m}{J_m+J_LL} \end{bmatrix} \begin{bmatrix} 0 \\ v_a \end{bmatrix} \tag{53}$$

Table 5 Parameters of the DC motor

Parameter	Value
J_m	0.02215 kg m^2
J_L	0.01 kg m^2
B_L	0.002953 Nm s
R	11.2 Ohms
L	0.1215 H
K_m	0.5161 Nm/A
Nominal speed	1,750 RPM

and

$$Y = Cx \tag{54}$$

where x_1 is the angular velocity of the motor, x_2 is the armature current and C is a 2×2 identity matrix. The motor parameters are shown in Table 5.

Solving the LMI (48) for P by an optimization algorithm, a matrix F can be found from M in order to obtain the gain matrices for the PID controller. The gain matrices of the PID anti windup controller can be obtained by (55).

$$F = \begin{bmatrix} 800 & 0 & 1 & 0 \\ 0 & 800 & 0 & 1 \end{bmatrix} \tag{55}$$

From F the following PID anti windup controller parameters are found

$$\begin{aligned} F_1 &= \begin{bmatrix} -200 & 0 \\ 0 & -200 \end{bmatrix} \\ F_2 &= \begin{bmatrix} 2000 & 0 \\ 0 & 2000 \end{bmatrix} \\ F_3 &= \begin{bmatrix} 1000 & 0 \\ 0 & 1000 \end{bmatrix} \end{aligned} \tag{56}$$

The gain matrices when there is no anti windup compensation are the following:

$$\begin{aligned} F_1 &= \begin{bmatrix} -100000 & 0 \\ 0 & -100000 \end{bmatrix} \\ F_2 &= \begin{bmatrix} 2000 & 0 \\ 0 & 2000 \end{bmatrix} \\ F_3 &= \begin{bmatrix} 1000 & 0 \\ 0 & 1000 \end{bmatrix} \end{aligned} \tag{57}$$

The matrix F for the H_∞ controller are given by:

$$F = \begin{bmatrix} -4.0825 & 0 & 0 & 0 \\ 0 & -4.0825 & 0 & 0 \end{bmatrix} \times 10^8 \tag{58}$$

and the following gain matrices for the H_∞ anti windup controllers are given by:

$$\begin{aligned} F_1 &= \begin{bmatrix} -4.0825 & 0 \\ 0 & -4.0825 \end{bmatrix} \times 10^8 \\ F_2 &= \begin{bmatrix} 2000 & 0 \\ 0 & 2000 \end{bmatrix} \\ F_3 &= \begin{bmatrix} 1000 & 0 \\ 0 & 1000 \end{bmatrix} \end{aligned} \tag{59}$$

and the compensator gain Δ is given by:

$$\Delta = \begin{bmatrix} 2 & 0 \\ 0 & 2 \end{bmatrix} \tag{60}$$

The main idea of this example is to keep the nominal angular velocity(1,750 RPM) or (183.26 rad/s) while applying a disturbance torque of (100 Nm) at 0 s, so the anti windup PID controller must be able to keep this velocity even when an external disturbance is applied on the model.

In Fig. 18 the angular velocity of the DC motor in three cases; with anti windup, no anti windup and H_∞ anti windup PID controllers are shown; where in the H_∞ and standard static feedback anti windup controller better results were obtained with smaller settling time, smaller steady state error and smaller overshoot in comparison when there is not anti windup compensation. For these reasons, the anti windup controllers and compensator are better than the uncompensated controller version.

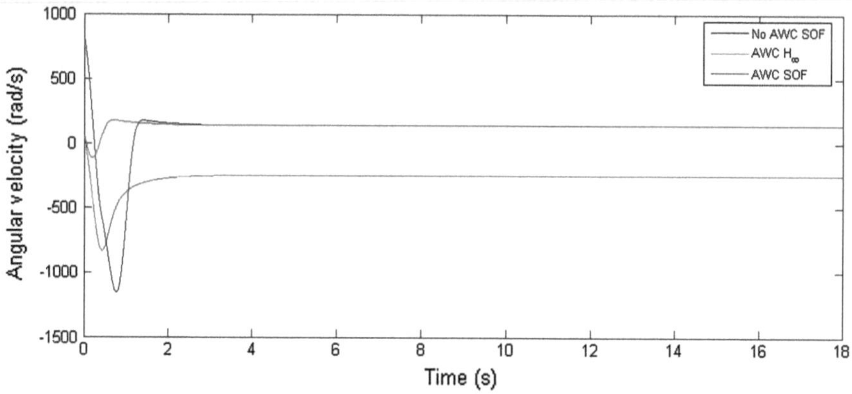

Fig. 18 Angular velocities with AWC, H_∞ and no AWC

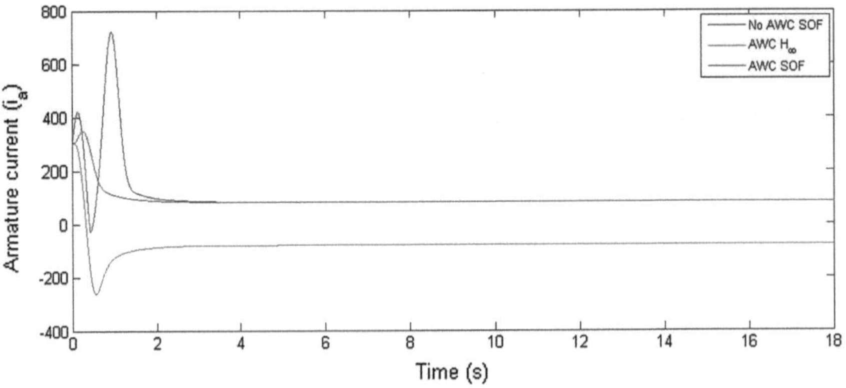

Fig. 19 Armature current i_a

It is clear when an anti windup compensator is implemented the performances is improved significantly, this means, that the settling time, steady state error and overshoot are smaller than the performance indexes of the uncompensated system.

In Fig. 19 the armature current i_a is depicted for the three cases of static output feedback (SOF) controller. It can be noticed that even than in the standard and H_∞ SOF the armature current is greater in comparison when no compensation is implemented in the SOF controller. These results are obtained due to the better performance of the standard and H_∞ SOF in comparison with the uncompensated controller, so more control effort is necessary in order to obtain an acceptable performance.

In Fig. 20 the input voltage of the DC motor (field voltage) is depicted where the input voltage of the non compensated systems increases to higher values than the compensated control systems. The input voltage for the uncompensated controller raises to higher values due to the windup effects that increases the integral action, similar to the SISO case, deteriorating the system performance.

The anti windup PID compensator by SOF improves the system response and performance significantly due to the compensator added to the MIMO PID controller, The windup effects are suppressed by the compensator action, reducing the integral action when the input of the system is saturated.

Finally, in Fig. 21 the mechanical torque of the DC motor is depicted where this variable reaches the final value of 100 Nm, which is the value of the disturbance input applied to the motor at 0 s while keeping the desired nominal velocity.

In this section the design of an anti windup PID controller for MIMO system by standard and H_∞ SOF is explained in order to obtain a suitable controller that eliminates the unwanted effects yielded by windup. As it occurs in the SISO case, the windup phenomena occurs when the input of the system is saturated increasing the integrator action, this effects damage the system performance, specially, it yields higher overshoot and longer settling times.

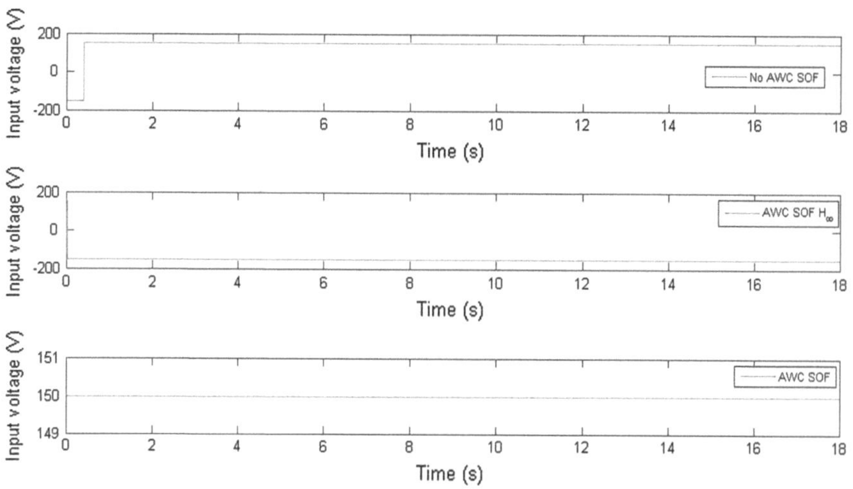

Fig. 20 Input voltage

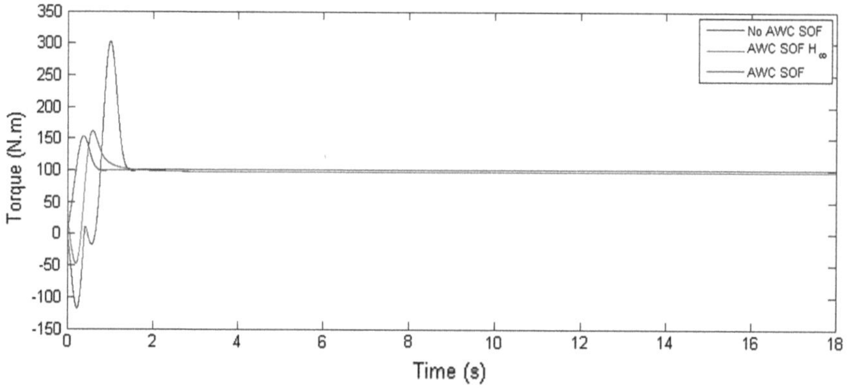

Fig. 21 Mechanical torque

In this section the design of two control strategy to deal with windup reducing the integral action and improving the system performance significantly. The saturation nonlinearity is implemented by the describing function method simplifying the design of the proposed control strategies. It is confirmed by an example, that the standard anti windup SOF controller yields better results than the uncompensated systems in which the system output is deteriorated by the windup effect. The standard and H_∞ SOF PID controllers are a perfect option for the control and compensation of saturated or constrained input MIMO systems.

In the next section, the MIMO counterpart of the control strategy presented in this section is shown. A discrete time MIMO system is obtained by a static output feedback, a PID compensator is selected similar as the continuous counterpart. In this section it is shown that an efficient control strategy is developed for the suppression of the unwanted effects yielded by windup, and the controller synthesis is done by a different saturation model.

6 Anti Windup Control of Discrete MIMO Systems by Static Output Feedback (SOF)

In this section the derivation of an anti windup PID controller for discrete MIMO system is proposed. The main idea behind this controller is to design an anti windup controller/compensator that minimizes the windup effects when the input of the system saturates producing an increasing of the integral action that deteriorates the system performance. The controller design for this kind of systems consist in deriving a static output feedback (SOF) control law, similar as the continuous time counterpart (Bateman and Zongli 2002; Kwan Ho et al. 2006; Matsuda and Ohse 2006) and then the SOF gain is obtained by solving the LMI's as an optimization problem.

In order to achieve suitable control gains for the PID controller, it is necessary to implement a saturation model (Li-Sheng et al. 2004; Zongli and Liang 2006; Shuping and Boukas 2009) where sufficient conditions are established in order to solve the LMI's by a convex optimization problem (Shuping and Boukas 2009). For the AWC design it is necessary to add a back calculation loop which consists in the difference between the non saturated and saturated input signal, similar as the continuous time counterpart, to reduce the effects of windup when the input system saturates. Then using the saturation model (Li-Sheng et al. 2004) this nonlinearity form is implemented to obtain the respective LMI's solved by a convex optimization problem. Beside from the standard solution of static output feedback controllers (SOF) a h_∞ SOF controller synthesis is obtained by solving the required LMI's (Lim and Lee 2008). In this section it is proved that a discrete time PID controller can be obtained by a static output feedback control law, simplifying the anti windup controller design and then the PID controller gains can be found by solving the linear matrix inequalities for SOF and H_∞ SOF.

As occurs in the continuous time case, there are several numerical methods to solve discrete time SOF problems by LMI's so with this method an optimal solution of the LMI's can be found. By implementing the appropriate LMI's and the saturation nonlinearity model an optimal solution can be found by any of the algorithm found in literature such as (Matsuda and Ohse 2006) for continuous time and (Kwan Ho et al. 2006) for discrete time systems. The proposed anti windup PID controller is designed taking into account the stability properties and characteristic of the closed loop system and for the H_∞ SOF problem the robustness of the closed loop system improves the system performance and reduces the deterioration of the system operation when a reference signal needs to be tracked.

This section is divided in two subsections, where in the first part the design of a PID anti windup controller is derived by adding a back calculation signal to the controller and converting the anti windup PID controller in a static output feedback problem and then this problem is solved by LMI's. Another anti windup PID controller is designed by a H_∞ synthesis where the stability and robustness of the system is considered, then the closed loop system is robust when unmodeled dynamics and disturbances are found in the system. Finally, an illustrative example is explained in the last subsection where the PID anti windup controller for a DC motor is shown, where the main objective is to maintain a constant nominal angular velocity by following a desired profile torque. With the theoretical background and the illustrative example shown in this section a complete demonstration of a PID anti windup control strategy for discrete time system is shown where the stability and robustness condition are met by selecting an appropriate static output feedback controller.

6.1 PID Anti Windup Controller Design for MIMO Discrete Time Systems

Consider the PID antiwindup controller shown in Fig. 22.

Where $G(z)$ is the discrete time transfer matrix, and v is the back calculation input signal of the anti windup PID controller. Consider the transfer matrix $G(z)$ in state space form

$$\begin{aligned} x(k+1) &= Ax(k) - B\sigma(u(k)) \\ y(k) &= Cx(k) \end{aligned} \tag{61}$$

where $x \in \Re^m$, A is a $\Re^{m\times m}$ matrix x is a $\Re^m$ vector, B is a $\Re^{n\times m}$ and C is a $\Re^{l\times m}$ matrix. $m > 0$ denotes the number of states, $n > 0$ is the number of inputs, l is the number of outputs and $\sigma(\cdot)$ is the saturation input. Consider the following anti windup PID controller given by:

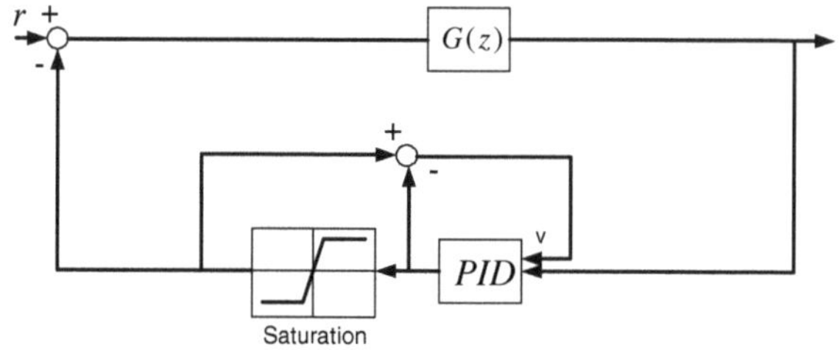

Fig. 22 Anti windup PID controller

$$\begin{aligned} x(k+1) &= Ax(k) - B\sigma(u(k)) \\ y(k) &= Cx(k) \end{aligned} \tag{62}$$

Consider the following PID controller with anti windup compensation (Lim and Lee 2008):

$$\begin{aligned} \sum_{i=0}^{k-1} y(k+1) &= y(k) + \sum_{i=0}^{k-1} y(k) - \Gamma(\sigma(u(k)) - u(k)) \\ u(k) &= -k_p y(k) - k_I \sum_{i=0}^{k-1} y(i) - k_D \Delta y(k) \end{aligned} \tag{63}$$

Due to $\sum_{i=0}^{k-1} y(k+1) - \sum_{i=0}^{k-1} y(k) = y(k)$ with $y(0) = 0$ subtracting the correction signal $\Gamma(\sigma(u(k)) - u(k))$ as done in the continuous time case explained in the previous section (Cao et al. 2002). Where k_p, k_I, k_D are diagonal matrices for the proportional, integral and derivative parts of the PID controller, Γ is a positive definite matrix and $\Delta y(k) = y(k) - y(k-1)$. In order to design the PID controller the following augmented variables are introduced to represent (62) and (63)

$$\begin{aligned} x_a(k) &= \begin{bmatrix} x(k) \\ \sum_{i=0}^{k-1} y(k) \\ y(k-1) \end{bmatrix} \\ y_a(k) &= \begin{bmatrix} y(k) \\ \sum_{i=0}^{k-1} y(k) \\ \Delta y(k) \end{bmatrix} \end{aligned} \tag{64}$$

Then the augmented system is (Lim and Lee 2008):

$$\begin{bmatrix} x(k+1) \\ \sum_{i=0}^{k-1} y(k+1) \\ y(k) \end{bmatrix} = \begin{bmatrix} A & 0 & 0 \\ C & I & 0 \\ C & 0 & 0 \end{bmatrix} \begin{bmatrix} x(k) \\ \sum_{i=0}^{k-1} y(k) \\ y(k-1) \end{bmatrix} + \begin{bmatrix} -B \\ -\Gamma \\ 0 \end{bmatrix} \sigma(u(k)) + \begin{bmatrix} 0 \\ \Gamma \\ 0 \end{bmatrix} u(k) \tag{65}$$

$$y_a(k) = \begin{bmatrix} C & 0 & 0 \\ 0 & I & 0 \\ C & 0 & -I \end{bmatrix} x_a(k) \tag{66}$$

where

$$\begin{aligned}
A_{cl} &= \begin{bmatrix} A & 0 & 0 \\ C & I & 0 \\ C & 0 & 0 \end{bmatrix} \\
B_{cl} &= \begin{bmatrix} -B \\ -\Gamma \\ 0 \end{bmatrix} \\
C_{cl} &= \begin{bmatrix} C & 0 & 0 \\ 0 & I & 0 \\ C & 0 & -I \end{bmatrix} \\
D_{cl} &= \begin{bmatrix} 0 \\ \Gamma \\ 0 \end{bmatrix} \\
F &= \begin{bmatrix} -k_p & -k_I & -k_D \end{bmatrix}
\end{aligned} \tag{67}$$

The saturation nonlinearity can be modeled by the following definition (Li-Sheng et al. 2004; Zongli and Liang 2006; Shuping and Boukas 2009).

Definition 1 Let $F, H \in \Re^{m \times n}$ be given. For $x \in \Re^n$, if $\|Hy_a\|_\infty \leq 1$ then $\sigma(Fy_a) \in co\left\{E_j F y_a + E_j^- H y_a : j \in [1, 2]\right\}$ where $co\{\cdot\}$ denotes the convex hull and $E_j^- = I - E_j$ where E_j is the set of $m \times m$ diagonal matrices where all their elements are 1 or 0. With Definition 1 (65) can be transformed into:

$$x_a(k+1) = \Phi x_a(k) \tag{68}$$

where:

$$\Phi = A_{cl} + B_{cl} E_j F C_{cl} + B_{cl} E_j^- H C_{cl} + D_{cl} F C_{cl} \tag{69}$$

Considering definition 1 and the following Lyapunov function:

$$V(k) = x_a^T(k) P x_a(k) \tag{70}$$

where P is a positive definite function. The derivative of (70) is given by:

$$\begin{aligned}
\Delta V(k) &= V(k+1) - V(k) \\
\Delta V(k) &= x_a^T(k) \Phi^T P \Phi x_a(k) - x_a^T(k) P x_a(k)
\end{aligned} \tag{71}$$

where in LMI equivalent is given by

$$x_a^T(k)\Phi^T P\Phi x_a(k) - x_a^T(k)Px_a(k) < 0 \tag{72}$$

For the static output feedback problem the following LMI must be solved for P and F (Mayer et al. 2013)

$$\begin{bmatrix} P^{-1} & \Phi \\ \Phi^T & P \end{bmatrix} > 0 \tag{73}$$

with $P^{-1} > 0$ and for the H_∞ controller synthesis the following LMI must be solved considering

$$\|T_{2\omega}(z)\|_\infty < \gamma \tag{74}$$

where γ is a robustness parameter that indicates the disturbance rejection of the system and $T_{2\omega}(z)$ is the discrete time transfer function of the closed loop system (Kwan Ho et al. 2006). Then by solving for F and P in the following LMI the H_∞ static output feedback PID controller can be obtained.

$$\begin{bmatrix} P & 0 & \Phi^T P & C_{cl}^T \\ 0 & \gamma I & 0 & 0 \\ P\Phi & 0 & P & 0 \\ C_{cl} & 0 & 0 & \gamma I \end{bmatrix} > 0 \tag{75}$$

With these explanations the PID controller gains can be obtained by any of the SOF controllers. In the next subsection an illustrative example is shown to evince the numerical simulation of an anti windup controller for a DC motor in discrete time.

6.2 Example 4

In this subsection a DC motor is stabilized with a PID anti windup controller in MIMO form. The same DC motor model of example 3 is considered in this section, so the following discretized state space model of the DC motor is obtained with a sampling period $T = 0.1$ s

$$x(k+1) = \begin{bmatrix} 2.774 & 0.1749 \\ -1.993 & 0.9169 \end{bmatrix} x(k) + \begin{bmatrix} 0.1749 & 0.1609 \\ -0.08306 & 2.153 \end{bmatrix} \sigma(u(k)) \tag{76}$$

$$y(k) = \begin{bmatrix} 1 & 0 \\ 0 & 1 \end{bmatrix} x(k) \tag{77}$$

where $x(k) = [x_1(k), x_2(k)]^T$ and $x_1(k)$ is the angular velocity and $x_2(k)$ is the armature current $u(k)$ is the input voltage.

For the discrete time SOF the following values of F and the gain matrices k_p, k_I and k_D are obtained by solving the LMI (73) with a Γ value of

$$\Gamma = \begin{bmatrix} 0.002 & 0 \\ 0 & 0.002 \end{bmatrix} \tag{78}$$

$$F = \begin{bmatrix} -3.1623 & 0 & -3.1623 & 0 & -3.1623 & 0 \\ 0 & -3.1623 & 0 & -3.1623 & 0 & -3.1623 \end{bmatrix} \times 10^8 \tag{79}$$

and the following PID controller gain matrices are:

$$\begin{aligned} k_p &= \begin{bmatrix} -3.1623 & 0 \\ 0 & -3.1623 \end{bmatrix} \times 10^8 \\ k_D &= \begin{bmatrix} -1 & 0 \\ 0 & -1 \end{bmatrix} \times 10^8 \\ k_I &= \begin{bmatrix} -50 & 0 \\ 0 & -50 \end{bmatrix} \times 10^8 \end{aligned} \tag{80}$$

The same PID controller gains are implemented for the system when there is no anti windup compensation. In the case of the PID anti windup controller by H_∞ synthesis the following matrix F is obtained by solving the LMI shown in (75)

$$F = \begin{bmatrix} -3.1604 & 0 & -3.1604 & 0 & -3.1604 & 0 \\ 0 & -3.1604 & 0 & -3.1604 & 0 & -3.1604 \end{bmatrix} \times 10^8 \tag{81}$$

$$\begin{aligned} k_p &= \begin{bmatrix} -3.1604 & 0 \\ 0 & -3.1604 \end{bmatrix} \times 10^8 \\ k_D &= \begin{bmatrix} -1.5802 & 0 \\ 0 & -1.5802 \end{bmatrix} \times 10^8 \\ k_I &= \begin{bmatrix} -1.5802 & 0 \\ 0 & -1.5802 \end{bmatrix} \times 10^8 \end{aligned} \tag{82}$$

With these results, a numerical simulation of the DC motor with anti windup and no anti windup compensation was done with the PID anti windup controller gain matrices achieving the following outcome.

In Fig. 23 the respective angular velocities when a anti windup and no anti windup controllers are implemented in the feedback loop of the DC motor, the system is stabilized at the nominal speed 1,750 RPM (183.26 rad/s) when a

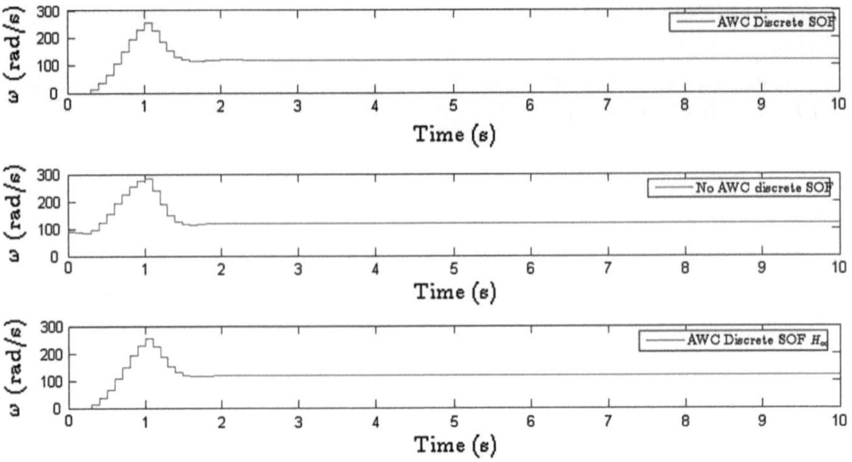

Fig. 23 Angular velocities of the DC motor

disturbance torque of 10 Nm is applied to the motor. It is verified that in the case when a SOF and H_∞ SOF the performance of the system is better than when no anti windup controller is implemented. The overshoot is smaller and a small settling time is obtained in the first mentioned cases, in comparison when only a PID controller, with no AW compensation, is implemented. This fact occurs due to the better controller and compensation characteristics when a SOF and H_∞ SOF PID anti windup compensators are implemented.

In Fig. 24 the input voltages of the DC motors in the three cases are depicted, where the voltages for the SOF and H_∞ SOF yields a more regular results than when no anti windup compensation is implemented. It can be noticed also that the input voltages are greater in the first cases than when no AWC is implemented, but

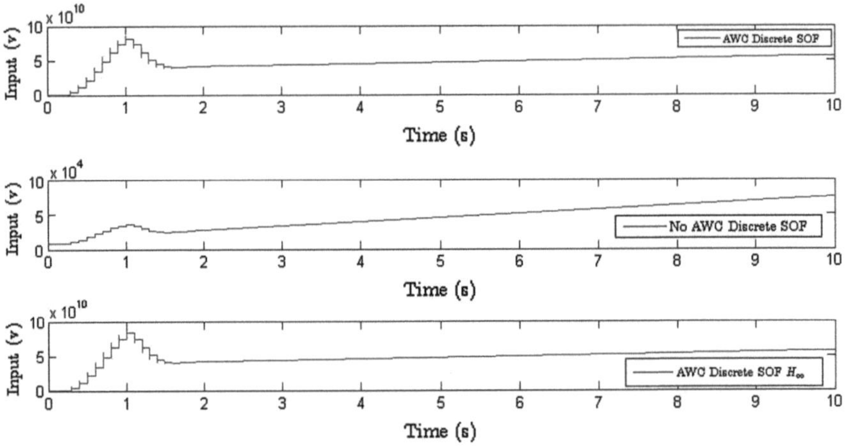

Fig. 24 Input voltage of the DC motor

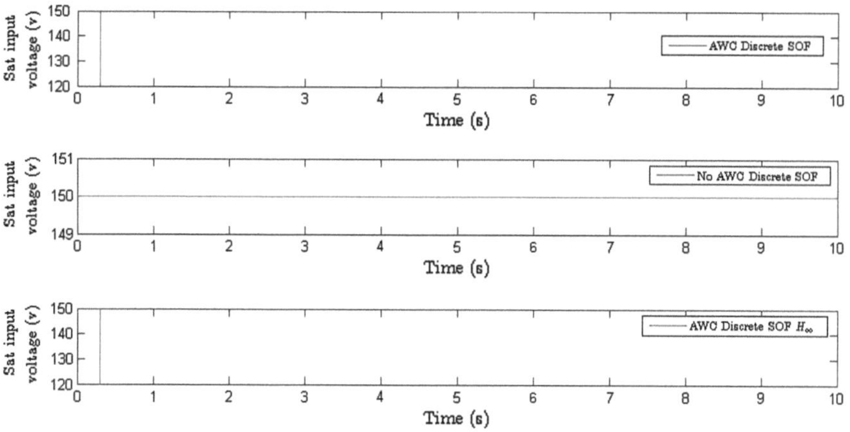

Fig. 25 DC motor saturated input voltages (v)

these values reach a steady state value in comparison with the increasing values when no AWC is implemented.

In Fig. 25 the saturated input voltages are obtained where as explained before, these values affects the system performance when the actuator, or in this case the DC motor input, is saturated due to the physical limits and properties of this model.

In Fig. 26 the armature currents of the DC motor are shown for the three cases in which the expected values are reached when a disturbance input torque is applied to the model. In Fig. 27 the mechanical torque of the model is depicted in the three cases, so as it is noticed the final value of 10 Nm, which is the value of the disturbance torque applied at 0 s is reached in the three cases but with a higher undershoot when there is no anti windup compensation.

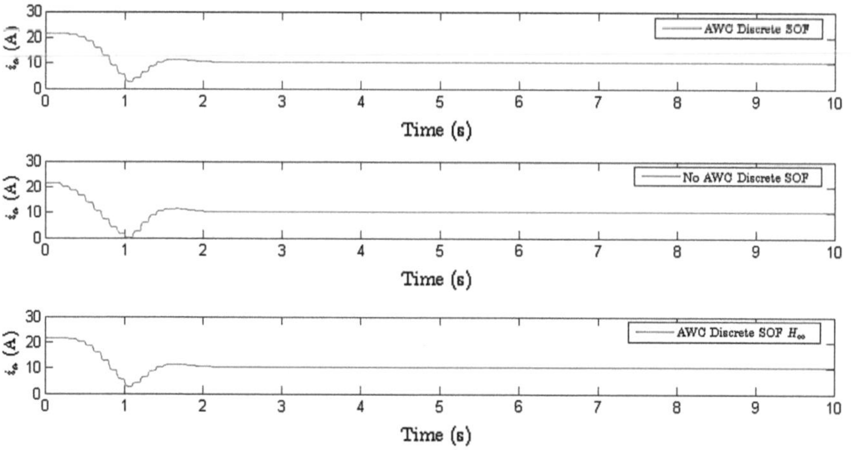

Fig. 26 DC motor armature current i_a

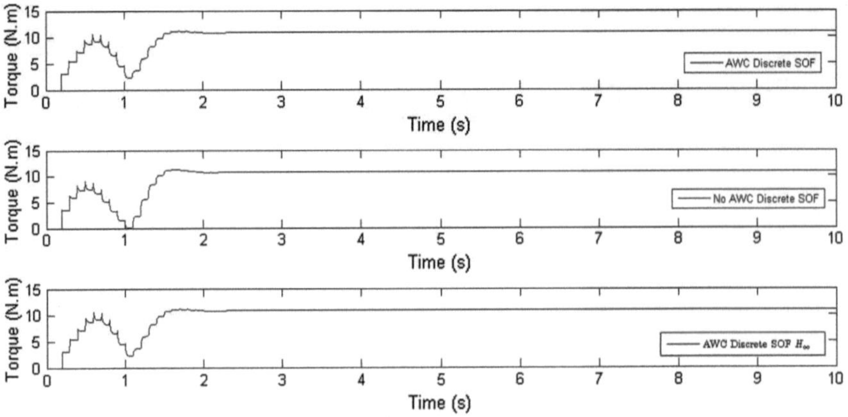

Fig. 27 DC motor torque

With the results obtained in this section, it is proved that a PID anti windup controller by SOF and H_∞ SOF can be derived for the control of MIMO systems when the inputs are saturated or constrained. In the following section a discussion of all the proposed anti windup controllers derived in this chapter are shown in order to analyze the advantages and disadvantages of these control techniques.

7 Discussion

In this chapter four anti windup controllers for SISO and MIMO continuous and discrete time systems are shown. For the first two cases, the anti windup controllers derived in the respective sections, it is shown that when the plant system is represented by a first order plus time delay model, the PID anti windup internal model controller techniques can be achieved meeting the internal stability and robustness requirements. A back calculation filter was implemented in order to suppress the integrator action when the input of the system is saturated, this compensation strategy reduces the increasing of the integrator output when the system saturates while ensuring the system stability.

It is proved that the advantages of the anti windup SISO continuous time systems is that they reduce the unwanted effects produced by windup, obtaining better results such as small overshoot and small settling time. It is proved that when a PID controller is implemented with no anti windup back calculation the performance of the system deteriorates due to the windup phenomena.

In the case of the discrete time SISO systems, a similar approach such as the SISO continuous time system where a back calculation filter is implemented to improve and avoid the deterioration of the system performance. The advantages of this control strategies is that the overshoot, settling time and other characteristics of the system performances are improved due to the filter included in the additional loop reduces the integral action when the input of the system saturates. Similar to the continuous time counterpart, an internal model controller (IMC) is implemented with an additional loop which includes a filter as a compensator while the internal stability and robustness of the system are ensured to guarantee an appropriate system performance.

In the case of the design of an anti windup controller for continuous and discrete time MIMO systems, an anti windup compensation approach is implemented where the saturated and non saturated signals are added by a feedback loop to the PID controller to eliminate the windup phenomena when the input of the system is saturated, as occurs in SISO systems. For continuous and discrete time systems, an anti windup PID controller by static output feedback was implemented, converting the PID controller into an static output feedback law and adding the difference of the saturated and non saturated input to compensate the windup effects. The advantages of the PID anti windup controller for continuous time systems, is that the system performance is not deteriorated by the influence of windup when the input is constrained or saturated, improving the controller action and preventing a high overshoot, settling time and other performance properties. In the case of the anti windup control of discrete time systems, the same advantages and properties are proved theoretically similar as the continuous time case; the system performance is improved by the addition of the difference of the saturated and no saturated signal that improves the system performance avoiding the deterioration of the system output by decreasing the integral action when the input of the system are constrained or saturated.

8 Conclusions

In this chapter some anti windup control strategies for SISO and MIMO system for discrete and continuous time models are shown. In the SISO cases an internal model anti windup PID controller is implemented by a back calculation algorithm in order to suppress the unwanted effects yielded by windup, when the system input is saturated and the integral action of the PID controller is increased. It was proved that the system performance is improved by the implementation of the back calculation loop that includes a compensator filter. The performance of the system with anti windup and no anti windup compensation was tested, and it was proved that in the first case the system performance is not deteriorated producing a smaller overshoot and settling time reducing the integral action in the discrete and continuous time SISO systems.

In the MIMO cases, a continuous and discrete time anti windup PID controllers are implemented in order to eliminates the performance deterioration caused by the integral action of the controller when the input of the system is saturated. It is proved that in the discrete and continuous time cases, the PID controllers can be achieved by a static output feedback control law (SOF) and by a H_∞ controller synthesis ensuring the system stability and robustness properties of the closed loop system. The PID controller gain matrices, in both cases, are found by solving the respective linear matrix inequalities LMI's in order to obtain the required gains to stabilize the system. In both cases it is proved that the system performance is not deteriorated by the windup phenomena when the input of the system is constrained or saturated, then in comparison when only a PID controller is implemented, the settling time and overshoot are smaller due to the anti windup characteristics of the PID controller.

Appendix 1

In this appendix the internal model PID controller, explained in Sect. 3 the gain and time constants are found with the following equations.

Define:

$$D(s) = ((\lambda s + 1)^r - P_{1A}(s))/s \tag{83}$$

and

$$K_p = p_{1m}(0) \tag{84}$$

Then the following gain and time constants are obtained using (15) with the following equations of the function $f(s)$ and its derivatives (Lee et al. 1998):

$$f(0) = \frac{1}{K_p D(0)} \tag{85}$$

where $D(0)$ is

$$D(0) = r\lambda - \dot{P}_{1A}(0) \tag{86}$$

the derivatives of $D(0)$, $\dot{D}(0)$ and $\ddot{D}(0)$ are shown in (Lee et al. 1998). Then the derivative of $f(0)$ is given by:

$$\begin{aligned} \dot{f}(0) = {} & \left(\frac{K(\alpha + \beta \Delta_\phi) a_1}{a_0 - \alpha - \beta \Delta_\phi} - \frac{K(\alpha + \beta \Delta_\phi) a_0 a_1}{(a_0 - \alpha - \beta \Delta_\phi)^2} - \frac{K(\alpha + \beta \Delta_\phi) a_0 \tau}{a_0 - \alpha - \beta \Delta_\phi} \right) (r\lambda + \theta)^{-1} K_p^{-2} \\ & + \frac{1/2 r(r-1)\lambda^2 - 1/2\theta^2}{K_p (r\lambda + \theta)^2} \end{aligned} \tag{87}$$

$$\ddot{f}(0) = \dot{f}(0) \left(\left(\frac{\ddot{p}_{1m}(0) D(0) + 2\dot{p}_{1m}(0) \dot{D}(0) + K_p \ddot{D}(0)}{\dot{p}_{1m}(0) D(0) + K_p \dot{D}(0)} \right) + 2\dot{f}(0)/f(0) \right) \tag{88}$$

Appendix 2

In this appendix the internal model PID controller, explained in Sect. 4 the gain and time constant are found and shown in the following equations. Consider the following representation in Taylor series of the digital PID controller (31) based on the analog controller design shown in (Lee et al. 1998)

$$G_c(s) = \frac{f(z)}{z-1} = \frac{1}{z-1}\left(f(1) + f'(1)(z-1) + \frac{f''(1)}{2}(z-1)^2 + \cdots\right) \tag{89}$$

Due to $G_c(s) = \frac{f(z)}{z-1}$ the following equation can be considered:

$$D(z) = \frac{(z-\alpha) - P^*_{\gamma A}(1-\alpha)z}{z-1} \tag{90}$$

because of (30) can be represented by:

$$G_c(z) = \frac{(1-\alpha)zP^{*-1}_{\gamma M}}{(z-\alpha) - P^*_{\gamma A}(1-\alpha)z} \tag{91}$$

The design procedure of the discrete time SISO controller is similar to the continuous time SISO case, (Lee et al. 1998) where (90) can be represented by:

$$D(z) = \frac{N(z)}{z-1} \tag{92}$$

where

$$N(z) = (z-\alpha) - P^*_{\gamma A}(1-\alpha)z \tag{93}$$

Then by the Taylor series expansion of $D(z)$ the following equation is obtained:

$$D(z) = \frac{1}{z-1}\left(N(1) + N'(1)(z-1) + \frac{N''(1)}{2}(z-1)^2 + \frac{N'''(1)}{6}(z-1)^3 + \cdots\right) \tag{94}$$

Considering that $N(1) = 0$, (94) becomes in:

$$D(z) = N'(1) + \frac{N''(1)}{2}(z-1) + \frac{N'''(1)}{6}(z-1)^2 + \cdots \tag{95}$$

Expanding $D(z)$ in Taylor series expansion as an only term, the following result is obtained:

$$D(z) = D(1) + D'(1)(z-1) + \frac{D''(1)}{2}(z-1)^2 + \cdots \tag{96}$$

Then associating the similar terms of (95) and (96) the following values for $D(1)$ and its derivatives are obtained:

$$\begin{aligned} D(1) &= N'(1) \\ D'(1) &= N''(1)/2 \\ D''(1) &= N'''(1)/3 \end{aligned} \tag{97}$$

the values of $D(1)$ and its derivatives can be found by:

$$D(1) = 1 + (N-1)(1-\alpha) \tag{98}$$

$$D'(1) = (-N(N-1)(1-\alpha))/2 \tag{99}$$

$$D''(1) = ((N+1)N(N-1)(1-\alpha))/3 \tag{100}$$

Then the values for $f(1)$ and its derivatives are found by (Lee et al. 1998):

$$f(1) = \frac{1}{(p^*_{\gamma M}(1)/(1-\alpha))D(1)} \tag{101}$$

$$f'(1) = -\frac{(p'^*_{\gamma M}(1)/(1-\alpha))D(1) + (p^*_{\gamma M}(1)/(1-\alpha))D'(1)}{((p^*_{\gamma M}(1)/(1-\alpha))D(1))^2} \tag{102}$$

$$f''(1) = f'(1)\left(\frac{(p''^*_{\gamma M}(1)/(1-\alpha))D(1) + 2(p'^*_{\gamma M}(1)/(1-\alpha))D'(1) + (p^*_{\gamma M}(1)/(1-\alpha))D''(1)}{(p'^*_{\gamma M}(1)/(1-\alpha))D(1) + (p^*_{\gamma M}(1)/(1-\alpha))D'(1)}\right) + 2f'^2(1)/f(1) \tag{103}$$

With $f(1)$ and its respective derivatives, the parameters of the digital PID controllers can be found using (32).

References

Baheti, R. S. (1989). Simple anti-windup controllers. In *American Control Conference*, (pp. 1684–1686) June 21–23, 1989.

Bateman, A., & Zongli, L. (2002). An analysis and design method for discrete-time linear systems under nested saturation. *IEEE Transactions on Automatic Control, 47*(8), 1305–1310.

Bohn, C., & Atherton, D. P. (1995). An analysis package comparing PID anti-windup strategies. *Control Systems Magazine, IEEE, 15*(2), 34–40.

Cao, Y.-Y., Lin, Z., & Ward, D. (2002). An antiwindup approach to enlarging domain of attraction for linear systems subject to actuator saturation. *IEEE Transactions on Automatic Control, 47* (1), 140–145.

Chen, Y. S., Tsai, J. H., Shieh, L., & Moussighi, M. (2003). Digital redesign of anti-wind-up controller for cascaded analog system. *ISA Transactions, 42*(1), 73–88.

Cockbum, J. C., & Bailey, F. N. (1991). Loop gain-phase shaping design of SISO robust controllers having mixed uncertainty. In *American Control Conference*, (pp. 1981–1986).

Doyle III, F. J. (1999). An anti-windup input/output linearization scheme for SISO systems. *Journal of Process Control, 9*(3), 213–220.

Fujimori, A. (2004). Optimization of static output feedback using substitutive LMI formulation. *IEEE Transactions on Automatic Control, 49*(6), 995–999.

He, Y., & Wang, Q.-G. (2006). An improved ILMI method for static output feedback control with application to multivariable PID control. *IEEE Transactions on Automatic Control, 51*(10), 1678–1683.

Henrion, D., Tarbouriech, S., & Garcia, G. (1999). Output feedback robust stabilization of uncertain linear systems with saturating controls: An LMI approach. *IEEE Transactions on Automatic Control, 44*(11), 2230–2237.

Kwan Ho, L., Joon-Hwa, L., & Wook-Hyun, K. (2006). Sufficient LMI conditions for H infinity output feedback stabilization of linear discrete-time systems. *IEEE Transactions on Automatic Control, 51*(4), 675–680.

Lambeck, S., & Sawodny, O. (2004). Design of anti-windup-extensions for digital control loops. *Proceedings of the* 2004 *American Control Conference*, (pp. 5309–5314). 2004.

Lee, Y., Park, S., Lee, M., & Coleman, B. (1998). PID controller tuning for desired closed-loop responses for SI/SO systems. *AIChE Journal, 44*(1), 106–115.

Lim, J. S., & Lee, Y. I. (2008). Design of discrete-time multivariable PID controllers via LMI approach. *International Conference on Control, Automation and Systems (ICCAS 2008)*, (pp. 1867–1871).

Li-Sheng, H., Biao, H., & Yong-Yan, C. (2004). Robust digital model predictive control for linear uncertain systems with saturations. *IEEE Transactions on Automatic Control, 49*(5), 792–796.

Matsuda, Y., & Ohse, N. (2006). An approach to synthesis of low order dynamic anti-windup compensators for multivariable PID control systems with input saturation. *International Joint Conference SICE-ICASE* 2006, (pp. 988–993). doi:10.1109/SICE.2006.315736.

Mayer, S., Dehnert, R., & Tibken, B. (2013). Controller synthesis of multi dimensional, discrete LTI systems based on numerical solutions of linear matrix inequalities. American *Control Conference (ACC)* (pp. 2386–2391). June 17–19, 2013.

Morales, R. M., Heath, W. P., & Li, G. (2009). Robust anti-windup against LTI uncertainty using frequency dependent IQCs. ICCAS-SICE (pp. 3329–3334), August 18–21, 2009.

Morari, M., & Zafiriou, E. (1989). *Robust Process Control*. New Jersey: Prentice Hall.

Neto, A. T., & Kucera, V. (1991). Stabilization via static output feedback. In: *Proceedings of the* 30*th IEEE Conference on Decision and Control* 1991 (pp. 910–913). doi:10.1109/CDC.1991.261451.

Rehan, M., Khan, A. Q., Abid, M., & Iqbal, N. (2013). Anti-windup-based dynamic controller synthesis for nonlinear systems under input saturation. *Applied Mathematics and Computation, 220*(1), 382–393.

Saeki, M., & Wada, N. (1996). Design of anti-windup controller based on matrix inequalities. In: *Proceedings of the 35th IEEE Conference on Decision and Control* 1996 (pp. 261–262). doi:10.1109/CDC.1996.574310.

Shamsuzzoha, M., & Lee, M. (2007). IMC-PID controller design for improved disturbance rejection of time-delayed processes. *Industrial and Engineering Chemistry Research, 46*(7), 712–749.

Shuping, M., & Boukas, E. K. (2009). Stability and H infinity control for discrete-time singular systems subject to actuator saturation. In: *American Control Conference ACC* '09, (pp. 1244–1249). doi:10.1109/ACC.2009.5159906.

Taylor, J. H., & O'Donnell, J. R. (1990). Synthesis of nonlinear controllers with rate feedback via sinusoidal-input describing function methods. In: *American Control Conference* (pp. 2217–2222).

Tu, Y.-W., & Ho, M.-T. (2011). Synthesis of low-order anti-windup compensators for PID control. In: 2011 8th *Asian Control Conference (ASCC)*, (pp. 1437–1442).

Wittenmark, B. (1989). Integrators, nonlinearities, and anti-reset windup for different control structures. In: *American Control Conference*, (pp. 1679–1683).

Wu, F., Lin, Z., & Zheng, Q. (2005). Output feedback stabilization of linear systems with actuator saturation. *Proceedings of the 2005 American Control Conference*, (pp. 3385–3390). doi:10.1109/ACC.2005.1470494.

Zhang, J., Wu, J., Zhang, Y., & Hopwood, F. (2011). Design and applications of an optimal anti-windup digital controller using scalar sign function approach. In *IEEE International Conference on Control Applications* (*CCA*) (pp. 94–101). doi:10.1109/CCA.2011.6044418.

Zongli, L., & Liang, L. (2006). Set invariance conditions for singular linear systems subject to actuator saturation. In *Chinese Control Conference* (*CCC*) (pp. 2070–2074). doi:10.1109/CHICC.2006.280919.

A Hybrid Global Optimization Algorithm: Particle Swarm Optimization in Association with a Genetic Algorithm

M. Andalib Sahnehsaraei, M.J. Mahmoodabadi, M. Taherkhorsandi, K.K. Castillo-Villar and S.M. Mortazavi Yazdi

Abstract The genetic algorithm (GA) is an evolutionary optimization algorithm operating based upon reproduction, crossover and mutation. On the other hand, particle swarm optimization (PSO) is a swarm intelligence algorithm functioning by means of inertia weight, learning factors and the mutation probability based upon fuzzy rules. In this paper, particle swarm optimization in association with genetic algorithm optimization is utilized to gain the unique benefits of each optimization algorithm. Therefore, the proposed hybrid algorithm makes use of the functions and operations of both algorithms such as mutation, traditional or classical crossover, multiple-crossover and the PSO formula. Selection of these operators is based on a fuzzy probability. The performance of the hybrid algorithm in the case of solving both single-objective and multi-objective optimization problems is evaluated by utilizing challenging prominent benchmark problems including FON, ZDT1, ZDT2, ZDT3, Sphere, Schwefel 2.22, Schwefel 1.2, Rosenbrock, Noise, Step, Rastrigin, Griewank, Ackley and especially the design of the parameters of linear feedback control for a parallel-double-inverted pendulum system which is a complicated, nonlinear and unstable system. Obtained numerical results in comparison to the outcomes of other optimization algorithms in the literature demonstrate the

M. Andalib Sahnchsaraci
Department of Mechanical Engineering, Faculty of Engineering,
The University of Guilan, Rasht, Iran
e-mail: morteza_andalib@yahoo.com

M.J. Mahmoodabadi · S.M. Mortazavi Yazdi
Department of Mechanical Engineering, Sirjan University of Technology, Sirjan, Iran
e-mail: mahmoodabadi@sirjantech.ac.ir

S.M. Mortazavi Yazdi
e-mail: m.mechanic88@yahoo.com

M. Taherkhorsandi (✉) · K.K. Castillo-Villar
Department of Mechanical Engineering, The University of Texas at San Antonio,
San Antonio, TX 78249, USA
e-mail: m.taherkhorsandi@gmail.com; milad.taherkhorsandi@utsa.edu

K.K. Castillo-Villar
e-mail: krystel.castillo@utsa.edu

Q. Zhu and A.T. Azar (eds.), *Complex System Modelling and Control Through Intelligent Soft Computations*, Studies in Fuzziness and Soft Computing 319,
DOI 10.1007/978-3-319-12883-2_2

efficiency of the hybrid of particle swarm optimization and genetic algorithm optimization with regard to addressing both single-objective and multi-objective optimization problems.

Keywords Particle swarm optimization · Genetic algorithm optimization · Single-objective problems · Multi-objective problems · State feedback control · Parallel-double-inverted pendulum system

1 Introduction

Optimization is the selection of the best element from some sets of variables with a long history dating back to the years when Euclid conducted research to gain the minimum distance between a point and a line. Today, optimization has an extensive application in different branches of science, e.g. aerodynamics (Song et al. 2012), robotics (Li et al. 2013; Cordella et al. 2012), energy consumption (Wang et al. 2014), supply chain modeling (Yang et al. 2014; Castillo-Villar et al. 2014) and control (Mahmoodabadi et al. 2014a; Wang and Liu 2012; Wibowo and Jeong 2013). Due to the necessity of addressing a variety of constrained and unconstrained optimization problems, many changes and novelties in optimization approaches and techniques have been proposed during the recent decade. In general, optimization algorithms are divided into two main classifications: deterministic and stochastic algorithms (Blake 1989). Due to employing the methods of successive search based upon the derivative of objective functions, deterministic optimization algorithms are appropriate for convex, continuous and differentiable objective functions. On the other hand, stochastic optimization techniques are applicable to address most of real optimization problems, which are heavily nonlinear, complex and non-differentiable. In this regard, a great number of studies have recently been devoted to stochastic optimization algorithms, especially, genetic algorithm optimization and particle swarm optimization.

The genetic algorithm, which is a subclass of evolutionary algorithms, is an optimization technique inspired by natural evolution, that is, mutation, inheritance, selection and crossover to gain optimal solutions. Lately, it was enhanced by using a novel multi-parent crossover and a diversity operator instead of mutation in order to gain quick convergence (Elsayed et al. 2014), utilizing it in conjunction with several features selection techniques, involving principle components analysis, sequential floating, and correlation-based feature selection (Aziz et al. 2013), using the controlled elitism and dynamic crowding distance to present a general algorithm for the multi-objective optimization of wind turbines (Wang et al. 2011), and utilizing a real encoded crossover and mutation operator to gain the near global optimal solution of multimodal nonlinear optimization problems (Thakur 2014). Particle swarm optimization is a population-based optimization algorithm mimicking the behavior of social species such as flocking birds, swimming wasps, school fish, among others. Recently, its performance was enhanced by using a multi-stage clustering procedure splitting the particles of the main swarm over a number of sub-swarms based upon the values of objective functions and the particles positions (Nickabadi et al. 2012), utilizing multiple ranking criteria to define three global bests of the swarm as well as employing fuzzy variables to evaluate the objective function and constraints of the problem (Wang and Zheng 2012), employing an innovative method to choose the global and personal best positions to enhance the rate of convergence and diversity of solutions (Mahmoodabadi et al. 2014b), and using a self-clustering algorithm to divide the particle swarm into multiple tribes and choosing appropriate evolution techniques to update each particle (Chen and Liao 2014).

Lately, researchers have utilized hybrid optimization algorithms to provide more robust optimization algorithms due to the fact that each algorithm has its own advantages and drawbacks and it is not feasible that an optimization algorithm can address all optimization problems. Particularly, Ahmadi et al. (2013) predicted the power in the solar stirling heat engine by using neural network based on the hybrid of genetic algorithm and particle swarm optimization. Elshazly et al. (2013) proposed a hybrid system which integrates rough set and the genetic algorithm for the efficient classification of medical data sets of different sizes and dimensionalities. Abdel-Kader (2010) proposed an improved PSO algorithm for efficient data clustering. Altun (2013) utilized a combination of genetic algorithm, particle swarm optimization and neural network for the palm-print recognition. Zhou et al. (2012) designed a remanufacturing closed-loop supply chain network based on the genetic particle swarm optimization algorithm. Jeong et al. (2009) developed a hybrid algorithm based on genetic algorithm and particle swarm optimization and applied it for a real-world optimization problem. Mavaddaty and Ebrahimzadeh (2011) used the genetic algorithm and particle swarm optimization based on mutual information for blind signals separation. Samarghandi and ElMekkawy (2012) applied the genetic algorithm and particle swarm optimization for no-wait flow shop problem with separable setup times and make-span criterion. Deb and Padhye (2013) enhanced the performance of particle swarm optimization through an algorithmic link with genetic algorithms. Valdez et al. (2009) combined particle swarm optimization and genetic algorithms using fuzzy logic for decision making. Premalatha and Natarajan (2009) applied discrete particle swarm optimization with genetic algorithm operators for document clustering. Dhadwal et al. (2014) advanced particle swarm assisted genetic algorithm for constrained optimization problems. Bhuvaneswari et al. (2009) combined the genetic algorithm and particle swarm optimization for alternator design. Jamili et al. (2011) proposed a hybrid algorithm based on particle swarm optimization and simulated annealing for a periodic job shop scheduling problem. Joeng et al. (2009) proposed a sophisticated hybrid of particle swarm optimization and the genetic algorithm which shows robust search ability regardless of the selection of the initial population and compared its capability to a simple hybrid of particle swarm optimization and the genetic algorithm and pure particle swarm optimization and pure the genetic algorithm. Castillo-Villar et al. (2012) used genetic algorithm optimization and simulated annealing for a model of supply-chain design regarding the cost of quality and the traditional manufacturing and distribution costs. Talatahari and Kaveh (2007) used a hybrid of particle swarm optimization and ant colony optimization for the design of frame structures. Thangaraj et al. (2011) presented a comprehensive list of hybrid algorithms of particle swarm optimization with other evolutionary algorithms e.g. the genetic algorithm. Valdez et al. (2011) utilized the fuzzy logic to combine the results of the particle swarm optimization and the genetic algorithm. This method has been performed on some single-objective test functions for four different dimensions contrasted to the genetic algorithm and particle swarm optimization, separately.

For the optimum design of traditional controllers, the evolutionary optimization techniques are appropriate approaches to be used. To this end, Fleming and Purshouse (2002) is an appropriate reference to overview the application of the evolutionary algorithms in the field of the design of controllers. In particular, the design of controllers in Fonseca and Fleming (1994) and Sanchez et al. (2007) was formulated as a multi-objective optimization problem and solved using genetic algorithms. Furthermore, in Ker-Wei and Shang-Chang (2006), the sliding mode control configurations were designed for an alternating current servo motor while a particle swarm optimization algorithm was used to select the parameters of the controller. Also, PSO was applied to tune the linear control gains in Gaing (2004) and Qiao et al. (2006). In Chen et al. (2009), three parameters associated with the control law of the sliding mode controller for the inverted pendulum system were properly chosen by a modified PSO algorithm. Wai et al. (2007) proposed a total sliding-model-based particle swarm optimization to design a controller for the linear induction motor. More recently, in Gosh et al. (2011), an ecologically inspired direct search method was applied to solve the optimal control problems with Bezier parameterization. Moreover, in Tang et al. (2011), a controllable probabilistic particle swarm optimization (CPPSO) algorithm was applied to design a memory-less feedback controller. McGookin et al. (2000) optimized a tanker autopilot control system using genetic algorithms. Gaing (2004) used particle swarm optimization to tune linear gains of the proportional-integral-derivative (PID) controller for an AVR system. Qiao et al. (2006) tuned the proportional-integral (PI) controller parameters for doubly fed induction generators driven by wind turbines using PSO. Zhao and Yi (2006) proposed a GA-based control method to swing up an acrobot with limited torque. Wang et al. (2006) designed a PI/PD controller for the non-minimum phase system and used PSO to tune the controller gains. Sanchez et al. (2007) formulated a classical observer-based feedback controller as a multi-objective optimization problem and solved it using genetic algorithm. Mohammadi et al. (2011) applied an evolutionary tuning technique for a type-2 fuzzy logic controller and state feedback tracking control of a biped robot (Mahmoodabadi et al. 2014c). Zargari et al. (2012) designed a fuzzy sliding mode controller with a Kalman estimator for a small hydro-power plant based on particle swarm optimization. More recently, a two-stage hybrid optimization algorithm, which involves the combination of PSO and a pattern search based method, is used to tune a PI controller (Puri and Ghosh 2013).

In this chapter, a hybrid of particle swarm optimization and the genetic algorithm developed previously by authors (Mahmoodabadi et al. 2013) is described and used to design the parameters of the linear state feedback controller for the system of a parallel-double-inverted pendulum. In elaboration, the used operators of the hybrid algorithm include mutation, crossover of the genetic algorithm and particle swarm optimization formula. The classical crossover and the multiple-crossover are two parts of the crossover operator. A fuzzy probability is used to choose the particle swarm optimization and genetic algorithm operators at each iteration for each particle or chromosome. The optimization algorithm is based on the non-dominated sorting concept. Moreover, a leader selection method based upon particles density and a dynamic elimination method which confines the numbers of non-dominated solutions are utilized to present a high convergence and uniform spread. Single and multi-objective problems are utilized to assess the capabilities of the optimization algorithm. By using the same benchmarks, the results of simulation are contrasted to the results of other optimization algorithms. The structure of this chapter is as follows. Section 2 presents the genetic algorithm and its details including the crossover operator and the mutation operator. Particle swarm optimization and its details involving inertia weight and learning factors are provided in Sect. 3. Section 4 states the mutation probabilities at each iteration which is based on fuzzy rules. Section 5 includes the architecture, the pseudo code, the parameter settings, and the flow chart of the single-objective and multi-objective hybrid optimization algorithms. Furthermore, the test functions and the evolutionary trajectory for the algorithms are provided in Sect. 5. State feedback control for linear systems is presented in Sect. 6. Section 7 presents the state space representation, the block diagram, and the Pareto front of optimal state feedback control of a parallel-double-inverted pendulum. Finally, conclusions are presented in Sect. 8.

2 Genetic Algorithm

The genetic algorithm inspired from Darwin's theory is a stochastic algorithm based upon the survival fittest introduced in 1975 (Holland 1975).

Genetic algorithms offer several attractive features, as follows:

- An easy-to-understand approach that can be applied to a wide range of problems with little or no modification. Other approaches have required substantial alteration to be successfully used in applications. For example, the dynamic programming was applied to select the number, location and power of the lamps along a hallway in such a way that the electrical power needed to produce the required illuminance will be minimized. In this method, significant alternation is needed since the choice of the location and power of a lamp affect the decisions made about previous lamps (Gero and Radford 1978).
- Genetic algorithm codes are publicly available which reduces set-up time.
- The inherent capability to work with complex simulation programs. Simulation does not need to be simplified to accommodate optimization.

- Proven effectiveness in solving complex problems that cannot be readily solved with other optimization methods. The mapping of the objective function for a day lighting design problem showed the existence of local minima that would potentially trap a gradient-based method (Chutarat 2001). Building optimization problems may include a mixture of a large number of integer and continuous variables, non-linear inequality and equality constraints, a discontinuous objective function and variables embedded in constraints. Such characteristics make gradient-based optimization methods inappropriate and restrict the applicability of direct-search methods (Wright and Farmani 2001). The calculation time of mixed integer programming (MIP), which was used to optimize the operation of a district heating and cooling plant, increases exponentially with the number of integer variables. It was shown that it takes about two times longer than a genetic algorithm for a 14 h optimization window and 12 times longer for a 24 h period (Sakamoto et al. 1999), although the time required by MIP was sufficiently fast for a relatively simple plant to make on-line use feasible.
- Methods to allow genetic algorithms to handle constraints that would make some solutions unattractive or entirely infeasible.

Performing on a set of solutions instead of one solution is one of notable abilities of stochastic algorithms. Thus, at first, initial population consisting of a random set of solutions is generated by the genetic algorithm. Each solution in a population is named an individual or a chromosome. The size of population (N) is the number of chromosomes in a population. The genetic algorithm has the capability of performing with coded variables. In fact, the binary coding is the most popular approach of encoding the genetic algorithm. When the initial population is generated, the genetic algorithm has to encode the whole parameters as binary digits. Hence, while performing over a set of binary solutions, the genetic algorithm must decode all the solutions to report the optimal solutions. To this end, a real-coded genetic algorithm is utilized in this study (Mahmoodabadi et al. 2013). In the real coded genetic algorithm, the solutions are applied as real values. Thus, the genetic algorithm does not have to devote a great deal of time to coding and decoding the values (Arumugam et al. 2005). Fitness which is a value assigned to each chromosome is used in the genetic algorithm to provide the ability of evaluating the new population with respect to the previous population at any iteration. To gain the fitness value of each chromosome, the same chromosome is used to obtain the value of the function which must be optimized. This function is the objective function. Three operators, that is, reproduction, crossover and mutation are employed in the genetic algorithm to generate a new population in comparison to the previous population. Each chromosome in new and previous populations is named offspring and parent, correspondingly. This process of the genetic algorithm is iterated until the stopping criterion is satisfied and the chromosome with the best fitness in the last generation is proposed as the optimal solution. In the present study, crossover and mutation are hybridized with the formula of particle swarm optimization (Mahmoodabadi et al. 2013). The details of these genetic operators are elaborated in the following sections.

2.1 The Crossover Operator

The role of crossover operator is to generate new individuals, that is, offspring from parents in the mating pool. Afterward, two offspring are generated based upon the selected parents and will be put in the place of the parents. Moreover, this operator is used for a number of pair of parents to mate (Chang 2007). This number is calculated by using the formula as $\frac{P_{tc} \times N}{2}$, where P_{tc} and N denote the probability of traditional crossover and population size, correspondingly. By regarding $\overrightarrow{x_i}(t)$ and $\overrightarrow{x_j}(t)$ as two random selected chromosomes in such a way that $\overrightarrow{x_i}(t)$ has a smaller fitness value than $\overrightarrow{x_j}(t)$, the traditional crossover formula is as follows

$$\begin{aligned}\overrightarrow{x_i}(t+1) &= \overrightarrow{x_i}(t) + \gamma_1(\overrightarrow{x_i}(t) - \overrightarrow{x_j}(t)) \\ \overrightarrow{x_j}(t+1) &= \overrightarrow{x_j}(t) + \gamma_2(\overrightarrow{x_i}(t) - \overrightarrow{x_j}(t))\end{aligned} \tag{1}$$

where γ_1 and $\gamma_2 \in [0, 1]$ represent random values. When Eq. (1) is calculated, between $\vec{x}(t)$ and $\vec{x}(t+1)$, whichever has the fewer fitness should be chosen. Another crossover operator called multiple-crossover operator is employed in this paper (Mahmoodabadi et al. 2013). This operator was presented in (Ker-Wei and Shang-Chang 2006) for the first time. The multiple-crossover operator consists of three chromosomes. The number of $\frac{P_{mc} \times N}{3}$ chromosomes is chosen for adjusting in which P_{mc} denotes the probability of multiple-crossover. Furthermore, $\overrightarrow{x_i}(t)$, $\overrightarrow{x_j}(t)$ and $\overrightarrow{x_k}(t)$ denote three random chosen chromosomes in which $\overrightarrow{x_i}(t)$ has the smallest fitness value among these chromosomes. Multiple-crossover is computed as follows

$$\begin{aligned}\overrightarrow{x_i}(t+1) &= \overrightarrow{x_i}(t) + \lambda_1(2\overrightarrow{x_i}(t) - \overrightarrow{x_j}(t) - \overrightarrow{x_k}(t)) \\ \overrightarrow{x_j}(t+1) &= \overrightarrow{x_j}(t) + \lambda_2(2\overrightarrow{x_i}(t) - \overrightarrow{x_j}(t) - \overrightarrow{x_k}(t)) \\ \overrightarrow{x_k}(t+1) &= \overrightarrow{x_k}(t) + \lambda_3(2\overrightarrow{x_i}(t) - \overrightarrow{x_j}(t) - \overrightarrow{x_k}(t))\end{aligned} \tag{2}$$

where λ_1, λ_2, and $\lambda_3 \in [0, 1]$ represent random values. When Eq. (2) is computed, between $\vec{x}(t)$ and $\vec{x}(t+1)$, whichever has the fewer fitness should be selected.

2.2 The Mutation Operator

According to the searching behavior of GA, falling into the local minimum points is unavoidable when the chromosomes are trying to find the global optimum solution. In fact, after several generations, chromosomes will gather in several areas or even just in one area. In this state, the population will stop progressing and it will become unable to generate new solutions. This behavior could lead to the whole population being trapped in the local minima. Here, in order to allow the chromosomes' exploration in the area to produce more potential solutions and to explore new regions of the parameter space, the mutation operator is applied. The role of this operator is to change the value of the number of chromosomes in the population. This number is calculated via $P_m \times N$, in which P_m and N are the probability of mutation and population size, correspondingly. In this regard, a variety in population and a decrease in the possibility of convergence toward local optima are gained through this operation. By regarding a randomly chosen chromosome, the mutation formula is obtained as (Mahmoodabadi et al. 2013):

$$\vec{x_i}(t+1) = \vec{x}_{\min}(t) + \upsilon(\vec{x}_{\max}(t) - \vec{x}_{\min}(t)) \tag{3}$$

in which $\vec{x_i}(t)$, $\vec{x}_{\max}(t)$ and $\vec{x}_{\min}(t)$ present the randomly chosen chromosome, upper bound and lower bound with regard to search domain, correspondingly and $\upsilon \in [0, 1]$ is a random value. When Eq. (3) is calculated, between $\vec{x}(t)$ and $\vec{x}(t+1)$, whichever has the fewer fitness should be chosen.

The second optimization algorithm used for the hybrid algorithm is particle swarm optimization and this algorithm and its details involving the inertia weight and learning factors will be presented in the following section.

3 Particle Swarm Optimization (PSO)

Particle swarm optimization introduced by Kennedy and Eberhart (1995) is a population-based search algorithm based upon the simulation of the social behavior of flocks of birds. While this algorithm was first used to balance the weights in neural networks (Eberhart et al. 1996), it is now a very popular global optimization algorithm for problems where the decision variables are real numbers (Engelbrecht 2002, 2005).

In particle swarm optimization, particles are flying through hyper-dimensional search space and the changes in their way are based upon the social-psychological tendency of individuals to mimic the success of other individuals. Here, the PSO operator adjusted the value of positions of particles which are not chosen for genetic operators (Mahmoodabadi et al. 2013). In fact, the positions of these particles are adjusted based upon their own and neighbors' experience. $\vec{x_i}(t)$ represents the position of a particle and it is adjusted through adding a velocity $\vec{v_i}(t)$ to it, that is to say:

$$\vec{x_i}(t+1) = \vec{x_i}(t) + \vec{v_i}(t+1) \tag{4}$$

The socially exchanged information is presented by a velocity vector defined as follows:

$$\vec{v_i}(t+1) = W\vec{v_i}(t) + C_1 r_1(\vec{x}_{pbest_i} - \vec{x_i}(t)) + C_2 r_2(\vec{x}_{gbest} - \vec{x_i}(t)) \tag{5}$$

where C_1 represents the cognitive learning factor and denotes the attraction that a particle has toward its own success. C_2 is the social learning factor and represents the attraction that a particle has toward the success of the entire swarm. W is the inertia weight utilized to control the impact of the previous history of velocities on the current velocity of a given particle. $\vec{x}_{pbest_i}$ denotes the personal best position of the particle *i*. $\vec{x}_{gbest}$ represents the position of the best particle of the entire swarm. $r_1, r_2 \in [0, 1]$ are random values. Moreover, in this paper, a uniform probability distribution is assumed for all the random parameters (Mahmoodabadi et al. 2013). The trade-off between the global and local search abilities of the swarm is adjusted by using the parameter W. An appropriate value of the inertia weight balances between global and local search abilities by regarding the fact that a large inertia weight helps the global search and a small one helps the local search. Based upon experimental results, linearly decreasing the inertia weight over iterations enhances the efficiency of particle swarm optimization (Eberhart and Kennedy 1995). The particles approach to the best particle of the entire swarm ($\vec{x}_{gbest}$) via using a small value of C_1 and a large value of C_2. On the other hand, the particles converge into their personal best position ($\overrightarrow{x}_{pbest_i}$) through employing a large value of C_1 and a small value of C_2. Furthermore, it was obtained that the best solutions were gained via a linearly decreasing C_1 and a linearly enhancing C_2 over iterations (Ratnaweera et al. 2004). Thus, the following linear formulation of inertia weight and learning factors are utilized as follows:

$$W_1 = W_1 - (W_1 - W_2) \times \left(\frac{t}{maximum\ iteration}\right) \tag{6}$$

$$C_1 = C_{1i} - (C_{1i} - C_{1f}) \times \left(\frac{t}{maximum\ iteration}\right) \tag{7}$$

$$C_2 = C_{2i} - (C_{2i} - C_{2f}) \times \left(\frac{t}{maximum\ iteration}\right) \tag{8}$$

in which W_1 and W_2 represent the initial and final values of the inertia weight, correspondingly. C_{1i} and C_{2i} denote the initial values of the learning factors C_1 and C_2, correspondingly. C_{1f} and C_{2f} represent the final values of the learning factors C_1 and C_2, respectively. t is the current number of iteration and *maximum iteration* is the maximum number of allowable iterations. The mutation probabilities at each iteration which is based on fuzzy rules will be presented in the next section.

4 The Mutation Probabilities Based on Fuzzy Rules

The mutation probability at each iteration is calculated via using the following equation:

$$P_m = \zeta_m \times Limit \tag{9}$$

in which ζ_m is a positive constant. *Limit* represents the maximum number of iteration restraining the changes in positions of the particles or chromosomes of the entire swarm or population. Equations (10) and (11) are utilized to compute other probabilities, as follows:

$$P_{tc} = \xi_{tc} \times P_g \tag{10}$$

$$P_{mc} = \xi_{mc} \times P_g \tag{11}$$

in which ξ_{tc} and ξ_{mc} are positive constants. P_g denotes a fuzzy variable and its membership functions and fuzzy rules are presented in Fig. 1 and Table 1.

The inference result P_g of the consequent variable can be computed via employing the min-max-gravity method, or the simplified inference method, or the product-sum-gravity method (Mizumoto 1996).

Single objective and multi-objective hybrid algorithms based on particle swarm optimization and the genetic algorithm and the details including the pseudo code, the parameter settings will be presented in the following section. These algorithms will be evaluated with many challenging test functions.

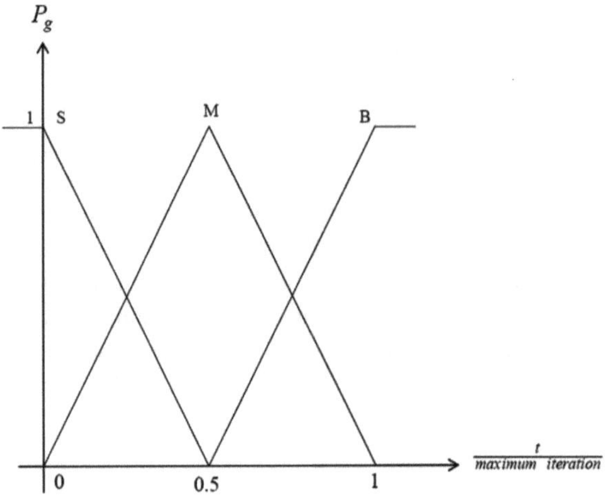

Fig. 1 Membership functions of fuzzy variable P_g

Table 1 Fuzzy rules of fuzzy variable P_g

Antecedent variable	Consequence variable
S	0.0
M	0.5
B	1.0

5 Optimization

Optimization is mathematical and numerical approaches to gain and identify the best candidate among a range of alternatives without having to explicitly enumerate and evaluate all possible alternatives (Ravindran et al. 2006). While maximum or minimum solutions of objective functions are the optimal solutions of an optimization problem, optimization algorithms are usually trying to address a minimization problem. In this regard, the goal of optimization is to gain the optimal solutions which are the points minimizing the objective functions. Based upon the number of objective functions, an optimization problem is classified as single-objective and multi-objective problems. This study uses both single-objective and multi-objective optimization algorithms to evaluate the capabilities of the hybrid of particle swarm optimization and the genetic algorithm (Mahmoodabadi et al. 2013). To this end, challenging benchmarks of the field of optimization are chosen to evaluate the optimization algorithm. The hybrid of particle swarm optimization and the genetic algorithm is applied to these benchmarks and the obtained results are compared to the obtained results of running a number of similar algorithms on the same benchmark problems.

5.1 Single-Objective Optimization

5.1.1 Definition of Single-Objective Optimization Problem

A single-objective optimization problem involves just one objective function as there are many engineering problems where designers combine several objective functions into one. Each objective function can include one or more variables. A single-objective optimization problem can be defined as follow:

$$\begin{aligned}&\text{Find } \overrightarrow{x}^{*} = [x_1^*, x_2^*, \ldots, x_n^*] \in R^n\\&\text{To minimize } f(\vec{x})\end{aligned}$$

By regarding p equality constraints $g_i(\vec{x}) = 0,\ i = 1, 2, \ldots, p$; and q inequality constraints $h_j(\vec{x}) \leq 0,\ j = 1, 2, \ldots, q$, where $\vec{x}$ represents the vector of decision variables and $f(\vec{x})$ denotes the objective function.

5.1.2 The Architecture of the Algorithm of Single-Objective Optimization

In this section, a single-objective optimization algorithm is used which is based on a hybrid of genetic operators and PSO formula to update the chromosomes and particle positions (Mahmoodabadi et al. 2013). In elaboration, the initial population is randomly produced. At each iteration, the inertia weight, the learning factors, and the operators probabilities are computed. Afterward, $\vec{x}_{pbest_i}$ and $\vec{x}_{gbest}$ are specified when the fitness values of all members are evaluated. The genetic algorithm operators, that is, traditional crossover, multiple-crossover and mutation operators are used to adjust the chromosomes (which are chosen randomly). Each chromosome is a particle and the group of chromosome is regarded as a swarm. Furthermore, the chromosomes which are not chosen for genetic operations will be appointed to particles and improved via PSO. Until the user-defined stopping criterion is satisfied, this cycle is repeated. Figures 2 and 3 illustrate the pseudo code and flow chart of the technique respectively.

5.1.3 The Results of Single-Objective Optimization

To evaluate the performance of the hybrid of particle swarm optimization and the genetic algorithm, nine prominent benchmark problems are utilized regarding a single-objective optimization problem. Essential information about these functions is summarized in Table 2 (Yao et al. 1999). Some of these functions are unimodal and the others are multimodal. Unimodal functions have only one optimal point while multimodal functions have some local optimal points in addition to one global optimal point.

The hybrid of particle swarm optimization and the genetic algorithm is applied to all the test functions with 30 dimensions ($n = 30$) (Mahmoodabadi et al. 2013). The mean and standard deviation of the best solution for thirty runs are presented in Table 4. In this regard, the results are contrasted to the results of three other algorithms [i.e., GA with traditional crossover (Chang 2007), GA with multiple-crossover (Chang 2007; Chen and Chang 2009), standard PSO (Kennedy and Eberhart 1995)]. Table 3 illustrates the list of essential parameters to run these

```
Initialize population and determine the algorithm configuration of the hybrid method.
While stopping criterion is satisfied
    Determine P_tc, P_mc, P_m based on the hybrid formulations.
    Find fitness values of each member and store x_pbest_i and x_gbest
    If rand < P_tc
        Select randomly two chromosomes from population and update them using traditional crossover operator;
    Elseif rand < P_mc
        Select randomly three chromosomes from population and update them using multiple-crossover operator;
    Elseif rand < P_m
        Select randomly a chromosome from population and update it using mutation operator;
    Else
        Select randomly a particle from swarm and update its position based on PSO formula;
End.
```

Fig. 2 The pseudo code of the hybrid algorithm for single-objective optimization

Fig. 3 The flow chart of the hybrid algorithm for single-objective optimization

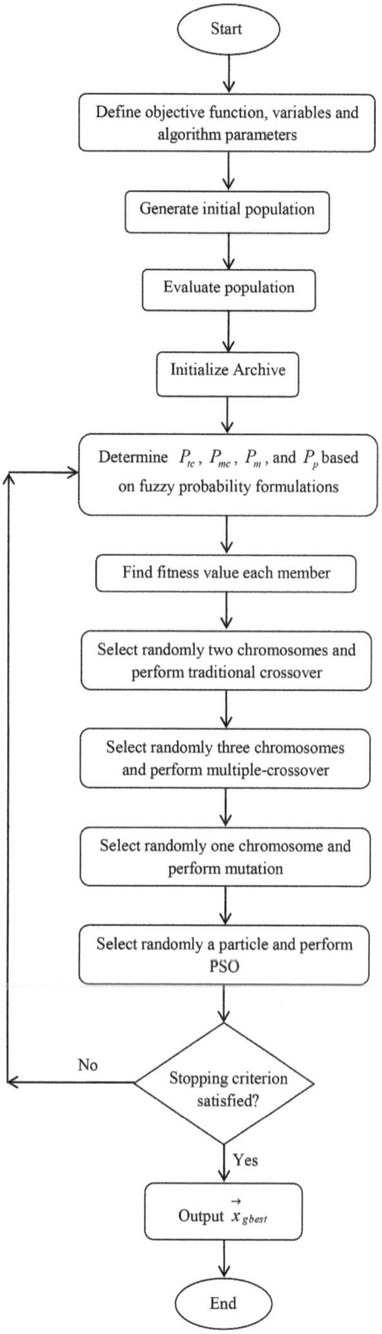

Table 2 Single-objective test functions

Name (comment)	Formula: $f(x)$	Search domain
Sphere (unimodal)	$\sum_{i=1}^{n} x_i^2$	$[-100, 100]^n$
Schwefel 2.22 (unimodal)	$\sum_{i=1}^{n} \lvert x_i \rvert + \prod_{i=1}^{n} \lvert x_i \rvert$	$[-10, 10]^n$
Schwefel 1.2 (unimodal)	$\sum_{i=1}^{n} \left(\sum_{j=1}^{i} x_j \right)^2$	$[-100, 100]^n$
Rosenbrock (unimodal)	$\sum_{i=1}^{n-1} \left[100 \left(x_{i+1} - x_i^2 \right)^2 + (x_i - 1)^2 \right]$	$[-30, 30]^n$
Noise (unimodal)	$\sum_{i=1}^{n} i x_i^4 + random[0, 1)$	$[-1.28, 1.28]^n$
Step (unimodal)	$\sum_{i=1}^{n} (\lfloor x_i + 0.5 \rfloor)^2$	$[-100, 100]^n$
Rastrigin (multimodal)	$\sum_{i=1}^{n} (x_i^2 - 10\cos(2\pi x_i) + 10)$	$[-5.12, 5.12]^n$
Griewank (multimodal)	$\frac{1}{4000} \sum_{i=1}^{n} x_i^2 - \prod_{i=1}^{n} \cos\left(\frac{x_i}{\sqrt{i}} \right) + 1$	$[-600, 600]^n$
Ackley (multimodal)	$20 + e - 20\exp\left(-0.2\sqrt{\frac{1}{n}\sum_{i=1}^{n} x_i^2} \right) - \exp\left(\frac{1}{n}\sum_{i=1}^{n} \cos(2\pi x_i) \right)$	$[-32, 32]^n$

Table 3 The parameter settings of optimization algorithms

Algorithm	Parameter
GA (traditional crossover)	$P_r = 0.2,\ P_c = 0.4,\ P_m = 0.1,\ S = 0.05$, tournament method for selection
GA (multiple-crossover)	$P_r = 0.2,\ P_c = 0.4,\ P_m = 0.1,\ S = 0.05$, tournament method for selection
Standard PSO	$W = 0.9,\ C_1 = C_2 = 2$
The hybrid algorithm	$W_1 = 0.9,\ W_2 = 0.4,\ C_{1i} = C_{2f} = 2.5,\ C_{1f} = C_{2i} = 0.5,$ $\zeta_m = 0.001, \xi_{tc} = \xi_{mc} = 0.2$

Table 4 The comparison results among single-objective optimization algorithms for the Sphere function

	GA (traditional crossover)	GA (multiple-crossover)	PSO (standard)	The hybrid algorithm
Mean	3.28×10^{-14}	2.56×10^{-18}	2.25×10^{-98}	1.58×10^{-119}
Standard deviation	4.58×10^{-14}	1.34×10^{-17}	6.54×10^{-98}	8.58×10^{-119}

algorithms. The population size and maximum iteration are set at 20 and 10,000, accordingly.

By contrasting the results of GA with traditional crossover, GA with multiple-crossover, standard PSO, and the hybrid of particle swarm optimization and the genetic algorithm (Tables 4, 5, 6, 7, 8, 9, 10, 11 and 12), it can be found that the hybrid algorithm has a superior performance with respect to other optimization algorithms. Moreover, the hybrid algorithm presents the best solutions in all test functions except Schwefel 2.22, in which the PSO algorithm has the best solution

Table 5 The comparison results among single-objective optimization algorithms for the Schwefel 2.22 function

	GA (traditional crossover)	GA (multiple-crossover)	PSO (standard)	The hybrid algorithm
Mean	1.83×10^{-6}	8.16×10^{0}	3.15×10^{-26}	9.11×10^{-25}
Standard deviation	4.32×10^{-6}	$3.13 \times 10^{+1}$	1.73×10^{-25}	4.99×10^{-24}

Table 6 The comparison results among single-objective optimization algorithms for the Schwefel 1.2 function

	GA (traditional crossover)	GA (multiple-crossover)	PSO (standard)	The hybrid algorithm
Mean	$7.74 \times 10^{+2}$	$3.14 \times 10^{+2}$	6.73×10^{-5}	2.14×10^{-11}
Standard deviation	$4.16 \times 10^{+2}$	$1.99 \times 10^{+2}$	9.40×10^{-5}	6.91×10^{-11}

Table 7 The comparison results among single-objective optimization algorithms for the Rosenbrock function

	GA (traditional crossover)	GA (multiple-crossover)	PSO (standard)	The hybrid algorithm
Mean	$1.33 \times 10^{+2}$	$8.25 \times 10^{+1}$	$2.04 \times 10^{+1}$	5.34×10^{-1}
Standard deviation	$1.32 \times 10^{+2}$	$5.51 \times 10^{+1}$	$2.53 \times 10^{+1}$	1.38×10^{0}

Table 8 The comparison results among single-objective optimization algorithms for the Noise function

	GA (traditional crossover)	GA (multiple-crossover)	PSO (standard)	The hybrid algorithm
Mean	5.10×10^{-2}	8.64×10^{-2}	1.06×10^{0}	3.38×10^{-3}
Standard deviation	1.88×10^{-2}	2.48×10^{-2}	3.12×10^{-1}	1.47×10^{-3}

Table 9 The comparison results among single-objective optimization algorithms for the Step function

	GA (traditional crossover)	GA (multiple-crossover)	PSO (standard)	The hybrid algorithm
Mean	$2.10 \times 10^{+2}$	$3.40 \times 10^{+1}$	1.00×10^{-1}	0
Standard deviation	$3.27 \times 10^{+2}$	$1.32 \times 10^{+2}$	3.05×10^{-1}	0

Table 10 The comparison results among single-objective optimization algorithms for the Rastringin function

	GA (traditional crossover)	GA (multiple-crossover)	PSO (standard)	The hybrid algorithm
Mean	$7.54 \times 10^{+1}$	$1.52 \times 10^{+2}$	$4.11 \times 10^{+1}$	2.15×10^{-3}
Standard deviation	$1.90 \times 10^{+1}$	$4.37 \times 10^{+1}$	9.19×10^{0}	1.18×10^{-2}

Table 11 The comparison results among single-objective optimization algorithms for the Griewank function

	GA (traditional crossover)	GA (multiple-crossover)	PSO (standard)	The hybrid algorithm
Mean	7.16×10^{-1}	4.86×10^{-1}	2.15×10^{-2}	2.03×10^{-2}
Standard deviation	2.87×10^{0}	1.4478×10^{0}	2.31×10^{-2}	2.21×10^{-2}

Table 12 The comparison results among single-objective optimization algorithms for the Ackley function

	GA (traditional crossover)	GA (multiple-crossover)	PSO (standard)	The hybrid algorithm
Mean	$1.57 \times 10^{+1}$	$1.80 \times 10^{+1}$	2.81×10^{-1}	4.30×10^{-8}
Standard deviation	1.13×10^{0}	6.23×10^{-1}	5.92×10^{-1}	9.90×10^{-8}

but the result is very close to the results of the hybrid method. Figures 4, 5, 6, 7, 8, 9 and 10 illustrate the evolutionary traces of some test functions of Table 2. In these figures, the mean best values are gained for thirty runs. The maximum iteration, population size, and dimension are set at 1,000, 10 and 50, respectively. In these figures, the vertical axis is the value of the best function found after each iteration of the algorithms and the horizontal axis is the iteration. By comparing these figures, it is obtained that the combination of the traditional crossover, multiple-crossover and mutation operator can enhance the performance of particle swarm optimization.

5.2 Multi-objective Optimization

5.2.1 Definition of Multi-objective Optimization Problem

In most of real problems, there is more than one objective function required to be optimized. Furthermore, most of these functions are in conflict with each other. Hence, there is not just one solution for the problem and there are some optimal

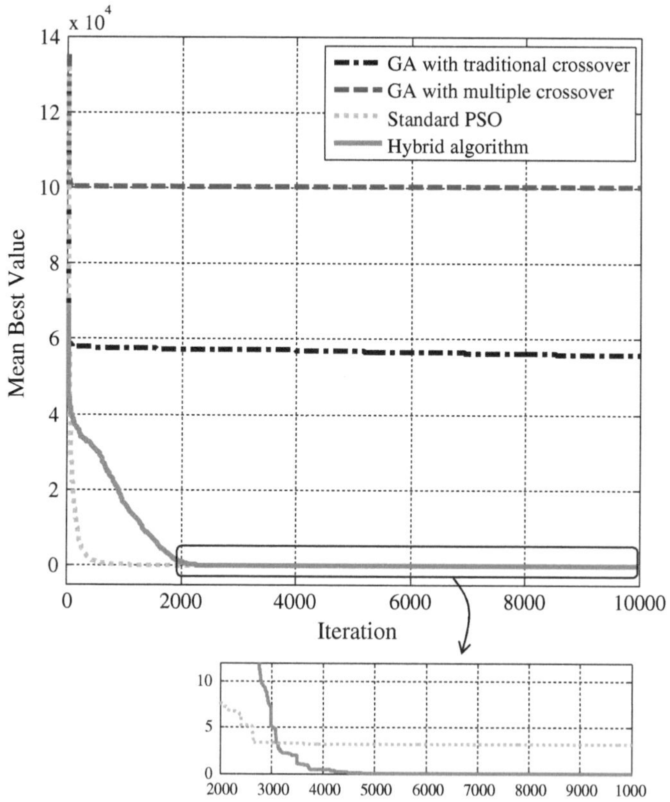

Fig. 4 The evolutionary trajectory of the single-objective optimization algorithm on the Sphere test function

solutions which are non-dominated with respect to each other and the designer can use each of them based upon the design criteria.

$$\text{Find } \vec{x}^* = [x_1^*, x_2^*, \ldots, x_n^*] \in R^n$$
$$\text{To minimize } \vec{f}(\vec{x}) = [f_1(\vec{x}), f_2(\vec{x}), \ldots, f_m(\vec{x})] \in R^m$$

By regarding p equality constraints $g_i(\vec{x}) = 0, \quad \mathrm{i} = 1, 2, \ldots, \mathrm{p}$ and q inequality constraints $h_j(\vec{x}) \leq 0, \quad \mathrm{j} = 1, 2, \ldots, \mathrm{q}$, where $\vec{x}$ represents the vector of decision variables and $\vec{f}(\vec{x})$ denotes the vector of objective functions.

As it is mentioned earlier, there is not one unique optimal solution for multi-objective problems. There exists a set of optimal solutions called Pareto-optimal solutions. The following definitions are needed to describe the concept of optimality (Deb et al. 2002).

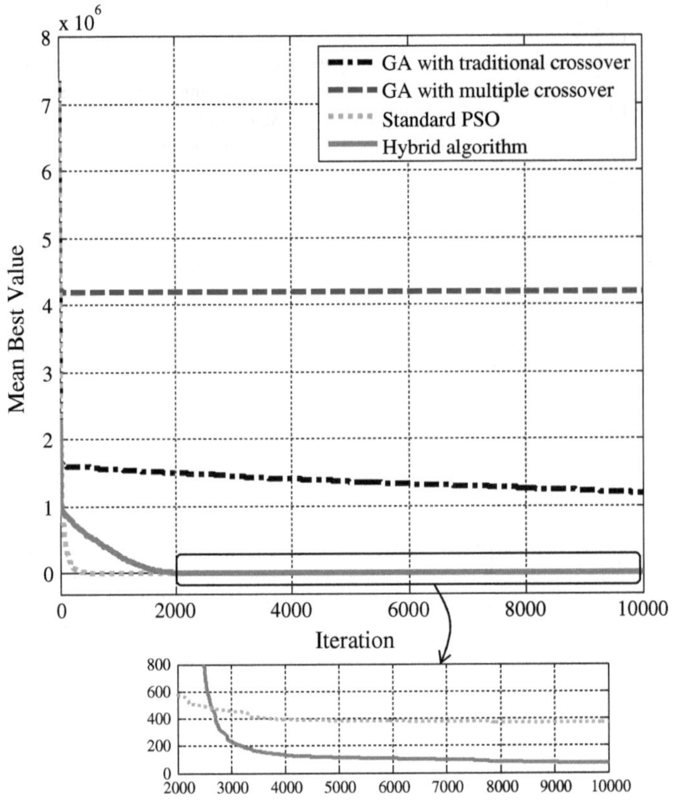

Fig. 5 The evolutionary trajectory of the single-objective optimization algorithm on the Rosenbrock test function

Definition 1 Pareto Dominance: It says that the vector $\vec{u} = [u_1, u_2, \ldots, u_n]$ dominates the vector $\vec{v} = [v_1, v_2, \ldots, v_n]$ and it illustrates $\vec{u} \prec \vec{v}$, if and only if:

$$\forall i \in \{1, 2, \ldots, \mathrm{n}\} : u_i \leq v_i \wedge \exists\ j \in \{1, 2, \ldots \mathrm{n}\} : u_j < v_j$$

Definition 2 Non-dominated: A vector of decision variables $\vec{x} \in X \subset R^{\mathrm{n}}$ is non-dominated, if there is not another $\vec{x}' \in X$ which dominates $\vec{x}$. That is to say that $\forall \vec{x} \in X,\ \nexists \vec{x}' \in X, \vec{x} \neq \vec{x}' : \vec{f}(\vec{x}') \prec \vec{f}(\vec{x})$, where $\vec{f} = \{f_1, f_2, \ldots, f_m\}$ denotes the vector of objective functions.

Definition 3 Pareto-optimal: the vector of decision variables $\vec{x}^* \in X \subset R^{\mathrm{n}}$,where X is the design feasible region, is Pareto-optimal if this vector is non-dominated in X.

Definition 4 Pareto-optimal set: In multi-objective problems, a Pareto-optimal set or in a more straightforward expression, a Pareto set denoted by P^* consists of all Pareto-optimal vectors, namely:

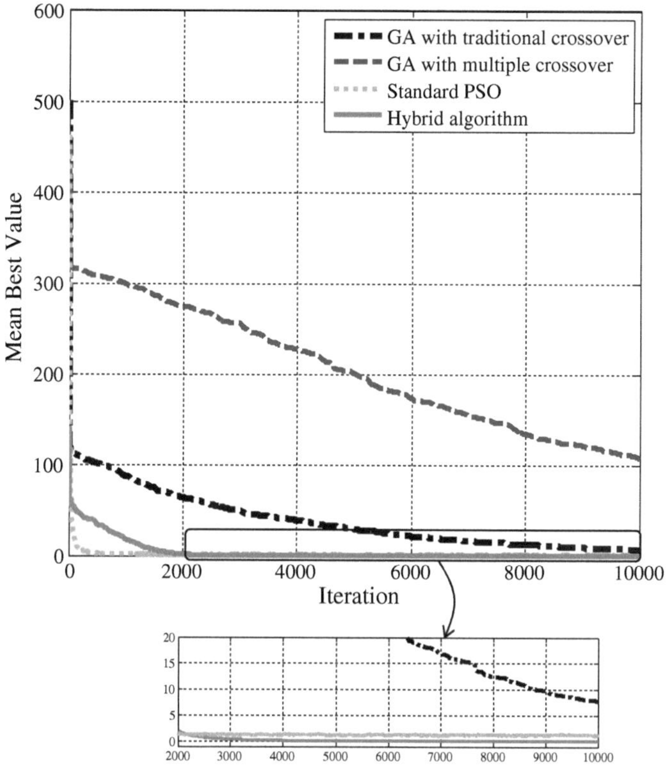

Fig. 6 The evolutionary trajectory of the single-objective optimization algorithm on the Noise test function

$$P^* = \{\vec{x} \in X | \vec{x} \, \text{is Pareto-optimal}\}$$

Definition 5 Pareto-optimal front: The Pareto-optimal front or in a more straightforward expression, Pareto front PF^* is defined as:

$$PF^* = \{\vec{f}(\vec{x}) \in R^m | \vec{x} \in P^*\}.$$

5.2.2 The Structure of the Hybrid Algorithm for Multi-objective Optimization

It is necessary to make modifications to the original scheme of PSO in finding the optimal solutions for multi-objective problems. In the single-objective algorithm of PSO, the best particle of the entire swarm ($\vec{x}_{gbest}$) is utilized as a leader. In the multi-objective algorithm, each particle has a set of different leaders that one of them is chosen as a leader. In this book paper, a leader selection method based upon density

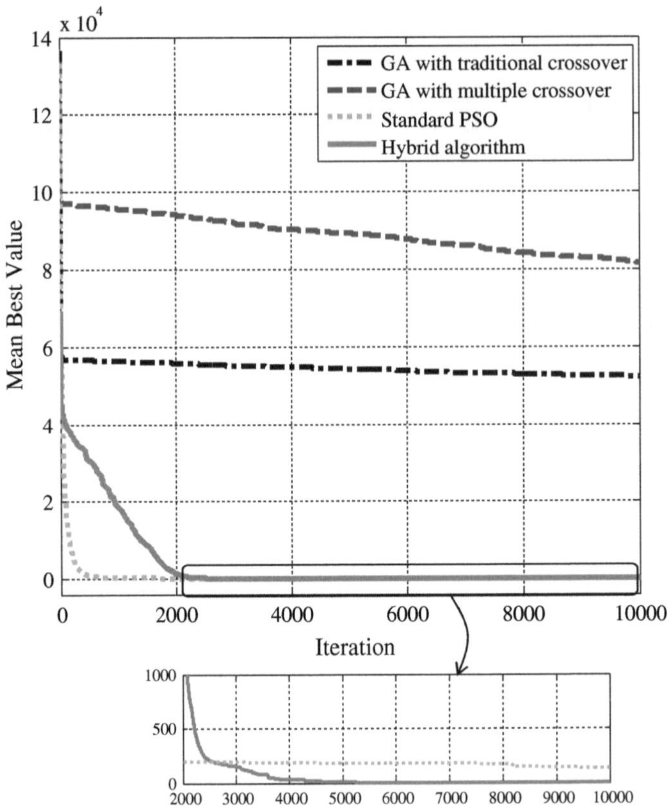

Fig. 7 The evolutionary trajectory of the single-objective optimization algorithm on the Step test function

measures is used (Mahmoodabadi et al. 2013). To this end, a neighborhood radius $R_{neighborhood}$ is defined for the whole non-dominated solutions. Two non-dominated solutions are regarded neighbors in case the Euclidean distance of them is less than $R_{neighborhood}$. Based upon this radius, the number of neighbors of each non-dominated solution is computed in the objective function domain and the particle having fewer neighbors is chosen as leaders. Furthermore, for particle i, the nearest member of the archive is devoted to $\overrightarrow{x}_{pbest_i}$. At this stage, a multi-objective optimization algorithm using the hybridization of genetic operators and PSO formula can be presented (Mahmoodabadi et al. 2013). In elaboration, the population is randomly generated. Once the fitness values of all members are computed, the first archive can be produced. The inertia weight, the learning factors and operator's probabilities are computed at each iteration. The genetic operators, that is, mutation operators, traditional crossover and multiple-crossover are utilized to change some chromosomes selected randomly. Each chromosome corresponds to a particle in it and the group of chromosome can be regarded as a swarm. On the other hand, the

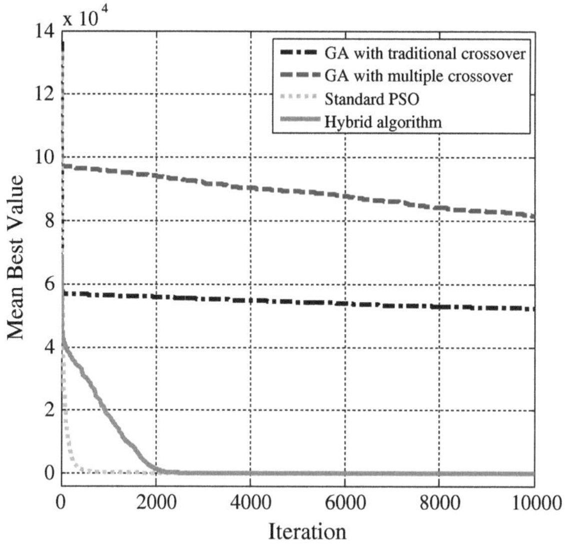

Fig. 8 The evolutionary trajectory of the single-objective optimization algorithm on the Rastringin test function

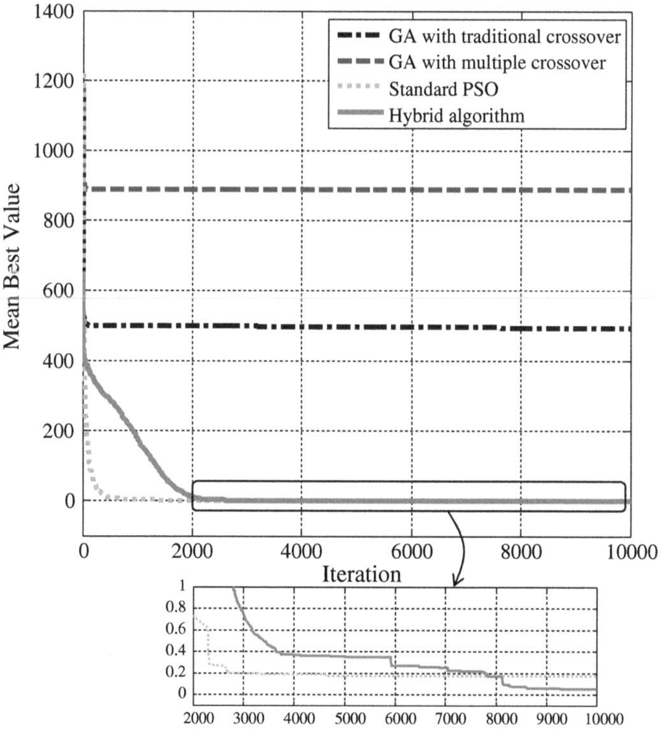

Fig. 9 The evolutionary trajectory of the single-objective optimization algorithm on the Griewank test function

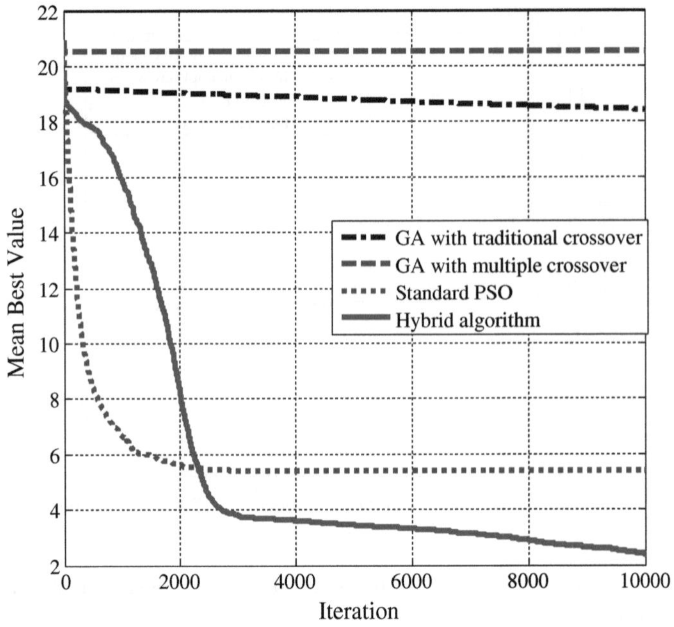

Fig. 10 The evolutionary trajectory of the single-objective optimization algorithm on the Ackley test function

chromosomes which are not chosen for genetic operations are enhanced via particle swarm optimization. Then, the archive is pruned and updated. This cycle is repeated until the user-defined stopping criterion is met. Figure 11 illustrates the flow chart of this algorithm.

The set of non-dominated solutions is saved in a different location named archive. If all of the non-dominated solutions are saved in the archive, the size of archive enhances rapidly. On the other hand, since the archive must be updated at each iteration, the size of archive will expand significantly. In this respect, a supplementary criterion is needed that resulted in saving a bounded number of non-dominated solutions. To this end, the dynamic elimination approach is utilized here to prune the archive (Mahmoodabadi et al. 2013). In this method, if the Euclidean distance between two particles is less than $R_{\text{elimination}}$ which is the elimination radius of each particle, then one of them will be eliminated. As an example, it is illustrated in Fig. 12. To gain the value of $R_{\text{elimination}}$, the following equation is utilized:

$$R_{elimination} = \begin{cases} \frac{t}{\alpha \times maximum\ iteration} & if \left(\frac{t}{\beta}\right) = fix\left(\frac{t}{\beta}\right) \\ 0 & else \end{cases} \tag{12}$$

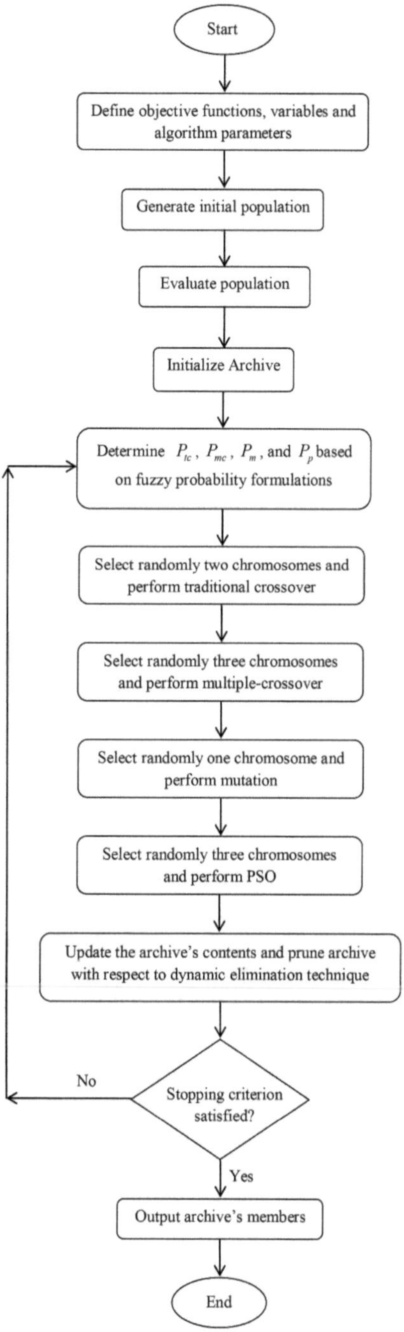

Fig. 11 The flow chart of the hybrid algorithm for multi-objective optimization problems

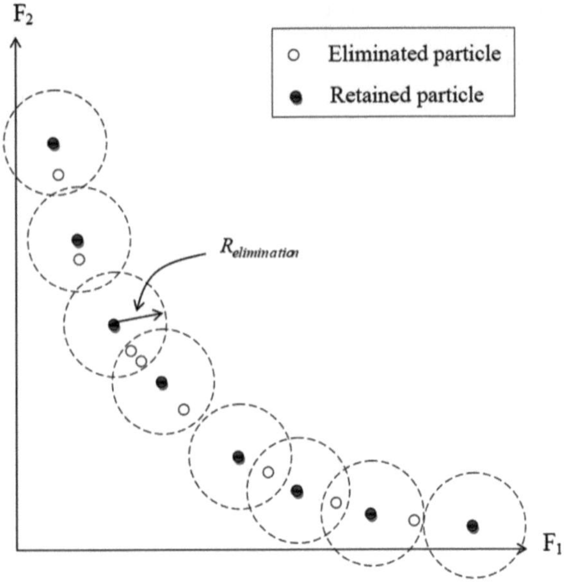

Fig. 12 The particles located in another particle's $R_{elimination}$ will be removed using the dynamic elimination technique

In which, t stands for the current iteration number and *maximum iteration* is the maximum number of allowable iterations. α and β are positive constants regarding as $\alpha = 100$ and $\beta = 10$.

5.2.3 Results for Multi-objective Optimization

Five multi-objective benchmark problems are regarded which have similar features such as the bounds of variables, the number of variables, the nature of Pareto-optimal front and the true Pareto optimal solutions. These problems which are unconstrained have two objective functions. The whole features of these algorithms are illustrated in Table 13. The contrast of the true Pareto optimal solutions and the results of the hybrid algorithm is illustrated in Figs. 13, 14, 15, 16 and 17. As it is obtained, the hybrid algorithm can present a proper result in terms of converging to the true Pareto optimal and gaining advantages of a diverse solution set.

In this comparison, the capability of the hybrid algorithm is contrasted to three prominent optimization algorithms, that is, NSGA-II (Deb et al. 2002), SPEA (Zitzler and Theile 1999) and PAES (Knowles and Corne 1999) with respect to the same test functions. Two crucial facts considered here are the diversity solution of the solutions with respect to the Pareto optimal front and the capability to gain the Pareto optimal set. Regarding these two facts, two performance metrics are utilized in evaluating each of the above-mentioned facts.

Table 13 Multi-objective test functions

Name	Variable bounds (dimension n)	Formula	Optimal solutions	Comments
SCH	$[-10^3, 10^3]n$ $(n = 1)$	$f_1(x) = x^2$ $f_2(x) = (x-2)^2$	$x \in [0, 2]$	Convex
FON	$[-4, 4]^n$ $(n = 3)$	$f_1(x) = 1-\exp(-\sum_{i=1}^{3} (x_i - \frac{1}{\sqrt{3}})^2)$ $f_2(x) = 1 - \exp(-\sum_{i=1}^{3} (x_i + \frac{1}{\sqrt{3}})^2)$	$x_1 = x_2 = x_3$ $\in [-\frac{1}{\sqrt{3}}, \frac{1}{\sqrt{3}}]$	Nonconvex
ZDT1	$[0, 1]^n$ $(n = 30)$	$f_1(x) = x_1, f_2(x) = g(x)[1-\sqrt{x_1/g(x)}]$ $g(x) = 1 + 9(\sum_{i=2}^{n} x_i) / (n-1)$	$x_1 \in [0, 1], x_i = 0$ $i = 2, \ldots, n$	Convex
ZDT2	$[0, 1]^n$ $(n = 30)$	$f_1(x) = x_1, f_2(x) = g(x)[1-(x_1 / g(x))^2]$ $g(x) = 1 + 9(\sum_{i=2}^{n} x_i) / (n-1)$	$x_1 \in [0, 1], x_i = 0$ $i = 2, \ldots, n$	Nonconvex
ZDT3	$[0, 1]^n$ $(n = 30)$	$f_1(x) = x_1, f_2(x) = g(x)[1-\sqrt{x_1/g(x)} - \frac{x_1}{g(x)} \sin(10\pi x_1)]$ $g(x) = 1 + 9(\sum_{i=2}^{n} x_i) / (n-1)$	$x_1 \in [0, 1], x_i = 0$ $i = 2, \ldots, n$	Convex, disconnected

Fig. 13 The non-dominated solutions of the hybrid method for the SCH test function

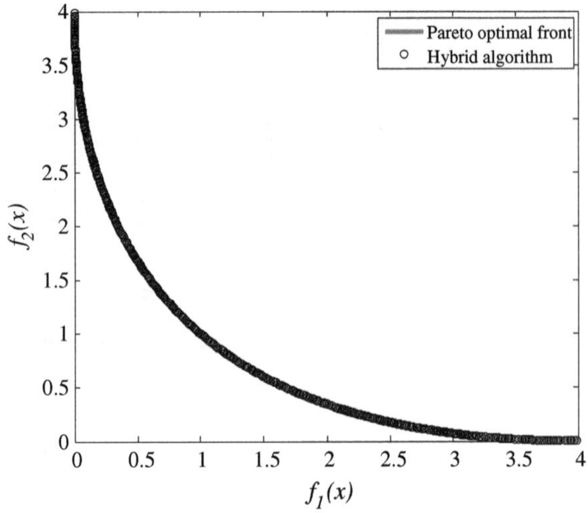

Fig. 14 The non-dominated solutions of the hybrid method for the FON test function

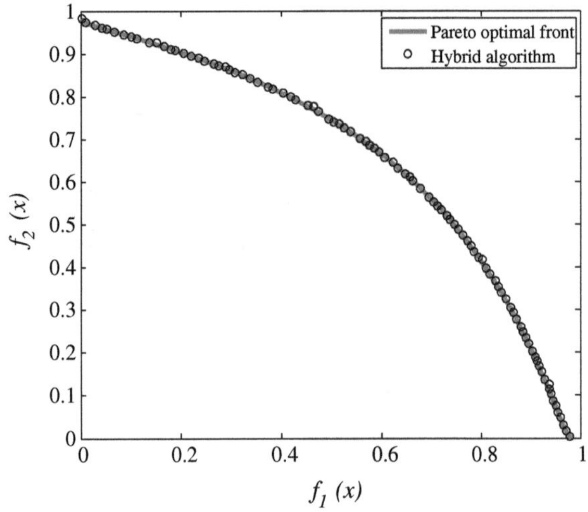

(1) A proper indication of the gap between the non-dominated solution members and the Pareto optimal front is gained by means of the metric of distance (Υ) (Deb et al. 2002) as follows:

$$\Upsilon = \sum_{i=1}^{n} d_i^2 \tag{13}$$

where n is the number of members in the set of non-dominated solutions and d_i is the least Euclidean distance between the member i in the set of non-dominated

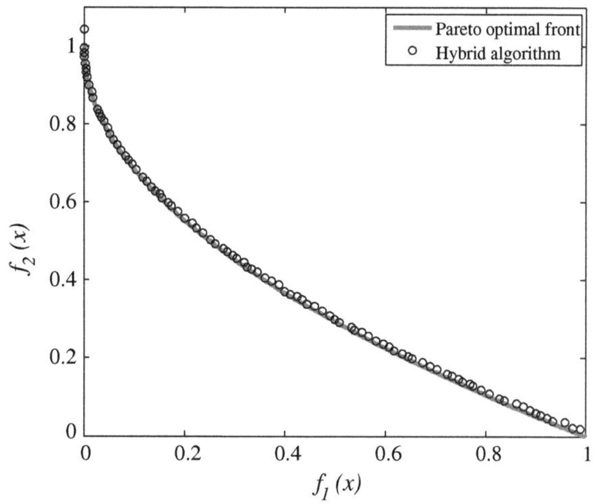

Fig. 15 The non-dominated solutions of the hybrid method for the ZDT1 test function

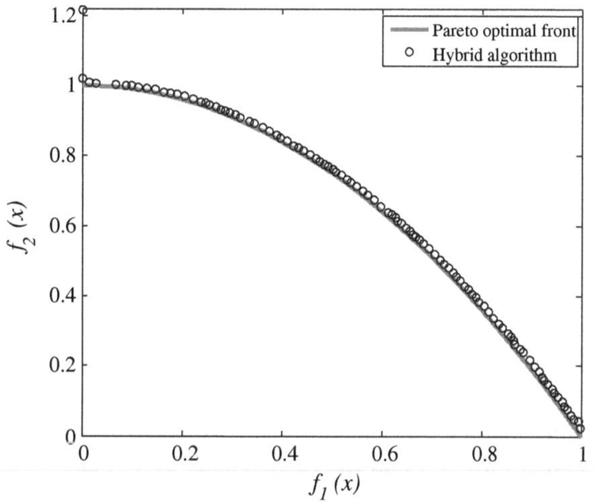

Fig. 16 The non-dominated solutions of the hybrid method for the ZDT2 test function

solutions and Pareto optimal front. If all members in the set of non-dominated solutions are in Pareto optimal front then $\Upsilon = 0$.

(2) The metric of diversity (Δ) (Deb et al. 2002) measures the extension of spread achieved among non-dominated solutions, which is given as

$$\Delta = \frac{d_f + d_l + \sum_{i=1}^{n-1} \left|d_i - \bar{d}\right|}{d_f + d_l + (n-1)\bar{d}} \tag{14}$$

In this formula, d_f and d_l denote the Euclidean distance between the boundary solutions and the extreme solutions of the non-dominated set, n stands for the

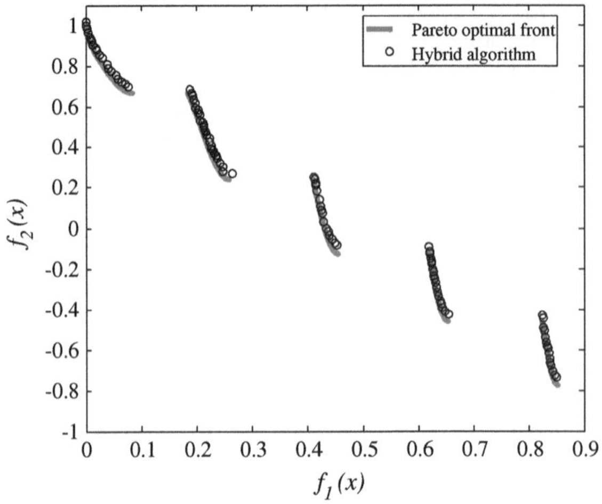

Fig. 17 The non-dominated solutions of the hybrid method for the ZDT3 test function

number of members in the set of non-dominated solutions, d_i is the Euclidean distance between consecutive solutions in the gained non-dominated set, and $\bar{d} = \frac{\sum_{i=1}^{n-1} \mathrm{d_i}}{n-1}$.

For the most extent spread set of non-dominated solutions $\Delta = 0$

The performance of the hybrid algorithm comparing to NSGA-II (Deb et al. 2002), SPEA (Zitzler and Theile 1999), and PAES (Knowles and Corne 1999) algorithms is illustrated in Tables 14, 15, 16, 17 and 18.

Based on the results of Tables 14, 15, 16, 17 and 18, the hybrid algorithm has very proper Δ values for all test functions excluding ZDT2. While NSGA-II presents proper Δ results for all test functions except ZDT3, the approaches SPEA and PAES do not illustrate proper performance in the diversity metric. The hybrid algorithm presents acceptable results for the convergence metric in all test functions. On the other hand, NSGA-II ZDT3 function, SPEA for FON function, and PAES for FON and ZDT2 functions do not show proper performance.

The hybrid optimization algorithm is used to design the parameters of state feedback control for linear systems. In the following section, state space representation and the control input of state feedback control for linear systems will be presented.

Table 14 The results of the comparison of multi-objective optimization algorithms for the SCH test function

Metrics		NSGA-II	SPEA	PAES	The hybrid algorithm
Δ	Mean	4.77×10^{-1}	1.02×10^{0}	1.06×10^{0}	6.00×10^{-1}
	Standard deviation	3.47×10^{-3}	4.37×10^{-3}	2.86×10^{-3}	1.81×10^{-2}
ϒ	Mean	3.39×10^{-3}	3.40×10^{-3}	1.31×10^{-3}	3.22×10^{-3}
	Standard deviation	0	0	3.00×10^{-6}	1.35×10^{-4}

Table 15 The results of the comparison of multi-objective optimization algorithms for the FON test function

Metrics		NSGA-II	SPEA	PAES	The hybrid algorithm
Δ	Mean	3.78×10^{-1}	7.92×10^{-1}	1.16×10^{0}	5.90×10^{-1}
	Standard deviation	6.39×10^{-4}	5.54×10^{-3}	8.94×10^{-3}	3.60×10^{-2}
ϒ	Mean	1.93×10^{-3}	1.25×10^{-1}	1.51×10^{-1}	1.56×10^{-3}
	Standard deviation	0	3.80×10^{-5}	9.05×10^{-4}	1.71×10^{-4}

Table 16 The results of the comparison of multi-objective optimization algorithms for the ZDT1 test function

Metrics		NSGA-II	SPEA	PAES	The hybrid algorithm
Δ	Mean	3.90×10^{-1}	7.84×10^{-1}	1.22×10^{0}	6.55×10^{-1}
	Standard deviation	1.87×10^{-3}	4.44×10^{-3}	4.83×10^{-3}	4.91×10^{-2}
ϒ	Mean	3.34×10^{-2}	1.79×10^{-3}	8.20×10^{-2}	8.16×10^{-3}
	Standard deviation	4.75×10^{-3}	1.00×10^{-6}	8.67×10^{-3}	2.73×10^{-3}

Table 17 The results of the comparison of multi-objective optimization algorithms for the ZDT2 test function

Metrics		NSGA-II	SPEA	PAES	The hybrid algorithm
Δ	Mean	4.30×10^{-1}	7.55×10^{-1}	1.16×10^{0}	9.57×10^{-1}
	Standard deviation	4.72×10^{-3}	4.52×10^{-3}	7.68×10^{-3}	3.20×10^{-2}
ϒ	Mean	7.23×10^{-2}	1.33×10^{-3}	1.26×10^{-1}	3.04×10^{-2}
	Standard deviation	3.16×10^{-2}	0	3.68×10^{-2}	1.84×10^{-2}

Table 18 The results of the comparison of multi-objective optimization algorithms for the ZDT3 test function

Metrics		NSGA-II	SPEA	PAES	The hybrid algorithm
Δ	Mean	7.38×10^{-1}	6.72×10^{-1}	7.89×10^{-1}	6.28×10^{-1}
	Standard deviation	1.97×10^{-2}	3.58×10^{-3}	1.65×10^{-3}	5.30×10^{-2}
ϒ	Mean	1.14×10^{-1}	4.75×10^{-2}	2.38×10^{-2}	8.88×10^{-3}
	Standard deviation	7.94×10^{-3}	4.70×10^{-5}	1.00×10^{-5}	6.97×10^{-3}

6 State Feedback Control for Linear Systems

The vector state equation can be utilized for a continuous time system, as follows:

$$\dot{\boldsymbol{x}}(t) = \boldsymbol{A}\boldsymbol{x}(t) + \boldsymbol{B}\boldsymbol{u}(t) + \boldsymbol{B}_{v}\boldsymbol{v}(t) \tag{15}$$

where $\boldsymbol{x}(t) \in \mathbb{R}^n$ stands for the state vector, $\dot{\boldsymbol{x}}(t)$ denotes the time derivative of state vector and $\boldsymbol{u}(t) \in \mathbb{R}^m$ is the input vector. The disturbance $\boldsymbol{v}(t)$ is assumed to be a deterministic nature. Furthermore, $\boldsymbol{A} \in \mathbb{R}^{n \times n}$ is the system or dynamic matrix, $\boldsymbol{B} \in \mathbb{R}^{n \times m}$ is the input matrix and $\boldsymbol{B}_v$ is the disturbance matrix. Measurements are made on this system which can be either the states themselves or linear combinations of them:

$$\boldsymbol{y}(t) = \boldsymbol{C}\boldsymbol{x}(t) + \boldsymbol{w}(t) \tag{16}$$

where $\boldsymbol{y}(t) \in \mathbb{R}^r$ is the output vector and $\boldsymbol{C} \in \mathbb{R}^{r \times n}$ is the output matrix. The vector $\boldsymbol{w}(t)$ stands for the measurement disturbance.

In order to establish linear state feedback around the above system, a linear feedback law can be applied as follows:

$$\boldsymbol{u}(t) = -\boldsymbol{K}\boldsymbol{x}(t) + \boldsymbol{r}(t) \tag{17}$$

In this formula, $\boldsymbol{K} \in \mathbb{R}^{m \times n}$ stands for a feedback matrix (or a gain matrix). $\boldsymbol{r}(t)$ denotes the reference input vector of the system having dimensions the same as the input vector $\boldsymbol{u}(t)$. The resulting feedback system is a full state feedback system due to measuring all of the states. To design the state feedback controller with an optimal control input and minimum error, the hybrid optimization algorithm is applied and the optimal Pareto front of the controller is shown in the following section.

7 Pareto Optimal State Feedback Control of a Parallel-Double-Inverted Pendulum

The model of a parallel-double-inverted pendulum system is presented in this section. The work deals with the stabilization control of a system which is a complicated nonlinear and unstable high-order system. Figure 18 illustrates the mechanical structure of the inverted pendulum. According to the figure, the cart is moving on a track with two pendulums hinged and balanced upward by means of a DC motor. In addition, the cart has to track a (varying) reference position. This system includes two pendulums and one cart. The pendulums are attached to the cart. While the cart is moving, the system has to be controlled in such a way that pendulums are placed in desired angels. The dynamic equations of the system are as follows:

$$I_1\ddot{\varphi} + C_1\dot{\varphi} - a_3 \sin\varphi + a_1\ddot{x}\cos\varphi = 0 \tag{18}$$

$$I_2\ddot{\alpha} + C_2\dot{\alpha} - a_4 \sin\alpha + a_2\ddot{x}\cos\alpha = 0 \tag{19}$$

$$M\ddot{x} + f_r\dot{x} + a_1\left(\ddot{\varphi}\cos\varphi - \dot{\varphi}^2\sin\varphi\right) + a_2\left(\ddot{\alpha}\cos\alpha - \dot{\alpha}^2\sin\alpha\right) = u \tag{20}$$

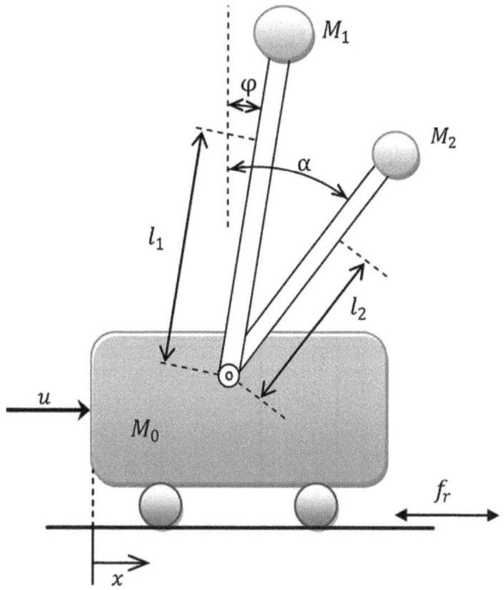

Fig. 18 The system of a parallel-double-inverted pendulum

where

$$I_1 = I_1' + M_1 l_1^2, I_2 = I_2' + M_2 l_2^2, \tag{21}$$

$$M = M_0 + M_1 + M_2 \tag{22}$$

$$a_1 = M_1 l_1,\ a_2 = M_2 l_2,\ a_3 = M_1 l_1 g,\ a_4 = M_2 l_2 g \tag{23}$$

where x is the position of the cart, $\dot{x}$ is the velocity of the cart, φ stands for the angular velocity of the first pendulum with respect to the vertical line, $\dot{\varphi}$ is the angular velocity of the first pendulum, α is the angular position of the second pendulum, $\dot{\alpha}$ represents the angular velocity of the second pendulum, M_1 is the mass of the first pendulum, M_2 is the mass of second pendulum, M_0 is the mass of the cart, l_1 denotes the length of the first pendulum with respect to its center, l_2 stands for the length of the second pendulum with respect to its center, f_r is the friction coefficient of the cart with ground, I_1' is the inertia moment of the first pendulum with respect to its center, I_2' represents the inertia moment of the second pendulum with respect to its center, C_1 is the angular frictional coefficient of the first pendulum, C_2 stands for the angular frictional coefficient of the second pendulum, and u is the control effort.

To obtain the state space representations of the dynamic equations, the state space variables are defined as $x = [x_1, x_2, x_3, x_4, x_5, x_6]^T$. This vector includes the position of the cart, the velocity of the cart, the angular position and velocity of the first pendulum, the angular position and velocity of the second pendulum. After linearization around the equilibrium point $x_a = [x_1, 0, 0, 0, 0, 0]^T$, the state space representation is obtained according to Eq. (25).

$$\begin{bmatrix} \dot{x}_1 \\ \dot{x}_2 \\ \dot{x}_3 \\ \dot{x}_4 \\ \dot{x}_5 \\ \dot{x}_6 \end{bmatrix} = \begin{bmatrix} 0 & 1 & 0 & 0 & 0 & 0 \\ 0 & \frac{f_r I_1 I_2}{P_2} & -\frac{-a_1 a_3 I_2}{P_2} & -\frac{a_1 C_1 I_2}{P_2} & -\frac{-a_2 a_4 I_1}{P_2} & -\frac{a_2 C_2 I_1}{P_2} \\ 0 & 0 & 0 & 1 & 0 & 0 \\ 0 & -\frac{a_1 f_r I_2}{P_1} & \frac{a_3(a_2^2 - m I_2)}{P_1} & -\frac{c_1(a_2^2 - m I_2)}{P_1} & -\frac{a_1 a_2 a_4}{P_1} & -\frac{a_1 a_2 a_4}{P_1} \\ 0 & 0 & 0 & 0 & 0 & 1 \\ 0 & -\frac{a_2 f_r I_1}{P_2} & \frac{-a_1^2 a_2 a_3 I_2}{P_3} & \frac{a_1 a_2 C_1}{P_2} & \frac{a_1^3 a_4 I_2 - a_1 a_4 m I_1 I_2}{P_3} & \frac{-a_1^3 C_2 I_2 + a_1 C_2 m I_1 I_2}{P_3} \end{bmatrix} \begin{bmatrix} x_1 \\ x_2 \\ x_3 \\ x_4 \\ x_5 \\ x_6 \end{bmatrix} + \begin{bmatrix} 0 \\ -\frac{I_1 I_2}{P_2} \\ 0 \\ \frac{a_1 I_2}{P_1} \\ 0 \\ \frac{a_2 I_1}{P_2} \end{bmatrix} u \tag{25}$$

where

$$P_1 = a_1^2 I_2 - I_1(-a_2^2 + m I_2) \tag{26}$$

$$P_2 = a_2^2 I_1 + a_1^2 I_2 - m I_1 I_2 \tag{27}$$

$$P_3 = a_1 I_2 (a_2^2 I_1 + a_1^2 I_2 - m I_1 I_2) \tag{28}$$

The block diagram of the linear state feedback controller to control the parallel-double-inverted pendulum is illustrated in Fig. 19. The control effort of the state feedback controller is obtained as follows

$$\begin{aligned} u = {} & K_1 (x_1 - x_{1,d}) + K_2 (x_2 - x_{2,d}) + K_3 (x_3 - x_{3,d}) \\ & + K_4 (x_4 - x_{4,d}) + K_5 (x_5 - x_{5,d}) + K_6 (x_6 - x_{6,d}) \end{aligned} \tag{29}$$

where $x_d = [x_{1,d}, x_{2,d}, x_{3,d}, x_{4,d}, x_{5,d}, x_{6,d}]^T$ is the vector of the desired states and $K = [K_1, K_2, K_3, K_4, K_5, K_6]$ is the vector of design variables obtained via the optimization algorithm. The boundaries of the system are:

The boundary of the control effort is $|u| \leq 20\ [N]$
The boundary of the length of x_1, x_3 and x_5 are $|x_1| \leq 0.5\ [m]$, $|x_3| \leq 0.174\ [rad]$, $|x_5| \leq 0.174\ [rad]$.

The initial state vector, final state vector, and the boundaries of design variables are as follows. Furthermore, the values of the parameters of the system of a parallel-double-inverted pendulum are presented in Table 19.

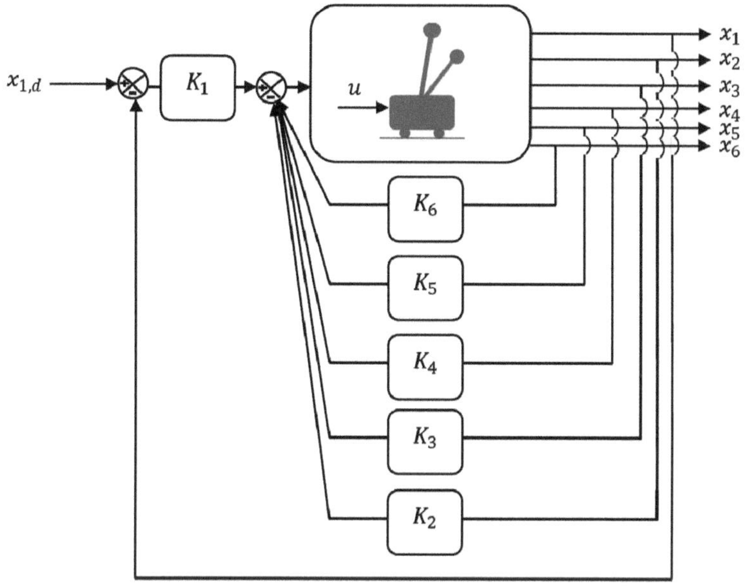

Fig. 19 The block diagram of the linear state feedback controller for a parallel-double-inverted pendulum for $\boldsymbol{x_d} = [\boldsymbol{x_{1,d}}, \boldsymbol{0}, \boldsymbol{0}, \boldsymbol{0}, \boldsymbol{0}, \boldsymbol{0}]^T$

Table 19 The values of the parameters of the system of a parallel-double-inverted pendulum

m_0	4.2774 kg
m_1	0.3211 kg
m_2	0.2355 kg
l_1	0.3533 m
l_2	0.0963 m
Θ_1	0.072 kg m^2
Θ_2	0.0044 kg m^2
F_r	10 Kg/s
C_1	0.023 Kg m^2/s
C_2	0.00145 Kg m^2/s

$$x_0 = [0, 0, 0, 0, 0, 0]^{\mathrm{T}} \tag{30}$$

$$x_d = [0.2, 0, 0, 0, 0, 0]^{\mathrm{T}} \tag{31}$$

$$50 \le K_1 \le 150 \tag{32}$$

$$150 \le K_2 \le 250 \tag{33}$$

$$14,900 \le K_3 \le 15,700 \tag{34}$$

$$3,000 \le K_4 \le 4,000 \tag{35}$$

$$-14,000 \le K_5 \le -12,000 \tag{36}$$

$$-3,000 \le K_6 \le -1,500 \tag{37}$$

In this problem, the objective functions of the multi-objective optimization algorithm are

F_1 = the sum of settling time and overshoot of the cart;
F_2 = the sum of settling time and overshoot of the first pendulum + the sum of settling time and overshoot of the second pendulum;

These objective functions have to be minimized simultaneously. The Pareto front of the control of the system of the parallel-double-inverted pendulum obtained via multi-objective hybrid of particle swarm optimization and the genetic algorithm is shown in Fig. 20. In Fig. 20, points A and C stand for the best sum of settling time and overshoot of the cart and the sum of settling time and overshoot of the first and second pendulums, respectively. It is clear from this figure that all the optimum design points in the Pareto front are non-dominated and could be chosen by a designer as optimum linear state feedback controllers. It is also clear that choosing a better value for any objective function in a Pareto front would cause a worse value for another objective. The corresponding decision variables (vector of linear state

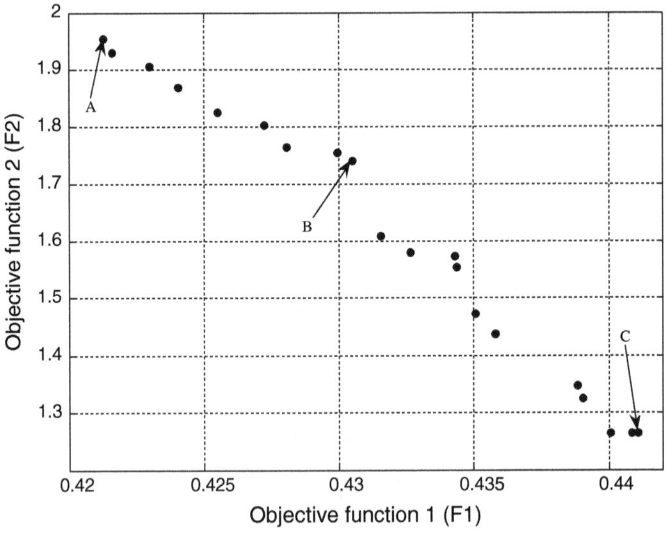

Fig. 20 Pareto front of multi-objective hybrid of particle swarm optimization and the genetic algorithm for the control of the system of the parallel-double-inverted pendulum

Table 20 The values of the parameters of the system of the parallel-double-inverted pendulum

	K_1	K_2	K_3	K_4	K_5	K_6	F_1	F_2
Point A	99.8110	224.5012	14,984.2210	3,967.5551	−12,480.5041	−2,502.5520	0.4213	1.9578
Point B	100.3566	225.5012	14,975.5521	3,968.2250	−12,404.6652	−2,493.3324	0.4305	1.7422
Point C	99.0315	224.5646	14,999.1835	3,966.4309	−12,391.4606	−2,491.0947	0.4411	1.2622

feedback controllers) of the Pareto front shown in Fig. 20 are the best possible design points. Moreover, if any other set of decision variables is selected, the corresponding values of the pair of those objective functions will locate a point inferior to that Pareto front. Indeed, such inferior area in the space of two objectives is top/right side of Fig. 20. Thus, the Pareto optimum design method causes to find important optimal design facts between these two objective functions. From Fig. 20, point B is the point which demonstrates such important optimal design facts. This point could be the trade-off optimum choice when considering minimum values of both sum of settling time and overshoot of the cart and sum of settling time and overshoot of the first and second pendulums. The values of the design variables obtained for three design points are illustrated in Table 20. The control effort, the angle of the first pendulum, the angle of the second pendulum, and the position of the cart are illustrated in Figs. 21, 22, 23 and 24. By regarding these figures, it can be concluded that the point A has the best time response (overshoot plus settling time) of the cart and the worst time responses (overshoot plus settling time) of the pendulums while point C has the best time responses of pendulums and the worst time response of the cart.

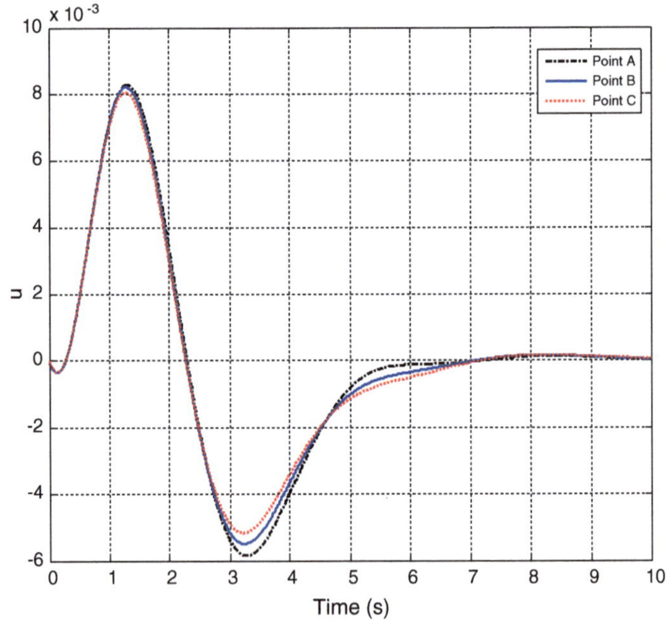

Fig. 21 The control effort of the system of the parallel-double-inverted pendulum for design points of the Pareto front of multi-objective hybrid of particle swarm optimization and the genetic algorithm

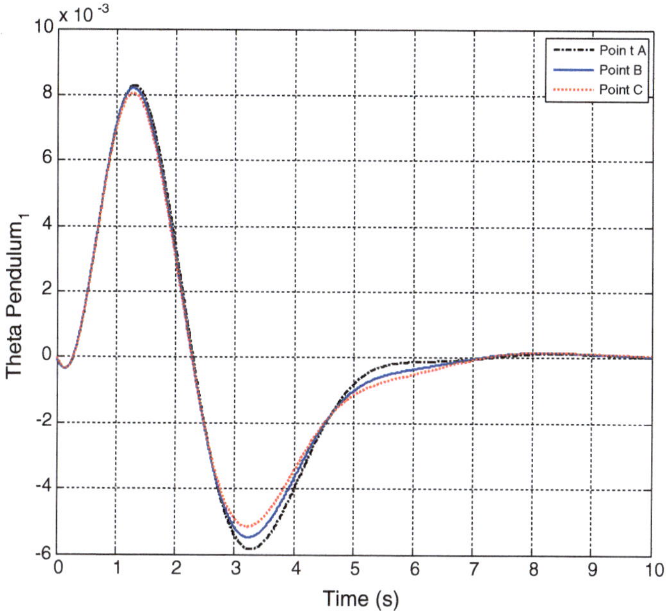

Fig. 22 The angle of the first pendulum of the system of the parallel-double-inverted pendulum for design points of the Pareto front of multi-objective hybrid of particle swarm optimization and the genetic algorithm

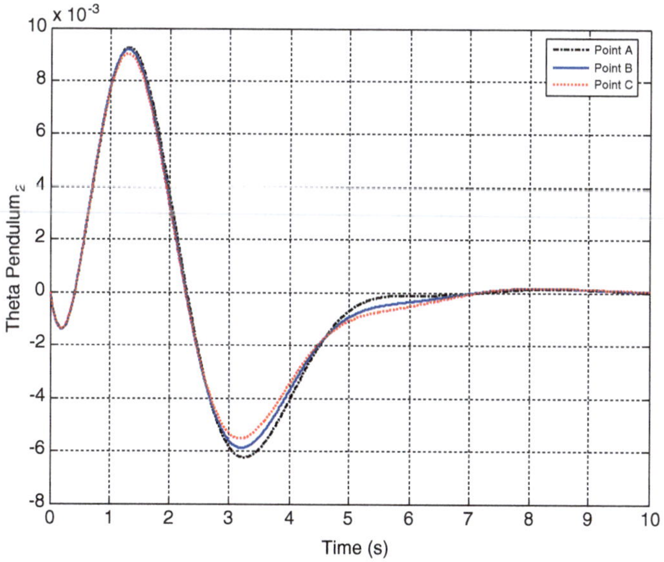

Fig. 23 The angle of the second pendulum of the system of the parallel-double-inverted pendulum for design points of the Pareto front of multi-objective hybrid of particle swarm optimization and the genetic algorithm

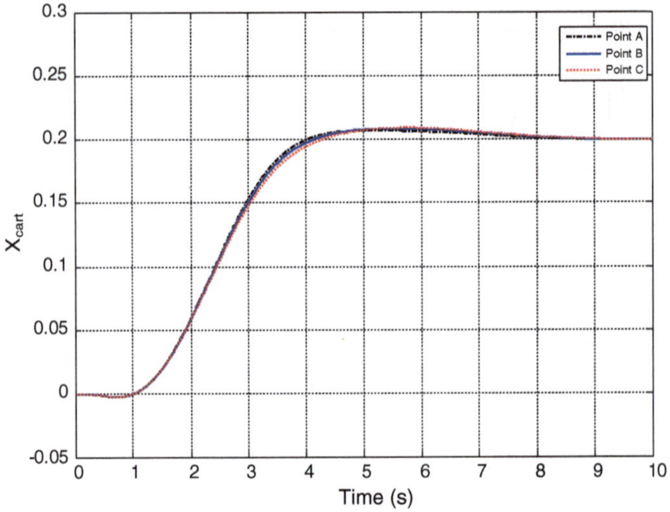

Fig. 24 The position of the cart of the system of the parallel-double-inverted pendulum for design points of the Pareto front of multi-objective hybrid of particle swarm optimization and the genetic algorithm

8 Conclusions

In this work, a hybrid algorithm using GA operators and PSO formula was presented via using effectual operators, for example, traditional and multiple-crossover, mutation and PSO formula. The traditional and multiple-crossover probabilities were based upon fuzzy relations. Five prominent multi-objective test functions and nine single-objective test functions were used to evaluate the capabilities of the hybrid algorithm. Contrasting the results of the hybrid algorithm with other algorithms demonstrates the superiority of the hybrid algorithm with regard to single and multi-objective optimization problems. Moreover, the hybrid optimization algorithm was used to obtain the Pareto front of non-commensurable objective functions in designing parameters of linear state feedback control for a parallel-double-inverted pendulum system. The conflicting objective functions of this problem were the sum of settling time and overshoot of the cart and the sum of settling time and overshoot of the first and second pendulums. The hybrid algorithm could design the parameters of the controller appropriately in order to minimize both objective functions simultaneously.

Acknowledgments The authors would like to thank the anonymous reviewers for their valuable suggestions that enhance the technical and scientific quality of this paper.

References

Abdel-Kader, R. F. (2010). Generically improved PSO algorithm for efficient data clustering. In *The 2010 Second International Conference on Machine Learning and Computing (ICMLC)*, February 9–11, 2010, Bangalore (pp. 71–75). doi:10.1109/ICMLC.2010.19.

Ahmadi, M. H., Aghaj, S. S. G., & Nazeri, A. (2013). Prediction of power in solar stirling heat engine by using neural network based on hybrid genetic algorithm and particle swarm optimization. *Neural Computing and Applications, 22*(6), 1141–1150.

Altun, A. A. (2013). A combination of genetic algorithm, particle swarm optimization and neural network for palmprint recognition. *Neural Computing and Applications, 22*(1), 27–33.

Aziz, A. S. A., Azar, A. T., Salama, M. A., Hassanien, A. E., & Hanafy, S. E. O. (2013). In *The 2013 Federated Conference on Computer Science and Information Systems (FedCSIS)*, September 8-11, 2013, Kraków (pp. 769–774).

Bhuvaneswari, R., Sakthivel, V. P., Subramanian, S., & Bellarmine, G. T. (2009). Hybrid approach using GA and PSO for alternator design. In *The 2009. SOUTHEASTCON '09. IEEE Southeastcon*, March 5–8, 2009, Atlanta (pp. 169–174). doi:10.1109/SECON.2009.5174070.

Blake, A. (1989). Comparison of the efficiency of deterministic and stochastic algorithms for visual reconstruction. *IEEE Transactions on Pattern Analysis and Machine Intelligence, 11*(1), 2–12.

Castillo-Villar, K. K., Smith, N. R., & Herbert-Acero, J. F. (2014). Design and optimization of capacitated supply chain networks including quality measures. *Mathematical Problems in Engineering, 2014*, 17, Article ID 218913. doi:10.1155/2014/218913.

Castillo-Villar, K. K., Smith, N. R., & Simonton, J. L. (2012). The impact of the cost of quality on serial supply-chain network design. *International Journal of Production Research, 50*(19), 5544–5566.

Chang, W. D. (2007). A multi-crossover genetic approach to multivariable PID controllers tuning. *Expert Systems with Applications, 33*(3), 620–626.

Chen, J. L., & Chang, W. D. (2009). Feedback linearization control of a two link robot using a multi-crossover genetic algorithm. *Expert Systems with Applications, 36*(2), 4154–4159.

Chen, C.-H., & Liao, Y.- Y. (2014). Tribal particle swarm optimization for neurofuzzy inference systems and its prediction applications. *Communications in Nonlinear Science and Numerical Simulation, 19*(4), 914–929.

Chen, Z., Meng, W., Zhang, J., & Zeng, J. (2009). Scheme of sliding mode control based on modified particle swarm optimization. *Systems Engineering-Theory & Practice, 29*(5), 137–141.

Chutarat, A. (2001). Experience of light: The use of an inverse method and a genetic algorithm in day lighting design. *Ph.D. Thesis*, Department of Architecture, MIT, Massachusetts, USA.

Cordella, F., Zollo, L., Guglielmelli, E., & Siciliano, B. (2012). A bio-inspired grasp optimization algorithm for an anthropomorphic robotic hand. *International Journal on Interactive Design and Manufacturing (IJIDeM), 6*(2), 113–122.

Deb, K., Pratap, A., Agarwal, S., & Meyarivan, T. (2002). A fast and elitist multiobjective genetic algorithm: NSGA-II. *IEEE Transactions on Evolutionary Computation, 6*(2), 182–197.

Deb, K., & Padhye, N. (2013). Enhancing performance of particle swarm optimization through an algorithmic link with genetic algorithms. *Computational Optimization and Applications, 57*(3), 761–794.

Dhadwal, M. K., Jung, S. N., & Kim, C. J. (2014). Advanced particle swarm assisted genetic algorithm for constrained optimization problems. *Computational Optimization and Applications, 58*(3), 781–806.

Eberhart, R., Simpson, P., & Dobbins, R. (1996). *Computational intelligence PC tools*. Massachusetts: Academic Press Professional Inc.

Eberhart, R. C., Kennedy, J. (1995). A new optimizer using particle swarm theory. In The Proceedings of the Sixth International Symposium on Micro Machine and Human Science, October 4–6, 1995, Nagoya (pp. 39–43). doi:10.1109/MHS.1995.494215.

Elsayed, S. M., Sarker, R. A., & Essam, D. L. (2014). A new genetic algorithm for solving optimization problems. *Engineering Applications of Artificial Intelligence, 27*, 57–69.

Elshazly, H. I., Azar, A. T., Hassanien, A. E., & Elkorany, A. M. (2013). Hybrid system based on rough sets and genetic algorithms for medical data classifications. *International Journal of Fuzzy System Applications (IJFSA), 3*(4), 31–46.

Engelbrecht, A. P. (2002). *Computational intelligence: An introduction*. New York: Wiley.

Engelbrecht, A. P. (2005). *Fundamentals of computational swarm intelligence*. New York: Wiley.

Fleming, P. J., & Purshouse, R. C. (2002). Evolutionary algorithms in control systems engineering: A survey. *Control Engineering Practice, 10*(11), 1223–1241.

Fonseca, C. M., & Fleming, P. J. (1994). Multiobjective optimal controller design with genetic algorithms. In The International Conference on Control, March 21–24, 1994, Coventry (pp. 745–749). doi:10.1049/cp:19940225.

Gaing, Z. L. (2004). A particle swarm optimization approach for optimum design of PID controller in AVR system. *IEEE Transactions on Energy Conversion, 19*(2), 384–391.

Gero, J., & Radford, A. (1978). A dynamic programming approach to the optimum lighting problem. *Engineering Optimization, 3*, 71–82.

Gosh, A., Das, S., Chowdhury, A., & Giri, R. (2011). An ecologically inspired direct search method for solving optimal control problems with Bezier parameterization. *Engineering Applications of Artificial Intelligence, 24*(7), 1195–1203.

Holland, J. H. (1975). *Adaptation in natural and artificial systems: An introductory analysis with applications to biology, control, and artificial intelligence*. Ann Arbor, Michigan: University of Michigan Press.

Jamili, A., Shafia, M. A., & Tavakkoli-Moghaddam, R. (2011). A hybrid algorithm based on particle swarm optimization and simulated annealing for a periodic job shop scheduling problem. *The International Journal of Advanced Manufacturing Technology, 54*(1–4), 309–322.

Jeong, S., Hasegawa, S., Shimoyama, K., & Obayashi, A. (2009). Development and investigation of efficient GA/PSO-hybrid algorithm applicable to real-world design optimization. *IEEE Computational Intelligence Magazine, 4*(3), 36–44.

Kennedy, J., & Eberhart, R. C. (1995). Particle swarm optimization. In The IEEE International Conference on Neural Networks, November/December, 1995, Perth (pp. 1942–1948). doi:10.1109/ICNN.1995.488968.

Ker-Wei, Y., & Shang-Chang, H. (2006). An application of AC servo motor by using particle swarm optimization based sliding mode controller. In The IEEE International Conference on Systems, Man and Cybernetics, October 8-11, 2006, Taipei (pp. 4146–4150). doi:10.1109/ICSMC.2006.384784.

Knowles, J., & Corne, D. (1999). The Pareto archived evolution strategy: A new baseline algorithm for multiobjective optimization. In The Proceedings of the 1999 Congress on Evolutionary Computation, July, 1999, Washington (pp. 98–105). doi:10.1109/CEC.1999.781913.

Li, Z., Yang, K., Bogdan, S., & Xu, B. (2013). On motion optimization of robotic manipulators with strong nonlinear dynamic coupling using support area level set algorithm. *International Journal of Control, Automation and Systems, 11*(6), 1266–1275.

Mahmoodabadi, M. J., Safaie, A. A., Bagheri, A., & Nariman-zadeh, N. (2013). A novel combination of particle swarm optimization and genetic algorithm for pareto optimal design of a five-degree of freedom vehicle vibration model. *Applied Soft Computing, 13*(5), 2577–2591.

Mahmoodabadi, M. J., Bagheri, A., Arabani Mostaghim, S., & Bisheban, M. (2011). Simulation of stability using Java application for Pareto design of controllers based on a new multi-objective particle swarm optimization. *Mathematical and Computer Modelling, 54*(5–6), 1584–1607.

Mahmoodabadi, M. J., Momennejad, S., & Bagheri, A. (2014a). Online optimal decoupled sliding mode control based on moving least squares and particle swarm optimization. *Information Sciences, 268*, 342–356.

Mahmoodabadi, M. J., Taherkhorsandi, M., & Bagheri, A. (2014b). Optimal robust sliding mode tracking control of a biped robot based on ingenious multi-objective PSO. *Neurocomputing, 124*, 194–209.

Mahmoodabadi, M. J., Taherkhorsandi, M., & Bagheri, A. (2014c). Pareto design of state feedback tracking control of a biped robot via multiobjective PSO in comparison with sigma method and genetic algorithms: Modified NSGAII and MATLAB's toolbox. *The Scientific World Journal, 2014*, 8, Article ID 303101.

Mavaddaty, S., & Ebrahimzadeh, A. (2011). Blind signals separation with genetic algorithm and particle swarm optimization based on mutual information. *Radioelectronics and Communications Systems, 54*(6), 315–324.

McGookin, E. W., Murray-Smith, D. J., Li, Y., & Fossen, T. I. (2000). The optimization of a tanker autopilot control system using genetic algorithms. *Transactions of the Institute of Measurement and Control, 22*(2), 141–178.

Mizumoto, M. (1996). Product-sum-gravity method = fuzzy singleton-type reasoning method = simplified fuzzy reasoning method. In The Proceedings of the Fifth IEEE International Conference on Fuzzy Systems, September 8–11, 1996, New Orleans (pp. 2098–2102). doi:10.1109/FUZZY.1996.552786.

Nickabadi, A., Ebadzadeh, M. M., & Safabakhsh, R. (2012). A competitive clustering particle swarm optimizer for dynamic optimization problems. *Swarm Intelligence, 6*(3), 177–206.

Premalatha, K., & Natarajan, A. M. (2009). Discrete PSO with GA operators for document clustering. *International Journal of Recent Trends in Engineering, 1*(1), 20–24.

Puri, P., & Ghosh, S. (2013). A hybrid optimization approach for PI controller tuning based on gain and phase margin specifications. *Swarm and Evolutionary Computation, 8*, 69–78.

Qiao, W., Venayagamoorthy, G. K., & Harley, R. G. (2006). Design of optimal PI controllers for doubly fed induction generators driven by wind turbines using particle swarm optimization. In The International Joint Conference on Neural Networks, Vancouver (pp. 1982–1987). doi:10.1109/IJCNN.2006.246944.

Ratnaweera, A., Halgamuge, S. K., & Watson, H. C. (2004). Self-organizing hierarchical particle swarm optimizer with time-varying acceleration coefficient computation. *IEEE Transactions on Evolutionary Computation, 8*(3), 240–255.

Ravindran, A., Ragsdell, K. M., & Reklaitis, G. V. (2006). *Engineering optimization: Method and applications* (2nd ed.). New Jersey: Wiley.

Sakamoto, Y., Nagaiwa, A., Kobayasi, S., & Shinozaki, T. (1999). An optimization method of district heating and cooling plant operation based on genetic algorithm. *ASHRAE Transaction, 105*, 104–115.

Samarghandi, H., & ElMekkawy, T. Y. (2012). A genetic algorithm and particle swarm optimization for no-wait flow shop problem with separable setup times and makespan criterion. *The International Journal of Advanced Manufacturing Technology, 61*(9–12), 1101–1114.

Sanchez, G., Villasana, M., & Strefezza, M. (2007). Multi-objective pole placement with evolutionary algorithms. *Lecture Notes in Computer Science, 4403*, 417–427. doi:10.1007/978-3-540-70928-2_33.

Arumugam, M. S., Rao, M. V. C., & Palaniappan, R. (2005). New hybrid genetic operators for real coded genetic algorithm to compute optimal control of a class of hybrid systems. *Applied Soft Computing, 6*(1), 38–52.

Song, K. S., Kang, S. O., Jun, S. O., Park, H. I., Kee, J. D., Kim, K. H., et al. (2012). Aerodynamic design optimization of rear body shapes of a sedan for drag reduction. *International Journal of Automotive Technology, 13*(6), 905–914.

Talatahari, S., & Kaveh, A. (2007). A discrete particle swarm ant colony optimization for design of steel frames. *Asian Journal of Civil Engineering (Building and Housing), 9*(6), 563–575.

Tang, Y., Wang, Z., & Fang, J. (2011). Controller design for synchronization of an array of delayed neural networks using a controllable probabilistic PSO. *Information Sciences, 181*(20), 4715–4732.

Thakur, M. (2014). A new genetic algorithm for global optimization of multimodal continuous functions. *Journal of Computational Science, 5*(2), 298–311.

Thangaraj, R., Pant, M., Abraham, A., & Bouvry, P. (2011). Particle swarm optimization: Hybridization perspectives and experimental illustrations. *Applied Mathematics and Computation, 217*(12), 5208–5226.

Valdez, F., Melin, P., & Castillo, O. (2011). An improved evolutionary method with fuzzy logic for combining particle swarm optimization and genetic algorithm. *Applied Soft Computing, 11* (2), 2625–2632.

Valdez, F., Melin, P., & Castillo, O. (2009). Evolutionary method combining particle swarm optimization and genetic algorithms using fuzzy logic for decision making. In The IEEE International Conference on Fuzzy Systems, August 20–24, 2009, Jeju Island (pp. 2114–2119). doi:10.1109/FUZZY.2009.5277165.

Wai, R. J., Chuang, K. L., & Lee, J. D. (2007). Total sliding-model-based particle swarm optimization controller design for linear induction motor. In The IEEE Congress on Evolutionary Computation, September 25–28, 2007, Singapore (pp. 4729–4734). doi:10.1109/CEC.2007.4425092.

Wang, H.-B., & Liu, M. (2012). Design of robotic visual servo control based on neural network and genetic algorithm. *International Journal of Automation and Computing, 9*(1), 24–29.

Wang, J. S., Zhang, Y., & Wang, W. (2006). Optimal design of PI/PD controller for non-minimum phase system. *Transactions of the Institute of Measurement and Control, 28*(1), 27–35.

Wang, K., & Zheng, Y. J. (2012). A new particle swarm optimization algorithm for fuzzy optimization of armored vehicle scheme design. *Applied Intelligence, 37*(4), 520–526.

Wang, L., Wang, T.-G., & Luo, Y. (2011). Improved non-dominated sorting genetic algorithm (NSGA)-II in multi-objective optimization studies of wind turbine blades. *Applied Mathematics and Mechanics, 32*(6), 739–748.

Wang, Q., Liu, F., & Wang, X. (2014). Multi-objective optimization of machining parameters considering energy consumption. *The International Journal of Advanced Manufacturing Technology, 71*(5–8), 1133–1142.

Wibowo, W. K., & Jeong, S.-K. (2013). Genetic algorithm tuned PI controller on PMSM simplified vector control. *Journal of Central South University, 20*(11), 3042–3048.

Wright, J., & Farmani, R. (2001). The simultaneous optimization of building fabric construction, HVAC system size, and the plant control strategy. In The Proceedings of the 7th IBPSA Conference: Building Simulation, Rio de Janeiro, August, 2001 (Vol. 2, pp. 865–872).

Yang, Y., Wang, L., Wang, Y., Bi, Z., Xu, Y., & Pan, S. (2014). Modeling and optimization of two-stage procurement in dual-channel supply chain. *Information Technology and Management, 15*(2), 109–118.

Yao, X., Lin, Y., & Lin, G. (1999). Evolutionary programming made faster. *IEEE Transactions on Evolutionary Computation, 3*(2), 82–102.

Zargari, A., Hooshmand, R., & Ataei, M. (2012). A new control system design for a small hydropower plant based on particle swarm optimization-fuzzy sliding mode controller with Kalman estimator. *Transactions of the Institute of Measurement and Control, 34*(4), 388–400.

Zhao, D., & Yi, J. (2006). GA-based control to swing up an acrobot with limited torque. *Transactions of the Institute of Measurement and Control, 28*(1), 3–13.

Zhou, X. C., Zhao, Z. X., Zhou, K. J., & He, C. H. (2012). Remanufacturing closed-loop supply chain network design based on genetic particle swarm optimization algorithm. *Journal of Central South University, 19*(2), 482–487.

Zitzler, E., & Thiele, L. (1999). Multi-objective evolutionary algorithms: A comparative case study. *IEEE Transactions on Evolutionary Computation, 3*(4), 257–271.

Fuzzy Adaptive Controller for a DFI-Motor

Naâmane Bounar, Abdesselem Boulkroune and Fares Boudjema

Abstract This chapter mainly deals with the fuzzy adaptive backstepping control (FABC) design of a doubly-fed induction motor (DFI-Motor). The proposed controller guarantees speed tracking and reactive power regulation at stator side. The DFI-Motor is controlled by acting on the rotor winding and its stator is directly connected to the grid. In the controller designing, a state-all-flux DFI-Motor model with stator voltage vector oriented reference frame is exploited. Our approach is based on the decomposition of the motor model in two coupled subsystems; the stator flux and the speed-rotor flux subsystems. Under some considerations on the system model, the DFI-Motor unity power factor control and speed tracking problem is transferred to the rotor flux control problem. In our control approach, the unknown load torque is estimated on-line by a suitable adaptive law and the nonlinear functions appearing in the tracking errors dynamics and uncertainties are reasonably approximated by adaptive fuzzy systems. A rigorous stability analysis based on Lyapunov theory is performed to guarantee that the complete control system is asymptotically stable. Furthermore, numerical simulation results are provided to verify the effectiveness of the proposed FABC approach.

Keywords Doubly-fed induction motor · Backstepping approach · Fuzzy adaptive control · Complex nonlinear systems

N. Bounar · A. Boulkroune (✉)
LAJ, University of Jijel, BP 98, Ouled Aissa 18000, Jijel, Algeria
e-mail: boulkroune2002@yahoo.fr

N. Bounar
e-mail: bounar18@yahoo.fr

F. Boudjema
LCP, Ecole Nationale Polytechnique (ENP), 10 Av. Hassen Badi, BP 182 Algiers, Algeria
e-mail: fboudjema@yahoo.fr

Q. Zhu and A.T. Azar (eds.), *Complex System Modelling and Control Through Intelligent Soft Computations*, Studies in Fuzziness and Soft Computing 319,
DOI 10.1007/978-3-319-12883-2_3

1 Introduction

The doubly fed induction machine (DFIM) is a wound rotor asynchronous machine; this form of drive is widely used in many industrial plants, for example pumps, compressors and fans. The DFIM has some distinct advantages over the conventional squirrel-cage machine. The DFIM can be fed and controlled from either or both the stator and the rotor windings. Sub and super-synchronous speeds are possible and the system can be used in generator or motor operation like a DC motor (Morel et al. 1998). In motor operation, two solutions are possible, namely: the machine can be supplied by one converter (at the rotor) or by two converters (one at the stator and one at the rotor). The advantage of the first solution is that the power electronic equipment only has to handle a fraction ($\sim$30 %) of the total system power. This allows the minimizing of converter size and therefore a decreased price of the whole system (Morel et al. 1998). However, the disadvantage in terms of cost of the second solution can be compensated by the best control performances of the powered systems (Brown et al. 1992). In the DFI-Motor operation, the inherent instability due of the double feeding requires a performing control to achieve a good stability and to obtain a high dynamic behavior. Different strategies were proposed in the literature to solve the DFI-Motor control problem. Most of the control strategies are established on the vector control based on the flux orientation that offers the decoupled control of the active and reactive powers (Bogalecka and Kzeminski 1993; Drid et al. 2005; Hopfensperger et al. 2000; Leonhard 1997; Morel et al. 1998; Peresada et al. 2003, 1999; Wang and Ding 1993). Therefore, most of the reported control approaches are based on exact knowledge of the DFI-Motor nonlinear model. Then, the control performance of the DFI-Motor is still influenced by the uncertainties, such as parameter variations, external disturbance and unmodeled dynamics, etc.

In electric motor drives and motion control, the fuzzy controller is considered as a promising alternative for conventional control methods in the control of complex nonlinear plants (Ghamri et al. 2007). The fuzzy controller is applied to static power converters, DC and induction motors. It has been reported that fuzzy controllers are more robust to system parameter changes and have better disturbance rejection. The main advantage of fuzzy control as compared to conventional control resides in the fact that no mathematical model of the plant is required and the human experience can be implanted in the controller as fuzzy rules. However, classical fuzzy controllers (i.e. the non-adaptive fuzzy controllers) can not adapt themselves to changes in their environment or in operating conditions. Then, it is necessary to add some form of adaptation that updates the controller parameters in order to maintain and improve the control performance in wide range of changing conditions Lee (1990); Li and Lau (1989). Using fuzzy systems for approximating of the nonlinear uncertain functions, adaptive fuzzy controllers for inductions motors (IM) have been developed in Agamy et al. (2004), Lin et al. (2002), Youcef and Wahba (2009).

Therefore, the motivation of this chapter is the design of a nonlinear controller for DFI-Motor drives which guarantees speed tracking and reactive power regulation at stator side. The DFI-Motor configuration taken in this work uses one converter in the rotor and the stator is directly connected to the line grid. Our approach is based on the decomposition of the machine model in two coupled subsystems; the stator flux and the speed-rotor flux subsystems. First, the stator voltage vector oriented reference frame is adopted, and the stator reactive power regulation purpose is converted into a stator flux regulation problem. In fact, the time varying stator flux vector is required to be orthogonal to line voltage. In fact, the d-axis component of rotor flux appears as the control input for the stator flux subsystem. Then, with an appropriate choice of the stator flux reference and a strict control of d-axis component of rotor flux to a suitable value, the stator flux error dynamics become linear and exponentially stable independently of the speed dynamics. Consequently, the DFI-Motor stator unity power factor control and the speed tracking problems are converted into a rotor flux control problem. The controller design is based on combination of sliding-mode control, fuzzy control and adaptive backstepping control approaches. The adaptive fuzzy systems are used to reasonably approximate the unknown nonlinear functions appearing in the DFI-Motor model and the tracking errors dynamics and the uncertainties. While, the sliding-mode control is used to effectively compensate for the unavoidable fuzzy approximation error. The adaptive laws, which are used to estimate on-line the load torque and the fuzzy parameters, are derived in the sense of Lyapunov stability theorem. Briefly, the nonlinear control approach described in this paper has the following important advantages:

- The motor-generated torque becomes linear with respect to system control states.
- The rotor flux can be easily regulated in order to increase the machine efficiency.
- The system robustness can be achieved against the uncertain parameters of DFI-Motor (rotor resistance, stator resistance), perturbations (i.e. the unknown load torque), functional uncertainties, etc.
- The controller design does not strongly depend on the model of DFI-Motor.

Moreover, to the authors' best knowledge, there is no result reported in the literature on the fuzzy adaptive control design for doubly-fed induction machine. It is worth noting that the design of the adaptive control based on state-all flux model, for a DFI-Motor controlled by acting on the rotor winding and with a stator which is directly connected to the grid, is very challenge.

This chapter is organized as follows: Section 2 introduces the state-all-flux DFI-Motor model. In Sect. 3, the DFI-Motor control problem is presented. In Sect. 4, the fuzzy logic system used for approximating the unknown nonlinear function is described. In Sect. 5, the proposed fuzzy adaptive backstepping controller (FABC) is presented. In Sect. 6, the effectiveness of our FABC for a DFI-Motor is demonstrated via some simulations results. Conclusions are drawn in Sect. 7.

2 The DFI-Motor Model

The Concordia and Park transformation's application to the traditional *abc* DFI-Motor model allows to write a dynamic model in a *d-q* synchronous reference frame as follows

$$\begin{aligned}\frac{d\varphi_{sd}}{dt} &= -\frac{R_s}{L_s\sigma}\varphi_{sd} + \frac{R_sM}{L_sL_r\sigma}\varphi_{rd} + \omega_s\varphi_{sq} + u_{sd}\\ \frac{d\varphi_{sq}}{dt} &= -\frac{R_s}{L_s\sigma}\varphi_{sq} + \frac{R_sM}{L_sL_r\sigma}\varphi_{rq} - \omega_s\varphi_{sd} + u_{sq}\\ \frac{d\varphi_{rd}}{dt} &= -\frac{R_r}{L_r\sigma}\varphi_{rd} + \frac{R_rM}{L_sL_r\sigma}\varphi_{sd} + \omega_r\varphi_{rq} + u_{rd}\\ \frac{d\varphi_{rq}}{dt} &= -\frac{R_r}{L_r\sigma}\varphi_{rq} + \frac{R_rM}{L_sL_r\sigma}\varphi_{sq} - \omega_r\varphi_{rd} + u_{rq}\end{aligned} \tag{1}$$

Stator and rotor flux equations are

$$\begin{aligned}\varphi_{sd} &= L_s i_{sd} + M i_{rd}\\ \varphi_{sq} &= L_s i_{sq} + M i_{rq}\\ \varphi_{rd} &= L_r i_{rd} + M i_{sd}\\ \varphi_{rq} &= L_r i_{rq} + M i_{sq}\end{aligned} \tag{2}$$

The mechanical equation is given by

$$J\frac{d\Omega}{dt} = \Gamma_e - \Gamma_l - k_f\Omega \tag{3}$$

The electromagnetic torque is given by

$$\Gamma_e = \frac{pM}{L_sL_r\sigma}\left(\varphi_{sq}\varphi_{rd} - \varphi_{sd}\varphi_{rq}\right) \tag{4}$$

where

s, r	Rotor and stator indices
d, q	Synchronous reference frame
α, β	Stationary reference frame
R, L, M	Resistance, inductance and mutual in ductance
u, i, φ	Voltage, current and flux
θ_s, θ_r	Stator and rotor electrical angles
θ, Ω	Rotor mechanical position and speed
$\omega_s = \frac{d\theta_s}{dt}$, $\omega_r = \frac{d\theta_r}{dt}$, $\omega = \frac{d\theta}{dt}$	Electrical frequencies of stator, rotor and shaft
Γ_l, Γ_e	Load and electromagnetic torque
J, p	Inertia, number of pole pairs
$\sigma = 1-(M^2/L_sL_r)$	Leakage coefficient

In a DFI-Motor, the combined effect of the stator and rotor currents produces a fundamental flux that is sinusoidally distributed around the air gap and that rotates at frequency proportional to the stator supply frequency. For all speed ranges the stator and the rotor angular frequencies are related to the shaft mechanical speed by $\omega_s = \omega_r + \omega$.

Expressions of stator and rotor active and reactive powers are respectively given by

$$\begin{aligned} P_s &= u_{sd} i_{sd} + u_{sq} i_{sq} \\ Q_s &= u_{sq} i_{sd} + u_{sd} i_{sq} \\ P_r &= u_{rd} i_{rd} + u_{rq} i_{rq} \\ Q_r &= u_{rq} i_{rd} + u_{rd} i_{rq} \end{aligned} \tag{5}$$

In the following section, the control objective of the DFI-Motor will be discussed.

3 DFI-Motor Control Objective

First, we suppose that the stator flux vector is aligned with d-axis as shown in Fig. 1. In the stationary frame abc, the component n of the stator voltage equation is given by

$$u_{sn} = R_s i_{sn} + \frac{d\varphi_{sn}}{dt} \tag{6}$$

By neglecting the stator resistance (Hopfensperger et al. 2000), (6) can be written as

$$u_{sn} \approx \frac{d\varphi_{sn}}{dt} \tag{7}$$

This equation demonstrates that the stator voltage vector is $\frac{\pi}{2}$ in advance of the stator flux. Then, in the chosen reference frame, we can write

$$\begin{aligned} u_{sd} &= 0 \\ u_{sq} &= u_s \end{aligned} \tag{8}$$

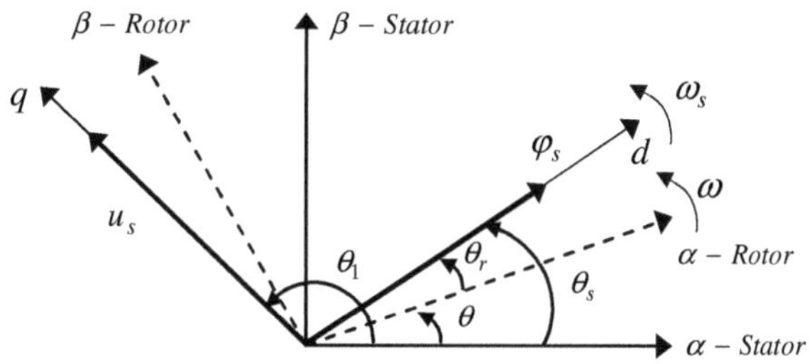

Fig. 1 Reference frames and angles for DFI-Motor

The stator is directly connected to the grid, then, the stator electrical angle θ_s is calculated only with the grid voltage.

$$\theta_s = \theta_1 - \frac{\pi}{2} \tag{9}$$

where $\theta_1 = \arctan\left(u_{s\beta}/u_{s\alpha}\right)$ is the stator voltage vector angle in the stationary reference frame *abc* as shown in Fig. 1.

Our control objective is the design of a controller for the DFI-Motor which ensures reactive power regulation at stator side and speed tracking reference with unknown load torque. It will be demonstrated that the stator-side reactive power regulation problem can be formalized as the requirement to guarantee that the line voltage vector and the stator flux vector are orthogonal.

Considering the stator equations expressed in terms of stator fluxes and currents in the line voltage reference frame

$$\begin{aligned} \dot{\varphi}_{sd} &= -R_s i_{sd} + \omega_s \varphi_{sq} \\ \dot{\varphi}_{sq} &= -R_s i_{sq} - \omega_s \varphi_{sd} + u_s \end{aligned} \tag{10}$$

From the second equation of (5), the unity power factor objective is equivalent to $i_{sd} = 0$. In steady-state condition, all the derivatives are zero. According to the first equation of (10), $\varphi_{sq} = 0$ is necessary to ensure $i_{sd} = 0$. Then, the stator-side unity power factor control is reformulated as a stator flux orientation control objective (the stator flux vector is required to be orthogonal to line voltage vector).

In the following section, the fuzzy logic system used to approximate the uncertain functions will be described in detail.

4 Description of the Fuzzy Logic System

The basic configuration of a fuzzy logic system consists of a fuzzifier, some fuzzy IF-THEN rules, a fuzzy inference engine and a defuzzifier, as shown in Fig. 2. The fuzzy inference engine uses the fuzzy IF-THEN rules to perform a mapping from an input vector $x^T = [x_1, x_2, \ldots, x_n] \in R^n$ to an output $\hat{f} \in R$. The *i*th fuzzy rule is written as

$$R^{(i)}: \text{ if } x_1 \text{ is } A_1^i \text{ and} \ldots \text{and } x_n \text{ is } A_n^i \text{ then } \hat{f} \text{ is } f^i \tag{11}$$

where $A_1^i, A_2^i, \ldots,$ and A_n^i are fuzzy sets and f^i is the fuzzy singleton for the output in the *i*th rule. By using the singleton fuzzifier, product inference, and center-average defuzzifier, the output of the fuzzy system can be expressed as follows:

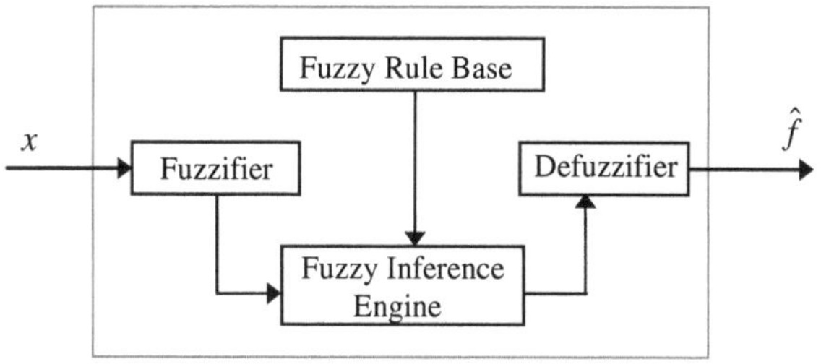

Fig. 2 The basic configuration of a fuzzy logic system

$$\begin{aligned}\hat{f}(x) &= \frac{\sum_{i=1}^{m} f^i \left(\prod_{j=1}^{n} \mu_{A_j^i}(x_j)\right)}{\sum_{i=1}^{m} \left(\prod_{j=1}^{n} \mu_{A_j^i}(x_j)\right)} \\ &= \theta^T \psi(x)\end{aligned} \tag{12}$$

where $\mu_{A_j^i}(x_j)$ is the degree of membership of x_j to A_j^i, m is the number of fuzzy rules, $\theta^T = [f^1, f^2, \ldots, f^m]$ is the adjustable parameter vector (composed of consequent parameters), and $\psi^T = [\psi^1 \psi^2 \ldots \psi^m]$ with

$$\psi^i(x) = \frac{\left(\prod_{j=1}^{n} \mu_{A_j^i}(x_j)\right)}{\sum_{i=1}^{m} \left(\prod_{j=1}^{n} \mu_{A_j^i}(x_j)\right)}$$

being the *fuzzy basis function* (*FBF*). Throughout the paper, it is assumed that the FBFs are selected so that there is always at least one active rule (Wang 1994), i.e. $\sum_{i=1}^{m} \left(\prod_{j=1}^{n} \mu_{A_j^i}(x_j)\right) > 0$.

It is worth noting that the fuzzy system (12) is commonly used in control applications. Following the universal approximation results (Wang 1994; Azar 2010a, b, 2012), the fuzzy system (12) is able to approximate any nonlinear smooth function $f(x)$ on a compact operating space to an arbitrary degree of accuracy. Of particular importance, it is assumed that the structure of the fuzzy system (i.e. the pertinent inputs, the number of membership functions for each input and the number of rules) and the membership function parameters are properly specified beforehand. The consequent parameters θ are then determined by appropriate adaptation algorithms.

In the following section, the proposed fuzzy adaptive backstepping controller will be presented.

5 Design of the Fuzzy Adaptive Backstepping Control

In this section, the stator flux subsystem control is designed in order to achieve asymptotic alignment of the stator flux vector with the d-axis of the line voltage vector reference frame, consequently, the stator voltage and flux vectors become orthogonal.

Introduce flux stator tracking errors as

$$\widetilde{\varphi}_{sd} = \varphi_{sd} - \varphi_s^*, \quad \widetilde{\varphi}_{sq} = \varphi_{sq} \tag{13}$$

where φ_s^* is the d-axis flux reference trajectory.

Using (8), the stator flux dynamic equations in (1) can be written in error form as

$$\begin{aligned} \dot{\widetilde{\varphi}}_{sd} &= -a_1 \widetilde{\varphi}_{sd} - a_1 \varphi_s^* + a_2 \varphi_{rd} + \omega_s \widetilde{\varphi}_{sq} - \dot{\varphi}_s^* \\ \dot{\widetilde{\varphi}}_{sq} &= -a_1 \widetilde{\varphi}_{sq} + a_2 \varphi_{rq} - \omega_s \widetilde{\varphi}_{sd} - \omega_s \varphi_s^* + u_s \end{aligned} \tag{14}$$

where $a_1 = R_s/L_s\sigma$, $a_2 = R_sM/L_rL_s\sigma$.

To realize the required stator flux orientation, the d-axis component of rotor flux φ_{rd} can be considered as control input in (14), and should be

$$\varphi_{rd} = \frac{1}{a_2}\left(a_1\varphi_s^* + \dot{\varphi}_s^*\right) \tag{15}$$

with the d-axis stator flux reference computed from the second equation of (14)

$$\varphi_s^* = \frac{1}{\omega_s}\left(u_s + a_2\varphi_{rq}\right) \tag{16}$$

Using (15) and (16), (14) becomes

$$\begin{aligned} \dot{\widetilde{\varphi}}_{sd} &= -a_1 \widetilde{\varphi}_{sd} + \omega_s \widetilde{\varphi}_{sq} + a_2(\varphi_{rd} - \varphi_{rd}^*) \\ \dot{\widetilde{\varphi}}_{sq} &= -a_1 \widetilde{\varphi}_{sq} - \omega_s \widetilde{\varphi}_{sd} \end{aligned} \tag{17}$$

However, in a DFI-Motor, the rotor flux is not available as control input and φ_{rd} in (15) can only represent the d-axis rotor flux reference φ_{rd}^* for the real flux φ_{rd}. The rotor voltages u_{rd} and u_{rq} are the only physical available control inputs of DFI-Motor. From (17), one concludes that the dynamic of the stator flux is exponentially stable (i.e. $\lim_{t\to\infty} \varphi_{sd} = \varphi_s^*$ and $\lim_{t\to\infty} \varphi_{sq} = 0$) provided that $\lim_{t\to\infty} \varphi_{rd} = \varphi_{rd}^*$.

Remark 1 From (17) and (15), it can be concluded that in the steady state (φ_s^* constant), $\varphi_{rd} = \frac{a_1\varphi_s^*}{a_2} = \frac{L_r}{M}\varphi_s^*$.

Now, it is required to design a control law (rotor voltages u_{rd} and u_{rq}) which guarantees that $\lim_{t\to\infty} \varphi_{rd} = \varphi_{rd}^*$ and $\lim_{t\to\infty} \Omega = \Omega^*$. Then, we will consider the reduced order DFI-Motor model represented by the rotor flux and speed equations.

$$\begin{aligned}\dot{x}_1 &= a_5(x_4x_3 - x_5x_2) - a_6x_1 - a_7\Gamma_l \\ \dot{x}_2 &= -a_3x_2 + a_4x_4 - \omega_r x_3 + \delta_1(x_1, x_2) + u_1 \\ \dot{x}_3 &= -a_3x_3 + a_4x_5 + \omega_r x_2 + \delta_2(x_3, x_2) + u_2\end{aligned} \tag{18}$$

with $x_1 = \Omega, x_2 = \varphi_{rq}$, $x_3 = \varphi_{rd}, x_4 = \varphi_{sq}$, $x_5 = \varphi_{sd}, u_1 = u_{rq}$, $u_2 = u_{rd}, a_3 = R_r/L_r\sigma$, $a_4 = R_rM/L_rL_s\sigma$, $a_5 = pM/JL_rL_s\sigma$, $a_6 = k_f/J$ and $a_7 = 1/J$.where δ_i $(i = 1, 2)$ are the unknown uncertainties and perturbations that can be naturally generated from the parameter variations.

Backstepping design procedure (Krstic et al. 1995) is used here for the construction of the FABC which guarantees asymptotic tracking of rotor speed and rotor flux reference signals. Then, the variables to be controlled in the model (18) are the rotor speed (x_1) and the rotor flux (x_2, x_3).

Step 1. For a continuous bounded reference signal x_{1d}, we define the tracking error e_1 as follows

$$e_1 = x_1 - x_{1d} \tag{19}$$

Its derivative $\dot{e}_1$ is given by

$$\dot{e}_1 = \dot{x}_1 - \dot{x}_{1d} \tag{20}$$

From the first subsystem of (18), we can write

$$\dot{e}_1 = a_5x_4x_3 - a_5x_5x_2 - a_6x_1 - a_7\Gamma_l - \dot{x}_{1d} \tag{21}$$

Choose $a_5x_5x_2$ as a virtual control to stabilize e_1 and select υ_1 as a desired reference signal for $a_5x_5x_2$

$$\upsilon_1 = a_5x_4x_{3d} + c_1e_1 - a_6x_{1d} - \dot{x}_{1d} - a_7\Gamma_l \tag{22}$$

where $c_1 > 0$ is a design constant.
However, the exact value of the external load torque Γ_l in (22) is generally difficult to be known in advance for practical applications. Then, it cannot be used in the virtual control signal. We can select the new virtual control as follows

$$\upsilon_2(z_0) = a_5x_4x_{3d} + c_1e_1 - a_6x_{1d} - \dot{x}_{1d} - a_7\hat{\Gamma}_l \tag{23}$$

where $\hat{\Gamma}_l$ is the estimate of Γ_l and $z_0 = [x_1, x_4, \hat{\Gamma}_l]^T$.
This leads to the following dynamics

$$\dot{e}_1 = a_5 x_4 e_3 - e_2 - (c_1 + a_6)e_1 - a_7 \widetilde{\Gamma}_l \tag{24}$$

where $\widetilde{\Gamma}_l = \Gamma_l - \hat{\Gamma}_l$ is the load torque estimation error, and e_2 is the tracking error of the variable $a_5 x_5 x_2$.

$$e_2 = a_5 x_5 x_2 - \upsilon_2 \tag{25}$$

Consider the following Lyapunov function candidate for the e_1-subsystem

$$V_1 = \frac{1}{2}\left(e_1^2 + \frac{1}{\gamma}\widetilde{\Gamma}_l^2\right) \tag{26}$$

where $\gamma > 0$ is a design constant.
By assuming that the load torque is slowly time-varying ($\dot{\Gamma}_l = 0$), the time-derivative of (26) along (24) is given by

$$\dot{V}_1 = -e_1 e_2 + a_5 x_4 e_3 e_1 - (c_1 + a_6)e_1^2 - \widetilde{\Gamma}_l\left(a_7 e_1 + \frac{1}{\gamma}\dot{\hat{\Gamma}}_l\right) \tag{27}$$

If the load torque adaptation law is designed as

$$\dot{\hat{\Gamma}}_l = \beta \widetilde{\Gamma}_l - \gamma a_7 e_1 \tag{28}$$

where $\beta > 0$ is a design parameter.
Then, (27) can be written as

$$\dot{V}_1 = -e_1 e_2 + a_5 x_4 e_3 e_1 - (c_1 + a_6)e_1^2 - \frac{\beta}{\gamma}\widetilde{\Gamma}_l^2 \tag{29}$$

The next step consists in stabilizing the tracking error e_2.

Step 2. The time-derivative of (25) is given by

$$\dot{e}_2 = a_5 x_5 \dot{x}_2 + a_5 \dot{x}_5 x_2 - \dot{\upsilon}_2 \tag{30}$$

From the second subsystem of (1), (18) and (23), we can write

$$\begin{aligned}\dot{e}_2 = {} & f_1(z_1) + e_1 + (a_5 a_2 x_2 - a_5 x_5 \omega_r - a_5 c_1 x_4)e_3 \\ & - a_7(\beta + c_1)\hat{\Gamma}_l + a_5 x_5 u_1\end{aligned} \tag{31}$$

with

$$\begin{aligned}f_1(z_1) = {} & -e_1 - a_5 a_3 x_5 x_2 + a_5 a_4 x_5 x_4 - a_5 x_5 \omega_r x_{3d} - a_5 a_1 x_5 x_2 \\ & + a_5 \omega_s x_4 x_2 + a_5 a_1 x_{3d} x_4 + a_5 \omega_s x_{3d} x_5 - a_5 x_{3d} u_s - a_5 x_4 \dot{x}_{3d} \\ & + a_6 \dot{x}_{1d} + \ddot{x}_{1d} + c_1(e_2 + (c_1 + a_6)e_1) - a_7^2 \gamma e_1 \\ & + a_7(\beta + c_1)\Gamma_l + a_5 x_5 \delta_1(x_1, x_2)\end{aligned}$$

where $z_1 = [x_1, x_2, x_4, x_5, v_2, \Gamma_l]^T$ and e_3 is the tracking error of x_3. It is given by

$$e_3 = x_3 - x_{3d} \tag{32}$$

The desired signal x_{3d} is given by the expression (15), i.e. $x_{3d} = \frac{1}{a_2}\left(a_1\varphi_s^* + \dot{\varphi}_s^*\right)$.
The uncertain continuous function $f_1(z_1)$ can be approximated by the fuzzy system (12) as follows

$$\hat{f}_1(z_1, \theta_1) = \theta_1^T \psi_1(z_1) \tag{33}$$

where $\psi_1(z_1)$ is the FBF vector, which is fixed a priori by the designer, and θ_1 is the adjustable parameter vector of the fuzzy system. Furthermore, the functions $f_1(z_1)$ can be approximated optimally (Wang 1993, 1994) as follows

$$\begin{aligned} f_1(z_1) &= \hat{f}_1(z_1, \theta_1^*) + \varepsilon_1(z_1) \\ &= \theta_1^{*T} \psi_1(z_1) + \varepsilon_1(z_1) \end{aligned} \tag{34}$$

where θ_1^* is the optimal parameter vector and $\varepsilon_1(z_1)$ is the unavoidable fuzzy approximation error which is generally assumed to be bounded (Boulkroune et al. 2008, 2009, 2010a, b; Wang 1993, 1994) as follows

$$|\varepsilon_1(z_1)| \leq \bar{\varepsilon}_1, \quad \forall z_1 \in \Omega_{z_1}$$

where $\bar{\varepsilon}_1$ is an unknown constant.
Since the input vector $z_1 = [x_1, x_2, x_4, x_5, v_2, \Gamma_l]^T$ is not available, it must be replaced by its estimate $\hat{z}_1 = [x_1, x_2, x_4, x_5, v_2, \hat{\Gamma}_l]^T$ in (33). Thus, the fuzzy system (33) used to approximate $f_1(z_1)$ is replaced by the following fuzzy system:

$$\hat{f}_1(\hat{z}_1, \theta_1) = \theta_1^T \psi_1(\hat{z}_1) \tag{35}$$

From (33–35), we have

$$\begin{aligned} f_1(z_1) &= f_1(z_1) - \hat{f}_1(z_1, \theta_1^*) + \hat{f}_1(z_1, \theta_1^*) - \hat{f}_1(\hat{z}_1, \theta_1^*) + \hat{f}_1(\hat{z}_1, \theta_1^*) \\ &= \hat{f}_1(\hat{z}_1, \theta_1^*) + f_1(z_1) - \hat{f}_1(z_1, \theta_1^*) + \hat{f}_1(z_1, \theta_1^*) - \hat{f}_1(\hat{z}_1, \theta_1^*) \\ &= \theta_1^{*T} \psi(\hat{z}_1) + \varepsilon_1(z_1) + [\theta_1^{*T} \psi_1(z_1) - \theta_1^{*T} \psi_1(\hat{z}_1)] \\ &= \theta_1^{*T} \psi(\hat{z}_1) + \vartheta_1(z_1, \hat{z}_1) \end{aligned} \tag{36}$$

where $\vartheta_1(z_1, \hat{z}_1) = \varepsilon_1(z_1) + [\theta_1^{*T}\psi_1(z_1) - \theta_1^{*T}\psi_1(\hat{z}_1)]$ is the approximation error. Notice that $\vartheta_1(z_1, \hat{z}_1)$ has an upper bound, i.e. $|\vartheta_1(z_1, \hat{z}_1)| \leq \kappa_1^*$ with κ_1^* is an unknown positive constant (Boulkroune et al. 2008).
To stabilise the dynamics (31), the following fuzzy adaptive controller is proposed

$$u_1 = \frac{1}{a_5x_5}\left(a_7(\beta + c_1)\hat{\Gamma}_l - \theta_1^T\psi_1(\hat{z}_1) - \lambda_1 e_2 - \frac{\kappa_1^2 e_2}{\kappa_1|e_2| + \sigma_1\, e^{-\sigma_2 t}}\right) \tag{37}$$

where σ_1 and $\sigma_2 > 0$ are small design constants and λ_1 is a positive design constant and κ_1 is the estimate of the unknown bound κ_1^*.

Remark 2 The magnetising flux x_5 must be non-zero (remanence flux).
Replacing (37) into (31) and using (36) yields

$$\begin{aligned}\dot{e}_2 &= e_1 + (a_5a_2x_2 - a_5x_5\omega_r - a_5c_1x_4)e_3 - \widetilde{\theta}_1^T\psi_1(\hat{z}_1)\\ &\quad + \vartheta_1(z_1, \hat{z}_1) - \lambda_1 e_2 - \frac{\kappa_1^2 e_2}{\kappa_1|e_2| + \sigma_1\, e^{-\sigma_2 t}}\end{aligned} \tag{38}$$

where $\widetilde{\theta}_1 = \theta_1 - \theta_1^*$ is the parameter error vector.
Multiplying (38) by e_2, we get

$$\begin{aligned}e_2\dot{e}_2 &= e_1e_2 + (a_5a_2x_2 - a_5x_5\omega_r - a_5c_1x_4)e_2e_3 - e_2\widetilde{\theta}_1^T\psi_1(\hat{z}_1)\\ &\quad + e_2\vartheta_1(z_1, \hat{z}_1) - \lambda_1 e_2^2 - \frac{\kappa_1^2 e_2^2}{\kappa_1|e_2| + \sigma_1\, e^{-\sigma_2 t}}\\ &\leq e_1e_2 + (a_5a_2x_2 - a_5x_5\omega_r - a_5c_1x_4)e_2e_3 - e_2\widetilde{\theta}_1^T\psi_1(\hat{z}_1) + \kappa_1^*|e_2|\\ &\quad - \lambda_1 e_2^2 - \frac{\kappa_1^2 e_2^2}{\kappa_1|e_2| + \sigma_1\, e^{-\sigma_2 t}}\\ &= e_1e_2 + (a_5a_2x_2 - a_5x_5\omega_r - a_5c_1x_4)e_2e_3 - e_2\widetilde{\theta}_1^T\psi_1(\hat{z}_1)\\ &\quad - \widetilde{\kappa}_1|e_2| - \lambda_1 e_2^2 + \sigma_1\, e^{-\sigma_2 t}\end{aligned} \tag{39}$$

where $\widetilde{\kappa}_1 = \kappa_1 - \kappa_1^*$ is the parameter error.
Define a Lyapunov function candidate for the (e_1, e_2)-subsystem as follows

$$V_2 = V_1 + \frac{1}{2}e_2^2 + \frac{1}{2\gamma_1}\widetilde{\theta}_1^T\widetilde{\theta}_1 + \frac{1}{2\eta_1}\widetilde{\kappa}_1^2 \tag{40}$$

where γ_1 and $\eta_1 > 0$ are design constants.
Take the derivative of V_2 with respect to time and using (39) and (29), one can obtain

$$
\begin{aligned}
\dot{V}_2 &= \dot{V}_1 + e_2\dot{e}_2 + \frac{1}{\gamma_1}\,\widetilde{\theta}_1^T\dot{\theta}_1 + \frac{1}{\eta_1}\,\widetilde{\kappa}_1\dot{\kappa}_1 \\
&\le (a_5a_2x_2 - a_5x_5\omega_r - a_5c_1x_4)e_2e_3 + a_5x_4e_1e_3 - (c_1 + a_6)e_1^2 - \frac{\beta}{\gamma}\widetilde{\Gamma}_l^2 \\
&\quad - e_2\widetilde{\theta}_1^T\psi_1(\widetilde{z}_1) - \widetilde{\kappa}_1|e_2| - \lambda_1e_2^2 + \sigma_1e^{-\sigma_2 t} + \frac{1}{\gamma_1}\,\widetilde{\theta}_1^T\dot{\theta}_1 + \frac{1}{\eta_1}\,\widetilde{\kappa}_1\dot{\kappa}_1 \\
&= (a_5a_2x_2 - a_5x_5\omega_r - a_5c_1x_4)e_2e_3 + a_5x_4e_1e_3 - (c_1 + a_6)e_1^2 - \frac{\beta}{\gamma}\widetilde{\Gamma}_l^2 \\
&\quad - \lambda_1e_2^2 + \sigma_1\,e^{-\sigma_2 t} + \frac{1}{\gamma_1}\,\widetilde{\theta}_1^T\left[\dot{\theta}_1 - \gamma_1e_2\psi_1(\hat{z}_1)\right] + \frac{1}{\eta_1}\,\widetilde{\kappa}_1[\dot{\kappa}_1 - \eta_1|e_2|]
\end{aligned} \tag{41}
$$

If the adaptation laws are designed as

$$\dot{\theta}_1 = \gamma_1e_2\psi_1(\hat{z}_1) \tag{42}$$

$$\dot{\kappa}_1 = \eta_1|e_2| \tag{43}$$

Then, (41) can be expressed as follows

$$
\begin{aligned}
\dot{V}_2 \le\, & (a_5a_2x_2 - a_5x_5\omega_r - a_5c_1x_4)e_2e_3 + a_5x_4e_1e_3 \\
& - (c_1 + a_6)e_1^2 - \frac{\beta}{\gamma}\widetilde{\Gamma}_l^2 - \lambda_1e_2^2 + \sigma_1\,e^{-\sigma_2 t}
\end{aligned} \tag{44}
$$

In the next step, we try to stabilize the tracking error e_3.

Step 3. At this step, we will construct the control law u_2. The time-derivative of (32) is given by

$$\dot{e}_3 = -a_3x_3 + a_4\,x_5 + \omega_rx_2 + \delta_2(x_3, x_2) + u_2 - \dot{x}_{3d} \tag{45}$$

We can rewrite (45) as follows

$$\dot{e}_3 = -(a_5a_2x_2 - a_5x_5\omega_r - a_5c_1x_4)e_2 - a_5x_4e_1 + f_2(z_2) + u_2 \tag{46}$$

with

$$
\begin{aligned}
f_2(z_2) = & (a_5a_2x_2 - a_5x_5\omega_r - a_5c_1x_5)e_2 + a_5x_4e_1 - a_3x_3 \\
& + a_4\,x_5 + \omega_rx_2 + \delta_2(x_3, x_2) - \dot{x}_{3d}
\end{aligned}
$$

where $z_2 = [x_1, x_2, x_3, x_4, x_5]^T$.

The uncertain continuous function $f_2(z_2)$ can be approximated by the fuzzy system (12) as follows

$$\hat{f}_2(z_2, \theta_2) = \theta_2^T \psi_2(z_2) \tag{47}$$

where $\psi_2(z_2)$ is the FBF vector, which is fixed a priori by the designer, and θ_2 is the adjustable parameter vector of the fuzzy system. Furthermore, the functions $f_2(z_2)$ can be approximated optimally (Wang 1993, 1994) as follows

$$\begin{aligned} f_2(z_2) &= \hat{f}_2(z_2, \theta_2^*) + \varepsilon_2(z_2) \\ &= \theta_2^{*T} \psi_2(z_2) + \varepsilon_2(z_2) \end{aligned} \tag{48}$$

where θ_2^* is the optimal parameter vector and $\varepsilon_2(z_2)$ is the unavoidable fuzzy approximation error which is assumed to be bounded (Boulkroune et al. 2008, 2009, 2010a, b; Wang 1993, 1994) as follows

$$|\varepsilon_2(z_2)| \le \bar{\varepsilon}_2, \quad \forall z_2 \in \Omega_{z_2},$$

where $\bar{\varepsilon}_2$ is an unknown constant.
From (47) and (48), we have

$$\begin{aligned} f_2(z_2) &= f_2(z_2) - \hat{f}_2(z_2, \theta_2^*) + \hat{f}_2(z_2, \theta_2^*) \\ &= \hat{f}_2(z_2, \theta_2^*) + f_2(z_2) - \hat{f}_2(z_2, \theta_2^*) \\ &= \theta_2^{*T} \psi_2(z_2) + \varepsilon_2(z_2) \end{aligned} \tag{49}$$

To stabilise the dynamics (46), the following fuzzy adaptive controller is proposed

$$u_2 = -\theta_2^T \psi_2(z_2) - \lambda_2 e_3 - \frac{\kappa_2^2 e_3}{\kappa_2 |e_3| + \sigma_3 e^{-\sigma_4 t}} \tag{50}$$

where σ_3 and $\sigma_4 > 0$ are small design constants, λ_2 is a positive design constant and κ_2 is the estimate of the unknown bound $\kappa_2^* = \bar{\varepsilon}_2$.
Replacing (50) into (46) and using (49) yields

$$\begin{aligned} \dot{e}_3 = &-(a_5 a_2 x_2 - a_5 x_5 \omega_r - a_5 c_1 x_4) e_2 - a_5 x_4 e_1 - \widetilde{\theta}_2^T \psi_2(z_2) \\ &+ \varepsilon_2(z_2) - \lambda_2 e_3 - \frac{\kappa_2^2 e_3}{\kappa_2 |e_3| + \sigma_3 e^{-\sigma_4 t}} \end{aligned} \tag{51}$$

where $\widetilde{\theta}_2 = \theta_2 - \theta_2^*$ is the parameter error vector.
Multiplying (51) by e_3, we get

$$
\begin{aligned}
e_3\dot{e}_3 &= -(a_5a_2x_2 - a_5x_5\omega_r - a_5c_1x_4)e_2e_3 - a_5x_4e_1e_3 \\
&\quad - e_3\widetilde{\theta}_2^T\psi_2(z_2) + e_3\varepsilon_2(z_2) - \lambda_2 e_3^2 - \frac{\kappa_2^2 e_3^2}{\kappa_2|e_3| + \sigma_3\, e^{-\sigma_4 t}} \\
&\leq -(a_5a_2x_2 - a_5x_5\omega_r - a_5c_1x_4)e_2e_3 - a_5x_4e_1e_3 \\
&\quad - e_3\widetilde{\theta}_2^T\psi_2(z_2) + \kappa_2^*|e_3| - \lambda_2 e_3^2 - \frac{\kappa_2^2 e_3^2}{\kappa_2|e_3| + \sigma_3\, e^{-\sigma_4 t}} \\
&= -(a_5a_2x_2 - a_5x_5\omega_r - a_5c_1x_4)e_2e_3 - a_5x_4e_1e_3 \\
&\quad - e_3\widetilde{\theta}_2^T\psi_2(z_2) - \lambda_2 e_3^2 - \widetilde{\kappa}_2|e_3| + \sigma_3\, e^{-\sigma_4 t}
\end{aligned}
\tag{52}
$$

where $\widetilde{\kappa}_2 = \kappa_2 - \kappa_2^*$.
Define a Lyapunov function candidate as follows

$$
V_3 = V_2 + \frac{1}{2}e_3^2 + \frac{1}{2\gamma_2}\widetilde{\theta}_2^T\widetilde{\theta}_2 + \frac{1}{2\eta_2}\widetilde{\kappa}_2^2
\tag{53}
$$

where γ_2 and $\eta_2 > 0$ are design constants.
Take the derivative of V_3 with respect to time and using (52) and (44), one can obtain

$$
\begin{aligned}
\dot{V}_3 &= \dot{V}_2 + e_3\dot{e}_3 + \frac{1}{\gamma_2}\widetilde{\theta}_2^T\dot{\theta}_2 + \frac{1}{\eta_2}\widetilde{\kappa}_2\kappa_2 \\
&\leq -(c_1 + a_6)e_1^2 - \frac{\beta}{\gamma}\widetilde{\Gamma}_l^2 - \lambda_1 e_2^2 - e_3\widetilde{\theta}_2^T\psi_2(z_2) - \lambda_2 e_3^2 \\
&\quad - \widetilde{\kappa}_2|e_3| + \sigma_1 e^{-\sigma_2 t} + \sigma_3 e^{-\sigma_4 t} + \frac{1}{\gamma_2}\widetilde{\theta}_2^T\dot{\theta}_2 + \frac{1}{\eta_2}\widetilde{\kappa}_2\dot{\kappa}_2 \\
&= -(c_1 + a_6)e_1^2 - \frac{\beta}{\gamma}\widetilde{\Gamma}_l^2 - \lambda_1 e_2^2 - \lambda_2 e_3^2 + \sigma_1 e^{-\sigma_2 t} + \sigma_3 e^{-\sigma_4 t} \\
&\quad + \frac{1}{\gamma_2}\widetilde{\theta}_2^T[\dot{\theta}_2 - \gamma_2 e_3\psi_2(z_2)] \\
&\quad + \frac{1}{\eta_2}\widetilde{\kappa}_2[\dot{\kappa}_2 - \eta_2|e_3|]
\end{aligned}
\tag{54}
$$

If the adaptation laws are designed as

$$
\dot{\theta}_2 = \gamma_2 e_3\psi_2(z_2)
\tag{55}
$$

$$
\dot{\kappa}_2 = \eta_2|e_3|
\tag{56}
$$

Then, (54) becomes

$$\dot{V}_3 = \leq -(c_1 + a_6)e_1^2 - \frac{\beta}{\gamma}\widetilde{\Gamma}_l^2 - \lambda_1 e_2^2 - \lambda_2 e_3^2 + \sigma_1 e^{-\sigma_2 t} + \sigma_3 e^{-\sigma_4 t} \tag{57}$$

One can write (57) as follows

$$\dot{V}_3 \leq -\lambda\|E\|^2 + \varsigma(t) \tag{58}$$

where $\lambda = \min\left\{(c_1 + a_6), \frac{\beta}{\gamma}, \lambda_2, \lambda_3\right\}$, $E = \left[e_1, e_2, e_3, \widetilde{\Gamma}_l\right]^T$, and $\varsigma(t) = \sigma_1 e^{-\sigma_2 t} + \sigma_3 e^{-\sigma_4 t}$.

Note that $\varsigma(t)$ verifies the following nice properties:

- $\varsigma(t) \in L_\infty$ and $\lim_{t\to\infty} \varsigma(t) = 0$
- $\varsigma(t) \in L_2$

Those properties will be exploited later in the stability analysis.

5.1 Study of the Tracking Error Convergence

The study of the asymptotic convergence of tracking errors is divided into three parts.

5.1.1 Proof of the Boundedness and Square Integrability of the Tracking Errors

By inequality (58), $\dot{V}_3$ can be rewritten as $\dot{V}_3 \leq -\lambda\|E\|^2 + \sigma_1 + \sigma_3$. Choosing $\lambda > \frac{\sigma_1 + \sigma_3}{\chi^2}$ for any small $\chi > 0$, there exists a constant λ_0 such that $\dot{V}_3 \leq -\lambda_0\|E\|^2 < 0$ for all $\|E\| > \chi$. Thus, there is a $T > 0$, such that $\|E\| \leq \chi$ for all $t \geq T$. This implies that the tracking errors are uniformly ultimately bounded (UUB), i.e. $(e_1, e_2, e_3, \widetilde{\Gamma}_l) \in L_\infty$ (Khalil 2001). According to the standard Lyapunov theorem, we conclude that $\widetilde{\theta}_1, \widetilde{\kappa}_1, \widetilde{\theta}_2$ and $\widetilde{\kappa}_2$ are all UUB. The boundedness of $\theta_1, \kappa_1, \theta_2$ and κ_2 is respectively established from that $\widetilde{\theta}_1, \widetilde{\kappa}_1, \widetilde{\theta}_2$ and $\widetilde{\kappa}_2$. Also, From (58) and since $\varsigma(t) \in L_2$, one can easily show that $(e_1, e_2, e_3, \widetilde{\Gamma}_l) \in L_2$.

5.1.2 Proof of $(\dot{e}_1, \dot{e}_2, \dot{e}_3, \dot{\tilde{\Gamma}}_l) \in L_\infty$ and the Boundedness of All Signals in the Closed Loop

Because $e_1, e_3 \in L_\infty$ and $x_{1d}, x_{3d} \in L_\infty$, therefore $x_1, x_3 \in L_\infty$. From (14), one can write the dynamics of the tracking errors of the stator fluxes as follows:

$$\dot{e}_5 = -a_1 e_5 + \omega_s e_4 + e_3$$
$$\dot{e}_4 = -a_1 e_4 - \omega_s e_5$$

with $e_4 = \widetilde{\phi}_{sq}$ and $e_5 = \widetilde{\phi}_{sd}$.

From those dynamics and since $e_3 \in L_\infty$, we can easily prove the boundedness of e_4, e_5 and x_4. From $x_4, x_{3d}, e_1, \dot{x}_{1d}, x_1 \in L_\infty$, it can be concluded that $\upsilon_2 \in L_\infty$ based on (23). Because $x_2 = (e_2 + \upsilon_2)/a_5 x_5$, $e_2, \upsilon_2 \in L_\infty$, $x_5 > 0$, we can show that $x_2 \in L_\infty$. The boundedness of φ^*_{sd} and x_5 follows that of x_2 and e_5. Due to the boundedness of $x_1, x_2, x_3, x_4, x_5, \hat{\Gamma}_l$ and since $\theta_1, \kappa_1, \theta_2, \kappa_2 \in L_\infty$, we can conclude that the controls (u_1 and u_2)are also bounded. The boundedness of states, reference signals, tracking errors and adaptation parameters implies the boundedness of $\dot{e}_1, \dot{e}_2, \dot{e}_3, \dot{\tilde{\Gamma}}_l$ (i.e. this implies that $(\dot{e}_1, \dot{e}_2, \dot{e}_3, \dot{\tilde{\Gamma}}_l) \in L_\infty$.)

5.1.3 Proof of the Asymptotic Convergence of the Tracking Errors

Because $(e_1, e_2, e_3, \widetilde{\Gamma}_l) \in L_\infty \cap L_2$ and $(\dot{e}_1, \dot{e}_2, \dot{e}_3, \dot{\tilde{\Gamma}}_l) \in L_\infty$, and using Barbalat's lemma (Khalil 2001), we can conclude that all tracking errors and the estimation error $\widetilde{\Gamma}_l$ converge asymptotically to zero, despite the presence of the uncertainties and perturbations.

5.2 An Implementable Version of the Load Torque Estimator

Now, let us consider the load torque adaptation law (28) that can be written in the following form

$$\dot{\hat{\Gamma}}_l = \beta \Gamma_l - \beta \hat{\Gamma}_l - \gamma a_7 e_1 \tag{59}$$

As the actual load torque Γ_l is unknown, the first equation in (18) will be used to compute its value. Consequently, Γ_l is given by

$$\Gamma_l = -\frac{(\dot{x}_1 + a_5 x_5 x_2 - a_5 x_4 x_3 + a_6 x_1)}{a_7} \tag{60}$$

which leads to

$$\dot{\hat{\Gamma}}_l = -\frac{\beta}{a_7}(\dot{x}_1 + a_5x_5x_2 - a_5x_4x_3 + a_6x_1) - \beta\hat{\Gamma}_l - \gamma a_7 e_1 \tag{61}$$

It is worth noticing that because of the integral structure of the adaptation law (61), this updating law is implementable despite the presence of the time derivative $\dot{x}_1$. To show that, let's rewrite the adaptation law as

$$\hat{\Gamma}_l = \hat{\Gamma}_l(0) - \frac{\beta}{a_7}(x_1(t) - x_1(0)) + \int_0^t h(\tau)d\tau \tag{62}$$

where

$$h = -\left(\beta\hat{\Gamma}_l + \gamma a_7 e_1 + \frac{\beta}{a_7}(a_5x_5x_2 - a_5x_4x_3 + a_6x_1)\right) \tag{63}$$

Consequently, the load torque adaptation law can be computed without the need of using $\dot{x}_1$.

Remark 3 From (59), we can rewrite $\dot{\tilde{\Gamma}}_l = -\beta\tilde{\Gamma}_l + \gamma a_7 e_1$, this equation can be seen as a standard disturbance observer. In fact, if e_1 converges to zero, then $\tilde{\Gamma}_l$ also converges to zero. Consequently, $\hat{\Gamma}_l$ converges to Γ_l.

To summary, Fig. 3 shows the block diagram of our FABC proposed. The overall scheme of the controlled DFI-Motor is depicted in Fig. 4 in which the stator is directly connected to the grid, and the DIF-Motor is controlled by acting on the rotor winding.

In the following section, the effectiveness of the proposed FABC will be illustrated via some simulations results.

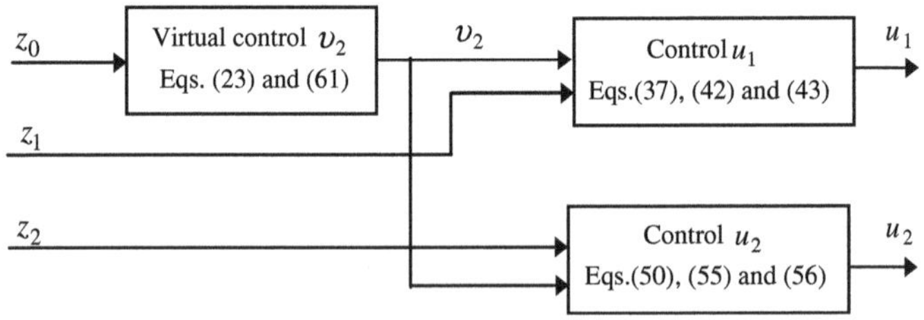

Fig. 3 The proposed FABC

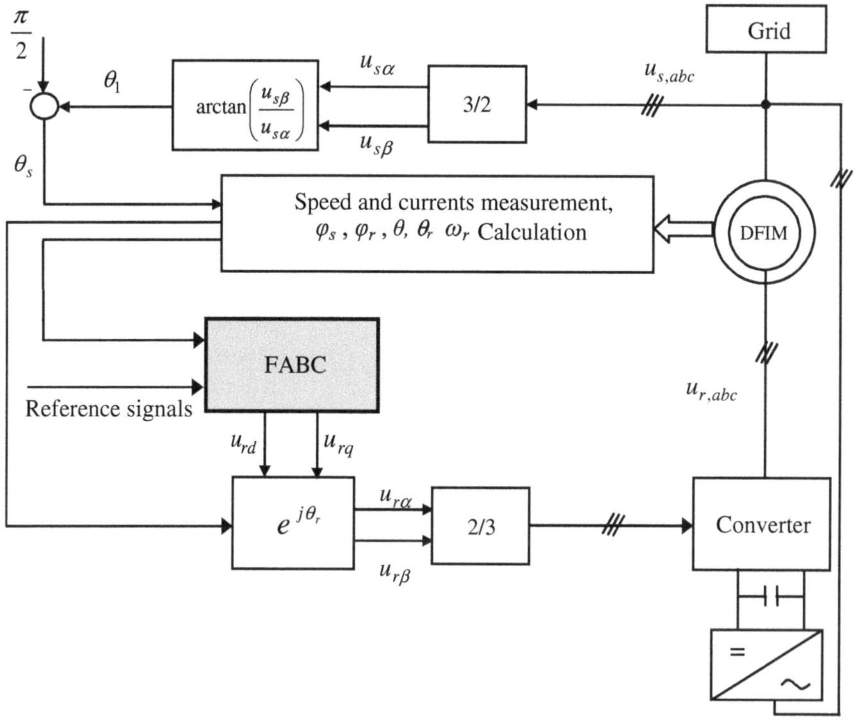

Fig. 4 The overall control scheme of the DFI-Motor

6 Simulation Results

In order to investigate the control system effectiveness, a numerical simulation has been realized with a 4 kW DFI-Motor. Table 1 summarizes the DFI-Motor's parameters along with their respective values (Vidal 2004). The performances of the control scheme are evaluated in terms of response to speed variation, sensitivity to external disturbances and robustness against machine parameters variations. The design parameters are selected as: $\gamma_1 = 0.001, \beta = 200, \lambda_1 = 200, \gamma_1 = 100, \eta_1 = 0.05, \lambda_2 = 200, \gamma_2 = 1{,}000, \eta_2 = 0.1, \sigma_1 = \sigma_3 = 0.1, \sigma_2 = \sigma_4 = 0.1$. The initial conditions are chosen as: $\kappa_1(0) = \kappa_1(0) = 0.2$, and $\theta_{1i}(0) = \theta_{2i}(0) = 0$. The unknown uncertainties and perturbations are selected as: $\delta_1(x_1, x_2) = 3x_2$ and $\delta_2(x_3, x_2) = 4x_2 + 2x_2$.

The fuzzy system $\theta_1^T \psi_1(\hat{z}_1)$ has the vector $[x_1, x_2, x_4, x_5, v_2, \hat{\Gamma}_l]^T$ as input, while the fuzzy system $\theta_2^T \psi_2(z_2)$ has the state vector $[x_1, x_2, x_3, x_4, x_5]^T$ as input. For each variable of the entries of these fuzzy systems, as in (Boulkroune et al. 2008), we define three (one triangular and two trapezoidal) membership functions uniformly distributed on the intervals $[-0.5, 1.5]$ for x_2, x_3, x_4 and x_5, $[-150, 150]$ for x_1, $[-2, 2]$ for υ_2, and $[-150, 150]$ for $\hat{\Gamma}_l$.

Table 1 DFI-Motor Parameters

Parameter	Value
Rated power	$P_n = 4$ kW
Stator—rotor voltages	$u_s = 400$ V
Stator—rotor currents	$I_s = 8.4$ A, $I_r = 19$ A
Synchronous speed	$\omega_{sn} = 2\pi 50$ Hz
Stator resistance	$R_s = 1.3740\ \Omega$
Rotor resistance	$R_r = 0.1000\ \Omega$
Stator inductance	$L_s = 0.2241$ H
Rotor inductance	$L_r = 0.0287$ H
Mutual inductance	$M = 0.0740$ H
Inertia	$J = 0.01862$ Nm/rad/s^2
Friction coefficient	$k_f = 0.01400$ Nm.s/rad
Pole pairs	p = 2

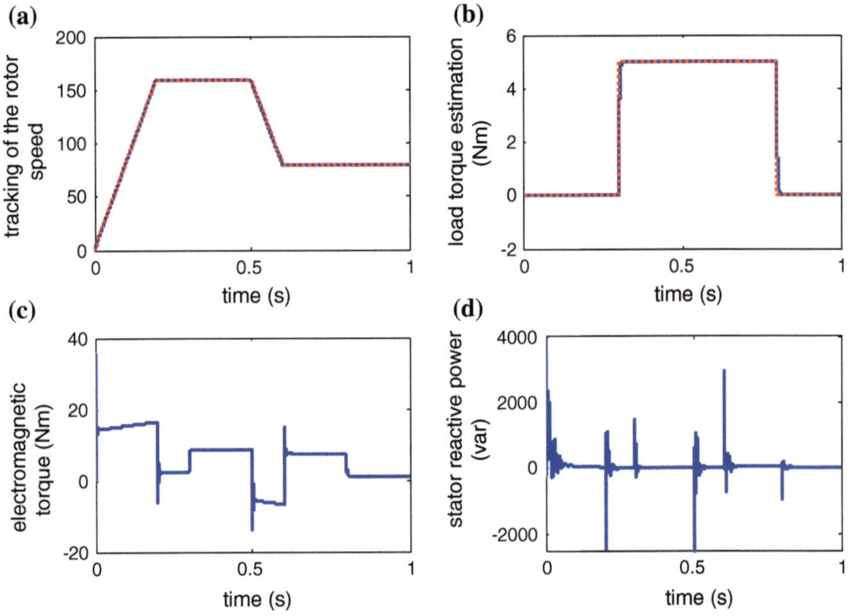

Fig. 5 Simulation results: **a** Tracking of the rotor speed: x_1 (*solid line*) and x_{1d} (*dotted line*). **b** Estimation of the load torque: the estimate $\hat{\Gamma}_l$ (*solid line*) and the actual value Γ_l (*dotted line*). **c** Electromagnetic torque. **d** Stator reactive power

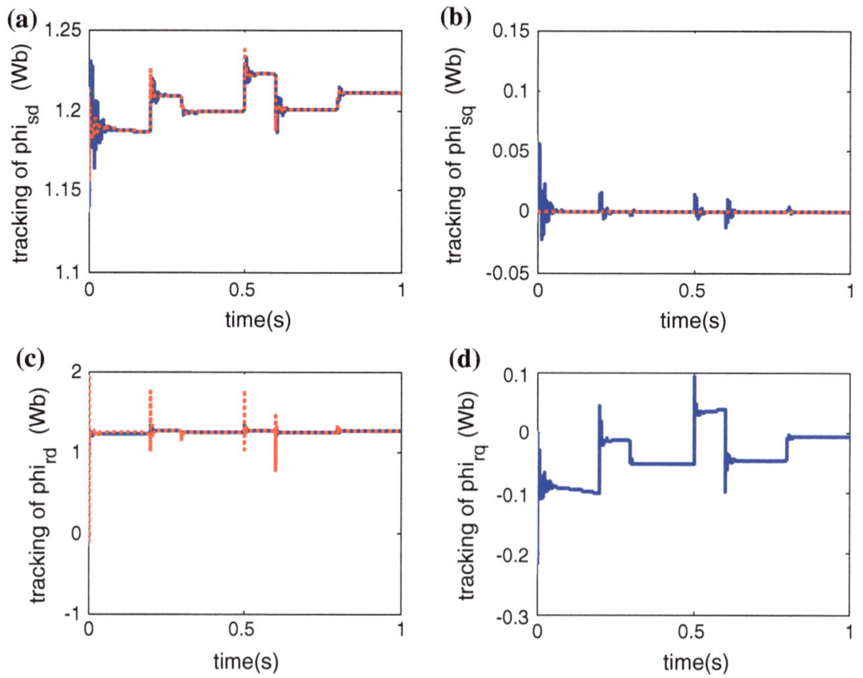

Fig. 6 Flux responses of the DFI-Motor: **a** Tracking of φ_{sd}: φ_{sd} (*solid line*) and φ_{sd}^* (*dotted line*). **b** Tracking of φ_{sq}: φ_{sq} (*solid line*) and φ_{sq}^* (*dotted line*). **c** Tracking of φ_{rd}: φ_{rd} (*solid line*) and φ_{rd}^* (*dotted line*). **d** Response of φ_{rq}

The simulation results of the proposed FABC system are depicted in Figs. 5, 6 and 7. From these simulation results, we can clearly see that a satisfactory behavior of the mechanical speed with regard to the imposed speed profile is obtained without the knowledge of the load torque. Moreover, the load torque estimator gives a correct estimation for the actual load torque.

We can observe clearly that the flux responses respect the imposed constraints. So, after transient, the stator and the rotor fluxes recover respectively their reference signals. Consequently, the flux orientation objective is guaranteed, and the stator reactive power is equal zero in steady-state operation. Also, the results show quickness of transients, good robustness and insensitivity in the face of the uncertainties.

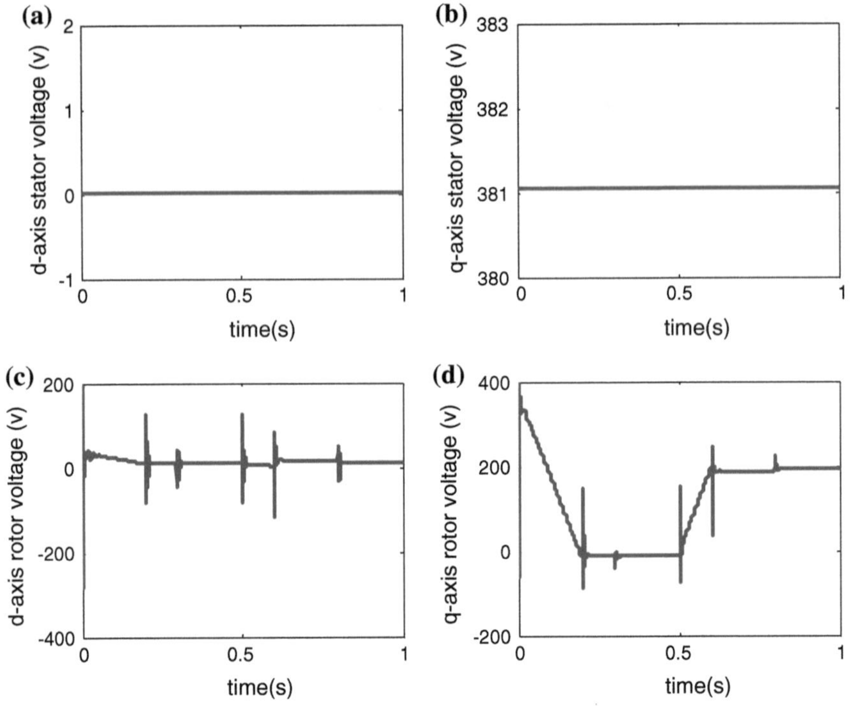

Fig. 7 Voltages applied to the DFI-Motor

7 Conclusion

In this chapter, a new fuzzy adaptive backstepping controller has been developed for a DFI-Motor. A Lyapunov approach has been adopted to derive the parameter adaptation laws and prove the stability of the control system as well as the asymptotic convergence of the underlying tracking and estimation errors to zero. Simulation results show clearly the effectiveness of this control approach. In spite of the presence of the model uncertainties, the dynamic behavior of the DFI-Motor presents high performances in terms of the speed and the load torque tracking accuracy, satisfactory flux control and consequently, stator reactive power regulation to zero in steady-state. It is worth noting that the control methodology proposed here can be easily extended to any other high performance electric drives. In our future work, one will address the experimental implementation of this proposed control scheme and the design of a speed sensorless controller.

References

Agamy, M., Youcef, H. A., & Sebakhy, O. A. (2004). Adaptive fuzzy variable structure control of induction motors. In *The* 2011 *Canadian Conference of Computer and Electrical Engineering* May 2–5, 2004, Ontario, (pp. 89–94). doi:10.1109/CCECE.2004.1344964.

Azar, A. T. (2010a). Adaptive neuro-fuzzy systems. In A. T. Azar (Ed.), *Fuzzy systems*. Vienna Austria: INTECH.

Azar, A. T. (2010b). *Fuzzy systems*. Vienna, Austria: INTECH.

Azar, A. T. (2012). Overview of type-2 fuzzy logic systems. *International Journal of Fuzzy System Applications, 2*(4), 1–28.

Bogalecka, E., & Kzreminski, Z. (1993). Control system of a doubly-fed induction machine supplied by current controlled voltage source inverter. In *The Sixth International Conference on Electrical Machines and Drives* September 1993, (pp. 168–172).

Boulkroune, A., Tadjine, M., M'saad, M., & Farza, M. (2008). How to design a fuzzy adaptive control based on observers for uncertain affine nonlinear systems. *Fuzzy Sets and Systems, 159* (8), 926–948.

Boulkroune, A., Tadjine, M., M'saad, M., & Farza, M. (2009). Adaptive fuzzy controller for non-affine systems with zero dynamics. *International Journal of Systems Science, 40*(4), 367–382.

Boulkroune, A., M'saad, M., & Chekireb, H. (2010a). Design of a fuzzy adaptive controller for MIMO nonlinear time-delay systems with unknown actuator nonlinearities and unknown control direction. *Information Sciences, 180*(24), 5041–5059.

Boulkroune, A., Tadjine, M., M'saad, M., & Farza, M. (2010b). Fuzzy adaptive controller for mimo nonlinear systems with known and unknown control direction. *Fuzzy Sets and Systems, 161*(6), 797–820.

Brown, G. M., Szabados, B., Hoolboom, G. J., & Poloujadoff, M. E. (1992). High-power cycloconverter drive for double-fed induction motors. *IEEE Transactions on Industrial Electronics, 39*(3), 230–240.

Drid, S., Tadjine, M., & Nait-Saïd, M. S. (2005). Nonlinear feedback control and torque optimization of a doubly-fed induction motor. *Journal of Electrical Engineering, 56*(3–4), 57–63.

Ghamri, A., Benchouia, M. T., Benbouzid, M. E. H., Golea, A., & Zouzou, S. E. (2007). Simulation and control of AC/DC converter and induction machine speed using adaptive fuzzy controller. In *The* 2007 *International Conference on Electrical Machines and Systems* Oct 8–11, 2007, Seoul, (pp. 539–542).

Hopfensperger, B., Atkinson, D. J., & Lakin, R. A. (2000). Stator-flux-oriented control of a doubly-fed induction machine without position encoder. *IEE Proceedings-Electric Power Applications, 147*(4), 241–250.

Khalil, H. (2001). *Nonlinear Systems* (3rd Edition). Prentice Hall.

Krstic, M., Kananellakopoulos, I., & Kokotovic, p. (1995). *Nonlinear and adaptive control design*. Hoboken: Wiley-Interscience.

Lee, C. C. (1990). Fuzzy logic in control system: fuzzy logic controller, part I and part II. *IEEE Transactions on Systems, Man and Cybernetics, 20*(2), 404–435.

Leonhard, W. (1997). *Control of electrical drives*. Berlin: Springer.

Li, Y. F., & Lau, C. C. (1989). Development of fuzzy algorithm for servo systems. *IEEE Control Systems Magazine, 9*(3), 65–72.

Lin, F. J., Shen, P. H., & Hsu, S. P. (2002). Adaptive sliding mode control for linear induction motor drive. *IEE Proceedings-Electric Power Applications, 149*(3), 184–194.

Morel, L., Godfroid, H., Mirzaian, A., & Kauffmann, J. M. (1998). Double-fed induction machine: converter optimization and field oriented control without position sensor. *IEE Proceedings-Electric Power Applications, 145*(4), 360–368.

Peresada, S., Tilli, A., & Tonielli, A. (1999). Dynamic output feedback linearizing control of a doubly-fed induction motor. In *The* 1999 *IEEE International Symposium on Industrial Electronics* (*ISIE*) (pp. 1256-1260), July 12–16, 1999, Bled. doi:10.1109/ISIE.1999.796880.

Peresada, S., Tilli, A., & Tonielli, A. (2003). Indirect stator flux-oriented output feedback control of a doubly-fed induction machine. *IEEE Transactions on Control Systems Technology, 11*(6), 875–888.

Vidal, P. E. (2004). Commande Non-Linéaire d'une Machine Asynchrone à Double Alimentation. PhD Thesis, Institut National Polytechnique de Toulouse.

Wang, L. X. (1993). Stable adaptive fuzzy control of nonlinear systems. *IEEE Transactions on Fuzzy Systems, 1*(2), 146–155.

Wang, L. X. (1994). *Adaptive fuzzy systems and control: design and stability analysis*. Englewood Cliffs, NJ: Prentice-Hall.

Wang, S., & Ding, Y. (1993). Stability analysis of field oriented doubly-fed machine drive based on computer simulation. *Electric Machines and Power Systems, 21*(1), 11–24.

Youcef, H. A., & Wahba, M. A. (2009). Adaptive fuzzy MIMO control of induction motors. *Expert Systems with Applications, 36*(3), 4171–4175.

Expert-Based Method of Integrated Waste Management Systems for Developing Fuzzy Cognitive Map

Adrienn Buruzs, Miklós F. Hatwágner and László T. Kóczy

Abstract Movement towards more sustainable waste management practice has been identified as a priority in the whole of EU. The EU Waste Management Strategy's requirements emphasize waste prevention; recycling and reuse; and improving final disposal and monitoring. In addition, in Hungary the national waste strategy requires an increase in the household waste recycling and recovery rates. Integrated waste management system (IWMS) can be defined as the selection and application of suitable and available techniques, technologies and management programs to achieve waste management objectives and goals. In this paper, the concept of 'key drivers' are defined as factors that change the status quo of an existing waste management system in either positive or negative direction. Due to the complexity and uncertainty occurring in sustainable waste management systems, we propose the use of fuzzy cognitive map (FCM) and bacterial evolutionary algorithm (BEA) methods to support the planning and decision making process of integrated systems, as the combination of the FCM and BEA seem to be suitable to model complex mechanisms such as IWMS. Since the FCM is formed for a selected system by determining the concepts and their relationships, it is possible to quantitatively simulate the system considering its parameters. The goal of optimization was to find such a connection matrix for FCM that makes possible to generate the

A. Buruzs (✉)
Department of Environmental Engineering, Széchenyi István University,
Egyetem tér 1, Győr 9026, Hungary
e-mail: buruzs@sze.hu

M.F. Hatwágner
Department of Information Technology, Széchenyi István University,
Egyetem tér 1, Győr 9026, Hungary
e-mail: miklos.hatwagner@sze.hu

L.T. Kóczy
Department of Telecommunications and Media Informatics,
Budapest University of Technology and Economics,
Magyar Tudósok körútja 2, Budapest 1117, Hungary
e-mail: koczy@tmit.bme.hu; koczy@sze.hu

Q. Zhu and A.T. Azar (eds.), *Complex System Modelling and Control Through Intelligent Soft Computations*, Studies in Fuzziness and Soft Computing 319,
DOI 10.1007/978-3-319-12883-2_4

most similar time series. This way a more objective description of IWMS can be given. While the FCM model represents the IWMS as a whole, BEA is used for parameter optimization and identification. Based on the results, in the near future we intend to apply the systems of systems (SoS) approach to regional IWMS.

Keywords Integrated waste management system • Sustainability factors • Fuzzy cognitive map • Bacterial evolutionary algorithm • Optimization

1 Introduction

Waste is one of the most visible environmental problems in the world. Increasing population, changing consumption patterns, economic development, urbanization and industrialization result in the increased generation of solid waste and a diversification of the types of the waste. Waste management is an umbrella term that refers to a host of interlinked activities such as reduction, recycling, collection, transportation, processing, disposal, and monitoring of waste materials.

The European Landfill Directive (1999/31/EC) and the Packaging and Packaging Waste Directive (94/62/EC) aim to reduce the amount of biodegradable municipal waste going to landfill. In addition, in Hungary the national waste strategy requires an increase in the household waste recycling and recovery rates. Movement towards more sustainable waste management practice has been identified as a priority in the whole of EU (Phillips et al. 1999). The EU Waste Management Strategy's requirements emphasize waste prevention; recycling and reuse; and improving final disposal and monitoring. As a consequence, the so-called waste hierarchy has become a major guiding principle for waste management policies (Demirbas 2011).

Despite the progress that the EU and Hungary have made, the volume of most waste streams continues to rise. By 2020 the waste generation is expected to be doubled (den Boer and Lager 2007). The main approach to solid waste management in Hungary is unfortunately still landfilling. The expected new measures require the development of different alternatives to improve the long-term performance and the sustainability of the current waste management systems in order to reach the targets set (Bovea and Powell 2006).

Waste management in Hungary is primarily controlled through legal regulations. Legal provisions determine technical requirements for waste management, the applicable economic incentives and sanctions, the responsibilities of the waste generators and managers of waste as well as the licensing and supervisory duties of the authorities. In Hungary, the local government is entrusted with the task of waste management services.

In Hungary, huge waste management projects were in progress in the last years, many forming part of a waste management mega-project of about 3 Mio EUR (10 billion HUF). These projects are designed to establish EU-compliant waste management in extant nationwide garbage disposal sites by introducing a selective waste collection strategy and optimizing logistics systems.

Increasing environmental concerns, legislative and public pressures have led to the evaluation of the perspectives of other treatment processes and technologies. However, technical and economic approaches towards designing solid waste management systems should not be considered as the only possible solution. While some research is dedicated to the physical management of municipal solid waste, relatively little attention has been paid to the larger context necessary for sustainable waste treatment. Integrated municipal solid waste management can be defined as the selection and application of suitable techniques, technologies and management programs to achieve waste management objectives and goals (Tanskanen 2000). Sustainable waste management provides a comprehensive inter-disciplinary framework for addressing the problems of managing municipal solid waste (Kurian 2006).

Systems with source control can avoid many problems of the processing technology by respecting different qualities and quantities of the waste streams, by treating them appropriately for reuse and recycling. Sustainable waste management means less reliance on landfill and greater amounts of recycling and composting (Demirbas 2011; Graymore et al. 2008). The purpose of this paper is to describe and model the sustainability elements of IWMS on regional level.

This chapter is structured as follows. Section 2 describes the history and background of sustainable waste management and introduces the driving factors of the IWMS. Section 3 presents the methodological approach of the simulations by two computational intelligence tools: fuzzy cognitive map (FCM) and bacterial evolutionary algorithm (BEA). Section 4 presents the results of the simulations. Finally, a summary is given in Sect. 5 which concludes in answering the question about the ranking of sustainability factors in waste management.

2 History and Background

The IWMS has to be an economically affordable, environmentally effective and socially acceptable system. Among others, it includes the practical aspects of waste management (i.e. transport, treatment and disposal) and the attitudes of citizens (how they feel about source separation, recycling, incineration, etc.). The evolution of waste management from truck and dump, to the highly integrated systems requires an investment of both time and resources (Wilson et al. 2001).

2.1 The History of Waste Management

Numerous studies introduce the history of waste management. According to Shmeleva and Powell (2006), until the 1960s municipal waste management was concentrated only on the collection and transportation of waste from households to the disposal facilities without any separation, which in the majority of cases were local dumps or landfills. Processes were planned or optimised merely on the basis of efficiency in terms of costs. Environmental effects were only marginally taken into account. In a second phase, waste treatment and landfilling technologies were improved. After Hung et al. (2007), in the 1970s, the goals of the municipal waste management systems were simply to optimize waste collection routes for vehicle or to locate appropriate transfer stations. In the 1980s, the focus was extended to encompass municipal waste management on a system level, minimizing the costs. This was the first time that the aspect of waste as a resource was taken into consideration. Complex waste management systems were first introduced and further developed from the 1980s onward. In the 1990s, specific treatment technologies for several types of waste were introduced, together with advanced landfill technologies (Salhofer et al. 2007). With the transition from waste management to materials management, tools are needed that consider all aspects and effects of waste management (Wilson et al. 2001).

2.2 The Development of Methods

Many environmental problems would benefit from models based on experts' knowledge (Özesmi and Özesmi 2004), among them IWMS modelling as well. Several models have been developed in recent decades to support decision making in IWMS to monitor present conditions, to assess future risks and to visualize alternative futures (Hung et al. 2007; Papageorgiou and Kontogianni 2012). According to Hung et al. (2007), Salhofer et al. (2007) and Tanskanen (2000), early waste management models developed during the 1960s and 1970s focused on studying individual functional elements, i.e. optimizing waste collection routes for vehicles or locating appropriate transfer stations. In the 1980s, the investigation was extended to encompass waste management on the system level, minimizing waste treatment costs. In the 1990s, the waste management models focused principally on economic (e.g. system cost and system benefit), environmental (air emission, water pollution) and technological (the maturity of technology) aspects. An environmental impact assessment model, the life cycle assessment (LCA) is also often used to aid the decision-making in waste management. Numerous studies applied the LCA method to evaluate the environmental impact of waste treatment alternatives. In several strategic planning models, both costs and emissions of waste management systems have been included in the research. In some models, the whole life cycle of products has been studied instead of only the waste management system when searching for environmentally optimal waste management strategies.

The increasing demand for types of models which combine environmental, economic and further aspects (like social, technological aspects) has led to the development of a latest generation of computerised models, which are similar to the LCA-based models, but include additional cost effects and/or social effects. In this case, cost effects can be regarded as an additional impact category. Examples of this type of models are GABI and Umberto (Kalakula et al. 2014), well known computerised tools especially in the German speaking community. From both methodological and practical point of view, it is a complex task to compare alternatives with respect to environmental effects, costs and social aspects. In most cases, the antagonistic targets of cost minimisation, reduction of environmental effects and high convenience for the user (mainly of the waste collection scheme) cannot be met by one single scenario. It is increasingly likely that a scenario in which high costs are linked with high environmental standards and high convenience will be involved, whereas low-cost scenarios prove to be less environmentally friendly or less convenient.

2.3 The Evolution of Factors

In the preliminaries of this research we investigated the conditions and driving factors of sustainability of IWMS and determined its main aspects based on various authors. The concept of 'key drivers' are defined as factors that change the status quo of an existing waste management system (in either positive or negative direction), be it legislation that encourages an integrated approach to waste management or change of public perception in an IWMS. A large body of literature on factors that influence municipal waste management systems is available. According to the development of methods investigating urban waste management systems, the number of factors influencing system element increased dramatically worldwide. In the 1990s, the factors considered in municipal waste management models were principally economic (e.g. system cost and system benefit), environmental (air emission, water pollution) and technological (the maturity of technology) (Salhofer et al. 2007). In the late 1990s, to compare different waste treatment and disposal scenarios, and rank them (from the 'best' to the 'worst'), the authors (Haastrup et al. 1989; Maniezzo et al. 1998; Tanskanen 2000) investigated technical data (number of treatment/disposal technologies and available plants, relative capacities, geographical data), social progress (demography), environmental aspects (protection of the environment, use of natural resources, greenhouse gas load, acid load) and economic variables (maintenance of economic activity) (Phillips et al. 1999). In some studies (Kurian 2006; Shmeleva and Powell 2006; van de Klundert and Anschutz 1999) examining the situation of waste management in the developing countries, the authors introduced six principles: technical/operational, environmental, financial, socio-economic, institutional/administrative and policy/legal ones.

In the early 2000s, the development of factors continued. In the European Union (Wilson et al. 2001), the role of policy, management and institutional structure (local and regional politics and planning strategy); operational demands (infrastructure and waste disposal, security, waste stream composition and change); economic and financial factors (available funding and subsidies, cost of current system and other option); legislation (prescriptive or enabling legislation, international, national and regional legislation) and social considerations (public opinion and support) came to the front. In the middle of the 2000s, more factors and subsystem elements were involved in the newly developed methods, such as savings from energy generation (Bovea and Powell 2006), habitats diversity (Salhofer et al. 2007), and also the social factors such as human well-being and motivation received bigger attention (since the separation of waste is undertaken by the inhabitants of a considered city, the citizens' behaviour is the key influencing factor) (den Boer and Lager 2007), life-cycle analysis for production and consumption of energy and full-cost accounting (Thorneloe et al. 1999). In some cases, the weight of factors is determined by stakeholders using questionnaires to obtain stakeholder opinions to develop fuzzy criteria weights (Hung et al. 2007).

Over recent years, the method of development of factors and subsystem elements has been refined. In the developing countries where the realization of sustainable waste management is still an urgent challenge, researches (Kurian 2006; McBean et al. 2005; Jadoon et al. 2014; Worku and Muchie 2012) focus on among others the involvement and participation of all the stakeholders, features of existing infrastructure, seasonal and daily variations of waste generation, etc. Therefore the key factors here are: environmental (regulations, standards, monitoring and enforcement); policy (guidance with long-term view in allocating resources, poor awareness about the benefits of proper waste management); public (participation in decision-making, the income of households, family size, education, profession); NGOs (mobilizing community); private sector (searching and implementing appropriate actions); media (environmental awareness, focus on real local priorities); scientific community (focus on needs of vulnerable population and communication); financial (institutions supporting environmentally sound developments); technical (presence/lack of infrastructural capacity, failure to adequately utilize modern waste management and processing technology, the absence of an integrated waste management system).

The so called horizontal factors describe the processes of interchanges between different waste types (shifts between residual waste, bulky waste, recyclable waste and illegally disposed waste), and vertical factors are due to changes of the total sum of all waste streams depending on demographic, economic, social and technical development (mass-related data and monetary data) (Beigl et al. 2008).

On the basic of the above review, we can conclude that there is a wide consensus in the related literature that a typical IWMS includes at least the following six key factors: environmental, economic, social, institutional, legal and technical. These factors are the 'key drivers' of a sustainable IWMS that determine why the system operates as it does (den Boer and Lager 2007; Langa et al. 2006; Morrissey and Browne 2004; Wilson et al. 2001; van de Klundert and Anschutz 1999; Thorneloe et al. 1999).

In Table 1 the main factors and some examples of their respective subsystems are introduced.

Table 1 'Key drivers' of IWMS and their respective subsystems

Factors	Subsystem elements
Environmental factors	Emissions; climate change; land use; recovery and recycling targets; depletion of natural resources; human toxicity
Economic factors	Efficiency at subsystem level; efficiency at system level; available funding/subsidies; equity; system costs and revenues; pricing system for waste services, secondary materials market
Social factors	Public opinion; public participation in the decision making process; risk perception; employment; local demographics—population density, household size and household income; public resistance (NIMBY—not in my backyard, LULU—locally unacceptable land use)
Institutional factors	Local and regional politics and planning; managerial conditions and future directions; institutional and administrative structure of waste management
Legal factors	Relevant legislation (international, national, regional and municipal)
Technical factors	Collection and transfer system; treatment technologies; waste stream composition and change

We have accepted this approach as well-founded; however, some of the results of our present research motivate us to re-validate the inputs by the stakeholders in a later phase of the investigation.

As a result of the incompleteness and multiple uncertainties occurring in sustainable waste management systems, we propose the use of FCM to support the planning and decision making process. It is obvious that uncertainties involved with waste management represent vagueness rather than probability. Fuzzy sets and fuzzy logic are suitable to construct a formal description and a mathematically manageable model of systems and processes with such uncertainties. By observation of the model and its time dependent behaviour we determined under what conditions the long-term sustainability of a regional waste management system could be ensured. In this paper, we introduce a model of waste management which investigates the six most common factors—environmental, economic, social, technical, legal and institutional aspects. The next section introduces the approach applied for the modelling.

3 The Methodological Approach

In the development of the FCM, in the first step of the design process the number and features of constituting factors were determined by the relevant literature, as it was mentioned beforehand. These six concepts are supposed to be combined all together in a single system, with mutual interactions.

Modern technological systems are complex and they are usually comprised of a large number of interacting and coupling entities that are called subsystems and/or components. These systems have nonlinear behaviour and cannot simply be derived from summation of analyzed individual component behaviour (Stylos and Groumpos 2004). Feedback mechanisms are important in the analysis of vulnerability and resilience of social-ecological-technical systems. But how to evaluate systems with direct feedbacks has been a great challenge. FCM was derived from the fusion of fuzzy logic and theory of cognitive maps. Kosko (1986) developed the fuzzy signed directed graphs with feedback in order to represent knowledge in a comprehensive way. Since the FCM is formed for a selected system by determining the concepts and their relationships, it is possible to quantitatively simulate the system considering its parameters. It has to be noted however, that a FCM is suitable for short term time series analysis and prediction. A FCM is a dynamic modelling tool in which the resolution of the system representation can be increased by applying a further mapping. The resulting fuzzy model can be used to analyze, simulate, and test the influence of parameters and predict the behaviour of the system (Papageorgiou and Kontogianni 2012).

According to Papageorgiou and Kontogianni (2012), the design of a FCM is a process that heavily relies on the input from experts and/or stakeholders. This methodology extracts the knowledge from the stakeholders and exploits their experience on the system's model and behaviour. A FCM is fairly simple and easy to understand for the participants. With the use of a participatory process it should be ensured that different interests are used to build up synergies as well as partnerships and hence find sustainable solutions as a joint decision (Malena 2004). Even though, the cognitive nature of a FCM makes it inevitably a subjective representation of the system. The model is not arbitrary as it is built carefully and reflexively with stakeholders (Isak et al. 2009).

On the basis of a FCM's development, during the first step in the designing process, the number and features of concepts are determined by a group of experts. After the identification of the main factors affecting the topic under investigation, each stakeholder is asked to describe the existence and type of the causal relationships among these factors and then assesses the strength of these causal relationships using a predetermined scale, capable to describe any kind of relationship between two factors, positive and negative.

Starting from the primary elements of a FCM, the *i*th concept denotes a state, a procedure, an event, a variable or an input of the system and is represented by C_i ($i = 1, 2, \ldots, n$). Another component of a FCM is the directed edge which connects the concepts i and j. Each edge includes a weight w_{ij} which represents the causality between concepts C_i and C_j. The values of the concepts are within the range [0, 1], while the values of the weights belong to the interval [−1, 1]. A positive value of the weight w_{ij} indicates that an increase (decrease) in the value of concept C_i results to an increment (decrement) of the concept's value C_j. Similarly, a negative weight w_{ij} indicates that an increase (decrease) in the value of concept C_i results to a decrement (increment) of the concept's value C_j, while a zero weight denotes the absence of relationship between C_i and C_j (Fig. 1). Considering the interrelations between the concepts of a FCM, the corresponding adjacency matrix can easily be formed.

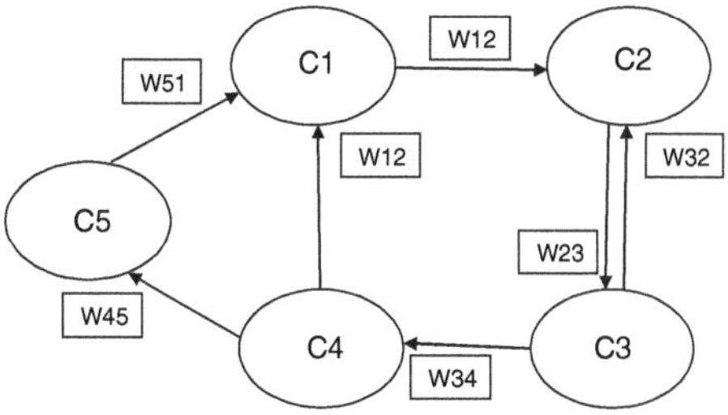

Fig. 1 The symbolic graph of a fuzzy cognitive map

Usually it is accepted that causality is not self reflexive, i.e., a concept cannot cause itself, which means that the weight matrix always has '0-s' in its diagonal (Carvalho 2010). Otherwise the component would grow without limits.

The description of the inference mechanism, which represents the behaviour of the physical system, lies in the interpretation of FCM's mathematical formulation. After the initialization of the FCM and the determination of concept activation values by experts, concepts are ready to interact. As it is obvious, the activation of a concept influences the values of concepts that are connected to it. At each step of interaction (simulation step), every concept acquires a new value that is calculated according to equations (Eqs. 1 and 2) and the interaction between concepts continues until a fix equilibrium is reached; a limit cycle is reached; or a chaotic behaviour is observed (Ketipi et al. 2010).

The mathematical description of our FCM system is a simple loop:

$$V_{k+1} = f(N \cdot V_k) \tag{1}$$

where V_k is the state k of the system; N is the matrix of the system which contains the weight $w_{i,j}$, and

$$f(x) = \frac{1}{1 + e^{-\lambda x}} \tag{2}$$

where $\lambda > 0$ determines the steepness of the of the continuous function f.

We have conducted an online survey where each stakeholder was asked to describe the existence and type of the causal relationships among the determined factors and then to assess the strength of these using a predetermined simple scale, capable to describe any kind of relationship between a pair of factors, both positive and negative ones. It was helpful to draw a guideline in order to describe the terms of concepts and the basics of the development of a FCM before starting with the survey. The questionnaire guideline functioned as a support material in answering the questions. In order to conduct the survey and to draw a suitable FCM, we needed to apply the steps of designing the process. At first, we explained to the participant what a FCM is, what its elements are and what our aim is with the results. As the interviewees understood the underlying basic information, they were able to assess the value of the connections. Thus, from each interviewee theoretically a different hypothetical FCM could be established. The 75 individual maps were however merged into a representative, collective map. In this phase we were primarily interested in investigating how the stakeholders perceive the future prospects of the IWMS.

3.1 Fuzzy Cognitive Map

In the next two chapters the applied Computational Intelligence Tool Kit will be briefly described.

As mentioned above, the FCM is a very convenient and simple tool for modelling complex systems. It is rather popular due to its simplicity and user friendliness. According to Stach et al. (2005), human experts are generally rather subjective and can handle only relatively simple networks therefore there is an urgent need to develop methods for automated generation of FCM models. The present research deploys the FCM and applies the BEA for parameter optimization.

An FCM is a fuzzy graph structure representing causal reasoning. Causality is represented here as a fuzzy relation of causal concepts. The FCM may be used for dynamic modelling of systems. The FCM approach uses nodes corresponding to the factors and edges for their interactions, to model different aspects in the behaviour of the system. These factors interact with each other in the FCM simulation, presenting the dynamics of the original system (Stylos and Groumpos 2004). The FCM has been described as the combination of neural networks and fuzzy logic. Thus, learning techniques and algorithms can be borrowed and utilized in order to train the FCM and adjust the weights of its interconnections (Stylos et al. 1997).

We have to mention here, that optimization algorithms (e.g. BEA) can be considered as machine learning algorithms in the sense that the optimized FCM parameters (the λ parameter of the threshold function and the weights of the connection matrix) result in the most realistic description of the examined system (IWMS in this case). This chapter does not deal with the learning of a huge amount of data. The goal of this study is to optimize the parameters of FCM first, then to compare the time series generated by this FCM with the time series given in the literature. Thus the words 'learning' and 'optimization' are used as synonyms in this paper. If optimization is considered as a kind of learning, the performance index, learning set and test set can also be identified.

- The performance index corresponds to the objective function (the difference between the time series generated by FCM using the optimized parameter values and the time series given in the literature).
- The time series given in literature can be considered as the training set.

The information collected from the above mentioned survey generates the test set.

3.2 The Bacterial Evolutionary Algorithm

Optimisation problems often arise in our everyday life. For example, the shape of cars are optimised to lower the drag co-efficient hereby lowering CO_2 emission as well, the workflows of factories are optimised to shorten the manufacturing time. In our case, optimisation helped us to get to know better the operation of waste management systems and the relationships between its factors.

IWMS are represented by FCM in this study. The time series of the state of IWMS factors (technical, environmental, economic, etc.) were already known from literature. The goal of optimisation was to find such a connection matrix for FCM that makes possible to generate the most similar time series. (Every optimisation problem can be considered as a search problem; during optimisation the best solution have to be found in the specified search space.) This way a more objective description of IWMS can be given. The elements of the connection matrix represent the strength of connections between the factors of IWMS.

This optimisation problem is quite complex. Because six main factors were established in the IWMS, the connection matrix contains $6 \times 6 = 36$ elements. Fortunately, the main diagonal always contains only zeroes; hence it is sufficient to handle only 30 variables. Because the λ value of the FCM had to be determined also, the objective function to optimise had 31 variables.

In such a complex case different kind of evolutionary algorithms are often used as an adequate solution to the problem because of their favourable properties. Our FCM-based model was optimised with BEA because our previous experiences and results with various benchmark data sets revealed that BEA and Bacterial Memetic Algorithms (BMA) were among the most efficient evolutionary algorithms (Balázs et al. 2010a, b). This was especially true for the variants equipped with the most appropriate and suitable operators. Several papers presented comparisons of these algorithms with other evolutionary and population based heuristics, e.g. when the goal was fuzzy rule-based learning of various physical models (Dányádi et al. 2010a; Balázs et al. 2010b), or when the Permutation Flow Shop Problem had to be optimised under certain conditions (Balázs et al. 2012).

The early algorithms of evolutionary computation (Bäck et al. 1997; Engelbrecht 2007) appeared in the 1960s. The name "evolutionary" refers to the main idea of these algorithms. They try to imitate some ideas appeared in Darwin's theory (Darwin 1859), e.g. natural selection, reproduction to create better solutions of problems. Historically, the problem they were researched and developed to solve for was different, and the details of these algorithms also differ. The best known four tends were Genetic Programming, Genetic Algorithm, Evolution Strategy, and Evolutionary Programming. During the next decades several newer evolutionary or related algorithms were also developed, e.g. particle-swarm optimisation, ant-colony optimisation.

Despite of this diversity, the above mentioned algorithms are similar in many looks. One common property is that these algorithms use a list of possible solutions, and iteratively improve the elements of it. They contain operators for exploration (to randomly explore the various points of the search space) and exploitation (to combine the already available good sub-solutions to get even better solution of the problem).

The genetic algorithm (GA) (Goldberg 1989) was developed by John Holland and his students. They created this algorithm originally to examine to properties of selection and adaptation (Holland 1975), to give a mathematical description of these phenomena, and to model it with computers. Later it became a popular optimisation technique.

The list of possible, candidate solutions is called population in GA's methodology. The elements of the population are called individuals or phenotypes. Every individual has a set of properties, the chromosome or genotype. The individuals can be represented e.g. with bit strings or floating point numbers as well, depending on the specific problem. GA works in the following way (see Fig. 2). At first, it

Fig. 2 Flow chart of the GA

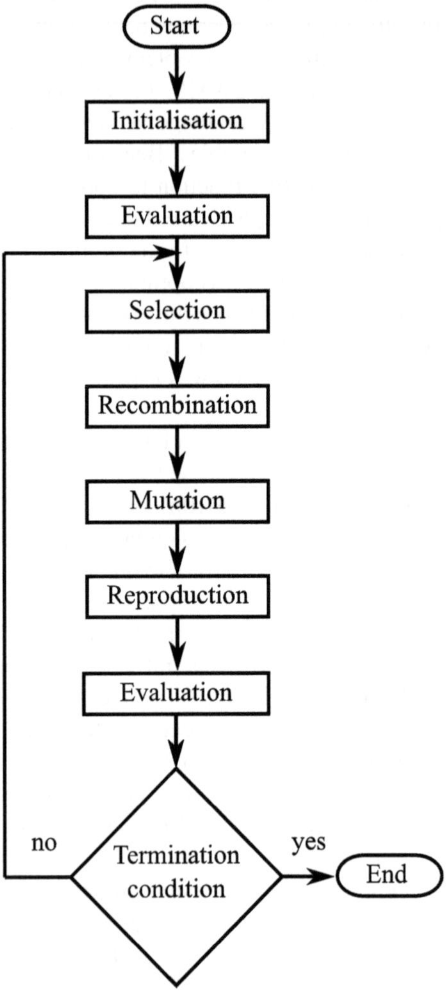

initializes the population. In most cases, the individuals are random generated, but sometimes a priori information about the problem helps to create near-optimal start population. In the next step GA calculates the values of the objective function for every individual. These values are the basis of fitness value calculation. In the most straightforward case, the fitness value and the objective value of the same individual is the same, but some advanced technique (ranking, scaling) helps to prevent some undesirable effects (primarily premature convergence) during optimisation. The next generation of the population contains new individuals (temporary population) created by recombination (combination of the parents' genetic data) and mutation (random modification of the child's chromosome with small probability). The parent individuals are selected with fitness-proportional selection. The reproduction step forms the next generation of the population using the current and the temporary population. Next, the whole process starts again from the evaluation of the objective function values. GA iteratively runs this process until some termination condition is fulfilled.

A special descendant of GA was suggested by Nawa et al. at the late 1990s. Because this algorithm was inspired by the evolution of bacteria, they named it pseudo-bacterial genetic algorithm (PBGA) (Nawa et al. 1997). PBGA uses bacterial mutation instead of GA's mutation operator. Some years later PBGA was further improved, and BEA (Nawa and Furuhashi 1998, 1999) was born. BEA includes a new gene transfer operator instead of crossover and contains bacterial mutation as well.

BEA was originally developed to determine and optimise the parameters of fuzzy rule bases made for solving general approximation and optimisation problems. The algorithms can be used in other engineering applications as well. BEA has some positive properties contrary to GA, e.g. simpler, shorter implementation; the gene transfer operator ensures the survival of the fittest bacteria without additional operator (called elitism in GA).

The exhaustive review of BEA can be found e.g. in Nawa and Furuhashi (1999), thus we will give only a short introduction here. Similarly to GA, BEA also uses a record of possible solutions. These candidate solutions are often called bacteria as well. The bacteria together form the population. The repeated utilization of bacterial mutation and gene transfer results in a series of generations. When some kind of termination condition is fulfilled, the best bacterium of the last population is accepted as the result of the optimization (see Fig. 3).

Bacterial mutation (Fig. 4) optimizes all the bacteria individually. The mutation functions in the following way. At first, K clones (exact copies) are generated for every bacterium. All genes of the bacteria are mutated during mutation in random order. In each step of it, exactly one gene at a specified position is modified randomly in every clone. If a better gene value (allele) has been found, it is copied into the other clones, too. On the end of mutation, if the objective value of the best clone is better than the value of the original bacterium, the bacterium is replaced with this clone.

Fig. 3 Flow chart of the BEA

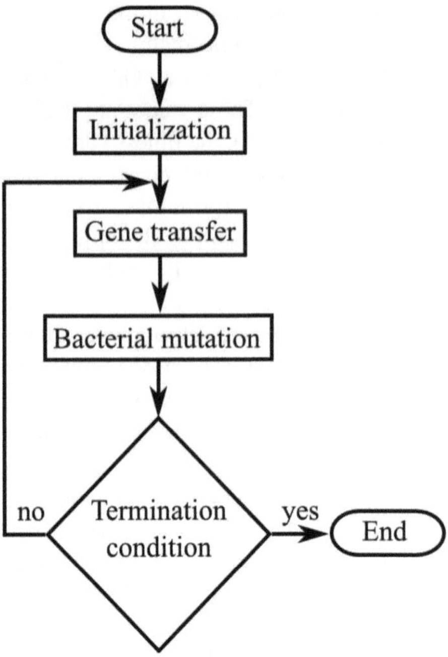

Gene transfer (Fig. 5) operates with the ordered list of bacteria. The so called superior half of the population contains the bacteria with better objective values. The other bacteria are the members of the inferior half. The operator repeats T times the following: after the selection of exactly one bacterium from the superior half and one from the inferior half, it selects a portion of the genes of the superior bacterium and copies it into the inferior bacterium. The objective value of the modified bacterium must be re-evaluated, and the whole population has to be re-sorted, too. Depending on the objective value of the modified bacterium it may get into the superior half.

GA and BEA are global optimisation techniques and provide near-optimal, approximate solution to the specific problem. They can be used even if the objective function is noisy, nonlinear, high dimensional, multimodal or non-continuous. The derivatives of the objective function is not needed thus it does not cause a problem if it is unknown or does not exists.

The original BEA was applied to a wide range of problems, e.g. to solve bin packing problem (Dányádi et al. 2010b) or a special kind of the travelling salesman problem (Botzheim et al. 2009b). BEA was improved several times during the past years. Several results are collected in Botzheim et al. (2009a). An important milestone of the research was the creation of the bacterial memetic algorithm (BMA) (Botzheim et al. 2009a). It extends the two main operators of BEA with a

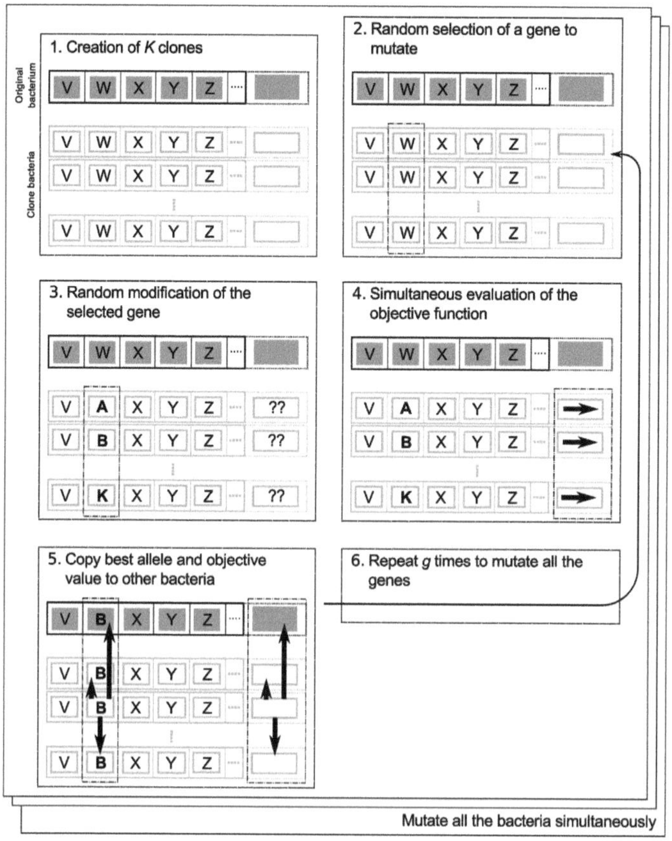

Fig. 4 Bacterial mutation

local search step. The usage of the gradient-based local search algorithm (Levenberg-Marquart, LM) increased the convergence speed. The modified version of BMA (Gál et al. 2008) is able to handle the knot order violation of LM, and uses a more efficient operator execution order as well. Other researchers tried to shorten the optimisation speed with modified, parallel gene transfer operator (Hatwágner and Horváth 2011, 2012a). BEA lacks of an operator that maintains the genetic diversity in the successive populations and this problem become more important if the operators are parallelised. A possible solution is described in Hatwágner and Horváth (2012a). The following section describes the results of the simulation.

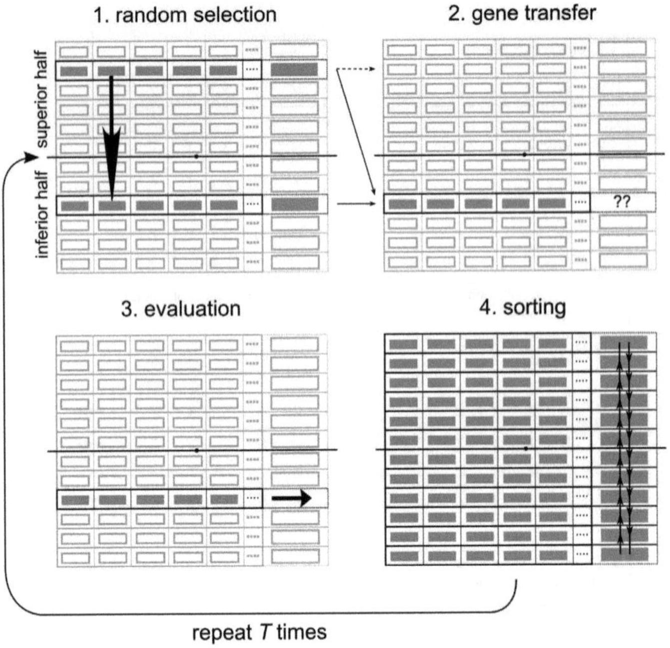

Fig. 5 Gene transfer

4 Results

In the first simulation our starting point was a fixed connection matrix. In this approach we studied the changes of the importance values of the factors over time.

The second experiment was about parameter identification using BEA. The connection matrix of FCM was determined so that the difference between the original time series of concepts given in literature and the generated ones using this matrix shall be as small as possible.

4.1 Results with the FCM Simulation

The goal of this first investigation (Buruzs et al. 2013b) was to assess the sustainability of the IWMS by investigating the FCM method with a holistic approach. First, the input data are presented here, then the experience obtained during the simulation and finally the results are introduced. The model consists of the expert system database which is based on human expert experience and knowledge obtained from the questionnaires (N = 75), namely, the initial draft connection matrix is the data gathered and averaged from the survey process shown in Table 2.

Table 2 The initial draft of the connection matrix

	C1	C2	C3	C4	C5	C6
C1	0	0.8	0.6	0.6	0.4	0.4
C2	0.6	0	0.6	0.6	0.4	0.4
C3	0.8	0.6	0	0.6	0.4	0.4
C4	0.4	0.6	0.4	0	0.4	0.4
C5	0.6	0.6	0.4	0.6	0	0.6
C6	0.4	0.4	0.4	0.4	0.4	0

This model includes the identification of concept nodes and relationships among them (Fig. 6).

The matrix presented above indicated that each node of the FCM is connected to each other node and the algorithm was used to set up values of connections.

The factors in the matrix are represented as follows:

- C1: technical factor (collection, transport, treatment methods, etc.)
- C2: environmental factor (emission of pollution, depletion of resources, human toxicity, etc.)
- C3: economic factor (subsidies, efficiency at system/subsystem level, economic sound and continuous operation, coverage of all after case expenses, etc.)
- C4: social factor (involving local need and requirements, minimizing public health risk, providing employment, etc.)
- C5: legal factor (EU packaging directive, EU landfill directive, waste hierarchy, national, regional and local regulations)
- C6: institutional factor (involvement of stakeholders, existence of feedback mechanisms of citizens, organisational structure, etc.)

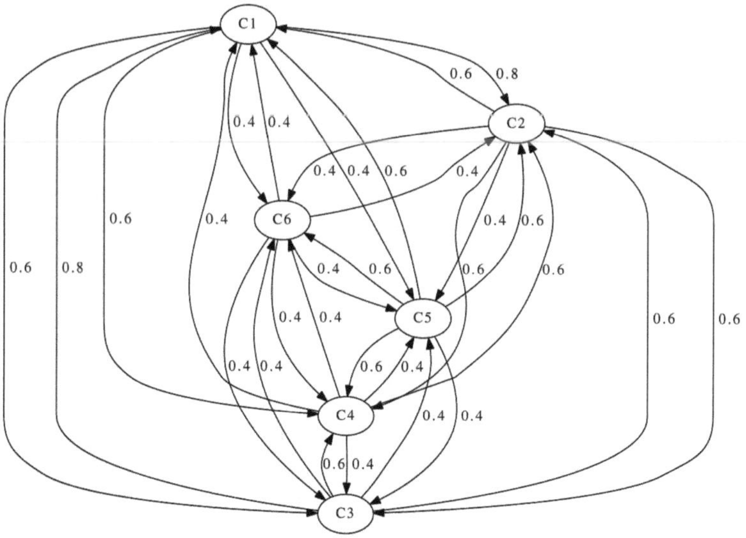

Fig. 6 The initial fuzzy cognitive maps

The other input data set was the range of historical data consisting of sequences of the state vectors. According to Demirbas (2011, den Boer and Lager 2007, Graymore et al. 2008, Langa et al. 2006, Morrissey and Browne 2004, Wilson et al. 2001, van de Klundert and Anschutz 1999, Thorneloe et al. 1999), the trend of the studied factors was assessed by values between 0 and 1 from the 1980s to the 2010s. The sequences of the state vector were designed on the basis of the literature and therefore it may be assumed that they specify soundly the role of the factors according to changes in the legislation, the available techniques, the social attitude, the state of the environment and the economic and institutional context as a time series (see Table 3, columns t_0–t_4).

During the simulation, we selected various values for λ in order to see how the parameter influenced the results of the simulation. The simulation was always started with the input of the above data. The simulation resulted in somewhat different iterations according to the value of λ. We scaled the initial state of the system in the [0, 1] interval and we used this model and ran the simulation for 10 iteration cycles. The results are presented below.

From Fig. 7 it can be observed that the system converges to an equilibrium state which is robust to the initial state variation however, the values of λ are different in each simulation. The estimated optimal value of λ may be determined by comparing the obtained results with the expert system database.

It may be observed that in the FCM model all factors converge rather fast to a steady state. After the first five iterations the transient behaviour seems to end and the FCM approaches an obviously stable state where each concept assumes a constant value (a 'plateau', depending on λ, between 0.5 and 0.9). While the qualitative behaviour of the simulation result is virtually independent from the steepness, the actual constant values to which the concept influence state converges are more or less similar, thus after normalization, the results are very consistent.

The initial states of the factors are known from Table 2. The final states of the concepts computed for each λ are shown in Table 4.

The average results of the simulation with different λ values are presented in the last column of Table 3. As IWMS are sophisticated and complex systems, priorities and targets need to be set up at the early stage of planning and implementation. The technical, environmental, economic, legal, social and institutional factors need to be balanced to attain sustainable waste management (Kurian 2006). Assuming, that the

Table 3 The sequences of the state vectors

	t_0	t_1	t_2	t_3	t_4	FCM averages
Technical	0.20	0.35	0.60	0.75	0.80	0.80
Environmental	0.15	0.20	0.40	0.60	0.80	0.71
Economic	0.10	0.15	0.30	0.50	0.70	0.62
Social	0.10	0.15	0.20	0.40	0.60	0.56
Legal	0.10	0.30	0.50	0.70	0.80	0.71
Institutional	0.10	0.20	0.30	0.50	0.60	0.58

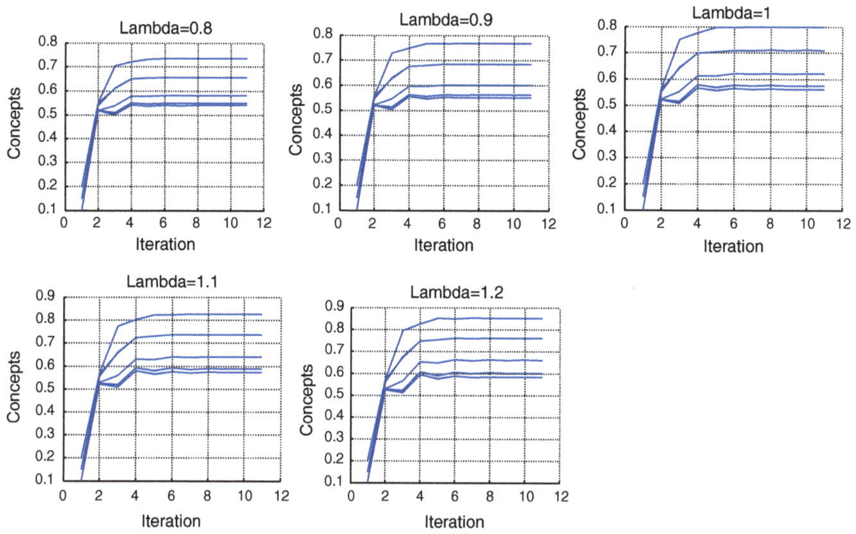

Fig. 7 The model simulation with λ = 0.8; 0.9; 1; 1.1 and 1.2

Table 4 The final state of the concepts computed for each λ

	λ					Normalized values λ = 0.8–1.2
	0.8	0.9	1	1.1	1.2	
C1	0.74	0.77	0.80	0.83	0.85	0.87–1.00
C2	0.66	0.68	0.71	0.74	0.76	0.78–0.89
C3	0.58	0.60	0.62	0.64	0.66	0.68–0.78
C4	0.54	0.55	0.56	0.57	0.59	0.64–0.69
C5	0.66	0.68	0.71	0.74	0.76	0.78–0.89
C6	0.55	0.56	0.58	0.59	0.60	0.65–0.71

initial values are estimated more or less correctly by the experts, we might conclude the following main statement of this part of the research: the ranking of the factors below influencing the sustainability of the waste management systems shows the way how the roles and weights of the factors should be considered within an IWMS in order to ensure environmental efficiency, economical affordability and social acceptability, this way providing a comprehensive interdisciplinary framework for addressing all problems of managing urban solid waste.

1. C1 (technical factor),
2. C2 (environmental factor) and C5 (legal factor),
3. C3 (economic factor),
4. C6 (institutional factor), and
5. C4 (social factor).

On the basis of this investigation, the priority sequence of the factors in the waste management systems on regional level might be declared.

The first or most important issue is how and what materials are managed, treated and disposed of (features of the collection, transfer and treatment systems, e.g. materials recovery, organic material treatment, thermal treatment, and final disposal). As second and third, the environmental and legal factors both have the same weight. These concern the state of the environment (pollution in the different areas, liveability of the settlements), and the relevant legislation (e.g. prescriptive or enabling legislation; EU, national, and municipal level legislation; legal definition of municipal solid waste). Then, they are followed by the economic issues of the system (system costs and revenues, available funding, etc.), and the institutional factor such as stakeholder involvement, accountability, professionalism and transparency. Finally, the list closes with the social factor where the main issue is to accept the IWMS and to participate in its activities (selection, collection), to minimize the risks to public health, adapting the system to the local demands and requirements and to willingness and ability to pay. However, the public plays an important role in sustainable waste management for which awareness on waste reduction, segregation and recycling need to be enhanced.

We set up the FCM model of the IWMS, and implemented its structure in a way that its parameters and weights were flexibly variable. As the data were obtained from a wide scope of experts, we are convinced that by using the proposed new approach sustainable waste management systems may be directly planned and established, at least in any more or less closed geographical area.

Even though the FCM model was proposed for the integrated analysis of the sustainability factors of IWMS on regional level, the validity of the method is depending on the reliability of the input data. In the first approach, we carried out an online survey where each stakeholder was asked to describe the existence and type of the causal relationships among the factors and then to assess the strength of these using a predetermined simple scale. In order to support their work we sent out a guideline to describe the terms of concepts and the basics of the development of an FCM before starting with the survey. This questionnaire guideline functioned as a support material in answering the questions. In this case, the interviewees had to rely only on the available information and had no chance to clarify uncertainties with the researchers.

In order to enhance the efficiency and pragmatics of this research, further, to establish a more suitable FCM model, we recently organized a workshop with the participation of waste management experts from all areas. During the workshop we explained to the participants what an FCM was, what its elements were and what our aim was with the results thus obtained. In this situation, if any issues were raised, we were able to explain the topics in more detail. As the participants were able to understand the underlying basic information, they could assess the values assigned to the connections more thoroughly. So, we assume, the difference between the two expert knowledge extraction methods (online survey and workshop with personal presence) influences the input data reliability essentially.

4.2 *Identification of the Elements of the Connection Matrix Using BEA*

In our second part of our research (Buruzs et al. 2013a), the model uses two different sets of input data. The sources of these two sets are different; one set is based on observations that may be considered more or less objective; observations on the trend of the studied factors in the time period from the 1980s till the 2010s. It is obvious that measuring the mutual influence of various factors within a complex phenomenon, like waste management is not easy. Nevertheless, it might be assumed that the time series published in the related literature (Demirbas 2011; den Boer and Lager 2007; Graymore et al. 2008; Langa et al. 2006; Morrissey and Browne 2004; Wilson et al. 2001; van de Klundert and Anschutz 1999; Thorneloe et al. 1999) is based on a consensus concerning the interrelationship of the concepts playing a determinative role in the procedure of waste management, thus these values are widely supported by independent observations and manually calculated partial models. In this research, the following data will be considered ‘objective’, even though they are not obtained by ‘measurements’ of some automatic machinery, but by the observation and evaluation of humans involved in the management of the procedure. It must be clearly understood that our learning model is based on these ‘objective’ data and therefore it makes it unnecessary to consult continuously the experts in order to obtain again and again up-to-date but entirely subjective data.

Nevertheless, in order to speed up the learning procedure, and to some extent, out of scientific curiosity, we used the data collected from the above mentioned survey. It must be stressed that results of these questionnaires (which were compared, and the medium values selected for each matrix element as the ‘typical subjective values’ of the given influence) were used only as initial values for the learning procedure, under the assumption that starting with more or less realistic values would speed up the convergence of the matrix to the stable ‘objective’ values. It turned out during the optimization that the convergence speed is quite high with randomly generated start population as well, thus prudent composition of the bacteria in the first generation was not an important issue. It is nevertheless interesting to compare the ‘subjective’ mutual influence values obtained from the questionnaires and the ‘objective’ matrix obtained from the time series observed starting with the data from the 1980s. On the basis of the gathered data we constructed the initial draft of the connection matrix, including identification of concept nodes and their mutual relationships represented by the graph edges.

Simulation in this context consisted of computing the states of the system described by the state vector over a number of successive iterations. In every iteration cycle the state vector specifies the current values of all factors (the nodes) in a particular moment. The values of the given states (nodes) are obtained from the preceding iteration values of all the nodes, which exert influence on the given node through cause-effect relationship. The transformation function is used to confine the weighted sum to the range set to [0, 1]. This normalization hinders the absolute quantitative analysis, but allows the comparison between nodes, which are attached by fuzzy activity degrees (defined as 'active': 1, 'inactive': 0 or 'active to a certain degree': values between 0 and 1), see Stach et al. (2005).

During the optimisation of our FCM with BEA, forced mutation (Hatwágner and Horváth 2012a) was used to increase the otherwise very low value of genetic diversity, to speed up computations in this manner. Forced mutation is a simple and easily implementable operator that slightly modifies some bacteria in the population if they seem very similar (typically in the final generations of the optimization). Forced mutation was applied in all subsequent generations after gene transfer.

The value of λ used by the transformation function was represented by the first gene of the bacteria. The following 30 genes corresponded to the elements of the 6 × 6 connection matrix (without the elements of the main diagonal, which were not stored).

The FCM determined the values of the factors in the subsequent iterations using the connection matrix. The goal of using the BEA was to find a connection matrix that minimizes a difference between the state values obtained from the literature (see Table 3) and the generated values of the factors. This difference *d* is expressed in Eq. 3.

$$d = \sum_{t=1}^{6} \left([c_i]_t - [\hat{c}_i]_t\right)^2 \tag{3}$$

where $[c_i]_t$ denotes the real and $[\hat{c}_i]_t$ the calculated values of factors.

The results of the optimization are contained in the connection matrix presented in Table 5. Here λ = 1, which resulted in *d* = 0.727 between the obtained and the experts suggested state vectors. It is rather surprising how far the interrelation coefficients obtained by automatic learning (based on the more or less objective data of the time series observed) are from the coefficients calculated from the median of the experts' questionnaires! We have no doubt that the matrix obtained by learning is rather independent from subjective elements, especially as it resulted from data obtained throughout a relatively long observation period. The fact that expert opinions differ so much from the objective reality definitely poses a question how deep the insight of waste management experts may be wherever the system on hand is constituted from a set of complex technical and social subsystems consisting of several mutually influencing (and rather fluctuating) factors.

Table 5 The resulting optimized connection matrix

	C1	C2	C3	C4	C5	C6
C1	0.00	−0.39	1.00	−1.00	1.00	0.75
C2	0.21	0.00	1.00	−1.00	1.00	−1.00
C3	−0.72	1.00	0.00	−1.00	1.00	−1.00
C4	−1.00	0.38	−1.00	0.00	1.00	−1.00
C5	1.00	1.00	0.75	−1.00	0.00	−1.00
C6	−1.00	0.82	−1.00	−1.00	1.00	0.00

Table 6 The time series predictions

	t_1'	t_2'	t_3'	t_4'	Normalized values
C1	0.53	0.68	0.69	0.72	0.74–1.00
C2	0.51	0.56	0.62	0.65	0.71–0.90
C3	0.48	0.45	0.50	0.52	0.67–0.72
C4	0.44	0.33	0.34	0.34	0.61–0.47
C5	0.56	0.62	0.70	0.72	0.78–1.00
C6	0.46	0.39	0.41	0.42	0.67–0.72

Despite the fact that a waste management system consists of only six main factors, it is obvious now that overviewing the whole procedure properly needs an approach based on the systems of systems concept (Perusich 2010). This approach is namely suitable to handle problems with essentially different type system components' where interoperability and seamless interfacing is necessary. The application of this approach then easily leads to unexpected emerging phenomena—such as e.g. the surprising values in the resulting connection matrix. The results obtained by the FCM model are unambiguously such emerging features that will necessarily lead to re-evaluation of the knowledge and views of experts dealing with waste management.

While in this approach we tried to find to optimize parameters with the help of the BEA and thus obtained a single set of results for the connection matrix in an alternative research (Buruzs et al. 2013b) we found that results obtained with various, non-optimal steepness values λ, the results differed essentially only in the scaling. All estimated time series predictions converged to essentially the same limit values (Table 6).

Findings from these simulations are not surprising. From the above table (Table 7), it can be stated, that the ranking of the factors influencing the sustainability of the waste management systems is similar in both simulations. This is the way how the roles and weights of the factors should be considered within an IWMS in order to ensure overall efficiency. In the next part, we summarize the results and give a short overview about the future research intentions.

Table 7 The normalized values of the two simulations

	Normalized values (FCM simulation)	Normalized values (BEA optimisation)
C1	0.87–1.00	0.74–1.00
C2	0.78–0.89	0.71–0.90
C3	0.68–0.78	0.67–0.72
C4	0.64–0.69	0.61–0.47
C5	0.78–0.89	0.78–1.00
C6	0.65–0.71	0.67–0.72

5 Summary

For large and complex systems it is extremely difficult to describe the entire system by a precise mathematical model (Jamshidi 2009). IWMSs are real and important elements of our everyday life therefore problems generated from these systems are real problems. From the unexpected results obtained, from the fact that the mutual influence matrix calculated from the observation data is so thoroughly different from the matrix given by the experts, the obvious question raises whether the approach and the objective results are mathematically stable and reliable enough in terms of the uncertainty of the observed values.

6 Further Research

Based on the above results, in the near future we intend to apply the systems of systems (SoS) approach to regional IWMS.

A system is a collection of main factors and their interrelationships gathered together to form a whole, greater than the sum of the parts (Boardman and Sauer 2009). The knowledge necessary for managing complex projects, for the development of complicated systems, has not kept pace with the increasing complexity and integration of these projects themselves. This increased complexity has permitted some to establish distinctions among system projects and to propose a framework of systems called system of systems (SoS).

Despite the fact that a waste management system consists of only six main factors, it is obvious now that overviewing the whole procedure properly needs an approach based on the systems of systems concept (Buruzs et al. 2013a). The challenge with the SoS emerges in the interoperability and interfacing of the component systems.

SoS integration is a method to pursue development, integration, interoperability, and optimization of systems to enhance performance, but it definitely needs a view that includes all views of the disciplines associated with the constituent systems. This can guarantee that among subsystems of different types and with various influence surfaces complete interoperability and seamless interfacing could be provided, and thus a deeply justifiable and relevant hierarchical adaptive FCM network model of IWMS can be established that may be used for actually determining the optimal inputs belonging to any intended change in the sustainable states while adequately predicting any unexpected emerging phenomena as well.

Acknowledgments The authors would like to thank to the National Science Research Fund (OTKA) K105529, K108405, the Social Renewal Operational Programme (TÁMOP) 4.1.1.C-12/1/KONV-2012-0017 grant for the support of the research.

References

Bäck, T., Fogel, D. B., & Michalewicz, Z. (1997). *Handbook of evolutionary computation*. London: IOP Publishing and Oxford University Press.

Balázs, K., Botzheim, J., & Kóczy, L. T. (2010a). Comparison of various evolutionary and memetic algorithms. *Integrated Uncertainty Management and Applications, Advances in Intelligent and Soft Computing, 68*, 431–442.

Balázs, K., Horváth, Z., & Kóczy, L. T. (2012). Different chromosome based evolutionary approaches for the permutation flow shop problem. *Acta Polytechnica Hungarica, 2*(2), 115–138.

Balázs, K., Kóczy, L. T., & Botzheim, J. (2010b). Comparative investigation of various evolutionary and memetic algorithms. In I. J. Rudas., J. Fodor., & J. Kacprzyk (Eds.), *Computational intelligence in engineering, studies in computational intelligence* (vol 313, pp. 129–140). Berlin: Springer.

Beigl, P., Lebersorger, S., & Salhofer, S. (2008). Modelling municipal solid waste generation: a review. *Journal of Waste Management, 28*, 200–214.

Botzheim, J., Cabrita, C., Kóczy, L. T., & Ruano, A. E. (2009a). Fuzzy rule extraction by bacterial memetic algorithms. *International Journal of Intelligent Systems, 24*(3), 312–339.

Botzheim, J., Földesi P., & Kóczy L. T. (2009b). Solution for fuzzy road transport travelling salesman problem using eugenic bacterial memetic algorithm. In *Proceedings of IFSA/ EUSFLAT Conference '2009* (pp. 1667–1672).

Bovea, M. D., & Powell, J. C. (2006). Alternative scenarios to meet the demands of sustainable waste management. *Journal of Environmental Management, 79*, 115–132.

Buruzs, A., Hatwágner, M. F., Pozna, R. C., & Kóczy, L. T. (2013b). Advanced learning of fuzzy cognitive maps of waste management by bacterial algorithm. In *Proceedings of IFSA World Congress and NAFIPS Annual Meeting* (pp. 890–895). IEEE.

Buruzs, A., Pozna, R. C., & Kóczy, L. T. (2013a). Developing fuzzy cognitive maps for modelling regional waste management systems. In Y, Tsompanakis. (Ed.), *Proceedings of the Third International Conference on Soft Computing Technology in Civil, Structural and Environmental Engineering*, Paper 19. Stirlingshire, UK: Civil-Comp Press doi:10.4203/ccp.103.19.

Carvalho, J. P. (2010). On the semantics and the use of fuzzy cognitive maps in social sciences. *Soft Computing in the Humanities and Social Science, 214*, 6–19.

Council Directive 1999/31/EC of April 26 1999 on the landfill of waste.

Dányádi, Zs, Balázs, K., & Kóczy, L. T. (2010a). A comparative study of various evolutionary algorithms and their combinations for optimizing fuzzy rule-based inference systems. *Scientific Bulletin of "Politechnica" University of Timisoara, Romania, Transactions on Automatic Control and Computer Science 55*(69), 247–254.

Dányádi, Zs., Földesi, P., & Kóczy, L. T. (2010b). A fuzzy bacterial evolutionary solution for three dimensional bin packing problems. *Acta Technica Jaurinensis, Series Logistica, 3*(3), 325–334.

Darwin, C. R. (1859). *The origin of species*. London: John Murray.

Demirbas, A. (2011). Waste management, waste resource facilities and waste conversion processes. *Journal of Energy Conservation and Management, 52*(2), 1280–1287.

den Boer, E., & Lager, J. (2007). LCA-IWM: A decision support tool for sustainability assessment of waste management systems. *Journal of Waste Management, 27*(8), 1032–1045.

Engelbrecht, A. P. (2007). *Computational intelligence: An introduction*. England: Wiley.

European Parliament and Council Directive 94/62/EC of December 20, 1994 on packaging and packaging waste.

Gál, L., Botzheim, J., & Kóczy, L. T. (2008). Modified bacterial memetic algorithm used for fuzzy rule base extraction. In *Proceedings of the 5th International of Conference on Soft Computing as Transdisciplinary Science and Technology* (pp. 425–431). USA:ACM.

Goldberg, D. E. (1989). Genetic algorithms in search, optimization, and machine learning. Boston: Addison-Wesley Publishing Company, Inc.

Graymore, M. L. M., Sipe, N. G., & Rickson, R. E. (2008). Regional sustainability: how useful are current tools of sustainability assessment at the regional scale? *Journal of Ecological Economics, 67*(3), 362–372.

Haastrup, P., Maniezzo, V., Mattarelli, M., Rinaldi, F. M., Mendes, I., & Paruccini, M. (1989). A decision support system for urban waste management. *European Journal of Operational Research, 109*(2), 330–341.

Hatwágner, F. M., & Horváth, A. (2011). Parallel gene transfer operations for the bacterial evolutionary algorithm. *Acta Technica Jaurinensis, 4*(1), 89–112.

Hatwágner, F. M., & Horváth, A. (2012a). Comparative analysis of parallel gene transfer operators in the bacterial evolutionary algorithm. *Acta Polytechnica Hungarica, 9*(4), 65–84.

Hatwágner, F. M., & Horváth, A. (2012b). Maintaining genetic diversity in bacterial evolutionary algorithm. *ANNALES Universitatis Scientiarum Budapestinensis de Rolando Eötvös Nominatae Sectio Computatorica 37*, 175–194.

Holland, J. H. (1975). *Adaptation in natural and artificial systems*. Ann Arbor: The University of Michigan Press.

Hung, M.-L., Ma, H.-W., & Yang, W.-F. (2007). A novel sustainable decision making model for municipal solid waste management. *Journal of Waste Management, 27*(2), 209–219.

Isak, K. G. Q., Wildenberg, M., Adamescu, M., Skov, F., De Blust, G., & Varjopuro, R. (2009). A long-term biodiversity, ecosystem and awareness research network manual for applying fuzzy cognitive mapping—experiences from ALTER-Net. Project no. GOCE-CT-2003-505298, ALTER-Net Deliverable type: Report, WPR6-2009-02—Deliverable 4.R6.D2.

Jadoon, A., Batool, S. A., & Chaudhry, M. N. (2014). Assessment of factors affecting household solid waste generation and its composition in Gulberg town, Lahore, Pakistan. *Journal of Mater Cycles Waste Management, 16*(1), 73–81. doi:10.1007/s10163-013-0146-5.

Jamshidi, M. (Ed.). (2009). *Systems of system engineering. Innovation for the 21th century*, 480 p. Wiley: Hoboken, ISBN 978-0-470-19590-1.

Kalakula, S., Malakulb, P., Siemanondb, K., & Gania, R. (2014). Integration of life cycle assessment software with tools for economic and sustainability analyses and process simulation for sustainable process design. *Journal of Cleaner Production, 17*, 98–109.

Ketipi, M. K., Koulouriotis, D. E., Karakasis, E. G., Papakostas, G. A., & Tourassis, V. D. (2010). A flexible nonlinear approach to represent cause–effect relationships in FCMs. *Journal of Applied Soft Computing, 12*(12), 3757–3770.

Kosko, B. (1986). Fuzzy cognitive maps. *International Journal of Man-Machine Studies, 24*(1), 65–75.

Kurian, J. (2006). Stakeholder participation for sustainable waste management. *Journal of Habitat International, 30*(4), 863–871.

Langa, D. L., Binderb, C. R., Stauffachera, M., Zieglera, C., Schleiss, K., & Scholz, R. W. (2006). Material and money flows as a means for industry analysis of recycling schemes. A case study of regional bio-waste management. *Journal of Resources, Conservation and Recycling, 49*(06), 159–190.

Malena, C. (2004). Strategic partnership: challenges and best practices in the management and governance of multi-stakeholder partnerships involving UN and civil society actors. Background paper prepared by for the multi-stakeholder workshop on partnerships and UN-civil society relations, Pocantico, New York.

Maniezzo, V., Mendes, I., & Paruccini, M. (1998). Decision support for siting problems. *Journal of Decision Support Systems, 23*(3), 273–284.

McBean, E. A., del Rosso, E., & Rovers, F. A. (2005). Improvements in financing for sustainability in solid waste management. *Journal of Resources, Conservation and Recycling, 43*(4), 391–401.

Morrissey, A. J., & Browne, J. (2004). Waste management models and their application to sustainable waste management. *Journal of Waste Management, 24*(3), 297–308.

Nawa, N. E., & Furuhashi, T. (1998). A study on the effect of transfer of genes for the bacterial evolutionary algorithm. In L. C., Jain., & R. K., Jain (Eds.), *Second international conference on knowledge-based intelligent electronic system* Adelaide, Australia (pp. 585–590).

Nawa, N. E., & Furuhashi, T. (1999). Fuzzy system parameters discovery by bacterial evolutionary algorithm. *IEEE Transactions on Fuzzy Systems, 7*(5), 608–616.

Nawa, N. E., Hashiyama, T., Furuhashi, T., & Uchikawa, Y. (1997). Fuzzy logic controllers generated by pseudo-bacterial genetic algorithm. In *Proceedings of the IEEE International Conference of Neural Networks (ICNN97)* Houston, USA (pp. 2408–2413).

Özesmi, U., & Özesmi, S. L. (2004). Ecological models based on people's knowledge: A multi-step fuzzy cognitive mapping approach. *Journal of Ecological Modelling, 176*(15), 3–64.

Papageorgiou, E., & Kontogianni, A. (2012). Using fuzzy cognitive mapping in environmental decision making and management: A methodological primer and an application. In S. Young (Ed.), *International perspectives on global environmental change*. InTech doi: 10.5772/29375 ISBN: 978-953-307-815-1.

Perusich, K. (2010). System diagnosis using fuzzy cognitive maps. *cognitive maps*. InTech ISBN: 978-953-307-044-5.

Phillips, P. S., Read, A. D., Green, A. E., & Bates, M. P. (1999). UK waste minimisation clubs: A contribution to sustainable waste management. *Journal of Resources, Conservation and Recycling, 27*(3), 217–247.

Salhofer, S., Wassermann, G., & Binner, E. (2007). Strategic environmental assessment as an approach to assess waste management systems. Experiences from an Austrian case study. *Journal of Environmental Modelling and Software, 22*(5), 610–618.

Boardman J., & Sauser, B. (2009). System of systems—The meaning of. In *Proceeding of the 2006 IEEE/SMC International Conference on System of Systems Engineering*, Los Angeles, CA, USA.

Shmeleva, S. E., & Powell, J. R. (2006). Ecological–economic modelling for strategic regional waste management systems. *Journal of Ecological Economics, 59*(1), 115–130.

Stach, W., Kurgan, L., Pedrycz, W., & Reformat, M. (2005). Genetic learning of fuzzy cognitive maps. *Journal of Fuzzy Sets and Systems, 153*(3), 371–401.

Stylos, C. D., Georgopoulos, V. C., & Groumpos, P. P. (1997). The use of fuzzy cognitive maps in modelling systems. In *Proceedings of 5th IEEE Mediterranean Conference on Control and Systems*, Paphos, Cyprus.

Stylos, D., & Groumpos, P. P. (2004). Modelling complex systems using fuzzy cognitive maps. *IEEE Transactions on Systems, Man, and Cybernetics—Part A: Systems and Humans, 34*(1), 155–162.

Tanskanen, J.-H. (2000). Strategic planning of municipal solid waste management. *Journal of Resources, Conservation and Recycling, 30*(2), 111–133.

Thorneloe, S. A., Weitz, K., Barlaz, M., & Ham, R. K. (1999). Tools for determining sustainable waste management through application of life-cycle assessment: Update on U.S. Research. In *Proceedings of 7th International Waste Management and Landfill Symposium vol V*, pp. 629–636.

van de Klundert, A., & Anschutz, J. (1999). Integrated sustainable waste management: The selection of appropriate technologies and the design of sustainable systems is not (only) a technological issue. *CEDARE/IETC inter-regional workshop on technologies for sustainable waste management*, pp. 1–17 Alexandria, Egypt.

Wilson, E. J., McDougall, F. R., & Willmore, J. (2001). Euro-Trash: Searching Europe for a more sustainable approach to waste management. *Journal of Resources Conservation and Recycling, 31*(4), 327–346.

Worku Y., & Muchie, M. (2012). An attempt at quantifying factors that affect efficiency in the management of solid waste produced by commercial businesses in the city of Tshwane, South Africa. *Journal of Environmental and Public Health 2012*, 12 p, Article ID 165353. doi:10. 1155/2012/165353 (research article).

Leukocyte Detection Through an Evolutionary Method

Erik Cuevas, Margarita Díaz and Raúl Rojas

Abstract Classical image processing methods often face great difficulties while dealing with images containing noise and distortions. Under such conditions, the use of soft computing approaches has been recently extended to address challenging real-world image processing problems. The automatic detection of Leukocytes or White Blood Cells (WBC) still remains as an unsolved issue in medical imaging. The analysis of WBC images has engaged researchers from fields of medicine and image processing alike. Since WBC can be approximated by an ellipsoid form, an ellipse detector algorithm may be successfully applied in order to recognize such elements. This chapter presents an algorithm for the automatic detection of leukocytes embedded into complicated and cluttered smear images that considers the complete process as a multi-ellipse detection problem. The approach, which is based on the Differential Evolution (DE) algorithm, transforms the detection task into an optimization problem whose individuals represent candidate ellipses. An objective function evaluates if such candidate ellipses are actually present in the edge map of the smear image. Guided by the values of such function, the set of encoded candidate ellipses (individuals) are evolved using the DE algorithm so that they can fit into the leukocytes which are enclosed within the edge map of the smear image. Experimental results from white blood cell images with a varying range of complexity are included to validate the efficiency of the proposed technique in terms of its accuracy and robustness.

Keywords Leukocyte detection · Image processing · WBC image analysis · Differential evolution · Evolutionary algorithms · Metaheuristics

E. Cuevas (✉)
Departamento de Electrónica, CUCEI, Universidad de Guadalajara,
Guadalajara, Mexico
e-mail: erik.cuevas@cucei.udg.mx

M. Díaz
División de Biotecnología y Salud, ITESM, Campus Guadalajara, Zapopan, Mexico

R. Rojas
Institut Für Informatik, Freie Universität Berlin, Berlin, Germany

Q. Zhu and A.T. Azar (eds.), *Complex System Modelling and Control Through Intelligent Soft Computations*, Studies in Fuzziness and Soft Computing 319,
DOI 10.1007/978-3-319-12883-2_5

1 Introduction

Soft computing has emerged as a powerful tool for information processing, decision making and knowledge management. The techniques of soft computing have been successfully developed in areas such as neural networks, fuzzy systems and evolutionary algorithms. It is predictable that in the near future soft computing will play a more important role in tackling several engineering problems. Image processing is a very important research area. Classical image processing methods often face great difficulties while dealing with images containing noise and distortions. Under such conditions, the use of computational intelligence approaches has been recently extended to address challenging real-world image processing problems.

On the other hand, medical image processing has become more and more important in diagnosis with the development of medical imaging and computer technique. Huge amounts of medical images are obtained by X-ray radiography, CT and MRI. They provide essential information for efficient and accurate diagnosis based on advance computer vision techniques (Zhuang and Meng 2004; Scholl et al. 2011).

White Blood Cells (WBC) also known as leukocytes play a significant role in the diagnosis of different diseases. Although computer vision techniques have successfully contributed to generate new methods for cell analysis, which in turn, have lead into more accurate and reliable systems for disease diagnosis. However, high variability on cell shape, size, edge and localization, complicates the data extraction process. Moreover, the contrast between cell boundaries and the image's background may vary due to unstable lighting conditions during the capturing process.

Many works have been conducted in the area of blood cell detection. In Wang and Chu (2009) a method based on boundary support vectors is proposed to identify WBC. In such approach, the intensity of each pixel is used to construct feature vectors whereas a Support Vector Machine (SVM) is used for classification and segmentation. By using a different approach, Wu et al. in 2006, developed an iterative Otsu method based on the circular histogram for leukocyte segmentation. According to such technique, the smear images are processed in the Hue-Saturation-Intensity (HSI) space by considering that the Hue component contains most of the WBC information. One of the latest advances in white blood cell detection research is the algorithm proposed by Wang et al. in 2007, which is based on the fuzzy cellular neural network (FCNN). Although such method has proved successful in detecting only one leukocyte in the image, it has not been tested over images containing several white cells. Moreover, its performance commonly decays when the iteration number is not properly defined, yielding a challenging problem itself with no clear clues on how to make the best choice.

Since white blood cells can be approximated with an ellipsoid form, computer vision techniques for detecting ellipses may be used in order to recognize them. Ellipse detection in real images is an open research problem since long time ago. Several approaches have been proposed which traditionally fall under three categories: Symmetry-based, Hough transform-based (HT) and Random sampling.

In symmetry-based detection (Muammar and Nixon 1989; Atherton and Kerbyson 1993), the ellipse geometry is taken into account. The most common elements used in ellipse geometry are the ellipse center and axis. Using these elements and edges in the image, the ellipse parameters can be found. Ellipse detection in digital images is commonly solved through the Hough Transform (Fischer and Bolles 1981). It works by representing the geometric shape by its set of parameters, then accumulating bins in the quantized parameter space. Peaks in the bins provide the indication of where ellipses may be. Obviously, since the parameters are quantized into discrete bins, the intervals of the bins directly affect the accuracy of the results and the computational effort. Therefore, for fine quantization of the space, the algorithm returns more accurate results, while suffering from large memory loads and expensive computation. In order to overcome such a problem, some other researchers have proposed other ellipse detectors following the Hough transform principles by using random sampling. In random sampling-based approaches (Shaked et al. 1996; Xu et al. 1990), a bin represents a candidate shape rather than a set of quantized parameters, as in the HT. However, like the HT, random sampling approaches go through an accumulation process for the bins. The bin with the highest score represents the best approximation of an actual ellipse in the target image. McLaughlin's work (Han et al. 1993) shows that a random sampling-based approach produces improvements in accuracy and computational complexity, as well as a reduction in the number of false positives (non existent ellipses), when compared to the original HT and the number of its improved variants.

As an alternative to traditional techniques, the problem of ellipse detection has also been handled through optimization methods. In general, they have demonstrated to give better results than those based on the HT and random sampling with respect to accuracy and robustness (Ayala-Ramirez et al. 2006). Such approaches have produced several robust ellipse detectors using different optimization algorithms such as Genetic algorithms (GA) (Lutton and Martinez 1994; Yao et al. 2005) and Particle Swarm Optimization (PSO) (Cheng et al. 2009).

Although detection algorithms based on optimization approaches present several advantages in comparison to traditional approaches, they have been scarcely applied to WBC detection. One exception is the work presented by Karkavitsas and Rangoussi, in 2005 that solves the WBC detection problem through the use of GA. However, since the evaluation function, which assesses the quality of each solution, considers the number of pixels contained inside of a circle with fixed radius, the method is prone to produce misdetections particularly for images that contained overlapped or irregular WBC.

In this work, the WBC detection task is approached as an optimization problem and the differential evolution algorithm is used to build the ellipsoidal approximation. Differential Evolution (DE), introduced by Storn and Price, in 1995, is a novel evolutionary algorithm which is used to optimize complex continuous nonlinear functions. As a population-based algorithm, DE uses simple mutation and crossover operators to generate new candidate solutions, and applies one-to-one competition scheme to greedily decide whether the new candidate or its parent will survive in the next generation. Due to its simplicity, ease of implementation, fast convergence, and robustness, the DE algorithm has gained much attention, reporting a wide range of successful applications in the literature (Babu and Munawar 2007; Mayer et al. 2005; Kannan et al. 2003; Chiou et al. 2005; Cuevas et al. 2010).

This chapter presents an algorithm for the automatic detection of blood cell images based on the DE algorithm. The proposed method uses the encoding of five edge points as candidate ellipses in the edge map of the smear. An objective function allows to accurately measure the resemblance of a candidate ellipse with an actual WBC on the image. Guided by the values of such objective function, the set of encoded candidate ellipses are evolved using the DE algorithm so that they can fit into actual WBC on the image. The approach generates a sub-pixel detector which can effectively identify leukocytes in real images. Experimental evidence shows the effectiveness of such method in detecting leukocytes despite complex conditions. Comparison to the state-of-the-art WBC detectors on multiple images demonstrates a better performance of the proposed method.

The main contribution of this study is the proposal of a new WBC detector algorithm that efficiently recognize WBC under different complex conditions while considering the whole process as an ellipse detection problem. Although ellipse detectors based on optimization present several interesting properties, to the best of our knowledge, they have not yet been applied to any medical image processing up to date.

This chapter is organized as follows: Sect. 2 provides a description of the DE algorithm while in Sect. 3 the ellipse detection task is fully explained from an optimization perspective within the context of the DE approach. The complete WBC detector is presented in Sect. 4. Section 5 reports the obtained experimental results whereas Sect. 6 conducts a comparison between state-of-the-art WBC detectors and the proposed approach. Finally, in Sect. 7, some conclusions are drawn.

2 Differential Evolution Algorithm

In the proposed approach, the problem of WBC detection is faced as an optimization problem. As optimization tool, the differential evolution (DE) algorithm is used to solve the detection problem. In this section, the main characteristics of DE are discussed.

DE algorithm is a simple and direct search algorithm which is based on population and aims for optimizing global multi-modal functions. DE employs the mutation operator as to provide the exchange of information among several solutions.

There are various mutation base generators to define the algorithm type. The version of DE algorithm used in this work is known as rand-to-best/1/bin or "DE1" (Storn and Price 1995). DE algorithms begin by initializing a population of N_p and D-dimensional vectors considering parameter values that are randomly distributed between the pre-specified lower initial parameter bound $x_{j,\text{low}}$ and the upper initial parameter bound $x_{j,\text{high}}$ as follows:

$$x_{j,i,t} = x_{j,\text{low}} + \text{rand}(0,1) \cdot (x_{j,\text{high}} - x_{j,\text{low}}); \\ j = 1,2,\ldots,D; \quad i = 1,2,\ldots,N_p; \quad t = 0. \tag{1}$$

The subscript t is the generation index, while j and i are the parameter and particle indexes respectively. Hence, $x_{j,i,t}$ is the jth parameter of the ith particle in generation t. In order to generate a trial solution, DE algorithm first mutates the best solution vector $\mathbf{x}_{best,t}$ from the current population by adding the scaled difference of two vectors from the current population.

$$\mathbf{v}_{i,t} = \mathbf{x}_{best,t} + F \cdot (\mathbf{x}_{r_1,t} - \mathbf{x}_{r_2,t}); \\ r_1, r_2 \in \{1,2,\ldots,N_p\} \tag{2}$$

with $\mathbf{V}_{i,t}$ being the mutant vector. Indices r_1 and r_2 are randomly selected with the condition that they are different and have no relation to the particle index i whatsoever (i.e. $r_1 \neq r_2 \neq i$). The mutation scale factor F is a positive real number, typically less than one. Figure 1 illustrates the vector-generation process defined by Eq. (2).

In order to increase the diversity of the parameter vector, the crossover operation is applied between the mutant vector $\mathbf{v}_{i,t}$ and the original individuals $\mathbf{X}_{i,t}$. The result is the trial vector $\mathbf{u}_{i,t}$ which is computed by considering element to element as follows:

$$u_{j,i,t} = \begin{cases} v_{j,i,t}, & \text{if rand(0,1)} \le CR \text{ or } j = j_{\text{rand}}, \\ x_{j,i,t}, & \text{otherwise.} \end{cases} \tag{3}$$

with $j_{\text{rand}} \in \{1,2,\ldots,D\}$. The crossover parameter ($0.0 \le CR \le 1.0$) controls the fraction of parameters that the mutant vector is contributing to the final trial vector.

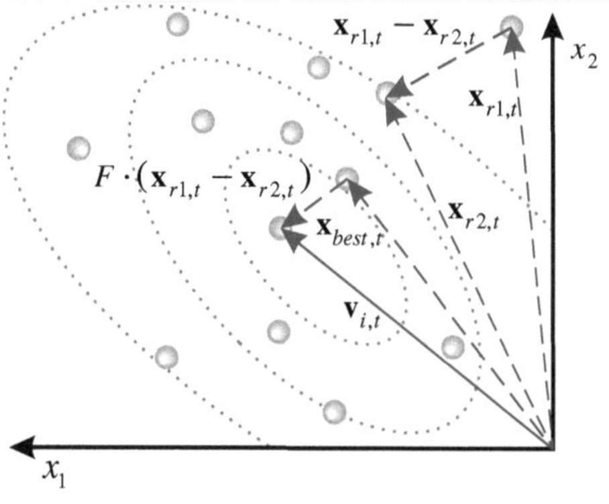

Fig. 1 Two-dimensional example of an objective function showing its contour lines and the process for generating **v** in scheme DE/best/l/exp from vectors of the current generation

In addition, the trial vector always inherits the mutant vector parameter according to the randomly chosen index j_{rand}, assuring that the trial vector differs by at least one parameter from the vector to which it is compared ($\mathbf{x}_{i,t}$).

Finally, a greedy selection is used to find better solutions. Thus, if the computed cost function value of the trial vector $\mathbf{u}_{i,t}$ is less or equal than the cost of the vector $\mathbf{x}_{i,t}$, then such trial vector replaces $\mathbf{x}_{i,t}$ in the next generation. Otherwise, $\mathbf{x}_{i,t}$ remains in the population for at least one more generation:

$$\mathbf{x}_{i,t+1} = \begin{cases} \mathbf{u}_{i,t}, & \text{if } f(\mathbf{u}_{i,t}) \leq f(\mathbf{x}_{i,t}), \\ \mathbf{x}_{i,t}, & \text{otherwise.} \end{cases} \tag{4}$$

Here, $f(\cdot)$ represents the objective function. These processes are repeated until a termination criterion is attained or a predetermined generation number is reached.

3 Ellipse Detection Using DE

Since WBC can be approximated by an ellipsoid form, an ellipse detector algorithm may be successfully applied in order to recognize such elements. In this section, the problem of ellipse detection is translated to an optimization task.

3.1 Data Preprocessing

In order to detect ellipse shapes, candidate images must be preprocessed first by an edge detection algorithm which yields an edge map image. Then, the (x_i, y_i) coordinates for each edge pixel p_i are stored inside the edge vector $P = \{p_1, p_2, \ldots, p_{N_p}\}$, with N_p being the total number of edge pixels.

3.2 Individual Representation

Just as a line requires two points to completely define its characteristics, an ellipse is defined by five points. Therefore, each candidate solution E (ellipse candidate) considers five edge points to represent an individual. Under such representation, edge points are selected following a random positional index within the edge array P. This procedure will encode a candidate solution as the ellipse that passes through five points p_1, p_2, p_3, p_4 and p_5 ($E = \{p_1, p_2, p_3, p_4, p_5\}$). Thus, by substituting the coordinates of each point of E into Eq. 5, we gather a set of five simultaneous equations which are linear in the five unknown parameters a, b, f, g and h.

$$ax^2 + 2hxy + by^2 + 2gx + 2fy + 1 = 0 \tag{5}$$

Considering the configuration of the edge points shown by Fig. 2, the ellipse center (x_0, y_0), the radius maximum (r_{max}), the radius minimum (r_{min}) and the ellipse orientation (θ) can be calculated as follows:

$$x_0 = \frac{hf - bg}{C}, \tag{6}$$

$$y_0 = \frac{gh - af}{C}, \tag{7}$$

$$r_{max} = \sqrt{\frac{-2\Delta}{C(a + b - R)}}, \tag{8}$$

$$r_{min} = \sqrt{\frac{-2\Delta}{C(a + b + R)}}, \tag{9}$$

$$\theta = \frac{1}{2}\arctan\left(\frac{2h}{a - b}\right) \tag{10}$$

where

$$R^2 = (a - b)^2 + 4h^2, \; C = ab - h^2 \text{ and } \Delta = \det\left(\begin{vmatrix} a & h & g \\ h & b & f \\ g & f & 1 \end{vmatrix}\right). \tag{11}$$

3.3 Objective Function

Optimization refers to choosing the best element from one set of available alternatives. In the simplest case, it means to minimize an objective function or error by systematically choosing the values of variables from their valid ranges. In order to calculate the error produced by a candidate solution E, the ellipse coordinates are calculated as a virtual shape which, in turn, must also be validated, i.e. if it really exists in the edge image. The test set is represented by $S = \{s_1, s_2, \ldots, s_{N_s}\}$, where N_s are the number of points over which the existence of an edge point, corresponding to E, should be tested.

The set S is generated by the Midpoint Ellipse Algorithm (MEA) (Bresenham 1987) which is a searching method that seeks required points for drawing an ellipse. For any point (x, y) lying on the boundary of the ellipse with a, h, b, g and f, it does satisfy the equation $f_{ellipse}(x, y) \cong r_{max}x^2 + r_{min}y^2 - r_{max}^2 r_{min}^2$, where r_{max} and r_{min}

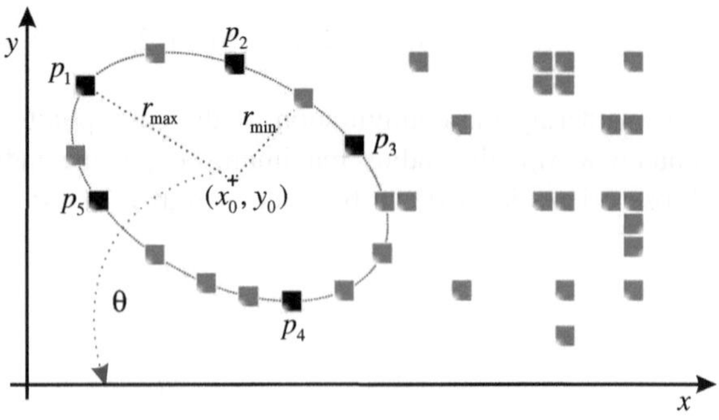

Fig. 2 Ellipse candidate (individual) built from the combination of points p_1, p_2, p_3, p_4 and p_5

represent the semi-major and semi-minor axis, respectively. However, MEA avoids computing square root calculations by comparing the pixel separation distances. A method for direct distance comparison is to test the halfway position between two pixels (sub-pixel distance) to determine if this midpoint is inside or outside the ellipse boundary. If the point is in the interior of the ellipse, the ellipse function is negative. Thus, if the point is outside the ellipse, the ellipse function is positive. Therefore, the error involved in locating pixel positions using the midpoint test is limited to one-half the pixel separation (sub-pixel precision). To summarize, the relative position of any point (*x*, *y*) can be determined by checking the sign of the ellipse function:

$$f_{ellipse}(x,y)\begin{cases} <0 & \text{if } (x,y) \text{ is inside the ellipse boundary} \\ =0 & \text{if } (x,y) \text{ is on the ellipse boundary} \\ >0 & \text{if } (x,y) \text{ is outside the ellipse boundary} \end{cases} \quad (12)$$

The ellipse-function test in Eq. 12 is applied to mid-positions between pixels nearby the ellipse path at each sampling step. Figure 3a, b show the midpoint between the two candidate pixels at sampling position. The ellipse is used to divide the quadrants into two regions the limit of the two regions is the point at which the curve has a slope of −1 as shown in Fig. 3.

In MEA the computation time is reduced by considering the symmetry of ellipses. Ellipses sections in adjacent octants within one quadrant are symmetric with respect to the dy/dy = −1 line dividing the two octants. These symmetry conditions are illustrated in Fig. 4. The algorithm can be considered as the quickest providing a sub-pixel precision (Van Aken 2005). However, in order to protect the MEA operation, it is important to assure that points lying outside the image plane must not be considered in S.

The objective function *J*(*E*) represents the matching error produced between the pixels *S* of the ellipse candidate *E* and the pixels that actually exist in the edge image, yielding:

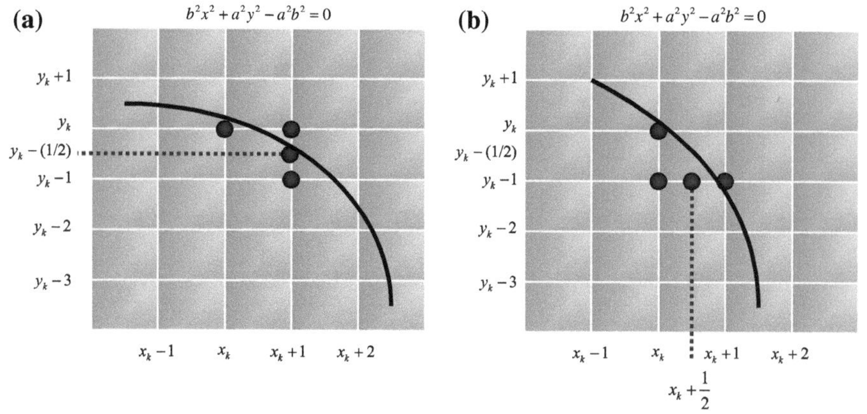

Fig. 3 **a** Symmetry of the ellipse: an estimated one octant which belong to the first region where the slope is greater than −1, **b** in this region the slope will be less than −1 to complete the octant and continue to calculate the same so the remaining octants

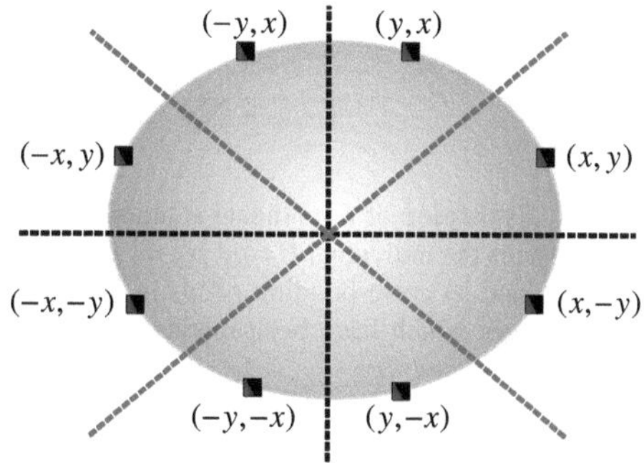

Fig. 4 Midpoint between candidate pixels at sampling position x_k along an *elliptical* path

$$J(E) = 1 - \frac{\sum_{v=1}^{Ns} G(x_v, y_v)}{Ns} \tag{13}$$

where $G(x_i, y_i)$ is a function that verifies the pixel existence in (x_v, y_v), with $(x_v, y_v) \in S$ and N_s being the number of pixels lying on the perimeter corresponding to E currently under testing.

Hence, function $G(x_v, y_v)$ is defined as:

$$G(x_v, y_v) = \begin{cases} 1 & \text{if the pixel } (x_v, y_v) \text{ is an edge point} \\ 0 & \text{otherwise.} \end{cases} \tag{14}$$

A value of $J(E)$ near to zero implies a better response from the “ellipsoid” operator. Figure 5 shows the procedure to evaluate a candidate action E with its representation as a virtual shape S. Figure 5a shows the original edge map, while Fig. 5b presents de individual $E = \{p_1, p_2, p_3, p_4, p_5\}$ to be evaluated. Figure 5c,

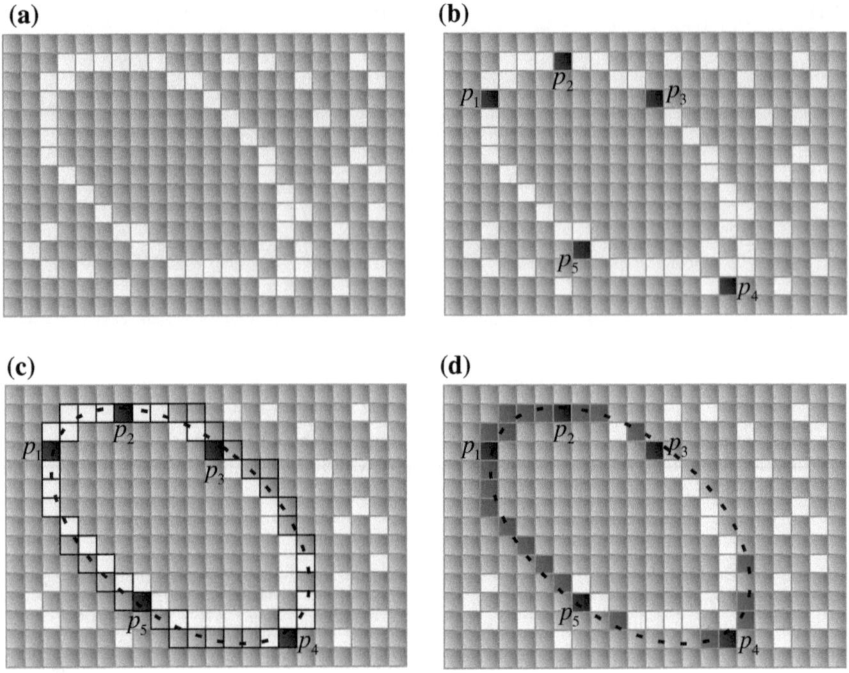

Fig. 5 Evaluation of a candidate solution E: the image in (**a**) shows the original edge map while (**b**) presents the individual $E = p_1, p_2, p_3, p_4$ and p_5 to be evaluated. The image (**c**) shows the virtual shape S and its corresponding pixels in *bold line*. The image in (**d**) shows coincidences between both images which have been marked by *darker* pixels while the virtual shape is also depicted through a *dashed line*

presents the generated virtual shape drawn from points p_1, p_2, p_3, p_4 and p_5. Finally, Fig. 5d shows the virtual shape S compared to the original image, point by point, in order to find coincidences between virtual and edge points.

The virtual shape S (Fig. 5c), obtained by MEA, gathers 47 points ($N_s = 47$) with only 25 of them existing in both images (shown as darker points in Fig. 5d) and yielding: $\sum_{v=1}^{Ns} G(x_v, y_v) = 25$, therefore $J(E) = 0.255$. This value indicates that the virtual shape S obtained, has a considerable number of coincidences with the edge map.

3.4 Implementation of DE for Ellipse Detection

The ellipse detector algorithm based on DE can be summarized in the following steps:

Step 1:	Set the DE parameters $F = 0.25$ and $CR = 0.8$
Step 2:	Initialize the population of m individuals $\mathbf{E}^k = \{E_1^k, E_2^k, \ldots, E_m^k\}$ where each decision variable p_1, p_2, p_3, p_4 and p_5 of E_a^k is set randomly within the interval $[1, N_p]$. All values must be integers. Considering that $k = 0$ and $a \in (1, 2, \ldots, m)$
Step 3:	Evaluate the objective value $J(E_a^k)$ for all m individuals, and determining the $E^{best,k}$ showing the best fitness value, such that $E^{best,k} \in \{\mathbf{E}^k\} \mid J(E^{best,k}) = \min\{J(E_1^k), J(E_2^k), \ldots, J(E_m^k)\}$

Step 4:	Generate the trial population $\mathbf{T} = \{T_1, T_2, \ldots, T_m\}$:
	
	for (i=1; $i<m+1$; i++)
	do r_1=floor(rand(0,1)$\cdot m$); while ($r_1 = i$);
	do r_2=floor(rand(0,1)$\cdot m$); while (($r_2 = i$) or ($r_2 = r_1$));
	jrand=floor($5 \cdot$ rand(0,1));
	
	for (j=1; $j<6$; j++) // generate a trial vector
	if (rand(0,1)$<=CR$ or j=jrand)
	$T_{j,i} = E_j^{best,k} + F \cdot (E_{j,r_1}^k - E_{j,r_2}^k)$;
	else
	$T_{j,i} = E_{j,i}^k$;
	end if
	end for
	end for

Step 5:	Evaluate the fitness values $J(T_i)$ ($i \in \{1, 2, \ldots, m\}$) of all trial individuals. Check all individuals. If a candidate parameter set is not physically plausible, i.e. out of the range $[1, N_p]$, then an exaggerated cost function value is returned. This aims to eliminate "unstable" individuals.

Step 6:	Select the next population $\mathbf{E}^{k+1} = \{E_1^{k+1}, E_2^{k+1}, \ldots, E_m^{k+1}\}$:
	
	for (i=1; $i<m+1$; i++)
	if $(J(T_i) < J(E_i^k))$
	$E_i^{k+1} = T_i$
	else
	$E_i^{k+1} = E_i^k$
	end if
	end for

Step 7:	If the iteration number (NI) is met, then the output $E^{best,k}$ is the solution (an actual ellipse contained in the image), otherwise go back to Step 3

4 The White Blood Cell Detector

In this section, the complete WBC detection strategy is described. Such a strategy combines a segmentation method with the ellipse detection approach presented in Sect. 3.

4.1 Image Preprocessing

To employ the proposed detector, smear images must be preprocessed to obtain two new images: the segmented image and its corresponding edge map. The segmented image is produced by using a segmentation strategy whereas the edge map is generated by a border extractor algorithm. Such edge map is considered by the objective function to measure the resemblance of a candidate ellipse with an actual WBC.

The goal of the segmentation strategy is to isolate the white blood cells (WBC's) from other structures such as red blood cells and background pixels. Information of color, brightness and gradients are commonly used within a thresholding scheme to generate the labels to classify each pixel. Although a simple histogram thresholding can be used to segment the WBC's, at this work the Diffused Expectation-Maximization (DEM) has been used to assure better results (Boccignone et al. 2004).

DEM is an Expectation-Maximization (EM) based algorithm which has been used to segment complex medical images (Boccignone et al. 2007). In contrast to classical EM algorithms, DEM considers the spatial correlations among pixels as a part of the minimization criteria. Such adaptation allows to segment objects in spite of noisy and complex conditions. The method models an image as a finite mixture, where each mixture component corresponds to a region class and uses a maximum likelihood approach to estimate the parameters for each class, via the expectation maximization (EM) algorithm, which is coupled to anisotropic diffusion over classes in order to account for the spatial dependencies among pixels.

For the WBC's segmentation, it has been used the implementation of DEM provided in (2012). Since the implementation allows to segment gray-level images and color images, it can be used for operating over all smear images with no regard about how each image has been acquired. The DEM has been configured considering three different classes ($K = 3$), $g(\nabla h_{ik}) = |\nabla h_{ik}|^{-9/5}$, $\lambda = 0.1$ and $m = 10$ iterations. These values have been found as the best configuration set according to (Boccignone et al. 2004).

As a final result of the DEM operation, three different thresholding points are obtained: the first corresponds to the WBC's, the second to the red blood cells whereas the third represents the pixels classified as background. Figure 6b presents the segmentation results obtained by the DEM approach employed at this work considering the Fig. 6a as the original image.

Once the segmented image has been produced, the edge map is computed. The purpose of the edge map is to obtain a simple image representation that preserves object structures. The DE-based detector operates directly over the edge map in order to recognize ellipsoidal shapes. Several algorithms can be used to extract the edge map; however, at this work, the morphological edge detection procedure (Gonzalez and Woods 1992) has been used to accomplish such a task. Morphological edge detection is a traditional method to extract borders from binary images in which original images (I_B) are eroded by a simple structure element (I_E) composed by a matrix-template of 3×3 with all its values equal to one. Then, the

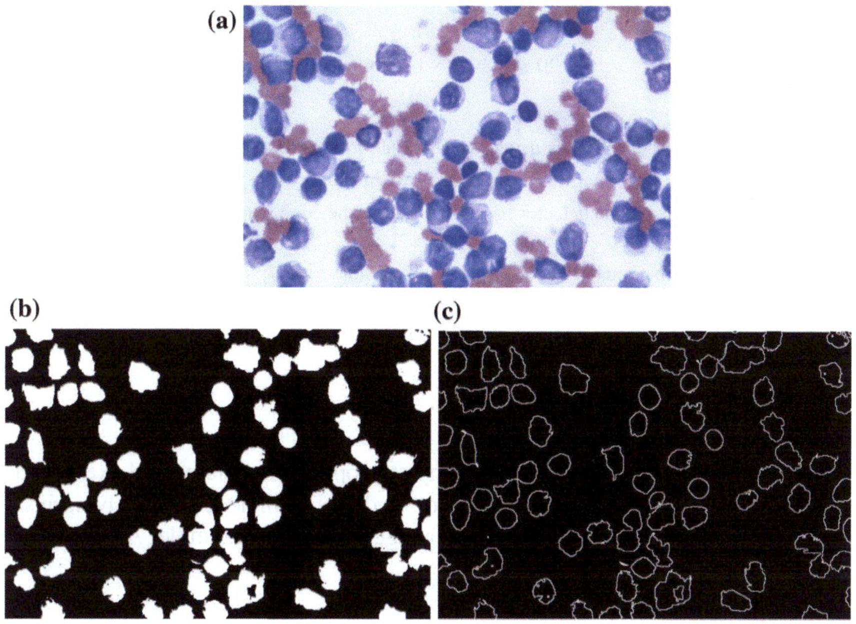

Fig. 6 Preprocessing process. **a** original smear image, **b** segmented image obtained by DEM and **c** the edge map obtained by using the morphological edge detection procedure

eroded image is inverted ($\bar{I}_E$) and compared with the original image ($\bar{I}_E \wedge I_B$) in order to detect pixels which are present in both images. Such pixels compose the computed edge map from I_B. Figure 6c shows the edge map obtained by using the morphological edge detection procedure.

4.2 Ellipse Detection Approach

The edge map is used as input image for the ellipse detector presented in Sect. 3. Table 1 presents the parameter set that has been used in this work for the DE algorithm after several calibration examples have been conducted. The final configuration matches the best possible calibration proposed in (Wang and Huang 2010), where it has been analyzed the effect of modifying the DE-parameters for several generic optimization problems. The population-size parameter (m = 20) has been selected considering the best possible balance between convergence and computational overload. Once it has been set, such configuration has been kept for all test images employed in the experimental study.

Under such assumptions, the complete process to detect WBC's is implemented as follows:

Table 1 DE parameters used for leukocytes detection in medical images

m	F	CR	NI
20	0.25	0.80	200

Step 1:	Segment the WBC's using the DEM algorithm (described in 4.1)
Step 2:	Get the edge map from the segmented image
Step 3:	Start the ellipse detector based in DE over the edge map while saving best ellipses (Sect. 3)
Step 4:	Define parameter values for each ellipse that identify the WBC's

4.3 Numerical Example

In order to present the algorithm's step-by-step operation, a numerical example has been set by applying the proposed method to detect a single leukocyte lying inside of a simple image. Figure 7a shows the image used in the example. After applying

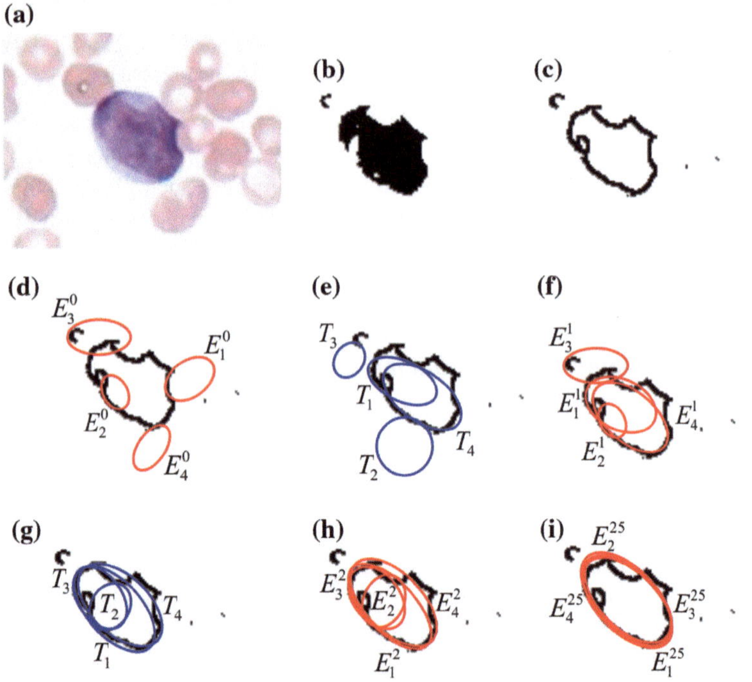

Fig. 7 Detection numerical example: **a** the image used as example. **b** Segmented image. **c** Edge map. **d** Initial particles $\mathbf{E}^0$. **e** Trial elements **T** produced by the DE operators. **f** New population $\mathbf{E}^1$. **g** Trial elements produced considering $\mathbf{E}^1$ as input population. **h** New population $\mathbf{E}^2$. **i** Final particle configuration after 25 iterations

the threshold operation, the WBC is located besides few other pixels which are merely noise (see Fig. 7b). Then, the edge map is subsequently computed and stored pixel by pixel inside the vector P. Figure 7c shows the resulting image after such procedure.

The DE-based ellipse detector is executed using information of the edge map (for the sake of easiness, it only considers a population of four particles). Like all evolutionary approaches, DE is a population-based optimizer that attacks the starting point problem by sampling the search space at multiple, randomly chosen, initial particles. By taking five random pixels from vector P, four different particles are constructed. Figure 7d depicts the initial particle distribution $\mathbf{E}^0 = \{E_1^0, E_2^0, E_3^0, E_4^0\}$. By using the DE operators, four different trial particles $\mathbf{T} = \{T_1, T_2, T_3, T_4\}$ (ellipses) are generated, their locations are shown in Fig. 7e. Then, the new population $\mathbf{E}^1$ is selected considering the best elements obtained among the trial elements $\mathbf{T}$ and the initial particles $\mathbf{E}^0$. The final distribution of the new population is depicted in Fig. 7f. Since the particles E_2^0 and E_2^0 hold (in Fig. 7f) a better fitness value ($J(E_2^0)$ and $J(E_3^0)$) than the trial elements T_2 and T_3, they are considered as particles of the final population $\mathbf{E}^1$. Figure 7g, h present the second iteration produced by the algorithm whereas Fig. 6i shows the population configuration after 25 iterations. From Fig. 7i, it is clear that all particles have converged to a final position which is able to accurately cover the WBC.

5 Experimental Results

In this section, the experimental results are presented. Several tests have been developed in order to evaluate the performance of the WBC detector. It was tested over microscope images from blood-smears holding a 960 × 720 pixel resolution. They correspond to supporting images on the leukemia diagnosis. The images show several complex conditions such as deformed cells and overlapping with partial occlusions. The robustness of the algorithm has been tested under such demanding conditions. All the experiments has been developed using an Intel Core i7-2600 PC, with 8 GB in RAM.

Figure 8a shows an example image employed in the test. It was used as input image for the WBC detector. Figure 8b presents the segmented WBC's obtained by the DEM algorithm. Figure 8c, d present the edge map and the white blood cells after detection, respectively. The results show that the proposed algorithm can effectively detect and mark blood cells despite cell occlusion, deformation or overlapping. Other parameters may also be calculated through the algorithm: the total area covered by white blood cells and relationships between several cell sizes.

Other example is presented in Fig. 9. It represents a complex example with an image showing seriously deformed cells. Despite such imperfections, the proposed approach can effectively detect the cells as it is shown in Fig. 9d.

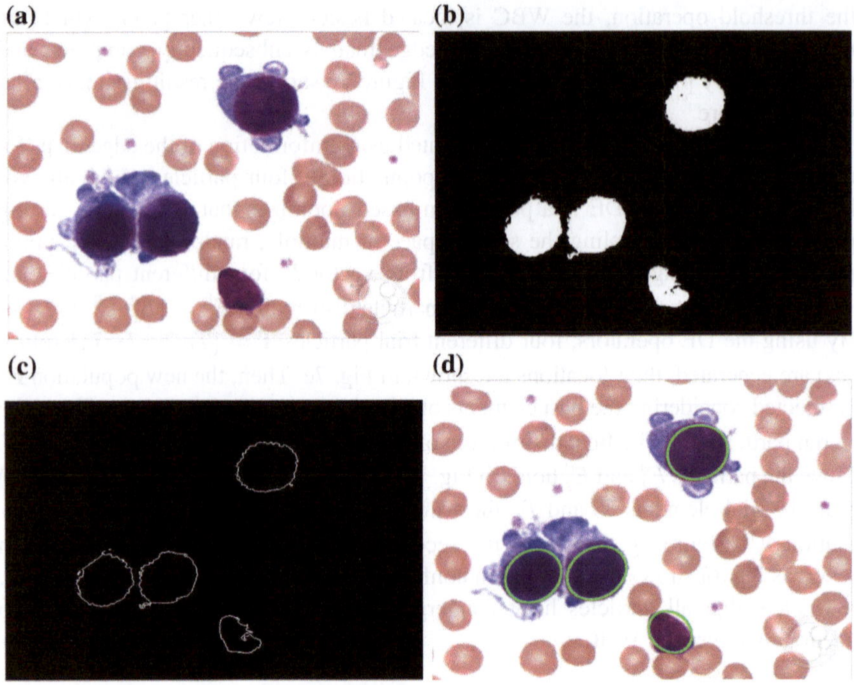

Fig. 8 Resulting images of the first test after applying the WBC detector: **a** original image, **b** image segmented by the DEM algorithm, **c** edge map and **d** the *white* detected blood cells

6 Comparisons to Other Methods

In this section, a comprehensive set of smear-blood test images is used to test the performance of the proposed approach. We have applied the proposed DE-based detector to test images in order to compare its performance to other WBC detection algorithms such as the Boundary Support Vectors (BSV) approach (Wang and Chu 2009), the iterative Otsu (IO) method (Wu et al. 2006), the Wang algorithm (Wang et al. 2007) and the Genetic algorithm-based (GAB) detector (Karkavitsas and Rangoussi 2005). In all cases, the algorithms are tuned according to the value set which is originally proposed by their own references.

6.1 Detection Comparison

To evaluate the detection performance of the proposed detection method, Table 2 tabulates the comparative leukocyte detection performance of the BSV approach, the IO method, the Wang algorithm, the BGA detector and the proposed method, in terms of detection rates and false alarms. The experimental data set includes 50 images which are collected from the ASH Image Bank (http://imagebank.hematology.org/).

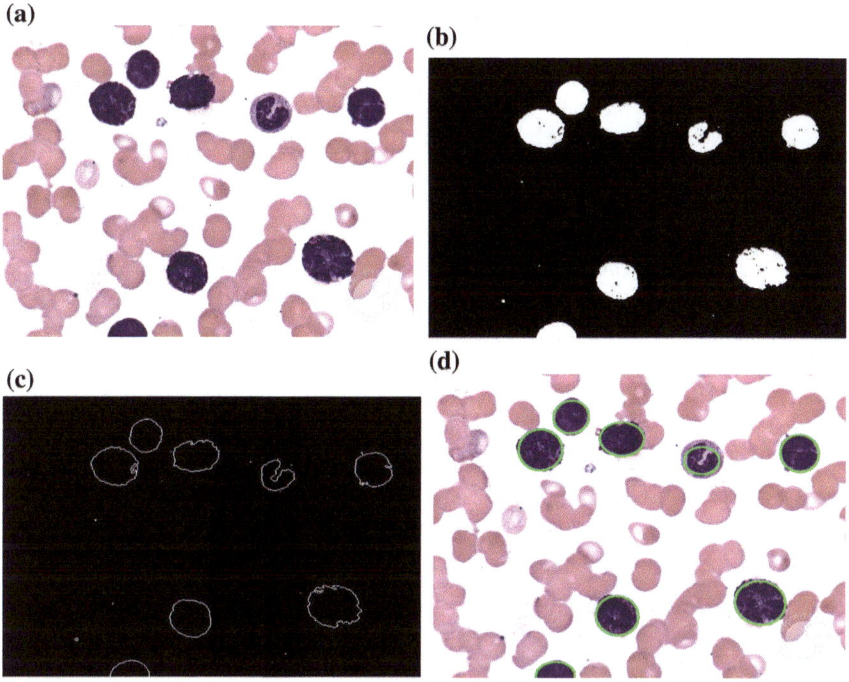

Fig. 9 Resulting images of the second test after applying the WBC detector: **a** original image, **b** image segmented by the DEM algorithm, **c** edge map and **d** the white detected blood cells

Such images contain 517 leukocytes (287 bright leukocytes and 230 dark leukocytes according to smear conditions) which have been detected and counted by a human expert. Such values act as ground truth for all the experiments. For the comparison, the detection rate (DR) is defined as the ratio between the number of leukocytes correctly detected and the number leukocytes determined by the expert. The false alarm rate (FAR) is defined as the ratio between the number of non-leukocyte objects that have been wrongly identified as leukocytes and the number leukocytes which have been actually determined by the expert.

Experimental results show that the proposed DE method, which achieves 98.26 % leukocyte detection accuracy with 2.71 % false alarm rate, is compared favorably against other WBC detection algorithms, such as the BSV approach, the IO method, the Wang algorithm and the BGA detector.

6.2 Robustness Comparison

Images of blood smear are often deteriorated by noise due to various sources of interference and other phenomena that affect the measurement processes in imaging

Table 2 Comparative leukocyte detection performance of the BSV approach, the IO method, the Wang algorithm, the BGA detector and the proposed DE method over the data set which contains 30 images and 426 leukocytes

Leukocyte type	Method	Leukocytes detected	Missing	False alarms	DR (%)	FAR (%)
Bright leukocytes **(287)**	BSV [3]	130	157	84	45.30	29.27
	IO [4]	227	60	73	79.09	25.43
	Wang [5]	231	56	60	80.49	20.90
	GAB [16]	220	67	22	76.65	7.66
	DE-based	281	6	11	97.91	3.83
Dark leukocytes **(230)**	BSV [3]	105	125	59	46.65	25.65
	IO [4]	183	47	61	79.56	26.52
	Wang [5]	196	34	47	85.22	20.43
	GAB [16]	179	51	23	77.83	10.00
	DE-based	227	3	3	98.70	1.30
Overall **(517)**	BSV [3]	235	282	143	45.45	27.66
	IO [4]	410	107	134	79.30	25.92
	Wang [5]	427	90	107	82.59	20.70
	GAB [16]	399	118	45	77.18	8.70
	DE-based	508	9	14	98.26	2.71

and data acquisition systems. Therefore, the detection results depend on the algorithm's ability to cope with different kinds of noises. In order to demonstrate the robustness in the WBC detection, the proposed DE approach is compared to the BSV approach, the IO method, the Wang algorithm and the BGA detector under noisy environments. In the test, two different experiments have been studied.

The first inquest explores the performance of each algorithm when the detection task is accomplished over images corrupted by Salt and Pepper noise. The second experiment considers images polluted by Gaussian noise. Salt and Pepper and Gaussian noise are selected for the robustness analysis because they represent the most compatible noise types commonly found in images of blood smear (Landi and Piccolomini 2012). The comparison considers the complete set of 50 images presented in Sect. 6.1 containing 517 leukocytes which have been detected and counted by a human expert.

The added noise is produced by MatLab©, considering two noise levels of 5 and 10 % for Salt and Pepper noise whereas $\sigma = 5$ and $\sigma = 10$ are used for the case of Gaussian noise. Such noise levels, according to (Tapiovaara and Wagner 1993), correspond to the best trade of between detection difficulty and the real existence in medical imaging. If higher noise levels are used then the detection process would be unnecessarily complicated without representing a feasible image condition.

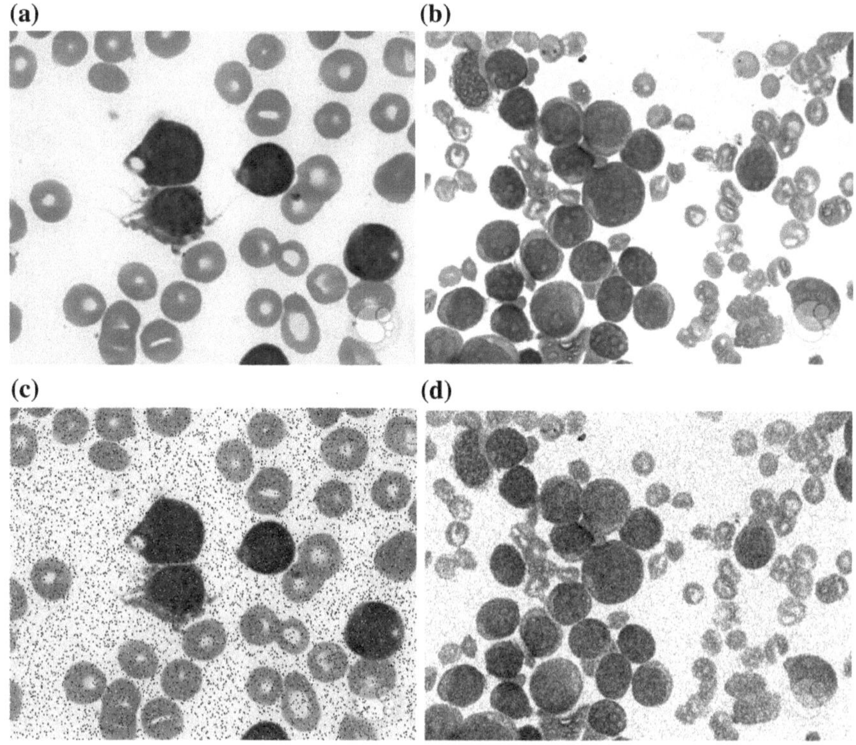

Fig. 10 Examples of images included in the experimental set for robustness comparison. **a–b** Originals images. **c** Image contaminated with 10 % of Salt and Pepper noise and **d** image polluted with $\sigma = 10$ of Gaussian noise

Figure 10 shows two examples of the experimental set. The outcomes in terms of the detection rate (DR) and the false alarm rate (FAR) are reported for each noise type in Tables 3 and 4. The results show that the proposed DE algorithm presents the best detection performance, achieving in the worst case a DR of 89.55 and 91.10 %, under contaminated conditions of Salt and Pepper and Gaussian noise, respectively. On the other hand, the DE detector possesses the least degradation performance presenting a FAR value of 5.99 and 6.77 %.

6.3 Stability Comparison

In order to compare the stability performance of the proposed method, its results are compared to those reported by Wang et al. (2007), in which is considered as an accurate technique for the detection of WBC.

Table 3 Comparative WBC detection among methods that considers the complete data set of 30 images corrupted by different levels of Salt and Pepper noise

Noise level	Method	Leukocytes detected	Missing	False alarms	DR (%)	FAR (%)
5 % Salt and Pepper noise 517 leukocytes	BSV [3]	185	332	133	34.74	26.76
	IO [4]	311	206	106	63.38	24.88
	Wang [5]	250	176	121	58.68	27.70
	GAB [16]	298	219	135	71.83	24.18
	DE-based	482	35	32	91.55	7.04
10 % Salt and Pepper noise 517 leukocytes	BSV [3]	105	412	157	20.31	30.37
	IO [4]	276	241	110	53.38	21.28
	Wang [5]	214	303	168	41.39	32.49
	GAB [16]	337	180	98	65.18	18.95
	DE-based	463	54	31	89.55	5.99

Table 4 Comparative WBC detection among methods that considers the complete data set of 30 images corrupted by different levels of Gaussian noise

Noise level	Method	Leukocytes detected	Missing	False alarms	DR (%)	FAR (%)
$\sigma = 5$ Gaussian noise 517 leukocytes	BSV [3]	214	303	98	41.39	18.95
	IO [4]	366	151	87	70.79	16.83
	Wang [5]	358	159	84	69.25	16.25
	GAB [16]	407	110	76	78.72	14.70
	DE-based	487	30	21	94.20	4.06
$\sigma = 10$ Gaussian noise 517 leukocytes	BSV [3]	162	355	129	31.33	24.95
	IO [4]	331	186	112	64.02	21.66
	Wang [5]	315	202	124	60.93	23.98
	GAB [16]	363	154	113	70.21	21.86
	DE-based	471	46	35	91.10	6.77

The Wang algorithm is an energy-minimizing method which is guided by internal constraint elements and influenced by external image forces, producing the segmentation of WBC's at a closed contour. As external forces, the Wang approach uses edge information which is usually represented by the gradient magnitude of the image. Therefore, the contour is attracted to pixels with large image gradients, i.e. strong edges. At each iteration, the Wang method finds a new contour configuration which minimizes the energy that corresponds to external forces and constraint elements.

In the comparison, the net structure and its operational parameters, corresponding to the Wang algorithm, follow the configuration suggested in Wang et al. (2007) while the parameters for the DE-based algorithm are taken from Table 1.

Figure 11 shows the performance of both methods considering a test image with only two white blood cells. Since the Wang method uses gradient information in order to appropriately find a new contour configuration, it needs to be executed iteratively in order to detect each structure (WBC). Figure 11b shows the results after the Wang approach has been applied considering only 200 iterations. Furthermore, Fig. 11c shows results after applying the DE-based method which has been proposed in this chapter.

The Wang algorithm uses the fuzzy cellular neural network (FCNN) as optimization approach. It employs gradient information and internal states in order to find a better contour configuration. In each iteration, the FCNN tries, as contour points, different new pixel positions which must be located nearby the original contour position. Such fact might cause the contour solution to remain trapped into a local minimum. In order to avoid such a problem, the Wang method applies a considerable number of iterations so that a near optimal contour configuration can be found. However, when the number of iterations increases the possibility to cover other structures increases too. Thus, if the image has a complex background (just as smear images do) or the WBC's are too close, the method gets confused so that finding the correct contour configuration from the gradient magnitude is not easy. Therefore, a drawback of Wang's method is related to its optimal iteration number

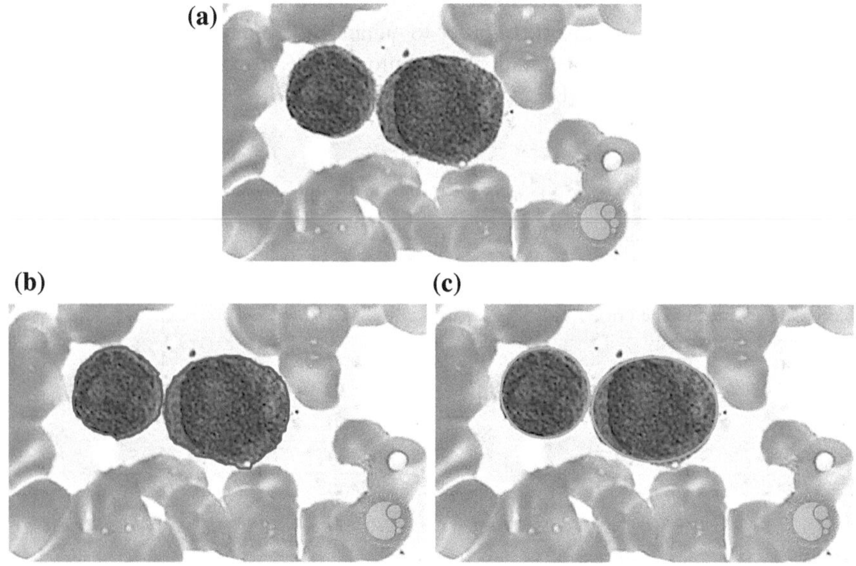

Fig. 11 Comparison of the DE and the Wang's method for white blood cell detection in medical images. **a** Original image. **b** Detection using the Wang's method, **c** detection after applying the DE method

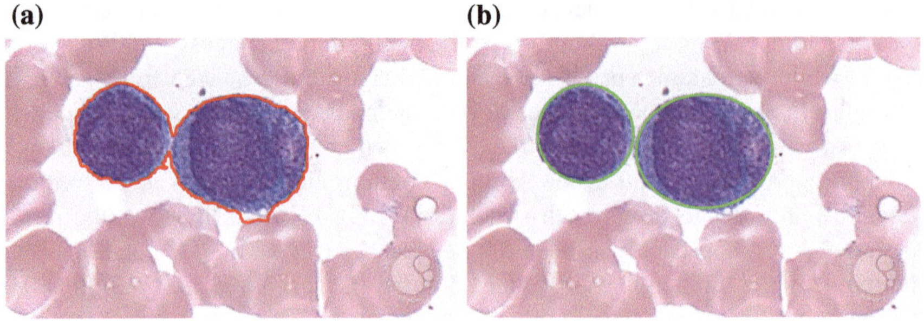

Fig. 12 Result comparison for the white blood cells detection showing (**a**) Wang's algorithm after 400 cycles and (**b**) DE detector method considering 1,000 cycles

(instability). Such number must be determined experimentally as it depends on the image context and its complexity. Figure 12a shows the result of applying 400 cycles of the Wang's algorithm while Fig. 12b presents the detection of the same cell shapes after 1,000 iterations using the proposed algorithm. From Fig. 12a, it can be seen that the contour produced by Wang's algorithm degenerates as the iteration process continues, wrongly covering other shapes lying nearby.

In order to compare the accuracy of both methods, the estimated WBC area which has been approximated by both approaches, is compared to the actual WBC size considering different degrees of evolution i.e. the cycle number for each algorithm. The comparison considers only one WBC because it is the only detected shape in the Wang's method. Table 5 shows the averaged results over twenty repetitions for each experiment. In order to enhance the analysis, Fig. 13 illustrates the Error-percentage versus Iterations evolution from an extended data set which has been compiled from Table 5.

Table 5 Error in cell's size estimation after applying the DE algorithm and the Wang's method to detect one leukocyte embedded into a blood-smear image. The error is averaged over twenty experiments

Algorithm	Iterations	Error (%)
Wang	30	88
	60	70
	200	1
	400	121
	600	157
DE-based	30	24.30
	60	7.17
	200	2.25
	400	2.25
	600	2.25

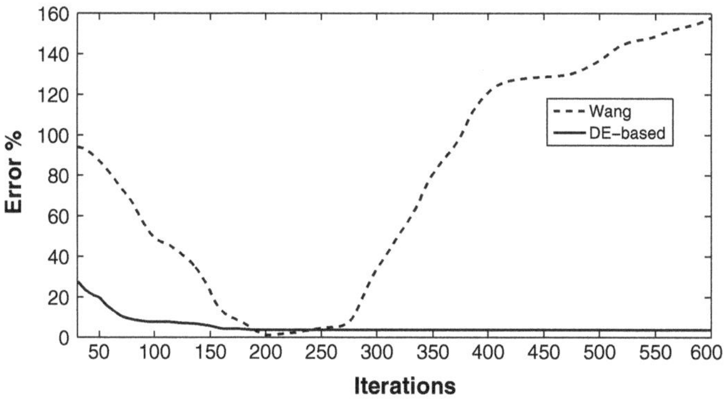

Fig. 13 Error percentage versus iterations evolution from an extended data set from Table 5

7 Conclusions

In this chapter, an algorithm for the automatic detection of blood cell images based on the DE algorithm has been presented. The approach considers the complete process as a multiple ellipse detection problem. The proposed method uses the encoding of five edge points as candidate ellipses in the edge map of the smear. An objective function allows to accurately measure the resemblance of a candidate ellipse with an actual WBC on the image. Guided by the values of such objective function, the set of encoded candidate ellipses are evolved using the DE algorithm so that they can fit into actual WBC on the image. The approach generates a sub-pixel detector which can effectively identify leukocytes in real images.

The performance of the DE-method has been compared to other existing WBC detectors such as the Boundary Support Vectors (BSV) approach (Wang and Chu 2009), the iterative Otsu (IO) method (Wu et al. 2006), the Wang algorithm (Wang et al. 2007) and the Genetic algorithm-based (GAB) detector (Karkavitsas and Rangoussi 2005) considering several images which exhibit different complexity levels. Experimental results demonstrate the high performance of the proposed method in terms of detection accuracy, robustness and stability.

References

Atherton, T., & Kerbyson, D. (1993). Using phase to represent radius in the coherent circle Hough transform. In *IEE Colloquium on the Hough Transform* (pp. 1–4), May 7 1993, IEEE.

Ayala-Ramirez, V., Garcia-Capulin, C. H., Perez-Garcia, A., & Sanchez-Yanez, R. E. (2006). Circle detection on images using genetic algorithms. *Pattern Recognition Letters, 27*(6), 652–657.

Babu, B., & Munawar, S. (2007). Differential evolution strategies for optimal design of shell-and-tube heat exchangers. *Chemical Engineering Science, 62*(14), 3720–3739.

Boccignone, G., Ferraro, M., & Napoletano, P. (2004). Diffused expectation maximisation for image segmentation. *Electron Letters, 40*(18), 1107–1108.

Boccignone, G., Napoletano, P., Caggiano, V., & Ferraro, M. (2007). A multi-resolution diffused expectation–maximization algorithm for medical image segmentation. *Computers in Biology and Medicine, 37*(1), 83–96.

Bresenham, J. E. (1987). A linear algorithm for incremental digital display of circular arcs. *Communications of the ACM, 20*(2), 100–106.

Cheng, H. D., Guo, Y., & Zhang, Y. (2009). A novel Hough transform based on eliminating particle swarm optimization and its applications. *Pattern Recognition, 42*(9), 1959–1969.

Chiou, J., Chang, C., & Su, C. (2005). Variable scaling hybrid differential evolution for solving network reconfiguration of distribution systems. *IEEE Transactions on Power Systems, 20*(2), 668–674.

Cuevas, E., Zaldivar, D., & Pérez-Cisneros, M. (2010). A novel multi-threshold segmentation approach based on differential evolution optimization. *Expert Systems with Applications, 37*(7), 5265–5271.

DEM: Diffused expectation maximization function for image segmentation. (2012). Version 1.0. http://www.mathworks.com/matlabcentral/fileexchange/37197-dem-diffused-expectation-maximisation-for-image-segmentation.

Fischer, M., & Bolles, R. (1981). Random sample consensus: A paradigm to model fitting with applications to image analysis and automated cartography. *CACM, 24*(6), 381–395.

Gonzalez, R. C., & Woods, R. E. (1992). *Digital image processing*. Reading, MA: Addison Wesley.

Han, J., Koczy, L., & Poston, T. (1993). Fuzzy Hough transform. In *Proceedings 2nd International Conference on Fuzzy Systems, San Francisco, California* (Vol. 2, pp. 803–808), March 28–April 01 1993. doi:10.1109/FUZZY.1993.32.7545.

Kannan, S., Slochanal, S. M. R., & Padhy, N. (2003). Application and comparison of metaheuristic techniques to generation expansion planning problem. *IEEE Transactions on Power Systems, 20*(1), 466–475.

Karkavitsas, G., & Rangoussi, M. (2005). Object localization in medical images using genetic algorithms. *International Journal of Medical, Dentistry, Pharmaceutical, Health Science and Engineering, 1*(2), 6–9.

Landi, G., & Piccolomini, E. L. (2012). An efficient method for nonnegatively constrained total variation-based denoising of medical images corrupted by Poisson noise. *Computerized Medical Imaging and Graphics, 36*(1), 38–46.

Lutton, E., & Martinez, P. (1994). A genetic algorithm for the detection of 2D geometric primitives in images. In *Proceedings of the 12th International Conference On Pattern Recognition, Jerusalem, Israel* (Vol. 1, pp. 9–13, 526–528), October 1994. doi:10.1109/ICPR.1994.576345.

Mayer, D., Kinghorn, B., & Archer, A. (2005). Differential evolution—An easy and efficient evolutionary algorithm for model optimization. *Agricultural Systems, 83*(3), 315–328.

Muammar, H., & Nixon, M. (1989). Approaches to extending the Hough transform. In *Proceedings International Conference on Acoustics, Speech and Signal Processing ICASSP-89, Glasgow* (Vol. 3, pp. 23–26, 1556–1559), May 1989. doi:10.1109/ICASSP.1989.266739.

Scholl, I., Aach, T., Deserno, T. M., & Kuhlen, T. (2011). Challenges of medical image processing. *Computer Science Research and Development., 26*(1–2), 5–13.

Shaked, D., Yaron, O., & Kiryati, N. (1996). Deriving stopping rules for the probabilistic Hough transform by sequential analysis. *Computer Vision Image Understanding, 63*(3), 512–526.

Storn, R., & Price, K. (1995). Differential evolution—A simple and efficient adaptive scheme for global optimization over continuous spaces. Technical Report No. TR-95-012, International Computer Science Institute, Berkley (CA).

Tapiovaara, M., & Wagner, R. (1993). SNR and noise measurements for medical imaging: I. A practical approach based on statistical decision theory. *Physics in Medicine and Biology, 38* (1), 71–92.

Van Aken, J. R. (2005). Efficient ellipse-drawing algorithm. *IEEE Computer Graphics and Applications, 4*(9), 24–35.

Wang, M., & Chu, R. (2009). A novel white blood cell detection method based on boundary support vectors. *Proceedings of the 2009 IEEE International Conference on Systems, Man, and Cybernetics, San Antonio, TX, USA* (pp. 2595–2598), October 11–14, 2009. DOI:10.1109/ICSMC.2009.5346736.

Wang, L., & Huang, F. (2010). Parameter analysis based on stochastic model for differential evolution algorithm. *Applied Mathematics and Computation, 217*(7), 3263–3273.

Wang, S., Korris, F. L., & Fu, D. (2007). Applying the improved fuzzy cellular neural network IFCNN to white blood cell detection. *Neurocomputing, 70*(7–9), 1348–1359.

Wu, J., Zeng, P., Zhou, Y., & Oliver, C. (2006). A novel color image segmentation method and its application to white blood cell image analysis. In *8th International Conference on Signal Processing, Beijing, China* (Vol. 2, pp. 16–20, 16–20), November 2006. DOI:10.1109/ICOSP.2006.345700.

Xu, L., Oja, E., & Kultanen, P. (1990). A new curve detection method: Randomized Hough transform (RHT). *Pattern Recognition Letters, 11*(5), 331–338.

Yao, J., Kharma, N., & Grogono, P. (2005). A multi-population genetic algorithm for robust and fast ellipse detection. *Pattern Analysis Applications, 8*(1–2), 149–162.

Zhuang, X., & Meng, Q. (2004). Local fuzzy fractal dimension and its application in medical image processing. *Artificial Intelligence in Medicine, 32*(1), 29–36.

PWARX Model Identification Based on Clustering Approach

Zeineb Lassoued and Kamel Abderrahim

Abstract This chapter addresses the problem of clustering based procedure for the identification of PieceWise Auto-Regressive eXogenous (PWARX) models. In order to overcome the main drawbacks of the existing methods such as their sensitivity to poor initializations and the existence of outliers, we propose the use of the Chiu's clustering algorithm and the Density-Based Spatial Clustering of Applications with Noise (DBSCAN) algorithm. A comparative study of the two proposed approaches with the k-means method is achieved in simulation. The results of experimental validation are also presented to illustrate the effectiveness of the proposed methods.

Keywords Identification · PWARX systems · Clustering approach · Chiu's algorithm · DBSCAN algorithm · Experimental validation

1 Introduction

Hybrid systems are heterogeneous dynamical systems that arise out of the interaction of continuous and discrete dynamics. The continuous behavior is the fact of the natural evolution of the physical process whereas the discrete behavior can be due to the presence of switches, operating phases, transitions, computer program codes, etc. These hybrid dynamics characterize the behavior of a broad class of physical systems, for example, the real-time control systems where physical processes are controlled by embedded controllers. The notion of hybrid system can also be used to represent complex nonlinear continuous systems. In fact, the operating range of a nonlinear system can be decomposed into a group of operating

Z. Lassoued · K. Abderrahim (✉)
Numerical Control of Industrial Processes, National School of Engineers of Gabes, University of Gabes, St. Omar Ibn-Khattab, 6029 Gabes, Tunisia
e-mail: kamelabderrahim@yahoo.fr

Z. Lassoued
e-mail: zeineb.lassoued1@gmail.com

Q. Zhu and A.T. Azar (eds.), *Complex System Modelling and Control Through Intelligent Soft Computations*, Studies in Fuzziness and Soft Computing 319,
DOI 10.1007/978-3-319-12883-2_6

point. For each operation point, we associate a simple sub-model (linear or affine) with it. Indeed, a complex system can be modeled as a hybrid system switching between simple sub-models.

This chapter addresses the problem of identification of hybrid systems represented by piecewise autoregressive models with exogenous input (PWARX). This problem consists in building mathematical models of hybrid systems from observed input-output data. The PWARX models have attracted a considerable attention in recent years, since they provide an efficient solution for modeling a wide range of engineering applications (Roll et al. 2004; Nakada et al. 2005; Wen et al. 2007; Xu et al. 2012). In addition, these models are able to approximate any nonlinear system with arbitrary accuracy (Lin and Unbehauen 1992). Moreover, the PWA model can be considered as a generic representation for other hybrid models such as jump linear models (JL models) (Vidal et al. 2002), Markov jump linear models (MJL models) (Doucet et al. 2001), mixed logic dynamical models (MLD models) (Bemporad et al. 2000), max-min-plus-scaling systems (MMPS models) (De Schutter and Van den Boom 2000), linear complementarity models (LC models) (Vander-Schaft and Schumacher 1998), extended linear complementarity models (ELC models) (De Schutter and De Moor 1999). In fact, the transfer of the results of PWARX models to other classes of hybrid systems is insured thanks to the properties of equivalence of PWARX models (Heemels et al. 2001). The PWARX models are obtained by decomposing the regression domain into a finite number of non-overlapping convex polyhedral regions and by associating a simple linear model with each region. Consequently, two main problems must be considered for the identification of PWARX models: one is the estimation of the parameters of the sub-models and two is the determination of the hyperplanes defining the partitions of the state-input regression. Consequently, the identification of PWARX models is one of the most difficult problems that represent an area of research where considerable work has been done in the last decade. In fact, numerous solutions have been proposed in the literature for the identification of the PWARX models such as the clustering-based solution (Ferrari-Trecate et al. 2003), the Bayesian solution (Juloski et al. 2005), the bounded-error solution (Bemporad et al. 2005), the greeting solution (Bemporad et al. 2003), the sparse optimization solution (Bako 2011; Bako and Lecoeuche 2013), and so on. The sparse solutions do not smooth out the effect of the measurement noise. Then, they often fail in real time applications since the measurement data are usually contaminated by an unknown additional noise. The greedy algorithms are very time consuming since they involve the solution of NP-hard problems. In addition, it can cause a loss of information because it sometimes fails to associate data to the appropriate regressors. The Bayesian approach assumes that the probability density functions of the unknown parameters of the system are known a priori. Otherwise, it requires an additional sequential processing to improve the identification results. The clustering solution is based on a simple and instructive procedure. It does not require a priori knowledge of the system. Therefore, only the clustering approach is considered in this chapter. This solution consists of three main steps, which are data classification, parameter estimation and region reconstruction. It is easy to remark that the performance of this approach depends on the efficiency of the used classification algorithm (Lassoued and Abderrahim 2013a, b, c, d, 2014a, b). The early methods have favored the simplicity of implementation. In fact, they present several drawbacks, which can be summarized as follows:

- Most of them are based on the optimization of nonlinear criteria. Consequently, they may converge to local minima in the case of poor initializations.
- Their performances degrade in the case of the presence of outliers in the data to be classified.
- Most of them assume that the number of sub-models is a priori known.

To overcome these problems, we have proposed the use of other clustering algorithms such as Chiu's method (Chiu 1997) and Density Based Spatial Clustering of Applications with Noise (DBSCAN) method (Chaitali 2012; Sander et al. 1998). This choice is justified by the fact that these algorithms automatically generate the number of models. In addition, they are characterized by their robustness to the classification of noisy measurements that containing also outliers.

This chapter is organized as follows. Section 2 presents the assumptions for PWARX model identification. In Sect. 3, we recall the main steps of the identification of PWARX systems based on clustering algorithm and its main drawbacks. Section 4 proposes two solutions to overcome the main problems of the existing methods. In Sect. 5, we present three simulation examples in order to illustrate the performance of the proposed solutions and to compare their efficiency with the modified k-means method. Section 6 proposes an application of the developed approach to an olive oil esterification reactor.

2 Piecewise Affine System Identification

Consider a discrete-time PieceWise Auto-Regressive eXogenous model (PWARX) with input $u(k) \in \mathbb{R}$, output $y(k) \in \mathbb{R}$ defined in the bounded polyhedron regressor space $H \subset \mathbb{R}^d$ ($d = n_a + n_b + 1$). The system is decomposed in s different modes $\{H_i\}_{i=1}^{s}$, in each one an ARX model is associated:

$$y(k) = f(\varphi(k)) + e(k). \tag{1}$$

f is a piecewise affine function defined by:

$$f(\varphi) = \begin{cases} \theta_1^T \bar{\varphi} & if\ \varphi \in H_1 \\ \vdots & \\ \theta_s^T \bar{\varphi} & if\ \varphi \in H_s \end{cases} \tag{2}$$

where

$$\overline{\varphi} = \begin{bmatrix} \varphi^T & 1 \end{bmatrix}^T. \tag{3}$$

$e(k)$ is the additive noise and $\varphi(k)$ is the regressor vector, containing past input and output observations, defined as:

$$\varphi(k) = \begin{bmatrix} y(k-1)\ldots y(k-n_a) & u(k-1)\ldots u(k-n_b) \end{bmatrix}^T. \tag{4}$$

$\theta_i \in R^{d+1}$ is the parameter vector, valid in H_i, defined as follows:

$$\theta_i^T = \begin{bmatrix} a_1 & a_2 & \ldots & a_{n_a} & b_1 & b_2 & \ldots & b_{n_b} & g \end{bmatrix} \tag{5}$$

where a_i and b_i are the coefficients of the model related respectively to the output and the input data, while n_a and n_b are the model orders. g is the independent affine coefficient.

Problem statement
Given input-output data generated by a PWARX system, we are interested simultaneously in identifying the number of submodels s, the parameter vectors $\{\theta_i\}_{i=1}^{s}$ and the partitions $\{H_i\}_{i=1}^{s}$ taking into account the following assumptions:

- The orders n_a and n_b of the system are known.
- The noise $e(k)$ is assumed to be a Gaussian process independent and identically distributed with zero mean and finite variance σ^2.
- The regions $\{H_i\}_{i=1}^{s}$ are the polyhedral partitions of a bounded domain $H \subset \mathbb{R}^d$ such that:

$$\begin{cases} \bigcup_{i=1}^{s} H_i = H \\ H_i \bigcap H_j = \emptyset \quad \forall i \neq j \end{cases} \tag{6}$$

3 Clustering Based PWARX Identification

The main steps of the clustering-based approach for the identification of PWARX models can be summarized as follows: constructing small data set from the initial data set, estimating a parameter vector for each small data set, classifying the parameter vectors in s clusters, classifying the initial data set and estimating the s sub-models with their partitions.

1. Form $\{\varphi(k), y(k)\}_{k=1}^{N}$ from the given dataset $S = (u(k), y(k)),\ k = 1, \ldots, N$
2. Create local datasets C_k and identify the local parameter vectors θ_k

(a) Choose n_ρ, the cardinality of data points to be contained in C_k, randomly.
(b) For each dataset $\varphi(k), y(k)$, build C_k containing $\{\varphi(k), y(k)\}$ and its $(n_\rho - 1)$ nearest neighbors satisfying:

$$\left\|\varphi(k) - \breve{\varphi}\right\|^2 \leq \|\varphi(k) - \hat{\varphi}\|^2, \quad \forall (\hat{\varphi}, \hat{y}) \notin C_k. \tag{7}$$

(c) Determine θ_k for each data in $C_k, k = 1, \ldots, N$ using the least square method.

$$\theta_k = (\phi_k^T \phi_k)^{-1} \phi_k^T Y_k. \tag{8}$$

where

$$\phi_k = \left[\overline{\varphi}\,(t_k^1) \ldots \overline{\varphi}\,(t_k^{n_\rho})\right]^T,$$

$$Y_k = \left[y\,(t_k^1) \ldots y\,(t_k^{n_\rho})\right]^T.$$

and $(t_k^1, \ldots, t_k^{n_\rho})$ are the indexes of the elements belonging in C_k

3. Cluster the local parameter vectors $(\theta_k,\ k = 1, \ldots, N)$ into s disjoint clusters while determining the value of s by using a suitable classification technique.
4. Identify the final models $\{\theta_i\}_{i=1}^s$.
5. Estimate the polyhedral partitions $\{H_i\}_{i=1}^s$ i.e. estimate the hyperplanes separating H_i from H_j, $i \neq j$. This is a standard pattern recognition/classification problem that can be solved by several established techniques. The most common technique is the Support Vector Machines (SVM) (Wang 2005; Duda et al. 2001).

The classification of data represents the main step for PWARX system identification because a successful identification of models' parameters and hyperplanes depends on the correct data classification. For the sake of simplicity, the early approaches use classical clustering algorithms for the data classification such as k-means algorithms.

However, these algorithms present several drawbacks. In fact, they may converge to local minima in the case of poor initializations because they are based on the minimization of non linear criterion. Furthermore, their performances degrade in the case of the presence of outliers in the data to be classified. In addition, most of them assume that the number of sub-models is a priori known.

4 The Proposed Clustering Techniques

In order to improve the identification results we propose the use of other classification algorithms such as Chiu's alogorithm and DBSCAN algorithm.

4.1 The Chiu's Clustering Technique

The Chiu's clustering method is a modified form of the Mountain method for cluster estimation (Chiu 1994). Each data point is considered as a potential cluster center instead of considering it as a grid point. This method is very advantageous compared with the Mountain method:

- The number of points to be evaluated is equal to the number of data points.
- It does not need to specify a grid solution which trades off between the accuracy and the computational complexity.
- It improves the computational efficiency and robustness of the original method.

Chiu's classification method consists in computing a potential value for each point of the data set based on its distances to the other data points and consider each data point as a potential cluster center. The point having the highest potential value is chosen as the first cluster center. The key idea in this method is that once the first cluster center is chosen, the potential of all other points is reduced according to their distance from the cluster center. All the points which are close to the first cluster center will have greatly reduced potentials. The next cluster center take then the highest remaining potential value. The procedure for determining a new center and updating other potentials is executed until a predefined condition is reached. This condition depends on the minimum value of the potentials or the required number of clusters which are reached.

This method consists in computing a potential value for each point $(\theta_i,\ i = 1,\ldots,N)$, based on its distances to the other data points and consider each data point as a potential cluster center. The potential is computed using the following expression:

$$P_i = \sum_{j=1}^{N} e^{-\frac{4}{r_a^2}\left\|\theta_i-\theta_j\right\|^2}. \tag{9}$$

The potential of each local parameter is a function of the distance from this parameter to all the other local parameters. Thus, a local parameter with many neighboring local parameters will have the highest potential value. The constant r_a is the radius defining the neighborhood which can be determined by the following expression:

$$r_a = \frac{\alpha}{N}\sum_{i=1}^{N}\frac{1}{n_\rho}\sum_{j=1}^{n_\rho}\left\|\theta_i - \theta_j\right\|. \tag{10}$$

where α can be chosen as follows $0<\alpha<1$.

Equation (9) can be exploited to eliminate the outliers. As this equation attribute to the outliers a low potential, we can fix a threshold γ under which the local parameters are not accepted and then removed from the data set. This threshold is described by the following equation:

$$\gamma = \min(P) + \beta(\max(P) - \min(P)). \tag{11}$$

where P is the vector containing the potentials P_i such that $P = [P_1, \ldots, P_N]$ and β is a parameter chosen as $0 < \beta < 1$.

The elimination of outliers reduces the parameter vectors to $(\theta_i,\ i = 1, \ldots, N')$ $(N' < N)$. Then, from this new data set, we select the data point with the highest potential value as the first cluster center.

Let θ_1^* be this first center and P_1^* be its potential. The other potentials P_j, $(j = 1, \ldots, N')$ are then updated using this expression:

$$P_i \Leftarrow P_i - P_1^* e^{-\frac{4}{r_b^2}\left\|\theta_i - \theta_1^*\right\|^2}. \tag{12}$$

Expression (13) allows to associate lower potentials to the local parameters close to the first center. Consequently, this choice guaranties that these parameters are not selected as cluster centers in the next step. The parameter r_b is a positive constant that must be chosen larger than r_a to avoid obtaining cluster centers which are too close to each other. The constant r_b is computed using this formula:

$$r_b = \frac{\alpha}{N} \sum_{i=1}^{N} \max_{j=1\,:\,n_\rho} \left(\left\|\theta_i - \theta_j\right\|\right). \tag{13}$$

In general after obtaining the kth cluster center, the potential of every local parameter is updated by the following formula:

$$P_i \Leftarrow P_i - P_k^* e^{-\frac{4}{r_b^2}\left\|\theta_i - \theta_k^*\right\|^2}. \tag{14}$$

where P_k^* and θ_k^* are respectively the potential and the center of the kth local parameter.

The number of sub-models s is a parameter that we would like to determine. Therefore, we have developed some criteria for accepting or rejecting the cluster centers as it is explained in the algorithm of the next section.

To search the elements belonging to each cluster, we compute the distance between the estimated output and the real one and classify $\varphi(k)$ within the cluster which has the minimum distance.

$$\arg\ \min\left(\theta_i^T \overline{\varphi}_k - y_k\right), \quad i = 1, \ldots, s. \tag{15}$$

The Chiu's clustering technique can be summarized by the following algorithm:

Algorithm 1: Identification algorithm

Data: Dispose of $\{\theta_i\}_{i=1}^{N}$ from a given data set (φ_i, y_i)
Main steps:
- Compute P_i for every $\{\theta_i\}_{i=1}^{N}$ according to (9)
- Determine the filtered local parameters $\{\theta_i\}_{i=1}^{N'}$, $(N' < N)$
- Compute the first cluster center θ_1^* from (9)

repeat
  Compute the other cluster centers according to the updated potential formula (14)
  if $P_k^* > \gamma$ **then**
    Compute $V(c)$ such as:

$$V(c) = \|\theta_k^* - \theta_c^*\|, \quad c = 1, ..., k-1. \tag{16}$$

    where $\theta_k{}^*$ is the current cluster center and θ_c^*, $c = 1, ..., k-1$ are the last selected ones.
    if $V(c) > \varepsilon$, $c = 1, ..., k-1$ **then**
      accept $\theta_k{}^*$ as a cluster center and continue
    else
      reject $\theta_k{}^*$ and compute a new potential
    end
  else
    reject $\theta_k{}^*$ and break
  end
until $V(c) \leq \varepsilon, \quad c = 1, ...k-1$;
Result: Determination of the number of clusters s and the parameters $\{\theta_i\}_{i=1}^{s}$

While ε is a small parameter characterizing the minimum distance between the new cluster center and the existing ones.

4.2 The DBSCAN Clustering Technique

The Density-Based Spatial Clustering of Applications with Noise (DBSCAN) algorithm is a pioneer algorithm of density-based clustering (Chaitali 2012; Sander et al. 1998). This algorithm is based on the concepts of density-reachability and density-connectivity. These concepts depend on two input parameters: epsilon (ε) and (*MinPts*).

- ε: is the radius around an object that defines its ε-neighborhood.
- *MinPts*: is the minimum number of points.

For a given object *q*, when the number of objects within the ε-neighborhood is at least *MinPts*, then *q* is defined as a core object. All objects within its ε-neighborhood are said to be directly density-reachable from *q*.

In general, an object *p* is considered density-reachable if it is within the ε-neighborhood of an object that is directly density-reachable or just density-reachable from q. The objects p and q are said to be density-connected if there exist an object g that both p and q are density-reachable from.

The DBSCAN algorithm define then a cluster as the set of objects in a data set that are density-connected to a particular core object. Any object that is not part of a cluster is categorized as noise. For a given data set $S = \{\theta_k\}_{k=1}^{N}$, ε and *MinPts* as inputs, the ε-neighborhood of a point θ_i is defined as:

$$N_\varepsilon(\theta_i) = \{\theta_j \in S;\ \|\theta_i - \theta_j\| \leq \varepsilon\} \tag{17}$$

The DBSCAN constructs clusters by checking the ε-neighborhood of each object in the data set. If the cardinal of the ε-neighborhood (denoted by cN_ε) of an object θ_k contains more than *MinPts*, a new cluster is created having θ_k as core. The DBSCAN then iteratively collects directly density-reachable objects from these core objects. The process terminates when no new objects can be added to any cluster. The main steps of this algorithm can be summarized as follows:

Algorithm 2: DBSCAN algorithm

Data: Define the input parameters: $S = \{\theta_k\}_{k=1}^{N}$, $MinPts$ and ϵ
Main steps:
for k=*1:N* **do**
    if θ_k *is not in a cluster* **then**
        Compute $N_\epsilon(\theta_k)$
        if $cN_\epsilon(\theta_k) < MinPts$ **then**
            Mark θ_k as noise
        else
            cluster-label=cluster-label+1
            for j =*1:*$cN_\epsilon(\theta_k)$ **do**
                Mark all point in $N_\epsilon(\theta_k)$ with the current cluster label
            end
            Lend $N_\epsilon(\theta_k)$ to the Seed list $L_S = [L_S \quad N_\epsilon(\theta_k)]$
            while L_S *is not empty* **do**
                $\theta_r = L_S(1)$
                Compute $N_\epsilon(\theta_r)$
                if $cN_\epsilon(\theta_r) \geq MinPts$ **then**
                    for o=*1:*$cN_\epsilon(\theta_r)$ **do**
                        if θ_o *is not in a cluster nor marked as noise* **then**
                            Mark θ_o with the current cluster-label
                            Lend $N_\epsilon(\theta_r)$ to the Seed list $L_S = [L_S \quad N_\epsilon(\theta_r)]$
                            $L_S(1) = [\ \ ]$
                        end
                    end
                else
                    $L_S(1) = [\ \ ]$
                end
            end
        end
    end
end
Result: Determination of the number of clusters s and the parameters $\{\theta_i\}_{i=1}^{s}$

5 Simulation Examples

In this section, we aim at illustrating the performance of the proposed methods with three simulation examples. First of all, we take an academic PWARX model where the proposed methods are compared with the well known k-means one (Ferrari-Trecate et al. 2001, 2003). After that, a nonlinear model is considered to show the efficiency of the proposed methods in approximating any nonlinear model. Finally, a pH neutralization process is simulated in order to prove their ability to model complex systems and to determine the number of sub-models.

5.1 Quality Measures

To achieve the purpose of these simulations, we consider the following quality measures (Juloski et al. 2006):

- The maximum of relative error of parameter vectors is defined by

$$\Delta_\theta = \max_{i=1,\ldots,s} \frac{\left\|\theta_i - \overline{\theta}_i\right\|_2}{\left\|\overline{\theta}_i\right\|_2} \tag{18}$$

  where $\overline{\theta}_i$ and θ_i are the true and the estimated parameter vectors for sub-model i, respectively. The identified model is deemed acceptable if Δ_θ is small or close to zero.
- The averaged sum of the squared residuals is defined by

$$\sigma_e^2 = \frac{1}{s}\sum_{i=1}^{s} \frac{SSR_i}{|D_i|} \tag{19}$$

  where $SSR_i = \sum_{(y(k),\varphi(k))\in D_i} (y(k) - [\varphi(k)'1]\theta_i)^2$ and $|D_i|$ is the cardinality of cluster D_i.
  The identified model is considered acceptable if σ_e^2 is small and/or close to the expected noise variance of the true system.
- The percentage of the output variation that is explained by the model is defined by

$$FIT = 100 \cdot \left(1 - \frac{\|\hat{y} - y\|_2}{\|y - \overline{y}\|_2}\right) \tag{20}$$

  where $\hat{y}$ and y are the estimated and the real outputs' vectors, respectively, and $\overline{y}$ is the mean value of y.
  The identified model is considered acceptable if FIT is close to 100.

- The relative error expressed in percentage (%) is given by:

$$e_r(k) = 100 \cdot \frac{|y(k) - \hat{y}(k)|}{|y(k)|} \tag{21}$$

where $\hat{y}(k)$ and $y(k)$ are the estimated and the real outputs at time k. The identified model is considered acceptable if e_r is close to 0 %.

5.2 Identification Results of a PWARX Model

Consider the following PWARX model (Boukharouba 2011):

$$y(k) = \begin{cases} [0.4 \quad 0.5 \quad 0.3]\bar{\varphi}(k) + e(k) & \text{if } \varphi(k) \in H_1, \\ [-0.7 \quad 0.6 \quad -0.5]\bar{\varphi}(k) + e(k) & \text{if } \varphi(k) \in H_2, \\ [0.4 \quad -0.2 \quad -0.2]\bar{\varphi}(k) + e(k) & \text{if } \varphi(k) \in H_3, \end{cases} \tag{22}$$

$$\begin{aligned} H_1 &= \left\{\varphi \in \Re^2 : [1 \quad 0.3 \quad 0]\overline{\varphi} \geq 0 \text{ and } [0 \quad 0.5 \quad 0]\overline{\varphi} > 0\right\} \\ H_2 &= \left\{\varphi \in \Re^2 : [1 \quad 0.3 \quad 0]\overline{\varphi} \leq 0 \text{ and } [1 \quad -0.3 \quad 0]\overline{\varphi} < 0\right\} \\ H_3 &= \left\{\varphi \in \Re^2 : [1 \quad -0.3 \quad 0]\overline{\varphi} \geq 0 \text{ and } [0 \quad 0.5 \quad 0]\overline{\varphi} \leq 0\right\} \end{aligned} \tag{23}$$

where $s = 3$, $n_a = 1$, $n_b = 1$, and $\varphi(k) = [y(k-1) \quad u(k-1)]^T$ is the regressor vector.

System (22) is simulated using an input signal $u(k)$ and a noise signal $e(k)$ which are normal distributions with variances respectively 0.5 and 0.05. The output $y(k)$ is presented in Fig. 1.

Table 1 presents the estimated parameter vectors obtained with the proposed methods and the k-means one.

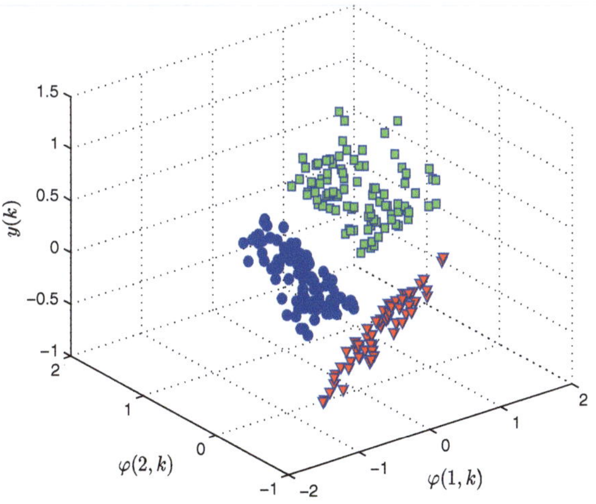

Fig. 1 The real output of the system (*squares*: output of sub-model 1, *triangles*: output of sub-model 2 and *dots*: output of sub-model 3)

Table 1 Estimated parameters

	True values	Chiu ($n_\rho = 20$)	DBSCAN ($n_\rho = 21$)	k-means ($n_\rho = 7$)
θ_1	$\begin{bmatrix} 0.4 \\ 0.5 \\ 0.3 \end{bmatrix}$	$\begin{bmatrix} 0.4046 \\ 0.5138 \\ 0.2919 \end{bmatrix}$	$\begin{bmatrix} 0.4054 \\ 0.4903 \\ 0.2992 \end{bmatrix}$	$\begin{bmatrix} 0.4064 \\ 0.5464 \\ 0.2598 \end{bmatrix}$
θ_2	$\begin{bmatrix} -0.7 \\ 0.6 \\ -0.5 \end{bmatrix}$	$\begin{bmatrix} -0.6179 \\ 0.5336 \\ -0.4740 \end{bmatrix}$	$\begin{bmatrix} -0.7369 \\ 0.6675 \\ -0.5239 \end{bmatrix}$	$\begin{bmatrix} -0.6955 \\ 0.5903 \\ -0.4939 \end{bmatrix}$
θ_3	$\begin{bmatrix} 0.4 \\ -0.2 \\ -0.2 \end{bmatrix}$	$\begin{bmatrix} 0.4015 \\ -0.2071 \\ -0.2042 \end{bmatrix}$	$\begin{bmatrix} 0.4679 \\ -0.1977 \\ -0.2298 \end{bmatrix}$	$\begin{bmatrix} 0.4792 \\ -0.2101 \\ -0.2406 \end{bmatrix}$

After obtaining the estimated parameter vectors, we apply the SVM algorithm in order to estimate the regions. We can then attribute each parameter vector to the corresponding region where it is defined. The estimated outputs obtained with three algorithms are presented in Fig. 2.

Table 2 presents the quality measures (18), (19) and (20) of the two proposed methods and the k-means method. The obtained results prove the efficiently of the proposed methods compared with the existing method (k-means).

5.3 Identification Results of a Nonlinear Model

Consider the nonlinear system described by the following equation (Lai et al. 2010):

$$\begin{aligned} y(k) = & \frac{1.5y(k-1)y(k-2)}{1+y^2(k-1)+y^2(k-2)} + \sin(y(k-1)+y(k-2)) \\ & + u(k-1) + 0.8u(k-2) \end{aligned} \tag{24}$$

This nonlinear system can be modeled by a PWARX model of the form (Lai 2011):

$$y(k) = \begin{cases} \theta_1^T \overline{\varphi}(k) & if\ \varphi \in H_1 \\ \vdots & \\ \theta_s^T \overline{\varphi}(k) & if\ \varphi \in H_s \end{cases} \tag{25}$$

where

$$\varphi(k) = [y(k-1),\ y(k-2),\ u(k-1),\ u(k-2)]^T \tag{26}$$

$$\overline{\varphi} = \begin{bmatrix} \varphi^T & 1 \end{bmatrix}^T. \tag{27}$$

θ_i are the parameter vectors and s is the number of submodels to be determined. u (k) is a random input in the range of [−2, 2].

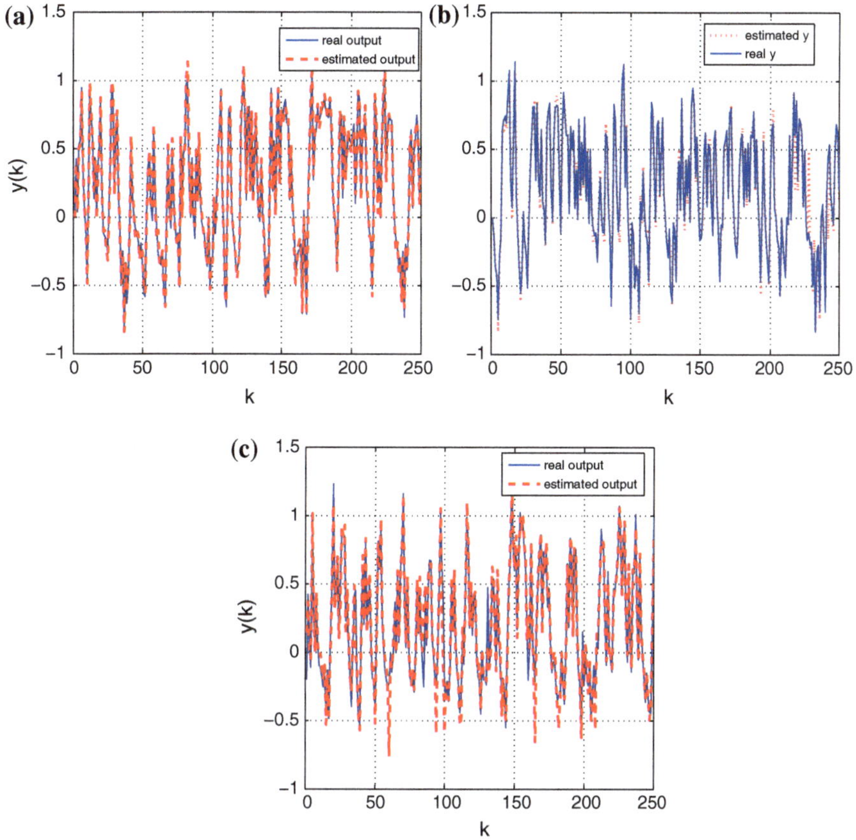

Fig. 2 The estimated outputs **a** with Chiu, **b** with DBSCAN, and **c** with k-means

Table 2 Validation results

Quality measures	Chiu	DBSCAN	k-means
Δ_θ	0.0876	0.1514	0.1828
σ_e^2	0.0023	0.0097	0.0109
FIT	89.4191	79.0183	74.195

For the DBSCAN based method, the choice of the synthesis parameters n_ρ, *MinPts* and ε is as follows:

$$\begin{cases} n_\rho = 20; \\ MinPts = 35; \\ \varepsilon = 0.85 \end{cases}$$

For the Chiu clustering algorithm, we have only one synthesis parameter: $n_\rho = 17$.

The number of submodels s depends on the initial parameters chosen. With the parameters described above, we obtain $s = 6$.

The parameter vectors are presented in Tables 3 and 4.

Figures 3 and 4 illustrate the outputs and the relative error signals of the two proposed methods.

In Table 5, the *FIT* is computed for the identification and the validation with the two proposed methods. The obtained results are very satisfactory and show that the performances of the two methods are close.

5.4 Identification Results of a PH Neutralisation Process

5.4.1 Process Description

The 'neutralization' is used to describe the reaction result between an acid and a base in which the properties of H^+ and OH^- that characterized the acid and base will be destroyed or neutralized. In fact, the ions H^+ and OH^- will be combined to form the water molecule H_2O. The resulting solution produced by the reaction is

Table 3 Estimated parameter vectors with the Chiu's method

Parameters vector	Estimated values with Chiu
θ_1	$[1.0996\ \ 0.6338\ \ 1.0259\ \ 0.8247\ \ 0.0762]^T$
θ_2	$[-0.7769\ \ -1.0611\ \ 1.0016\ \ 0.8446\ \ -2.2539]^T$
θ_3	$[0.5066\ \ 0.9219\ \ 0.9781\ \ 0.7686\ \ 0.1886]^T$
θ_4	$[-0.6576\ \ -0.4256\ \ 0.9388\ \ 0.8097\ \ -1.4943]^T$
θ_5	$[-0.3013\ \ -0.5044\ \ 1.0644\ \ 0.6904\ \ 1.7459]^T$
θ_6	$[-0.8306\ \ -0.5365\ \ 1.0030\ \ 0.7557\ \ 2.7559]^T$

Table 4 Estimated parameter vectors with the DBSCAN method

Parameters vector	Estimated values with DBSCAN
θ_1	$[0.5134\ \ 0.4871\ \ 1.0377\ \ 0.8082\ \ 0.0187]^T$
θ_2	$[-0.5974\ \ -0.3021\ \ 0.8905\ \ 0.7355\ \ 2.1231]^T$
θ_3	$[0.7044\ \ 0.0158\ \ 1.0153\ \ 0.8255\ \ 0.1468]^T$
θ_4	$[-0.6958\ \ -0.6557\ \ 1.0714\ \ 0.7435\ \ -1.5669]^T$
θ_5	$[0.3111\ \ 0.0277\ \ 0.9738\ \ 0.7616\ \ -0.3855]^T$
θ_6	$[-0.6177\ \ -0.3720\ \ 0.9606\ \ 0.7291\ \ -1.2358]^T$

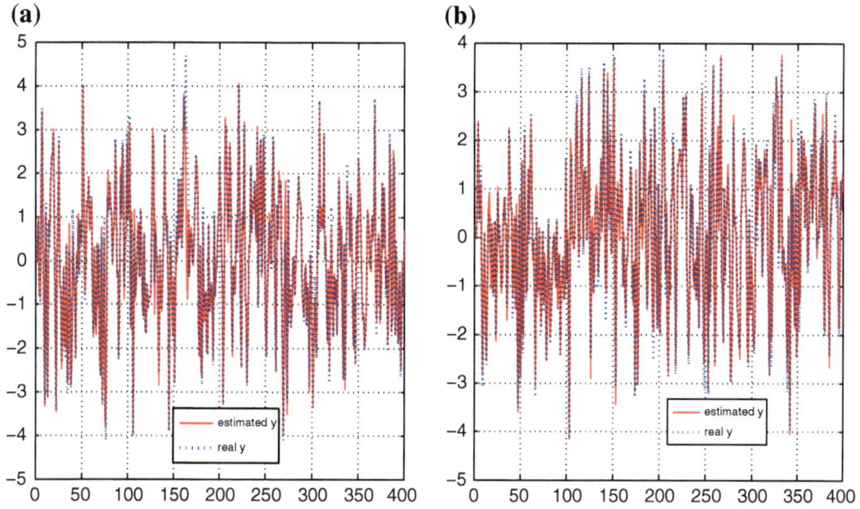

Fig. 3 The estimated outputs **a** with Chiu, and **b** with DBSCAN

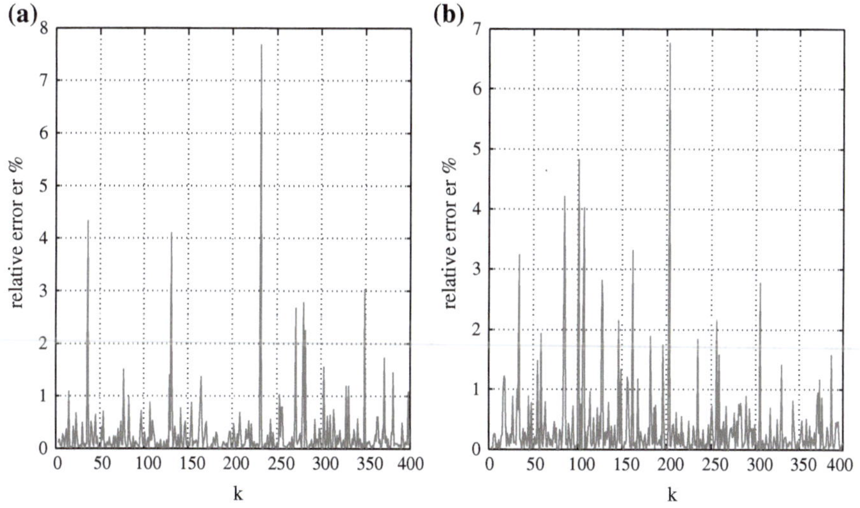

Fig. 4 The relative error **a** with Chiu, and **b** with DBSCAN

Table 5 Quality measures with the two proposed methods

	Chiu	DBSCAN
FIT for identification	78.0543	84.1135
FIT for validation	75.2050	74.1906

composed of a salt and water. The general formula for acid–base neutralization reactions can be written as:

$$acid + base \rightarrow salt + water \tag{28}$$

The process of pH neutralization (see Fig. 5) is constituted essentially of a treatment tank of cross sectional area A, a mixer, acid and base injection pipes, a pH probe, a level sensor to measure the level h in the tank and a discharge valve (Henson and Seborg 1994; Salehi et al. 2009). It consists of an acid stream q_1, buffer stream q_2 and base stream q_3 that are mixed in the tank. The effluent stream q_4 exits the tank via the discharge valve with an adjusted pH_m. The streams $\{q_i\}_{i=1}^{4}$ are characterized by the following parameters:

- $\{W_{ai}\}_{i=1}^{4}$ are the charge related quantities for $\{q_i\}_{i=1}^{4}$.
- $\{W_{bi}\}_{i=1}^{4}$ are the mass balance quantities for $\{q_i\}_{i=1}^{4}$.

The pH probe introduces a delay time τ in the measured pH_m value such as $pH_m = pH(t - \tau)$.

The objective of the pH neutralization process is to control the pH value of the effluent through manipulating the base flow rate q_3 while considering the acid flow rate q_1 and the buffer flow rate q_2 as disturbances.

The dynamic model of the neutralization process is developed as follows:

- The pH value of the obtained solution is derived from the conservation equations and equilibrium reactions as follows:

$$W_{a4} + \frac{K_w}{[H^+]} + W_{b4} \frac{\frac{K_{a1}}{[H^+]} + \frac{2K_{a1}K_{a2}}{[H^+]^2}}{1 + \frac{K_{a1}}{[H^+]} + \frac{K_{a1}K_{a2}}{[H^+]^2}} - [H^+] = 0. \tag{29}$$

Fig. 5 A pH neutralization process

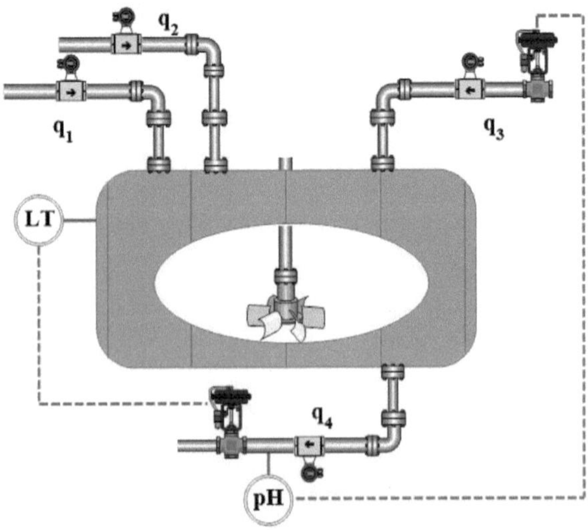

Knowing that

$$pH_m = -\log([H^+]) \tag{30}$$

$$K_w = [H^+][OH^-], \tag{31}$$

Equation (29) can be then rewritten as:

$$W_{a4} + 10^{pH_m-14} + W_{b4}\frac{1 + 2(10^{pH_m-pK_{a2}})}{1 + 10^{pH_m-pK_{a1}} + 10^{pH_m-pK_{a2}}} - 10^{-pH_m} = 0 \tag{32}$$

- The mass balance yields to:

$$A\frac{dh}{dt} = q_1 + q_2 + q_3 - q_4 \tag{33}$$

Taking into account that the exit flow rate $q_4 = C_v.h^{0.5}$, Eq. (33) becomes:

$$A\frac{dh}{dt} = q_1 + q_2 + q_3 - C_v \cdot h^{0.5} \tag{34}$$

where C_v is the constant valve coefficient.
- The differential equations of the effluent reaction invariants (W_{a4}, W_{b4}) can be determined as follows:

$$A\,h\frac{dW_{a4}}{dt} = q_1(W_{a1} - W_{a4}) + q_2(W_{a2} - W_{a4}) + q_3(W_{a3} - W_{a4}) \tag{35}$$

$$A\,h\frac{dW_{b4}}{dt} = q_1(W_{b1} - W_{b4}) + q_2(W_{b2} - W_{b4}) + q_3(W_{b3} - W_{b4}) \tag{36}$$

Nominal model parameters and operating conditions (Xiao et al. 2014) are given in Table 6.

The static nonlinearity of this process can be represented by the titration curve shown in Fig. 6 with a beginning pH of 2.7 and an ending pH of 10.7. A brief glance at the curve indicates that the process of pH neutralization is highly nonlinear.

Table 6 Operation parameters of the pH neutralization process

$q_1 = 16.6$ ml/s	$W_{a1} = 3 \times 10^{-3}$ mol/l
$q_2 = 0.55$ ml/s	$W_{a2} = -3 \times 10^{-2}$ mol/l
$q3 = 15.6$ ml/s	$W_{a3} = -3.05 \times 10^{-3}$ mol/l
$h = 14.0$ cm	$W_{b1} = 0$
$A = 207$ cm^2	$W_{b2} = 3 \times 10^{-2}$ mol/l
$C_v = 8.75$ ml/cm/s	$W_{b3} = 5 \times 10^{-5}$ mol/l
$pK_{a1} = 6.35$	$pH_4 = 7$
$pK_{a2} = 10.25$	$\tau = 0.5$

Fig. 6 The titration curve

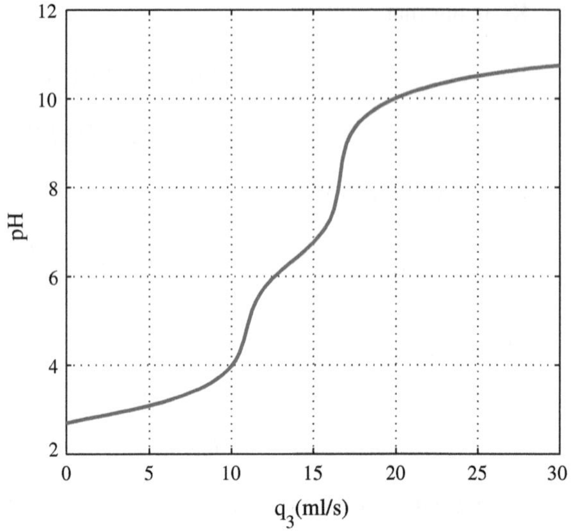

5.4.2 Structure Identification

It was mentioned that the early approaches of identification of pH neutralization process approximate this process around an operating range as a First Order Plus Delay Time model. Added to that, the evolution of the pH in Fig. 7, for a fixed values of the input q_3, is similar to a first order system response.

Therefore, we propose to represent the sub-models by a discrete first order plus dead time models ($n_a = 1$, $n_b = 2$) defined by:

Fig. 7 The pH evolution with different values of q_3

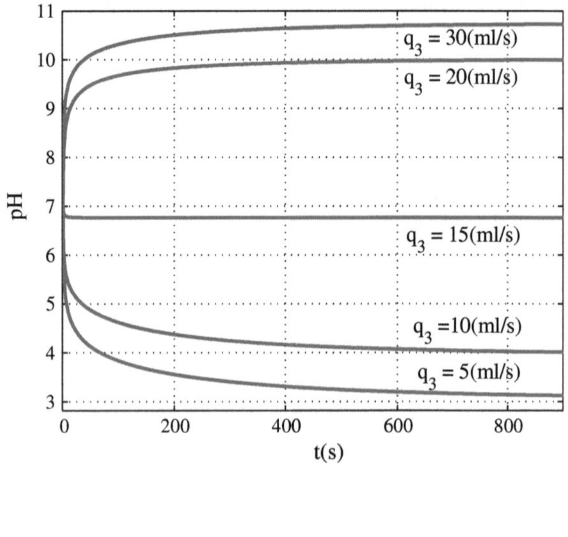

$$y(k) = \begin{cases} a_{1,1}y(k-1) + b_{1,1}u(k-1) + b_{1,2}u(k-2) \\ \quad \text{if } \varphi(k) \in H_1 \\ \vdots \\ a_{s,1}y(k-1) + b_{s,1}u(k-1) + b_{s,2}u(k-2) \\ \quad \text{if } \varphi(k) \in H_s \end{cases} \tag{37}$$

where the regressor vector is defined by:

$$\varphi(k) = [y(k-1),\ u(k-1),\ u(k-2)]^T$$

and the parameter vectors are denoted by:

$$\theta_i(k) = [a_{i,1}, b_{i,1}, b_{i,2}],\ i = 1, \ldots, s.$$

5.4.3 Input Design

The input design is an important aspect to be considered when implementing nonlinear system identification experiments. In fact, two main properties must be verified by this input in order to generate representative data measurements to be used in identification purpose. First, the input must be able to excitep the totality of dynamics range present in the system. Second, the used input signal must illustrate the response of the system to a range of amplitude changes since these models have nonlinear gains. For these reasons, we have considered the Multi-Sine sequence as input sequences to identify the ph neutralization process since it satisfies the above two conditions. It presents several frequencies and exhibits different amplitude changes. The dynamic of this input is defined according to the dominant time constant range of the process. The amplitudes are selected to cover the totality operating region around the nominal value of the base flow rate $q_3 = 15.6$ ml/s.

5.4.4 Results

The nonlinear model of the pH process defined by Eqs. (32), (34), (35) and (36) and the parameters of Table 6 is used to generate the output using a Multi-Sine excitation Sequence. The system output is corrupted by a Gaussian white noise with zero mean and standard deviation $\sigma = 0.001$ in order to simulate industrial situations where the obtained measurements are often noisy. The obtained input-output data illustrated in Fig. 8 are then divided into two parts. The first part is used for the identification and the second is considered for the validation purpose.

The number of neighboring is chosen $n_\rho = 85$ for the two methods. The DBSCAN approach uses the following synthesis parameters

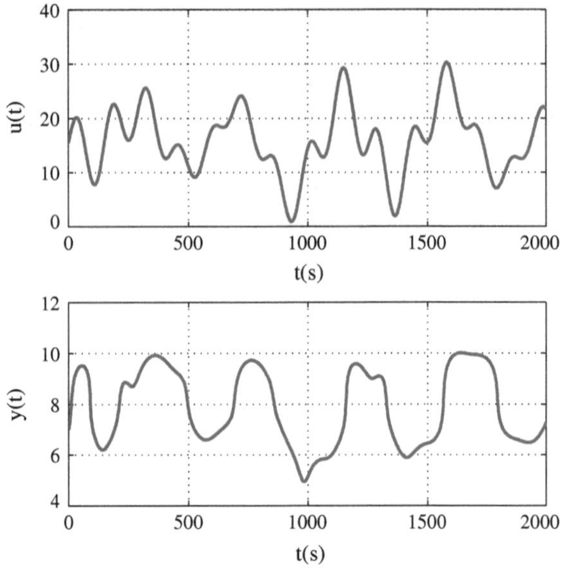

Fig. 8 The data of the multi-sine input

$$\begin{cases} MinPts = 40; \\ \varepsilon = 0.18 \end{cases}$$

The number of submodels obtained with these parameters is (s = 6). The parameter vectors are illustrated in Table 7.

The validation results and the estimated titration curves are presented respectively in Figs. 9 and 10 which shows that the obtained model gives good results in terms of dynamic and nonlinear gain of the pH process.

Now, we compare the performance of the two proposed methods using the quality measures (19), (20) and (21). The obtained results are summarized in Table 8 and Fig. 11.

Table 7 Estimated parameter vectors

Parameter vectors	Estimated values DBSCAN	Estimated values Chiu
θ_1	$[0.9906 \quad 0.0264 \quad -0.0218]^T$	$[0.9855 \quad 0.0179 \quad -0.0101]^T$
θ_2	$[0.9808 \quad 0.0287 \quad -0.0190]^T$	$[0.9692 \quad 0.0593 \quad -0.0423]^T$
θ_3	$[0.9763 \quad 0.0282 \quad -0.0152]^T$	$[1.0102 \quad -0.0216 \quad 0.0200]^T$
θ_4	$[1.0272 \quad 0.0343 \quad -0.0434]^T$	$[1.0351 \quad 0.0273 \quad -0.0550]^T$
θ_5	$[0.9968 \quad -0.0390 \quad 0.0432]^T$	$[0.9632 \quad -0.0289 \quad 0.0440]^T$
θ_6	$[0.9690 \quad -0.0404 \quad 0.0533]^T$	$[1.0021 \quad 0.0903 \quad -0.0910]^T$

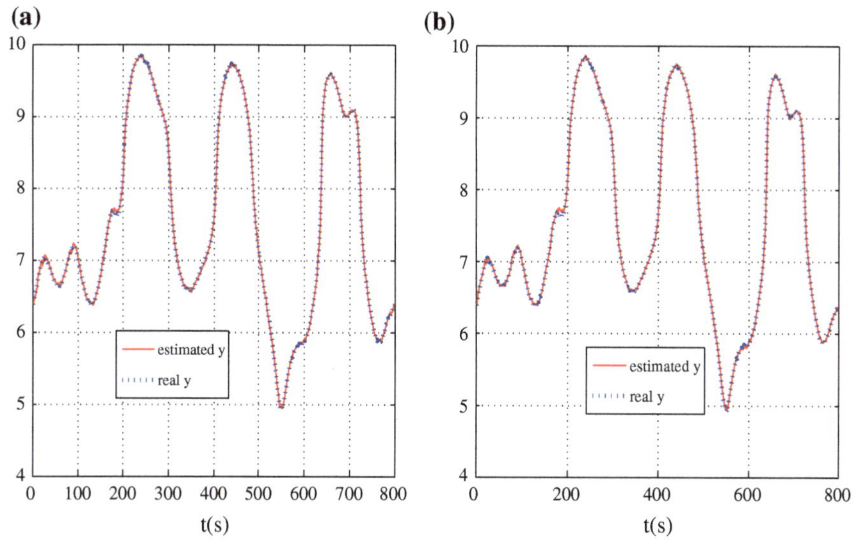

Fig. 9 The validation outputs **a** with Chiu, and **b** with DBSCAN

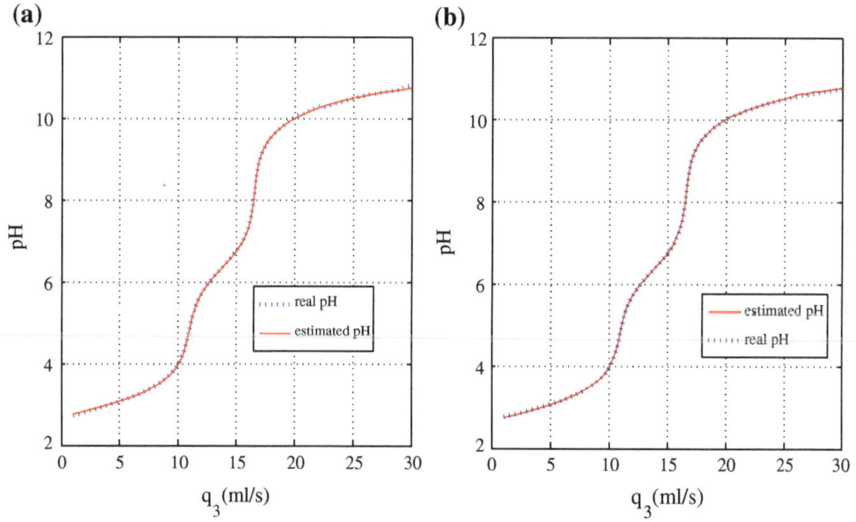

Fig. 10 The validation of the titration curve **a** with Chiu, and **b** with DBSCAN

Table 8 Quality measures with the two proposed methods

	Chiu	DBSCAN
FIT for identification	98.0203	97.8688
FIT for validation	98.1254	98.0405
σ_e^2	8.8552×10^{-4}	0.0013

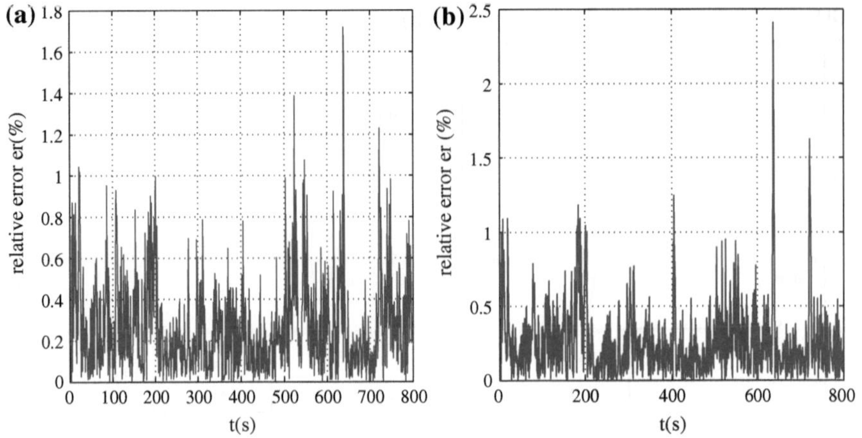

Fig. 11 Relative error for the two proposed methods **a** with Chiu, and **b** with DBSCAN

6 Experimental Example: A Semi-batch Reactor

6.1 Process Description

The olive oil esterification reactor produces ester with a very high added value which is used in fine chemical industry such as cosmetic products. The esterification reaction between vegetable olive oil with free fatty acid and alcohol, producing ester, is given by the following equation:

$$Acid + Alcohol \leftrightarrow Ester + Water. \tag{38}$$

The ratio of the alcohol to acid represents the main factor of this reaction because the esterification reaction is an equilibrium reaction i.e. the reaction products, water and ester, are formed when equilibrium is reached. In addition, the yield of ester may be increased if water is removed from the reaction. The removal of water is achieved by the vaporisation technique while avoiding the boiling of the alcohol. In fact, we have used an alcohol (1-butanol), characterized by a boiling temperature of 118 °C which is greater than the boiling temperature of the water (close to 100 °C). In addition, the boiling temperatures of the fatty acid (oleic acid) and the ester are close to 300 °C. Therefore, the boiling point of water may be provided by a temperature slightly greater than 100 °C.

The block diagram of the process is shown in Fig. 12. It is constituted essentially of:

- A reactor with double-jackets: It has a cylindrical shape manufactured in stainless steel. It is equipped with a bottom valve for emptying the product, an agitator, an orifice introducing the reactants, a sensor of the reaction mixture temperature, a pressure sensor and an orifice for the condenser. The double-

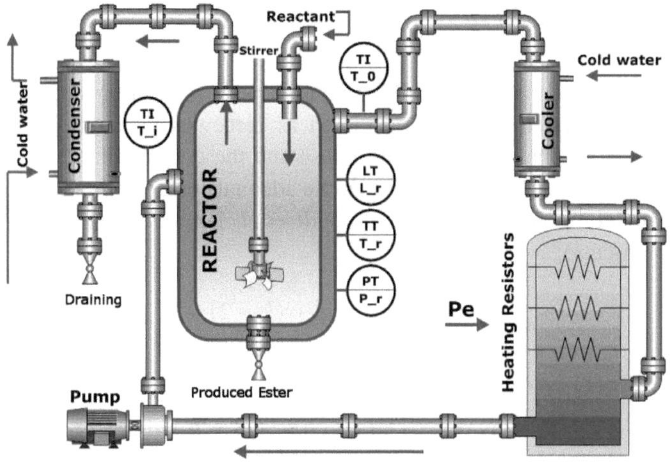

Fig. 12 Block diagram of the reactor

jackets ensure the circulation of a coolant fluid which is intended for heating or for cooling the reactor.

- A heat exchanger: It allows to heat or to cool the coolant fluid circulating through the reactor jacket. Heating is carried out by three electrical resistances controlled by a dimmer for varying the heating power. It is intended to achieve the required reaction temperature of the esterification. Cooling is provided by circulating cold water through the heat exchanger. It is used to cool the reactor when the reaction is completed.
- A condenser: It allows to condense the steam generated during the reaction. It plays an important role because it is also used to indicate the end of the reaction which can be deduced when no more water is dripping out of the condenser.
- A data acquisition card between the reactor and the calculator.

The ester production by this reactor is based on three main steps as illustrated in Fig. 13.

Fig. 13 Specific trajectory of the reactor temperature

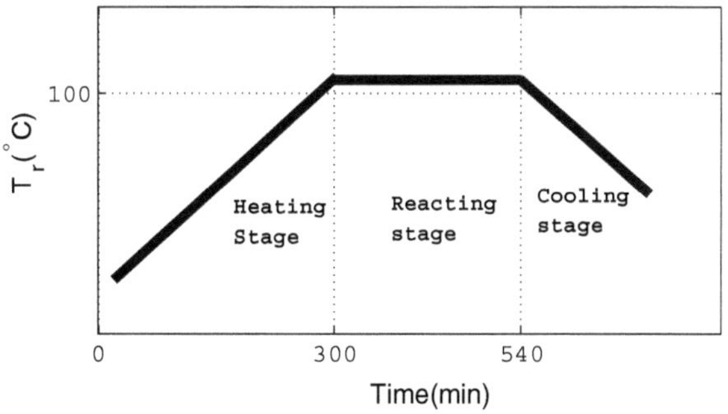

6.2 Experimental Results

The alternative of considering a PWA map is very interesting because the characteristic of the system can be considered as piecewise linear in each operating phase: the heating phase, the reacting phase and the cooling phase.

Previous works has demonstrated that the adequate estimated orders n_a and n_b of each sub-model are equal to two (Talmoudi et al. 2008). Thus, we can adopt the following structure:

$$y(k) = \begin{cases} -a_{1,1}y(k-1) - a_{1,2}y(k-2) + b_{1,1}u(k-1) \\ +b_{1,2}u(k-2) \quad if\ \varphi(k) \in H_1 \\ \vdots \\ a_{s,1}y(k-1) + a_{s,2}y(k-2) + b_{s,1}u(k-1) \\ +b_{s,2}u(k-2) \quad if\ \varphi(k) \in H_s \end{cases} \tag{39}$$

where the regressor vector is defined by:

$$\varphi(k) = [-y(k-1),\ -y(k-2),\ u(k-1),\ u(k-2)]^T$$

and the parameter vectors is denoted by:

$$\theta_i(k) = \left[a_{i,1},\ a_{i,2},\ b_{i,1},\ b_{i,2}\right], \quad i = 1, \ldots, s.$$

We have picked out some input-output measurements from the reactor in order to identify a model to this process. We have taken two measurement files, one for the identification having a length $N = 220$ and another one of length $N = 160$ for the validation.

The measurement file used in this identification is presented in Fig. 14.

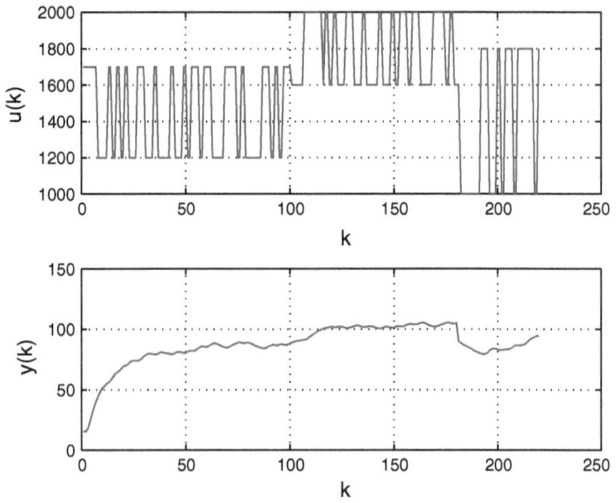

Fig. 14 The real input-output evolution

Table 9 Estimated parameter vectors with the proposed clustering techniques

Parameter vectors	Estimated parameters with Chiu	Estimated parameters with DBSCAN
θ_1	$\begin{bmatrix} -1.4256 \\ 0.4508 \\ 4.9853 \times 10^{-4} \\ 0.0010 \end{bmatrix}$	$\begin{bmatrix} -1.4404 \\ 0.4692 \\ 0.0003 \\ 0.0014 \end{bmatrix}$
θ_2	$\begin{bmatrix} -1.1604 \\ 0.2111 \\ 0.0015 \\ 0.0014 \end{bmatrix}$	$\begin{bmatrix} -1.1144 \\ 0.1772 \\ 0.0003 \\ 0.0032 \end{bmatrix}$
θ_3	$\begin{bmatrix} -1.0847 \\ 0.1490 \\ -3.9782 \times 10^{-4} \\ 0.0040 \end{bmatrix}$	$\begin{bmatrix} -1.0591 \\ 0.1304 \\ 0.0006 \\ 0.0034 \end{bmatrix}$

We apply the proposed identification procedures in order to represent the reactor by a PWARX model. The number of neighboring is chosen $n_\rho = 70$ with the two proposed techniques. Our purpose is to estimate the number of sub-models s, the parameter vectors $\theta_i(k)$, $i = 1, \ldots, s$ and the hyperplanes defining the partitions $\{H_i\}_{i=1}^{s}$.

The obtained results are as follows:

- The number of sub-models is $s = 3$.
- The parameter vectors $\theta_i(k)$, $i = 1, 2$ and 3 are illustrated in Table 9.

The attribution of every parameter vector to the submodel that has generated it is ensured by the SVM algorithm. The obtained outputs are then computed and they are represented in Fig. 15.

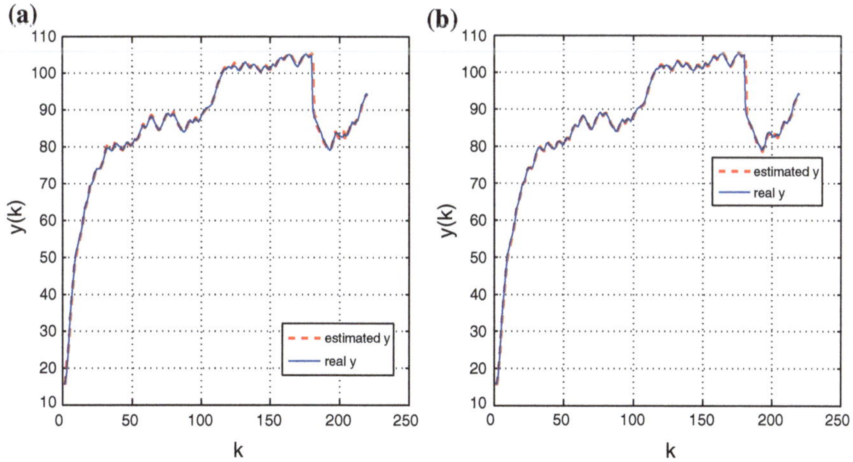

Fig. 15 Estimated outputs with two methods **a** with Chiu, and **b** with DBSCAN

To validate the obtained models, we have considered a new input-output measurement file having a length $N = 160$ shown in Fig. 16.

The real and the estimated validation outputs and the errors are presented in Fig. 17.

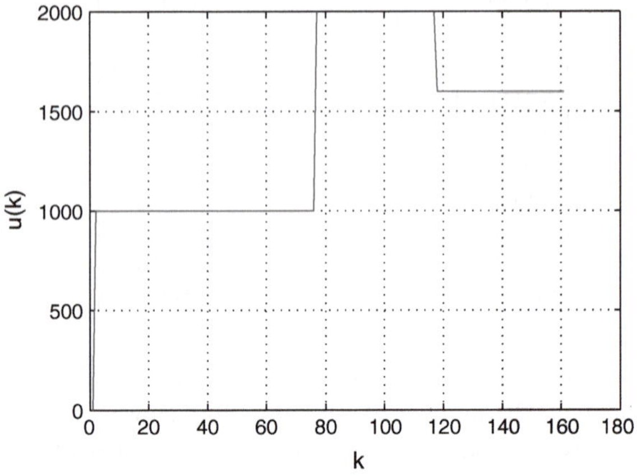

Fig. 16 The validation input

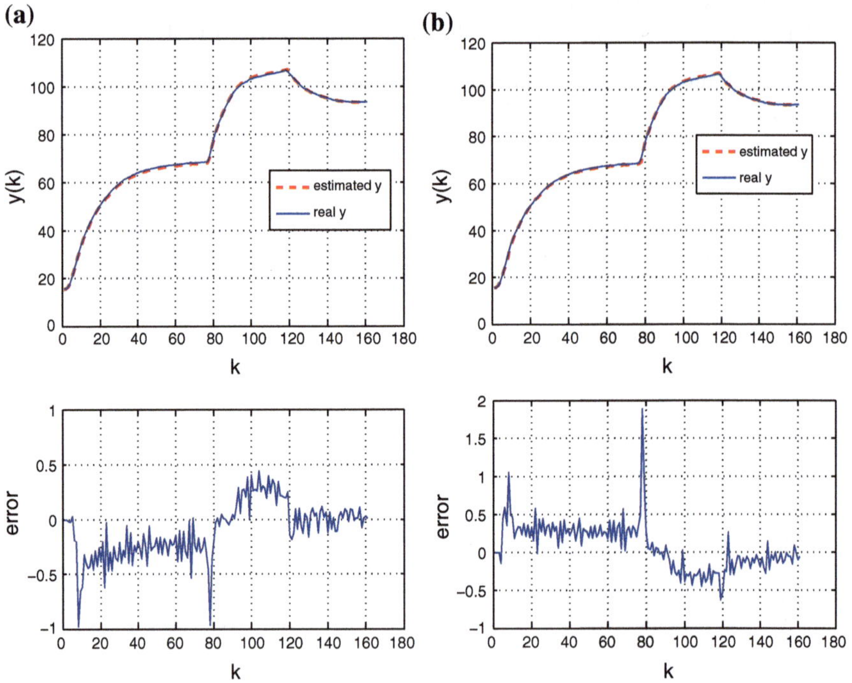

Fig. 17 Estimated validation outputs and the errors with two methods **a** with Chiu, and **b** with DBSCAN

7 Conclusion

In this chapter, we have considered only the clustering based procedures for the identification of PWARX systems. We focused on the most challenging step which is the task of data points classification. In fact, we have proposed the use of two clustering techniques which are the Chiu's clustering algorithm and the DBSCAN algorithm. These algorithms present several advantages. Firstly, they do not require any initialization so the problem of convergence towards local minima is overcome. Secondly, these algorithms are able to remove the outliers from the data set. Finally, our approaches generate automatically the number of sub-models. Numerical simulation results are presented to demonstrate the performance of the proposed approaches and to compare them with the k-means one. Also, an experimental validation with an olive oil reactor is presented to illustrate the efficiency of the developed methods.

References

Bako, L. (2011). Identification of switched linear systems via sparse optimization. *Automatica, 47*(4), 668–677.

Bako, L., & Lecoeuche, S. (2013). A sparse optimization approach to state observer design for switched linear systems. *Systems and Control Letters, 62*(2), 143–151.

Bemporad, A., Ferrari-Trecate, G., & Morari, M. (2000). Observability and controllability of piecewise affine and hybrid systems. *IEEE Transactions on Automatic Control, 45*(10), 1864–1876.

Bemporad, A., Garulli, A., Paoletti, S., & Vicino, A. (2003). A greedy approach to identification of piecewise affine models. In *Hybrid systems: Computation and control* (pp. 97–112). New York: Springer.

Bemporad, A., Garulli, A., Paoletti, S., & Vicino, A. (2005). A bounded-error approach to piecewise affine system identification. *IEEE Transactions on Automatic Control, 50*(10), 1567–1580.

Boukharouba, K. (2011). Modélisation et classification de comportements dynamiques des systemes hybrides. *Ph.D. thesis*, Université de Lille, France.

Chaitali, C. (2012). Optimizing clustering technique based on partitioning DBSCAN and ant clustering algorithm. *International Journal of Engineering and Advanced Technology (IJEAT), 2*(2), 212–215.

Chiu, S. (1994). Fuzzy model identification based on cluster estimation. *Journal of Intelligent and Fuzzy Systems, 2*(3), 267–278.

Chiu, S. (1997). Extracting fuzzy rules from data for function approximation and pattern classification. In D. Dubois, et al. (Eds.), *Chapter 9 in fuzzy information engineering: A guided tour of applications.* New York: Wiley.

De Schutter, B., & De Moor, B. (1999). The extended linear complementarity problem and the modeling and analysis of hybrid systems. In *Hybrid systems V* (pp. 70–85). New York: Springer.

De Schutter, B., & Van den Boom, T. (2000). On model predictive control for max-min-plus-scaling discrete event systems. Technical report, bds 00-04: Control Systems Engineering, Faculty of Information Technology and Systems, Delft University of Technology, The Netherlands.

Doucet, A., Gordon, N., & Krishnamurthy, V. (2001). Particle filters for state estimation of jump markov linear systems. *IEEE Transactions on Signal Processing, 49*(3), 613–624.

Duda, R. O., Hart, P. E., & Stork, D. G. (2001). *Pattern classification* (2nd ed.). New York: Wiley.

Ferrari-Trecate, G., Muselli, M., Liberati, D., & Morari, M. (2001). A clustering technique for the identification of piecewise affine systems. In *Hybrid systems: Computation and control* (pp. 218–231). New York: Springer.

Ferrari-Trecate, G., Muselli, M., Liberati, D., & Morari, M. (2003). A clustering technique for the identification of piecewise affine systems. *Automatica, 39*(2), 205–217.

Heemels, W. P., De Schutter, B., & Bemporad, A. (2001). Equivalence of hybrid dynamical models. *Automatica, 37*(7), 1085–1091.

Henson, M. A., & Seborg, D. E. (1994). Adaptive nonlinear control of a ph neutralization process. *IEEE Transactions on Control Systems Technology, 2*(3), 169–182.

Juloski, A., Weiland, S., & Heemels, W. (2005). A bayesian approach to identification of hybrid systems. *IEEE Transactions on Automatic Control, 50*(10), 1520–1533.

Juloski, A. L., Paoletti, S., & Roll, J. (2006). Recent techniques for the identification of piecewise affine and hybrid systems. In *Current trends in nonlinear systems and control* (pp. 79–99). New York: Springer.

Lai, C. Y. (2011). Identification and control of nonlinear systems using multiple models. *Ph.D. thesis.*

Lai, C. Y., Xiang, C., & Lee, T. H. (2010). Identification and control of nonlinear systems via piecewise affine approximation. In *The 49th IEEE Conference on Decision and Control (CDC)* (pp. 6395–6402).

Lassoued, Z., & Abderrahim, K. (2013a). A comparison study of some PWARX system identification methods. In *The 17th IEEE International Conference in System Theory, Control and Computing (ICSTCC)* (pp. 291–296).

Lassoued, Z., & Abderrahim, K. (2013b). A Kohonen neural network based method for PWARX identification. In *Adaptation and learning in control and signal processing, IFAC* (Vol. 11, pp. 742–747).

Lassoued, Z., & Abderrahim, K. (2013c). New approaches to identification of PWARX systems. *Mathematical Problems in Engineering.* http://dx.doi.org/10.1155/2013/845826.

Lassoued, Z., & Abderrahim, K. (2013d). A new clustering technique for the identification of PWARX hybrid models. In *The 9th IEEE Asian Control Conference (ASCC)* (pp. 1–6).

Lassoued, Z., & Abderrahim, K. (2014a). An experimental validation of a novel clustering approach to PWARX identification. *Engineering Applications of Artificial Intelligence, 28*, 201–209.

Lassoued, Z., & Abderrahim, K. (2014b). New results on PWARX model identification based on clustering approach. *International Journal of Automation and Computing, 11*(2), 180–188.

Lin, J., & Unbehauen, R. (1992). Canonical piecewise-linear approximations. *IEEE Transactions on Circuits and Systems I: Fundamental Theory and Applications, 39*(8), 697–699.

Nakada, H., Takaba, K., & Katayama, T. (2005). Identification of piecewise affine systems based on statistical clustering technique. *Automatica, 41*(5), 905–913.

Roll, J., Bemporad, A., & Ljung, L. (2004). Identification of piecewise affine systems via mixed-integer programming. *Automatica, 40*(1), 37–50.

Salehi, S., Shahrokhi, M., & Nejati, A. (2009). Adaptive nonlinear control of ph neutralization processes using fuzzy approximators. *Control Engineering Practice, 17*(11), 1329–1337.

Sander, J., Ester, M., Kriegel, H.-P., & Xu, X. (1998). Density-based clustering in spatial databases: The algorithm gdbscan and its applications. *Data Mining and Knowledge Discovery, 2*(2), 169–194.

Talmoudi, S., Abderrahim, K., Abdennour, R. B., & Ksouri, M. (2008). Multimodel approach using neural networks for complex systems modeling and identification. *Nonlinear Dynamics and Systems Theory, 8*(3), 299–316.

Vander-Schaft, A. J., & Schumacher, J. M. (1998). Complementarity modeling of hybrid systems. *IEEE Transactions on Automatic Control, 43*(4), 483–490.

Vidal, R., Chiuso, A., & Soatto, S. (2002). Observability and identifiability of jump linear systems. In *Proceedings of the 41st IEEE Conference on Decision and Control* (Vol. 4, pp. 3614–3619).

Wang, L. (2005). *Support vector machines: Theory and applications* (Vol. 177). New York: Springer.

Wen, C., Wang, S., Jin, X., & Ma, X. (2007). Identification of dynamic systems using piecewise-affine basis function models. *Automatica, 43*(10), 1824–1831.

Xiao, C., Xue, A., Peng, D., & Guo, Y. (2014). Modeling of ph neutralization process using fuzzy recurrent neural network and dna based nsga-ii. *Journal of the Franklin Institute, 351*(7), 3847–3864.

Xu, J., Huang, X., Mu, X., & Wang, S. (2012). Model predictive control based on adaptive hinging hyperplanes model. *Journal of Process Control, Elsevier, 22*(10), 1821–1831.

Supplier Quality Evaluation Using a Fuzzy Multi Criteria Decision Making Approach

Anjali Awasthi

Abstract Supplier quality evaluation is a vital decision for buyer organizations. Every year, several organizations suffer not only monetary loss but waste of materials, man power, resources etc. due to poor quality product and services leading to loss of clientele and market reputation. In this paper, we address the problem of supplier quality evaluation and propose a multi-perspective fuzzy multicriteria decision making approach based on fuzzy TOPSIS. Four perspectives namely product quality, process quality, service quality, and organizational quality are considered to identify the supplier quality evaluation criteria. Supplier quality evaluation is performed by a committee of decision making experts who use linguistic assessments to rank the criteria and the suppliers in the lack of quantitative information. These assessments are then combined through fuzzy TOPSIS to generate an overall performance score for each alternative. The alternative with the highest score is finally chosen and recommended for procurement. A practical application is provided. Sensitivity analysis and comparison with another approach called fuzzy SAW is performed for model verification and validation. The novelty of the proposed approach is supplier quality evaluation from multiple perspectives namely product quality, process quality, organizational quality, and service quality and ability to perform supplier quality evaluation under limited or lack of quantitative data.

Keywords Supplier quality evaluation · Multicriteria decision making · Fuzzy theory · TOPSIS · Multiple perspectives

---

A. Awasthi (✉)
CIISE, Concordia University, Montreal, Canada
e-mail: awasthi@ciise.concordia.ca

Q. Zhu and A.T. Azar (eds.), *Complex System Modelling and Control Through Intelligent Soft Computations*, Studies in Fuzziness and Soft Computing 319,
DOI 10.1007/978-3-319-12883-2_7

1 Introduction

Managing supplier quality is vital for efficient functioning of modern supply chains and achieving customer satisfaction. Poor quality of materials from suppliers can lead to high internal and external failure costs for organizations. To avoid these costs, organizations need to adopt a pro-active approach in managing product quality. This involves assuring quality in all stages of product development from conceptual design, procurement, processing, packaging until delivery to customer. It is useful to maintain and record product and service data to monitor changes in quality performance over time for selecting quality product and services from suppliers.

In literature, several studies have been reported by researchers on quality assurance in supply chain management. Foster (2007) presents a detailed study of approaches for managing quality in supply chains namely supplier quality management, voice of the customer, voice of the market, statistical process control, quality management standards, and six sigma. Flynn and Flynn (2005) study the implications of synergies between supply chain management and quality management. Their research emphasizes on developing cumulative capabilities with suppliers than trade-off orientations (usually for minimizing costs). In other words, seeing them as partners and not adversaries and involving them in product and process development. Bessant et al. (1994) investigate the role of greater cooperation and collaboration among organizations for managing successful total quality relationships in the supply chain. Casadesus and Castrao (2005) analyse the effect of ISO 9000 on improving quality in supply chain management and found it to be effective in areas of improving supplier relationships, customer satisfaction and reducing non-conformity costs. They also encourage enterprises to adopt modern process change initiatives such as business process redesign and enterprise resource planning (ERP) systems. Foster and Ogden (2008) present the differences in how operations and supply chain managers approach quality management. Their study found that operations managers tend to manage supply chain relationship through procedural methods such as ISO 9000 and supplier evaluation. Supply chain managers tend to adopt more collaborative approaches such as supplier development, awards, and complaint resolution processes. Trent et al. (1999) addresses the increasing importance of suppliers, particularly in supporting product and service quality requirements, and presents a series of questions concerning how well purchasing and sourcing activities contribute to total quality. Robinson and Malhotra (2005) discuss the parallels between quality management and supply chain management.

In all the above studies supplier quality management has been widely emphasized for quality management in supply chains. In this paper, we present a multi-criteria decision making approach based on fuzzy TOPSIS for evaluating supplier quality. The advantage of using fuzzy TOPSIS is that it distinguishes between Benefit (The more the better) and the Cost (The less the better) category criteria and selects solutions that are close to the positive ideal solutions and far from negative ideal solutions. Fuzzy set theory is used to model vagueness and uncertainty in decision making processes arising due to lack of quantitative or precise information (Zadeh 1965; Dubois and Prade 1982; Klir and Yuan 1995). The supplier quality is evaluated from multiple perspectives namely product, process, service and organizational. More details are provided in later sections.

The rest of the paper is organized as follows. In Sect. 2, we present related works on supplier quality management. Section 3 presents the proposed methodology for evaluating quality performance of suppliers. Section 4 presents a case study for the proposed approach. Section 5 presents the results discussion. In the sixth and the last section, we present the conclusions and future work.

2 Related Works

Existing studies on supplier quality evaluation can be mainly categorized into multicriteria decision making approaches, standards and certifications based approaches (ISO, Balridge etc.), data mining based approaches, total cost of ownership based approaches, and empirical studies.

Multicriteria decision making involves evaluating a set of alternatives by a committee of decision makers. Examples of multicriteria decision making techniques are AHP, TOPSIS, ELECTRE, PROMETHEE etc. Chin et al. (2006) propose an AHP based assessment system for supplier quality management. Pi and Low (2006) perform supplier evaluation and selection via Taguchi loss functions and AHP. de Boer et al. (1998) present outranking methods in support of supplier selection. Ho et al. (2011) present a combined QFD and AHP approach for strategic sourcing in manufacturing. Chin et al. (2006) propose an AHP based assessment system for supplier quality management. For a detailed review on multicriteria decision making techniques for supplier evaluation and selection, please refer to Chai et al. (2013) and Ho et al. (2010).

The standards and certifications based approaches rely on ISO, Malcolm Baldrige, Deming award etc. for supplier quality evaluation. Lee et al. (2003) propose an ISO 9001 system based multiobjective programming model for supplier quality evaluation. The appraisal factors include Quality management system audit, Product test, Percentage of workforce with a technical qualification, Process capability index and Annual training hours per employee. Grieco (1989) study the role of certification in assuring supplier quality and supported its role in increasing buyer-supplier partnership, trust and communication. Lee et al. (2008) propose an integrated model for supplier quality performance assessment in the semiconductor industry using ISO 9001 management framework, importance-performance analysis and Taguchi's signal-to-noise ratio techniques. Vokurka and Lummus (2003) study Baldrige awards for better supply chains and emphasize on information sharing, customized production, reducing the number of suppliers, flexibility, co-ordination and employee involvement and focus on end-users for supply chain excellence. Curkovic and Handfield (2006) study the use of ISO 9000 and Baldrige Award

Criteria in supplier quality evaluation in 314 North American organizations. Their results suggest that ISO 9000 registration focuses solely on reducing negative quality—defects and nonconformities and fails to measure key areas of quality management, including strategic planning, employee involvement, quality results, competitive benchmarking, and customer satisfaction. Based on these results, the implications for the design of supplier quality measurement and evaluation systems are discussed. Sroufe and Curkovic (2008) examine ISO 9000:2000 for supply chain quality assurance and show that ISO 9000:2000 has the potential, when used under the right circumstances, to improve QA across the supply chain.

The data mining based approaches have been often coupled with statistical quality control techniques for supplier quality evaluation. One such study is by Chen and Chen (2006) who use process capability index for supplier quality evaluation. Shu and Wu (2009) perform quality-based supplier selection and evaluation using process capability index based on fuzzy data. Wang et al. (2004) apply six-sigma for supplier development.

The total cost of ownership based approaches use various costs of quality namely prevention, appraisal, internal failure and external failure for supplier quality evaluation. Dogan and Aydin (2011) use Bayesian Networks and Total Cost of Ownership method for supplier selection analysis. Ramanathan (2007) perform supplier selection by integrating DEA with the approaches of total cost of ownership and AHP. Dogan and Sahin (2003) perform supplier selection using activity-based costing and fuzzy present-worth techniques. Chen and Weng (2002) use fuzzy approaches to evaluate quality improvement alternative based on quality costs. Bhutta and Huq (2002) compare total cost of ownership and analytical hierarchy process approaches for supplier selection problem. Chen and Yang (2003) propose a total costs based evaluation system of supplier quality performance.

The empirical studies rely on case studies, surveys, personal interviews, expert opinions for supplier quality assessment. Seth et al. (2006) propose a SERVQUAL (Parasumraman et al. 1988) type strategic planning tool to measure supplier service quality in supply chain. Watts and Hahn (1993) conduct an empirical study of supplier development programs and found that they are more prevalent than expected and that large companies are more likely to be involved. Dean and Kiu (2002) conduct a survey study for performance monitoring and quality outcomes in contracted services and found that organisations rely on inspections by their own employees or contractor checklists, but that these practices are in conflict with their views on best practice for performance monitoring.

Kuei et al. (2008) describe a strategic framework based on vision- and gap driven change for the development of supply chain quality management (SCQM). Four drivers for supply chain quality are identified namely supply chain competence, critical success factors (CSF), strategic components, and SCQ practices/activities/programmes. The dimensions of supply chain competence are quality product, delivery reliability, supplier/buyer trust, operational efficiency and delivery value/innovation to customer. The CSF include customer focus, quality of IT system, supplier relationship, externally focused process integration and supply chain quality leadership. The strategic components include quality management culture, technology management, supplier participation, supply chain configuration design, and strategic planning. The SCQ practices/activities/programs involve supplier/buyer quality meeting, quality data and reporting, supply chain quality office, supply chain optimisation, and policy deployment. Theodorakioglou et al. (2006) investigate supplier management and its relationship to buyers' quality management through a survey study in the Greek manufacturing industry. They found that intra-firm adoption of the quality philosophy can lead firms to better supplier management in the SCM context.

Forker (1997) conducted a survey study of 348 manufacturing firms to investigates factors affecting supplier quality performance and found that asset specificity and organizational efficiency at implementing total quality management hold great promise. Trent et al. (1999) propose steps for achieving world-class supplier quality. These include supply base optimization, supplier performance measurement, aggressive supplier improvement targets, performance improvement rewards, supplier certification, supplier performance development, and involvement of supplier in product and process design.

To deal with the lack of quantitative data for supplier quality evaluation, several authors have coupled fuzzy theory (Zadeh 1965) with the above proposed approaches. In fuzzy set theory, linguistic terms are used to represent decision maker preferences (Zimmermann 2001). For example, it is much easier to represent the quality performance of suppliers as good, very good, poor, very poor etc. than in numbers. Ku et al. (2010) present a fuzzy analytic hierarchy process and fuzzy goal programming based approach for global supplier selection. Liu et al. (2012) use axiomatic fuzzy set and TOPSIS methodology for supplier selection. Kumar and Mahapatra (2009) propose a fuzzy multi-criteria decision making approach for supplier selection involving weighted decision makers preferences through an AHP-TOPSIS like procedure. Kumar and Mahapatra (2011) propose a fuzzy multi-criteria decision-making approach for supplier selection in supply chain management in group decision making environment. Yang et al. (2008) study vendor selection by integrated fuzzy MCDM techniques considering independent and interdependent relationships. Dursun and Karsak (2013) use a QFD-based fuzzy MCDM approach for supplier evaluation.

3 Solution Approach

Our proposed solution approach for evaluating supplier quality involves following steps.

1. Selection of evaluation criteria for supplier quality assessment from multiple perspectives.
2. Evaluation and selection of best supplier(alternative) using selected quality criteria.

3. Sensitivity analysis to determine the influence of criteria weights on model results.
4. Results validation by comparison against another standard approach.

These steps are presented in detail as follows:

3.1 Criteria Selection

The first step involves selection of criteria for evaluating supplier quality. A comprehensive literature review review and discussion with departmental quality representatives (see the case study), and five academic supply chain experts is formed to identify the criteria for supplier quality evaluation. Four perspectives are used namely product quality, process quality, organizational quality and service quality. The product quality perspective takes into account the quality of the product whereas the process quality perspective takes into account the quality of processes used to generate the product. Six criteria are considered from product quality perspective namely customer satisfaction, environmental considerations, product features, documentation, product design, and conformance to standards. Five criteria are considered from process quality perspective namely technical capability, nonconformities/defects generated during production, statistical process control, and process capability index ($Cp > 1.33$). The organizational perspective refers to the incorporation of the quality in organization itself such as in employees and the service quality refers to the quality of service offered by the suppliers to the buyer organizations. Four criteria are considered from service quality perspective namely responsiveness, reliability, flexibility and handling of returned material and warranties and three from organizational perspective namely quality certifications, employee training, and management commitment. A total of 17 criteria are chosen for supplier quality evaluation from these four perspectives. Table 1 presents these criteria along with their categories.

It can be seen in Table 1 that but except the eighth criterion, the remaining criteria are the Benefit (B) category criteria that is the higher the value, the more preferable the alternative (supplier). The eleventh criteria has the cost (C) category that is, the lower the value the more preferable the alternative.

3.2 Alternatives Evaluation and Selection Using Fuzzy TOPSIS

The second step involves allocation of linguistic ratings to the 17 criteria chosen for evaluating supplier quality and to the potential alternatives (suppliers) with respect to each criteria. The decision making committee provides linguistic ratings using scales given in Table 3 to the criteria and using Table 2 to the alternatives. The

Table 1 Criteria for evaluating supplier quality

Perspective	Criteria (id)	Category
Product quality	Customer satisfaction (C1)	B
	Environmental considerations (C2)	B
	Product features (C3)	B
	Documentation (C4)	B
	Product design (C5)	B
	Conformance to standards (C6)	B
Process quality	Technical capability (C7)	B
	Nonconformities/Defects generated during production (C8)	**C**
	Statistical process control (C9)	B
	Process capability index (Cp > 1.33) (C10)	B
Service quality	Quality of service (responsiveness) (C11)	B
	Reliability (C12)	B
	Flexibility (C13)	B
	Handling of returned material and warranties (C14)	B
Organizational quality	Quality certifications (C15)	B
	Employee training (C16)	B
	Management commitment (C17)	B

B Benefit (the higher the better), **C** Cost (the lower the better)

Table 2 Linguistic ratings for alternatives

Linguistic term	Membership function
Very poor (VP)	(1, 1, 3)
Poor (P)	(1, 3, 5)
Fair (F)	(3, 5, 7)
Good (G)	(5, 7, 9)
Very good (VG)	(7, 9, 9)

Table 3 Linguistic ratings for criteria

Linguistic term	Membership function
Very low	(1, 1, 3)
Low	(1, 3, 5)
Medium	(3, 5, 7)
High	(5, 7, 9)
Very high	(7, 9, 9)

linguistics terms are then transformed to fuzzy triangular numbers (Azar 2010a, b). Then, fuzzy TOPSIS (Sect. 3) is applied to aggregate the criteria and the alternative ratings to generate an overall score for assessing supplier quality (alternatives). The alternative with the highest score is finally chosen.

3.2.1 Fuzzy TOPSIS

The fuzzy TOPSIS approach involves fuzzy assessments of criteria and alternatives in TOPSIS (Hwang and Yoon 1981). The TOPSIS approach chooses alternative that is closest to the positive ideal solution and farthest from the negative ideal solution. A positive ideal solution is composed of the best performance values for each criterion whereas the negative ideal solution consists of the worst performance values. The various steps of Fuzzy TOPSIS are presented as follows:

Step 1: Assignment of ratings to the criteria and the alternatives.

Let us assume there are j possible candidates (in our case suppliers) called $A = \{A_1, A_2, \ldots, A_j\}$ which are to be evaluated against n criteria, $C = \{C_1, C_2, \ldots, C_n\}$. The criteria weights are denoted by $w_i(i = 1, 2, \ldots, m)$. The performance ratings of decision maker $D_k(k = 1, 2, \ldots, K)$ for each alternative $A_j(j = 1, 2, \ldots, n)$ with respect to criteria $C_i(i = 1, 2, \ldots, m)$ are denoted by $\tilde{R}_k = \tilde{x}_{ijk}(i = 1, 2, \ldots, m; j = 1, 2, \ldots, n; k = 1, 2, \ldots, K)$ with membership function $\mu_{\tilde{R}_k}(x)$.

Step 2: Compute aggregate fuzzy ratings for the criteria and the alternatives.

If the fuzzy ratings of decision makers are described by triangular fuzzy number $\tilde{R}_k = (a_k, b_k, c_k), k = 1, 2, \ldots, K$, then the aggregated fuzzy rating is given by $\tilde{R} = (a, b, c), k = 1, 2, \ldots, K$ where;

$$a = \min_k\{a_k\}, b = \frac{1}{K}\sum_{k=1}^{K} b_k, c = \max_k\{c_k\}$$

If the fuzzy rating and importance weight of the kth decision maker are $\tilde{x}_{ijk} = (a_{ijk}, b_{ijk}, c_{ijk})$ and $\tilde{w}_{ijk} = (w_{jk1}, w_{jk2}, w_{jk3}), i = 1, 2, \ldots, m, j = 1, 2, \ldots n$ respectively, then the aggregated fuzzy ratings $(\tilde{x}_{ij})$ of alternatives with respect to each criteria are given by $\tilde{x}_{ij} = (a_{ij}, b_{ij}, c_{ij})$ where

$$a_{ij} = \min_k\{a_{ijk}\}, b_{ij} = \frac{1}{K}\sum_{k=1}^{K} b_{ijk}, c_{ij} = \max_k\{c_{ijk}\} \tag{2}$$

The aggregated fuzzy weights $(\tilde{w}_{ij})$ of each criterion are calculated as $\tilde{w}_j = (w_{j1}, w_{j2}, w_{j3})$ where

$$w_{j1} = \min_k\{w_{jk1}\}, w_{j2} = \frac{1}{K}\sum_{k=1}^{K} w_{jk2}, w_{j3} = \max_k\{c_{jk3}\} \tag{3}$$

Step 3: Compute the fuzzy decision matrix.

The fuzzy decision matrix for the alternatives ($\tilde{D}$) and the criteria ($\tilde{W}$) is constructed as follows:

$$\tilde{D} = \begin{matrix} \\ A_1 \\ A_2 \\ \dots \\ A_m \end{matrix} \begin{matrix} C_1 & C_2 & & C_n \\ \begin{bmatrix} \tilde{x}_{11} & \tilde{x}_{12} & \dots & \tilde{x}_{1n} \\ \tilde{x}_{21} & \tilde{x}_{22} & \dots & \tilde{x}_{2n} \\ \dots & \dots & \dots & \dots \\ \tilde{x}_{m1} & \tilde{x}_{m2} & \dots & \tilde{x}_{mn} \end{bmatrix} \end{matrix}, i = 1, 2, \dots, m;\ j = 1, 2, \dots, n \quad (4)$$

$$\tilde{W} = (\tilde{w}_1, \tilde{w}_2, \dots, \tilde{w}_n) \quad (5)$$

Step 4: Normalize the fuzzy decision matrix.

The raw data are normalized using linear scale transformation to bring the various criteria scales into a comparable scale. The normalized fuzzy decision matrix $\tilde{R}$ is given by:

$$\tilde{R} = [\tilde{r}_{ij}]_{mxn}, i = 1, 2, \dots, m;\ j = 1, 2, \dots, n \quad (6)$$

where:

$$\tilde{r}_{ij} = (\frac{a_{ij}}{c_j^*}, \frac{b_{ij}}{c_j^*}, \frac{c_{ij}}{c_j^*}) \quad \text{and} \quad c_j^* = \max_i c_{ij} \quad \text{(benefit criteria)} \quad (7)$$

$$\tilde{r}_{ij} = (\frac{a_j^-}{c_{ij}}, \frac{a_j^-}{b_{ij}}, \frac{a_j^-}{a_{ij}}) \quad \text{and} \quad a_j^- = \min_i a_{ij} \quad \text{(cost criteria)} \quad (8)$$

Step 5: Compute the weighted normalized matrix.

The weighted normalized matrix $\tilde{V}$ for criteria is computed by multiplying the weights ($\tilde{w}_j$) of evaluation criteria with the normalized fuzzy decision matrix $\tilde{r}_{ij}$.

$$\tilde{V} = [\tilde{v}_{ij}]_{mxn}, i = 1, 2, \dots, m;\ j = 1, 2, \dots, n \quad \text{where } \tilde{v}_{ij} = \tilde{r}_{ij}(.)\tilde{w}_j \quad (9)$$

Step 6: Compute the fuzzy ideal solution (FPIS) and fuzzy negative ideal solution (FNIS)

The FPIS and FNIS of the alternatives is computed as follows:

$$A^* = (\tilde{v}_1^*, \tilde{v}_2^*, \ldots, \tilde{v}_n^*) \text{ where } \tilde{v}_j^* = \max_i\{v_{ij3}\}, i = 1, 2, \ldots, m; j = 1, 2, \ldots, n \quad (10)$$

$$A^- = (\tilde{v}_1, \tilde{v}_2, \ldots, \tilde{v}_n) \text{ where } \tilde{v}_j^- = \min_i\{v_{ij1}\}, i = 1, 2, \ldots, m; j = 1, 2, \ldots, n \quad (11)$$

Step 7: Compute the distance of each alternative from FPIS and FNIS:

The distance (d_i^*, d_i^-) of each weighted alternative $i = 1, 2, \ldots, m$ from the FPIS and the FNIS is computed as follows:

$$d_i^* = \sum_{j=1}^{n} d_v(\tilde{v}_{ij}, \tilde{v}_j^*), i = 1, 2, \ldots, m \quad (12)$$

$$d_i^- = \sum_{j=1}^{n} d_v(\tilde{v}_{ij}, \tilde{v}_j^-), i = 1, 2, \ldots, m \quad (13)$$

where $d(\tilde{a}, \tilde{b}) = \sqrt{\frac{1}{3}\left[(a_1 - b_1)^2 + (a_2 - b_2)^2 + (a_3 - b_3)^2\right]}$ is the distance measurement between two fuzzy numbers $\tilde{a}$ and $\tilde{b}$.

Step 8: Compute the closeness coefficient (CC_i) of each alternative.

The closeness coefficient CC_i represents the distances to the fuzzy positive ideal solution (A*) and the fuzzy negative ideal solution (A−) simultaneously. The closeness coefficient of each alternative is calculated as:

$$CC_i = \frac{d_i^-}{d_i^- + d_i^*}, i = 1, 2, \ldots, m \quad (14)$$

Step 9: Rank the alternatives

In step 9, the different alternatives are ranked according to the closeness coefficient (CC_i) in decreasing order. The best alternative is closest to the FPIS and farthest from the FNIS.

3.3 Sensitivity Analysis

The third step involves conducting the sensitivity analysis. Sensitivity analysis addresses the question, "How sensitive is the overall decision to small changes in the individual weights assigned during the pair-wise comparison process?". This question can be answered by varying slightly the values of the weights and observing the effects on the decision. This is useful in situations where uncertainties exist in the definition of the importance of different factors. In our case, we will conduct sensitivity analysis to see the importance of criteria weights in evaluating supplier quality.

3.4 Results Validation

To validate the model results, we will compare the results of our study with another standard approach called Fuzzy Simple Aggregated Weighting (SAW). This method is defined as follows.

3.4.1 Fuzzy SAW

Let w_j represent the weight of the criteria j and x_{ij} represents the rating of alternative i against criteria j.

Step 1: Normalize the data using the following equations.

$$r_{ij} = \frac{x_{ij}}{\max_i\{x_{ij}\}} \quad \forall i = 1, 2, \ldots, m; j = 1, 2, \ldots, n \quad \text{(benefit type criteria)} \tag{15}$$

$$r_{ij} = \frac{\min_i\{x_{ij}\}}{x_{ij}} \forall i = 1, 2, \ldots, m; j = 1, 2, \ldots, n \quad \text{(cost type criteria)} \tag{16}$$

Step 2: Calculate the overall performance rating u_i for alternative i by aggregating the product of its various criteria values with their respective weights.

$$u_i = \sum_j w_j x_{ij} \tag{17}$$

Step 3: Select the alternative with the highest overall performance value $u_i \, \forall i = 1, 2, \ldots m$

4 Case Study

In this section, we present the application of the proposed approach on an oven manufacturing organization in Poland. The name of the company is not revealed for confidentiality reasons. Three experts from the organization are involved. One is manager in the production (D1) department and the other two are in purchasing (D2) and quality control (D3) department. Three suppliers (A1, A2, A3) were assessed for quality. The decision makers evaluated the three suppliers using the quality criteria provided in Table 1. The linguistic ratings for criteria are given using Table 3 and for the three suppliers using Table 2. These results can be found in Tables 4 and 5 respectively.

4.1 Fuzzy TOPSIS

Once the ratings are received, fuzzy TOPSIS is applied to generate overall performance scores for quality for the three suppliers. The aggregated fuzzy weights ($\tilde{w}_{ij}$) of each criterion are calculated using Eq. (3). For example, for criteria C1, the aggregated fuzzy weight is given by $\tilde{w}_j = (w_{j1}, w_{j2}, w_{j3})$ where

Table 4 Linguistic assessments for the 17 criteria

Criteria	Decision makers		
	D1	D2	D3
C1	H	M	VH
C2	VL	VH	L
C3	L	L	VH
C4	VL	H	VH
C5	VL	M	H
C6	M	VH	M
C7	H	5	H
C8	M	M	H
C9	VH	H	H
C10	M	H	H
C11	VH	H	H
C12	VH	H	H
C13	M	VH	M
C14	H	H	H
C15	L	H	VL
C16	M	VL	VH
C17	H	M	L

Table 5 Linguistic assessments for the three suppliers (alternatives)

Criteria	Alternatives								
	A1			A2			A3		
	D1	D2	D3	D1	D2	D3	D1	D2	D3
C1	H	VH	VL	VH	VH	M	VH	H	VL
C2	L	VL	VH	VL	VL	M	H	H	VL
C3	M	H	L	H	M	M	H	VL	VL
C4	VH	L	VH	M	M	M	L	VH	L
C5	VL	M	M	H	L	H	M	M	M
C6	M	VH	VH	M	M	VL	M	VL	VH
C7	VL	H	H	H	VL	VL	M	VH	L
C8	H	M	M	H	L	VH	H	M	L
C9	H	VL	VH	VH	H	L	L	VH	VL
C10	VL	M	VH	VL	VL	VL	H	VL	M
C11	L	H	M	VH	VH	M	L	VH	VH
C12	VL	L	L	VL	VL	M	M	VH	M
C13	VH	L	H	VL	H	M	L	H	VH
C14	H	H	H	L	M	M	M	VL	M
C15	VL	H	M	VH	VL	L	L	M	H
C16	H	VL	VL	M	L	VL	L	VL	H
C17	VL	VH	M	VH	VH	M	H	VH	H

$$w_{j1} = \min_k(5,3,7), w_{j2} = \frac{1}{3}\sum_{k=1}^{3} 7+5+9,$$

$$w_{j1} = \max_k(9,7,9)$$

$$= \tilde{w}_j = (3,7,9)$$

Likewise, we computed the aggregate weights for the remaining criteria. The aggregate weights of the 17 criteria are presented in Table 6.

The aggregate fuzzy weights of the three alternatives (suppliers) are computed using Eq. (2). For example, the aggregate rating for supplier A1 for criteria C1 using the rating of the three decision makers is computed as follows:

$$a_{ij} = \min_k(5,7,1), b_{ij} = \frac{1}{3}\sum_{k=1}^{3} 7+9+1,$$

$$c_{ij} = \max_k(9,9,3)$$

$$= (1, 5.667, 9)$$

Table 6 Aggregate fuzzy weights for 17 criteria

Criteria	Decision makers			Aggregate fuzzy weight
	D1	D2	D3	
C1	(5, 7, 9)	(3, 5, 7)	(7, 9, 9)	(3, 7, 9)
C2	(1, 1, 3)	(7, 9, 9)	(1, 3, 5)	(1, 4.334, 9)
C3	(1, 3, 5)	(1, 3, 5)	(7, 9, 9)	(1, 5, 9)
C4	(1, 1, 3)	(5, 7, 9)	(7, 9, 9)	(1, 5.667, 9)
C5	(1, 1, 3)	(3, 5, 7)	(5, 7, 9)	(1, 4.334, 9)
C6	(3, 5, 7)	(7, 9, 9)	(3, 5, 7)	(3, 6.334, 9)
C7	(5, 7, 9)	(5, 7, 9)	(5, 7, 9)	(5, 7, 9)
C8	(3, 5, 7)	(3, 5, 7)	(5, 7, 9)	(3, 5.667, 9)
C9	(7, 9, 9)	(5, 7, 9)	(5, 7, 9)	(5, 7.667, 9)
C10	(3, 5, 7)	(5, 7, 9)	(5, 7, 9)	(3, 6.334, 9)
C11	(7, 9, 9)	(5, 7, 9)	(5, 7, 9)	(5, 7.667, 9)
C12	(7, 9, 9)	(5, 7, 9)	(5, 7, 9)	(5, 7.667, 9)
C13	(3, 5, 7)	(7, 9, 9)	(3, 5, 7)	(3, 6.334, 9)
C14	(5, 7, 9)	(5, 7, 9)	(5, 7, 9)	(5, 7, 9)
C15	(1, 3, 5)	(5, 7, 9)	(1, 1, 3)	(1, 3.667, 9)
C16	(3, 5, 7)	(1, 1, 3)	(7, 9, 9)	(1, 5, 9)
C17	(5, 7, 9)	(3, 5, 7)	(1, 3, 5)	(1, 5, 9)

Likewise, the aggregate ratings for the alternatives (A1, A2, A3) with respect to the 17 criteria are computed. The aggregate fuzzy decision matrix for the alternatives is presented in Table 7.

Then, normalization of the fuzzy decision matrix of alternatives is performed using Eqs. (6), (7) and (8). For example, the normalized rating for alternative A1 for criteria C1 is given by:

$$c_j^* = \max_i(9,9,9) = 9$$

$$a_j^- = \min_i(1,3,1) = 1$$

Since C1 is a benefit (B) category criteria,

$$\tilde{r}_{ij} = (\frac{1}{9}, \frac{5.667}{9}, \frac{9}{9}) = (0.111, 0.629, 1)$$

Likewise, the normalized values of the three alternatives with respect to the 17 criteria are computed. The value of $a_j^- = 1$ and $c_j^* = 9$ for all criteria. The normalized fuzzy decision matrix for the three alternatives is presented in Table 8.

Next, the fuzzy weighted decision matrix for the three alternatives is constructed using Eq. (9). The $\tilde{r}_{ij}$ values from Table 8 and $\tilde{w}_j$ values from Table 6 are used to

Table 7 Aggregate fuzzy decision matrix for alternatives

Criteria	Alternatives			Min	Max
	A1	A2	A3		
C1	(1, 5.667, 9)	(3, 7.667, 9)	(1, 5.667, 9)	1	9
C2	(1, 4.333,9)	(1, 2.333, 7)	(1, 5, 9)	1	9
C3	(1, 5, 9)	(3, 5.667, 9)	(1, 3, 9)	1	9
C4	(1, 7, 9)	(3, 5, 7)	(1, 5, 9)	1	9
C5	(1, 3.667)	(1, 5.667, 9)	(3, 5, 7)	1	9
C6	(3, 7.667, 9)	(1, 3.667, 7)	(1, 5, 9)	1	9
C7	(1, 5, 9)	(1, 3, 9)	(1, 5.667, 9)	1	9
C8	(3, 5.667, 9)	(1, 6.333, 9)	(1, 5, 9)	1	9
C9	(1, 5.667, 9)	(1, 6.333, 9)	(1, 4.333, 9)	1	9
C10	(1, 5, 9)	(1, 1, 3)	(1, 4.333, 9)	1	9
C11	(1, 5, 9)	(3, 7.667, 9)	(1, 7, 9)	1	9
C12	(1, 2.333)	(1, 2.333, 7)	(3, 6.333, 9)	1	9
C13	(1, 6.333, 9)	(1, 4.333, 9)	(1, 6.333, 9)	1	9
C14	(5, 7, 9)	(1, 4.333, 7)	(1, 3.667, 7)	1	9
C15	(1, 4.333, 9)	(1, 4.333, 9)	(1, 5, 9)	1	9
C16	(1, 3, 9)	(1, 3, 7)	(1, 3.667, 9)	1	9
C17	(1, 5, 9)	(3, 7.667, 9)	(5, 7.667, 9)	1	9

Table 8 Normalized fuzzy decision matrix for alternatives

Criteria	A1	A2	A3
C1	(0.111, 0.629, 1)	(0.333, 0.851, 1)	(0.111, 0.629, 1)
C2	(0.111, 0.481, 1)	(0.111, 0.259, 0.778)	(0.111, 0.556, 1)
C3	(0.111, 0.556, 1)	(0.333, 0.629, 1)	(0.111, 0.333, 1)
C4	(0.111, 0.778, 1)	(0.333, 0.556, 0.778)	(0.111, 0.556, 1)
C5	(0.111, 0.523, 0.778)	(0.111, 0.629, 1)	(0.333, 0.556, 0.778)
C6	(0.333, 0.851, 1)	(0.111, 0.407, 0.778)	(0.111, 0.556, 1)
C7	(0.111, 0.556, 1)	(0.111, 0.333, 1)	(0.111, 0.629, 1)
C8	(0.111, 0.176, 0.333)	(0.111, 0.157, 1)	(0.111, 0.2, 1)
C9	(0.111, 0.629, 1)	(0.111, 0.703, 1)	(0.111, 0.481, 1)
C10	(0.111, 0.556, 1)	(0.111, 0.111, 0.333)	(0.111, 0.481, 1)
C11	(0.111, 0.556, 1)	(0.333, 0.851, 1)	(0.111, 0.778, 1)
C12	(0.111, 0.259, 0.556)	(0.111, 0.259, 0.778)	(0.333, 0.703)
C13	(0.111, 0.703, 1)	(0.111, 0.481, 1)	(0.111, 0.703, 1)
C14	(0.556, 1, 1)	(0.111, 0.481, 0.778)	(0.111, 0.407, 0.778)
C15	(0.111, 0.481, 1)	(0.111, 0.481, 1)	(0.111, 0.556, 1)
C16	(0.111, 0.333, 1)	(0.111, 0.333, 0.778)	(0.111, 0.407, 1)
C17	(0.111, 0.556, 1)	(0.333, 0.851, 1)	(0.556, 0.851, 1)

compute the fuzzy weighted decision matrix for the alternatives. For example, for alternative A1, the fuzzy weight for criteria C1 is given by:

$$\tilde{v}_{ij} = (0.111, 0.629, 1)(.)(3, 7, 9) = (0.333, 4.407, 9)$$

Likewise, the fuzzy weights of the three alternatives for the 17 criteria are computed (Table 9). Then, the fuzzy positive ideal solution (A*) and the fuzzy negative ideal solutions (A−) are computed using Eqs. (10) and (11) for the alternatives. For example, for criteria C1, $A^{-} = (0.333, 0.333, 0.333)$ and

Table 9 Weighted normalized alternatives, FPIS and FNIS

Criteria	A1	A2	A3	Min	Max	A−	A*
C1	(0.33, 4.40, 9)	(1, 5.963, 9)	(0.33, 4.407, 9)	0.33	9	(0.333, 0.333, 0.333)	(9, 9, 9)
C2	(0.11, 2.08, 9)	(0.11, 1.12, 7)	(0.111, 2.407, 9)	0.11	9	(0.111, 0.111, 0.111)	(9, 9, 9)
C3	(0.11, 2.78, 9)	(0.33, 3.14, 9)	(0.11, 1.667, 9)	0.11	9	(0.11, 0.111, 0.111)	(9, 9, 9)
C4	(0.11, 4.40, 9)	(0.33, 3.14, 7)	(0.11, 3.148, 9)	0.111	9	(0.111, 0.111, 0.111)	(9, 9, 9)
C5	(0.11, 2.27, 7)	(0.11, 2.72, 9)	(0.33, 2.407, 7)	0.111	9	(0.111, 0.111, 0.111)	(9, 9, 9)
C6	(1, 5.395, 9)	(0.33, 2.58, 7)	(0.33, 3.519, 9)	0.333	9	(0.333, 0.333, 0.333)	(9, 9, 9)
C7	(0.56, 3.89, 9)	(0.56, 2.33, 9)	(0.556, 4.407, 9)	0.556	9	(0.556, 0.556, 0.556)	(9, 9, 9)
C8	(0.33, 1.0, 3)	(0.33, .895, 9)	(0.333, 1.133, 9)	0.333	9	(0.333, 0.333, 0.333)	(9, 9, 9)
C9	(0.56, 4.82, 9)	(0.56, 5.39, 9)	(0.556, 3.691, 9)	0.556	9	(0.556, 0.556, 0.556)	(9, 9, 9)
C10	(0.33, 3.51, 9)	(0.33, .704, 3)	(0.333, 3.049, 9)	0.333	9	(0.333, 0.333, 0.333)	(9, 9, 9)
C11	(0.56, 4.29, 9)	(1.67, 6.53, 9)	(0.556, 5.963, 9)	0.556	9	(0.556, 0.556, 0.556)	(9, 9, 9)
C12	(0.56, 1.98, 5)	(0.56, 1.98, 7)	(1.67, 5.395, 9)	0.556	9	(0.556, 0.556, 0.556)	(9, 9, 9)
C13	(0.33, 4.45, 9)	(0.33, 3.04, 9)	(0.333, 4.457, 9)	0.333	9	(0.333, 0.333, 0.333)	(9, 9, 9)
C14	(2.78, 7, 9)	(0.56, 3.37, 7)	(0.556, 2.852, 7)	0.556	9	(0.556, 0.556, 0.556)	(9, 9, 9)
C15	(0.11, 1.76, 9)	(0.11, 1.76, 9)	(0.111, 2.037, 9)	0.111	9	(0.111, 0.111, 0.111)	(9, 9, 9)
C16	(0.11, 1.67, 9)	(0.11, 1.67, 7)	(0.111, 2.037, 9)	0.111	9	(0.111, 0.111, 0.111)	(9, 9, 9)
C17	(0.11, 2.78, 9)	(0.33, 4.25, 9)	(0.556, 4.259, 9)	0.111	9	(0.111, 0.111, 0.111)	(9, 9, 9)

$A^* = (9, 9, 9)$. Similar computations are performed for the remaining criteria. The results are presented in last two columns of Table 9.

Next, the distance d_v (.) of each alternative from the fuzzy positive ideal matrix (A*) and fuzzy negative ideal matrix (A−) are computed using Eqs. (12) and (13). For example, for alternative A1 and criteria C1, the distances $d_v(A_1, A^*)$ and $d_v(A_1, A^-)$ are computed as follows:

$$d_v(A_1, A^-) = \sqrt{\frac{1}{3}[(0.333 - 0.333)^2 + (4.407 - 0.333)^2 + (9 - 0.333)^2]} = 5.529$$

$$d_v(A_1, A^*) = \sqrt{\frac{1}{3}[(0.333 - 9)^2 + (4.407 - 9)^2 + (9 - 9)^2]} = 5.663$$

Likewise, we compute the distances for the remaining criteria for the three alternatives. The results are shown in Table 10.

Then, we compute the distances d_i^* and d_i^- using Eqs. (12) and (13). For example, for alternative A1 and criteria C1, the distances d_i^* and d_i^- are given by:

$$d_i^- = 5.529 + 5.257 + \cdots + 5.358 = 85.057$$
$$d_i^* = 5.663 + 6.502 + \cdots + 6.264 = 101.802$$

Table 10 Distance d_v (A_i, A^*) and d_v (A_i, A^-) for alternatives

Criteria	d−(min)			d*(max)		
	A1	A2	A3	A1	A2	A3
C1	5.529	5.979	5.529	5.663	4.940	5.663
C2	5.257	4.020	5.300	6.502	6.953	6.389
C3	5.358	5.425	5.210	6.264	6.038	6.653
C4	5.700	4.349	5.423	5.777	6.147	6.144
C5	4.168	5.350	4.194	6.540	6.281	6.392
C6	5.807	4.062	5.331	5.066	6.333	5.921
C7	5.241	4.982	5.359	5.699	6.212	5.550
C8	1.587	5.014	5.025	7.640	6.851	6.758
C9	5.464	5.619	5.201	5.438	5.301	5.759
C10	5.331	1.554	5.244	5.921	7.745	6.070
C11	5.324	6.007	5.789	5.591	4.467	5.181
C12	2.696	3.811	5.656	6.745	6.442	4.718
C13	5.541	5.244	5.541	5.650	6.070	5.650
C14	6.266	4.060	3.950	3.773	5.972	6.140
C15	5.220	5.220	5.251	6.617	6.617	6.519
C16	5.210	4.077	5.251	6.653	6.753	6.519
C17	5.358	5.665	5.669	6.264	5.703	5.591
Σ	85.057	80.439	88.924	101.802	104.824	101.616

Table 11 Closeness coefficient (CC_i) for alternatives

	Alternatives		
	A1	A2	A3
d_i^-	85.057	80.439	88.924
d_i^+	101.802	104.824	101.616
CC_i	0.455	0.434	0.467

The last row of Table 10 presents the distances d_i^* and d_i^- for the three alternatives. Now, using distances d_i^* and d_i^- (Eq. 14), we compute the closeness coefficient (CC_i) of the three alternatives. For example, for alternative A1, the closeness coefficient is given by:

$$CC_i = d_i^- / (d_i^- + d_i^+) = 85.057/(85.057 + 101.802) = 0.455$$

Likewise, CC_i for the other two alternatives are computed. The final results are shown in Table 11.

By comparing the CC_i values of the three alternatives (Table 12), we find that A3 (0.467) > A1 (0.455) > A1(0.434). Therefore, A3 is selected as the supplier with highest quality and is recommended for procurement of materials by the organization.

4.2 Sensitivity Analysis

To investigate the impact of criteria weights (denoted by W_{Ci} for criteria Ci where i = 1, 2,…,n) on the selection of highest quality suppliers, we conducted the sensitivity analysis. 23 experiments were conducted. In the first five experiments (#1–5), weights of all criteria are set equal to (1, 1, 3), (1, 3, 5), (3, 5, 7), (5, 7, 9) and (7, 9, 9) respectively. In experiments #6–22, the weight of each criteria is set as highest (7, 9, 9) one by one and the remaining criteria are set to the lowest value (1, 1, 3). The goal is to see which criteria is most important in influencing the decision making process. For example, in experiment #6, the criteria C1 has the highest weight = (7, 9, 9) whereas the remaining criteria have weight = (1, 1, 3). In experiment #23, the weight of the cost category criteria (C8) is set as lowest = (1, 1, 3), whereas the other criteria are set to highest weights = (7, 9, 9). The contrary of experiment #23 is experiment #16 where weight of C8 = (7, 9, 9) and weight of remaining criteria = (1, 1, 3). The details of the 23 experiments and their results are presented in Table 12.

Figure 1 presents the spider diagram for the results of the sensitivity analysis. It can be seen from Table 12 and Fig. 1 that out of 23 experiments, alternative A3 (Supplier 3) has scored highest in 22/23 experiments followed by supplier A1

Table 12 Experiments for sensitivity analysis

S. No.	Definition	Overall scores (CC_i)			Ranking
		A1	A2	A3	
1	$W_{C1–C17}$ = (1, 1, 3)	0.415	0.402	0.431	A3 > A1 > A2
2	$W_{C1–C17}$ = (1, 3, 5)	0.445	0.427	0.459	A3 > A1 > A2
3	$W_{C1–C17}$ = (3, 5, 7)	0.456	0.438	0.471	A3 > A1 > A2
4	$W_{C1–C17}$ = (5, 7, 9)	0.463	0.444	0.478	A3 > A1 > A2
5	$W_{C1–C17}$ = (7, 9, 9)	0.490	0.469	0.506	A3 > A1 > A2
6	W_{C1} = (7, 9, 9), $W_{C2–C17}$ = (1, 1, 3)	0.429	0.430	0.444	A3 > A2 > A1
7	W_{C2} = (7, 9, 9), $W_{C1,C3–C17}$ = (1, 1, 3)	0.425	0.400	0.441	A3 > A1 > A2
8	W_{C3} = (7, 9, 9), $W_{C1–C2,C4–C17}$ = (1, 1, 3)	0.427	0.422	0.435	A3 > A1 > A2
9	W_{C4} = (7, 9, 9), $W_{C1–C3,C5–C17}$ = (1, 1, 3)	0.434	0.414	0.441	A3 > A1 > A2
10	W_{C5} = (7, 9, 9), $W_{C1–C4,C6–C17}$ = (1, 1, 3)	0.421	0.417	0.441	A3 > A1 > A2
11	W_{C6} = (7, 9, 9), $W_{C1–C5,C7–C17}$ = (1, 1, 3)	0.441	0.404	0.441	A3 = A1 > A2
12	W_{C7} = (7, 9, 9), $W_{C1–C6,C8–C17}$ = (1, 1, 3)	0.427	0.409	0.444	A3 > A1 > A2
13	W_{C8} = (7, 9, 9), $W_{C1–C7,C9–C17}$ = (1, 1, 3)	0.394	0.404	0.432	A3 > A2 > A1
14	W_{C9} = (7, 9, 9), $W_{C1–C8,C10–17}$ = (1, 1, 3)	0.429	0.420	0.439	A3 > A1 > A2
15	W_{C10} = (7, 9, 9), $W_{C1–C9,C11–17}$ = (1, 1, 3)	0.427	0.380	0.439	A3 > A1 > A2
16	W_{C11} = (7, 9, 9), $W_{C1–C10,\ C12–17}$ = (1, 1, 3)	0.427	0.430	0.448	A3 > A2 > A1
17	W_{C12} = (7, 9, 9), $W_{C1–C11,C13–17}$ = (1, 1, 3)	0.405	0.400	0.451	A3 > A1 > A2
18	W_{C13} = (7, 9, 9), $W_{C1–C12,C14–17}$ = (1, 1, 3)	0.432	0.413	0.446	A3 > A1 > A2
19	W_{C14} = (7, 9, 9), $W_{C1–C13,C15–17}$ = (1, 1, 3)	0.452	0.407	0.431	A1 > A3 > A2
20	W_{C15} = (7, 9, 9), $W_{C1–C14,C16–17}$ = (1, 1, 3)	0.425	0.413	0.441	A3 > A1 > A2
21	W_{C16} = (7, 9, 9), $W_{C1–C15,\ C17}$ = (1, 1, 3)	0.421	0.402	0.437	A3 > A1 > A2
22	W_{C17} = (7, 9, 9), $W_{C1–C16}$ = (1, 1, 3)	0.427	0.430	0.462	A3 > A2 > A1
23	W_{C8} = (1, 1, 3), $W_{C1–C7,\ 9–17}$ = (7, 9, 9)	0.394	0.404	0.432	A3 > A2 > A1

(1/23) and supplier A2 (0/23). Therefore, we can conclude from these results that our decision process is relatively insensitive to criteria weights with supplier A3 emerging as winner with clear majority. Also, the sensitivities of supplier A1 and supplier A2 are relatively close as they differ by only one vote.

4.3 Results Validation Using Fuzzy SAW

Tables 13 and 14 present the crisp criteria weights for the 17 criteria and the 3 alternatives. The crisp value ($\bar{a}$) for a fuzzy triangular number $\tilde{a} = (a_1, a_2, a_3)$ is obtained using $\frac{a_1+4a_2+a_3}{6}$.

Table 15 presents the normalized weighted alternative scores for the three alternatives for the 17 criteria calculated using Eqs. (15-17). It can be seen that criteria C8 is a cost type criteria, therefore Eq. (16) will be used whereas for the rest of the criteria Eq. (15) will be used for normalization. The overall scores for the

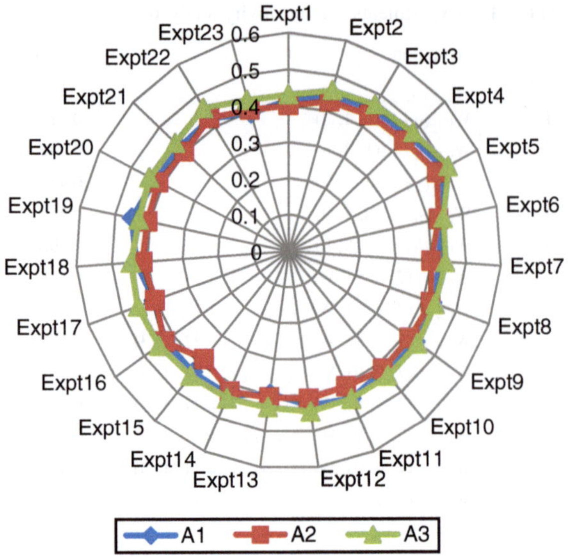

Fig. 1 CCi scores for the three suppliers

Table 13 Crisp weight values

Criteria	D1	D2	D3	Aggregate score	Crisp score
C1	H	M	VH	(3, 7, 9)	6.667
C2	VL	VH	L	(1, 4.333, 9)	4.556
C3	L	L	VH	(1, 5, 9)	5
C4	VL	H	VH	(1, 5.667, 9)	5.444
C5	VL	M	H	(1, 4.333, 9)	4.556
C6	M	VH	M	(3, 6.333, 9)	6.222
C7	H	H	H	(5, 7, 9)	7
C8	M	M	H	(3, 5.667, 9)	5.778
C9	VH	H	H	(5, 7.667, 9)	7.444
C10	M	H	H	(3, 6.333, 9)	6.222
C11	VH	H	H	(5, 7.667, 9)	7.444
C12	VH	H	H	(5, 7.667, 9)	7.444
C13	M	VH	M	(3, 6.333, 9)	6.222
C14	H	H	H	(5, 7, 9)	7
C15	L	H	VL	(1, 3.667, 9)	4.111
C16	M	VL	VH	(1, 5, 9)	5
C17	H	M	L	(1, 5, 9)	5

three alternatives calculated using Eq. (17) can be seen in the last row of Table 15. It can be seen that alternative A3 scores highest followed by A1 and A2. These results are in agreement with Fuzzy TOPSIS results and hence validate the proposed model for supplier quality evaluation.

Table 14 Aggregate and crisp alternative values

Criteria	Aggregate score			Crisp score		
	A1	A2	A3	A1	A2	A3
C1	(1, 5.667, 9)	(3, 7.667, 9)	(1, 5.667, 9)	5.444	7.111	5.444
C2	(1, 4.333, 9)	(1, 2.333, 7)	(1, 5, 9)	4.556	2.889	5
C3	(1, 5, 9)	(3, 5.667, 9)	(1, 3, 9)	5	5.778	3.667
C4	(1, 7, 9)	(3, 5, 7)	(1, 5, 9)	6.333	5	5
C5	(1, 3.667, 7)	(1, 5.667, 9)	(3, 5, 7)	3.778	5.444	5
C6	(3, 7.667, 9)	(1, 3.667, 7)	(1, 5, 9)	7.111	3.778	5
C7	(1, 5, 9)	(1, 3, 9)	(1, 5.667, 9)	5	3.667	5.444
C8	(3, 5.667, 9)	(1, 6.333, 9)	(1, 5, 9)	5.778	5.889	5
C9	(1, 5.667, 9)	(1, 6.333, 9)	(1, 4.333, 9)	5.444	5.889	4.556
C10	(1, 5, 9)	(1, 1, 3)	(1, 4.333, 9)	5	1.333	4.556
C11	(1, 5, 9)	(3, 7.667, 9)	(1, 7, 9)	5	7.111	6.333
C12	(1, 2.333, 5)	(1, 2.333, 7)	(3, 6.333, 9)	2.556	2.889	6.222
C13	(1, 6.333, 9)	(1, 4.333, 9)	(1, 6.333, 9)	5.889	4.556	5.889
C14	(5, 7, 9)	(1, 4.333, 7)	(1, 3.667, 7)	7	4.222	3.778
C15	(1, 4.333, 9)	(1, 4.333, 9)	(1, 5, 9)	4.556	4.556	5
C16	(1, 3, 9)	(1, 3, 7)	(1, 3.667, 9)	3.667	3.333	4.111
C17	(1, 5, 9)	(3, 7.667, 9)	(7.667, 9)	5	7.111	7.444

Table 15 Weighted crisp values for alternatives

Criteria	A1	A2	A3
C1	36.296	47.407	36.296
C2	20.753	13.160	22.778
C3	25	28.889	18.333
C4	34.481	27.222	27.222
C5	17.209	24.802	22.778
C6	44.246	23.506	31.111
C7	35	25.667	38.111
C8	1	0.981	1.156
C9	40.530	43.839	33.913
C10	31.111	8.296	28.345
C11	37.222	52.938	47.148
C12	19.024	21.506	46.320
C13	36.641	28.345	36.641
C14	49	29.556	26.444
C15	18.728	18.728	20.556
C16	18.333	16.667	20.556
C17	25	35.556	37.222
Total	489.580	447.067	494.933
		A3 > A1 > A2	

5 Discussion

In this chapter, we demonstrated successful application of fuzzy multicriteria decision making technique based on TOPSIS for supplier quality evaluation. The results were validated against fuzzy SAW and found to be consistent. However, it is difficult to generalize the results since the results of proposed technique is highly dependent on the data used i.e. number of experts and their rankings which is again dependent on their familiarity and experience with the subject. Therefore, interested readers are advised to interpret these results from methodological perspective and make a careful selection of the model parameters (number of experts, rankings of criteria and alternatives) for their respective works.

Secondly, the criteria proposed for supplier quality evaluation although considered from multiple perspectives are generic in nature to be able to be applied for majority of industries. However, for specific industries such as pharmaceuticals, defense etc. adding more/adapting the proposed criteria is recommended depending on the context under consideration.

Thirdly, fuzzy triangular numbers were used for modeling uncertainties in our study. There is also possibility of investigating other types of fuzzy numbers such as trapezoidal, intuitionistic and various defuzzification techniques to see their impact on final results.

Lastly, the sensitivity analysis showed that for the specific numerical application under study, no changes in the best supplier ranking was observed with variation in weights. The sensitivity analysis can be further extended by incorporating more experiments and/or linking with simulated data sets. There is also possibility of coupling uncertainty analysis with the same to test model robustness.

6 Conclusions and Future Works

This chapter presents a multi-criteria decision making approach based on fuzzy TOPSIS for evaluating supplier quality from multiple perspectives under partial or lack of quantitative information. Four perspectives namely product quality, process quality, organizational quality and service quality are adopted to retain 17 criteria for supplier quality evaluation. These criteria are Quality of service, Reliability, Quality certifications, Employee training, Management commitment, Flexibility, Quality of product, Technical Capability, Statistical process control, Environmental considerations, Nonconformities/defects generated during production, Process Capability Index, Product features, Documentation, Handling of returned material and warranties, Quality in Design of products, and Conformance to Quality standards. Fuzzy TOPSIS is used to aggregate the expert ratings and generate an overall performance score for measuring the quality performance of each alternative (supplier). The alternative with the highest score is finally chosen and is recommended for procurement. Sensitivity analysis to determine the influence of criteria weights on the decision making process. A case study is conducted and results validated against fuzzy SAW technique.

The strength of our approach is the ability to perform supplier quality evaluation under partial or lack of quantitative information and consideration of supplier quality evaluation from multiple perspectives. Since the decision making process is sensitive to the number of participants involved and their expertise with the subject, their selection should be carefully done.

The next step of our work addresses deals with criteria correlation in supplier quality evaluation and quality management in global supply chains.

References

Azar A. T. (2010a). Fuzzy systems. *IN-TECH*, Vienna, Austria. ISBN 978-953-7619-92-3.1.

Azar A. T. (2010b). Adaptive neuro-fuzzy systems. In A. T. Azar (Ed.) *Fuzzy systems*. IN-TECH, Vienna, Austria, ISBN 978-953-7619-92-3.

Bessant, J., Levy, P., Sang, B., & Lamming, R. (1994). Managing successful total quality relationships in the supply chain. *European Journal of Purchasing and Supply Management, 1* (1), 7–17.

Bhutta, K. S., & Huq, F. (2002). Supplier selection problem: A comparison of the total cost of ownership and analytical hierarchy process approaches. *Supply Chain Management, 7*(3), 126–135.

Casadesus, M., & Castrao, R. (2005). How improving quality improves supply chain management: Empirical study. *The TQM Magazine, 17*(4), 345–357.

Chai, J., Liu, J. N. K., & Ngai, E. W. T. (2013). Application of decision-making techniques in supplier selection: A systematic review of literature. *Expert Systems with Applications, 40*(10), 3872–3885.

Chen, K. S., & Chen, K. L. (2006). Supplier selection by testing the process incapability index. *International Journal of Production Research, 44*(3), 589–600. (1366–588X).

Chen, L.-H., & Weng, M.-C. (2002). Using fuzzy approaches to evaluate quality improvement alternative based on quality costs. *International Journal of Quality and Reliability Management, 19*(2), 122–136.

Chen, C.-C., & Yang, C.-C. (2003). Total-costs based evaluation system of supplier quality performance. *Total Quality Management and Business Excellence, 14*(3), 325–339.

Chin, K. S., Yeung, I., & Pun, K. F. (2006). Development of an assessment system for supplier quality management. *International Journal of Quality and Reliability Management, 23*(7), 743–765.

Curcovic, S., & Handfield, R. (2006). Use of ISO 9000 and baldrige award criteria in supplier quality evaluation. *The Journal of Supply Chain Management, 32*(2), 2–11.

de Boer, L., van der Wegen, L., & Telgen, J. (1998). Outranking methods in support of supplier selection. *European Journal of Purchasing and Supply Management, 4*, 109–118.

Dean, A. M., & Kiu, C. (2002). Performance monitoring and quality outcomes in contracted services. *International Journal of Quality and Reliability Management, 19*(4), 396–413.

Dogan, I., & Aydin, N. (2011). Combining bayesian networks and total cost of ownership method for supplier selection analysis. *Computers and Industrial Engineering, 61*(4), 1072–1085.

Dogan, I., & Sahin, U. (2003). Supplier selection using activity-based costing and fuzzy present-worth techniques. *Logistics Information Management, 16*(6), 420–426.

Dubois, D., Prade, H. (1982). The use of fuzzy numbers in decision analysis. In M. M. Gupta & E. Sanchez (Eds.), *Fuzzy Information and Decision Processes* (pp. 309−322). North-Holland: Amsterdam.

Dursun, M., & Karsak, E. E. (2013). A QFD-based fuzzy MCDM approach for supplier selection. *Applied Mathematical Modelling, 37*(8), 5864–5875.

Flynn, B., & Flynn, E. (2005). Synergies between supply chain management and quality management: Emerging implications. *International Journal of Production Research, 43*(16), 3421–3436.

Forker, L. B. (1997). Factors affecting supplier quality performance. *Journal of Operations Management, 15*(4), 243–269.

Foster, S. T. (2007). *Managing quality: Integrating the supply chain*. NJ: Prentice Hall.

Foster, S. T, Jr, & Ogden, J. (2008). On differences in how operations and supply chain managers approach quality management. *International Journal of Production Research, 46*(24), 6945–6961. (1366–588X).

Grieco, J. P. L. (1989). Why supplier certification and will it work? *Production and Inventory Management Review and APICS News, 9*(5), 38–42.

Ho, W., Dey, P. K., & Lockström, M. (2011). Strategic sourcing: A combined QFD and AHP approach in manufacturing. *Supply Chain Management: An International Journal, 16*(6), 446–461.

Ho, W., Xu, X., & Dey, P. K. (2010). Multi-criteria decision making approaches for supplier evaluation and selection: A literature review. *European Journal of Operational Research, 202* (1), 16–24.

Hwang, C. L., & Yoon, K. (1981). *Multiple attribute decision making: Methods and applications*. Berlin, Heidelberg, New York: Springer.

Klir, G. R., & Yuan, B. (1995). *Fuzzy sets and fuzzy logic theory and applications*. Upper Saddle River, NJ: Prentice-Hall.

Ku, C. Y., Chang, C. T., & Ho, H. P. (2010). Global supplier selection using fuzzy analytic hierarchy process and fuzzy goal programming. *Quality and Quantity, 44*, 623–640.

Kuei, C. H., Madu, C. N., & Lin, C. (2008). Implementing supply chain quality management. *Total Quality Management and Business Excellence, 19*(11), 1127–1141.

Kumar, S., & Mahapatra, S. S. (2009). A fuzzy multi-criteria decision making approach for supplier selection in supply chain management. *African journal of business management, 3*, 168–177.

Kumar, S., & Mahapatra, S. S. (2011). Supplier selection in supply chain management: A fuzzy multi-criteria decision-making approach. *International Journal of Services and Operations Management, 8*, 108–126.

Lee, M.-S., Lee, Y.-H., & Jeong, C.-S. (2003). A high-quality-supplier selection model for supply chain management and ISO 9001 system. *Production Planning & Control, 14*(3), 225–232.

Lee, Y.-C., Yen, T.-M., & Tsai, C.-H. (2008). The Study of an integrated rating system for supplier quality performance in the semiconductor industry. *Journal of Applied Science, 8*(3), 453–461.

Liu, X., Li, Y., & Chen, Y. (2012). Supplier selection using axiomatic fuzzy set and TOPSIS methodology in supply chain management. *Fuzzy Optimization and Decision Making, 11*, 147–176.

Parasuraman, A., Zeithml, V. A., & Berry, L. L. (1988). SERVQUAL: A multiple-item scale for measuring consumer perceptions of service quality. *Journal of Retail, 64*, 2–40. (Spring).

Pi, W.-N., & Low, C. (2006). Supplier evaluation and selection via Taguchi loss functions and an AHP. *International Journal of Advanced Manufacturing Technology, 27*, 625–630.

Ramanathan, R. (2007). Supplier selection problem: Integrating DEA with the approaches of total cost of ownership and AHP. *Supply Chain Management: An International Journal, 12*(4), 258–261.

Robinson, C. J., & Malhotra, M. K. (2005). Defining the concept of supply chain quality management and its relevance to academic and industrial practice. *International Journal of Production Economics, 96*, 315–337.

Seth, N., Deshmukh, S. G., & Vrat, P. (2006). SSQSC: A tool to measure supplier service quality in supply chain. *Production Planning and Control, 17*(5), 448–463.

Shu, M. H., & Wu, H. C. (2009). Quality-based supplier selection and evaluation using fuzzy data. *Computers and Industrial Engineering, 57*(3), 1072–1079.

Sroufe, R., & Curkovic, S. (2008). An examination of ISO 9000:2000 and supply chain quality assurance. *Journal of Operations Management, 26*(4), 503–520.

Theodorakioglou, Y., Gotzamani, K., & Tsiolvas, G. (2006). Supplier management and its relationship to buyers' quality management. *Supply Chain Management, 11*(2), 148–159.

Trent, R., Monczka, J., & Robert, M. (1999). Achieving world-class supplier quality. *Total Quality Management and Business Excellence, 10*(6), 927–938.

Vokurka, R., & Lummus, R. (2003). Better supply chains with Baldrige. *Quality Progress, 36*(4), 51–58.

Wang, F. K., Du, T., & Li, E. (2004). Applying six-sigma to supplier development. *Total Quality Management and Business Excellence, 15*(9), 1217–1229.

Watts, C. A., & Hahn, C. K. (1993). Supplier development programs: An empirical analysis. *International Journal of Purchasing and Materials Management, 29*(2), 10–17.

Yang, J. L., Chiu, H. N., Tzeng, G. H., & Yeh, R. H. (2008). Vendor selection by integrated fuzzy MCDM techniques with independent and interdependent relationships. *Information Sciences, 178*(21), 4166–4183.

Zadeh, L. A. (1965). Fuzzy set. *Information and Control, 8*(3), 338–353.

Zimmermann, H. J. (2001). *Fuzzy set theory and its applications* (4th ed.). Boston: Kluwer, Academic Publishers.

Concept Trees: Building Dynamic Concepts from Semi-structured Data Using Nature-Inspired Methods

Kieran Greer

Abstract This paper describes a method for creating structure from heterogeneous sources, as part of an information database, or more specifically, a 'concept base'. Structures called 'concept trees' can grow from the semi-structured sources when consistent sequences of concepts are presented. They might be considered to be dynamic databases, possibly a variation on the distributed Agent-Based or Cellular Automata models, or even related to Markov models. Semantic comparison of text is required, but the trees can be built more, from automatic knowledge and statistical feedback. This reduced model might also be attractive for security or privacy reasons, as not all of the potential data gets saved. The construction process maintains the key requirement of generality, allowing it to be used as part of a generic framework. The nature of the method also means that some level of optimisation or normalisation of the information will occur. This gives comparisons with databases or knowledge-bases, but a database system would firstly model its environment or datasets and then populate the database with instance values. The concept base deals with a more uncertain environment and therefore cannot fully model it beforehand. The model itself therefore evolves over time. Similar to databases, it also needs a good indexing system, where the construction process provides memory and indexing structures. These allow for more complex concepts to be automatically created, stored and retrieved, possibly as part of a more cognitive model. There are also some arguments, or more abstract ideas, for merging physical-world laws into these automatic processes.

Keywords Concept · Tree · Database · Self-organise · AI · Semi-structured · Semantic

K. Greer (✉)
Distributed Computing Systems, Office 2038, PO Box 1213, Belfast BT1 9JY, UK
e-mail: kgreer@distributedcomputingsystems.co.uk
URL: http://distributedcomputingsystems.co.uk

Q. Zhu and A.T. Azar (eds.), *Complex System Modelling and Control Through Intelligent Soft Computations*, Studies in Fuzziness and Soft Computing 319,
DOI 10.1007/978-3-319-12883-2_8

1 Introduction

The term 'concept base' has been used previously (Jarke et al. 1995; Zhao et al. 2007, for example) and has been adopted in Greer (2011) to describe a database of heterogeneous sources, representing information that has been received from the environment and stored in the database for processing. The key point is that the information received from one source does not have to be wholly consistent with information received from another source. The uncertain environment in which it operates, means that information can be much more fragmented, heterogeneous, or simply unrelated to other sources. This could be particularly true in a sensorised environment, when sensors provide a relatively small and specific piece of information. As the sensor-based information would be determined by the random and/or dynamic environment in which it operates, there can be much less cohesion between all of the different input sources. For example, event 1 triggers sensor A with value X and shortly afterwards, event 2 triggers sensor B with value Y. Later, event 1 again triggers sensor A with value X, but instead, event 3 occurs and triggers sensor B with value Z. While the nature of the data is much more random, statistical processes can still be used to try to link related pieces of information. This linked data can then represent something about the real world. The term 'concept' can be used to describe a single value or a complex entity equally and so the concept base can consistently store information from any kind of data source. Intelligent linking mechanisms can be used to try to turn the smaller, more simplistic and separate concepts into larger, more complex and meaningful ones. This is probably also more realistic in terms of what humans have to deal with, in our interaction with the real world.

While information might be input and stored in an ad hoc manner, it is probably the case that some level of structure must firstly be added to the information, before it can be processed, data mined, or reasoned over. When looking for patterns or meaningful relations; then if the data always appears to be random, it is more difficult to find the consistent relations and so a first stage that does this would always be required. This paper looks at a very generic and simplistic way of adding structure to the data, focusing particularly on using whatever existing structure there is, as a guide. Other statistical processes can then use the structure to try to generate some knowledge. Thinking of the sensors or data streams, for example—if it can be determined that concepts A and B usually occur together, while concepts C and D also occur together; knowledge might be able to tell us that when A-B occurs, C-D is likely to occur soon afterwards, or maybe should occur as something else. The current context is to extract this structure from textual information sources, but this is only an example of how the method would work. If consistent patterns can be found, they can be used to grow 'concept trees'. A concept tree is essentially and AND/OR graph of related concepts that grows naturally from the ordering that already exists in the data sources. This paper is concerned with describing the structure of these concept trees and how the process might work. Note that this is at the structure-creation level and not the knowledge-creation level just mentioned.

The rest of this paper is organised as follows: Section 2 describes what type of information might be received and why it can be useful. Section 3 gives examples of related work. Section 4 gives step-by-step examples of how the process might work. Section 5 tries to define the processes formally, as would be done for a database. Section 6 gives some suggestions, relating the structure more closely to nature or the physical world. Section 7 describes how this fits in with an earlier cognitive model and linking mechanisms research, while Sect. 8 gives some conclusions on the work.

2 Adding Structure to Semi-structured Data

Computers require some level of standardisation or structure, to allow them to process information correctly. The problem is therefore how to add this structure, to give the computer system a standardised global view over the data. Even the idea of structure is not certain and can be different for different scenarios. Therefore, obtaining the correct structure probably also means the addition of knowledge to the system. As described in the related work in Sect. 3, this type of modelling started with relational databases (Codd 1970), but then extended to semi-structured and even completely unstructured information. Wikipedia[1] explains that distinct definitions of these are not clear for the following reasons:

1. Structure, while not formally defined, can still be implied.
2. Data with some form of structure may still be characterised as unstructured if its structure is not helpful for the processing task at hand.
3. Unstructured information might have some structure (semi-structured) or even be highly structured but in ways that are unanticipated or unannounced.

The introduction of random events and time elements means that the data sources can also change (Zhang and Ji 2009), requiring statistical or semi-intelligent processes to recognise patterns that cannot be determined beforehand. This could result in a different type of modelling problem than for a traditional database. For one scenario, the designer creates a model of what he/she wishes to find out about and then dynamically adds specific data instances, as they occur, to try to confirm the model. For another scenario, the actual model itself is not known but is derived from an underlying theory. With the second situation, not only are the model values updated dynamically, but the model itself can change in a dynamic and unknown way.

---

[1] http://en.wikipedia.org/wiki/Unstructured_data.

2.1 Types of Data Input

With regard to the text sequences considered in this paper, Greer (2011) describes how a time element can be used to define sequences of events that might contain groups of concepts. A time stamp can record when the concept is presented to the concept base, with groups presented at the same time being considered to be related to each other. This is therefore built largely on the 'use' of the system, where these concept sequences could be recognised and learnt by something resembling a neural network, for example. The uncertainty of the real world would mean that concept sequences are unlikely to always be the same, and so key to the success is the ability to generalise over the data and also to accommodate a certain level of randomness or noise. The intention is that the neural network will be able to do this relatively well. It is also true that there is a lot of existing structure already available in information sources, but it might not be clear what the best form of that is. Online datasets, for example, can be continuous streams of information, defined by time stamps. While the data will contain structure, there is no clearly defined start or end, but more of a continuous and cyclic list of information, from which clear patterns need to be recognised.

As well as events, text might be presented in the form of static documents or papers that need to be classified. For the proposed system, there are some simple answers to the problem of how to recognise the existing structure. The author has also been working on a text-based processing application. One feature of the text processor is the ability to generate sorted list of words from whole text documents. Word lists can also appear as cyclic lists and patterns can again be recognised. This current section of text, for example, is a list of words with nested patterns. In that case, structure could be recognised as a sequence, ending when the word that started the sequence is encountered again. To sort the text, each term in the sequence could be assigned a count of the number of times it has occurred, as part of the sequence. How many times does 'tree' follow 'concept' for example, but a sequence can be more than one word deep. Sequences that contain the same words, or overlap, can be combined, to create the concept trees in the concept base. To select starting or base words, for example, a bag-of-words with frequency counts can determine the most popular ones. The decision might then be to read or process text sequences only if they start with these key words. Pre-formatting or filtering of the text can also be performed. Because this information would be created from existing text documents, the process would be more semantic and knowledge-based. This does not exclude the addition of a time element however and a global system would benefit from using all of these attributes.

The concept trees can then evolve, adding and updating branches as new information is received. Processing just a few test documents however shows that different word sorts of the original data will produce different sequences, from which these basic structures are built, so the decision of correct structure is still quite arbitrary. On the technical front, it might be more correct to always use complete lists of concepts, as they are presented or received, and then try to re-structure the trees that they get added to. Each tree should try to represent some sort of distinct entity and it would be desirable to have larger trees, as this indicates a greater level of coherence. The structure must also be as stable as possible however and so we could try to always add to a base set of concepts, so that the base always has the largest count values. Therefore a triangular structure is realised, with respect to count values, where the base has the largest count, narrowing to the branches. If this basic principle is broken, it might be an indication that the structure is incorrect. Additions to an existing tree should include additions from the base upwards when possible, with new concepts creating new branches if required. It should 'extend' the existing tree along the whole of one of its branches.

2.2 *Structure Justification*

An earlier paper (Greer 2011) gave a slightly philosophical argument that if two concepts are always used together, then at some level they are a single entity. This is a very general rule not related to any particular application, but describes how any sort of entity can be important based on its relevance to the scenario. Consider then the following made-up scenario: There is a farm with a fence in a field. A sheep comes up to the fence and jumps over it. Sensors in the field record this and send the information to the concept base. The concept base recognises the sheep and the fence objects and assigns them to be key concepts in the event. With our existing knowledge, we would always assign more importance to the sheep, but if we had never encountered either object, maybe the sheep and the fence would be equally important to ourselves as well. The scenario continues, where a cow comes up to the fence and jumps over it, then a chicken comes up to the fence and jumps over it. In this case, the fence now becomes the main and key concept. Without the fence, 'none' of the scenarios can occur. A count of these concepts would give the fence the largest total, again suggesting that it is the key concept. The process to combine these scenarios might then compare these stored events and decide that a concept tree with the fence at its base would be the most stable. This process is described further after the related work section, where the addition of existing knowledge is also suggested, to add a natural ordering to things.

3 Related Work

The related work section is able to include topics from both the information processing and AI areas. After introducing some standard data processing techniques and structures, some intelligent methods, relating to nature in particular, are described. It would also be an important topic for problems like data management in the business or online worlds, for example Blumberg and Atre (2003) or Karin et al. (2012). While concepts are the main focus of interest, combining service functionality is more important for the Internet or Cloud at the moment (Aslam et al. 2007; Atkinson et al. 2007, for example). The paper Carr et al. (2001) describes slightly earlier ideas about linked data and marking-up documents on the Internet. It notes how the lines between search and link, or web and database have become blurred and even just searching over metadata tags can be considered as a sort of database operation.

3.1 Ontologies and Semantics

A tree structure, or directed graph, is often used to model text sequences, because it allows for the reuse of sequence paths, extending from the same base. Ontologies are essentially definitions of domains that describe the concepts in that domain and how they relate to each other. A section from the book Greer (2008, Chap. 4) describes that ontologies can be used to represent a domain of knowledge, allowing a system to reason about the contents of that domain. The concepts are related through semantics, for example, 'a car is a vehicle'. For traditional constructions, relations can then be organised into hierarchical tree-like structures. The 'subclass' relation is particularly useful, where the previous example shows that a car is a subclass of a vehicle. There are different definitions of what an ontology is depending on what subject area you are dealing with. Gruber (1993) gives the following definition for the area of 'AI and knowledge representation', which is suitable for this work:

> An ontology is an explicit specification of a conceptualisation. The term is borrowed from philosophy, where an ontology is a systematic account of Existence. For knowledge-based systems, what 'exists' is exactly that which can be represented. When the knowledge of a domain is represented in a declarative formalism, the set of objects that can be represented is called the universe of discourse. This set of objects, and the describable relationships among them, are reflected in the representational vocabulary with which a knowledge-based program represents knowledge. Thus, we can describe the ontology of a program by defining a set of representational terms. In such an ontology, definitions associate the names of entities in the universe of discourse (e.g., classes, relations, functions, or other objects) with human-readable text describing what the names are meant to denote, and formal axioms that constrain the interpretation and well-formed use of these terms.

This is a desirable definition, but because a concept base is constructed slightly differently, the related ontology construction will also be slightly different. The additional knowledge that defines something like 'subclass' is not automatically present, where the system has to determine the correct position, relation and ordering for any concept, mostly from statistics. Because the knowledge is missing however, the relation must also be more simplistic and would probably normally just be 'related to'. It is also worth noting that the future vision of the Web would probably require distributed ontologies. Again from Greer (2008), the future Internet should maybe describe itself at a local level, with larger centralised representations being created by specific applications, based on the domains of information that they typically use. This would naturally happen as part of the Semantic Web. The construction of these ontologies will enable computers to autonomously search the Internet and interact with the services that are provided and is also part of knowledge management on the Internet. The book 'Towards the Semantic Web: Ontology-Driven Knowledge Management' (2003) discusses the ontology construction problem in relation to p2p networks and the Semantic Web. They note that for reasons of scalability, ontology construction must be automated, based on information extraction and natural language processing technologies. However, for reasons of quality, the process still requires a human in the loop, to build and manipulate ontologies. With a slightly reduced knowledge-level, it is intended that the concept base can construct itself almost completely autonomously, giving it a major advantage in this respect.

For dynamic or autonomic systems, the context in which the knowledge is used can become a critical factor. Context is an information space that can be modelled as a directed graph, rather like an ontology. Context allows both recognition and mapping of knowledge, by providing a structured and unified view of the world in which it operates (Coutaz et al. 2005). It is about evolving, structured and shared spaces that can change from one process to the next, or even through the duration of a single process. As such, the meaning of the knowledge will evolve over time. The key lies in providing an ontological foundation, an architectural foundation, and an approach to adaptation that all scale alongside the richness of the environment. Contexts are defined by a specific set of situations, roles, relations and entities. A shift in context corresponds to a change in the set of entities, a change in the set of possible relations between entities, or a change in the set of roles that entities may play. Unfortunately, the context adaptation cannot currently be carried out in a totally automatic way and a concept base would not really consider context in the first instance. It is constructed primarily through statistical counts, but groups of terms presented at the same time can provide some level of context.

By describing the domain in a standardised way, the programs that use the domain will be able to understand what the domain represents. Through this process, different programs on the Internet will be able to learn about each other and form useful associations with other programs that provide the information that they require. This will enrich the knowledge that they can provide, thus turning the Internet into a knowledge-based system, rather than primarily as a source for direct information retrieval. This is of course, a utopian idea that has many possibilities and may never be fully realised.

3.2 Dynamic Databases

As a concept base is a type of database, this is probably the first technology to look at, where the following text is also taken from the book Greer (2008, Chap. 3). Databases are the first kind of organised information system, where the first models were developed in the 1960s. The relational model proposed by Codd (1970) has become the de facto standard and contains a sound mathematical foundation, allowing for optimisation of the storage and retrieval processes. During the 1980s, research on databases focused on distributed models, in the 1990s object-oriented models and then in the 2000s on XML-based models. The distributed models were necessary because of the evolution of the Internet and networking, which meant that distributed sites of related information could now be linked up electronically. The object-oriented models then arose with the invention of object-oriented programming and the theory that object-based models are a preferable way to store and manipulate information. Then with the emergence of XML as the new format for storing and representing information, XML-based models also needed to be considered. While XML is now the de facto standard for describing text on the Internet, meaning that most textual information will soon be stored in that format, it has not replaced the relational model for specific modelling. Neither has the object-oriented model. The increased complexity of these models can make them more difficult to use in some cases, when the mathematical foundations of the relational model remains appealing.

3.3 New Indexing Systems

The recent problems that 'Big Data' provides, linking up mobile or Internet of Things with the Web, has meant that new database structures, or particularly, their indexing systems, have had to be invented. Slightly more akin to Object-oriented databases are new database versions such as NoSql and NewSql (Grolinger et al. 2013), or navigational databases.[2] As stated in Grolinger et al. (2013), the modern Web, with the introduction of mobile and sensor devices has led to the proliferation of huge amounts of data that can be stored and processed. While the relational model is very good for structured information on a smaller scale, it cannot cope with larger amounts of heterogeneous data. It is usually required to process full tables to answer a query. As stated in Grolinger et al. (2013), CAP (Gilbert and Lynch 2002) stands for 'consistence, availability and partition tolerance' and has been developed along-side Cloud Computing and Big Data. 'More specifically, the challenges of RDBMS in handling Big Data and the use of distributed systems techniques in the context of the CAP theorem led to the development of new classes of data stores called NoSQL and NewSQL.' They note that the consistency in CAP refers to having a single up-to-date instance of the data, whereas in RDBMs it means that the whole database is consistent. NoSql now has different meanings and might also be termed 'Not Only SQL'. It can use different indexing systems that might not even have an underlying schema. So it can be used to store different types of data structure, probably more as objects than tables. The database aspect however can try to provide an efficient indexing system, to allow for consistent search

[2] http://en.wikipedia.org/wiki/Navigational_database.

and retrieval over the distributed contents. There are different data models for implementing NoSql. 'Key-value stores' have a simple data model based on key-value pairs, which resembles an associative map or a dictionary. The key uniquely identifies the data value and is used to store and retrieve it from the data store. The data value can be of any type. In 'column-family stores' the data are stored in a column-oriented way. One example might be where the dataset consists of several rows, each of which is addressed by a unique row key, also known as a primary key. Each row is composed of a set of column families, and different rows can have different column families. Similarly to key-value stores, the row key resembles the key, and the set of column families resembles the value represented by the row key. However, each column family further acts as a key for the one or more columns that it holds, where each column consists of a name-value pair. 'Document stores' provide another derivative of the key-value store data model by using keys to locate documents inside the data store. Each document can be highly heterogeneous and so the store can provide the capability to index also on the document contents. 'Graph databases' originated from graph theory and use graphs as their data model. By using a completely different data model to the other 3 types, graph databases can efficiently store the 'relationships' between different data nodes. Graph databases are specialized in handling highly interconnected data and therefore are very efficient in traversing relationships between different entities. NewSql is based more on the relational model, where clients would interact in terms of table and relations. Its internal data model however might be different and there can be semi-relational models as well.

A navigational database is a type of database in which its records or objects are found primarily by following references from other objects. Navigational interfaces are usually procedural, though some modern systems like XPath (2014), can be considered to be simultaneously navigational and declarative. Navigational databases therefore use a tree indexing system and can fall under the graph-based category of NoSql. These graph-based databases therefore look more similar to a concept base or concept tree. While the problems of semi-structured or unstructured data remain, these new databases do offer general architectures and indexing systems. One criticism of graph-based ones however, is that they tend to lead to very messy sets of indexing links that do not have very much structure. This is possible for concept trees as well, but as the concept tree might have a more specific construction process, it can provide some kind of mathematical foundation to help with the organisation.

3.4 Semantic Environment

As well as a sensorised environment, a concept base is also closely related to the Web 3.0, that is, the Semantic Web (Berners-Lee et al. 2001) combined with Service Oriented Architectures (SOA) (OASIS 2014). This is because they can also produce individual pieces of semantic information dynamically and computer-to-computer processing likes to link these up. This would mean that real-time information retrieved from sensors, for example, could be combined with more knowledge-intensive, but static information provided by the Internet, to answer a wider variety of queries. A hierarchical structure is also appealing for reasons of organisation and search efficiency, and so as has been suggested previously by other researchers (Robinson and Indulska 2003), at least a shallow hierarchy would be useful. The largest network of information that we have at the moment is of course the Internet. This is composed of many individual Web sites that contain information by themselves. However, the only relation to other Web sites is through hyperlinks that are typically created by human users. This is really the only way to try and combine the information provided into a meaningful whole. To try and turn the Internet into a network of knowledge, the Semantic Web has thus been invented. With the Semantic Web, the programs that run on the Internet can describe themselves through metadata, which will allow other programs to look them up and be able to understand what they represent. Metadata is 'data about data' and provides extra descriptive information about the contents of a document or piece of information. If this information is available in a machine-readable format, then computer-to-computer interaction will be enabled as well as the typical human-to-computer interaction.

While the Internet is the main source for information, an evolving area is that of mobile devices, including the Pervasive sensorised (Hansmann 2003) or Ubiquitous computing (Greenfield 2006) environments. The mobile environment, by its very nature, is much more dynamic. The Internet contains static Web pages that once loaded will remain on a server, at a site from where they can be located. With mobile networks, devices may be continually moving and so they may connect and disconnect to a network at different locations. Ubiquitous computing is a model of human-to-computer interaction in which information processing has been integrated into everyday objects and activities. An example of this would be to embed sensors into our clothes, to identify us when we went to a particular location. This dynamism actually presents problems to a network that tries to organise through experience. The experience-based organisation requires some level of consistency to allow it to reliably build up the links, but if the structure constantly changes then this consistency may be lost. However, the mobile devices may be peripheral to the main knowledge content. They would be the clients that want to use the knowledge rather than the knowledge providers. For example, in the case of people wearing sensors, it would be the building that they entered that would learn from the sensor information and provide the knowledge, not the people themselves. The sensors would continually be bringing new information into the environment that would need to be processed and integrated. The paper Encheva (2011) also includes the ideas of concept stability and nesting, which are central to the whole problem. The following sections describe how the laws of nature have helped with building these complex structures.

3.5 Underlying Theories and the Natural World

If the model cannot be pre-defined, then it needs to be learned. To do this, the computer program needs to be given a set of rules to use as part of the construction process. For a generic solution, these rule sets are usually quite simplistic in nature. Again taken from Greer (2008, Chap. 1), Complex Adaptive Systems is a general term that would also comprise the sciences of bio-inspired computing. The term Complex Adaptive Systems (or complexity science), is often used to describe the loosely organised academic field that has grown up around the study of such systems. Complexity science encompasses more than one theoretical framework and is highly interdisciplinary, seeking the answers to some fundamental questions about living, adaptable and changeable systems. A Complex Adaptive System (for example, Holland 1995; Kauffman 1993) is a collection of self-similar agents interacting with each other. They are complex in that they are diverse and made up of multiple interconnected elements and adaptive in that they have the capacity to change and learn from experience. One definition of CAS by Holland (1995), one of the founders of this science, can also be found in Waldrop (1993) and is as follows:

> A Complex Adaptive System (CAS) is a dynamic network of many agents (which may represent cells, species, individuals, firms, nations) acting in parallel, constantly acting and reacting to what the other agents are doing. The control of a CAS tends to be highly dispersed and decentralised. If there is to be any coherent behaviour in the system, it has to arise from competition and cooperation among the agents themselves. The overall behaviour of the system is the result of a huge number of decisions made every moment by many individual agents.

The nature of the interactions between the individual entities is the key aspect that distinguishes such complex systems from complicated systems (Al-Obasiat and Braun 2007). A system is called complex if the interactions between its components are not predictable and if it has at least one or more of the following characteristics:

- It is non-linear.
- It is dynamic.
- It is time-variant.
- It is chaotic or stochastic.

All telecommunication networks possess one or more of these attributes. Complicated systems are an alternative type of complex system. However, while complicated systems interact in a predictable way, with CAS, the unpredictable interactions between individual components in the system give rise to 'emergent' behaviour. Emergence is the process of complex pattern formation from simpler rules. An emergent behaviour arises at the global or system level and cannot be predicted or deduced from observing the behaviour of the individual components in the lower-level entities. Because of external forces, concept trees would probably be classified as complex, because their construction is unpredictable.

3.5.1 Mathematical Theories

If one considers the natural world, then Cellular Automata might be thought to be relevant and at some level they provide the required mechanisms. There are different versions of Cellular Automata (Wolfram 1983, for example). They work using a localised theory and entropy (Shannon 1948) could be a key consideration for the structure that is described in the following sections. As described in Wikipedia[3]: In thermodynamics, entropy is commonly associated with the amount of order, disorder, and/or chaos in a thermodynamic system. For a modern interpretation of entropy in statistical mechanics, entropy is the amount of additional information needed to specify the exact physical state of a system, given its thermodynamic specification. If thought of as the number of microstates that the system can take; as a system evolves through exchanges with its environment, or outside reservoir, through energy, volume or molecules, for example; the entropy will increase to a maximum and equilibrium value. The information that specifies the system will evolve to the maximum amount. As the microstates are realised, the system achieves its minimum potential for change, or best entropy state. In information theory, entropy is a measure of the uncertainty in information content, or the amount of unpredictability in a random variable (Shannon 1948). As more certainty about the information source is achieved, the entropy (potential uncertainty) reduces, to a minimum and more balanced amount.

However, it would be difficult to map these types of state machine, or mini-computers, over to a process that is designed only to link up text, to create ontologies. Most distributed systems use some kind of localised theory as well, in any case. The reason for this section is the fact that the dynamic linking uses a basic association equation to create links and also, as described later, makes a decision about breaking a link and creating a new structure. To show their relation to distributed systems and nature, the following quote is from the start of the paper (Wolfram 1983).

> It appears that the basic laws of physics relevant to everyday phenomena are now known. Yet there are many everyday natural systems whose complex structure and behaviour have so far defied even qualitative analysis. For example, the laws that govern the freezing of water and the conduction of heat have long been known, but analysing their consequences for the intricate patterns of snowflake growth has not yet been possible. While many complex systems may be broken down into identical components, each obeying simple laws, the huge number of components that make up the whole system act together to yield very complex behaviour.

If we know what the underlying theory of the system is, then it can build itself in a distributed manner, even if we do not know what the eventual structure will be. Cellular Automata would be too rigid for a concept tree, as they can be created from a fixed grid structure with local interactions only; while a concept tree is required to

[3] http://en.wikipedia.org/wiki/Entropy, plus_(information_theory), _(statistical_thermo-dynamics), or _(order_and_disorder), for example.

create and move structures, as well as link up existing ones. Fractals (Mandelbrot 1983; Fractal Foundation 2014) are also important and cover natural systems and chaos theory. There are many examples of fractals in nature. Using a relatively simple 'feedback with minor change mechanism', the complex systems that actually exist can be created. As described in Wolfram (1983), automata and fractals share the feature of self-similarity, where portions of the pattern, if magnified, are indistinguishable from the whole. Tree and snowflake shapes can be created using fractals, for example. Fractals also show how well defined these natural non-bio processes are already. So automata would belong to the group called fractals and are created using the same types of recursive feedback mechanism. The construction of a concept tree would be a self-repeating process, but the created structures are not self-similar. However, they would result from same sort of simplistic feedback mechanism that these self-similar systems use.

Agent-Based modelling is another form of distributed and potentially intelligent modelling. Scholarpedia[4] notes that Agent-Based Models (ABM) can be seen as the natural extension of the Ising model (Ising 1925) or Cellular Automata-like models. It goes on to state that one important characteristic of ABMs, which distinguishes them from Cellular Automata, is the potential asynchrony of the interactions among agents and between agents and their environments. Also ABMs are not necessarily grid-based nor do agents 'tile' the environment. An introduction to ABM could be the paper Macal and North (2006). Agent-based models usually require the individual components to exhibit autonomous or self-controlled behaviour and to be able to make decision for themselves, sometimes pro-actively. While Cellular Automata would be considered too inflexible, agents would probably be considered as too sophisticated. Although as noted in Macal and North (2006), some modellers consider that any individual component can be an agent (Bonabeau 2001) and that its behaviour can be as simple as a reactive decision.

3.5.2 Biologically-Related

As Artificial Intelligence tries to do, there are clear comparisons with the natural world. Comparisons with or copying of the biological world happens often, but trying to copy the non-biological world is less common, at least in computer science. There are lots of processes or forces that occur in the non-biological world that have an impact on physical systems that get modelled. Trying to integrate, or find a more harmonious relationship between the two, could be quite an interesting topic and computer programs might even make the non-bio processes a bit more intelligent. It might currently have more impact in the field of Engineering and the paper Goel (2013) describes very clearly how important the biological designs are there. With relation to a concept base, a small example of this sort of thing is described in Sect. 6. As noted in Wolfram (1983) and other places, as the second

---

[4] Scholarpedia http://www.scholarpedia.org/article/Agent_based_modeling.

law of thermodynamics implies, natural systems tend towards maximum entropy, or a minimal and balanced energy state. While biological ones tend towards order as well, the non-biological ones tend towards disorder. Cellular Automata are therefore closer to biological systems, where a cross-over is required if non-biological systems are going to exhibit the same levels of intelligence. Although, something like wave motion in the sea shows a steady and consistent behaviour until the wave breaks. So the equations of wave motion can certainly work together in a consistent manner. Even the snowflake shows consistent behaviour for its growth stage, and so on. So with the non-biological systems, a consistent energy input can be the controlling mechanism that triggers specific mechanics. If this is lost or changes, the system can behave more chaotically and may have problems healing or fixing itself. Biological systems might be driven by something more than the specific mechanic, which allows for another level of—possibly overriding—control. Consider the re-join (self-heal?) capability of the concept tree later, in Sects. 4 and 5.

The Gaia theory (Lovelock and Epton 1975) should probably be mentioned. As stated in Wikipedia: 'The Gaia hypothesis, also known as Gaia theory or Gaia principle, proposes that organisms interact with their inorganic surroundings on Earth to form a self-regulating, complex system that contributes to maintaining the conditions for life on the planet'. So the inorganic elements of the planet have a direct effect on the evolution of the biological life. Maybe the inorganic mechanisms suggested here are for smaller individual events, than the more gradual self-regulation of a large global system. Although in that case, they could still be the cause for global changes.

4 Concept Tree Examples

The examples provided in this section show how the concept trees can be built from text sequences. They also describe some problems with the process and the proposed solutions. With these examples, it is more important to understand the general idea than consider them as covering every eventuality. Section 5 then tries to give a more formal definition, based on these examples. To show how a tree is created, consider the following piece of text: *The black cat sat on the mat. The black cat drank some milk.* If punctuation and common words are removed, this can result in the following two text sequences:

Black cat sat mat
Black cat drank milk

From this, as illustrated in Fig. 1, a tree can be built with the following counts. The base set of 'black cat' can be extended by either the set 'sat mat' or the set 'drank milk'. The base 'black cat' concept set has been found from a sort that starts with these terms, where combining the two sets of terms then reinforces the base.

It also appears when constructing these trees that sets of counts should in fact balance, unless additions with missing information are allowed. The counts for the

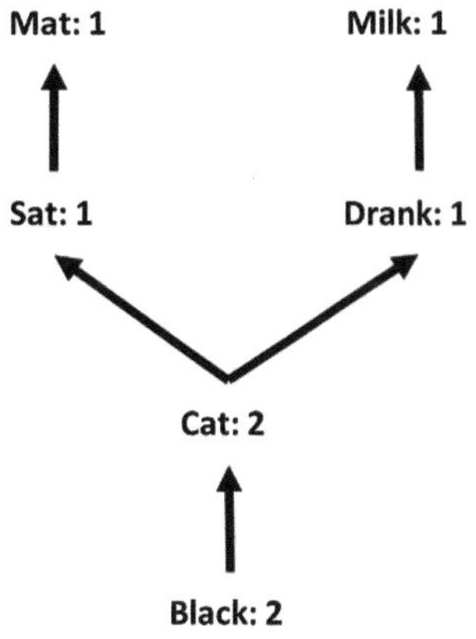

Fig. 1 Concept tree generated from two text sentences

immediate child nodes should add up to the count for the parent. If, for example, a new list starting with 'sat mat' was allowed to be added, it would only increment counts higher up the tree, altering the tree's balance. If this caused the triangular rule to be broken, a re-structuring process, starting where the count becomes larger again should 'prune' the tree and create a new one, with the more stable branch at its base. As will be suggested in the rules of Sect. 5, in fact, if the trees are always constructed from the base up, this particular problem will not exist.

4.1 Combining on Common Branches

If the following text was also stored in the concept base: *The thirsty boy drank some milk. The thirsty elephant drank some milk.* This could result in two more concept sequences:

Thirsty boy drank milk
Thirsty elephant drank milk

To add these to the concept base data structure, the process might firstly create two new trees, one starting with 'thirsty boy' and another with 'thirsty elephant'. However, the terms 'drank milk' have now become the most important overall and therefore should be at the base of a tree. Adding to Fig. 1, the 'drank milk' branch of the 'black cat' tree should be pruned and added with the other two 'drank milk' sequences, to start a new tree, with a count of 3, as shown in Fig. 2.

It would then be necessary to add links between the trees, where they were related. Links can be created using indexing or unique key values, for example. The structure of each tree can then develop independently and so long as they exist, any links between them will remain, giving some level of orderly navigation. So why separate the concepts and not just extend the 'black cat' tree? One reason is the new base concept sets of 'thirsty boy' and 'thirsty elephant', creating a new tree by

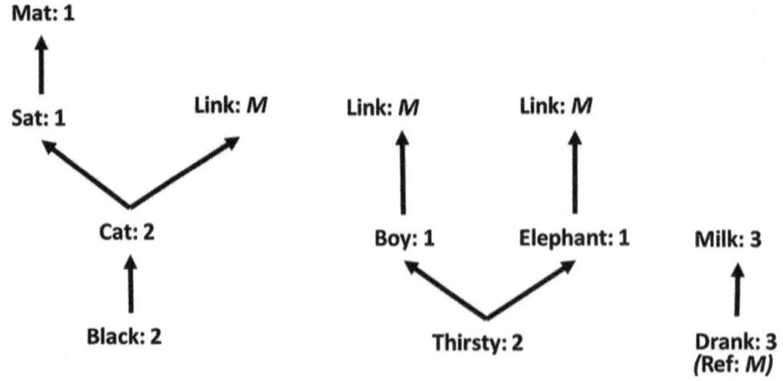

Fig. 2 Example of 'pruning' and optimising the structure when new trees are added

default. Separating and linking is also an optimisation or normalisation feature. If the process concludes that 'drank milk' is an important concept in its own right, it should not be duplicated in different places, but rather it should be stored and updated in one place and referenced to by other trees. The understanding of 'drinking milk' is the same for all 3 animals. Then of course, also the triangular count rule.

4.2 Combining with Unrelated Branches

Another situation would be if the concept base then receives, for example, 3 instances of: *The thirsty elephant drank milk and ate grass*. This would automatically add 'ate grass' to the 'drank milk' tree and the count would still be OK. The sequence 'ate grass' however, only relates to the original elephant branch in this case. There is no indication that the boy or cat ate grass. The final solution to this is another new tree and also a new indexing key, to link the elephant with both the 'milk' and the 'grass' trees. This might happen however after a stage of monitoring (see Sect. 4.3). The other tree branches keep the original index, where Fig. 3 shows what the new set of structures might look like. Note the way that existing links only

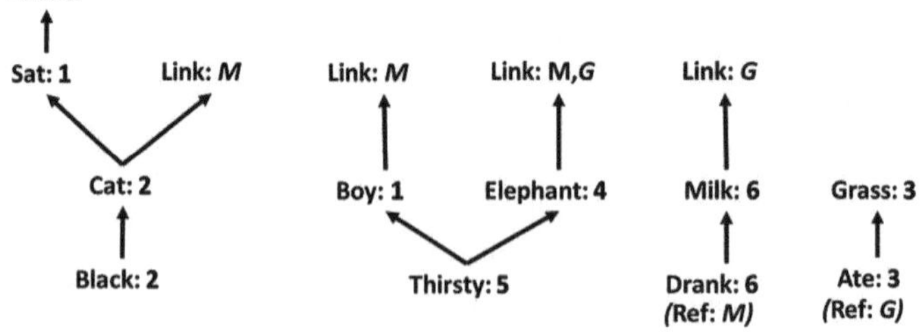

Fig. 3 New tree and key value changes the indexes

need to be traversed if the parent tree also has the key value; and it is more important to note the types of structure and links that get created, rather than the exact semantics of the whole process. For example, start with the first linking keyset, remove a link key if it gets used and only continue if the next link key is in the list.

Not shown in the figure, the key sets might possibly be similar to the NoSql column-family stores, with a primary key retrieving a set of secondary keys. The secondary keyset would then allow for the traversal of the linked database trees. The primary key can relate to an external entity that is interested in certain groups of concepts. As each tree is quite separate, the secondary keys could, for example, relate or point to sets of base tree concepts. In this case, graph-like navigation is also possible, as leaf nodes can link to other tree base nodes as well, where the secondary keyset helps to define the allowed starting paths.

4.2.1 Query Example

A query process is still required to access the tree contents and can be related to the keysets that exist. For example, if the cat also starts to eat grass, a primary key that points to the 'Thirsty' tree, might then include the 'Black' tree in its secondary keyset as well. The 'Cat' link now also includes the G graph link. Without any additional information, the traversal can return either the cat or the elephant for milk and grass. This might need to be specified, because another similarly indexed primary key could accept, cat, boy or elephant for just the milk concept. It might also be interesting to consider that the secondary keyset allows for the completion of circuits through the graph trees. For example, if the first primary key also includes the 'Ate' tree (and maybe 'Drank') in its secondary keyset and requires all indexed trees to be true; then only cat and elephant can be returned. This is just a possibility and is related to suggestions about intelligence in Sect. 7.2. It does however look like useful searches might require a set of conditions, but also allow for some automatic reasoning, where a query language is possible. For this example, it looks like a Horn clause could be used (Jarke et al. 1995; Greer 2011, for example).

4.3 Compound Counts

Another requirement is to break the structure up again. If the concept base, for example, received concept sequences of 'drank milk with a long trunk', with no relation to any of the animal tree base concepts; then the process should just add 'long trunk' to the 'drank milk' base. This is not necessarily incorrect because that specific information does not state that the cat or the boy do not have long trunks. If the information is then used and the cat or boy trees traversed, it will incorrectly return that they have long trunks. Some type of compound count can be used to check for this. A *positive:negative* compound count, for example, can indicate that the tree is possibly incorrect. If the negative count becomes too large, then the 'drank milk long trunk' tree can be split and a link added between them instead. The new link key could then get added to some primary keysets, to allow for its traversal. The trick is in being able to determine when the information is untrue. Does the information need to be returned and then evaluated as untrue, or can the update process allow for this to be recognised automatically? For example, a new text sequence of 'thirsty boy drank milk' is again added to the database, where it updates the related tree nodes. As it stops short in the 'drank milk' branch, any nodes further up that branch can have their negative count incremented instead. As the elephant tree reinforces the positive count here however, this is then an indication that the tree should be split, as the information is true sometimes and false other times. A break in any tree would automatically create a new indexing key as well. This would be sent to all related trees that can then decide what link keyset best relates to them—the one for just 'drank milk' or the one that also links to 'long trunk'. New entries can therefore be flagged in the first instance, until they become reliable.

There is also a process of reasoning and adjustment here, again over a period of time. Even if a tree branch is not proven to be incorrect, in some scenarios, a negative count might be required to indicate that part of a tree is no longer relevant to the current entity or scenario. More traditionally, a decay factor can be used to determine this. If, for example, the link is rarely used, its value decays until it is lost completely. If it is used so infrequently, then it might as well not be present, even if it is not false. So this is another alternative link update mechanism that could be added, but a compound key helps to decide to split rather than remove completely. With a single value that gets incremented or decremented, you have to judge how many times each has occurred. If there is a compound count, then this is clear and it is easy to tell if the branch is true as well as false.

4.4 Re-join or Multiple References

A final consideration might be the re-joining of one tree to another and also a problem with multiple reference links to the re-joined part. As the data can be random, it might initially be skewed in some way and force a tree to break into two trees. Over time this evens out and the original single tree can become correct again. This is determined by the fact that the counts become consistent with a single tree structure again. In that case, it would be possible to re-join the previous branch that is now a base, back onto the first tree. The only worry would be if there are also multiple references to the second tree that had a branch broken off. These references might not relate to the original tree as well. It might not be good practice to allow arbitrary references half-way up a tree and so if the previous branch has a different set of references now, then maybe it must stay as a base. Ideas here would therefore include transferring back only some of the new tree, while keeping the rest as the second tree, with its base. The next section tries to explain this again, but more formally. This might be the most ‘intelligent’ part of the process, as a re-join can be compared to a self-heal or fixing process. The non-bio systems would typically not do this and therefore continue to a more chaotic state.

5 Formal Specification

The concept tree idea, for a concept base, has a restricted construction process. It is based on a frequency count with a very strict rule about relative count sizes. It might therefore be possible to define the construction process more formally and even bring some standardisation or normalisation; where other similar techniques, such as Navigational Database or NoSql, are not able to. The following sets of declarations might be useful to standardising the process and bring added order to the structures. Initial tests have confirmed some of these rules, but are not variable enough to test all of the possible scenarios. Because the rules are more of a logical argument than a mathematical proof, they are listed as groups of points.

5.1 General

1. A concept tree can represent different types of entity. The entity however should be a whole concept. For example, it might be a single object in the real world, or a single action. Therefore, the base concepts in any tree are the ones that would be used first in any scenario.
2. Tree structures require that every child node has a count that is the same or less than its parent. This should always be the case if the linking integrity is maintained, unless branches are allowed to re-join.
3. Whenever possible, the process would prefer larger trees for the following reasons:
   (a) A larger tree has more meaning as a general concept and gives added confidence when reasoning over its group of nodes.
   (b) A larger tree gives more coherence to the concept base.
   (c) Larger trees mean less of a trend towards a chaotic structure.
4. Normalisation would like each concept to exist only once and so, also for this reason, the whole process tries to find what the main concepts are and place them as base concepts to trees. As with traditional databases, if a concept exists somewhere only once, then it only needs to be updated in one place. This is difficult or even impossible however, for every scenario:
   (a) If the concept gets used for different contexts, then its meaning and relation to other concepts changes slightly, when it needs different link sets.
   (b) For a distributed system over a large area, it might simply not be practical to have the concept at one place only and be able to find it.

   (c) Even if trees have similar branches, a link might be required if other factors do not allow a join.

5. Indexing and linking can use key sets, but it can also include graph-based navigations. This is because the structure is tree-based, with links between tree nodes defining concept sequences.

5.2 Truth Tests

1. For a tree to exist, every node in it must be true. That does not mean that every node is evenly used, but there should be no false information. This extends to being true for any entities that link to the tree or related sub-tree. If any part is false for any linking entity, then the tree needs to be split.
2. Note the difference between a part of a tree that is rarely used and a part or path that is false or untrue. Rarely used is OK, but untrue is not.
3. A set of links to a tree from different entities might make parts of the tree untrue, when it then needs to be split at the false branch. It might however be quite difficult to determine if a path is untrue, as information retrieval scenarios might mean that the path simply does not get traversed. So the type of count key can be important and the trick is to be able to recognise when tree paths are rarely used, or are simply false.
   (a) There could be a time-based degradation of a link, for example. If it degrades so much as to remove it, then it has never been used and so it is not relevant, even if it is not false.
   (b) Or possibly the counting mechanism's 'group:individual' count (Greer 2011), that reinforces a count, both for the concept group and the individual. This can determine when individual nodes no longer appear to be the same as others in the group.
   (c) Or there is a 'positive:negative' count, when the negative count can become too large.
   (d) There could also be a time-based count that measures when events happen at the same time. This is important for recognising when trees can be re-joined.

5.3 Tree Comparisons

1. Tree comparisons and updates are made using groups of concepts that represent individual input events. The event group is considered to be a complete entity itself and gets stored in the concept base as that. It is then also compared with the other structures as that, where it needs to match with existing tree paths in one of two ways:

   (a) If it matches exactly from the base of another tree up any branch, then it can be added to that tree.
   (b) If its' base matches to a different node of another tree, then a link between the two trees can be created.

2. If a smaller independent entity is added as a branch to a larger one, then it will not be possible to access it without going through the larger entity first. This means that the normal process of reconstruction will be to break at a tree branch and move to a tree base, with the other direction being used less often.
3. Re-structuring will therefore also prefer to link between trees than to re-join them permanently. This is because a link provides the appropriate navigation, while the base nodes still remain for each tree, allowing them to be accessed directly.
4. While linking is more practical, coherence would prefer permanent joins to create larger trees and so under the correct conditions, a join should be preferred, where any doubt would lead to a dynamic link instead. The re-joining process requires more intelligence, which may be why it would be a more difficult automatic process.

5.4 Linking or Joining

1. Any reinforcement of an existing tree, based on adding a new group of concepts, should always start from the base node.
   (a) If it would start part of the way up a tree, then the process should form a new tree instead.
   (b) Similarly, when a new path is added to an existing tree, it must start from the base only and traverse through any sub-path up to any leaf node. It can then extend that leaf node if required.
2. For a linking example, if we have two trees—tree 1 and tree 2, then:
   (a) The tree 2 is simply added as is, with a possible node link to tree 1 and subsequent events can try to change or combine the structures further.
   (b) This would be more in line with the general theory, but the idea of entropy or concept coherence would prefer the next scenario if possible.
3. For a permanent join example, if we have two trees—tree 1 and tree 2, then:
   (a) The tree 1 can be split at the node related to tree 2's base node and that branch combined with tree 2 for the new structure.
   (b) Less likely is a permanent join the other way, but it is still possible. For example, if tree 2 has a path from the base up that matches to a branch in tree 1. Then if the counts are OK, the path can be moved from tree 2 to tree 1.

4. Linking related nodes is always possible. Different keysets can then define, for different entities, whether they traverse all of the links or not.
5. Re-joining trees needs to consider the base entity links more.
   (a) Breaking off a branch from tree 1 to join to tree 2 at its base would be easier.
      (i) If tree 2 has all of the entity links that tree 1 has, then the join can be automatic.
      (ii) If tree 2 has additional entity links, then a compound count can be added, because it might still be unclear if the branches, new to some entities, are false.
      (iii) If the broken branch however is completely contained in the tree 2 path, then the join can be automatic.
   (b) Re-joining tree 2, or part of tree 2 from its base, to a tree 1 branch is more difficult.
      (i) If tree 2 does not have any entity links to its base, then it can be added to any other matching branch.
      (ii) If tree 2 has a different set of entity links, then its base must be accessible and so it cannot be removed.
      (iii) If tree 1 and tree 2 have the same set of entity links, then a join should be attempted. A check might be performed to determine if the two trees are always accessed together. If that is the case, then they can be joined over some common branch or node.
6. Unclear is when one branch has additional elements inside of it, so that the branch it is being compared with, would need to be extended internally and not at an edge. This is quite common with ontologies, for example. This might favour breaking the larger branch at the two points where the new nodes exist, creating 3 trees and linking between them. The first tree uses only two of the new trees while the other uses all 3.
7. A truth test might check if a join is preferable to a link, including branches not defined as false, but possibly now out of character and can be moved.
8. So there could be a statistical, or even a reasoning process that decides what join action to take and this could be different for different implementations.

6 Relation to Nature

This section gives some more comparisons with natural laws and is about trying to justify the proposed construction mechanism, by showing that it will give the best possible balance to a concept tree, with the minimum amount of additional intelligence or knowledge required. It is reasonable to think that in the random or chaotic world that we live in, there is no reason to always link from a larger 'measurement' to an equal or smaller one. This is however the main rule of the concept tree and there is some mathematical justification or foundation to it. Some of the evidence was found after the creation of the concept tree, more than the concept tree has been derived from it. However, if it can be used to support the general model or theory, then why not specify it here. The main point to note is the fact that base concepts should probably be the most frequently occurring ones statistically. That is probably a sound enough idea, based on statistics alone. If trying to compare to a real-world physical law, then if tree branches were allowed to become larger again, the tree would probably break at that place. This might be an opportunistic statement, but it is completely the idea behind the triangular counting rule. Other pieces of evidence that might provide support are listed in the following sections.

6.1 Problem Decomposition

Any sub-entity must be smaller than the entity it belongs to. This is particularly relevant to the process of problem decomposition that is used to solve large and difficult problems. The larger problem is broken down into smaller ones, until each smaller problem is simple enough to be solved. So this is another application of the natural ordering. It is also the case that you cannot be a sub-concept of something that does not exist. If thinking about Markov models, then one construction of these will count the number of occurrences, of transitions from one state to another. This process will necessarily require the 'from' state to exist first and therefore, if the model is tree-like without loops, each parent state must have a larger or the same count value as the following state, as part of the same rule. Similar to concept trees, Markov models have been used for text classification or prediction, as well as state-based models.

6.2 Clustering and Energy

Some of the research that has looked at clustering processes, for example, the single link theory (Sibson 1973) might provide support. This original theory proposed that any node should link to its closest neighbour. These small clusters could then link to their nearest neighbours in the next iteration, and so on. Therefore, through only one link from each group, at each iteration, larger clusters can eventually be formed. It is interesting to note that if there is a certain ordering of the nodes, this process will work particularly well. A measurement of closeness depends on what is being measured and also the evaluation criteria. However, suppose that spatial distance is the metric, where a line of evaluated nodes can only cluster with the node on either side—necessarily being the closest nodes. Consider the two sets of nodes, represented by Figs. 4 and 5. In these figures, each node value is represented by its height in the graph and each node position in the cluster space, is represented by its position in the graph.

In Fig. 4, the nodes have a perfect descending order. Using the process of linking to your closest neighbour, this could lead to the whole set of nodes creating a single cluster in one go. In Fig. 5, the perfect ordering is broken, where node 3 will link with node 2 only. This forces two clusters to be formed, or forces a break in the sequence. We also have the idea of a minimal energy, or entropy (Shannon 1948). This has already been used to cluster or sort text documents, for example Decision Trees (Quinlan 1986) and the principle of entropy can also be applied to a concept tree. If one considers the simplistic sorting mechanism in Fig. 4 again, it can be seen that the most efficient sort, causing the least amount of energy to move from one place to the next, is in fact the uniform decreasing of the entity lengths, from largest to smallest. If each energy change is 1 unit, then a total of 7 units are required. Any change in this order, for example Fig. 5, would require a larger amount of energy to traverse all of the entities—9 units in this case. As natural systems like lower energy states, a self-organising system might favour the lower energy state. This therefore supports the idea of not adding larger counts to smaller ones, because the required energy amount for the same entity set increases, as in Fig. 5. It could increase and then decrease uniformly, but in general, it would support the rule. Entropy also

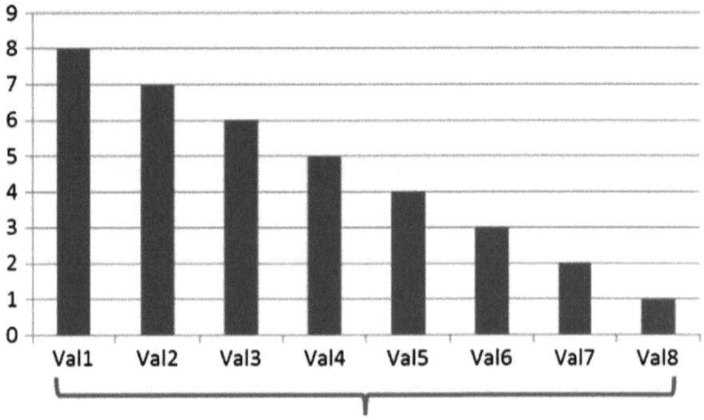

Fig. 4 Energy of 7 required to traverse all elements

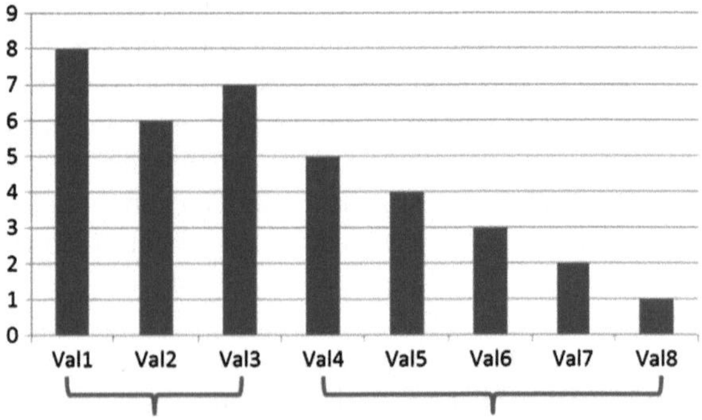

Fig. 5 A single change increases the energy amount to 9

deals with the problem of micro-states, where possibly Fig. 4 has only one and Fig. 5 has two, but Fig. 1 is better because the whole dataset is more coherent and it is already at its minimal state.

6.3 Language Structure

Another comparison should be with how we process natural language. Language is so fluent that there are not many restrictions on what can be said or written. As the concept tree is a simplified model of natural language however, it might allow some rules to be included. Generic or autonomous rules are desirable and also plausible. They might be thought of as an extra layer above the basic statistical counts that help to direct the initial structure. They would not be allowed to override the triangular count rule however. The ordering used in WordNet[5] (Fellbaum 1998; Miller 1995), for example, is the sort of ordering that would be useful. The base, for example, could typically be formed from nouns and verbs, with adjectives or adverbs forming mostly the leaf nodes or end branches. In a real-world sense, the descriptive words would possibly define specific instances of the more grounded noun or verb concept groups. For example, 'the black cat sat on the mat', gives a count of 1 initially to each concept and so the ordering before adding to a tree could be changed. The rule might state to add 'cat' at the base instead of 'black', as it is an object. Then possibly some sort of reverse polish notation 'cat—mat—sat' to push nouns down, or just 'cat—sat—mat'. So the exact language structure might get lost, but the associations will still exist and the rules will help to reconstruct text sequences from the tree. As another example, we can have a cat and a mat, but maybe only the 'black' cat sat on a 'red' mat, and so on. Descriptive nodes at the end would also help to relate the concept tree more closely with earlier work, as described in Sect. 7.

6.4 Natural Weight

The following, associated with size or weight, is possibly even more interesting. It would be a strange way of looking at ordering text, but it is again a physical-world rule being applied in a slightly different context and again relates to the idea of sub-concepts. Note that text often relates to real world objects and so its construction would have to be consistent with the physical world. In the real world, it is often the case that the largest and therefore heaviest entity, resides at the bottom of things. Putting a heavier object on-top of a lighter one is not often done and so there is a

[5] I have to note my recent interest in WordNet, although, most of the new theory here was formulated before that, with WordNet then supporting it.

natural order here. It might be possible to use this knowledge, as part of the tree structure, without requiring more sophisticated natural language or ontology understandings. For example, every event that takes place, takes place on planet earth. If we were creating a structured ontology, planet earth would be at the bottom. Then, for example, a car always drives on a road and so a road should link to a car, not the other way around. It might be the case that the car branch, when it gains relations to lots of other things, would be broken off to form a new base, but it still makes more sense to link from road to car and not car to road. So this ordering, based on some knowledge of relative size or use in the real world, might also become part of a structuring rule. It would be useful because the related context-specific information should not be very sophisticated and so it might be possible to apply the knowledge automatically again. We just need to know that there is a car and a road, for example. One could imagine a large database that stores different bands of entities, grouped simply by size or weight, that are not allowed to be ordered before/after another entity. It is not a typeOf or subClass relation, but a more functional one. Maybe something like relative use, but it is really only the ordering that is required. This fixed ordering would again be a secondary aid, where the statistical counts and dynamic relations of the parsed text would still have the most influence. The trees of Figs. 1, 2 and 3 might have their ordering changed slightly, for example, but the word groups and concept associations would still be determined by the dynamic text, not fixed knowledge. For example, the mat should probably be placed before the cat, when the cat branch could be broken off later. It might be 'mat − cat − black + sat', or something.

7 Relation to Earlier Work

This section is slightly different, looking at a specific cognitive model, rather than general theories. It is helpful for developing that cognitive model further and will hopefully add ideas for a more intelligent system, but can be skipped if the database model is specifically of interest. Earlier research by the author has looked at how a whole cognitive model might be developed from very simple mechanisms, such as stigmergic or dynamic links (Greer 2008, 2013b). The earlier work described how a reinforcement mechanism can be used to determine the reliability of linked source references in a linking structure. These links are created through user feedback only and are therefore very flexible, as the feedback can be much more variable than static rules can accommodate. User feedback adds the intelligence of the user, which a rule set might not contain. While concept trees are also built from user feedback, they are then constrained by pre-determined rules and knowledge. They are also more semantic, complementing the event instances of the earlier work. A concept tree could therefore be created from similar source types—sensor-based, dynamic input, specific concepts, but deal more with the existing structure than the events that created it. It is still possible to make comparisons with earlier work on a neural network model (Greer 2011) that clustered without considering semantics, but blindly presumed that input presented at the same time must simply be related. The original cognitive model did include an ontology or knowledge-base reference, to provide this type of support. Some comparisons with bio-related models in general can also be made.

7.1 Biological Comparisons

More recent work again (Greer 2013a) has put Hebb's well-known theory of 'neurons that wire together, fire together', into a computer model. It has added a mechanism using the idea that when they fire together they may be attracted to each other and grow links to join up. The rules of Sect. 5.4 (maybe point 5.b) fit in well with this as it suggests comparing the sets of input links to trees. If both trees have the same set of input links, then when these fire, both trees will be activated and can therefore decide to join up. It does not however suggest exactly how they might grow towards each other or combine in a biological model. The earlier cognitive model (Greer 2008, 2013b) defines a 3-layer architecture, where the bottom level links for optimisation purposes, the middle layer links to aggregate these pattern groups, while the top layer links to create higher-level concepts and trigger dynamic events. As the concept trees theory is more about aggregating and balance, over all of the data, it is more suited to this middle level. It has also been noted in the formal specification of Sect. 5 that a concept might be duplicated, simply because of the distributed nature of the system. This is also the case for the human brain, as it is known to duplicate information and the practical aspects of trying to access a particular brain region might make it easier to simply duplicate some information locally. Ideas of entropy and automatic monitoring can also be related to both the stigmergic/dynamic linking model (Greer 2008, Sect. 8.5) or the concept trees. As either system develops, it will tend towards some sort of fixed structure. This trend would then only be broken by a change in the input state. So, after the formations are created, any more dramatic changes might indicate a change in data, and so on. This idea probably applies to most dynamic systems with similar designs.

7.2 Higher-Level Concepts

There is a reference to a linking structure in Greer (2008, Sect. 9.3.7, Fig. 24) that describes how linked concepts might only be related or activated if they are assigned specific values. For example, if we have mother, son and uncle concepts linked, then it might only be true if the mother is called Susan, the son is called John and the uncle is called David. The idea of pushing the descriptive text to the leaf nodes, so as to represent specific instances, has been written about in Sect. 6.3. There is also a reference to another linking structure in Greer (2008, Sect. 9.3.7, Fig. 25) that tries to index different concept sets through unique keys. It has the

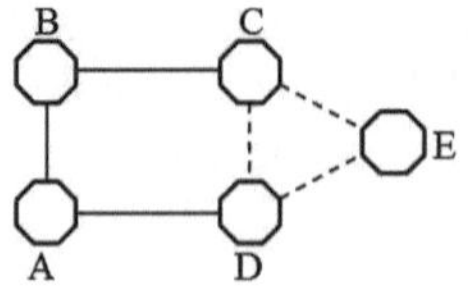

Fig. 6 Example network with two higher-level concepts A-B-C-D and C-D-E (Greer 2008)

same idea as the indexing system being used here and a diagram of this is shown in Fig. 6.

The nodes are meant to represent concepts and groups of them, higher-level concepts. However, because there can be overlap between concepts they can be grouped together, with different indexes defining each exact group. If concept trees were used, a tree consisting of A-B-C-D could link to a tree consisting of E only, for example. The ABCD tree would have a base node with some value and then branches, one of 'A to B' and one of 'C to D'. An event entity would then need to activate the base node of the tree and activate all of its branches, to realise the first concept group. To realise the second group, a different event entity would need to link to and activate both the ABCD tree and the E tree, at the same time. Then possibly and interestingly, can a link between the two trees themselves complete a circuit, to indicate the other concept group of CDE. If a link between the leaf D node and the base E tree node exists, for example? This might be a more dynamic model than the original design of Fig. 6 that considered fixed sets of (unique) reinforced links only. The key sets possibly sit on-top of the linking structure, where both can change dynamically. So there are two different possibilities for dynamic change, but with the new functionality, there are also other technical difficulties. The intention is that concept groups will represent something more meaningful and therefore can be used as part of a reasoning process. This paper would suggest that it is more of a memory structure, but with the same goal of defining higher-level concepts more accurately, to allow them to be reasoned over.

7.3 Complementary Structures

Dynamic links have therefore been used previously (Greer 2008, 2011) as part of a neural network, but the two techniques are probably compatible. It is curious that the knowledge-based concept tree, in relation to the 3-level cognitive model described in Greer (2008, Sect. 9.3.8 or 2013b), would be more closely associated with the first optimising level and the second aggregating level. It would create the base or bed of the system. The experience-based neural network would then be more closely associated with the third level. It would manipulate the knowledge (cleverly) through a dynamic, experience-based approach. Looking at the concept trees has actually helped to create a clearer picture and provide some more consistency over the whole model. If the concept trees are used to create pattern groups in the middle level, then it makes sense for them to have a main base concept that defines the tree, with branches or sub-groups that define its contents. It also makes sense for the construction process to start at the general base node and work through to smaller and more specific details at the leaf nodes. It also makes sense that it is more knowledge-based. The earlier neural network model (Greer 2011) also creates a hierarchical structure, but it was noted that the construction process there might start with the leaf or individual nodes that are then aggregated together into a main or higher-level concept. That neural network model was associated more with the third level of the cognitive model that deals more with dynamic events and triggers. If the concept groups there are based on events, then it could make sense that a reader of those would receive input as small events instances in time. Each event could be some knowledge, defined by some structure. The events would then be aggregated together into something more singular and maybe even learned. They are based on time and external forces, where learning and predicting is also important. But this then gives more sense to the architecture overall and allows for the two hierarchical construction directions to be OK. In a general sense, we already know this. As stated in Greer (2008, Sect. 4.8), with regard to service-based networks: different industries would prefer either a top-down or a bottom-up approach to organisation. Top-down starts with a central component and then adds to it when required. Bottom-up starts with simpler components and then combines them to provide the more complex organisation. If you want more control then a top-down approach is preferred. If you allow a more chaotic but independent organisation, then maybe bottom-up is preferred. It is the same argument for the cognitive model. Top-down relates to knowledge-based concept trees and in this context also, to small but specific entities. Bottom-up relates to the event-based clustering and also to self-organising these smaller structures. As an example, you could imagine a human seeing a tree and learning about its different components or varieties; but when out walking on a stormy day, learning in a different way to avoid falling branches when under a tree in high winds. Or following earlier papers' food examples, you could imagine a human tasting different food types and learning what they are made of; but when in a restaurant, selecting a menu based on the food types and discovering some new recipe through the experience.

8 Conclusions and Future Possibilities

This paper has introduced two new ideas of concept trees and concept bases. The concept base is a more general device that is the storage program for the trees. It is also responsible for sorting or creating the trees, and for managing the index and link sets. The concept trees are described in more detail and even formally defined. The counting rule that is introduced in this paper and probably a different construction method, make the concept tree a bit different to other graph-based techniques. The addition of some rules helps to standardise the construction process and give it some mathematical foundation. The idea of only allowing a narrowing structure with respect to count values is probably a good one, because it is statistically consistent and also consistent with the real world. Ideas from nature or the physical world support this and are interesting, but should probably not be taken too seriously. They could introduce a very light form of intelligence, although a light form of knowledge is required first. Any concept is allowed to be a main one and this is defined by an automatic count. The rule set can then give additional structure independently, but it is still the presented data that determines what trees get built. Problems with the process might include the creation of a long list of very short trees that represent nothing in particular by themselves. This then begins to look a bit like the standard memory storage on a computer, with pointers between pieces of memory linking them up. There is however the possibility of building larger more meaningful trees as well. A comparison, or relation, with Markov models has been introduced because they are known to work well and may exhibit the same statistical counting property.

The construction process builds hierarchies automatically and these can represent any type of concept. A slightly weaker idea is therefore to try to build service-based business processes or compositions in the same way, where the earlier stigmergic links were suggested for the same task. See, for example (Greer 2008, Sect. 7.3.2.1), or maybe (Aslam et al. 2007) or (Atkinson et al. 2007). While real-world concepts or natural language might be restricted by sets of relations that can justify the triangular counting mechanism, more complex business processes might not be. There is a difference between a sub-process and linking two independent processes. In that case, statistical counts would be used purely for reliability, but it is a known problem and several solutions that are at least semi-automatic, have already been suggested. It is worth noting that the count values could be used as probability values, or something similar, as each tree is a bit self-contained. If a particular structure was presented to a network and one of the concepts was missing, the system could try to calculate a probability value, indicating the confidence that the missing concept was in fact an error. This would be an automatic way to assign a value range to the stored data, for security reasons, or other. So concept trees can also be looked at in terms of automatically creating process hierarchies and really does span from the large Internet-based network to the small cognitive model.

Not every group of concepts should be added either and dynamic factors like reinforcement and time can also be considered. So while the construction process is automatic, a reasoning component might also make certain decisions. For example, does a link to another newly created tree actually apply to my instance? If a real tree is taken as the natural world model, and why not, then it obeys the rule that a heavier branch will cause a lighter one to snap. The new AI part then is the idea of an intelligent indexing and linking system, to keep consistency between the split trees. This means that even if the original structures disintegrate, while the natural world entity would tend to chaos, the linked elements will allow for traversal through specific channels and maintain the order. The question would be how efficient or accurate the structure can be.

The idea to use this as part of an indexing and memory structure is optional, but it would fit in well with the cognitive model written about in earlier papers (Greer 2008, 2011, 2013a, 2013b). The whole process could mostly be performed automatically, with minimum existing knowledge. The earlier model diagrams are relevant enough to be compared with the concept tree directly and even compliment it. This research is still a work in progress and the hope is to be able to provide more substantive results in the future.

Disclosure This paper is an updated version of a paper called 'Concept Trees: Indexing and Memory from Semi-Structured Data', originally published on DCS and Scribd, June 2012.

References

Al-Obasiat, Y. & Braun, R. (2007). A multi-agent flexible architecture for autonomic services and network management. In *IEEE/ACS International Conference on Computer Systems and Applications, AICCSA'07* (pp. 132–138). ISBN 1-4244-1031-2.

Aslam, M. A., Shen, J., Auer, S. & Herrmann, M. (2007). An integration life cycle for semantic web services composition. In *Proceedings of the 2007 11th International Conference on Computer Supported Cooperative Work in Design* (pp. 490–495).

Atkinson, C., Bostan, P., Hummel, O. & Stoll, D. (2007). A practical approach to web service discovery and retrieval. In *IEEE International Conference on Web Services (ICWS 2007).*

Berners-Lee, T., Hendler, J. & Lassila, O. (2001, May). The semantic web: A new form of web content that is meaningful to computers will unleash a revolution of new possibilities. *Scientific American.*

Blumberg, R. & Atre, S. (2003, February). The problem with unstructured data. *DM Review* (pp. 42–46).

Bonabeau, E. (2001). Agent-based modeling: Methods and techniques for simulating human systems. *Proceedings of the National Academy of Sciences, 99*(3), 7280–7287.

Carr, L., Hall, W., Bechhofer, S. & Goble, C. (2001). Conceptual linking: Ontology-based open hypermedia. In *WWW10* (pp. 334–342), Hong Kong.

Codd, E. F. (1970). A relational model of data for large shared data banks. *Communications of the ACM, 13*(6), 377–387.

Coutaz, J., Crowley, J. L., Dobson, S., & Garlan, D. (2005). Context is Key. *Communications of the ACM, 48*(3), 49–53.

Encheva, S. (2011). Lattices and patterns. In *Proceedings of the 10th WSEAS International Conference on Artificial Intelligence, Knowledge Engineering and Data Bases (AIKED'11)* (pp. 156–161), Cambridge, UK.

Fellbaum, C. (Ed.). (1998) *WordNet: An electronic lexical database*. Cambridge, MA: MIT Press.

Fractal Foundation. (2014). http://fractalfoundation.org/. Accessed 25 February 14.

Gilbert, S., & Lynch, N. (2002). Brewer's conjecture and the feasibility of consistent, available, partition-tolerant web services. *ACM SIGACT News, 33*(2), 51–59. doi:10.1145/564585.564601.

Goel, A. K. (2013). Biologically inspired design: A new program for computational sustainability. *IEEE Intelligent Systems, 28*(3), 80–84.

Greenfield, A. (2006). *Everyware: The dawning age of ubiquitous computing* (1st ed.). Berkeley, CA: New Riders Press. ISBN 0321384016.

Greer, K. (2008). Thinking networks—The large and small of it: Autonomic and reasoning processes for information networks. published with LuLu.com, 2008. ISBN 1440433275. Also available on Google books.

Greer, K. (2011). Symbolic neural networks for clustering higher-level concepts. *NAUN International Journal of Computers, 3*(5), 378–386 [extended version of the WSEAS/EUROPMENT International Conference on Computers and Computing (ICCC'11)].

Greer, K. (2013a). New ideas for brain modelling. Published on arXiv at http://arxiv.org/abs/1403.1080, also on Scribd.

Greer, K. (2013b). Turing: Then, now and still key. In: X-S. Yang (Ed.), *Artificial intelligence, evolutionary computation and metaheuristics (AIECM)—Turing 2012*. Studies in Computational Intelligence. Berlin: Springer.

Grolinger, K., Wilson, A. H., Tiwari, A., & Capretz, M. (2013). Data management in cloud environments: NoSQL and NewSQL data stores. *Journal of Cloud Computing: Advances, Systems and Applications, 2*(22), 1–24.

Gruber, T. (1993). A translation approach to portable ontology specifications. *Knowledge Acquisition, 5*, 199–220.

Hansmann, U. (2003). *Pervasive Computing: The mobile word*. Berlin: Springer. ISBN 3540002189.

Holland, J. (1995). *Hidden Order: How adaptation builds complexity*. Reading, MA: Perseus.

Ising, E. (1925). A contribution to the theory of ferromagnetism. *Zeitschrift für Physik, 31*(1), 253–258.

Jarke, M., Eherer, S., Gallersdorfer, R., Jeusfeld, M. A., & Staudt, M. (1995). ConceptBase—A deductive object base manager. *Journal on Intelligent Information Systems, 4*(2), 167–192.

Karin, M., Prasad, M. D., Atreyee, D., Ramanujam, H., Mukesh, M., Deepak, P., Reed, J. & Schumacher, S. (2012). Exploiting evidence from unstructured data to enhance master data management. In *Proceedings of the VLDB Endowment The 38th International Conference on Very Large Data Bases* (Vol. 5(12) pp. 1862–1873). Istanbul: Turkey.

Kauffman, S. A. (1993). *The origins of order: Self-organization and selection in evolution*. Oxford, UK: Oxford University Press.

Lovelock, J. & Epton, S. (1975). The quest for Gaia. *New Scientist Magazine*. Available on Google Books.

Macal, C. M. & North, M. J. (2006). Tutorial on agent-based modelling and simulation part 2: How to model with agents. In L. F. Perrone, F. P. Wieland, J. Liu, B. G. Lawson, D. M. Nicol, & R. M. Fujimoto. (Eds.), *Proceedings of the 2006 Winter Simulation Conference*.

Mandelbrot, B. B. (1983). *The fractal geometry of nature*. New York: Macmillan.

Miller, G. A. (1995). WordNet: A lexical database for English. *Communications of the ACM, 38* (11), 39–41.

OASIS. (2014). http://www.oasis-open.org. Accessed 25 Jan 2014.

Quinlan, J. R. (1986). Induction of decision trees. *Machine Learning, 1*, 81–106.

Robinson, R. & Indulska, J. (2003). Superstring: A scalable service discovery protocol for the wide-area Pervasive environment. In *The 11th IEEE International Conference on Networks, ICON2003* (pp. 699–704). ISSN 1531-2216, ISBN 0-7803-7788-5.

Shannon, C. E. (1948). A mathematical theory of communication (continued). *The Bell System Technical Journal, 27*(4), 623–656. ISSN 0005-8580.

Sibson, R. (1973). SLINK: An optimally efficient algorithm for the single-link cluster method. *The Computer Journal (British Computer Society), 16*(1), 30–34.

Towards the Semantic Web: Ontology-driven Knowledge Management. (2003). In John Davies, Dieter Fensel, Frank van Harmelen (Eds.), Wiley. ISBN 0470858079, 9780470858073

Waldrop, M. M. (1993). In L. Sternlieb (Ed.), *Complexity: The emerging science at the edge of order and chaos*.

Wolfram, S. (1983). Cellular Automata, *Los Alamos science*.

XPath. (2014). http://www.w3.org/TR/xpath/. Accessed 10 Mar 2014.

Zhang, Y., & Ji, Q. (2009). Efficient sensor selection for active information fusion. *IEEE Transaction on Systems, Man, and Cybernetics—Part B: Cybernetics, 10*(3), 719–728.

Zhao, J., Gao, Y., Liu, H., & Lu, R. (2007). Automatic construction of a lexical attribute knowledge base. In Z. Zhang & J. Siekmann (Eds.), *Proceedings of Second International Conference, KSEM 2007*, Melbourne, Australia (pp. 198–209). LNAI 4798 Berlin: Springer.

Swarm Intelligence Techniques and Their Adaptive Nature with Applications

Anupam Biswas and Bhaskar Biswas

Abstract Swarm based techniques have huge application domain covering multiple disciplines, which include power system, fuzzy system, forecasting, bio-medicine, sociological analysis, image processing, sound processing, signal processing, data analysis, process modeling, process controlling etc. In last two decades numerous techniques and their variations have been developed. Despite many variations are being carried out, main skeleton of these techniques remain same. With diverse application domains, most of these techniques have been modified to fit into a particular application. These changes undergo mostly in perspective of encoding scheme, parameter tuning and search strategy. Sources of real world problems are different, but their nature sometimes found similar to other problems. Hence, swarm based techniques utilized for one of these problems can be applied to others as well. As sources of these problems are different, applicability of such techniques are very much dependent on the problem. Same encoding scheme may not be suitable for the other similar kind of problems, which has led to development of problem specific encoding schemes. Sometimes found that, even though encoding scheme is compatible to a problem, parameters used in the technique does not utilized in favor of the problem. So, parameter tuning approaches are incorporated into the swarm based techniques. Similarly, search strategy utilized in swarm based techniques are also vary with the application domain. In this chapter we will study those problem specific adaptive nature of swarm based techniques. Essence of this study is to find pros and cons of such adaptation. Our study also aims to draw some aspects of such problem specific variations through which it can be predicted that what kind of adaptation is more convenient for any real world problem.

A. Biswas (✉) · B. Biswas
Department of Computer Science and Engineering, Indian Institute of Technology (BHU), Varanasi 221005, India
e-mail: anupam.rs.cse13@iitbhu.ac.in; abanumail@gmail.com

B. Biswas
e-mail: bhaskar.cse@iitbhu.ac.in

Q. Zhu and A.T. Azar (eds.), *Complex System Modelling and Control Through Intelligent Soft Computations*, Studies in Fuzziness and Soft Computing 319,
DOI 10.1007/978-3-319-12883-2_9

1 Introduction

Advancement of technology has led real world problems to become complex and more challenging. To acquire requisite quality of such advanced technology, associated problems needed to solve intelligently and efficiently. In these days intelligent techniques become very popular for solving such technology oriented problems. There are variety of ways to solve a single problem, so it is very crucial to decide exactly which cases an intelligent technique needs to be adopted. There are few aspects of such decision. Firstly, the problem in hand has to be feasible for an intelligent technique. There has to be a plug point in the problem where such techniques are to be plugged in. Secondly, even if problem is found suitable for intelligent techniques, the best possible technique has to choose from the archive of numerous techniques available. Lastly, which of the multiple versions of selected technique will be most suited for the problem has to be decided.

Before considering any intelligent technique, feasibility analysis of the problem as well as available techniques have to be carried out. Once suitable technique is found, arises another key issue in implementation of the technique with respect to the problem. As far as swarm intelligence techniques are concerned, mostly were developed for solving optimization problems. Again when we say optimization problem, it covers huge domain. There are different types of optimization problems and have special characteristics of each. Very basic notion of optimization is to find best possible solution from a set of solutions (referred as solution space) to any problem. Corresponding problem with solution space can be summarized with some functions, generally referred as objective function or fitness function. Some problems require constraints along with the functions to define solution space, that case problem is referred as constrained optimization problem. If problem is defined with linear objective functions and constraints, problem is called linear optimization problem, otherwise it is termed as nonlinear. Hence, optimization problems can be categorized as linear or nonlinear on the basis of linearity in the problem definition. Objective functions can be continuous or discrete, accordingly problems are referred as continuous and combinatorial optimization problem respectively. Most of the real world problems experience several constraints, sometime those constraints are defined with nonlinear functions. Often such problems require multiple objective functions to optimize with necessary constraints, referred as multi-objective optimization problem. Solution space may have several preferred solutions, each of them represents best solution and cannot be dominated by other. In this case we have best solution set instead of one best solution, such problems are referred as multi-modal optimization problems. Determination of possibles best solution set is very important to engineering problems, but due to constraints present in the problem, best solutions may not always be realized. Both single and multi-objective problem experiences such hurdles along with diverse constraints and linearity. Todays technology oriented problems become more complex with these issues associated with the problem.

Each kind of optimization problem can be solved very efficiently with some specific techniques. These techniques include mathematical approaches such as linear programming (Matoušek and Gäartner 2007; Todd 2002; Wiki 2014), non-linear programming (Borwein and Lewis 2010; Ruszczynski 2006) and iterative techniques (Wiki 2014) as well as numerous heuristic approaches such as Evolutionary Algorithms (Rechenberg 1994; Schwefel 1994; Yan et al. 2005), Genetic Algorithms (Deb et al. 2002; Holland 1975), Swarm Intelligence (Dorigo 1992; Karaboga 2005; Kennedy and Eberhart 1995; Rashedi et al. 2009; Shah-Hosseini 2008, 2009) and other nature inspired methods. In these days heuristic approaches have gained popularity specially Genetic Algorithms and Swarm Intelligence techniques. Applications of almost all fields utilize swarm intelligence techniques in their specific problems. As mentioned above, swarm intelligent techniques are utilized mainly for solving optimization problem. These techniques are inspired from chemical, biological and physical phenomenon of nature. Extensively used approaches in various applications include Particle Swarm Optimization (PSO) (Kennedy and Eberhart 1995), Ant Colony Optimization (ACO) (Dorigo 1992), Artificial Bee Colony (ABC) (Karaboga 2005), Gravitational Search Algorithm (GSA) (Rashedi et al. 2009) and Intelligent Water Drop (IWD) (Shah-Hosseini 2008, 2009). In this chapter we will try to cover these popular swarm based techniques in perspective of their applicability to numerous problems. These techniques have undergone several changes. We will study those changes with respect to their applications and try to draw key issues behind such changes. Variety of applications of different domains as well as frequent variation within the technique create chaos. Induced several confusions regarding selection of suitable technique for the application. We will try to address those issues and summarize them in a generalized manner for all the applications. We will also try to generalize issues regarding variations of techniques and their applicability throughout the chapter.

Rest of the chapter is organized as follows: Section 2 surveys various works which presents applications of swarm based techniques and extrapolates trade off between applications and swarm based techniques. Section 3 provides brief introduction to popular swarm based approaches and generalize these techniques into common framework. Section 4 addresses issues related objectives behind incorporation of swarm based approaches, probable plug points in any application and suitable problem types. Considering these constraints a generalized framework is presented for any application and such techniques. Section 5 presents various encoding schemes of applications to fit swarm based techniques and encoding related changes in techniques. Section 6 explains strategic changes of swarm based techniques. Section 7 illustrates parameter tuning related issues of swarm based techniques. Section 8 discusses various application related problems, advantages and difficulties. Finally, concluded in Sect. 9.

2 Related Work

Several application centric studies on Swarm Intelligent (SI) techniques have been done. Main focus of this brief survey is to cover the kind of works related to the application have been done rather than covering those works and draw application related issues. Numerous studies have been done related to application and SI techniques. Most of those studies were specific to a particular approach (Al Rashidi and El-Hawary 2009; Borwein and Lewis 2010; Chandrasekhar and Naga 2011; Janacik et al. 2013; Kameyama 2009; Kothari et al. 2012; Kulkarni and Venayagamoorthy 2011). Some studies are even more specific to single problem of one particular application domain (Al Rashidi and El-Hawary 2009; Chandrasekhar and Naga 2011; Kulkarni and Venayagamoorthy 2011). There are several applications in single domain which have utilized SI techniques (Al Rashidi and El-Hawary 2009; Kulkarni and Venayagamoorthy 2011).

Several surveys on various applications of SI techniques covering multiple domain have been done (Chandra Mohan and Baskaran 2012; Monteiro et al. 2012). Such studies are often done by considering a single variant of any SI technique such as multi-objective (Reyes-Sierra and Coello 2006), parallel versions (Chu et al. 2003; Schutte et al. 2004; Vanneschi et al. 2012) and so on to different application domains. With these studies one can have an intuitive idea about problems of various domains which could absorb such approach. Problem specific study, associated issues of the problem and study of solution to that problem which covers numerous techniques including SI techniques have often done (Al Rashidi and El-Hawary 2009; Kulkarni and Venayagamoorthy 2011). Such study gives an overall idea of various techniques to solve that problem, but internal matters of those techniques remain unclear, which is the source of confusion regarding which one of the available techniques will be better for the problem.

Generalized studies that covers all application domains where intelligence techniques are utilized, have not been done yet. Studies done earlier also not based on applicability. No clear idea has been come out from previous studies about the way which would be better for incorporating any intelligent techniques into an application. Numerous studies about applications and swarm techniques (Al Rashidi and El-Hawary 2009; Chandra Mohan and Baskaran 2012; Chandrasekhar and Naga 2011; Eslami et al. 2012; Kameyama 2009) have been done. But, study related to usability of any intelligent technique in an application, like what points have to be taken care of, which changes have to be done in application in order to fit the technique in hand has not been done yet. At present, a single approach have several variants and each variant has specific characteristics (Chandra Mohan and Baskaran 2012; Kameyama 2009; Monteiro et al. 2012). Suitable variants of any SI technique for any particular application has not been done yet. Problems faced (Yang et al. 2007) during implementation of SI techniques into an application and solutions to those problems are given. Whether those problems can be arise in other application or if arises whether same solution can be suitable for that application, such issues have not been surveyed yet. Off course, different variants of SI techniques proposed, addition of new parameters and tuning of various parameters have often been surveyed (Kameyama 2009), sometime such study extended to various applications.

Intuition behind introduction of a new technique to solve any problem is to apply in real life problems and solve associated problems more efficiently. Hence, generalized study would be more realistic and helpful to the society of diverse application domain rather than any particular application specific study. SI techniques in particular covers huge application domain. In this case, generalized study with comparative analogy to the applications of multiple domain will be more helpful. Implementer of any domain can have an intuitive idea about applicability of such techniques into the specific applications of any domain.

3 Swarm Intelligences

Swarm intelligent (SI) techniques are heuristic stochastic search processes. SI approaches can be generalized as follows: all approaches are initiated with a set of solutions called population, then in successive steps each candidate of the set learns collectively from other candidates and adapts itself in accordance to the solution space. Strategy incorporated and learning mechanism of these techniques mostly mimic the natural facts and phenomena. Such nature inspired mathematical models can be plugged into one framework. In this section we will brief popular SI techniques and try to wrap up in one generalized framework.

3.1 Particle Swarm Optimization

Particle Swarm Optimization (PSO) originally introduced by Kennedy and Eberhart in 1995 (Kennedy and Eberhart 1995). Basic intuition behind PSO was simulation of cooperative learning mechanism of bird's flocking. Flying birds in flock show learning through individual's experience and follow other. One of them leading the flock and other follows that leader. Once leader changes, immediately all other individual including previous leader begin following it. This process continues until reach their destination. Kennedy and Eberhart formulated this process into a mathematical model with two very simple equations. One of those equations was analogous to the position and other one was analogous to the velocity of bird or particle. Experience of individual particle was conserved as personal best i.e. any particle experienced best position so far. Experience of flock or swarm or population was conserved as global best i.e. the best position experienced by the swarm so far. These experiences were used to learn and control velocity of particle. Finally, particles moved to new position with learned velocity. So each particle in solution space are associated with position and velocity. Velocity and position equation proposed by Kennedy and Eberhart in original PSO are shown below:

$$V_i(t+1) = V_i(t) + C_1R_1[X_{pb}(t) - X_i(t)] + C_2R_2[X_{gb}(t) - X_i(t)] \quad (1)$$

$$X_i(t+1) = X_i(t) + V_i(t+1) \quad (2)$$

Here, t denotes discrete time steps, $X_i(t)$ denotes the position vector of particle i and $V_i(t)$ denotes the velocity vector of a particle in the solution space at time step t, $X_{pb}(t)$ be the position vector of particle i's personal best position so far and $X_{gb}(t)$ be the position vector of global best particle so far. R_1 and R_2 are the vectors of uniform random values in range (0, 1). Parameters C_1 and C_2 are the cognitive acceleration coefficient and social acceleration coefficient respectively. $X_i(t+1)$ and $V_i(t+1)$ denote new position and velocity vector at time step $t+1$ respectively.

This strategy can be easily fitted into the generalized SI process as shown in Fig. 1. Only requirement is to represent particle set as vectors of position and velocity. Once we have initial velocity and position of each individual particle, we can update iteratively with strategic Eqs. (1) and (2) until the attain desired approximated position. Actual strategy behind the PSO lies in these two equations. This basic version of PSO has been modified significantly over the years to improve performance, those have been studied in coming sections in perspective of applications.

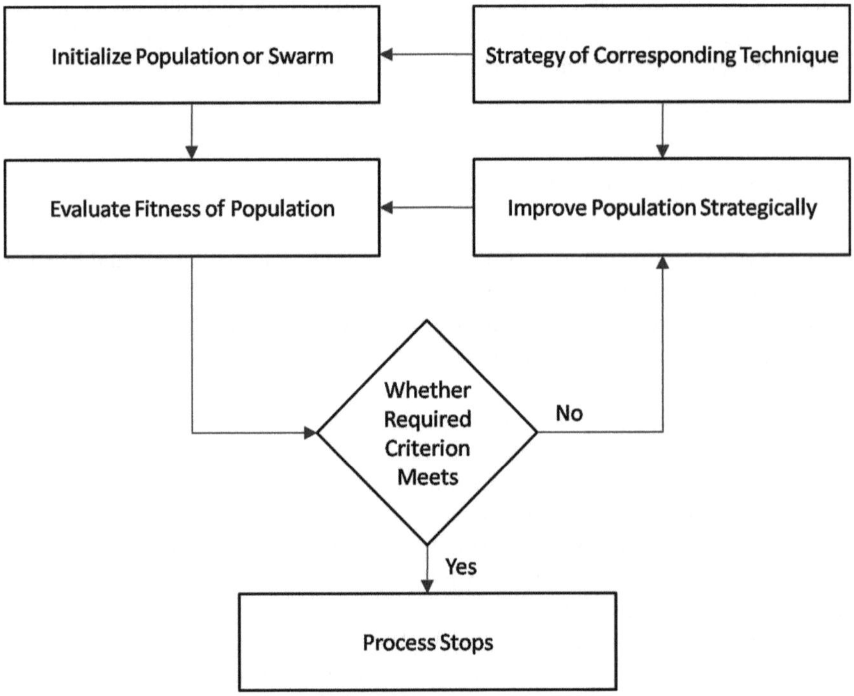

Fig. 1 Generalized flow diagram of swarm intelligence technique. Strategy of any SI technique can be inserted into the generalized process once population is initialized in accordance of the strategy

3.2 Ant Colony Optimization

Ant colony optimization (ACO) is inspired by foraging behavior of ants. Ants deposit pheromone to mark the favorable path and that path is followed by other member of the colony to collect foods. Over the time pheromone evaporates and hence, relative attraction to that specific path decreases. As much time an ant takes to travel to and fro from source to destination, proportionate amount of pheromone evaporates by that time. Summarily smaller path takes less time, which implies less evaporation, so density of pheromone becomes higher. Therefore, if one ant finds a good (i.e., short) path from the source to destination, other ants are more likely to follow that path. Artificial ant system has to do two main tasks: updating pheromone and selection of path with maximum pheromone densities to reach destination. To acquire ACO method, solution space is represented with graph. The pheromone τ_{ij}, associated with the edge joining nodes i and j, is updated as follows:

$$\tau_{ij} \leftarrow (1-\rho)\cdot\tau_{ij} + \sum_{k=1}^{m} \Delta\tau_{ij} \tag{3}$$

where, ρ is the evaporation rate, m is the number of ants, and $\Delta\tau_{ij}$ is the quantity of pheromone deposited on edge (i, j) by ant k. $\Delta\tau_{ij}$ is measured with a ratio Q/L_k if ant k used edge (i, j) in its tour, if not considered 0 value, where Q is a constant, and L_k is the cost of the tour constructed by ant k.

ACO adopts probabilistic approach to select path. This probability depends on priori desirability represented by attractiveness η_{ij} of the move and posteriori τ_{ij} of the move, indicating how desirable it has been in the past to make that particular move. When an ant k is at node i then the probability of going to node j is given by the equation below:

$$p_{ij}^{k} = \frac{\tau_{ij}^{\alpha}\cdot\tau_{ij}^{\beta}}{\sum_{j\in allowed_j}\tau_{ij}^{\alpha}\cdot\eta_{ij}^{\beta}} \tag{4}$$

Here, τ_{ij} is the amount of pheromone deposited for moving from node i to j, α is a parameter to control the influence of τ_{ij} which takes greater than equal to 0 value, β is another parameter to control the influence of η_{ij} which takes value greater than equal to 1. η_{ij} is the attractiveness of edge (i, j) which is considered as inverse of cost of the edge (i, j).

In each iteration ants add new transition to construct final solution and update pheromone level in the path. Once solutions are represented with graph and initialize ants accordingly, ACO can be fitted in general framework.

3.3 Artificial Bee Colony

Karaboga introduced Artificial Bee Colony (ABC) (Karaboga 2005) algorithm by modeling foraging behavior of bee. Artificial bee colony have groups of three kind of bees employed bee, onlooker bee and scout bee. Employed bee and onlooker bee are responsible for determining amount of food present in the food source. One employed bee is assigned for every source of food. Employed bee become scout if food source assigned to it is exhausted.

Basic strategy of ABC consists three steps: (1) move the employed and onlooker bees onto the food sources to determine food amounts, (2) determine scout bees from employed bees, and (3) direct scout bees to search new food source. Position of food source found by scout bees represents a possible solution to the problem that has to be optimized. Onlooker bees are placed probabilistically to each food source. As the quality of food source increases, the probability with which the food source chosen by onlooker bees is also increases. Natural scout bees have no guidance for exploring new food sources, but artificial bee can explore solution space as a task to find new food sources strategically. This has led to fast discovery of feasible solution that are represented by food sources. If within the predetermined number of trials a solution representing a food source does not improved, then that food source is abandoned by associated employed bee and converted to a scout.

This strategy can be incorporated into generalized SI framework by interpreting group of solutions as artificial bees. Divide population into three groups as mentioned above and perform task accordingly.

3.4 Gravitational Search Algorithm

Main inspiration of Gravitational search algorithm (GSA) was taken from Newton's law of universal gravitation and laws of motion. Every point in solution space is considered as a searching agent having mass. Searching agents of GSA are considered as objects and their performance is measured by their masses. Due to gravitational force each object attracts other objects. Object with lower masses will be attracted towards the object with heavier mass. Heavy masses represent comparatively better solution and moves slowly. Another fact of physics is also considered, according to which objects in space do not feel uniform gravitational force due to varying gravitational constant. Actual value of gravitational constant G depends on the age of the universe. So gravitational constant $G(t)$ at time t can be computed as follows:

$$G(t) = G(t_0) \times \left(\frac{t_0}{t}\right)^{\beta}, \quad \beta < 1 \tag{5}$$

Here, $G(t_0)$ is the value of the gravitational constant at the first cosmic quantum-interval of time t_0. This varying $G(t)$ is used to compute total force F_i^d exerted on agent i from direction d in the space. Due to this force agents are accelerated towards each other. Acceleration of any agent i having mass M_i at time t in direction d is computed with the equation presented below:

$$a_i^d(t) = \frac{F_i^d(t)}{M_i(t)} \tag{6}$$

The position velocity of each agent is updated with laws of motion as follows:

$$v_i^d(t+1) = R \times v_i^d(t) + a_i^d(t) \tag{7}$$

$$x_i^d(t+1) = x_i^d(t) + v_i^d(t+1) \tag{8}$$

Here, $x_i^d(t+1)$ and $v_i^d(t+1)$ are the position and velocity of agent i in direction d at time $t+1$. R is an uniformly distributed random variable within range $(0, 1]$. This strategy can be fitted into generalized framework of SI by considering population as set of searching agents.

3.5 Intelligent Water Drop

Intelligent Water Drops (IWD) algorithm was introduced by Shah-Hosseini (2008). IWD simulates flow of river water. It seems that natural river often follows favorable paths among lots of possible different paths on the ways from the source to destination. Paths through which water flows in rivers may have several twists and turns, but always chose best possible path. Intelligence behind those twists and turns are the key inspiration of the IWD algorithm. Those approximate best paths are resulted by the actions and reactions, which occur among the water drops and the water drops with the soil. Considering these aspects IWDs are created with two properties soil and velocity. Solution space is represented with a graph. IWDs are distributed over the graph and starts moving through edges. An IWD flows from a source to a destination. Initially IWDs have velocity but zero soil. During movement from one node to another, removes soil from the path and gain speed. Increment in velocity of an IWD is non-linearly and inversely proportional to the soil present in the path. Therefore, the IWD become faster in a path with less soil.

Three things happening during movement of IWDs in graphically represented solution space. Firstly, IWDs gain velocity and gather soil from the path they moved through. Secondly, proportionate amount of soil is removed from the path of the graph through which they move. Time factor is used during removal and addition of soil. Less the time to pass IWD through a path can remove larger amount of soil. The time is proportional to the velocity of the IWD and inversely proportional to the distance between the two nodes. Lastly, IWD needs to chose path to next node from the multiple paths. Mechanism to chose path is that the IWD prefers path with less soil. Hence, paths with less soil have higher chance to get chosen by an IWD.

The strategy looks very similar to ACO. Pheromones are deposited through the path an ant moves and with the time pheromone decreases as it evaporates. In case of IWD, soil is removed from the path when an IWD moves through it. Only difference is changes to the path is constant in case of ACO, whereas in case of IWD these change are dependent on velocity and soil gained by an IWD. This strategy can be fitted to the generalized SI framework once we have initial soil and velocity. In coming sections application and variation related issues with SI techniques explained above are addressed, and also extended to other remaining SI techniques in a generalized form.

4 Applicability of SI Techniques

An application is a composition of several sub-applications or modules. Each module may be an explicit and working application. Considering the feature selection problem of pattern recognition, it has several applications such as Medical disease diagnosis (Selvaraj and Janakiraman 2013), salient object detection (Singh et al. 2014) etc. Salient object detection can be used in surveillance systems (Graefe and Efenberger 1996), image retrieval (Amit 2002; Gonzalez and Woods 2002), advertising a design (Itti 2000) etc. All these abstract level applications in background require feature selection. Often intelligent techniques are used for feature selection (Selvaraj and Janakiraman 2013; Singh et al. 2014). Similarly, several real world applications at different level of abstraction require intelligent techniques in background to solve associated problems. Sometime, intelligent techniques are hybridized with classical methods (Ranaee et al. 2010). Hence, even if an application in hand not directly incorporate intelligent techniques in it, alternatively can be hybridized with classical approaches to improve efficiency.

Application systems may have several control parameters that decide overall behavior of the system. Behavior of any system can be depicted as function of control parameters of that system. To optimize such functions optimization techniques are utilized. Intuitively there may have two kinds of objectives behind using any optimization technique. First one is about finding optimal values of the function for the system. Second one is to find optimal settings of control parameters for the system. Both objectives are interrelated, as optimal value of the function implies optimal settings of control parameters. Normally it looks both are inseparable, but depending on how the application will going to be benefited with the optimization techniques, one can get clear indication about the notion of objective behind incorporation of such techniques. Both kind of objectives can be understood with very general application such as Traveling Salesman Problem (TSP) (Dorigo et al. 2006; Shah-Hosseini 2009). Normally, main aim of TSP is to find shortest route to visit all cities. It is clear that salesman would be benefited with shortest route obtained through any optimization technique. If a vendor gives salary to the salesman depending on the distance, then the route obtained is irreverent to the vendor. In this case simply the minimum distance matters, no matter what route salesman have followed. TSP fitness function is considered as sum of distances among all cities, here distance between any two cities acts as control parameter to the system. To utilize optimization technique, both vendor and salesman will consider same function as both needed minimum distance to cover all cities. But at the end, salesman uses the route obtained with minimum distance i.e. the control parameters and vendor uses minimum distance i.e. the function value.

Concept of background objective for incorporating SI techniques can be understood with other examples. Equalizer plays very important role in digital transmission system. Das et al. (2014) used Artificial Neural Network (ANN) for channel and co-channel equalizer. ANN has three main components i.e. weight, network topology and transfer function. ANN with specific topology and transfer function, performs well when weights are suitable. ANNs are trained to obtain suitable weight and this application used PSO to do so. This application is of second kind as it grasps trained ANN to equalizer not the fitness value. Similarly, application to feature selection mostly used features corresponding to the best accuracy (Ranaee et al. 2010; Selvaraj and Janakiraman 2013; Singh et al. 2014).

Whatever may be the intuition behind utilization of SI techniques finally the technique has to be plugged in somewhere in the application. Implementer has to detect such plug points in the application. Indication to those plug point can be obtained by analyzing problems of application. Those plug points may be unclear or dissimilar to the considered SI technique, in those cases encoding mechanism (described in Sect. 5) is used to unite. A generalized model to incorporate of SI technique in any application is presented in Fig. 2. SI techniques are very suitable for NP-hard problems. Those complex problems suffer exponential worst case

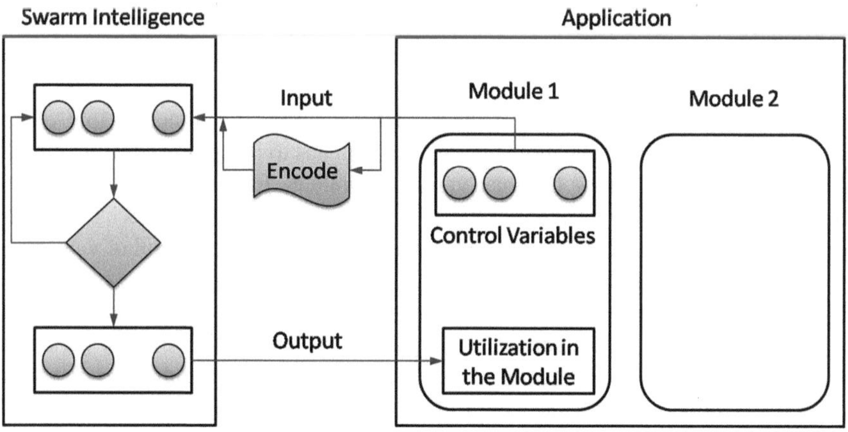

Fig. 2 View of an application with swarm intelligence technique

complexity. Hence, these problems require approximated solution which is relatively good. SI techniques resulted high quality solution to those problem (Dorigo et al. 2006; Shah-Hosseini 2009). Real life application often experiences such problem. In power system applications, problems such as optimal power flow is a NP-hard problem (Alzalg et al. 2011). As explained above, an application have several modules. Those modules may be of different kind of problems. Some of those problems may be of NP-hard category. Summarily, one can get an indication to adopt SI techniques to solve those problems. For example clustering is a NP-hard problem (Kulkarni and Venayagamoorthy 2011). Many application of various domains such as social network analysis (Honghao et al. 2013; Kumar and Jayaraman 2013), image processing (Hancer et al. 2012), wireless sensor network (Kulkarni and Venayagamoorthy 2011) used clustering for their respective problems. This is notable that all these applications adopt SI techniques related to clustering. Hence, detection of problem complexity can be used as good indicator to decide whether associated application needed to grasp SI technique to solve the problem. Objective behind such grasping may vary with problem to problem and can be decided on the basis of problem. As mentioned above, clear objective and problem complexity of associated problem can be handy to decide on applicability of SI techniques.

5 Encoding Schemes

SI techniques act as black box to the application and to implementers act as white box. These techniques take control parameters and fitness function of an application as input and gives optimal solutions as output. Irrespective of the application, optimization techniques will result solution to any input. Those results may be suitable for the application or may not be. It is the implementer who can judge and manipulate the inputs with respect to the considered technique. Results are very much dependent on representative inputs of the application to the SI technique. Hence, representation of control parameters of an application in terms of input to any SI technique is very important. Decision on such issues are very sensitive to the application.

SI techniques use inputs to evaluate fitness or objective function that effects control parameters of application and this function has to be optimized. Generally SI techniques use Boolean or Decimal values to evaluate fitness function. If application's control parameters are not in the same form as the considered technique, then control parameters have to be encoded so that it become required kind of values that are to be used by the technique. Otherwise, application control parameters can be used directly as input to the SI technique. The generalized mechanism of encoding scheme is presented in Fig. 3 with suitable examples. When application control parameter values are Boolean, actually in this case value of any control parameter indicates its selection if value is 1 and deselection if value is 0. During encoding into Decimal, control parameter values are taken together as Boolean string. All control parameter values can be considered at once to represent a single Boolean string or grouped into

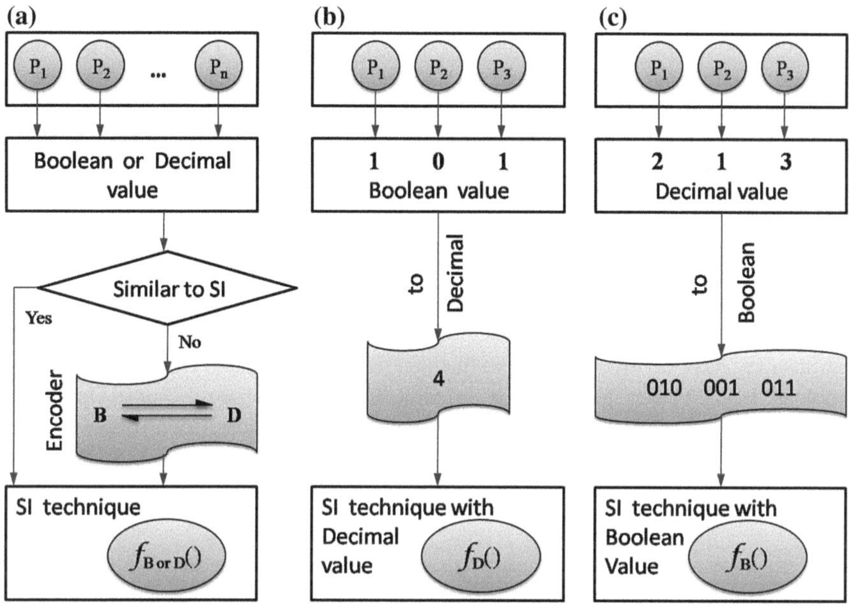

Fig. 3 Encoding mechanism of control parameters of any module of an application corresponding to the inputs of SI technique. Encoding of Boolean to Decimal or Decimal to Boolean is presented in (**a**), **b** shows Boolean to decimal encoding where SI technique uses decimal inputs to evaluate objective function, and **c** shows decimal to Boolean encoding where SI technique uses Boolean inputs to evaluate objective function

different strings. In part (b) of Fig. 3 shows a single string to encode into Decimal value. When application control parameter values are Decimal, they are either defined as continuous or discrete values within some range. Each individual value of control parameter is encoded into Boolean value of predefined bit size. In part (c) of Fig. 3 shows encoding into Boolean value of bit size 3. Each Boolean string is considered as different input to multi-dimensional objective function of SI technique or considered as single Boolean string for uni-dimensional objective function.

Despite of different encoding schemes, it is better to consider such a SI technique which uses similar values as control parameters of application to avoid possible pitfalls. Though such decisions are constraint to the application. Most of the SI techniques including PSO, ACO and GSA use Decimal representation. Even if encoding is not needed when inputs of SI technique and application control parameters are similar, results obtained may not be good. Hence, strategic changes (discussed in Sect. 6) to existing technique is adopted. In some cases require both kinds of encoding schemes. For example feature selection problem uses both Boolean and Decimal representation. Boolean representations are done by representing 1 if particular feature is selected and 0 otherwise. String of these binary values represented as individual particles or solutions to the SI techniques (Chakraborty and Chakraborty

2013; Selvaraj and Janakiraman 2013). However, assigning of weights to each feature is done through Decimal representation (Singh et al. 2014). Features in the vector indicate whether they are selected or not with predefined threshold values. Each vector represents a solution. Another example is in (Das et al. 2014), particles are considered as individual network, which comprises links and weights of these links between two neurons and transfer function. Presence and absence of link can be represented as Boolean, weights can be represented as Decimal and transfer function may give Boolean values. Hence author used both kinds of encoding to represent particles.

Improperly detected control parameters may not be suitable for the application and may show unexpected results. Even if suitable control parameters are indentified, it is a matter of judgment whether to encode as Decimal or Boolean. Sometimes, applications may have control parameters that can not be represented either as Decimal or Boolean. In such cases needed some functions to transform them either to Decimal or to Boolean. Moreover, some applications don't even have clear control parameters. Implementer needs to be identified suitable control parameters from the application in those cases.

6 Strategic Changes

Over the decades SI techniques have been undergone several strategic changes to improve efficiency, reliability, scalability and solution quality. Those changes were carried out irrespective of applications. Often used benchmark functions to test those techniques. But, when those improved variants are applied to some real life problems, found that they are not comply with the problem. It seems that they perform better on some specific problems. Even though performed well on benchmark functions, it cannot be generalize that same technique will perform better in all problems as stated in no free lunch (NFL) theorem (Wolpert and Macready 1997). Same method may show best result on some problem or even on standard benchmark functions but may show worst result in other problems. Hence, problem specific improvements were done to suit real life problem, but always have exceptions. For example PSO often found better to feature selection when hybridized with other approaches, but on the contrary hybridized ACO shows degraded results on the same problem (Kothari et al. 2012).

Introduction of new strategy to the existing technique have added extra overhead to the technique. Though shows better results than previous actual techniques, they become more complex. Implementer has to look over more constraints in order to realize actual variant of the technique. If we take a look at strategic changes of PSO, it has undergone several modifications to the PSO originally introduced by Kennedy and Eberhart (1995). Following the original version of PSO discrete variant was also introduced by Kennedy and Eberhart (1997). Concept of constriction factor was introduced by Cleric and Kennedy (2002). Fully informed PSO (FIPS) was proposed by Mendes (2004), where every particle is informed about others experience. This requires too high computation time. Linearly varying coefficients along with linearly varying weight introduced in Ratnaweera et al. (2004). Implantation of linear variation requires two parameters to define upper and lower limit. Hence, to perceive this concept requires six additional parameters, two for each of three core parameters. Similar parameter related issues can be noted in hybridized versions with Chemical Reaction Optimization (CRO) (Nguyen et al. 2014), which requires eight parameter to maintain, five for CRO and three for PSO. Orthogonal learning approach proposed by Zhan and his colleague (Zhan et al. 2011), where used concept of orthogonal experiment design (OED). Factor analysis part of this strategy consumes much computation time as compared to PSO without OED. However, this version have resulted high quality solutions. Recently, Ganapathy and his colleague proposed ortho-cyclic circle concept (Ganapathy et al. 2014) to PSO, which considers two level particles. Particles are identified as belonging to a group referred as circles. These circles are also considered as high level particles. Maintenance of two level particle and identification of circles has added extra burden and make concept little bit complex. Similar problem can be noted in case of ACO. Min-Max ant system (MMAX) (Stützle and Hoos 2000) added bound on pheromone deposit, which required to maintain two extra parameters. In rank-based ant system (ASrank) (Bullnheimer et al. 1997), solutions are ranked in accordance of their cost. Ants are allowed to deposit pheromones proportional to their derived solution cost. Ranking of ants based on the solution cost is a time consuming process.

Advancement of strategy of a technique implies accumulation of new concept, which may be complex in real sense. Basic and original version of SI techniques found to be simpler than advanced versions of the same method. Hence, instead of made strategic changes sometime parameters of existing SI technique are tuned to improve results. Parameter tuning schemes on various SI techniques are explained in Sect. 7. Moreover, Complex advance version cannot fit into the applications easily. Hence, basic versions are used enormously even though results are comparatively degraded than advanced versions. Methodology of SI technique (it may be advanced version or basic version) often found to be incompatible with applications, those cases the strategy has to be modified so that it comply with that application.

7 Parameter Tuning

SI techniques have specific parameters to perceive their strategic needs by tuning those parameters. Collective aim of such tuning is to improve efficiency, raise effectiveness, increase applicability of these techniques to practical problems. In perspective of applications, parameters of SI techniques are tuned to comply with application environment. Every application has its own environment and needs to solve associated problems with respect to those environment. Techniques incorporated (including SI techniques) to solve such problems have to be practised in terms of application environment. So, parameter values of any SI technique has to be tuned in accordance of environment of the application. Some application may have constraints for those cases require additional parameters to fit the technique in it. Hence, such application specific parameter tunings cannot be generalized for other applications.

To account maximum application domains for any SI technique, require generalized parameters settings which can be used in any application and also needs to ensure that those setting will result better. So, most of such tunings are carried out with benchmark functions to test tuned parameter. Benchmark function based parameter adjustments can be considered as generalized settings for applications of different domain.

Generalized parameter tuning can be of two kinds, constant value based tuning and strategic tuning. Universally used PSO have three parameters C_1, C_2, ω. Initial version of PSO had only two parameters C_1 and C_2. Later on suggested one more parameter because, uncontrolled velocity often led to move particles much ahead of optimal solution as expected, which implies divergence of particles from the objective, resulting slow convergence of the process to the optimal solution. Shi and Eberhart (1998a) has observed that velocity of particle has to be controlled in order to control search scope of particle. To overcome this problem they introduced new parameter inertia weight to control velocity of particle. Cleric and Kennedy has done similar control by introducing constriction factor c instead of inertia weight (Clerc and Kennedy 2002). They showed values of two coefficient have to be 1.4962, while inertia weight has to be 0.7968. Kennedy and his colleague (Kennedy et al. 2001) shows value of C_1 and C_2 has to be 2. Apart from these constant value based tuning, parameters of PSO also tuned strategically to act more friendly to the applications of different domains. Shi and Eberhart (1999) has varied inertia weight ω linearly. Sometime added new parameters which might be for realization of strategic tuning. For example, to vary ω of PSO there has to be upper bound ω_{up} and lower bound ω_{low}. These newly introduced parameter also has to be tuned to obtain better range of ω for varying linearly. Shi and Eberhart (1998b) showed values of ω_{up} and ω_{low} are 0.9 and 0.4 respectively. Hence, tuning of parameters not only done as constant value based or strategic tuning, but also can be done both simultaneously. Parameter values obtained through constant value based tuning may not suit some application, but strategic tuning can fit all application which utilizes SI technique.

8 Discussion

Adaptation have to be done both from application side as well as SI techniques to comply with each other. In order to fit SI techniques into diverse applications of different domains have to be generalized. Such generalization has to be done in SI techniques with respect to any application or benchmark problems. Generally, control parameters of application varies with different application domain and the environment of the application. SI techniques' parameters and strategies have to be adjusted in accordance of control parameters of any application. Hence, parameters and strategy of an SI technique have to be tuned in such a way so that it can be grasped in any application without too many alternations both in SI techniques as well as application control parameters.

Structure of most SI techniques are very similar and can be generalized in single framework. Most of SI techniques treat solutions in similar way, only difference is in their strategy. Generally, SI techniques represent solutions in Boolean string or Decimal string. Some of the SI techniques consider multi level solution representation. Hence, to incase SI techniques in a generalized framework, solution representation has to be done in perspective of strategy incorporated into the approach. Once presentation of initial swarm is done with respect to strategy, swarm can be updated strategically in a common loop by absorbing strategy of corresponding approach.

SI techniques always have to go hand in hand with its application to solve common problems. In order to mastery any of the SI techniques in applicability to applications of various domains, compatibility of its strategy with the application has to be increased. Applicability of any SI technique can be increased if its parameters and strategy are generalized. Major issues observed for generalization of SI techniques in perspective of applications can be stated as below:

- Diversity of applications.
- Numerous techniques available creating confusion.
- Constraints in applications.
- Encoding of application.
- Multiple variants of single technique.

Issues related to applicability of SI techniques to any application:

- Selection of suitable technique for an application is very important task and always have confusions. Generally, simpler and effective techniques are the good criterion for selection of SI technique.
- Normally have two objectives behind the utilization of any SI technique, either it is the objective function value or the control parameters.
- Application associated problems have to be encoded in perspective of the selected technique to realize problem in that technique.
- Problem characteristics have be similar to problems where selected technique performs comparatively better.
- During implementation may face confusion regarding which one to adapt, is it the application or technique.
- Adaptation of application in terms of technique selected will be the better option, but of course selected technique has to be good one.
- Parameters of selected technique have to be tuned with respect to application to get good quality solution.

Advantages of incorporation of SI techniques into applications:

- Concept of some techniques such as PSO are very simple and competitive to other approaches.
- Simpler techniques just need to identify application problem's control parameters in order to consume strategy of the technique.
- Techniques with less parameters and simple strategy can easily fit into variety of applications of different domain.
- Generally SI techniques requires only a few steps i.e. initialize swarm and iterate with corresponding strategy.
- SI provides framework to solve collective or distributed problems through a global view model with decentralized control.
- Can easily solve non-linear complex problems.
- Can easily solve hard problems such as NP-hard and NP-complete problems, shows high quality approximated solution.
- Problem can be solved without any guidance through local shared information and cooperative learning of SI technique.

Disadvantages of SI techniques in perspective of applications:

- Application complexity creates problems during absorption of SI techniques.
- Some advanced variants show very high quality solutions but technique become complex and difficult to implement.
- SI techniques always give approximated result and have no grantee of good solution.
- Most of the SI techniques have parameters which have to be tuned in order to get better result.

9 Conclusion

Swarm based techniques have been applied enormously in real life problems. Most of these techniques are inspired by nature and have a common aim to optimize. Technology advances have led to solve problems more efficiently and more effectively. To deal with the requirements of such advanced technology, associated problem has to solve intelligently. SI techniques have been one of the repository for such intelligence. To align with current technology, SI techniques have also been improved. Though these advanced variations shown improvement in quality of solutions, in parallel increases complexity of the approach and time requirement. Also, creates confusion regarding selection of one from these too many techniques along with their variants. Such application and SI techniques related issues have been addressed in this chapter. Generalization of SI techniques can be one of the solution to such issues, but due to constraints present at variety of applications this solution can not be materialized fully. Another solution being noticed to this problem is the utilization of simpler and relatively effective variant to applications, which not only meets quality requirements but also can easily deal with implementation related issues. Despite of numerous difficulties SI techniques gaining popularity and also consuming enormously in various applications.

References

Al Rashidi, M. R., & El-Hawary, M. E. (2009). A survey of particle swarm optimization applications in electric power systems. *IEEE Transactions on Evolutionary Computation, 13* (4), 913–918.

Alzalg, B., Anghel, C., Gan, W., Huang, Q., Rahman, M., & Shum, A. (2011). A computational analysis of the optimal power problem. In *Institute of Mathematics and its Application*. IMA Preprint Series 2396. University of Minnesota.

Amit, Y. (2002). *2D object detection and recognition: Models, algorithms, and networks*. Cambridge: MIT Press.

Borwein, J. M. & Lewis, A. S. (2010). *Convex analysis and nonlinear optimization: Theory and examples* (2nd ed.). Berlin, Springer.

Bullnheimer, B., Hartl, R. F., & Strauss, C. (1997). A new rank based version of the ant system. A computational study. *SFB Adaptive Information Systems and Modelling in Economics and Management Science, 7*, 25–38.

Chakraborty, B., & Chakraborty, G. (2013). Fuzzy consistency measure with particle swarm optimization for feature selection. In *2013 IEEE International Conference on Systems, Man, and Cybernetics (SMC)* (pp. 4311–4315), October 13–16, 2013, Manchester. IEEE. doi:10.1109/SMC.2013.735.

Chandra Mohan, B., & Baskaran, R. (2012). A survey: Ant colony optimization based recent research and implementation on several engineering domain. *Expert Systems with Applications, 39*(4), 4618–4627.

Chandrasekhar, U., & Naga, P. (2011). Recent trends in ant colony optimization and data clustering: A brief survey. In *2011 2nd International Conference on Intelligent Agent and Multi-agent Systems (IAMA)* (pp. 32–36), September 7–9, 2011, Chennai. IEEE. doi:10.1109/IAMA.2011.6048999.

Chu, S.-C., Roddick, J. F., & Pan, J.-S. (2003). Parallel particle swarm optimization algorithm with communication strategies. *Journal of Information Science and Engineering, 21*(4), 809–818.

Clerc, M., & Kennedy, J. (2002). The particle swarm-explosion, stability, and convergence in a multidimensional complex space. *IEEE Transactions on Evolutionary Computation, 6*(1), 58–73.

Das, G., Pattnaik, P. K., & Padhy, S. K. (2014). Artificial neural network trained by particle swarm optimization for non-linear channel equalization. *Expert Systems with Applications, 41*(7), 3491–3496.

Deb, K., Pratap, A., Agarwal, S., & Meyarivan, T. (2002). A fast and elitist multiobjective genetic algorithm: Nsga-ii. *IEEE Transactions on Evolutionary Computation, 6*(2), 182–197.

Dorigo, M. (1992). Optimization, learning and natural algorithms. *Ph.D. Thesis*, Politecnico di Milano, Italy.

Dorigo, M., Birattari, M., & Stutzle, T. (2006). Ant colony optimization. *Computational Intelligence Magazine, IEEE, 1*(4), 28–39.

Eslami, M., Shareef, H., Khajehzadeh, M., & Mohamed, A. (2012). A survey of the state of the art in particle swarm optimization. *Research Journal of Applied Sciences, Engineering and Technology, 4*(9), 1181–1197.

Ganapathy, K., Vaidehi, V., Kannan, B., & Murugan, H. (2014). Hierarchical particle swarm optimization with ortho-cyclic circles. *Expert Systems with Applications, 41*(7), 3460–3476.

Gonzalez, R. C., & Woods, R. E. (2002). *Digital image processing*. New Jersey: Prentice Hall.

Graefe, V., & Efenberger, W. (1996). A novel approach for the detection of vehicles on freeways by real-time vision. In *Proceedings of the 1996 IEEE Intelligent Vehicles Symposium* (pp. 363–368), September 19–20, 1996, Tokyo. IEEE. doi:10.1109/IVS.1996.566407.

Hancer, E., Ozturk, C., & Karaboga, D. (2012). Artificial bee colony based image clustering method. In *2012 IEEE Congress on Evolutionary Computation (CEC)* (pp. 1–5), June 10–15, 2012, Brisbane. IEEE. doi:10.1109/CEC.2012.6252919.

Holland, J. H. (1975). *Adaptation in natural and artificial systems: An introductory analysis with applications to biology, control, and artificial intelligence*. Michigan: University Michigan Press.

Honghao, C., Zuren, F., & Zhigang, R. (2013). Community detection using ant colony optimization. In *2013 IEEE Congress on Evolutionary Computation (CEC)* (pp. 3072–3078), June 20–23, 2013, Cancun. IEEE. doi:10.1109/CEC.2013.6557944.

Itti, L. (2000). Models of bottom-up and top-down visual attention. *PhD thesis*, California Institute of Technology.

Janacik, P., Orfanus, D., & Wilke, A. (2013). A survey of ant colony optimization-based approaches to routing in computer networks. In *2013 4th International Conference on Intelligent Systems Modelling and Simulation (ISMS)* (pp. 427–432), January 29–31, 2013, Bangkok. IEEE. doi:10.1109/ISMS.2013.20.

Kameyama, K. (2009). Particle swarm optimization-a survey. *IEICE Transactions on Information and Systems, 92*(7), 1354–1361.

Karaboga, D. (2005). An idea based on honey bee swarm for numerical optimization. Technical report, (Technical report-tr06), Erciyes university, Engineering Faculty, Computer Engineering Department.

Kennedy, J., & Eberhart, R. (1995). Particle swarm optimization. In *Proceedings of IEEE international conference on neural networks*, (Vol. 4, pp. 1942–1948), 27 Nov 1995–2001 Dec 1995, Perth. IEEE. doi:10.1109/ICNN.1995.488968.

Kennedy, J., & Eberhart, R. C. (1997). A discrete binary version of the particle swarm algorithm. In *Systems, Man, and Cybernetics, 1997. Computational Cybernetics and Simulation., 1997 IEEE International Conference on*, Vol. 5, October 12–15, 1997, Orlando, IEEE (pp. 4104–4108). doi:10.1109/ICSMC.1997.637339.

Kennedy, J. F., Kennedy, J., & Eberhart, R. C. (2001). *Swarm intelligence*. Los Altos: Morgan Kaufmann.

Kothari, V., Anuradha, J., Shah, S., & Mittal, P. (2012). A survey on particle swarm optimization in feature selection. In *Global Trends in Information Systems and Software Applications* (pp. 192–201). Berlin: Springer

Kulkarni, R. V., & Venayagamoorthy, G. K. (2011). Particle swarm optimization in wireless-sensor networks: A brief survey. *IEEE Transactions on Systems, Man, and Cybernetics Part C: Applications and Reviews, 41*(2), 262–267.

Kumar, G. K., & Jayaraman, V. (2013). Clustering of complex networks and community detection using group search optimization. *CoRR*, (abs/1307.1372).

Matoušek, J., & Gärtner, B. (2007). *Understanding and using linear programming*. 7th edition. Berlin: Springer.

Mendes, A. (2004). *Building generating functions brick by brick*. San Diego: University of California.

Monteiro, M. S., Fontes, D. B., & Fontes, F. A. (2012). Ant colony optimization: a literature survey. Technical report, Universidade do Porto, Faculdade de Economia do Porto.

Nguyen, T. T., Li, Z., Zhang, S., & Truong, T. K. (2014). A hybrid algorithm based on particle swarm and chemical reaction optimization. *Expert Systems with Applications, 41*(5), 2134–2143.

Ranaee, V., Ebrahimzadeh, A., & Ghaderi, R. (2010). Application of the pso–svm model for recognition of control chart patterns. *ISA Transactions, 49*(4), 577–586.

Rashedi, E., Nezamabadi-Pour, H., & Saryazdi, S. (2009). Gsa: A gravitational search algorithm. *Information Sciences, 179*(13), 2232–2248.

Ratnaweera, A., Halgamuge, S., & Watson, H. C. (2004). Self-organizing hierarchical particle swarm optimizer with time-varying acceleration coefficients. *IEEE Transactions on Evolutionary Computation, 8*(3), 240–255.

Rechenberg, I. (1994). Evolution strategy *Computational intelligence: Imitating life*, (pp. 147–159). Piscataway: IEEE Press.

Reyes-Sierra, M., & Coello, C. C. (2006). Multi-objective particle swarm optimizers: A survey of the state-of-the-art. *International journal of computational intelligence research, 2*(3), 287–308.

Ruszczyński, A. P. (2006). *Nonlinear optimization* (Vol. 13). NJ: Princeton University Press.

Schutte, J. F., Reinbolt, J. A., Fregly, B. J., Haftka, R. T., & George, A. D. (2004). Parallel global optimization with the particle swarm algorithm. *International Journal for Numerical Methods in Engineering, 61*(13), 2296–2315.

Schwefel, H.-P. (1994). On the evolution of evolutionary computation, *Computational intelligence: Imitating life*, (pp. 116–124). IEEE Press: Piscataway.

Selvaraj, G., & Janakiraman, S. (2013). Improved feature selection based on particle swarm optimization for liver disease diagnosis. In *Swarm, Evolutionary, and Memetic Computing* (pp. 214–225). Berlin: Springer.

Shah-Hosseini, H. (2008). Intelligent water drops algorithm: A new optimization method for solving the multiple knapsack problem. *International Journal of Intelligent Computing and Cybernetics, 1*(2), 193–212.

Shah-Hosseini, H. (2009). The intelligent water drops algorithm: A nature-inspired swarm-based optimization algorithm. *International Journal of Bio-Inspired Computation, 1*(1), 71–79.

Shi, Y., & Eberhart, R. (1998a). A modified particle swarm optimizer. In *The 1998 IEEE International Conference on Evolutionary Computation Proceedings, IEEE World Congress on Computational Intelligence* (pp. 69–73), May 4-9, 1998, Anchorage. IEEE. doi:10.1109/ICEC.1998.699146.

Shi, Y., & Eberhart, R. C. (1998b). Parameter selection in particle swarm optimization. In *Evolutionary Programming VII*, March 25-27, 1998, San Diego, California, USA, Springer (pp. 591–600). doi:10.1007/BFb0040810.

Shi, Y., & Eberhart, R. C. (1999). Empirical study of particle swarm optimization. In *Evolutionary Computation, 1999. CEC 99. Proceedings of the 1999 Congress on*, Vol. 3, July 6-9, 1999, Washington, IEEE. doi:10.1109/CEC.1999.785511.

Singh, N., Arya, R., & Agrawal, R. (2014). A novel approach to combine features for salient object detection using constrained particle swarm optimization. *Pattern Recognition, 47*(4), 1731–1739.

Stützle, T., & Hoos, H. H. (2000). Max–min ant system. *Future generation computer systems, 16* (8), 889–914.

Todd, M. J. (2002). The many facets of linear programming. *Mathematical Programming, 91*(3), 417–436.

Vanneschi, L., Codecasa, D., & Mauri, G. (2012). An empirical study of parallel and distributed particle swarm optimization. In *Parallel Architectures and Bioinspired Algorithms* (pp. 125–150). Berlin: Springer.

Wiki (2014). Mathematical optimization. http://en.wikipedia.org/wiki/Mathematical_optimization. Accessed 2014-02-30.

Wolpert, D. H., & Macready, W. G. (1997). No free lunch theorems for optimization. *IEEE Transactions on Evolutionary Computation, 1*(1), 67–82.

Yan, X.-S., Li, C., Cai, Z.-H., & Kang, L.-S. (2005). A fast evolutionary algorithm for combinatorial optimization problems. In *Proceedings of 2005 International Conference on Machine Learning and Cybernetics* (Vol. 6, pp. 3288–3292) August 18-21, 2005, Guangzhou. IEEE. doi:10.1109/ICMLC.2005.1527510.

Yang, B., Chen, Y., & Zhao, Z. (2007). Survey on applications of particle swarm optimization in electric power systems. In *IEEE International Conference on Control and Automation (ICCA 2007)* (pp. 481–486), May 30 2007–June 1 2007, Guangzhou. IEEE. doi:10.1109/ICCA.2007.4376403.

Zhan, Z.-H., Zhang, J., Li, Y., & Shi, Y.-H. (2011). Orthogonal learning particle swarm optimization. *IEEE Transactions on Evolutionary Computation, 15*(6), 832–847.

Signal Based Fault Detection and Diagnosis for Rotating Electrical Machines: Issues and Solutions

Andrea Giantomassi, Francesco Ferracuti, Sabrina Iarlori, Gianluca Ippoliti and Sauro Longhi

Abstract Complex systems are found in almost all field of contemporary science and are associated with a wide variety of financial, physical, biological, information and social systems. Complex systems modelling could be addressed by signal based procedures, which are able to learn the complex system dynamics from data provided by sensors, which are installed on the system in order to monitor its physical variables. In this chapter the aim of diagnosis is to detect if the electrical machine is healthy or a change is occurring due to abnormal events and, in addition, the probable causes of the abnormal events. Diagnosis is addressed by developing machine learning procedures in order to classify the probable causes of deviations from system normal events. This chapter presents two Fault Detection and Diagnosis solutions for rotating electrical machines by signal based approaches. The first one uses a current signature analysis technique based on Kernel Density Estimation and Kullback–Liebler divergence. The second one presents a vibration signature analysis technique based on Multi-Scale Principal Component Analysis. Several simulations and experimentations on real electric motors are carried out in order to verify the effectiveness of the proposed solutions. The results show that the proposed signal based diagnosis procedures are able to detect and diagnose different electric motor faults and defects, improving the reliability of electrical machines. Fault Detection and Diagnosis algorithms could be used not only with the fault diagnosis purpose but also in a Quality Control scenario. In fact, they can be

---

A. Giantomassi (✉) · F. Ferracuti · S. Iarlori · G. Ippoliti · S. Longhi
Dipartimento di Ingegneria dell'Informazione, Università Politecnica delle Marche,
Via Brecce Bianche, 60131 Ancona, Italy
e-mail: a.giantomassi@univpm.it

F. Ferracuti
e-mail: f.ferracuti@univpm.it

S. Iarlori
e-mail: s.iarlori@univpm.it

G. Ippoliti
e-mail: gianluca.ippoliti@univpm.it

S. Longhi
e-mail: sauro.longhi@univpm.it

Q. Zhu and A.T. Azar (eds.), *Complex System Modelling and Control Through Intelligent Soft Computations*, Studies in Fuzziness and Soft Computing 319,
DOI 10.1007/978-3-319-12883-2_10

integrated in test benches at the end or in the middle of the production line in order to test the machines quality. When the electric motors reach the test benches, the sensors acquire measurements and the Fault Detection and Diagnosis procedures detect if the motor is healthy or faulty, in this last case further inspections can diagnose the fault.

1 Introduction

Mathematical process models describe the relationship between input signals $u(k)$ and output signals $y(k)$ and are fundamental for model-based fault detection. In many cases the process models are not known at all or some parameters are unknown. Further, the models have to be rather precise in order to express deviations as results of process faults. Therefore, process-identification methods have to be applied frequently before applying any model-based fault detection method as stated in Giantomassi (2012). But also the identification method itself may be a source to gain information on, e.g. process parameters which change under the influence of faults. First publications on fault detection with identification methods are found in Isermann (1984) and Filbert and Metzger (1982).

For dynamic processes the input signals may be the normal operating signal or may be artificially introduced for testing. A considerable advantage of identification methods is that with only one input and one output signal several parameters can be estimated, which give a detailed picture on internal process quantities. The generated features for fault detection are then impulse response values in the case of correlation methods or parameter estimates [see Isermann (2006)].

On-line process monitoring with fault detection and diagnosis can provide range of processes, as stated in Cheng et al. (2008), Giantomassi et al. (2011) and Ferracuti et al. (2010, 2011). A large number of applications have been reviewed, e.g. Isermann and Balle (1997) and Patton et al. (2000). Venkatasubramanian et al. (2000a, b, c) published an article series reviewing monitoring methods with attention in the field of chemical processes. They classified the Fault Detection and Diagnosis methods as model-based, signal-based and knowledge-based. Signal-based approaches to fault detection and isolation (FDI) in large-scale process plants are consolidated and well studied, because for these processes the development of model-based FDI methods requires considerable and eventually too high effort, and moreover because a large amount of data is collected, as stated in Chiang et al. (2000) and Isermann (2006).

Fault detection and diagnosis (FDD) in industrial applications regards two important aspects: the FDD for the production plant and for the systems that work for the plant; among these systems, induction motors are the most important electrical machineries in many industrial applications, considering that, electric motors account about 65 % of energy use. In the field of operational efficiency, the monitoring activity of rotating electrical machines by fault detection and diagnosis is in-depth investigated: Benloucif and Balaska (2006), Ran and Penman (2008), Singh and Ahmed (2004), Taniguchi et al. (1999), Tavner (2008), Verucchi and Acosta (2008). Vibration analysis is widely accepted as a tool to detect faults of a machine since it is nondestructive, reliable and it permits continuous monitoring without stopping the machine [see Ciandrini et al. (2010), Gani and Salami (2002), Hua et al. (2009); Immovilli et al. (2010), Shuting et al. (2002); Zhaoxia et al. (2009)]. In particular analysing the vibration power spectrum it is possible to detect different faults that arise in rotating machines. In traditional machine vibration signature analysis (MVSA), the Fourier transform is used to determine the vibration power spectrum and the signature at different frequencies are identified and compared with those related to healthy motors to detect faults in the machine, as in Lachouri et al. (2008). The shortcoming of this approach is that the Fourier analysis is limited to stationary signals while vibrations are not stationary by its nature.

The use of Soft Computing methods is considered an important extension to the model-based approach Patton et al. (2000). It allows to improve residual generation in FDD when process signals show complex behaviours. Multi-scale principal component analysis (MSPCA) deals with processes that operate at different scales: events occurring at different localizations in time and frequency, stochastic processes and variables measured at different sampling rate, as reported in Bakshi (1998) and Li et al. (2000). PCA, treated in Jolliffe (2002) and Jackson (2003), decorrelates the variables by extracting a linear relationship in order to transform the multivariate space into a subspace which preserves maximum variance of the original space. Wavelets extract deterministic features and approximately decorrelate autocorrelated measurements. MSPCA combines these two techniques to extract maximum information from multivariate sensor data (Misra et al. 2002).

Rotating electrical machines are well known systems with accurate analytical models and extensive results in literature. Failure surveys, as Thomson and Fenger (2001), report that failures, in induction motors, are: stator related (38 %), rotor related (10 %), bearing related (40 %) and others (12 %). Fast and accurate diagnosis of incipient faults allows actions to protect the power system, the process leaded by the machine and the machine itself.

FDD techniques based on MVSA have received great attention in literature because by vibrations it is possible to identify directly mechanical faults regarding rotating electrical machines. In recent years, many methodologies have been developed to detect and diagnose mechanical faults of electrical machines by current measurements. In this context motor current signature analysis (MCSA) involves detection and identification of current signature patterns that are indicative of normal and abnormal motor conditions. However, the motor current is influenced by many factors such as electric supply, static and dynamic load conditions, noise, motor geometry and faults. In Chilengue et al. (2011) an artificial immune system approach is investigated for the detection and diagnosis of faults in the stator and rotor circuits of induction machines. The proposed technique measures the stator currents to compute its representation before and after a fault condition. These patterns are used to construct a characteristic image of the machine operating condition. Moreover MCSA procedures are used to detect and diagnose not only classic motor faults (i.e. rotor eccentricity), but also gear faults (i.e. tooth spall), as presented in Feki et al. (2013). Fault Tolerant Control (FTC) as well as robust control systems have been applied in electric drive systems Ciabattoni et al. (2011a, 2011b, 2014). In Abdelmadjid et al. (2013) a FTC procedure is proposed for stator winding fault of induction motors. It consists of an algorithm which can detect an incipient fault in closed loop and switches itself between a nominal control strategy for healthy condition and a robust control for faulty condition. Samsi et al. (2009) validated a technique, called Symbolic Dynamic Filtering (SDF), for early detection of stator voltage imbalance in three-phase induction motors that involves Wavelet Transform (WT) of current signals. In Baccarini et al. (2010) a sensor-less approach has been proposed to detect one broken rotor bar in induction motors. This method is not affected by load and other asymmetries. The technique estimates stator and rotor flux and analyses the differences obtained in torque. A new saturation model that explains the experimental data is investigated in Pedra et al. (2009). The model has three different saturation effects, which have been characterized in four induction motors.

As possible solutions of the FDI problem for electrical machines, two different approaches are proposed: the first one uses vibration signals provided by accelerometer sensors placed on the machine, and the second one uses current signals provided by inverters.

In the first solution, based on current signal analysis of rotating electrical machines, different algorithms are applied for FDD: PCA is used to reduce the three-phase current space in two dimensions. Then, Kernel Density Estimation (KDE) is adopted to estimate the probability density function (PDF) of each healthy and faulty motor, which are typical features that can be used to identify each fault [see Ferracuti et al. (2013a)]. Kullback–Leibler (K–L) divergence is used as a distance between two PDF obtained by KDE. K–L allows to identify the dissimilarity between two probability distributions (that can also be multidimensional): one is related to the modelled signatures and the other one is related to the acquired data samples. The classification of each motor condition is performed by K-L divergence.

In the second approach, based on vibration analysis of rotating electrical machines, MSPCA is applied for fault detection and diagnosis (Ferracuti et al. 2013b; Lachouri et al., 2008; Misra et al. 2002). Fault identification is evaluated by calculating the contributions of each variable in the principal component subspace and in the residual space. KDE, which allows to estimate the PDF of random variables is introduced, in Odiowei and Cao (2010), to improve fault detection and isolation. The contributions PDFs are estimated by KDE, the thresholds are computed for each signal in order to improve fault detection. Faults are classified by using the contribution plots by Linear Discriminant Analysis (LDA).

The proposed data-driven algorithms for FDD based on MVSA and MCSA are tested by several simulations and experimentations in order to verify the effectiveness of the proposed methodologies.

The chapter will be organized in the following sections. In Sect. 2 the FDD algorithm based on Motor Current Signature Analysis is discussed with focus on Quality Control scenario. Experimental tests on real motors are reported in Sect. 3. The FDI algorithm based on vibration signals is described in Sect. 4. Experimental tests on real motors are reported in Sect. 5. Comments on the performances of the proposed solutions are reported in Sect. 6.

2 Electric Motor FDD by MCSA in Quality Control Scenario

In industry, QC is a collection of methods that are able to improve the quality and efficiency in processes, productions and in many others industry aspects. In 1924, Walter Shewhart designed the first control chart and gave a rationale for its use in process monitoring and control (Stuart et al. 1995). The main concept of QC is the "proactiveness" that ensures the product quality, processes and signals monitoring to detect when they "go out of control". In the last years, manufacturing industries are paying attention and efforts for the introduction of QC in the production lines. Large volumes of low-tech products involve many investigations on the efficient introduction of QC in production lines.

One of the major problems, in which these manufacturing industries are involved, is the customers satisfaction, because they usually purchase a lot of products with some unwanted defective component. In order to satisfy customers, manufacturing industries carry out some spot checks at the end of production lines. This method does not ensure the quality of products and total defective products removal. A desirable QC solution for these manufacturing industries should be minimally invasive, effective and with a low payback period. In addition, tests should be performed in a systematic way using a low-cost system based on a reduced set of sensors embedded in the test bench.

The proposed FDD system acquires sensor measurements and detects defective products. Moreover, by isolating and identifying the defective type, the FDD procedure helps to estimate in which subprocess the defect is introduced and allows to remove the defective products, improving the processes quality. The tests, performed at the end of production lines, allow to improve the quality of processes as proactive measures for the QC methodology.

2.1 Recalled Results

In this section authors present the algorithms used to develop the FDD procedure. They extract patterns by current signals using PCA and KDE. Then K–L divergence compares these patterns to extract the motor health index.

2.1.1 Principal Component Analysis

PCA is a dimensionality reduction technique that produces a lower dimensional representation in a way that preserves the correlation structure between the process variables capturing the variability in the data (Jolliffe 2002). PCA rotates the original coordinate system along the direction of maximum variance. Considering a data matrix $\boldsymbol{X} \in \mathbb{R}^{N \times m}$ of N sample rows and m variable columns that are normalized to zero mean with mean values vector $\boldsymbol{\mu}$. The matrix $\boldsymbol{X}$ can be decomposed as follows:

$$\boldsymbol{X} = \hat{\boldsymbol{X}} + \tilde{\boldsymbol{X}}, \tag{1}$$

where $\hat{\boldsymbol{X}}$ is the projection on the Principal Component Subspace (PCS) S_d, and $\tilde{\boldsymbol{X}}$, the residual matrix, is the projection on the Residual Subspace (RS) S_r (see Misra et al. 2002). Defining the loading matrix $\boldsymbol{P}$, whose columns are the right singular vectors of $\boldsymbol{X}$, and selecting the columns of the loading matrix $\boldsymbol{P} \in \mathbb{R}^{m \times d}$, which correspond to the loading vectors associated with the first d singular values, it follows that:

$$\hat{\boldsymbol{X}} = \boldsymbol{X}\boldsymbol{P}\boldsymbol{P}^T \in S_d. \tag{2}$$

The residuals matrix $\tilde{\boldsymbol{X}}$, is the difference between the data matrix $\boldsymbol{X}$ and its projection into the first d principal components retained in the PCA model:

$$\tilde{\boldsymbol{X}} = \boldsymbol{X}(\boldsymbol{I} - \boldsymbol{P}\boldsymbol{P}^T) \in S_r, \tag{3}$$

therefore the residual matrix captures the variations in the observations space spanned by the loading vectors associated with the $r = m - d$ smallest singular values. The projections of the observations into the lower-dimensional space are contained in the score matrix:

$$\boldsymbol{T} = \boldsymbol{X}\boldsymbol{P} \in \mathbb{R}^{N \times d}. \tag{4}$$

Here, PCA is applied to the three-phase currents of induction motors in order to reduce the inputs space from the three original dimensions to two because the currents are highly correlated. Indeed for healthy motor, with three-phase without neutral connection, ideal conditions and a balanced voltage supply, the stator currents are given by Eq. (5), where i_a, i_b and i_c denote the three stator currents, $I_{\max}$ their maximum value, f their frequency, ϕ their phase angle and t the time. It is known that each stator current is given by the combination of the others:

$$\begin{cases} i_a(t) = I_{\max}\sin(2\pi ft - \phi) \\ i_b(t) = I_{\max}\sin(2\pi ft - 2\pi/3 - \phi) \\ i_c(t) = I_{\max}\sin(2\pi ft - 4\pi/3 - \phi). \end{cases} \quad (5)$$

The PCA transform (4), applied to the signals in Eq. (5), makes the smallest singular value equal to zero. This implies that the information of the principal component, captured by the smallest singular value is null, then the last principal component could be deleted and the original space reduced from three to two without losing information. This is justified by the fact that in Eq. (5), each stator current is perfectly correlated to the sum of the others. Adding Gaussian white noise, with standard deviation σ, to the stator current signals (Eq. 5), the smallest singular value will not be equal to zero, but it will depend by the ratio between $I_{\max}$ and σ.

2.1.2 Kernel Density Estimation

Given N independent and identically distributed (i.i.d.) random vectors $\boldsymbol{X} = [\boldsymbol{X}_1, \ldots, \boldsymbol{X}_N]$, where $\boldsymbol{X}_\text{i} = [X_{i1}, \ldots, X_{id}]$, whose distribution function $F(\boldsymbol{x}) = P[\boldsymbol{X} \leq \boldsymbol{x}]$ is absolutely continuous with unknown PDF $f(\boldsymbol{x})$. The estimated density at $\boldsymbol{x}$ is given by Parzen (1962):

$$\hat{f}(\boldsymbol{x}) = \frac{1}{N}\sum_{i=1}^{N} \frac{1}{|H|^d} K\left(\frac{\boldsymbol{x} - \boldsymbol{X}_i}{|H|^d}\right). \quad (6)$$

In the present study a two-dimensional Gaussian kernel function is used so d is 2 and a further simplification, which follows from the restriction of kernel bandwidth $H = \{h^2 I : h > 0\}$, leads to the single bandwidth estimator so the estimated density $\hat{f}(\boldsymbol{x})$ becomes:

$$\hat{f}(\boldsymbol{x}) = \frac{1}{N}\sum_{i=1}^{N} \frac{1}{(2\pi h^2)^{1/2}} \mathrm{e}^{-\frac{\|\boldsymbol{x}-\boldsymbol{X}_i\|^2}{2h^2}}. \quad (7)$$

where $\boldsymbol{x} \in \mathbb{R}^d$ whose size n_{grid} is the points number in which the PDF is estimated, accordingly to Wand and Jones (1994a). It is well known that the value of the bandwidth h and the shape of the kernel function are of critical importance as stated in Mugdadi and Ahmad (2004). In many computational-intelligence methods that employ KDE, the issue is to find the appropriate bandwidth h [see for example Comaniciu (2003), Mugdadi and Ahmad (2004), Sheather (2004)]. In the present work the Asymptotic Mean Integrated Squared Error (AMISE) with plug-in bandwidth selection procedure is used to choose automatically the bandwidth h [treated in Wand and Jones (1994b)]. In the proposed algorithm, KDE is used to model a specific pattern for each motor condition, indeed the features of the current signals are mapped in the two-dimensional principal component space, representing specific signatures of the motor conditions.

2.1.3 Kullback–Leibler Divergence

Given two continuous PDFs $f_1(\boldsymbol{x})$ and $f_2(\boldsymbol{x})$, a measure of "divergence" or "distance" between $f_1(\boldsymbol{x})$ versus $f_2(\boldsymbol{x})$ is given in Kullback and Leibler (1951), as:

$$I_{1:2}(X) = \int_{\mathbb{R}^d} f_1(\boldsymbol{x}) \log\frac{f_1(\boldsymbol{x})}{f_2(\boldsymbol{x})} d\boldsymbol{x}, \tag{8}$$

and between $f_2(\boldsymbol{x})$ versus $f_1(\boldsymbol{x})$ is given by:

$$I_{2:1}(X) = \int_{\mathbb{R}^d} f_2(\boldsymbol{x}) \log\frac{f_2(\boldsymbol{x})}{f_1(\boldsymbol{x})} d\boldsymbol{x}. \tag{9}$$

Therefore the K–L divergence between $f_1(\boldsymbol{x})$ and $f_2(\boldsymbol{x})$ is:

$$\begin{aligned} J(f_1;f_2) &= I_{1:2}(X) + I_{2:1}(X) \\ &= \int_{\mathbb{R}^d} (f_1(\boldsymbol{x}) - f_2(\boldsymbol{x})) \log\frac{f_1(\boldsymbol{x})}{f_2(\boldsymbol{x})} d\boldsymbol{x}. \end{aligned} \tag{10}$$

The above equation is known as the symmetric K–L divergence, which represents a non negative measure between two PDFs. In the present work d is 2 and a discrete form of K–L divergence is adopted:

$$J(f_1;f_2) = \sum_{i=1}^{n_{grid}} \sum_{j=1}^{d} (f_1(x_{ij}) - f_2(x_{ij})) \log\frac{f_1(x_{ij})}{f_2(x_{ij})}. \tag{11}$$

The K–L divergence allows to define a fault index: if f_Ω is the PDF in the PCs space estimated by KDE of the oncoming current measurements, the motor condition is that which minimizes the K–L divergence between f_Ω and f_i that is the ith PDF related to each motor condition:

$$c = arg \min_i J(f_\Omega;f_i), \tag{12}$$

where c is the classification output.

2.2 Developed Algorithm

The developed FDD procedure based on KDE consists of two stages: training and FDD monitoring. In the first, a KDE model is computed for each motor condition, in order to have one KDE model in the case of healthy motor and one for each faulty case. The training steps are summarized below:

T1. Stator current signals for each motor condition are acquired;
T2. Data are normalized;
T3. PCA transform (4) is applied to stator current signals, which are projected into the two-dimensional principal component space;
T4. The matrices $\boldsymbol{P}$ and $\boldsymbol{\mu}$ are stored;
T5. KDE is performed on the lower-dimensional principal components space (4) using a grid of n_{grid} points and a bandwidth h for the Gaussian kernel function (7);
T6. PDFs are estimated by KDE (7) and stored.

In diagnosis step, the models previously obtained are compared with the new data and a fault index is calculated. The diagnosis steps are summarized below:

D1. Stator current signals are acquired;
D2. Data are normalized;
D3. The matrices $\boldsymbol{P}$ and $\boldsymbol{\mu}$, previously computed (T4), are applied to signals;
D4. KDE is performed on the lower-dimensional principal component space (4) using the same points grid n_{grid} and bandwidth h used in the training step (T5);
D5. Symmetric K–L divergence (11) is computed between the estimated PDF by KDE (7) using the acquired current signals, and those stored in the training step (one for each condition) (T6);
D6. Diagnosis is evaluated using Eq. (12).

Faults are identified using Eq. (12) where f_Ω is the PDF, estimated by KDE, in the PCs space of the oncoming current measurements and f_i is the ith PDF related to each motor condition. K–L divergence is used as an input for fault decision algorithm allowing to take decision automatically on the operating state and condition of the machine and detecting any abnormal operating condition.

The next Section introduces the FDD experimental results of induction motors in order to show the proposed method performances.

3 Electric Motor FDD by MCSA: Results

In order to verify the effectiveness of the proposed methodology several simulations are carried out using one benchmark and some experimentations using real asynchronous motors. The benchmark uses a Time Stepping Coupled Finite Element-State Space modelling (FEM) approach to generate current signals for induction motors as described in Bangura et al. (2003). The simulation dataset consists of

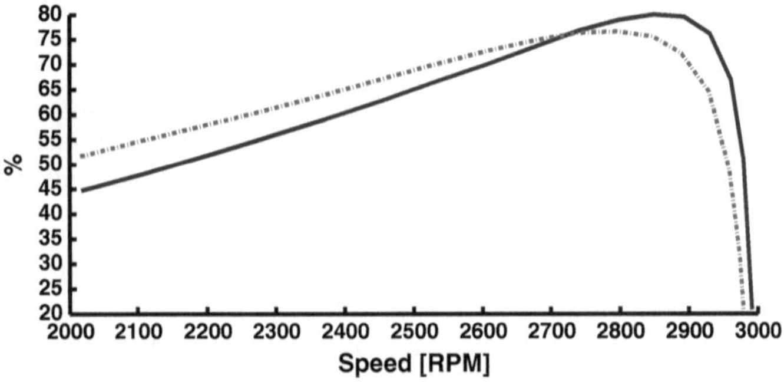

Fig. 1 Efficiency characterization of tested induction motors. *Blue solid line* refers the healthy motor, *red dashed line* refers to motor with defective rotor

twenty-one different motor conditions, which are: one healthy condition, ten broken bars conditions and ten broken connectors conditions. Twenty time series are generated for each motor condition. Each signal consists of 1,500 samples. The dataset can be download from UCR time series data mining archive in Keogh (2013). The characteristics of the three-phase induction motors are: 208 V input voltage, 60 Hz supply frequency, 34 rotor bars, 2 poles and power 1.2 hp. The sampling rate is 33.3 kHz and the processed data, for each test, are related to 0.3 s of acquisition. White noise with standard deviation $\sigma = 0.2$ is added to the simulated current signals. The results are the average of 200 Monte Carlo simulations where the training and testing data sets are randomly changed.

The real tests are carried out using three phase induction motors whose parameters are: 380 V input voltage, 60 Hz supply frequency, 0.75 kW power, 20 kHz sampling rate. Two different faults are tested: wrong rotor and cracked rotor. Wrong rotor refers to a non compliant rotor, in particular a single phase rotor is assembled instead of a three phase rotor. Ten motors are tested both for the healthy and faulty cases. The acquisition time is 14 s. The processed data, for each test, are related to 0.7 s of acquisition. In this case study the results are the average of 2,000 Monte Carlo simulations where the training and testing data sets are randomly changed. The motors, with a defective rotor installed, have about 3 % of efficiency drop at the operating point of 2,800 RPM, as shown in Fig. 1. So it is important to detect this defect in the energy efficiency context and QC.

3.1 Results and Discussion

The proposed approach processes the three-phase stator currents in order to perform defects detection and diagnosis as described in Sect. 2.2. The following two subsections show the results related to the two cases described previously. Figures 2, 3, 4 and 5 show the simulation and experimentation results. The classification accuracy is considered as an index to evaluate the performances of the proposed

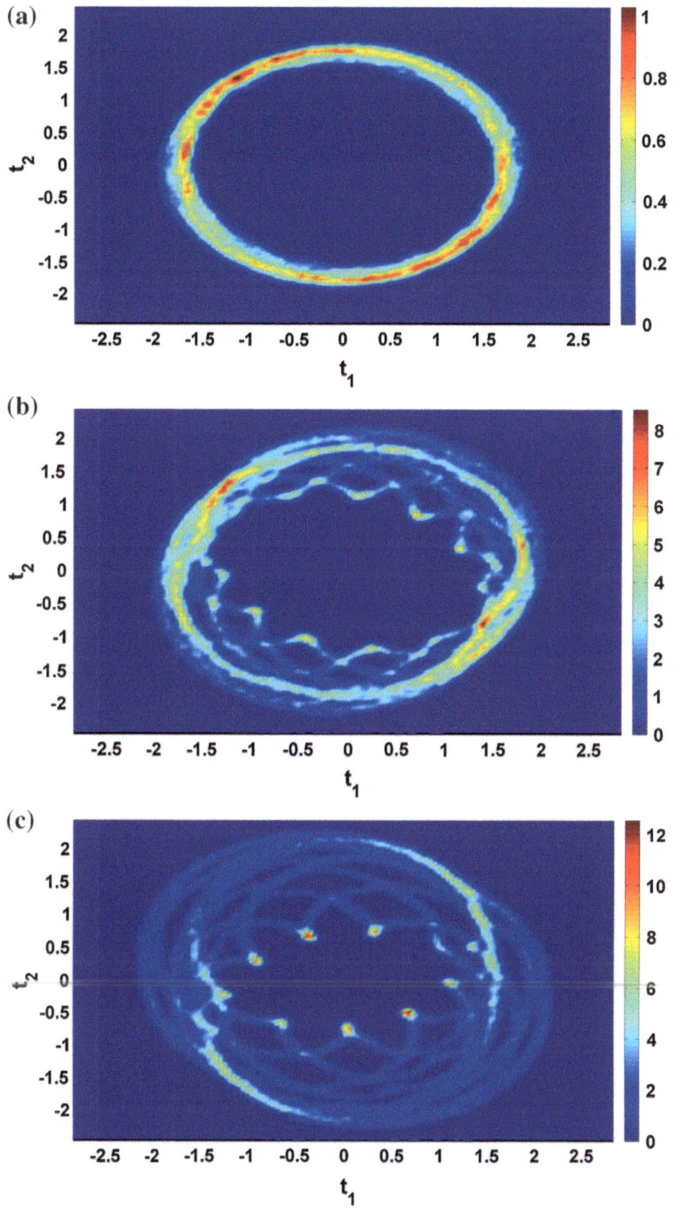

Fig. 2 Interpolated PDFs of a finite element motor in the two-dimensional principal component space estimated by KDE. **a** Healthy motor. **b** Motor with one broken bar. **c** Motor with one broken connector

algorithm as shown in Tables 1 and 2. This index is obtained using the probability distributions of the K-L distances of each class, approximated as normal distributions and estimated by Monte Carlo trials. The simulations are carried out changing

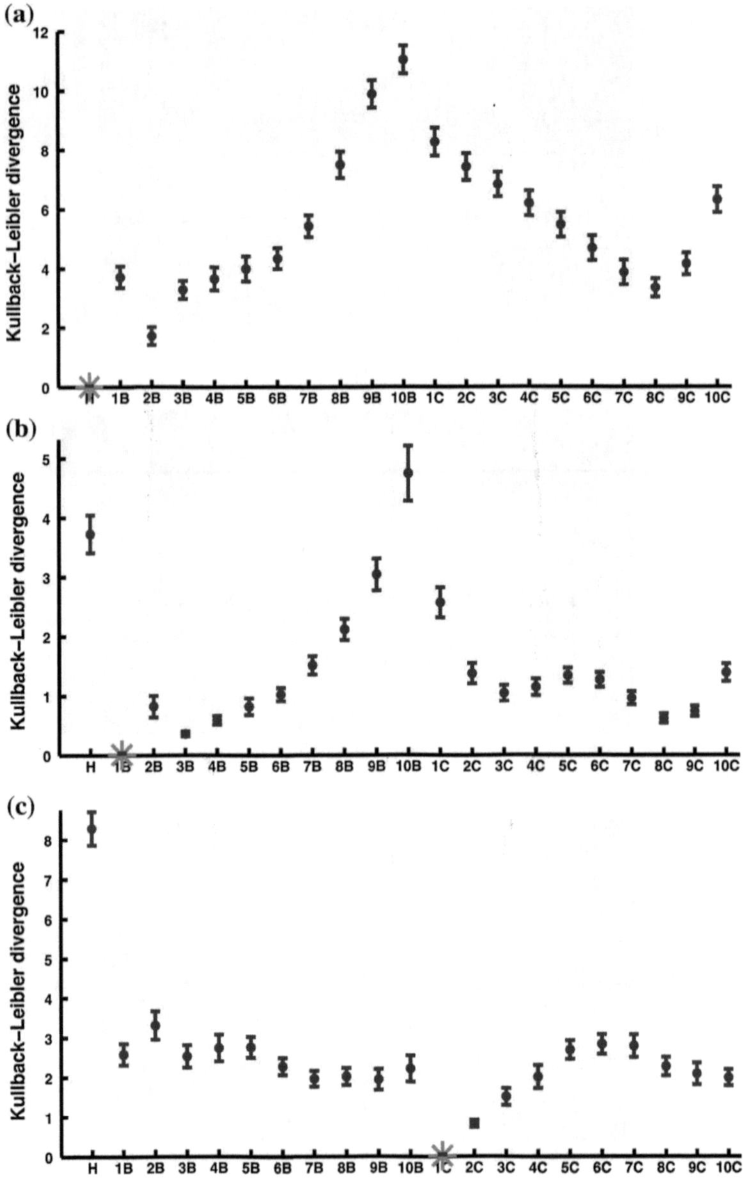

Fig. 3 K–L divergence in the case of a finite element motor. The *blue dots* are the mean, the *blue bars* are the four times standard deviation and the *red asterisks* are the classification output. Label H means healthy motor, labels 1–10B mean broken bars with the relative number, labels 1–10C mean broken connectors with relative number. **a** Healthy motor. **b** Motor with one broken bar. **c** Motor with one broken connector

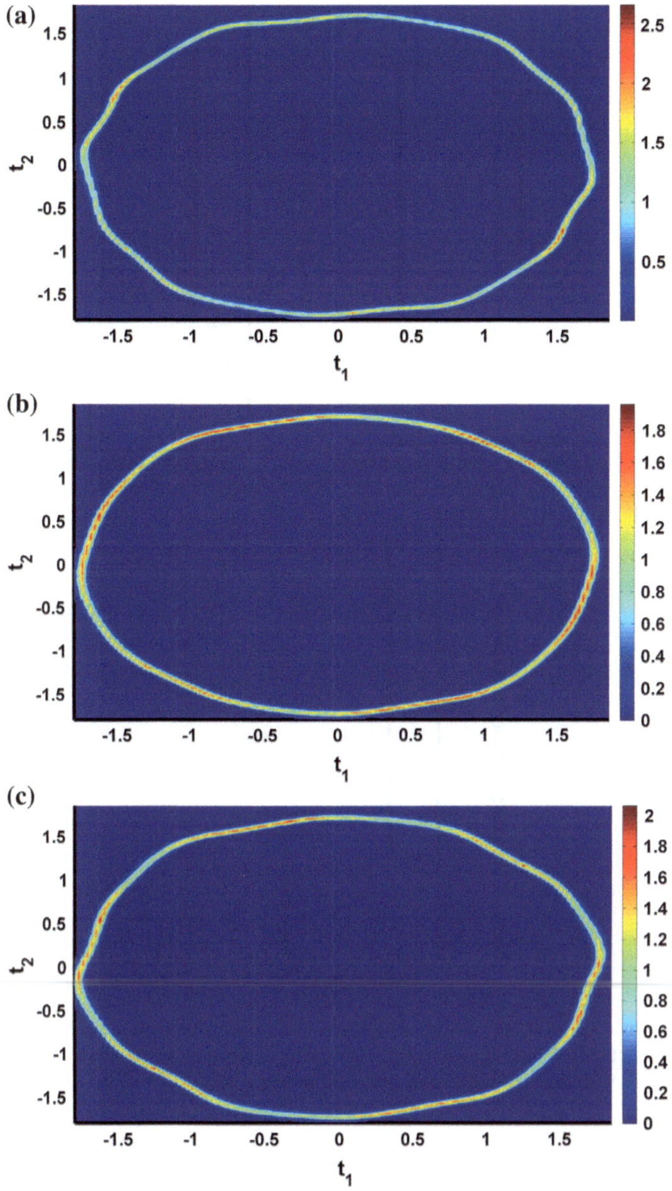

Fig. 4 Interpolated PDFs of real motors in the two-dimensional principal component space estimated by KDE. **a** Healthy motor. **b** Motor with cracked rotor. **c** Motor with wrong rotor

n_{grid}, the points number in which the PDF is estimated, and the current signals acquisition time in steady-state. Figures 3 and 5 show the K–L distances for all Monte Carlo trials. On each vertical line, the central dot is the mean and the

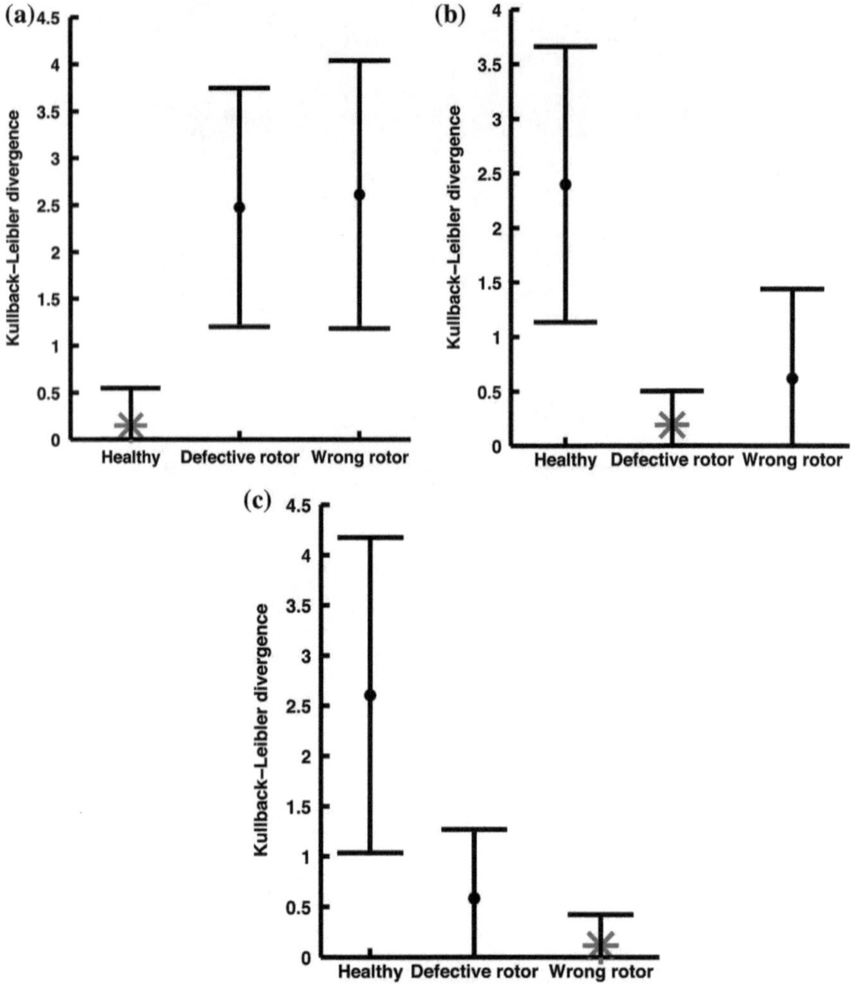

Fig. 5 K–L divergence in the case of real motors. The *blue dots* are the mean, the *blue bars* are the four times standard deviation and the *red asterisks* are the classification output. **a** Healthy motor. **b** Motor with cracked rotor. **c** Motor with wrong rotor

horizontal edges are the 4 times standard deviation. The figures show the results with $n_{grid} = 64 \times 64$ points and the acquisition time, for the benchmark and real motors, equals to 0.3 and 0.7 s respectively. This algorithm parameter setting guarantees better results for these cases taking into account the classification accuracy and the processing time. In the real motor the algorithm takes about 2.5 s for the classification output (Eq. 12): about 1 s to acquire the current signals, of which 0.25 s in transient state and 0.7 s in steady-state, and about 1.45 s to evaluate the PDF and the classification output (Eq. 12). Setting $n_{grid} = 32 \times 32$ points, the processing time is reduced to 1.5 s but decreasing the classification accuracy as

Table 1 Classification accuracy in the case of a finite element motor, changing n_{grid}, the points number in which the PDF is estimated, and the current signals acquisition time in steady-state

n_{grid}	128 × 128		64 × 64		32 × 32	
Acquisition time (s)	0.3	0.15	0.3	0.15	0.3	0.15
	%					
H	100	100	100	100	100	100
1B	100	100	100	100	100	100
2B	100	100	100	100	100	100
3B	100	100	100	100	100	99.93
4B	100	100	100	100	100	99.89
5B	100	100	100	99.96	100	99.70
6B	100	100	100	100	100	99.87
7B	100	99.98	100	99.94	100	98.19
8B	99.82	95.34	99.89	95.97	99.55	91.41
9B	100	99.74	100	99.53	100	98.56
10B	100	99.43	100	99.32	99.99	99.43
1C	100	100	100	100	100	99.99
2C	99.98	96.89	99.88	95.99	99.56	91.01
3C	99.88	91.71	99.72	95.74	98.48	93.75
4C	99.79	97.61	99.84	98.10	99.93	96.87
5C	99.98	98.96	99.99	98.49	99.96	96.37
6C	100	100	100	100	100	99.98
7C	100	100	100	100	100	99.94
8C	100	100	100	100	100	99.88
9C	100	99.99	100	100	100	99.47
10C	100	99.89	100	99.77	100	96.95
Mean	99.97	99.03	99.97	99.18	99.88	98.15

Label H means healthy motor, labels 1–10B mean broken bars with the relative number, labels 1–10C mean broken connectors with relative number

Table 2 Classification accuracy in the case of real motors, changing n_{grid}, the points number in which the PDF is estimated, and the current signals acquisition time in steady-state

n_{grid}	128 × 128			64 × 64			32 × 32		
Acquisition time (s)	0.7	0.5	0.3	0.7	0.5	0.3	0.7	0.5	0.3
	%								
Healthy	100	100	100	100	100	100	100	100	100
Cracked rotor	98.82	95.08	77.08	99.00	94.74	86.54	98.29	94.02	81.45
Wrong rotor	98.97	99.47	99.49	99.18	99.36	98.56	99.85	99.41	99.21
Mean	99.26	98.18	92.19	99.39	98.03	95.03	99.38	97.81	93.55

Motor conditions are: healthy, motor with cracked rotor and motor with wrong rotor shown in Tables 1 and 2. The tests are also performed for both cases using the asymmetric K–L divergence (Eq. 9). The results are comparable to those achieved with the symmetric K–L divergence described in the next subsections.

3.1.1 Broken Rotor Bars and Connectors Diagnosis

Figures 2a–c depict the patterns of a healthy motor, one broken bar and one broken connector conditions, respectively; these figures show how the PDFs, estimated by KDE in the principal component space, are used as the specific patterns for the motor conditions. The simulation results, given in Figs. 3a–c, show the faults diagnosis for broken rotor bars and connectors, setting $n_{grid} = 64 \times 64$ and the current signals acquisition time in steady-state condition is equal to 0.3 s. Figure 3 (a) shows the K–L divergence among the PDFs, estimated by KDE, of all motor conditions (i.e. healthy, from one to ten broken rotor bars and from one to ten broken connectors) and the PDF estimated by KDE from stator current signals of healthy motor. The results show as the minimum K–L distance is exactly the healthy condition. Figure 3b shows the K–L divergence among all PDFs and the PDF estimated from stator current signals affected by one broken rotor bar. In this case the graph shows as the minimum K–L distance is exactly the broken bar condition. The last graph, Fig. 3c, shows the one broken connector diagnosis. Even in this case the K–L divergence detects and identifies the fault, that is one broken connector. By Monte Carlo simulations, all fault types are diagnosed with 100 % accuracy hence the K–L divergence figures for the other faults are not reported. Moreover the classification accuracy is 100 % with acquisition time above 0.3 s for each fault, while below 0.3 s, the classification accuracy decreases as shown in Table 1.

3.1.2 Real Induction Motors Diagnosis

Figures 4a–c depict the patterns of three real motors: healthy, cracked and wrong rotor; these figures show as the PDFs, estimated by KDE in the principal component space, are different and therefore can be used as specific patterns for each motor condition. Experimental results given in Figs. 5a–c show the fault diagnosis for cracked and wrong rotors, setting $n_{grid} = 64 \times 64$ and the current signals acquisition time in steady-state is equal to 0.7 s. Figure 5a shows the K–L divergence among the PDFs, estimated by KDE, of all motor conditions (i.e. healthy, cracked and wrong rotors) and the PDF estimated by KDE from stator current signals of healthy motor. The results show as the minimum K–L distance is exactly the healthy condition. Figure 5b shows the K–L divergence among all PDFs and the PDF estimated from stator current signals where cracked rotors are diagnosed. In this case the graph shows as the minimum K–L distance is exactly the cracked rotor condition. The last graph, Fig. 5c, shows the wrong rotor diagnosis. Even in this case the K–L divergence detects and identifies the fault. By Monte Carlo simulations, all fault types are diagnosed with accuracy reported in Table 2. It can be noticed how the classification accuracy in the case of healthy motor is always 100 %, therefore the algorithm is able to detect if motors are healthy or if there are some faults or defects. In Figs. 5b and c the blue lines of motors with cracked and wrong rotor are never overlapped to the blue lines of healthy motors so, in these tests, the algorithm never confuses the cases of healthy motors from those not healthy.

The next Section describes the well-known MSPCA algorithm for fault detection and isolation based on vibration signals.

4 Electric Motor FDD by MVSA

In electric motors, faults and defects are often correlated to the vibration signals, which can be processed to model the motor behaviours by patterns that represent the normal and abnormal motor conditions. Vibration analysis is widely accepted as a tool to detect faults of a rotating machine since it is reliable, not destructive and it permits continuous monitoring without stopping the machine. A brief literature review is given by: Fan and Zheng (2007), Immovilli et al. (2010), Sawalhi and Randall (2008a, b), Tran et al. (2009), Yang and Kim (2006). In particular, it is possible to detect different faults by analysing the vibration power spectrum. Most common faults are unbalance and misalignment. Unbalance may be caused by poor balancing, shaft inflection (i.e. thermal expansion) and rotor distortion by magnetic forces (a well known problem in high power electrical machines). Misalignment may be caused by misaligned couplings, misaligned bearings or crooked shaft.

In order to model the vibration signals, MSPCA is taken into account, as presented in Bakshi (1998). MSPCA deals with processes that operate at different scales, and have contributions from:

- events occurring at different localizations in time and frequency;
- stochastic processes whose energy or power spectrum changes over time and/or frequency;
- variables measured at different sampling rate or containing missing data.

MSPCA transforms the process data information at different scales by WT. The information of each different scale is captured by PCA modelling. These patterns, which represent the process conditions, can be used to identify each fault and defect.

To detect the defects, a KDE algorithm is used on the PCA residuals, and the thresholds are computed for each sensors signal. It allows to identify if, for each wavelet scale, the signals are involved in the fault or not. When Gaussian assumption is not recognized, KDE method is a robust methodology to estimate numerically the PDF, by Odiowei and Cao (2010). Fault isolation is carried out by contribution plots, which is based on quantifying the contribution of each process variable to the single scores of the PCA. Diagnosis can be performed using the contribution plots because they represent the signatures of the rotating electrical machine conditions. The contributions are the inputs of a LDA classifier, which is a supervised machine learning algorithm used here to diagnose each motor defect. Several simulations are carried out using a benchmark provided by the Case Western Reserve University Bearing Data Center (2014).

4.1 Recalled Results

In this section authors present the algorithms used to develop the fault and defect diagnosis procedure. It extracts patterns by vibration signals using MSPCA and PCA contributions are used to diagnose each motor fault.

4.1.1 Principal Component Analysis

PCA is introduced in the Sect. 2.1.1, here an improved PCA fault detection index is described. A deviation of the new data sample $\boldsymbol{X}$ from the normal correlation could change the projections onto the subspaces, either S_d or S_r. Consequently, the magnitude of either $\tilde{\boldsymbol{X}}$ or $\hat{\boldsymbol{X}}$ could increase over the values obtained with normal data. The Square Prediction Error (SPE) is a statistic that measures lack of fit of a model to data. The SPE statistic is the difference, or residual, between a sample and its projection into the d components retained in the model. The description of the distribution of SPE is given in Jackson (2003):

$$SPE \equiv \|\tilde{\boldsymbol{X}}\|^2 = \|\boldsymbol{X}(\boldsymbol{I} - \boldsymbol{P}\boldsymbol{P}^{\boldsymbol{T}})\|^2. \tag{13}$$

The process is faultless if:

$$SPE \leq \delta^2 \tag{14}$$

where δ^2 is a confidence limit for SPE. A confidence limit expression for SPE, when $\boldsymbol{x}$ follows a normal distribution, is developed in Jackson and Mudholkar (1979), Misra et al. (2002) and Rodriguez et al. (2006). The fault detectability condition is given in Dunia and Joe Qin (1998) and recalled in the following. Defining:

$$\boldsymbol{X} = \boldsymbol{X}^* + \boldsymbol{f}\boldsymbol{\Xi}, \tag{15}$$

where the sample vector for normal operating conditions is denoted by $\boldsymbol{X}^*$, $\boldsymbol{f}$ represents the magnitude of the fault and Ξ is a fault direction vector. Necessary and sufficient conditions for detectability are:

- $\tilde{\Xi} = (\boldsymbol{I} - \boldsymbol{PP}^T)\Xi \neq 0$, with $\tilde{\Xi}$ the projection of Ξ on the residual subspace;
- $|\tilde{\boldsymbol{f}}| = |(\boldsymbol{I} - \boldsymbol{PP}^T)\boldsymbol{f}| > 2\delta$, with $\tilde{\boldsymbol{f}}$ the projection of $\boldsymbol{f}$ on the residual subspace.

The drawbacks of *SPE* index for fault detection are mainly two: the first is related to the assumption of normal distribution to estimate the threshold of this index, the second is that the *SPE* is a weighted sum, with unitary coefficients, of quadratic residues $\tilde{\boldsymbol{X}}_i$. To improve the fault detection, these two drawbacks are faced assuming that the process is faultless if, for each i:

$$\tilde{\boldsymbol{X}}_i^2 \leq \delta_i \quad i = 1, \ldots, m, \tag{16}$$

where δ_i is a confidence limit for $\tilde{\boldsymbol{X}}_i^2$. To estimate the confidence limit δ_i, even if the normality assumption of $\tilde{\boldsymbol{X}}_i^2$ is not valid, the solution is to estimate the PDF directly from $\tilde{\boldsymbol{X}}_i^2$ through a non parametric approach. In Yu (2011a, b) and Odiowei and Cao (2010), KDE is considered because it is a well established non parametric approach to estimate the PDF of statistical signals and evaluate the control limits. Assume y is a random variable and its density function is denoted by $p(y)$. This means that:

$$P(y < k) = \int_{-\infty}^{k} p(y)dy. \tag{17}$$

Hence, by knowing $p(y)$, an appropriate control limit can be given for a specific confidence bound α, using Eq. (17). Replacing $p(y)$, in Eq. (17), with the estimation of the probability density function of $\tilde{\boldsymbol{X}}_i^2$, called $\hat{p}(\tilde{\boldsymbol{X}}_i^2)$, the control limits will be estimated by:

$$\int_{-\infty}^{\delta_i} \hat{p}(\tilde{\boldsymbol{X}}_i^2)d\tilde{\boldsymbol{X}}_i^2 = \alpha. \tag{18}$$

Fault isolation and diagnosis are performed by the PCA contributions: defining the new observation vector $\boldsymbol{x}_j \in \mathbb{R}^m$, the total contribution of the ith process variable $\boldsymbol{X}_i$ is

$$CONT_i = \sum_{j=1}^{N} \tilde{x}_{ij}^2 \quad i = 1, \ldots, m. \tag{19}$$

4.1.2 Wavelet Transform

The Wavelet Transform (WT) is defined as the integral of the signal $f(t)$ multiplied by scaled, shifted version of basic wavelet function $\phi(t)$, that is a real valued function whose Fourier transform satisfies the admissibility criteria stated in Li et al. (1999). Then the wavelet transformation $c(\cdot,\cdot)$ of a signal $f(t)$ is defined as:

$$\begin{aligned} &c(a,b) = \int_{\mathbb{R}} f(t)\frac{1}{\sqrt{a}}\phi\left(\frac{t-b}{a}\right)dt \\ &a \in \mathbb{R}^{+} - \{0\} \\ &b \in \mathbb{R}, \end{aligned} \tag{20}$$

where a is the so-called scaling parameter, b is the time localization parameter. Both a and b can be continuous or discrete variables. Multiplying each coefficient by an appropriately scaled and shifted wavelet it yields the constituent wavelets of the original signal. For signals of finite energy, continuous wavelets synthesis provides the reconstruction formula:

$$f(t) = \frac{1}{K_\phi}\int_{\mathbb{R}}\int_{\mathbb{R}^{+}} c(a,b)\phi\left(\frac{t-b}{a}\right)\frac{da}{a^2}db \tag{21}$$

where:

$$K_\phi = \int_{-\infty}^{+\infty}\frac{|\hat{\phi}(\xi)|^2}{|\xi|}\mathrm{d}\xi \tag{22}$$

denotes a (Wavelet specific) normalization parameter in which $\hat{\phi}$ is the Fourier transform of ϕ. Mother wavelets must satisfy the following properties:

$$\int_{-\infty}^{+\infty}|\phi(t)|dt < \infty, \quad \int_{-\infty}^{+\infty}|\phi(t)|^2 dt = 1, \quad \int_{-\infty}^{+\infty}\phi(t)dt = 0. \tag{23}$$

To avoid intractable computations when operating at every scale of the Continuous WT (CWT), scales and positions can be chosen on a power of two, i.e. dyadic scales and positions. The Discrete WT (DWT) analysis is more efficient and accurate, as reported in Li et al. (1999) and Daubechies (1988). In this scheme a and b are given by:

$$a = a_0^j, \quad b = b_0 a_0^j k, \quad (j,k) \in \mathbb{Z}^2, \quad \mathbb{Z} := \{0, \pm 1, \pm 2, \cdots\}. \tag{24}$$

The variables a_0 and b_0 are fixed constants that are set, as in Daubechies (1988), to: $a_0 = 2$ and $b_0 = 1$. The discrete wavelet analysis can be described mathematically as:

$$\begin{aligned} &c(a,b) = c(j,k) = \sum_{n \in \mathbb{Z}^+} f(n)\phi_{j,k}(n), \\ &a = 2^j,\ b = 2^j k, \\ &j \in \mathbb{Z},\ k \in \mathbb{Z}, \end{aligned} \tag{25}$$

considering the simplified notation $f(n) = f(n \cdot t_c)$, $n \in \mathbb{Z}^+$ and t_c the sampling time, the discretization of continuous time signal $f(t)$ is considered. The inverse transform, also called discrete synthesis, is defined as:

$$f(n) = \sum_{j \in \mathbb{Z}} \sum_{k \in \mathbb{Z}} c(j,k)\phi_{j,k}(n). \tag{26}$$

In Mallat (1989), a signal is decomposed into various scales with different time and frequency resolutions, this algorithm is known as the multi-resolution signal decomposition. Defining:

$$\begin{aligned} \phi_{j,k}(n) &= 2^{-j/2}\phi(2^{-j}n - k), \\ \psi_{j,k}(n) &= 2^{-j/2}\psi(2^{-j}n - k), \\ V_j &= span\{\phi_{j,k}, k \in \mathbb{Z}\}, \\ W_j &= span\{\psi_{j,k}, k \in \mathbb{Z}\}, \end{aligned} \quad (j,k) \in \mathbb{Z}^2 \tag{27}$$

the wavelet function $\phi_{j,k}$, is the orthonormal basis of V_j and the orthogonal wavelet $\psi_{j,k}$, called scaling function, is the orthonormal basis of W_j. In Daubechies (1988) is shown that:

$$\begin{aligned} &V_j \perp W_j, \\ &V_m = W_{m+1} \oplus V_{m+1}. \end{aligned} \quad V_m, W_m \subset \mathbf{L}^2(\mathbb{R}) \tag{28}$$

Defining $f(n) = f$ as element of $V_0 = W_1 \oplus V_1$, f can be decomposed into its components along V_1 and W_1:

$$f = P_1 f + Q_1 f. \tag{29}$$

with P_j the orthogonal projection onto V_j and Q_j the orthogonal projection onto W_j. Defining $j \geq 1$ and $f(n) = c_n^0$, it results:

$$\begin{aligned} f(n) &= \sum\nolimits_{k \in \mathbb{Z}} c_k^1 \phi_{1,k}(n) + \sum_{k \in \mathbb{Z}} d_k^1 \psi_{1,k}(n), \\ c_k^1 &= \sum\nolimits_{n \in \mathbb{Z}} h(n-2k) c_n^0, \\ d_k^1 &= \sum\nolimits_{n \in \mathbb{Z}} g(n-2k) c_n^0, \\ h(n-2k) &= \langle \phi_{1,k}(n), \phi_{0,n}(n) \rangle, \\ g(n-2k) &= \langle \psi_{1,k}(n), \psi_{0,n}(n) \rangle. \\ &k,\ n \in \mathbb{Z}^2. \end{aligned} \tag{30}$$

where the terms g and h are high-pass and low-pass filter coefficients derived from the bases ψ and ϕ. Considering a dataset of N $(n = 1,\ldots,N)$ samples, and introducing a vector notation, c_k^1 and d_k^1 can be rewrite as Daubechies (1988):

$$\begin{aligned} \boldsymbol{c}^1 &= \boldsymbol{H}\boldsymbol{c}^0, \\ \boldsymbol{d}^1 &= \boldsymbol{G}\boldsymbol{c}^0, \end{aligned} \tag{31}$$

with

$$\boldsymbol{H} = \begin{bmatrix} h(0) & h(1) & \cdots & h(N) \\ h(-2) & h(-1) & \cdots & h(N-2) \\ \vdots & \vdots & \cdots & \vdots \\ h(-2k) & h(1-2k) & \cdots & h(N-2k) \end{bmatrix}, \tag{32}$$

$$\boldsymbol{G} = \begin{bmatrix} g(0) & g(1) & \cdots & g(N) \\ g(-2) & g(-1) & \cdots & g(N-2) \\ \vdots & \vdots & \cdots & \vdots \\ g(-2k) & g(1-2k) & \cdots & g(N-2k) \end{bmatrix}. \tag{33}$$

The procedure can be iterated obtaining:

$$\begin{aligned} c^j &= \boldsymbol{H}\boldsymbol{c}^{j-1}, \\ d^j &= \boldsymbol{G}\boldsymbol{d}^{j-1}. \end{aligned} \tag{34}$$

Then:

$$\begin{aligned} c^j &= \boldsymbol{H}_j c^0, \\ d^j &= \mathbf{G}_j \boldsymbol{d}^0, \end{aligned} \tag{35}$$

where $\boldsymbol{H}_j$ is obtained by applying the $\boldsymbol{H}$ filter j times, and $\boldsymbol{G}_j$ is obtained by applying the $\boldsymbol{H}$ filter $j-1$ times and the $\boldsymbol{G}$ filter once. Hence any signal may be decomposed into its contributions in different regions of the time-frequency space by projection on the corresponding wavelet basis function. The lowest frequency content of the signal is represented on a set of scaling functions. The number of wavelet and scaling function coefficients decreases dyadically at coarser scales due to the dyadic discretization of the dilation and translation parameters. The algorithms for computing the wavelet decomposition are based on representing the projection of the signal on the corresponding basis function as a filtering operation (Mallat 1989). Convolution with the filter $\boldsymbol{H}$ represents projection on the scaling function, and convolution with the filter $\boldsymbol{G}$ represents projection on a wavelet. Thus, the signal $f(n)$ is decomposed at different scales, the detail scale matrices and approximation scale matrices. Defining L the decomposition levels, the approximation scale $\boldsymbol{A}_L$

and the detail scales D_j, $j = 1, \ldots, L$ are the composition of c^j and d^j for every m variables of the data matrix X:

$$\begin{aligned} A_j &= [c_1^j, c_2^j, \ldots, c_m^j], \\ D_j &= [d_1^j, d_2^j, \ldots, d_m^j]. \end{aligned} \quad j = 1, \ldots, L \tag{36}$$

To select the wavelet decomposition level L it is considered the minimum number of decomposition levels, and used to obtain an approximation signal A_L so that the upper limit of its associated frequency band is under the fundamental frequency f, as described by the following condition Antonino-Daviu et al. (2006), Bouzida et al. (2011):

$$2^{-(L+1)} f_s < f. \tag{37}$$

where f_s is the sampling frequency of the signals and f is the fundamental frequency of the machine. From this condition, the decomposition level of the approximation signal is the integer L given by:

$$L = \lfloor \log_2(f_s/f) - 1 \rfloor. \tag{38}$$

4.2 MSPCA Formulation

WT and PCA can be combined to extract maximum information from multivariate sensor data. MSPCA can be used as a tool for fault detection and diagnosis by means of statistical indexes. In particular, faults are detected by using Eqs. 16 and 18 and the isolation is conducted by the contribution method (Eq. 19). In this way it is possible to detect which sensor is most affected by fault (see Misra et al. 2002). Two fundamental theorems exist for the MSPCA formulation, they assess that PCA assumptions remain unchanged under the Wavelet transformation. These theorems are useful to apply MSPCA methodology, as stated in Bakshi (1998).

Theorem 4.1 *Let* $W = \left[H_L', G_L', G_{L-1}', \ldots, G_1'\right]' \in \mathbb{R}^{N \times N}$ *the orthonormal matrix representing the orthonormal wavelet transformation operator containing the filter coefficients, the principal component loadings obtained by the PCA of* X *and* WX *are identical, whereas the principal component scores of* WX *are the wavelet transform of the scores of* X.

Theorem 4.2 *MSPCA reduces to conventional PCA if neither the principal components nor the wavelet coefficients at any scale are eliminated.*

The developed FDD MSPCA based procedure consists of two stages: in the first step, the faultless data are processed and a model of this data is built. MSPCA training steps are summarized below:

T1. Data are preprocessed;
T2. The Wavelet analysis is used, to refine the data, with a level of detail L which is chosen by Eq. (38);
T3. Normalize mean and standard deviation of detail and approximation matrices and apply PCA to the approximation matrix $\boldsymbol{A}_L$, of order L, and to the L detail matrices $\boldsymbol{D}_j$, where $j = 1, \ldots L$;
T4. The PCA transformation matrix $\boldsymbol{P}$ and the signal covariance matrix $\boldsymbol{S}$ are computed for each approximation and detail matrices;
T5. The $\tilde{\boldsymbol{X}}_i$ signals (Eq. 13) are computed, for each wavelet matrix;
T6. The δ_i thresholds are computed, for each detail matrix and for the approximation matrix of order L, using the KDE algorithm (Eq. 18) and a confidence bound α;

In the second step, the model previously obtained is on-line compared with the new data and a statistical index of failure is calculated. MSPCA diagnosis steps are summarized below:

D1. The previous steps, except the threshold computation step (T6), are repeated for each new dataset, the data are standardized as in the training step (T3) and the PCA and $\tilde{\boldsymbol{X}}_i$ signals are computed using the $\boldsymbol{P}$ and $\boldsymbol{S}$ matrices, obtained in the training step;
D2. **If** any of the $\tilde{\boldsymbol{X}}_i^2$ signals is over the thresholds δ_i, the fault is detected and the isolation is performed by the contributions, **else** the next data set is analysed [return to (D1)];
D3. Compute all the residual contributions, for each sensor, for all details and approximation matrices and isolate and diagnose the fault type.

The next Section introduces the FDD experimental results in order to show the MSPCA algorithm performances. Tests are carried out on real induction motors with different fault severity.

5 Electric Motor FDD by MVSA: Results

The diagnosis algorithm has been tested on the vibration signals provided by the Case Western Reserve University Bearing Data Center (2014). Experiments were conducted using a 2 hp Reliance Electric motor, and acceleration data was measured at locations near to and remote from the motor bearings. Motor bearings were seeded with faults using electro-discharge machining (EDM). Faults ranging from 0.007 in. to 0.040 in. of diameter were introduced separately at the inner raceway, rolling element (i.e. ball) and outer raceway. Faulty bearings were reinstalled into the test motor and vibration data was recorded for motor loads of 0–3 hp (motor speeds of 1,797–1,720 RPM). Accelerometers were placed at the 12 o'clock position at both the drive end and fan end of the motor housing. Digital data was collected at 12,000 samples per second. Experiments were conducted for both fan and drive end bearings with outer raceway faults located at 3 o'clock (directly in the load zone), at 6 o'clock (orthogonal to the load zone), and at 12 o'clock.

5.1 Results and Discussion

The proposed approach described in Sect. 4.2 has been tested using a Daubechies mother wavelet of order 15, defined *db*15 mother wavelet (defined kernel ϕ in Sect. 4.1.2). Since the motor rotation frequency is 30 Hz and the sampling frequency is 12 kHz, applying Eq. (38), the level of detail obtained is $L = 7$. The dimension of principal component subspace d, chosen by the Kaiser's rule, is described in Jolliffe (2002).

Incoming batch data samples are then fed into the MSPCA model and the PCA residual contributions are computed for the matrices $\boldsymbol{D}_j$, $j = 1, \ldots, L$, $\boldsymbol{A}_L$. In the following, these matrices are defined *scale matrices*, and they are compared with the respective thresholds. When, at any scale, the number of residual contribution samples over the thresholds is greater than $\alpha \cdot \gamma$, where α is the significance level used for the threshold δ_i calculation (stated in Sect. 4.2) and γ is a corrective index (fixed equal to 2), a fault is detected and the motor is considered faulty.

Once a fault is detected, the isolation and diagnosis tests are performed. At this step the PCA contributions are computed for each scale matrix. Fault isolation allows to detect which sensors are involved in the fault. By using several scales for the DWT analysis, it is possible to cluster the residual contributions of each scale and define a unique signature of the motor fault, as in a MVSA approach. More in detail, the signature of each fault is given by the contributions of each variable for each scale. The results are the average of 1,000 Monte Carlo simulations where the training and testing data sets are randomly changed.

Figures 6 and 7 show the residuals of the first accelerometer (i.e. placed at the drive end) for drive end bearing faults estimated by Eq. (16). The thresholds, drawn in dashed red line, are estimated by KDE (Eq. 18). While Fig. 6a shows the residuals for healthy motor, Fig. 6b, c show the residuals of rolling element and inner raceway faults respectively at the detail scales D_1 and D_4, which are, among all scales, the most affected by the faults.

Figures 7a–c show the residuals of outer raceway faults located at 3, 6, 12 o'clock respectively at the detail scales D_1, D_2 and D_4, which are, among all scales, the most affected by the faults. It can be noticed how the residuals are related to the fault type and so they can be exploited as signatures of the rotating electrical machine conditions.

Figures 8 and 9 show the contribution plots of each accelerometer at different scales for drive end bearing fault, particularly Figs. 8a–c show the contribution plots of healthy motor, rolling element and inner raceway faults while Figs. 9a–c

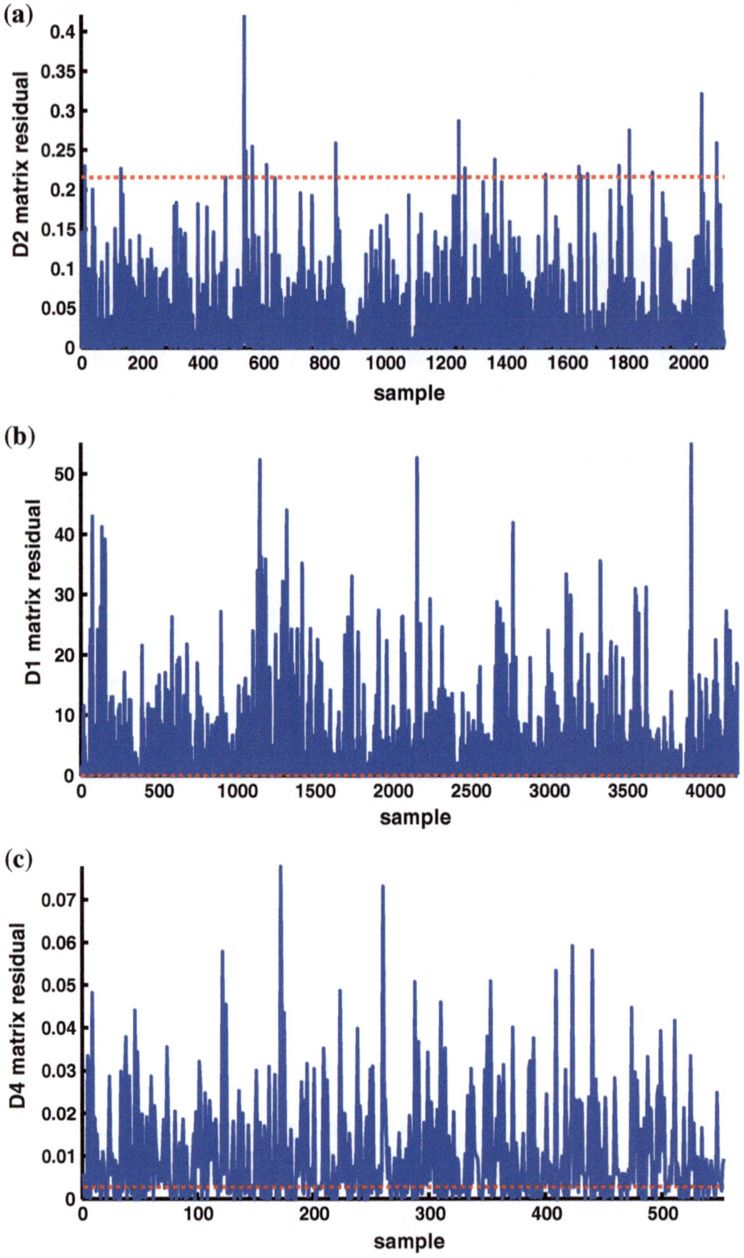

Fig. 6 Residuals of the accelerometer placed at the drive end. **a** Healthy motor at D_2 detail scale. **b** Rolling element fault of drive end bearing at D_1 detail scale. **c** Inner raceway fault of drive end bearing at D_4 detail scale

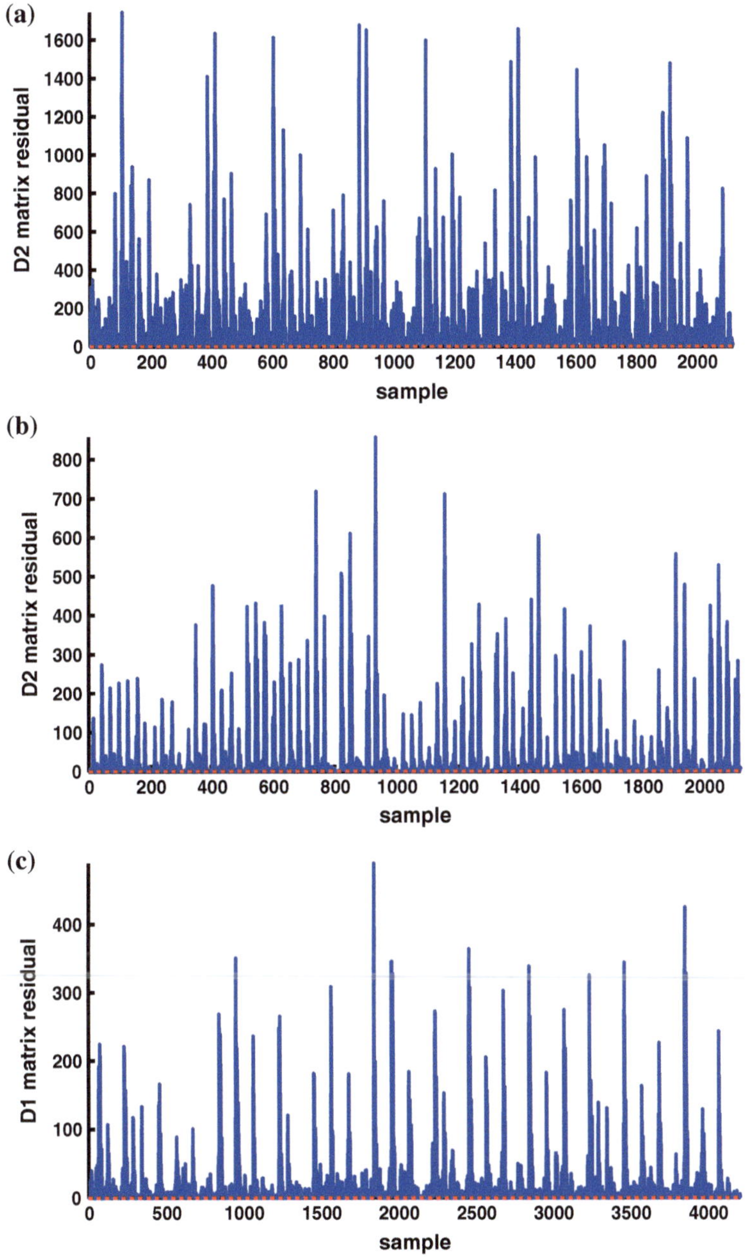

Fig. 7 Residuals of the accelerometer placed at the drive end. **a** Outer raceway fault located at 3 o'clock of drive end bearing at D_1 detail scale. **b** Outer raceway fault located at 6 o'clock of drive end bearing at D_2 detail scale. **c** Outer raceway fault located at 12 o'clock of drive end bearing at D_2 detail scale

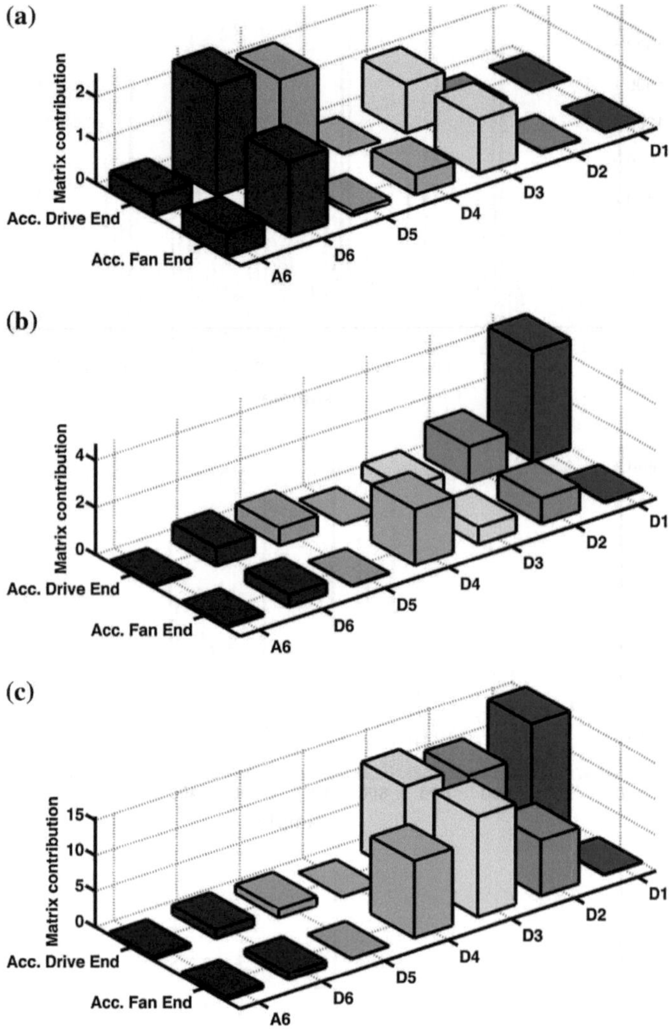

Fig. 8 Contribution plots. **a** Healthy motor. **b** Rolling element fault of drive end bearing. **c** Inner raceway fault of drive end bearing

show the contribution plots of outer raceway fault located at 3, 6, 12 o'clock respectively.

The contribution plots could be used as signatures of the electric motor conditions, so a supervised machine learning algorithm, with the PCA contributions as inputs, can be used to diagnose each motor fault. The Figs. 8 and 9 show that the identified signatures by PCA contributions are features for each fault. The accelerometers are involved in these signatures at different scales with different amplitudes. As shown by Figs. 8 and 9, all faults affect the accelerometer placed at the drive end.

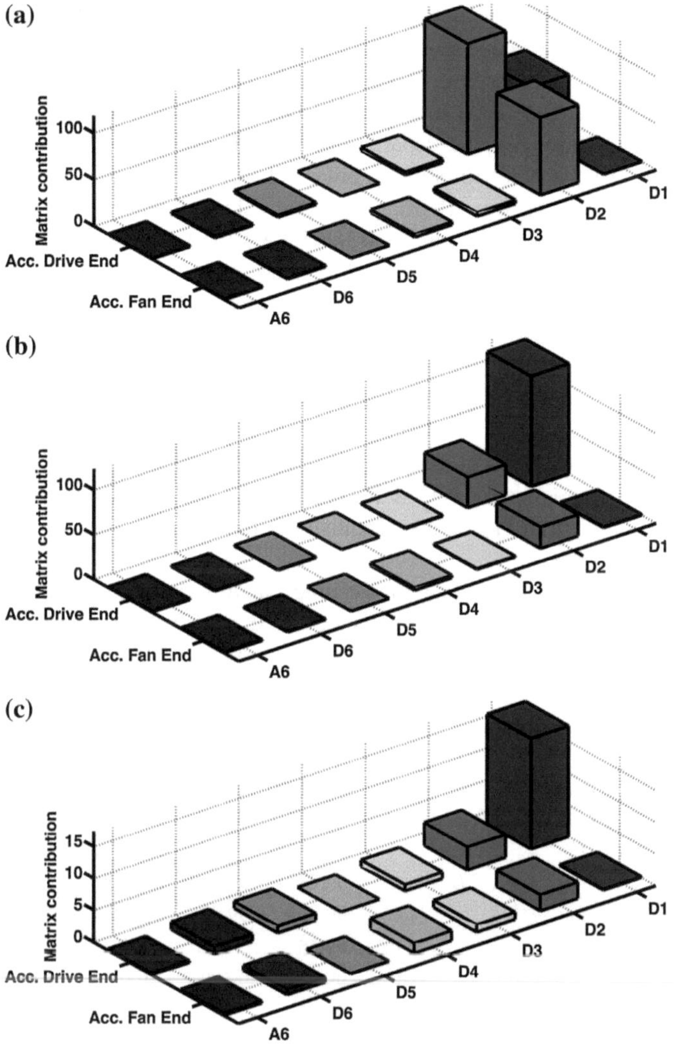

Fig. 9 Contribution plots. **a** Outer raceway fault located at 3 o'clock of drive end bearing. **b** Outer raceway fault located at 6 o'clock of drive end bearing. **c** Outer raceway fault located at 12 o'clock of drive end bearing

This points out that the contribution plots can be used to identify the sensors affected by the faults. Particularly, the fault isolation is performed by computing for each scale the average value: the sensor affected by the fault is that one with the highest average value. As shown in Figs. 8 and 9 the sensor most affected by the faults is the accelerometer placed at the drive end. The fault diagnosis is performed using the contribution plots because they are the signatures of the electric motor conditions and are features for each fault as shown in Figs. 8 and 9. Outer raceway fault located at 6 o'clock and outer raceway fault located at 12 o'clock of drive end bearing affect the contributions at the same scales (Figs. 9a–b), but the contribution amplitudes are different so they can be used to diagnose the faults.

In order to diagnose each motor fault LDA is used. It searches a linear transformation that maximizes class separability in a reduced dimensional space. LDA is proposed in Fisher (1936) for solving binary class problems. It is further extended to multi-class cases in Rao (1948). In general, LDA aims to find a subspace that minimizes the within-class scatter and maximizes the between-class scatter simultaneously. PCA contributions are used as features input to the LDA algorithm. Tables 3, 4 and 5 show the classification accuracy. The results are the average classification accuracy of each motor conditions (i.e. healthy motor, rolling element fault, inner raceway fault, outer raceway fault located at 3, 6 and 12 o'clock) and of 4 motor loads: from 0 to 3 hp (motor speeds of 1,797–1,720 RPM). Table 3 shows the classification accuracy at different wavelet decomposition level L and acquisition time of faults occurred at drive end bearing with fault diameter of 0.007 in. The classification accuracy is over 99 % for each level L and acquisition time, so a low wavelet decomposition level and acquisition time can be chosen to diagnose effectively this fault.

Table 4 shows the classification accuracy at different wavelet decomposition level L and acquisition time of faults occurred at drive end bearing with fault diameter of 0.021 in. The classification accuracy is over 99 % at each level L and acquisition time higher 0.3 s, so a low wavelet decomposition level and acquisition

Table 3 Average classification accuracy of drive end bearing fault with fault diameter of 0.007 in.

		Acquisition time s				
		0.1	0.2	0.3	0.5	0.7
		%				
L	1	99.23	99.96	99.98	100	100
	2	99.40	99.85	99.99	100	100
	3	99.55	99.95	100	99.99	100
	4	99.67	99.98	100	100	100
	5	99.47	99.95	100	100	100
	6	99.29	99.85	99.99	100	100

Table 4 Average classification accuracy of drive end bearing fault with fault diameter of 0.021 in.

		Acquisition time s				
		0.1	0.2	0.3	0.5	0.7
		%				
L	1	92.66	97.73	99.19	99.91	99.96
	2	94.76	98.36	99.51	99.75	99.88
	3	95.10	98.87	99.73	99.98	99.99
	4	94.64	98.39	99.21	99.81	99.97
	5	95.00	98.54	99.50	99.79	99.96
	6	94.38	98.25	99.35	99.83	99.88

Table 5 Average classification accuracy of fan end bearing fault with fault diameter of 0.007 in.

		Acquisition time s				
		0.1	0.2	0.3	0.5	0.7
		%				
L	1	73.61	82.50	89.43	93.04	94.31
	2	74.76	84.11	90.44	95.02	95.15
	3	83.08	92.54	94.82	98.49	98.60
	4	84.41	91.62	95.37	99.22	99.28
	5	84.41	90.73	95.56	98.90	98.76
	6	84.41	93.14	96.76	99.39	99.31

time of 0.3 s can be chosen to diagnose effectively this fault. Table 5 shows the classification accuracy at different wavelet decomposition level L and acquisition time of faults occurred at fan end bearing with fault diameter of 0.007 in. The classification accuracy is over 98 % for level $L = 3$ and acquisition time higher 0.5 s, so a wavelet decomposition level of 3 and acquisition time of 0.3 s can be chosen to diagnose effectively this fault.

6 Summary and Conclusions

This chapter addresses the modelling and diagnosis issues of rotating electrical machines by signal based solutions. With attention to real systems, two case studies related to rotating electrical machines are discussed. The first FDD solution uses PCA in order to reduce the three-phase current space in two dimensions. The PDFs of PCA-transformed signals are estimated by KDE. PDFs are the models that can be used to identify each fault. Diagnosis has been carried out using the K–L divergence, which measures the difference between two probability distributions. This divergence is used as a distance between signatures obtained by KDE. The second FDD solution uses MSPCA, KDE and PCA contributions to identify and diagnose the faults. Several experimentations on real motors are carried out in order to verify the effectiveness of the proposed methodologies. The first solution, based on current signals, has been tested on a motor modelled by FEM and real induction motors in order to diagnose broken rotor bars, broken connector, cracked and wrong rotor. The second solution, based on vibration signals, has been tested on a real induction motors in order to diagnose bearings faults: inner raceway, rolling element (i.e. ball) and outer raceway faults with different fault severities (i.e. diameter of 0.007 and 0.021 in.). Results show that the signal based solutions are able to model the fault dynamics and diagnose the motor conditions (i.e. healthy and faulty) and identify the faults.

References

Abdelmadjid, G., Mohamed, B. S., Mohamed, T., Ahmed, S., & Youcef, M. (2013). An improved stator winding fault tolerance architecture for vector control of induction motor: Theory and experiment. *Electric Power Systems Research, 104*, 129–137.

Antonino-Daviu, J., Riera-Guasp, M., Roger-Folch, J., Martnez-Giménez, F., & Peris, A. (2006). Application and optimization of the discrete wavelet transform for the detection of broken rotor bars in induction machines. *Applied and Computational Harmonic Analysis, 21*(2), 268–279.

Baccarini, L. M. R., Tavares, J. P. B., de Menezes, B. R., & Caminhas, W. M. (2010). Sliding mode observer for on-line broken rotor bar detection. *Electric Power Systems Research, 80*(9), 1089–1095.

Bakshi, B. R. (1998). Multiscale PCA with application to multivariate statistical process monitoring. *AIChE Journal, 44*(7), 1596–1610.

Bangura, J., Povinelli, R., Demerdash, N. A. O., & Brown, R. (2003). Diagnostics of eccentricities and bar/end-ring connector breakages in polyphase induction motors through a combination of time-series data mining and time-stepping coupled FE-state-space techniques. *IEEE Transactions on Industry Applications, 39*(4), 1005–1013.

Benloucif, M., & Balaska, H. (2006). Robust fault detection for an induction machine. In *Automation Congress, 2006. WAC '06. World* (pp. 1–6). Budapest.

Bouzida, A., Touhami, O., Ibtiouen, R., Belouchrani, A., Fadel, M., & Rezzoug, A. (2011). Fault diagnosis in industrial induction machines through discrete wavelet transform. *IEEE Transactions on Industrial Electronics, 58*(9), 4385–4395.

Case Western Reserve University Bearing Data Center. (2014). Bearing Data Center.

Cheng, H., Nikus, M., & Jamsa-Jounela, S. L. (2008). Evaluation of PCA methods with improved fault isolation capabilities on a paper machine simulator. *Chemometrics and Intelligent Laboratory Systems, 92*(2), 186–199.

Chiang, L. H., Russell, E. L., & Braatz, R. D. (2000). Fault diagnosis in chemical processes using Fischer discriminant analysis, discriminant partial least squares, and principal component analysis. *Chemometrics and Intelligent Laboratory Systems, 50*(2), 243–252.

Chilengue, Z., Dente, J., & Branco, P. C. (2011). An artificial immune system approach for fault detection in the stator and rotor circuits of induction machines. *Electric Power Systems Research, 81*(1), 158–169.

Ciabattoni, L., Corradini, M., Grisostomi, M., Ippoliti, G., Longhi, S., & Orlando, G. (2011a). Adaptive extended Kalman filter for robust sensorless control of PMSM drives. In *Proceedings of the IEEE Conference on Decision and Control* (pp. 934–939).

Ciabattoni, L., Corradini, M., Grisostomi, M., Ippoliti, G., Longhi, S., & Orlando, G. (2014). A discrete-time vs controller based on RBF neural networks for PMSM drives. *Asian Journal of Control, 16*(2), 396–408.

Ciabattoni, L., Grisostomi, M., Ippoliti, G., & Longhi, S. (2011b). Estimation of rotor position and speed for sensorless DSP-based PMSM drives. In *2011 19th Mediterranean Conference on Control and Automation, MED 2011* (pp. 1421–1426).

Ciandrini, C., Gallieri, M., Giantomassi, A., Ippoliti, G., & Longhi, S. (2010). Fault detection and prognosis methods for a monitoring system of rotating electrical machines. In *Industrial Electronics (ISIE), 2010 IEEE International Symposium on* (pp. 2085–2090).

Comaniciu, D. (2003). An algorithm for data-driven bandwidth selection. *IEEE Transactions on Pattern Analysis and Machine Intelligence, 25*(2), 281–288.

Daubechies, I. (1988). Orthonormal bases of compactly supported wavelets. *Communications on Pure and Applied Mathematics, 41*(7), 909–996.

Dunia, R., & Joe Qin, S. (1998). Joint diagnosis of process and sensor faults using principal component analysis. *Control Engineering Practice, 6*(4), 457–469.

Fan, Y., & Zheng, G. (2007). Research of high-resolution vibration signal detection technique and application to mechanical fault diagnosis. *Mechanical Systems and Signal Processing, 21*(2), 678–687.

Feki, N., Clerc, G., & Velex, P. (2013). Gear and motor fault modeling and detection based on motor current analysis. *Electric Power Systems Research, 95*, 28–37.

Ferracuti, F., Giantomassi, A., Iarlori, S., Ippoliti, G., & Longhi, S. (2013a). Induction motor fault detection and diagnosis using KDE and Kullback–Leibler divergence. In *Industrial Electronics Society, IECON 2013—39th Annual Conference of the IEEE* (pp. 2923–2928).

Ferracuti, F., Giantomassi, A., Ippoliti, G., & Longhi, S. (2010). Multi-scale PCA based fault diagnosis for rotating electrical machines. In *European workshop on advanced control and diagnosis, 8th ACD* (pp. 296–301). Ferrara, Italy.

Ferracuti, F., Giantomassi, A., & Longhi, S. (2013b). MSPCA with KDE thresholding to support QC in electrical motors production line. *In Manufacturing Modelling, Management, and Control, 7*, 1542–1547.

Ferracuti, F., Giantomassi, A., Longhi, S., & Bergantino, N. (2011). Multi-scale PCA based fault diagnosis on a paper mill plant. In *Emerging Technologies Factory Automation (ETFA), 2011 IEEE 16th Conference on* (pp. 1–8).

Filbert, D., & Metzger, L. (1982). Quality test of systems by parameter estimation. In *Proceedings of 9th IMEKO congress*. Berlin, Germany.

Fisher, R. A. (1936). The use of multiple measurements in taxonomic problems. *Annals of Eugenics, 7*(2), 179–188.

Gani, A., & Salami, M. (2002). Vibration faults simulation system (VFSS): A system for teaching and training on fault detection and diagnosis. In *Research and Development, 2002. SCOReD 2002. Student Conference on* (pp. 15–18).

Giantomassi, A. (2012). *Modeling estimation and identification of complex system dynamics*. LAP Lambert Academic Publishing.

Giantomassi, A., Ferracuti, F., Benini, A., Longhi, S., Ippoliti, G., & Petrucci, A. (2011). Hidden Markov model for health estimation and prognosis of turbofan engines. In *ASME 2011 International Design Engineering Technical Conferences and Computers and Information in Engineering Conference* (pp. 1–6).

Hua, L., Zhanfeng, L., & Zhaowei. (2009). Time frequency distribution for vibration signal analysis with application to turbo-generator fault diagnosis. In *Control and Decision Conference, 2009. CCDC '09. Chinese* (pp. 5492–5495). Guilin.

Immovilli, F., Bellini, A., Rubini, R., & Tassoni, C. (2010). Diagnosis of bearing faults in induction machines by vibration or current signals: A critical comparison. *IEEE Transactions on Industry Applications, 46*(4), 1350–1359.

Isermann, R. (1984). Process fault detection on modeling and estimation methods? A survey. *Automatica, 20*(4), 387–404.

Isermann, R. (2006). *Fault-diagnosis dystems*. Berlin: Springer.

Isermann, R., & Balle, P. (1997). Trends in the application of model-based fault detection and diagnosis of technical processes. *Control Engineering Practice, 5*(5), 709–719.

Jackson, J. E. (2003). *A user's guide to principal components* (Vol. 587). New York: Wiley-Interscience.

Jackson, J. E., & Mudholkar, G. S. (1979). Control procedures for residuals associated with principal component analysis. *Technometrics, 21*(3), 341–349.

Jolliffe, I. T. (2002). *Principal component analysis*. Berlin: Springer.

Keogh, E. (2013). UCR time series data mining archive.

Kullback, S., & Leibler, R. A. (1951). On information and sufficiency. *Annals of Mathematical Statistics, 22*(1), 79–86.

Lachouri, A., Baiche, K., Djeghader, R., Doghmane, N., & Ouhtati, S. (2008). Analyze and fault diagnosis by Multi-scale PCA. In *Information and communication technologies: From theory to applications, 2008. ICTTA 2008. 3rd International Conference on* (pp. 1–6).

Li, W., Yue, H. H., Valle-Cervantes, S., & Qin, S. J. (2000). Recursive PCA for adaptive process monitoring. *Journal of Process Control, 10*(5), 471–486.

Li, X., Dong, S., & Yuan, Z. (1999). Discrete wavelet transform for tool breakage monitoring. *International Journal of Machine Tools and Manufacture, 99*(12), 1944–1955.

Mallat, S. G. (1989). A theory for multiresolution signal decomposition: The wavelet representation. *IEEE Transactions on Pattern Analysis and Machine Intelligence, 11*(7), 674–693.

Misra, M., Yue, H., Qin, S., & Ling, C. (2002). Multivariate process monitoring and fault diagnosis by multi-scale PCA. *Computers and Chemical Engineering, 26*(9), 1281–1293.

Mugdadi, A., & Ahmad, I. A. (2004). A bandwidth selection for kernel density estimation of functions of random variables. *Computational Statistics and Data Analysis, 47*(1), 49–62.

Odiowei, P.-E., & Cao, Y. (2010). Nonlinear dynamic process monitoring using canonical variate analysis and kernel density estimations. *IEEE Transactions on Industrial Informatics, 6*(1), 36–45.

Parzen, E. (1962). On estimation of a probability density function and mode. *The Annals of Mathematical Statistics, 33*(3), 1065–1076.

Patton, R., Uppal, F., & Lopez-Toribio, C. (2000). Soft computing approaches to fault diagnosis for dynamic systems: a survey. In *4th IFAC Symposium on Fault Detection supervision and Safety for Technical Processes* (pp. 298–311). Budapest, Hungary.

Pedra, J., Candela, I., & Barrera, A. (2009). Saturation model for squirrel-cage induction motors. *Electric Power Systems Research, 79*(7), 1054–1061.

Ran, L., & Penman, J. (2008). *Condition monitoring of rotating electrical machines* (Vol. 56). London, United Kingdom: Institution of Engineering and Technology.

Rao, C. R. (1948). The utilization of multiple measurements in problems of biological classification. *Journal of the Royal Statistical Society—Series B, 10*(2), 159–203.

Rodriguez, P., Belahcen, A., & Arkkio, A. (2006). Signatures of electrical faults in the force distribution and vibration pattern of induction motors. *Electric Power Applications, IEE Proceedings, 153*(4), 523–529.

Samsi, R., Ray, A., & Mayer, J. (2009). Early detection of stator voltage imbalance in three-phase induction motors. *Electric Power Systems Research, 79*(1), 239–245.

Sawalhi, N., & Randall, R. (2008a). Simulating gear and bearing interactions in the presence of faults: Part I. the combined gear bearing dynamic model and the simulation of localised bearing faults. *Mechanical Systems and Signal Processing, 22*(8), 1924–1951.

Sawalhi, N., & Randall, R. (2008b). Simulating gear and bearing interactions in the presence of faults: Part II. simulation of the vibrations produced by extended bearing faults. *Mechanical Systems and Signal Processing, 22*(8), 1952–1966.

Sheather, S. J. (2004). Density estimation. *Statistical science* (pp. 588–597).

Shuting, W., Heming, L., & Yonggang, L. (2002). Adaptive radial basis function network and its application in turbine-generator vibration fault diagnosis. In *Power system technology, 2002. Proceedings of PowerCon 2002. International Conference on* (Vol. 3 pp. 1607–1610).

Singh, G., & Ahmed, S. A. K. S. (2004). Vibration signal analysis using wavelet transform for isolation and identification of electrical faults in induction machine. *Electric Power Systems Research, 68*(2), 119–136.

Stuart, M., Mullins, E., & Drew, E. (1995). Statistical quality control and improvement. *European Journal of Operational Research, 88*(2), 203–214.

Taniguchi, S., Akhmetov, D., & Dote, Y. (1999). Fault detection of rotating machine parts using novel fuzzy neural network. In *Systems, Man, and Cybernetics, 1999. IEEE SMC'99 Conference Proceedings. 1999 IEEE International Conference on*, (Vol. 1, pp. 365–369). Tokyo.

Tavner, P. (2008). Review of condition monitoring of rotating electrical machines. *IET Electric Power Applications, 2*(4), 215–247.

Thomson, W., & Fenger, M. (2001). Current signature analysis to detect induction motor faults. *Industry Applications Magazine, IEEE, 7*(4), 26–34.

Tran, V. T., Yang, B.-S., Oh, M.-S., & Tan, A. C. C. (2009). Fault diagnosis of induction motor based on decision trees and adaptive neuro-fuzzy inference. *Expert Systems with Applications, 36*(2, Part 1), 1840–1849.

Venkatasubramanian, V., Rengaswamy, R., Yin, K., & Kavuri, S. N. (2000a). A review of process fault detection and diagnosis part I: Quantitative model-based methods. *Computers and Chemical Engineering, 27*(3), 293–311.

Venkatasubramanian, V., Rengaswamy, R., Yin, K., & Kavuri, S. N. (2000b). A review of process fault detection and diagnosis part II: Qualitative models and search strategies. *Computers and Chemical Engineering, 27*(3), 313–326.

Venkatasubramanian, V., Rengaswamy, R., Yin, K., & Kavuri, S. N. (2000c). A review of process fault detection and diagnosis part III: Process history based methods. *Computers and Chemical Engineering, 27*(3), 327–346.

Verucchi, C. J., & Acosta, G. G. (2008). Fault detection and diagnosis techniques in induction electrical machines. *IEEE Transactions Latin America, 5*(1), 41–49.

Wand, M. P., & Jones, M. C. (1994a). *Kernel smoothing*. United Kingdom: Chapman and Hall CRC.

Wand, M. P., & Jones, M. C. (1994b). Multivariate plug-in bandwidth selection. *Computational Statistics, 9*, 97–116.

Yang, B.-S., & Kim, K. J. (2006). Application of Dempster-Shafer theory in fault diagnosis of induction motors using vibration and current signals. *Mechanical Systems and Signal Processing, 20*(2), 403–420.

Yu, J. (2011a). Bearing performance degradation assessment using locality preserving projections. *Expert Systems with Applications, 38*(6), 7440–7450.

Yu, J. (2011b). Bearing performance degradation assessment using locality preserving projections and Gaussian mixture models. *Mechanical Systems and Signal Processing, 25*(7), 2573–2588.

Zhaoxia, W., Fen, L., Shujuan, Y., & Bin, W. (2009). Motor fault diagnosis based on the vibration signal testing and analysis. In *Intelligent information technology application, 2009. IITA 2009. Third International Symposium on* (Vol. 2, pp. 433–436). Nanchang.

Modelling of Intrusion Detection System Using Artificial Intelligence—Evaluation of Performance Measures

Manojit Chattopadhyay

Abstract In recent years, applications of internet and computers are growing extremely used by many people all over the globe—so is the susceptibility of the network. In contrast, network intrusion and information security problems are consequence of internet application. The increasing network intrusions have placed people and organizations to a great extent at peril of many kinds of loss. With the aim to produce effectiveness and state-of-the-art concern, the majority organizations put their applications and service things on internet. The organizations are even investing huge money to care for their susceptible data from diverse attacks that they face. Intrusion detection system is a significant constituent to protect such information systems. A state-of-the-art review of the applications of neural network to Intrusion Detection System has been presented that reveals the positive trend towards applications of artificial neural network. Various other parameters have been selected to explore for a theoretical construct and identifying trends of ANN applications to IDS. The research also proposed an architecture based on Multi Layer Perceptron (MLP) neural network to develop IDS applied on KDD99 data set. Based on the identified patterns, the architecture recognized attacks in the datasets using the back propagation neural network algorithm. The proposed MLP neural network has been found to be superior when compared with Recurrent and PCA neural network based on the common measures of performance. The proposed neural network approach has resulted with higher detection rate (99.10 %), accuracy rate (98.89 %) and a reduced amount of execution time (11.969 s) and outperforms the benchmark results of six approaches from literature. Thus the analysis based on experimental outcomes of the MLP approach has established the robustness, effectiveness in detecting intrusion that can further improve the performance by reducing the computational cost without obvious deterioration of detection performances.

Keywords Artificial intelligence · Multilayer perceptron · Intrusion detection system · Detection rate · Fasle alarm · KDDCUP99

M. Chattopadhyay (✉)
Operations and Systems Area, Indian Institute of Management Raipur, GEC Campus, Sejbahar, Raipur 492015, Chhattisgarh, India
e-mail: mjc02@rediffmail.com

Q. Zhu and A.T. Azar (eds.), *Complex System Modelling and Control Through Intelligent Soft Computations*, Studies in Fuzziness and Soft Computing 319,
DOI 10.1007/978-3-319-12883-2_11

1 Introduction

In recent years, intrusion detection system (IDS) has attracted a great deal of concern and attention. The webopedia English Dictionary (http://www.webopedia.com/) defines intrusion detection system as "An intrusion detection system (IDS) inspects all inbound and outbound network activity and identifies suspicious patterns that may indicate a network or system attack from someone attempting to break into or compromise a system." (Heady et al. 1990). Heady et al. (1990) describe intrusion as "any set of actions that attempt to compromise the integrity, confidentiality or availability of a resource". Even after adopting various intrusion prevention techniques it is nearly impossible for an operational system to be completely secure (Lee et al. 1999). Therefore IDS are imperative to provide extra protection for being characterized as normal or legitimate behaviour of resources, models and techniques rather than to identify as abnormal or intrusive. The IDS has been formalized during the 1980s as a potential model (Denning 1987) to prevent the incident of unauthorized access to data (Eskin et al. 2002). During the last two decades has been categorized accepted definition of financial fraud, Wang et al. (2006) define it as "a deliberate act that is contrary to law, rule, or policy with intent to obtain unauthorized financial benefit."

Therefore due to the immense expansion of computer networks usage and the enormous increase in the number of applications running on top of it, network security is becoming more and more significant. As network attacks have increased in number and severity over the past few years, consequently Intrusion Detection Systems (IDSs) is becoming more important to detect anomalies and attacks in the network. Therefore, even with the most advanced protected environment, computer systems are still not 100 % secure.

In the domain of intrusion detection, there is a growing interest of the application and development of Artificial Intelligence (AI) based approach is (Laskov et al. 2005). AI and machine learning techniques were used to discover the underlying models from a set of training data. Commonly used methods were rule-based induction, classification and data clustering (Wu and Bunzhaf 2010). AI is a huge and sophisticated field still growing and certainly not optimized for network security. Definite effort will be required in AI to help its application to IDSs. Development on that face will take place more rapidly if the opportunity of using AI techniques in IDSs motivates more attention to the AI community. AI is a collection of approaches, which endeavors to make use of tolerance for imprecision, uncertainty and partial truth to achieve tractability, robustness and low solution cost. As AI techniques can also be used for computational intelligence, different computational intelligence approaches have been used for intrusion detection (Fuzzy Logic, Artificial Neural Networks, Genetic Algorithms) (Yao et al. 2005; Gong et al. 2005; Chittur 2001; Pan et al. 2003), but their potentials are still underutilized. Researcher are also using a term computational intelligence that deals with only numerical data to recognize patterns unlike that of artificial intelligence it has potential to computational adaptive, fault tolerant, maximizing speed, minimizing error rates corresponding to human performance (Bezdek 1994).

Wu and Banzhaf (2010) commented that the popular domain of AI is different from the CI. However there is neither full conformity on the exact nature of computational intelligence nor there is any far and wide established vision on which domain belong to CI: artificial neural networks, fuzzy sets, evolutionary computation, artificial immune systems, swarm intelligence, and soft computing. Majority of these approaches are able to process the information using either supervised or unsupervised learning algorithm. Supervised learning frequently constructs classifiers known as a function mapping data observations to matching class labels for misuse discovery from class-labeled training datasets. Classifiers are basically viewed. On the other hand, unsupervised learning is different from supervised learning due to non-availability of class-labeled data during the training stage and it works on based on similarities of data points. Therefore it becomes a more suitable approach to deal with anomaly detection.

Artificial Intelligence (AI) has recently been attracted significantly in the development of Intrusion Detection System (IDS) for anomaly detection, data reduction from the research community. Due to large trend of internet usage in the last decade in a more complex and un-trusted global internet environment, the information systems are inescapably uncovered to the growing threats. Intrusion Detection System is an approach use to respond to such threats. Diverse IDS techniques have been proposed, which identify and alarm for such threats or attacks. The Intrusion Detection System (IDS) generates huge amounts of alerts that are mostly false positives. The abundance of false positive alerts makes it difficult for the security analyst to identify successful attacks and to take remedial actions. Many of artificial intelligence approach have been used for classification, but they alone are incapable of dealing with new types of attack which are evolving due to the advent of real time data. To address with these new problems of networks, artificial intelligence based IDS are opening new research avenues. Artificial intelligence offers a vast range of techniques to classify these attacks. So to assist in categorizing the degree of the threat, different artificial intelligence techniques are used to classify the alerts, our research work will be based on analyzing the existing techniques and in the process identifying the best algorithm for the development of an efficient intrusion detection system.

The fundamental objectives of our contribution will be to explore for an optimal intrusion detection system model based on Artificial Intelligence techniques and evaluation perspective for performance of such predictive classification system. Therefore, the objective is basically to provide solutions in developing a complex system model. The principal chapter objectives of this research work can be summarized as:

1. Undertake detailed study on anomaly based intrusion detection systems.
2. Exploring the research trend for security challenges of ID based on anomaly detection after critical appraisal of the existing methodologies for intrusion detection system.

3. Propose a suitable methodology for anomaly detection using KDD99Cup Dataset. Specifically, the research work focuses on the followings:
   (a) To extract the data, normalize it and categorization of the attack based on numerical value
   (b) to develop an optimal neural network architecture of the anomaly detection for increase rate of correct classification of anomaly
   (c) to calculate the performance measure of the anomaly in IDSs result obtained after applying proposed supervised learning approach
   (d) to assess the predictive ability of the proposed neural network architecture

In the present research the intrusion detection has been considered as a binary classification problem and thus it is necessitated to highlight the back ground on the types of intrusion detection system in the next section.

1.1 Intrusion Detection

Intrusion detection mechanism can be divided into two broad categories (Anderson 1995; Tiwari 2002) (i) Misuse detection system (ii) Anomaly based detection.

The systems are described as below:

(i) Misuse detection system

It is perhaps the oldest and most frequent method and applies well-known knowledge of identified attack patterns to search for signatures, observe state transitions or employed at a mining system to classify potential attacks (Faysel and Haque 2010). The familiar attacks can be identified efficiently with a very low false alarm rate for which it is broadly applied in most of the commercial systems. As the attacks are frequently polymorph, and changed regularly therefore, misuse detection become unsuccessful due to unfamiliar attacks. This problem may be resolved by regularly updated knowledge base either through time consuming and laborious manual method or through automatic updating using supervised learning methods. However this becomes too costly to set up to perform labeling of each occurrence in the dataset as normal or a type of attack. Differently to deal with this problem is to apply the anomaly detection method as proposed by Denning (1987).

(ii) Anomaly based detection

Anomaly detection systems recognize difference from normal behaviour and alert to possible unknown or novel attacks lacking any past knowledge of them. It theorized that anomalous behavior is rare and dissimilar from normal behavior. Thus it is orthogonal to misuse detection (Wu and Banzhaf 2010). Anomaly detection can be of two types (Chebrolu et al. 2005): static and dynamic anomaly detection. In the first one it is assumed that the observed attack behavior is constant and the second one extracts pattern occasionally known as profiles from behavioral routine of end users, or usage history of networks/hosts.

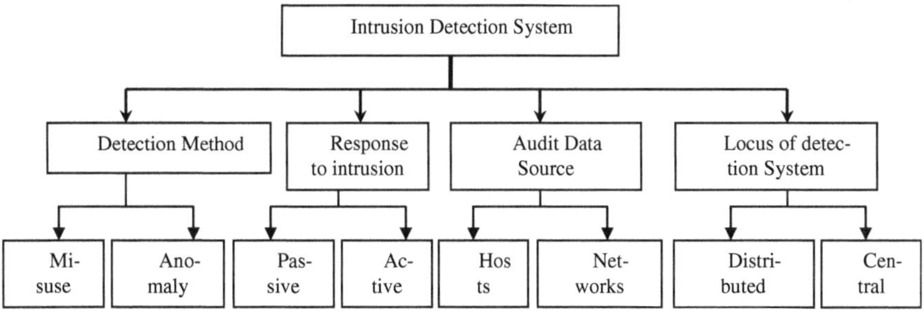

Fig. 1 Classification scheme of intrusion detection system taken from Wu and Banzhaf (2010)

Therefore, anomaly detection has the potential of identifying latest kind of attacks, and only necessitates normal data during generation of profiles. Though, the main intricacy involves in determining borders among normal and abnormal behaviors, as a result of the lack of abnormal examples during the learning stage. An additional complexity is to familiarizing itself to continually varying normal behavior, particularly for dynamic anomaly detection.

Additionally there are other features used to classify intrusion detection system approach, as shown in Fig. 1 (Wu and Banzhaf 2010).

One frequent method applied to identify intrusion detection is by classification defined as dividing the samples into distinct partition. The purpose of the classifier is not to investigate the data to determine interesting partition but also to settle on how new data will be classified. In intrusion detection, classification grouped the data records in a encoded classes applied as features to label each sample, discriminating elements fitting to anomaly or normal attack classes. However the classification has to be used with fine tuning approaches to decrease false positive rates. Thus intrusion detection is considered as a binary categorization problem (Liao and Vemuri 2002).

Artificial neural network is relatively new and emerging approach to easily deal with complex classification with much better precision and output and the conceptual background of different types of artificial neural network with diverse application domains explored in literature are discussed in the next section.

1.2 Artificial Neural Network

Artificial intelligence (AI) is an interdisciplinary domain exhibits human-like intelligence and demonstrated by hardware or software. The term AI was coined by McCarthy et al. (1955) and defined it as "the science and engineering of making intelligent machines" (McCarthy 2007). Artificial Neural Network (ANN) is massively parallel interconnections of simple neurons that act as a collective system (Haykin 2005). The ANNs mimic the human brain so as to perform intelligently. The major benefits include high computation rate due to their massive parallelism for which real time computation of large data sets become possible using proper hardware. The information is determined on connection weights between the layers. A processing unit consists of a learning rule and an activation function. The learning rule resolves the actual input of the node by mapping the output of all direct antecedent and extra external inputs onto a single input value. The activation function is then applied on the actual input and determines the output of the node. The output of the processing unit is also described as activation. In the Fig. 2 the two input nodes are shown in input layer, one output nodes is shown in output layer. Organizing the nodes in layers resulted in a layered network and the Fig. 2 shows in between input and output layers there are two hidden layers. The inputs to hidden and hidden to output nodes are connected by weight values that is initialized during the start of the training and a net input is calculated on which the activation function is applied to calculate the output. The multilayer perceptron has additional $L \geq 1$ hidden layers. The lth hidden layer consists of h(l) hidden units. MLP is applied to solve wide varieties of interdisciplinary problems like credit scoring (Khashei et al. 2013), medical (Peláez et al. 2014), food classification (Dębska and Guzowska-Świder 2011), forecasting (Valero et al. 2012), mechanical engineering (Hwang et al. 2010), production (Kuo et al. 2010) etc.

The next section will discuss specifically the various neural network approaches applied in the development of intrusion detection system.

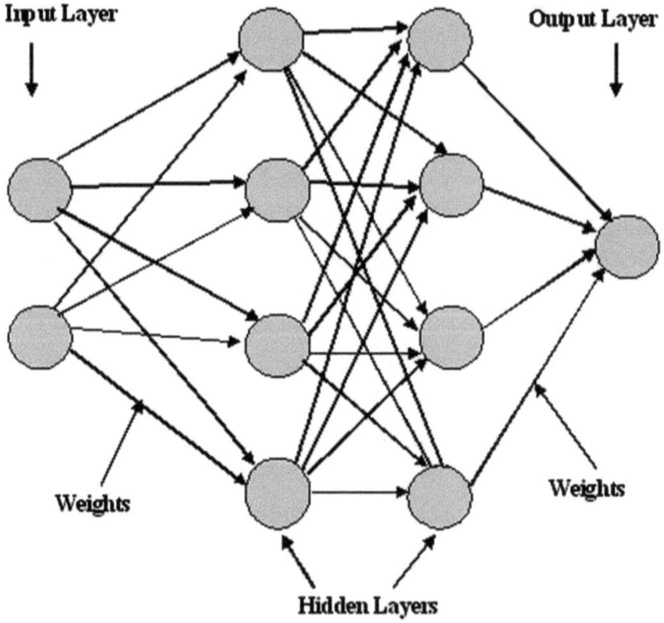

Fig. 2 ANN architecture

2 Survey on the AI Based Techniques Used for Intrusion Detection

Artificial neural network based intrusion detection system development is an important research trend in intrusion detection domain (Yang et al. 2013). Artificial Neural Network (ANN) has been used in the classification process of the system. The inputs of ANN are obtained from the features of packet headers, such as port number and IP number. The implemented embedded IDS has been first trained with training data. Then, packet classification has been performed in the real time and finally time of determining packet classes have been obtained (Tuncer and Tatar 2012). ANN has been shown to increase efficiency, by reducing the fault positive, and detection capabilities by allowing detection with partial available information on the network status (El Kadhi et al. 2012).

Different sizes of feed forward neural networks are compared for their evaluation performance using MSE. The generalization capacity of the trained network shows potential and the network is competent to predict number of zombies involved in a DDoS attack with very less test error (Gupta et al. 2012). Genetic Algorithm has successfully applied on NSL-KDD data set (Aziz et al. 2014). Research has revealed high accuracy and good detection rates but with moderate false alarm on novel attacks by the implementing Genetic Algorithms, Support Vector Machines, Neural Networks etc. (Abdel-Aziz et al. 2013; Zainaddin et al. 2013). In a research it is established that PSO outperforms GA both in population size and number of evolutions and can converge faster. Comparing PSO with some other machine learning algorithm it was found that PSO perform better in terms of detection rate, false alarm rate, and cost per example (Sheikhan and Sharifi 2013).

IDS development using Self Organization Map (SOM) neural network, has been successfully detected anomalies (Xiang et al. 2013). Comparative result analysis of SOM implementation based on several performance metrics revealed that detection rate for KDD 99 dataset was 92.37 %, while detection rate for NSL-KDD dataset was 75.49 % (Ibrahim et al. 2013).

ART2 neural network experiments with IDS demonstrated that the model effectively improved detection accuracy and decreased false alarm rate compared with the static learning intrusion detection method based on SVM (Liu 2013). Fuzzy adaptive resonance theory-based neural network (ARTMAP) has been used as a misuse detector (Sheikhan and Sharifi 2011).

In majority of the research ANNs has improved the performance of intrusion detection systems (IDS) when evaluated with traditional approaches. However for ANN-based IDS, detection precision, especially for low-frequent attacks, and detection stability are still required to be improved. FC-ANN approach, based on ANN and fuzzy clustering, has demonstrated to solve IDS that achieved higher detection rate, less false positive rate and stronger stability. Experimental outcomes on the KDD CUP 1999 dataset showed that FC-ANN approach outperforms BPNN and other well-known approaches like decision tree, the naive Bayes in terms of detection precision and detection stability (Wang et al. 2010).

Recurrent Neural Network out-performs Feed-forward Neural Network, and Elman Network for detecting attacks in a communication network (Anyanwu et al. 2011).

Theory and experiment show that Radial basis function network (RBFN) algorithm has better ability in intrusion detection, and can be used to improve the efficiency of intrusion detection, and reduce the false alarm rate (Peng et al. 2014). Binary Genetic Algorithm (BGA) as a feature extractor provide input for the classification task to a standard Multi-layer Perceptron (MLP) classifier that resulted with very high classification accuracy and low false positive rate with the lowest CPU time (Behjat et al. 2014).

Using k-means clustering, Naive Bayes feature selection and C4.5 decision tree classification for pinpointing cyber attacks resulted with a high degree of accuracy (Louvieris et al. 2013). Comparing the traditional BP networks and the IPSO-BPNN algorithm to simulate results of the KDD99 CUP data set with the intrusion detection system has demonstrated the BPN resulted with less time, better recognition rate and detection rate (Zhao et al. 2013).

Feizollah et al. (2014) evaluated five machine learning classifiers, namely Naive Bayes, k-nearest neighbour, decision tree, multi-layer perceptron, and support vector machine in wireless sensor network (WSN). A critical study has been made using genetic algorithm, artificial immune, and artificial neural network (ANN) based IDSs approaches (Yang et al. 2013).

A network IDS applied discretization with genetic algorithm (GA) as a feature selection to assess it's performance several classifiers algorithms like rules based classifiers (Ridor, Decision table), trees classifiers (REPTree, C 4.5, Random Forest) and Naïve bays classifier have been used on the NSL-KDD dataset (Aziz et al. 2012; Eid et al. 2013). Research revealed that discretization has a positive impact on the time to classify the test instances and is found to be an important factor for developing a real time network IDS.

Therefore only a detail analytical view on applications of neural network based intrusion detection system can quantitatively enlighten on the trend of kind of diverse research based on neural network as explored in literature.

2.1 Analysis of IDS Research Based on the Neural Network Algorithm

This paper provides a state-of-the-art review of the applications of neural network to IDS. The following query string has been searched using scopus search engine: (TITLE-ABS-KEY (intrusion detection system) AND SUBJAREA (mult OR ceng OR CHEM OR comp OR eart OR ener OR engi OR envi OR mate OR math OR phys) AND PUBYEAR > 1999) AND (neural network). It resulted with 2,185 articles and only the relevant information has been collected to interpret the significance of IDS research using neural network during the period 2000–2014. Figures 3, 4, 5, 6 and 7 organizes this review of the literature.

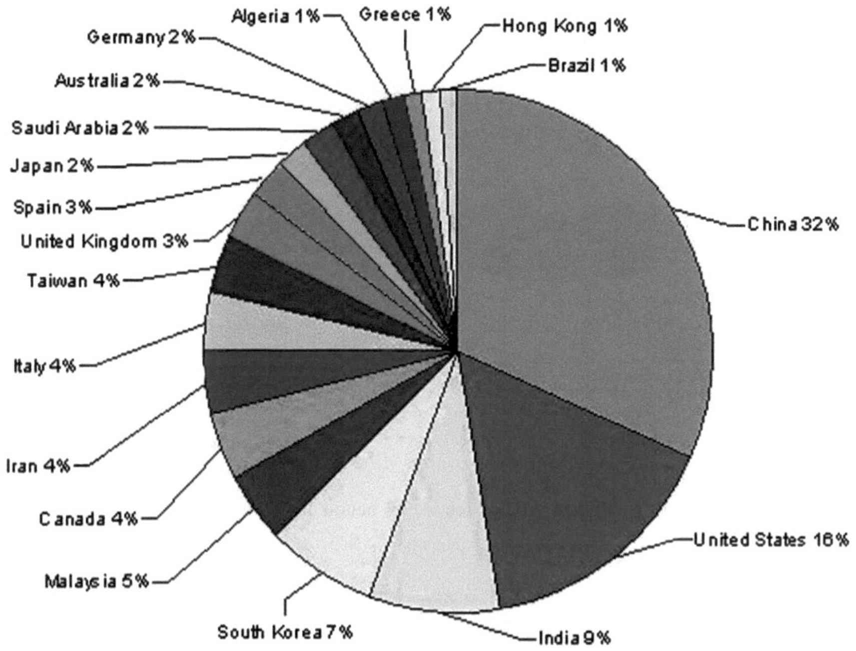

Fig. 3 Articles on neural network applied intrusion detection system development published by researcher from their affiliated country

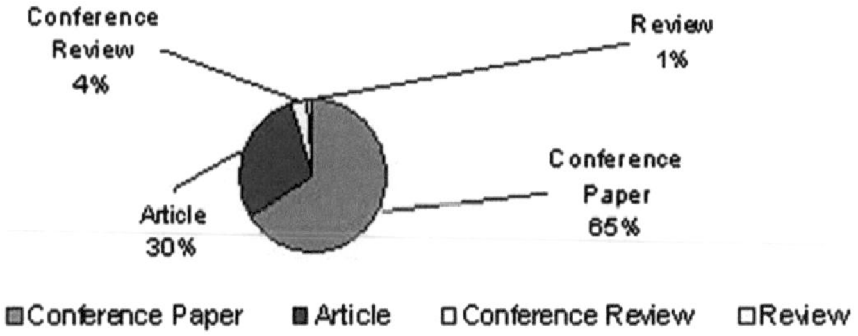

Fig. 4 Type of research documents published on neural network applied intrusion detection system development

Figures 3, 4, 5 and 6, dissects and organizes this review of the literature. For the classification of literature Fig. 3 shows the articles published by researcher from their affiliated country. It is shown in that China is leading (32 %) followed by USA (16 %) and India (9 %) as highest articles published by affiliated country.

The conference papers (65 %) are the major type of research documents followed by articles (30 %) as revealed by Fig. 4.

The Fig. 5 has not considered around 329 articles published in rest 141 journals having less than 7 articles published due to interpretability of this huge in formation

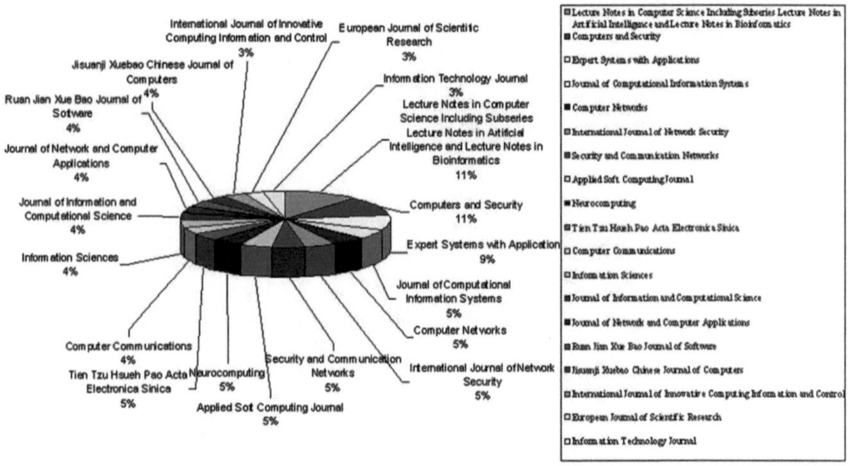

Fig. 5 Major journals published articles related to neural network applied intrusion detection system development

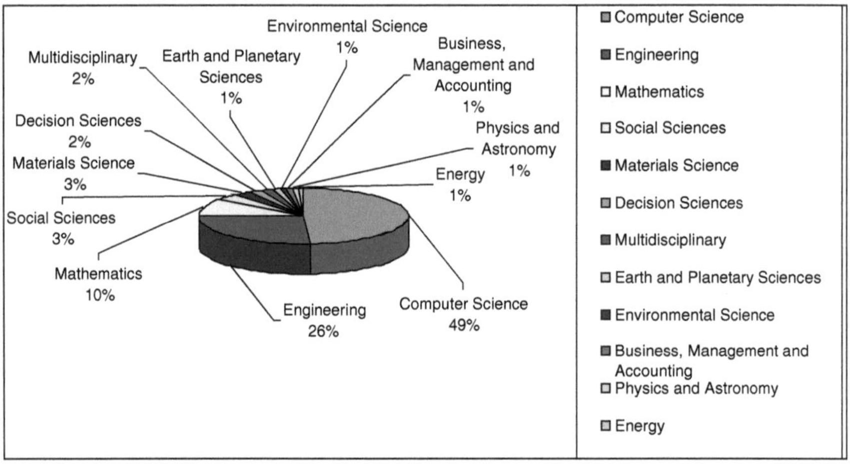

Fig. 6 Publication of neural network applied intrusion detection system development related articles in various domains

in a single graph. The figure depicts that Lecture Notes in Computer Science is the major journal publishing 25 articles on IDS based on neural network (11 %) followed by 24 articles in Computers and Security and 20 articles in Expert Systems with Applications (9 %) journals.

For more than 9 articles published in a domain are shown in the Fig. 6 to get information of different domains in which neural network based IDS articles are found. It is shown that computer science (49 %) is the major domain publishing 509

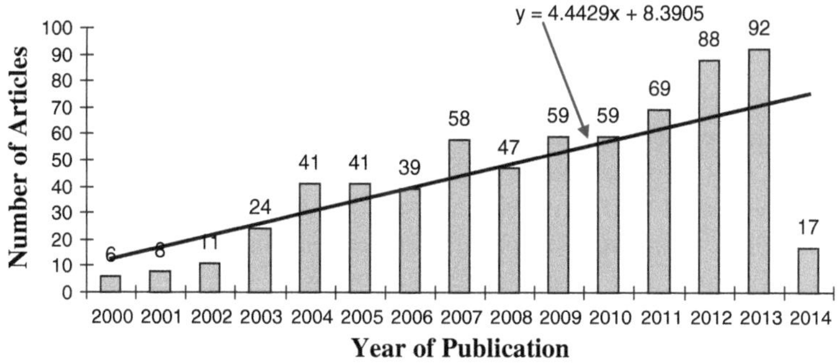

Fig. 7 Number of articles on neural network applied intrusion detection system development published during the year 2000–2014 and the research trend

articles followed by 273 articles in engineering (26 %) and 108 articles in mathematics (10 %).

In Fig. 7 the research trend based on the number of articles published between the years 2000–2013 has been shown to be increasing with R-squared value equals 0.9433 which is a good fit. The trend line in Fig. 7 for 2000–2014 is also increasing where the search on articles has been performed in February, 2014.

The next section has discussed the description of the data set applied in the development of the model for intrusion detection system.

3 KDD-99 Dataset

Mostly all the experiments on intrusion detection are done on KDDCUP'99 dataset, which is a subset of the 1998 DARPA Intrusion Detection Evaluation data set, and is processed, extracting 41 features from the raw data of DARPA 98 data set Stolfo et al. (2000) defined higher-level features that help in distinguishing between good normal connections from bad connections (attacks). This data can be used to test both host based and network based systems, and both signature and anomaly detection systems. A connection is a sequence of Transmission Control Protocol (TCP) packets starting and ending with well defined times, between which data flows from a source IP address to a target IP address under some well defined protocol. Each connection is labeled as normal, or as an attack, with exactly one specific attack type. Each connection record consists of about 100 bytes (https://kdd.ics.uci.edu/databases/kddcup99/kddcup99.html).

The data to be used in the model is organized and prepared to be used in the form of binary classification model. However the classification model needs to be evaluated based on certain metrics from their output results and discussed in the next section.

3.1 Evaluation Metrics

An elementary concern in the development of classification models is the evaluation of predictive accuracy (Guisan and Thuiller 2005; Barry and Elith 2006). The quantitative evaluation of the model is important as it helps in determining the ability of the model tom provide better solution for a specific problem and also assist in exploring the areas of model improvement. In the domain of binary predictions of anomaly and normal attacks, a confusion matrix (Table 1) known as contingency table or error matrix (Swets 1988) that represents the performance visualization of the predictive models of IDS that consists of two rows showing the actual class and two columns showing the predicted class. The aim is to check whether the system is confusing both classes. The IDSs are primarily distinguished binary classes: anomaly class (malicious, threats or abnormal data) and normal class (normal data points). Therefore, the proposed models generating normal-anomaly predictions of intrusion detection system are typically assessed in Table 1 through comparison of the predictions and developing a confusion matrix to predict the number of true positive (TP), false positive (FP), false negative (FN) and true negative (TN) cases. TP/TP+FN, is used as detection rate (DR) or sensitivity. It is also termed as recall in information retrieval Overall accuracy is a simple measure of accuracy that can be derived from the confusion matrix by calculating the proportion of correct prediction. Sensitivity is the proportion of observed normal attacks that are predicted as such, and therefore quantifies omission errors. Specificity is the proportion of observed anomaly attacks that are predicted as such, and therefore quantifies commission errors. Sensitivity and Specificity are independent of each other when compared across models. The most popular measure for the accuracy of yes–no predictions is Cohen's kappa (Shao and Halpin 1995; Segurado and Araujo 2004) which corrects the overall accuracy of model predictions by the expected random accuracy. The kappa statistic ranges from 0 to 1, where 1 indicates perfect agreement and values of zero indicate a performance no better than random (Cohen 1960). The principle benefits of kappa are for its simplicity and the reason that both commission and omission errors are accounted for in one parameter. In this paper we also introduced another measure known as the true skill statistic (TSS) for the performance of normal–anomaly classifier models, that still preserves the advantages of kappa.

In the next section a detail experiment and analysis demonstrated the efficacy of the proposed MLP in the development of IDS system based on the above discussed classification evaluation metrics.

Table 1 Confusion matrix

	Predicted class		
Actual class		Anomaly	Normal
	Anomaly	TP	FN
	Normal	FP	TN

4 Experiment and Analysis of Intrusion Detection System Based on MLP Algorithm

MLP is conceivably the most popular network architecture currently in use amongst the ANNs (Saftoiu et al. 2012). There are three layers of units: input layer, a hidden layer and an output layer in the architecture of MLP with feed-forward supervised learning. The proposed ANN architecture was implemented using the SPSS neural networks program using SPSS 16.0 (http://www-01.ibm.com/software/in/analytics/spss/downloads.html) in Windows XP environment. Neural Networks are nonlinear statistical data modeling approaches. ANNs can explore and extract nonlinear interactions among parameters to expose formerly unidentified associations among given input parameters and outcomes (Sall et al. 2007).

The Fig. 8 shows a feed forward architecture of the neural network because the connections in the network flow forward from the input layer to the output layer without any feedback loops. In this Fig. 8 the input layer contains the 39 predictors; one hidden layer contains unobservable nodes, or units. Based on some function of

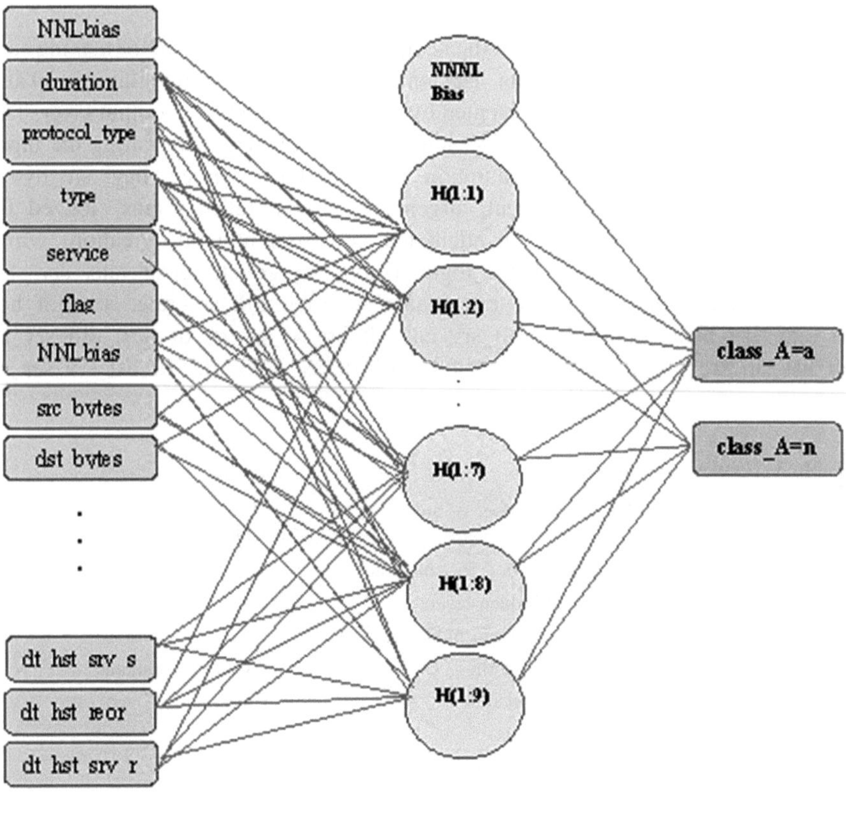

Fig. 8 Feedforward architecture with one hidden layer

Table 2 Case processing summary

		N	Percent (%)
Sample	Training	17,723	70.4
	Testing	7,468	29.6
Valid		25,191	100.0
Excluded		0	
Total		25,191	

the predictors represents the value of each hidden unit; that depends partly on the network type and on user-controllable condition. Anomaly and Normal from intrusion detection modeling point of view are being represented by the output layer as dependent variables. Since the class of response is a categorical variable with two classes, it is recoded as binary class variables. Each output node is some function of the hidden node that is also partly on the network type and on user-controllable condition. The proposed Multilayer Perceptron (MLP) model generates a predictive architecture for one dependent (target) variable to classify whether the attack class is anomaly or normal one.

In Table 2 the summary of case processing shows that 17,723 cases were assigned to the training sample and 7,468 to the testing sample.

Table 3 displays information on the neural network and is helpful for making sure that the specifications are accurate. The number of nodes in the input layer is 39 and similarly binary class out is represented by the two output units in output layer. The applied KDDCUP-99 dataset has 39 independent variables representing the input layer of the proposed model (duration, protocol_type, service, flag, src_bytes, dst_bytes, land, wrong_fragment, urgent, hot, num_failed_logins, logged_in, num_compromised, root_shell, su_attempted, num_root, num_file_creations, num_shells, num_access_files, is_guest_login, count, srv_count, serror_rate, srv_serror_rate, rerror_rate, srv_rerror_rate, same_srv_rate, diff_srv_rate, srv_diff_host_rate, dst_host_count, dst_host_srv_cnt, dt_hst_se_srv_rt, dt_host_diff_srv_rt, dt_hst_sm_src_prt_rt, dt_hst_srv_dif_ht_rt, dt_hst_seror_rt, dt_hst_srv_ser_rt,

Table 3 Network information

Input layer	Covariates number of units[a]	39 input variables from the KDDCUP 99 dataset
	Rescaling method for covariates	Standardized
Hidden layer(s)	Number of hidden layers	1
	Number of units in hidden layer 1[a]	9
	Activation function	Hyperbolic tangent
Output layer	Dependent variables	Class 1
	Number of units	2
	Activation function	Softmax
	Error function	Cross-entropy

[a] Excluding the bias unit

dt_hst_reor_rt, dt_hst_srv_rerr_rt). For the single default hidden layer there is 9 nodes. Rest of the information is default for the architecture.

Thus the developed IDS model required to be assessed based on performance measures using the model evaluation criteria discussed in the next section.

4.1 Measurement of Proposed Model Performance

In Table 4 the model summary shows information about the outcomes of training and applying the final network to the testing sample. Cross entropy error is shown since the output layer uses the softmax activation function using that the network tries to minimize the error during training. The confusion matrix provides the percentage of incorrect predictions. The execution of algorithm stopped when the maximum number of epochs reached and training has been completed ideally when the errors has converged.

In Table 5 the confusion matrix displays the useful outcomes of applying the network. For each case, the predicted response is anomaly if that cases's predicted pseudo-probability is greater than equal to 1 else it is normal attack. For each sample: Cells on the diagonal of the cross-classification of cases are correct predictions and off the diagonal of the cross-classification of cases are incorrect predictions.

Table 4 Model summary

Training	Cross entropy error	520.534
	Percent incorrect predictions	1.0 %
	Stopping rule used	1 consecutive step(s) with no decrease in error[a]
	Training time	00:00:11.969
Testing	Cross entropy error	331.762
	Percent incorrect predictions	1.3 %
Dependent variable: class		

[a] Error computations are based on the testing sample

Table 5 Confusion matrix

Sample	Observed	Predicted		
		a	n	Percent correct (%)
Training	a	8,142	73	99.1
	n	96	9,412	99.0
	Overall percent	46.5 %	53.5 %	99.0
Testing	a	3,484	43	98.8
	n	57	3,884	98.6
	Overall percent	47.4 %	52.6 %	98.7
Dependent variable: class				

Of the cases used to create the model, 9412 of the 9508 normal attacks are classified correctly (99 %) and 8142 of the 8215 anomaly attack types are classified correctly (99.1 %). Overall, 99.0 % of the training cases are classified correctly, corresponding to the 1 % incorrect shown in the Table 4 of model summary. Thus the model generates a better classification by correctly identifying a higher percentage of the cases. Classifications based upon the cases used to create the model tend to be too "optimistic" in the sense that their classification rate is inflated. The holdout sample facilitates to validate the model; here 98.8 % of these cases were correctly classified by the model. This suggests that, overall, the proposed model is in fact correct.

In Table 6 the model summary shows a couple of positive signs:

The percentage of incorrect predictions is roughly equal across training, testing, and holdout samples. The estimation algorithm stopped because the error did not decrease after a step in the algorithm. This further suggests that the original model did not over trained.

The confusion matrix in Table 7 shows that, the network does excellent at detecting anomaly than normal attacks. The detection rate and overall accuracy of

Table 6 Confusion matrix

Sample	Observed	Predicted		
		a	n	Percent correct (%)
Training	a	7,019	59	99.2
	n	70	7,981	99.1
	Overall percent	46.9 %	53.1 %	99.1
Testing	a	3,431	31	99.1
	n	53	4,044	98.7
	Overall percent	46.1 %	53.9 %	98.9
Holdout	a	1,190	12	99.0
	n	17	1,284	98.7
	Overall percent	48.2 %	51.8 %	98.8

Dependent variable: class

Table 7 Model summary

Training	Cross entropy error	389.173
	% Incorrect predictions	0.9 %
	Stopping rule used	1 consecutive step(s) with no decrease in error[a]
	Training time	00:00:25.563
Testing	Cross entropy error	246.806
	Percent incorrect predictions	1.1 %
Holdout	Percent incorrect predictions	1.2 %

Dependent variable: class

[a] Error computations are based on the testing sample

the testing outcomes have been calculated from Table 5 as 0.991045638, 0.988887419 respectively. Unfortunately, the single cutoff value (>zero) gives a very limited view of the predictive ability of the network, so it is not necessarily very useful for comparing competing networks rather focus should be on ROC curve

The Fig. 9 displays ROC curve that gives a visual display of the sensitivity and specificity for all possible cutoffs in a single plot, which is much cleaner and more powerful than a series of tables. The figure depicts here shows two curves, one for the category anomaly and one for the category normal. Since it is binary, the curves are symmetrical about a 45° line from the upper left corner of the chart to the lower right. This graph is based on the combination of training and testing samples.

The area under the curve is a numerical summary of the ROC curve, and the values in the table represent, for each category, the probability that the predicted pseudo-probability of being in that category is higher for a randomly chosen case in that category than for a randomly chosen case not in that category. In Table 8, for a randomly selected anomaly and randomly selected normal, there is a 0.999 probability that the model-predicted pseudo-probability of anomaly will be higher for the anomaly than for the normal. While the area under the curve is a useful one-statistic summary of the accuracy of the network, it is required to choose a specific criterion by which network intrusion is classified. The predicted-by-observed chart provides a visual start on this process (Fig. 10).

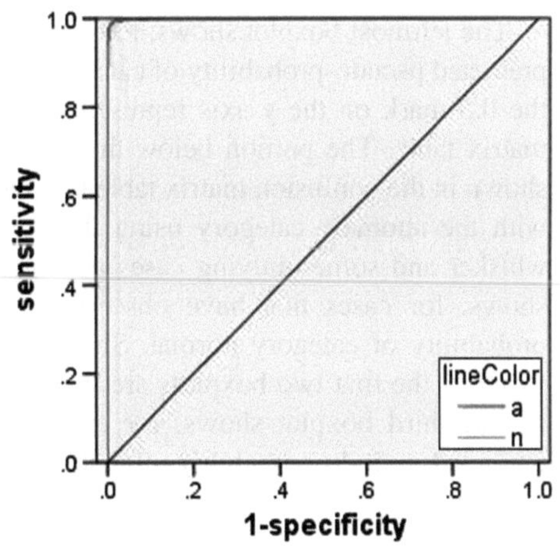

Fig. 9 ROC curve

Table 8 Area under the curve

		Area
Class	a	0.999
	n	0.999

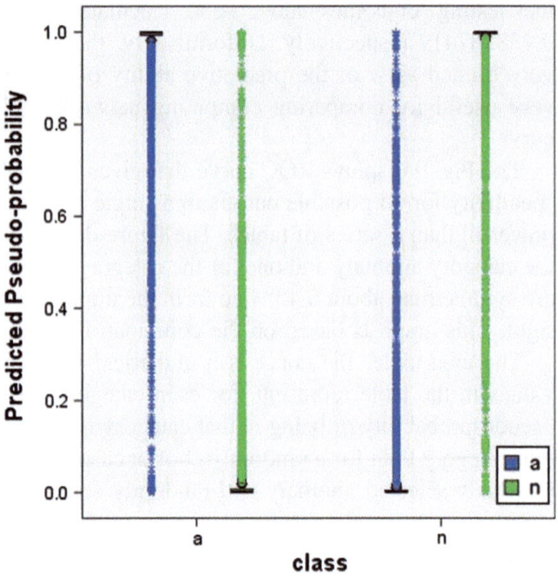

Fig. 10 Predicted-by-observed chart

For categorical dependent variables, the predicted-by-observed chart displays clustered boxplots of predicted pseudo-probabilities for the combined training and testing samples. The x axis corresponds to the observed response categories, and the legend corresponds to predicted categories.

The leftmost boxplot shows, for cases that have observed category anomaly, the predicted pseudo-probability of category anomaly. The portion of the boxplot above the 0.5 mark on the y axis represents correct predictions shown in the confusion matrix table. The portion below the 0.5 mark represents incorrect predictions. As shown in the confusion matrix table that the network is excellent at predicting cases with the anomaly category using the 0.5 cutoff, so only a portion of the lower whisker and some outlying cases are misclassified. The next boxplot to the right shows, for cases that have observed category anomaly, the predicted pseudo-probability of category normal. Since there are only two categories in the target variable, the first two boxplots are symmetrical about the horizontal line at 0.5.

The third boxplot shows, for cases that have observed category normal, the predicted pseudo-probability of category anomaly. It and the last boxplot are symmetrical about the horizontal line at 0.5. The last boxplot shows, for cases that have observed category normal, the predicted pseudo-probability of category normal. The portion of the boxplot above the 0.5 mark on the y axis represents correct predictions shown in the confusion matrix table. The portion below the 0.5 mark represents incorrect predictions. Remember from the confusion matrix table that the network predicts slightly more than half of the cases with the normal category using the 0.5 cutoff, so a good portion of the box is misclassified. Looking at the plot, it appears that by lowering the cutoff for classifying a case as normal from 0.5 to approximately 0.3—this is roughly the value where the top of the second box and the bottom of the fourth box are—that can increase the chance of correctly detecting possible intrusion without generating false alarm on normal attacks.

In the Fig. 11 cumulative gains chart demonstrates the percentage of the overall number of cases in a given category "gained" by targeting a percentage of the total number of cases. For example, the first point on the curve for the anomaly category is at (10, 20 %), meaning that if a dataset is scored with the network and sort all of the cases by predicted pseudo-probability of anomaly, it is expected that the top 10 % to contain approximately 20 % of all of the cases that actually take the category anomaly (attacks). Likewise, the top 20 % would contain approximately 45 % of the anomaly; the top 30 % of cases would contain 65 % of defaulters, and so on. If 100 % scored dataset is selected then all of the anomaly in the dataset will be obtained. The diagonal line is the "baseline" curve; if 10 % of scored dataset is selected at random, then it is expected to "gain" approximately 10 % of all of the cases that actually take the category anomaly. The farther above the baseline a curve lies, the greater the gain. The cumulative gains chart is used to help choose a classification cutoff by choosing a percentage that corresponds to a desirable gain, and then mapping that percentage to the appropriate cutoff value. What constitutes a "desirable" gain depends on the cost of Type I and Type II errors. That is, what is the cost of classifying a anomaly attack as a normal attack (Type I)? What is the cost of classifying a normal as a anomaly (Type II)? If any network parameter is the primary concern, then Type I error may be minimised; on the cumulative gains chart, this might correspond to generate alarm in the top 40 % of pseudo-predicted probability of anomaly, which captures nearly 90 % of the possible anomaly attacks

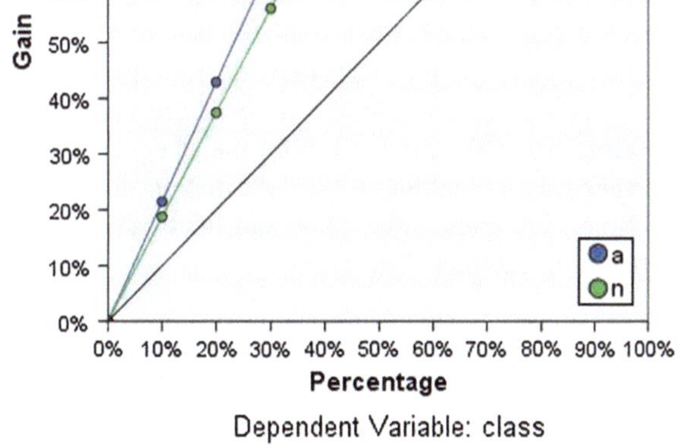

Fig. 11 Cumulative gains chart, dependent variable: class

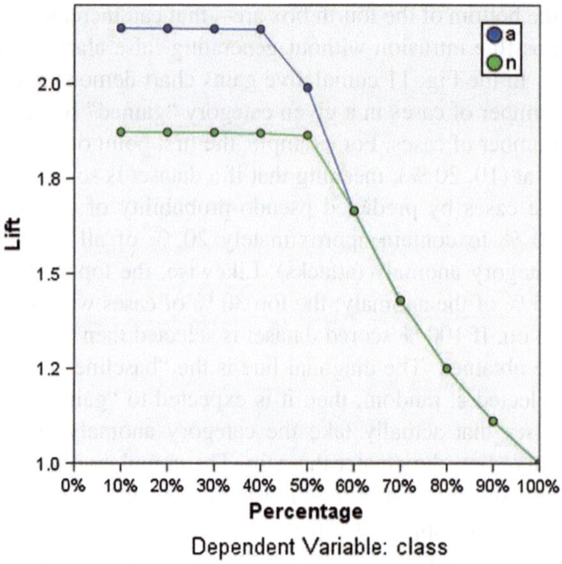

Fig. 12 Lift chart, dependent variable: class

Table 9 Comparison of three neural network architecture based on common measures of performance

	BPN	Recurrent	PCA
Sensitivity	0.9848	0.9650	0.9828
Specificity	0.9924	0.9306	0.9594
Overall accuracy	0.9889	0.9490	0.9719
Kappa statistic	0.9630	0.8970	0.9430
TSS	0.9772	0.8956	0.9422

but removes nearly half of total attacks. If a very large data set is the priority, then Type II error may be minimised. On the chart, this might correspond to rejecting the top 10 %, which captures 20 % of the anomaly and leaves most of KDD99 data set intact. Usually, both are major concerns, so a decision rule should have been chosen for classifying attacks that gives the best mix of sensitivity and specificity.

The lift chart in Fig. 12 is derived from the cumulative gains chart; the values on the y axis correspond to the ratio of the cumulative gain for each curve to the baseline. Thus, the lift at 10 % for the category Yes is 30 %/10 % = 3.0. It provides another way of looking at the information in the cumulative gains chart.

Note: The cumulative gains and lift charts are based on the combined training and testing samples (Table 9).

5 Conclusion

As described in the preceding section, MLP method has recognized them as a good choice for any existing intrusion detection system. This paper provides a state-of-the-art review of the applications of neural network to Intrusion Detection System. Following findings are significant in the research review of IDS:

- Artificial neural network based intrusion detection system development is an important research trend in intrusion detection domain.
- China has shown to be significantly contributing (32 %) followed by USA (16 %) and India (9 %) in terms of publication by affiliated country.
- The conference paper (65 %) has recognized as the major type of research documents followed by articles (30 %).
- Lecture Notes in Computer Science has emerged as leading journal that published 25 articles on IDS based on neural network (11 %) followed by 24 articles in Computers and Security and 20 articles in Expert Systems with Applications (9 %) journals.
- Undoubtedly the computer science (49 %) is shown to be the major domain publishing 509 articles followed by 273 articles in engineering (26 %) and 108 articles in mathematics (10 %).
- The current research trend based on the number of articles published between the years 2000–2013 has been shown to be increasing with R-squared value equals 0.9433 as a good fit. The trend line for 2000–2014 is also shown to be increased.

In this research, we have proposed architecture based on Multi Layer Perceptron neural network. The model builds the intrusion detection system learnt from the patterns of KDD99 data set. Based on the identified patterns, the architecture recognized attacks in the datasets using the back propagation neural network algorithm. The proposed neural network approach resulted with higher detection rate, a reduced amount of execution time. We continue our work in this direction in order to build an efficient intrusion detection model. When the proposed Back propa gation neural network approach is compared with the other two approaches: Recurrent and PCA neural network based on the common measures of performance it is clearly visible as shown in Fig. 13 to outperform the performances of the other two approaches. Further work will be undertaken to increase the performance of the intrusion detection model and reduce the false alarm and efficiently handle the identification of correct anomaly dynamically.

Since the goal of this research was also to evaluate the performance of our proposed approach by comparing with other six approaches available in literature in terms of three measures of performance: detection rate, accuracy rate and computation time of the intrusion detection (Table 10). The comparative research findings from Table 10 has revealed that the proposed approach has succeeded in achieving increased rate of anomaly detection, reduced false alarm and at the same time minimal execution time for the development of intrusion detection system. In KPCA and SVM approach, the accuracy rate is 99.2 % (98.89 % accuracy in case of

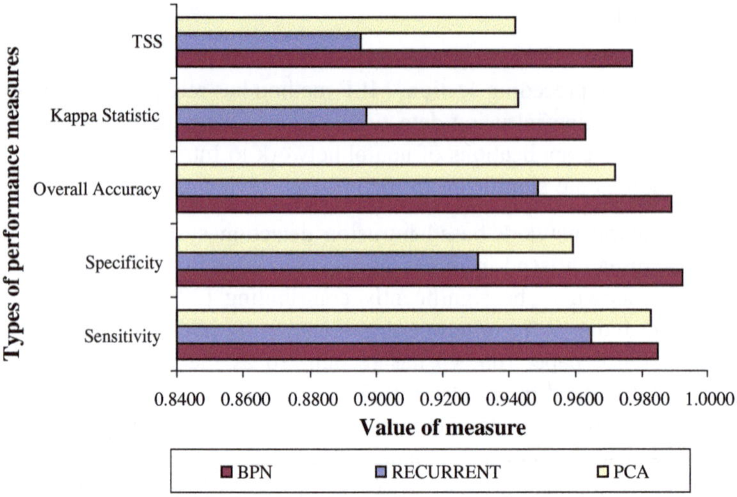

Fig. 13 Comparison of the measure of performance for three neural network architecture applied to develop intrusion detection system

Table 10 Comparative performance of literature available approaches used with proposed multilayer perceptron approach based on detection rate, accuracy and computation time

Approach used	References	Detection rate testing %	Accuracy testing %	Computation time (s)	Dataset used in experiment
KPCA and SVM	Kuanf et al. (2012)	–	99.2 (training: 99.975)	407.918466	KDD dataset 6000 sample-4000 for training, 2000 for testing (Han 2012)
Resilient back propagation neural network	Naoum et al. (2012)	94.7	–	–	KDD dataset (Naoum et al. 2005)
Decision tree based light weight intrusion detection using wrapper approach	Sivatha Sindhu et al. (2012)	98.38	–	–	KDD dataset (Sivatha Sindhu et al. 2012)
Neural network	Devaraju and Ramakrishnan (2011)	–	97.5	–	KDD dataset (Kuanf et al. 2012)
BPNN	Mukhopadhyay et al. (2011)	–	–	–	KDD dataset (Mukhopadhyay et al. 2011)
SOM	Ibrahim et al. (2013)	92.37		–	KDD 99
Our proposed approach[a]	–	99.10	98.89	11.969	KDD 20 % dataset

[a] The data is taken from 70 to 30 dataset as it is giving better detection rate

proposed approach) however detection rate is unknown with a larger computation time. In the rest of the result of Table 10 the detection rate and computation time of the proposed MLP approach are superior.

The future work should be directed towards developing hybrid neural network to increase the efficiency of intrusion detection and to deal the dynamic large data stream to secure from network intrusion.

References

Anderson, J. (1995). *An introduction to neural networks*. Cambridge: MIT Press.

Anyanwu, L. O., Keengwe, J., & Arome, G. A. (2011). Scalable intrusion detection with recurrent neural networks. *International Journal of Multimedia and Ubiquitous Engineering, 6*(1), 21–28.

Aziz, A. S. A., Azar, A. T., Hassanien, A. E., & Hanafy, S. E. O. (2012). Continuous features discretization for anomaly intrusion detectors generation. In *The 17th Online World Conference on Soft Computing in Industrial Applications* (*WSC17*), December 10–21.

Aziz, A. S. A., Azar, A. T., Hassanien, A. E., & Hanafy, S. E. O. (2014). Continuous features discretization for anomaly intrusion detectors generation. In *Soft computing in industrial applications* (pp. 209–221). Switzerland: Springer International Publishing.

Abdel-Aziz, A. S., Hassanien, A. E., Azar, A. T., & Hanafi, S. E. O. (2013). Machine learning techniques for anomalies detection and classification. *Advances in security of information and communication networks* (pp. 219–229). Berlin Heidelberg: Springer.

Barry, S., & Elith, J. (2006). Error and uncertainty in habitat models. *Journal of Applied Ecology, 43*(3), 413–423.

Behjat, A. R., Vatankhah, N., & Mustapha, A. (2014). Feature subset selection using genetic algorithm for intrusion detection system. *Advanced Science Letters, 20*(1), 235–238.

Bezdek, J. C. (1994). *What is computational intelligence? Computational intelligence imitating life* (pp. 1–12). New York: IEEE Press.

Chebrolu, S., Abraham, A., & Thomas, J. P. (2005). Feature deduction and ensemble design of intrusion detection systems. *Computers and Security, 24*(4), 295–307.

Chittur, A. (2001). Model generation for an intrusion detection system using genetic algorithms. High School Honors Thesis, Ossining High School. In Cooperation with Columbia Univ. Accessed on November 27, 2013.

Cohen, J. (1960). A coefficient of agreement for nominal scales. *Educational and Psychological Measurement, 20*(1), 37–46.

Dębska, B., & Guzowska-Świder, B. (2011). Application of artificial neural network in food classification. *Analytica Chimica Acta, 705*(1), 283–291.

Denning, D. E. (1987). An intrusion-detection model. *IEEE Transactions on Software Engineering, 13*(2), 222–232.

Devaraju, S., & Ramakrishnan, S. (2011). Performance analysis of intrusion detection system using various neural network classifiers. In Recent Trends in Information Technology (ICRTIT), June 2011 International Conference on (pp. 1033–1038). IEEE.

Eid, H. F., Azar, A. T., & Hassanien, A. E. (2013, January). Improved real-time discretize network intrusion detection system. In *Proceedings of seventh international conference on bio-inspired computing: theories and applications* (*BIC-TA 2012*) (pp. 99–109). India: Springer.

El Kadhi, N., Hadjar, K., & El Zant, N. (2012). A mobile agents and artificial neural networks for intrusion detection. *Journal of Software, 7*(1), 156–160.

Eskin, E., Arnold, A., Prerau, M., Portnoy, L., & Stolfo, S. (2002). A geometric framework for unsupervised anomaly detection. *Applications of data mining in computer security* (pp. 77–101). US: Springer.

Faysel, M. A., & Haque, S. S. (2010). Towards cyber defense: research in intrusion detection and intrusion prevention systems. *IJCSNS International Journal of Computer Science and Network Security, 10*(7), 316–325.

Feizollah, A., Anuar, N. B., Salleh, R., Amalina, F., Ma'arof, R. U. R., & Shamshirband, S. (2014). A study of machine learning classifiers for anomaly-based mobile Botnet detection. *Malaysian Journal of Computer Science, 26*(4), 251–265.

Gong, R. H., Zulkernine, M., & Abolmaesumi, P. (2005, May). A software implementation of a genetic algorithm based approach to network intrusion detection. In *Sixth international conference on software engineering, artificial intelligence, networking and parallel/distributed computing, 2005 and first ACIS international workshop on self-assembling wireless networks (SNPD/SAWN 2005)* (pp. 246–253). IEEE.

Guisan, A., & Thuiller, W. (2005). Predicting species distribution: Offering more than simple habitat models. *Ecology Letters, 8*(9), 993–1009.

Gupta, B. B., Joshi, R. C., & Misra, M. (2012). ANN based scheme to predict number of Zombies in a DDoS attack. *IJ Network Security, 14*(2), 61–70.

Han, L. (2012). Research of K-MEANS algorithm based on information Entropy in Anomaly Detection. In Multimedia Information Networking and Security (MINES), November 2012 Fourth International Conference on (pp. 71-74). IEEE.

Haykin, S. (2005). *Neural networks a comprehensive foundation*. New Delhi: Pearson Education.

Heady R., Luger G., Maccabe A., & Servilla M. (1990, August). The architecture of a network level intrusion detection system. Technical report, Computer Science Department, University of New Mexico.

Hwang, R. C., Chen, Y. J., & Huang, H. C. (2010). Artificial intelligent analyzer for mechanical properties of rolled steel bar by using neural networks. *Expert Systems with Applications, 37* (4), 3136–3139.

Ibrahim, L. M., Basheer, D. T., & Mahmod, M. S. (2013). A comparison study for intrusion database (Kdd99, Nsl-Kdd) based on self organization map (SOM) artificial neural network. *Journal of Engineering Science and Technology, 8*(1), 107–119.

Khashei, M., Rezvan, M. T., Hamadani, A. Z., & Bijari, M. (2013). A bi-level neural-based fuzzy classification approach for credit scoring problems. *Complexity, 18*(6), 46–57.

Kuanf, F., Xu, W., Zhang, S., Wang,Y., & Liu, K. (2012). A novel Approach of KPCA and SVM for Intrusion Detection, *Journal of Computational Information Systems*, pp 3237–3244.

Kuo, R. J., Wang, Y. C., & Tien, F. C. (2010). Integration of artificial neural network and MADA methods for green supplier selection. *Journal of Cleaner Production, 18*(12), 1161–1170.

Laskov, P., Düssel, P., Schäfer, C., & Rieck, K. (2005). Learning intrusion detection: Supervised or unsupervised? In *Image analysis and processing—ICIAP 2005* (pp. 50–57). Berlin Heidelberg: Springer.

Lee, W., Stolfo, S. J., & Mok, K. W. (1999). A data mining framework for building intrusion detection models. In *Proceedings of the 1999 IEEE symposium on security and privacy* (pp. 120–132). IEEE.

Liao, Y., & Vemuri, V. R. (2002). Use of K-nearest neighbor classifier for intrusion detection. *Computers and Security, 21*(5), 439–448.

Liu, J. (2013). An adaptive intrusion detection model based on ART2 neural network. *Journal of Computational Information Systems, 9*(19), 7775–7782.

Louvieris, P., Clewley, N., & Liu, X. (2013). Effects-based feature identification for network intrusion detection. *Neurocomputing, 121*, 265–273.

McCarthy, J., Minsky, M. L., Rochester, N., & Shannon, C. E. (1955). A proposal for the dartmouth summer research project on artificial intelligence, August 31, 1955. *AI Magazine, 27*(4), 12.

McCarthy, J. (2007). What is artificial intelligence. url: http://www-formal.stanford.edu/jmc/whatisai.html. (accessed on 22 November 2013)

Mukhopadhyay, I., Chakraborty, M., Chakrabarti, S., & Chatterjee, T. (2011). Back propagation neural network approach to Intrusion Detection System. In *Recent Trends in Information Systems (ReTIS), December 2011 International Conference* on (pp. 303–308). IEEE.

Naoum, R. S., Abid, N. A., Al-Sultani, Z. N. (2005) "An enhanced Resilient backpropagation artificial neural network for Intrusion detection", International Journal of Computer Science and Network Security, 2005, *12*(3), 11–16.

Pan Z., Chen, S., Hu, G., & Zhang, D. (2003). Hybrid neural network and C4.5 for misuse detection. In *Proceedings of the second international conference on machine learning and cybernetics* (Vol. 4, pp. 2463–2467). IEEE.

Peláez, J. I., Doña, J. M., Fornari, J. F., & Serra, G. (2014). Ischemia classification via ECG using MLP neural networks. *International Journal of Computational Intelligence Systems, 7*(2), 344–352.

Peng, Y., Wang, Y., Niu, Y., & Hu, Q. (2014). Application study on intrusion detection system using IRBF. *Journal of Software, 9*(1), 177–183.

Saftoiu, A., Vilmann, P., Gorunescu, F., Janssen, J., Hocke, M., & Larsen, M., et al. (2012). Efficacy of an artificial neural network-based approach to endoscopic ultrasound elastography in diagnosis of focal pancreatic masses. *Clinical Gastroenterology Hepatology, 10*(1), 84–90.

Sall, J., Creighton, L., & Lehman, A. (2007). Safari tech books online. JMP start statistics a guide to statistics and data analysis using JMP. *SAS press series* (4th edn.). Cary, N.C.: SAS Pub.

Segurado, P., & Araujo, M. B. (2004). An evaluation of methods for modelling species distributions. *Journal of Biogeography, 31*(10), 1555–1568.

Shao, G., & Halpin, P. N. (1995). Climatic controls of eastern North American coastal tree and shrub distributions. *Journal of Biogeography*, 1083–1089.

Sheikhan, M., & Sharifi Rad, M. (2011). Intrusion detection improvement using GA-optimized fuzzy grids-based rule mining feature selector and fuzzy ARTMAP neural network. *World Applied Sciences Journal, 14*, 772–781.

Sheikhan, M., & Sharifi, Rad M. (2013). Using particle swarm optimization in fuzzy association rules-based feature selection and fuzzy ARTMAP-based attack recognition. *Security and Communication Networks, 6*(7), 797–811.

Sivatha Sindhu, S. S., Geetha, S., & Kannan, A. (2012). Decision tree based light weight intrusion detection using a wrapper approach. *Expert Systems with applications, 39*(1), 129–141.

Stolfo, S. J., Fan, W., Lee, W., Prodromidis, A., & Chan, P. K. (2000). Cost-based modeling for fraud and intrusion detection: Results from the JAM project. In *Proceedings of the DARPA information survivability conference and exposition, 2000 (DISCEX'00)* (Vol. 2, pp. 130–144). IEEE.

Swets, J. A. (1988). Measuring the accuracy of diagnostic systems. *Science, 240*(4857), 1285–1293.

Tiwari, P. (2002). Intrusion detection. Technical Report, Department of Electrical Engineering, Indian Institute of Technology, Delhi.

Tuncer, T., & Tatar, Y. (2012). Implementation of the FPGA based programmable embedded intrusion detection system. *Journal of the Faculty of Engineering and Architecture of Gazi University, 27*(1), 59–69.

Valero, S., Senabre, C., López, M., Aparicio, J., Gabaldon, A., & Ortiz, M. (2012). Comparison of electric load forecasting between using SOM and MLP neural network. *Journal of Energy and Power Engineering, 6*(3), 411–417.

Wang, G., Hao, J., Ma, J., & Huang, L. (2010). A new approach to intrusion detection using artificial neural networks and fuzzy clustering. *Expert Systems with Applications, 37*(9), 6225–6232.

Wang, J. H., Liao, Y. L., Tsai, T. M., & Hung, G. (2006). Technology-based financial frauds in Taiwan: Issues and approaches. In *SMC* (pp. 1120–1124).

Wu, S. X., & Banzhaf, W. (2010). The use of computational intelligence in intrusion detection systems: A review. *Applied Soft Computing, 10*(1), 1–35.

Xiang, Z., Zhu, J., Han, W., & Ding, J. (2013). On the capability of SOINN based intrusion detection systems. *Journal of Computational Information Systems, 9*(3), 941–949.

Yang, S., Yang, Y., Shen, Q., & Huang, H. (2013). A method of intrusion detection based on semi-supervised GHSOM. In *Jisuanji Yanjiu yu Fazhan/Computer Research and Development. Jisuanji Yanjiu yu Fazhan/Computer Research and Development, November 2013* (Vol. 50(11), pp. 2375–2382).

Yao, J. T., Zhao, S. L., & Saxton, L. V. (2005). A study on fuzzy intrusion detection. In B. V. Dasarathy (Ed.), In *Proceedings of SPIE vol. 5812, data mining, intrusion detection, information assurance, and data networks security*, 28 March–1 April 2005 (pp. 23–30). Orlando, Florida, USA, Bellingham, WA: SPIE.

Zainaddin, A., Asyiqin, D., & Mohd Hanapi, Z. (2013). Hybrid of fuzzy clustering neural network over NSL dataset for intrusion detection system. *Journal of Computer Science, 9*(3), 391–403.

Zhao, Y., Zha, Y., & Zha, X. (2013). Network intrusion detection based on IPSO-BPNN. *Information Technology Journal, 12*(14), 2719–2725.

Enhanced Power System Security Assessment Through Intelligent Decision Trees

Venkat Krishnan

Abstract Power system security assessment involves ascertaining the post-contingency security status based on the pre-contingency operating conditions. A system operator accomplishes this by the knowledge of critical system attributes which are closely tied to the system security limits. For instance, voltage levels, reactive power reserves, reactive power flows are some of the attributes that drive the voltage stability phenomena, and hence provide easy guidelines for the operators to monitor and maneuver the highly stressed power system to a secure state. With tremendous advancements in computational power and machine learning techniques, there is increased ability to produce security guidelines that are highly accurate and robust under a wide variety of system conditions. Particularly, the decision trees, a data mining tool, has lend itself well in extracting highly useful and succinct knowledge from a very large repository of historical information. The most vital and sensitive part of such a decision tree based security assessment is the stage of training database generation, a computationally intensive process which involves sampling many system operating conditions and performing power system contingency assessment simulations on them. The classification performance of operating guidelines under realistic testing scenarios depend heavily on the quality of the training database used to generate the decision trees. So the primary objective of this chapter is to develop an improvised database generation process that creates a satisfactory training database by sampling the most influential operating conditions from the input operating parameter state space prior to the stage of power system contingency simulation. Embedding such intelligence to the system scenario sampling process enhances the information content in the training database, while minimizing the computing requirements to generate it. This chapter will clearly explain and demonstrate the process of identifying such high information contained sampling space and the advantage of deriving security guidelines from decision trees that exclusively use such an enhanced training database.

Keywords Security assessment · Operating guidelines · Decision trees · Intelligent training set · Monte Carlo simulation · Importance sampling

V. Krishnan (✉)
Department of Electrical and Computer Engineering,
Iowa State University, 1124 Coover Hall, Ames, IA 50011, USA
e-mail: vkrish@iastate.edu

Q. Zhu and A.T. Azar (eds.), *Complex System Modelling and Control Through Intelligent Soft Computations*, Studies in Fuzziness and Soft Computing 319,
DOI 10.1007/978-3-319-12883-2_12

1 Introduction

Traditionally, power system reliability assessments and planning involve deterministic techniques and criteria, which are being used in practical applications even now, such as WECC/NERC disturbance-performance table for transmission planning (WECC 2003; Abed 1999). But the drawback with deterministic criteria is that they do not reflect the stochastic or probabilistic nature of the system in terms of load profiles, component availability, failures etc. (Billinton et al. 1997). Therefore the need to incorporate probabilistic or stochastic techniques to assess power system reliability and obtain suitable indices or guidelines for planning has been recognized by the power system planners and operators; and several such techniques have been developed (Beshir 1999; Chowdhury and Koval 2006; Li and Choudhury 2007; Wan et al. 2000; Xiao and McCalley 2007).

In this regard, Monte Carlo simulation (MCS) methods lend themselves well by simulating the actual analytical process with randomness in system states (Billinton and Li 1994). In this way, several system effects or process including nonelectrical factors such as weather uncertainties can be included in a study based on appropriate parameter's probability distributions. Figure 1 shows an overview of MCS based security assessment methodology, which involves two major tasks: database generation approach and machine learning analysis.

The database generation approach involves the following steps:

- *Random Sampling*: Operating parameters (load, unit availability, circuit outages, etc.) are randomly selected as per a distribution (e.g., uniform, Gaussian, exponential, empirical etc.). This process is generally known as Monte Carlo sampling. Using the generated samples, various base cases are formed.

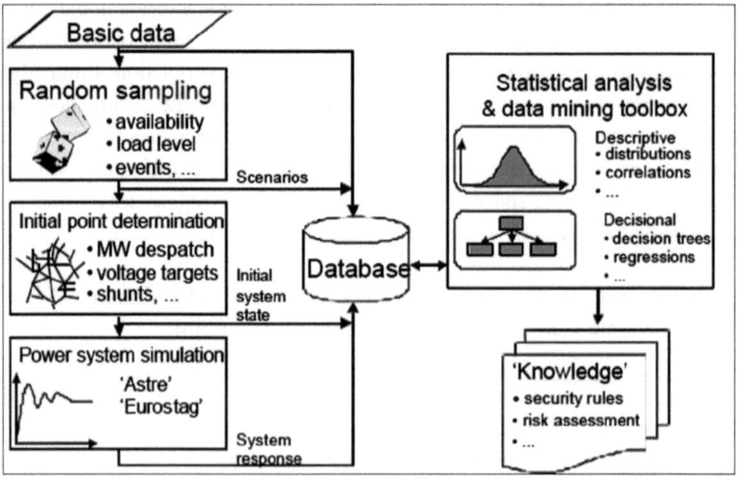

Fig. 1 Probabilistic reliability assessment based on MCS and data mining (Henry et al. 2004b)

- *Optimal power flow*: Initial states for every case is obtained using OPF
- *Contingency assessment*: Using steady-state or time-domain tools contingency events are simulated, and post-contingency performance measures are obtained.

The machine learning methods (Wehenkel 1998; Witten and Frank 2000) are used to extract a high level information, or knowledge from a huge database containing post-contingency responses obtained from the database generation step. These machine learning or data mining techniques are broadly classified as:

- *Unsupervised learning*: Those methods which do not have a class or target attribute. For example, association rule mining can be used to find the correlation between various attributes. Clustering methods such as k-means, EM etc. are generally used to discover classes.
- *Supervised learning*: Those methods that have a class or target attribute, such as classification, numerical prediction etc., and use the other attributes (other observable variables) to classify or predict class values of scenarios. For example, naïve bayes, decision trees, instance based learning, neural network, support vector machine, regression etc.

Among these, decision tree based inductive learning method serves as an attractive option for preventive-control approach in power system security assessment (Zhou et al. 1994; Wehenkel 1997; Zhou and McCalley 1999; Niimura et al. 2004; Wehenkel et al. 2006). It identifies key pre-contingency attributes that influence the post-contingency stability phenomena and provides the corresponding acceptable scenario thresholds. Based on it, security rule or guidelines are developed, which can be deductively applied to ascertain any new pre-contingency scenario's post-contingency performance. Information required for building decision tree are:

- A training set, containing several pre-contingency attributes with known class values
- The classification variable (i.e., class attribute with class values such as "secure" or "insecure"), which could be based on post-contingency performance indices
- An optimal branching rule, i.e., a rule to find critical attribute
- A stopping rule, such as maximum tree length or minimum instances

The aim of inducing a decision tree is to obtain a model that classifies new instances well and produces simple to interpret rules. Ideally we would like to get the best model that has no diversity (impurity), i.e., all instances within every branch of the tree belong to the same class. But due to many other uncertainties or interactions that have not been accounted for in the model, there would be some impurity (i.e., non-homogeneous branch) at most of the levels. So the goal is to select attributes at every level of branching such that impurity is reduced. There are many measures of impurity, which are generally used as optimal branching criteria to select the best attribute for splitting. Some of those are entropy, information gain, Gini index, gain ratio etc.

Classification accuracy and error rates are used as the performance measures of a decision tree. There are two kinds of errors: *false alarms*—acceptable cases classified as unacceptable; and *risks*—unacceptable cases as acceptable. Errors can be calculated by testing the obtained decision model on the training set, which is usually an over-estimate. There are training set sampling methods such as holdout procedures, cross-validation, bootstrap etc. (Witten and Frank 2000) to make the error estimation unbiased. It is even better if the testing is performed using an independent test dataset. There are numerous references that explain the process of building a decision tree from a database with algorithms such as ID3, J48 etc. CART, Answer Tree, Orange, WEKA etc. are some software available for building decision trees.

Many utilities have taken and are continuing to take a serious interest in implementing learning algorithm such as decision tree in their decision making environment. French transmission operator RTE has been using decision tree based security assessment methods to define operational security rules, especially regarding voltage collapse prevention (Lebrevelec et al. 1998, 1999; Schlumberger et al. 1999, 2002; Pierre et al. 1999; Martigne et al. 2001; Paul and Bell 2004; Henry et al. 1999, 2004a, 2006; Cholley et al. 1998). They provide operators a better knowledge of the distance from instability for a post-contingency scenario in terms of pre-contingency conditions, and thus save a great amount of money by preserving the reliability while enabling more informed operational control closer to the stability limits. So the central topic of this chapter will be: *what is the significant component of this decision tree induction process, and how to improve it for the betterment of the planning solutions that are needed under realistic operating conditions?*

The remaining parts of this chapter are organized as follows. Section 2 provides the background of this work in terms of motivation behind this research, related past work, and the objective of this work. Section 3 describes the concept of "information content" in the context of this work. Section 4 presents the technical approach of the proposed high information contained training database generation. Section 5 demonstrates the application in deriving operational rules for voltage stability problem in Brittany region of RTE's system, and presents results and discussions. Section 6 presents conclusions and future research directions.

2 Motivation, Related Work, and Objective

The most vital and sensitive part of MCS based reliability studies is the stage of database generation. The confidence we will have in the results generally reflects the confidence we have in the set of system states generated. The generated database does influence the classification performance of the decision tree against realistic scenarios, selection of critical attributes and their threshold values, and size of the operating rules.

Generally a uniform or random sampling of system states is carried out by varying parameters such as load level, unit availability, exchanges at the boarders, component availability etc. according to their independent probability distributions obtained from projected historical data (Henry et al. 1999, 2004b; Paul and Bell 2004; Lebrevelec et al. 1999; Senroy et al. 2006). Then, various scenarios are simulated for a pre-specified set of contingencies. This stage is generally very tedious and time consuming, as there could be a tremendously large number of combinations of variables [about 5,000–15,000 samples for a statistically valid study (Henry et al. 2004b)]. Therefore, the challenge of producing high information content training database at low computational cost needs to be addressed (Cutsem et al. 1993; Jacquemart et al. 1996; Wehenkel 1997; Dy-Liacco 1997).

In the open literature, there are re sampling methods to retain only the most important instances from an already generated training database (Jiantao et al. 2003; Foody 1999) for classification purposes. But such methods involve huge computational cost in first generating a training database, then identifying the most influential instances, and if need be, generate more of such instances. Genc et al. (2010) proposed an iterative method to a priori identify the most influential region in the operating parameter state space, and then enrich the training database with more instances from the identified high information content region for enhancing classification performance. In this case, the method proposed to identify the high information content region involves heavy computational cost when the dimension of the operating parameter space increases, even beyond 10 parameters.

This chapter proposes to develop an efficient sampling method to generate influential operating conditions that captures high information content for better classification and also reduces computing requirements. In short, the objective is to maximize information content in the training database, while minimizing computing requirements to generate it. This efficient sampling is constructed using the Monte Carlo Variance Reduction (MCVR) techniques. Among the mostly used MCVR methods, control variate and antithetic variate take advantage of the correlation between certain random variables to obtain variance reduction in statistical estimation studies. Stratification method and importance sampling method re-orient the way the random numbers are generated, i.e., alters the sampling distribution (Ripley 1987; Thisted 1988). The proposed efficient sampling method is constructed using the importance sampling method for its ability to bias the Monte Carlo sampling towards the influential region identified a priori; and generate samples within the influential region preserving the original relative likelihood of the operating conditions.

In order to sample the most influential operating conditions, the influential region must be first traced; which requires that the operating parameter state space be characterized with respect to post-contingency performance. A straight forward way to perform state space characterization is to divide the N-dimensional hypercube, where N is the number of selected operating parameters, into M smaller hypercubes, select the center point of each of the M smaller hypercubes and perform an assessment to identify post-contingency performance (NM contingency simulations). But for large N, there is a curse of dimensionality, resulting in very large computational cost. This work proposes a computationally efficient method based on Latin Hypercube sampling (LHS) to characterize the operational parameter space.

The next section introduces the concept of "high information content" and the measure that can be used to quantify it.

3 High Information Content Region

The decision tree learning algorithm requires a database that has good representation of all the class values, so that it can effectively classify new instances and not overlook the less representative classes. So, for a two-class problem, a good representation of operating conditions on both sides of the class boundary is required. Also, not every operating condition on both sides of the class boundary contributes equally to the operating rule derivation process. This is further demonstrated using Fig. 2 with the help of its four parts a-d, which explain the importance of sampling the most influential operating conditions for the purpose of rule making. For instance, consider sampling some operating conditions defined in terms of variations in Loads A and B as shown in Fig. 2a. Perform contingency analysis to find the post-contingency voltage stability performance (*yellow dots* have acceptable, and *red dots* have unacceptable performances). A suitable rule can be defined by line R that effectively partitions the operating region with acceptable post contingency performance from unacceptable performance. We refer to this line as the security boundary. Now, if more operating conditions are sampled as shown in Fig. 2b, the samples drawn near to the security boundary influences the rule making process more than the samples away from the boundary. This is evident from the consequent rule change (shifting line R) that is necessary as shown in Fig. 2c. So it is very essential that the database contains operating conditions nearer to the security boundary with finer granularity, since they convey more information on the variability of the performance measure, which thereby enables a clear cut decision making on the acceptability of any operating condition. Furthermore, if the some of the operating conditions with unacceptable performance near the rule line R in Fig. 2c are less likely to occur in reality, then the rule line R may be shifted slightly upwards to exploit more operating conditions for economic reasons, as shown in Fig. 2d. Hence the desired influential operating conditions are obtained by sampling according to the probability distribution of the boundary region, which is the shaded region in Fig. 2d where there is a high uncertainty in the acceptability of any operating condition. This will also ensure a very good representation of both the classes in the database at a reduced computational cost compared to sampling from the entire operational parameter state space probability distribution.

In this work Entropy, the most commonly used information theoretic measure for the information contained in a distribution, is used to quantify information content in a database (Unger et al. 1990). It is a function of class proportions, when

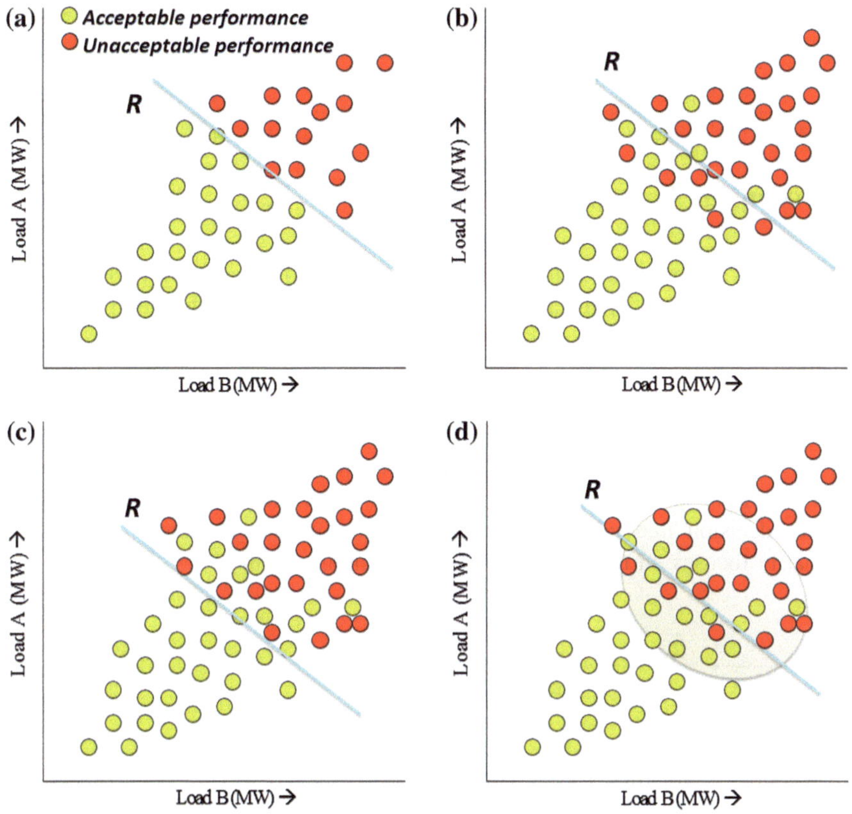

Fig. 2 High information content region

operating conditions are sampled according to their probability distribution. Entropy is given by Eq. (1)

$$Entropy(S) = \sum_{i=1}^{c} -p_i \log_2 p_i \tag{1}$$

where, S is training data, c is the number of classes, and pi is the proportion of S classified as class i. Given that the security boundary generally falls in the lower probability region of the operating parameter state space, a database containing samples within the boundary region has the maximum entropy, produced at reduced computational cost. This is the central principle that is used to devise the efficient training database generation approach proposed in this chapter.

The following section will delineate a technical approach that will be used in this chapter to devise the efficient sampling method to generate the high information contained training database. Later in the numerical results section, the entropy measure introduced in this section will be used to measure the information content in the training database used for producing the decision trees.

4 Technical Approach

The overall flowchart of risk-based planning approach is shown by Fig. 3, along with the proposed efficient sampling approach. The proposed algorithm consists of two stages, where stage I utilizes a form of stratified sampling to approximately identify the boundary region and stage II utilizes importance sampling to bias the sampling towards the boundary region. The database generation is performed for every critical contingency or a group of critical contingencies screened, as depicted by the left-side loop. The right-side loop feeds back information about the region of sampling state space requiring more emphasis in the training database, in order to reduce decision tree misclassifications and improve the accuracy. This chapter primarily focuses on the proposed efficient sampling method.

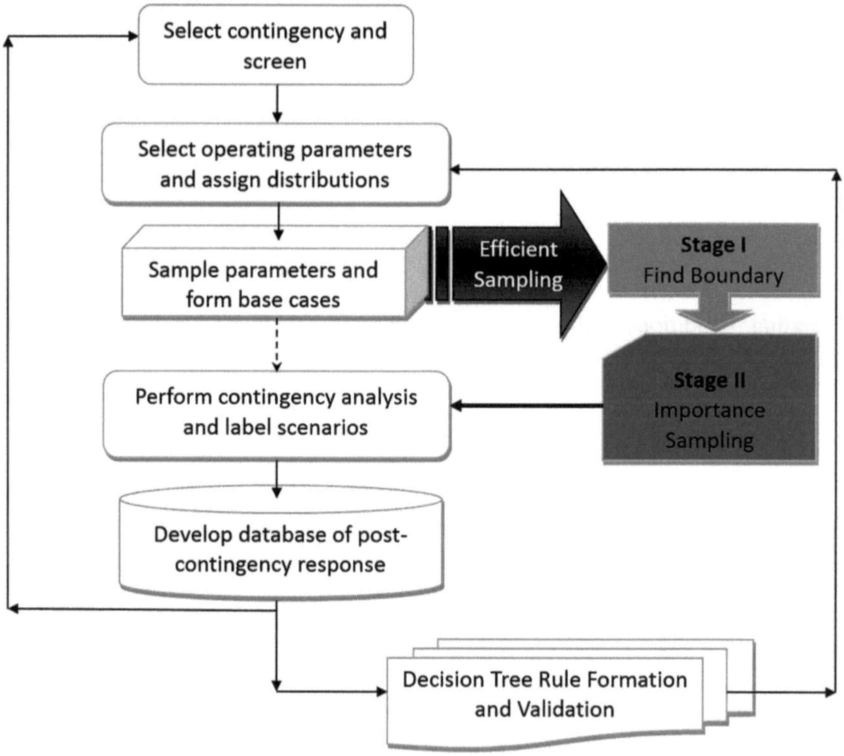

Fig. 3 Proposed approach

4.1 Stage I—Boundary Region Identification

This section develops a LHS method that uses linear sensitivity information to trace the boundary region in a computationally effective manner.

The sampling procedure is computationally very burdensome for a very large dimensional sampling space, especially if the individual load's mutual correlation information is taken into account. So, in order to provide a more reasonable sampling space which would reduce the computation, typically a very strong assumption is made that all loads vary in proportion to the total (also known as homothetic load distribution), so that the load at any bus i maintains a constant percentage of the total load, i.e., $P_{Li} = (P_{Li0}/P_{T0})\,P_T$, where P_{Li0} and P_{T0} are the bus i load and total load in the reference or base case; and P_{Li} and P_T are for any new loading scenario. In the language of voltage stability analysis, these assumptions amount to defining a particular stress direction through the space of possible load increases. Therefore, when a single stress direction is assumed, the uncertainty in load can be simply expressed in terms of the total system load (P_T). So in this case, the sampling is performed only in the univariate space of total system load (P_T) to identify the boundary region.

Generally, this assumption of individual loads having a homothetic distribution along the most probable stress direction is typically done in studies to reduce the computational burden. However, in reality the individual loads may vary along multiple stress directions each having substantial likelihood, and therefore confining to a single stress direction may result in incomplete characterization of the entire load state space. So it is important to consider the multivariate distribution of loads to capture the boundary region effectively. Otherwise, single stress direction assumption will identify only some portion of boundary, and consequently the rules derived from such a database may face challenges when applied to realistic operating conditions. Through the stratified sampling stage (LHS is one kind of stratified sampling), we would want to obtain the boundary region in the multi-dimensional load sampling state space, and then apply the importance sampling to bias the sampling towards this boundary region, which would capture maximum information content including the relative likelihood of sampled operating conditions.

In order to accomplish this, it is necessary to capture inter-load correlations from historical information while sampling from multivariate load distribution to create the training database, where such finer details will have crucial impact in a decision tree's ability to find rules suitable for realistic scenarios. While we can be assured of more information content from this approach, it is likely to increase computing requirements; especially for boundary region identification using stratified sampling. Singh and Mitra (1997) proposed a state space pruning method to identify the important region in a discrete parameter space composed of generation levels and transmission line capacities under a single load level for system adequacy assessment. Yu and Singh (2004) proposed self-organized mapping together with MCS to characterize the transmission line space. Dobson and Lu (2002) proposed a direct and iterative method to find the closest voltage collapse point with reduced computation in the hyperspace defined by loads. But the method's applicability to a specific distribution of loading conditions in the hyperspace was not shown, and doubts were also cast over its applicability to a large power system with dimension of the hyperspace in 100 s, as will be dealt in the case study of this chapter.

This chapter proposes a sampling space characterization method that uses Latin hypercube sampling (LHS) of homothetic stress directions and linear sensitivities, which promises to reduce the computational requirements. Using this approach the multivariate load state space for a given historical distribution is quickly characterized, under various combinations of Static Var Compensator (SVC) and generator unavailability states. The boundary identification method is described in Sect. 4.1.1, while the stress direction sampling approach (central piece of the proposed state space characterization method) is described in Sect. 4.1.2.

4.1.1 Fast Boundary Region Identification in Multivariate Space

For voltage stability related problems, voltage stability margin (VSM) can be used as the performance measure and hence voltage stability margin sensitivities (Greene et al. 1997; Long and Ajjarapu, 1999; Krishnan et al. 2009) with respect to operational parameters such as individual loads, generator availability, etc. can be used to identify the boundary. VSM is defined as the amount of additional load in a specific pattern of load increase (also termed as stress direction) that would cause voltage instability. It is computed using the continuation power flow (CPF) method. The assumption of a stress direction is important to perform CPF for identifying the voltage collapse point in that direction. Figure 4 depicts existence of several homothetic stress directions for load increase in the two dimensional space defined by loads A and B. The line $Load_A + Load_B = C$ defines various basecases with different inter-node repartitions among loads A and B for the same system load C. These basecases define various homothetic stress directions in the state space, as shown by the various lines from the origin.

The same concept of multiple stress directions is shown in a 3-D load space in the left-hand side of Fig. 5. CPF is performed on these basecases along their

Fig. 4 Multiple homothetic stress directions

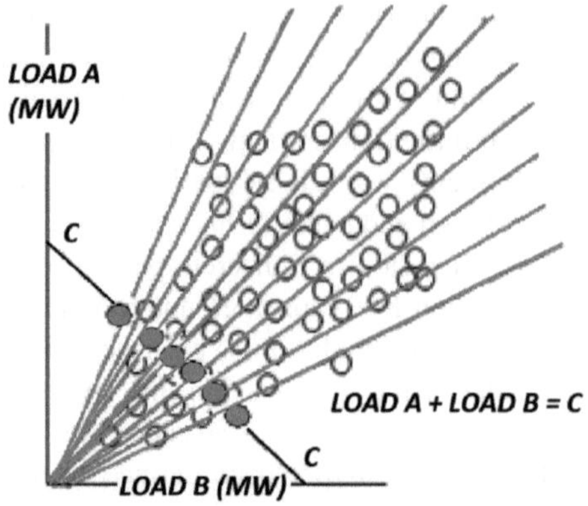

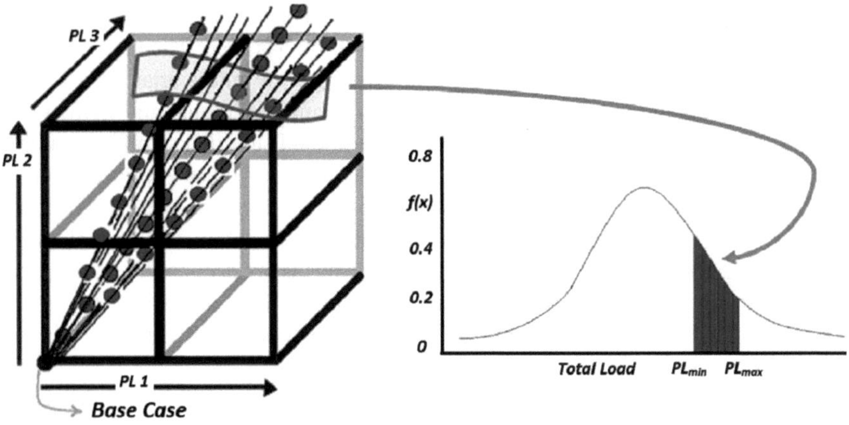

Fig. 5 Multiple homothetic stress directions in 3-D and boundary identification

intrinsic load increase directions, and the maximum loadability along each stress direction is computed. From these the boundary limits, $\{P_{Lmin}, P_{Lmax}\}$, in the total system load space is found, as shown in the right-hand side of Fig. 5. This limit in the hyperspace is subject to variation due to the influence of discrete variables such as SVC and generator unavailability states. The effect of these two variables is estimated using VSM sensitivities with respect to real and reactive power injections along every stress direction, and is given by the Eq. (2). Usage of such linear sensitivities significantly reduces the computational burden in characterizing a multi-dimensional operational parameter state space.

$$\Delta P_{L}^{svc} = Q_{svc}^{*} \cdot dVSMdQ_{svc} \quad (2)$$

where ΔP_{L}^{svc} is the change in boundary limit in a particular stress direction due to the influence of SVC unavailability, Q_{svc}^{*} is the amount of unavailable SVC reactive power at the collapse point, and $dVSMdQ_{svc}$ is the linear sensitivity of voltage stability margin with respect to reactive power injection at the SVC node, which is computed as a by-product of CPF study in that particular stress direction.

Finally, the boundary limits in the total system load space is identified, subject to these discrete variable influences. The key in realizing the computational benefit in boundary region identification lies in the manner in which the multiple homothetic stress directions are sampled from the historical data.

4.1.2 Sampling Homothetic Stress Directions Using Latin Hypercube Method

Latin Hypercube Sampling (LHS) is very prevalently used in Monte Carlo based reliability studies in many fields. LHS of multivariate distribution is performed by dividing every variable forming the multivariate distribution into k equiprobable intervals, and sampling once from each interval of the variable. Then these samples are paired randomly to form k random vectors from the multivariate distribution. Figure 6 depicts the stratified sampling in both forms, traditional and LHS, where the difference is in the pairing process. In the traditional stratified sampling, samples from every interval of variable i is paired with every other samples from all intervals of variable j; whereas in the LHS, one sample from an interval of variable i is paired only once with any one of the sample from an interval of variable j. The pairing in LHS can also be done in such a way as to account for the mutual correlation of the variables by preserving their rank correlation (Wyss and Jorgensen 1998), and hence capturing the inter-dependence structure of the multivariate distribution.

Similarly, LHS of homothetic stress directions is performed from historical data by dividing the load stress factor variables into k equidistant intervals (i.e., equal width; a modification to traditional LHS that partitions into equiprobable intervals), sampling once from each interval of the variable, and pairing them preserving their rank correlation, to form k homothetic stress directions. Figure 7 shows (a) traditional stratified sampling and (b) LHS of homothetic stress directions in 3-dimensional state space. In the case of LHS, for k intervals per dimension, irrespective of state space size the uniform stratification of stress direction is achieved with k samples; compared to the stratified sampling that produces k^{n-1} samples for k intervals per dimension, in a state space of dimension n. The ideal number of k is found in an incremental fashion until there is no further improvement in the boundary limits. Hence computation to find the boundary region can be decreased drastically by using the proposed method based on LHS of stress directions and linear sensitivities.

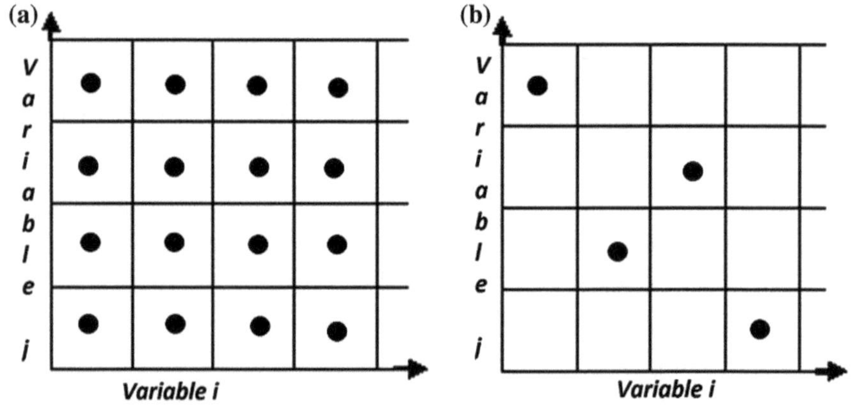

Fig. 6 Stratified sampling—**a** traditional, **b** LHS

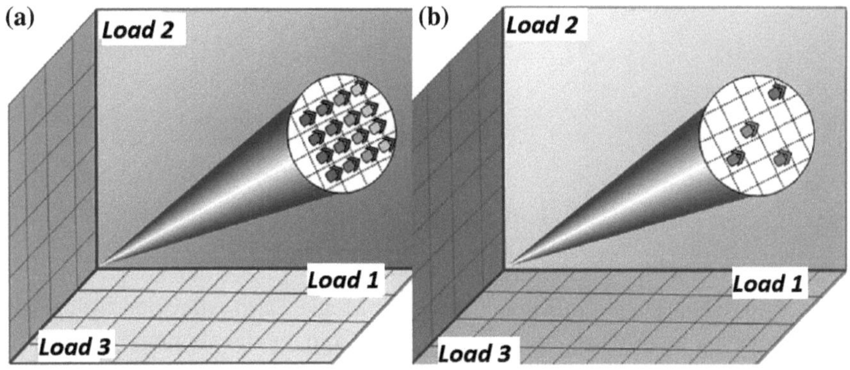

Fig. 7 Sampling homothetic stress directions for boundary identification. **a** Traditional stratified sampling. **b** Latin hypercube sampling

4.2 Stage II—Importance Sampling

Once the boundary region has been identified, the next step is to sample operating conditions from that. This section describes the central concept behind embedding such intelligence in the sampling approach.

The standard Monte Carlo sampling approach draws values for each parameter in proportion to the assigned distribution. Given the previous knowledge of the boundary region from Stage I, biasing the sampling process towards the boundary region can be implemented using the importance sampling method, which helps in maximizing the information content. In this study, the inter-load correlations are captured in the sampling process using copulas (Papaefthymiou and Kurowicka 2009), unlike many studies that approximate the inter-load correlations using multivariate Normal distribution for computational purposes. Copulas are generated based on non-parametric historical load distribution, and it enables sampling realistic scenarios.

4.2.1 Importance Sampling Variance Reduction

In risk-based security planning studies, the quantity of interest is probability of unacceptable performance, i.e., $P(Y \sim \textit{unacceptable events})$ (Billinton and Li 1994).

$$P(Y < t) = \int_{-\infty}^{t} f(y)dy \tag{3}$$

where, $y = t$ denotes the threshold performance such that $y < t$ is unacceptable performance. The indicator function $I(y)$ denoting region of interest $h(y)$ is defined as,

$$h(y) = I(Y < t) = \begin{cases} 1 \text{ if } Y < t \\ 0 \text{ if } Y \geq t \end{cases} \tag{4}$$

and hence,

$$P(Y < t) = \int_{-\infty}^{\infty} h(y)f(y)dy = E(h(Y)) = \sum_{i=1}^{n} h(y_i) \tag{5}$$

The above expectation function gives crude Monte Carlo estimation (Rubinstein 1981), where *yi* are Monte Carlo samples taken from the distribution *f*(*y*), the post-contingency performance index probability distribution. This estimation has a variance associated with it, as the quantity $h(y_i)$ varies with y_i. Importance sampling attempts to reduce the variance of the crude Monte Carlo estimator by changing the distribution from which the actual sampling is carried out. Suppose it is possible to find a distribution *g*(*y*) such that it is proportional to *h*(*y*) *f*(*y*), then the variance of estimation can be reduced by reformulating the expectation function as,

$$P(Y < t) = \int_{-\infty}^{\infty} h(y)f(y)\frac{g(y)}{g(y)}dy = E\left(\frac{h(Y)f(Y)}{g(Y)}\right) = \sum_{i=1}^{n} \frac{h(y_i)f(y_i)}{g(y_i)} \tag{6}$$

where y_i are Monte Carlo samples drawn from the distribution *g*(*y*). This ensures the quantity $\{h(y_i)f(y_i)/g(y_i)\}$ is almost equal for all y_i. In effect, by choosing the sampling distribution *g*(*y*) this way, the probability mass is redistributed according to the relative importance of *y*, measured by the function |*h*(y)| *f*(y) (Ripley 1987).

4.2.2 Proposed Efficient Sample Generation

The property of importance sampling to bias the sampling using an importance function *g*(*y*) towards an area of interest, as discussed above is used to generate influential operating conditions from operational state space, *X*. The joint probability distribution of the operational parameter space *f*(*x*) can be obtained from historical data (Rencher 1995). Once we have *a priori* information about *f*(*x*), stage-I operation provides the region in *X* through which the boundary most likely occurs and therefore identifies approximately the *x*-space in which we want to bias the sample generation. The region of interest for sampling is defined using the indicator function *h*(*X*), where S is the boundary region.

$$h(X) = I(X \in S) = \begin{cases} 1 \text{ if } Y(X) \in S \\ 0 \text{ if } Y(X) \notin S \end{cases} \tag{7}$$

In a univariate case, we can define it as $S = \{x : x_1 \leq x \leq x_2\}$, as shown in Fig. 8. The importance function or the sampling distribution *g*(*x*) can be constructed

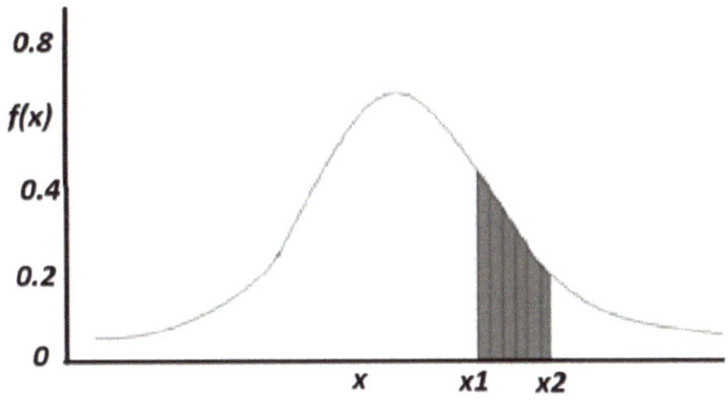

Fig. 8 Boundary region in the univariate operating parameter distribution $f(x)$

proportional to $|h(x)|\ f(x)$, i.e., focusing on the region S of $f(x)$. In general, the importance sampling density can be expressed as,

$$g(x) = p * f_1(x) \ * \ \mathrm{I}(x \in S) + (1 - p) * f_2(x) \ * \ \mathrm{I}(x \notin S) \tag{8}$$

where p controls the biasing satisfying the probability condition $p \le 1$, $f_1(x)$ is the probability density function of the boundary region, and $f_2(x)$ is the probability distribution function of the region outside boundary. We can adopt a composition algorithm to generate samples from the distribution $g(x)$ (Devroye 1986; Gentle 1998). Setting $p = 0.75$, 75 % of the points can be expected from region S, thereby performing an upward scaling of the distribution $f(x)$ towards the boundary region.

In the multivariate case, sampling techniques such as copulas or LHS or sequential conditional marginal sampling (SCMS) (Papaefthymiou and Kurowicka 2009; Hormann et al. 2004) is used to generate correlated multivariate random vectors from non-parametric distributions $f_1(x)$ and $f_2(x)$. The SCMS method is time consuming and requires a lot of memory usage for storing the entire historical data, while LHS and copulas are relatively faster and consume less memory since they work only with non-parametric marginal distributions and correlation data. We use copulas for their simpler and elegant approach in handling any non-parametric marginal distributions

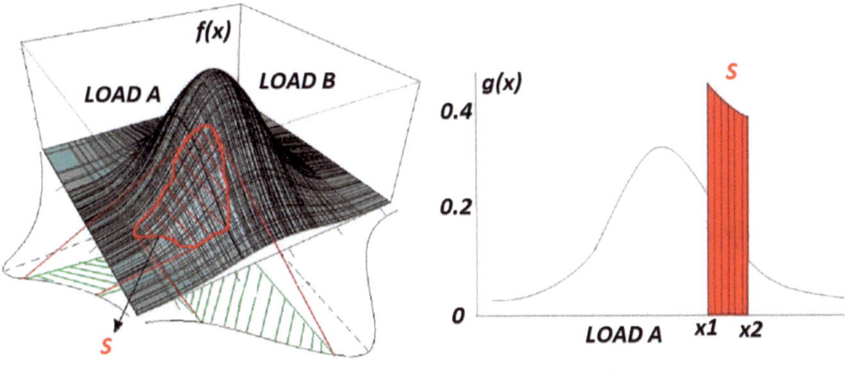

Fig. 9 Importance sampling: upward scaling of boundary region probability

and inter-dependencies. Again setting $p = 0.75$, 75 % of the points is expected from N-dimensional boundary region S, as the probability distribution is altered to produce more samples from S. Figure 9 depicts the probability reorientation by importance sampling process towards the boundary region in a 2-dimensional state space. The parameter p serves as a sliding parameter that controls the extent of biasing between a completely operational study with $p = 1$ to investment planning study with $p = 0$.

5 Case Study

The proposed sampling approach is applied for a decision tree based security assessment study for deriving operating rules against voltage stability issues on SEO region (*Système Eléctrique Ouest*, West France, Brittany), a voltage security-limited region of the French EHV system containing 5,331 buses with 432 generators supplying 83,782 MW. Figure 10 shows 400 kV network of the French system, where it can be seen that the Brittany region (highlighted in pink) is pretty weakly interconnected. During winter periods, when demand peaks, the system gets close to voltage collapse limits. Moreover the local production capabilities being far lower than the local consumption, it puts the EHV grid under pressure as the needed power comes from remote location, eventually leading to cascading phenomenon at the sub voltage levels. The busbar fault at 225 kV Cordemais bus is the most credible contingency in the Brittany region during winter period.

So in order to avoid the risk of collapse situations under such contingency events, the operator may have to resort to expensive preventive measures such as starting up close yet expensive production units. It is therefore very important to assess the risks of a network situation correctly considering uncertainties in operating conditions and obtain operating rules built with decision trees, that aid to take right decision at right time.

Section 5.1 describes the study specifications in terms of historical data used in this study, the sampling parameters and assumptions, and tools and methods used to perform power system assessments. Section 5.2 provides the numerical illustration, presenting the systematic application of stages 1 and 2 of the efficient sampling approach in Sects. 5.2.1 and 5.2.2 respectively, and finally discussing the results from the proposed method and their significance in Sect. 5.2.3 in terms of operating rule's classification accuracy and economic benefits.

5.1 Study Specifications

Data preparation: The historical database of French EHV power grid system for the study is extracted from records made every 15 s on the network by SCADA. The load in the SEO region starts to increase at the end of October, as the winter comes closer, and decreases in February. The heavily loaded period is the winter, during

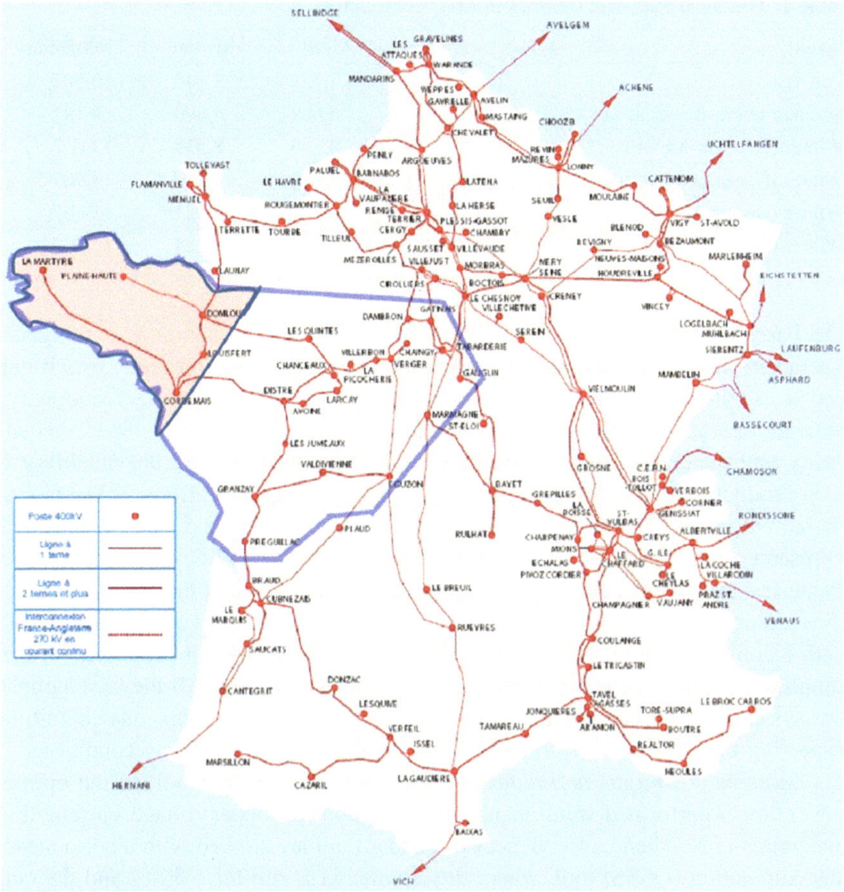

Fig. 10 French 400 kV network with SEO and Brittany highlighted

December, January, and February months. A lot of loads were shed in the month of January under stressful conditions motivated by economic and reliability considerations for system operation. The loading pattern over the year changes depending upon various factors such as, if it is winter or summer, week or week-end, day or night, peak-hours or off peak hours etc. Typically, the load is heavier during the daytime of weekdays in winter, as shown by the statistics in Table 1. Therefore, these heavily loaded periods are the most constraining in terms of voltage, and the study focuses on them for generating samples of operating conditions. Therefore, MCS is not performed on the entire year distribution, but only on those relevant periods that impact the considered stability problem.

Sampling: The pre-contingency operating conditions are generated from a base case, by considering random changes of key parameters. The basecase of SEO network considered corresponds to 2006/2007 winter, with the variable part of the

Table 1 Historical load data statistics in MW—year 2007

Period	Mean	Median	Maximum
Full year	7,729	7,640	13,607
Summer (June–September)	6,609	6,600	9,182
Winter (October–March)	8,585	8,539	13,607
Winter (December–February)	9,290	9,307	13,607
Winter (December–February)—weekdays	9,758	9,823	13,607
Winter (December–February)—week 8 h to 22 h	10,350	10,284	13,607

total baseload amounting to about 13,500 MW. The most constraining contingency is a busbar fault in the Brittany area that trips nearby generation units, which may lead to a voltage collapse under extreme conditions. The parameters sampled to generate operating conditions are variable part of total SEO load, SVC unavailability and generator group unavailability in Brittany area. The unavailability of main production units, consisting of nuclear groups at Civaux, Blayais, St-Laurent, Flamanville, and Chinon are sampled such that each of these 5 unavailabilities is represented in 1/6th of the total basecases. The unavailabilities of two SVCs at Plaine-Haute and Poteau-Rouge are sampled such that 25 % of the cases have both, 25 % do not have both and 50 % have only one of them. The variable part of total load, a continuous multivariate parameter, is sampled using our proposed efficient sampling method. The power factor of loads is kept constant. All the load samples are systematically combined with SVC and generator group unavailabilities respecting their respective sampling laws to form various operating conditions.

Contingency analysis and database generation: For each condition, an optimal power flow is performed, minimizing the production cost under voltage, current, flow constraints in N. Then consequences of busbar fault are studied with a quasi steady state simulation (QSSS) tool, where the simulation is run for 1,500 s and the contingency is applied at 900 s. Scenarios are characterized as unacceptable if any of SEO EHV bus voltage falls below 0.8 p.u or the simulation does not converge. Then a learning dataset is formed using pre-contingency attributes of every scenario (sampled at 890 s of QSSS) that drives voltage stability phenomenon, such as voltages, active/reactive power flows, productions etc., and their respective classifications. Then security rules are produced using decision tree to detect a probable voltage collapse situation contingent upon the severe event. An independent test set is used to validate the tree.

The software tools used in the study are:

1. ASSESS—Special platform for statistical and probabilistic analyses of power networks (Available at: http://www.rte-france.com/htm/an/activites/assess.jsp)
2. TROPIC—Optimal Power Flow tool, embedded with ASSESS, to create initial base cases
3. ASTRE—Simulating slow dynamic phenomena (QSSS), embedded with ASSESS
4. SAS—Statistical analysis and database processing
5. ORANGE, WEKA—Decision tree tools

5.2 Numerical Illustration

One of the major significances of this case study, apart from the demonstration of efficient training database generation for decision trees, is the consideration of system load with non-parametric multivariate distribution including the mutual correlation or inter-load dependencies. The multivariate load distribution is comprised of 640 load buses, on which the two-stage efficient sampling process is performed to generate influential operating conditions for preparing training database.

5.2.1 Stage-I: Fast Boundary Region Identification

There are 24 combinations of discrete parameters as shown in Table 2.

For the first combination, with no component unavailability, initial basecases are formed based on the sampled *k* homothetic stress directions using LHS. Then CPF

Table 2 Boundary identification under component combinations

S. No	SVC cases	Generator cases	P_L^{SEO}min (MW)	P_L^{SEO}max (MW)
1	None	None	11,627	12,700
2	None	Blayais	11,507	12,580
3	None	Chinon	11,474	12,547
4	None	Civaux	11,515	12,529
5	None	Flamanville	11,476	12,506
6	None	St-Laurent	11,490	12,562
7	Plaine-Haute	None	11,618	12,691
8	Plaine-Haute	Blayais	11,498	12,571
9	Plaine-Haute	Chinon	11,465	12,538
10	Plaine-Haute	Civaux	11,506	12,520
11	Plaine-Haute	Flamanville	11,467	12,497
12	Plaine-Haute	St-Laurent	11,481	12,553
13	Plaine-Rouge	None	11,608	12,681
14	Plaine-Rouge	Blayais	11,488	12,561
15	Plaine-Rouge	Chinon	11,455	12,528
16	Plaine-Rouge	Civaux	11,496	12,510
17	Plaine-Rouge	Flamanville	11,457	12,487
18	Plaine-Rouge	St-Laurent	11,471	12,543
19	Both	None	11,599	12,672
20	Both	Blayais	11,479	12,552
21	Both	Chinon	11,446	12,519
22	Both	Civaux	11,487	12,501
23	Both	Flamanville	11,448	12,478
24	Both	St-Laurent	11,462	12,534
	Boundary		**11,446**	**12,700**

Table 3 Incremental estimation of *k*

k	P_L^{SEO}min (MW)	P_L^{SEO}max (MW)	Gap (MW)
5	12,500	12,700	200
8	11,627	12,500	873
12	12,000	12,700	700
15	11,627	12,700	1,073
20	11,627	12,650	1,023
25	11,627	12,700	1,073

is performed to characterize the load state space and find the boundary limits of total SEO load $\{P_L^{SEO}\text{min}, P_L^{SEO}\text{max}\}$, which is found to be {11,627, 12,700} MW as shown in Table 2. The margin sensitivities are also computed along every *k* stress directions, which are used to estimate the change in boundary limits due to the influence of component combination change. Table 2 shows the estimated boundary limits for all the remaining combinations. The final boundary limits are estimated as 11,446 and 12,700 MW.

Table 3 shows the process of estimating *k* for LHS in an incremental fashion. Beyond *k* = *15*, the boundary region is identified fairly consistently. The Expectation-Maximization algorithm based clustering method, when applied to historical record of stress directions, optimally grouped the stress directions into 21 clusters. This information is useful to quickly zero in on the ideal value for *k*.

Figure 11 shows the boundary characterization from a simulation performed for 24,000 operating conditions with randomly selected combinations of discrete

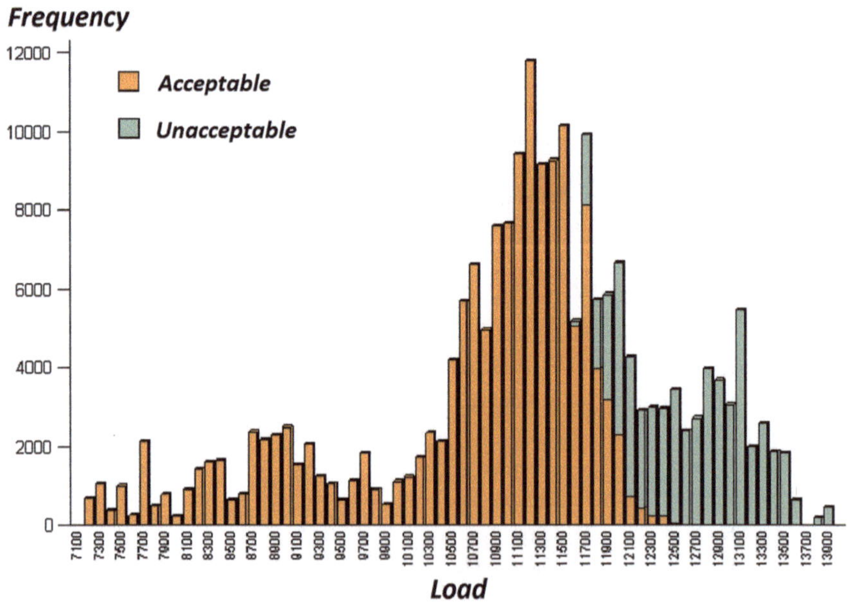

Fig. 11 Boundary characterization in total SEO load state space

parameters and loads. The boundary region (where both acceptable and unacceptable performances occur) begins approximately at around 11,500–11,700 MW and ends at around 12,500–12,700 MW. Therefore, these simulation results verify the ability of the proposed stage-I method to estimate the boundary region in the multidimensional operating parameter state space at a highly reduced computing requirements (i.e., about 20 CPF computations, compared to 24,000 simulations for Fig. 11).

5.2.2 Stage-II: Importance Sampling

Many MCS studies in the past have assumed a multivariate normal distribution of load data (Wan et al. 2000). But in this study, importance sampling is performed on actual empirical non-parametric distribution obtained from the projected historical data of loads. Figure 12 shows three marginal load distributions among the 640 load vectors that make up the multivariate historical data. It is seen that the multivariate distribution is made up of marginal distributions that are not exactly normal, but by visual inspection some looks close to normal, some uniform, some discrete and so on. So a multivariate Normality assumption will give misleading results.

Furthermore, these marginal distributions are not independent to model them separately as a group of normal, uniform and discrete distributions respectively; but they are mutually correlated, and the sampling method must preserve their inter-dependencies or correlations while sampling. So considering both the non-parametric nature of the marginal distributions and their mutual correlations, the whole sampling task becomes very challenging. Therefore, as mentioned in the Sect. 4.2.2, copulas are used that could efficiently work with multiple non-parametric marginal distributions and their mutual correlation (rank correlation) to produce correlated multivariate random vectors from original multivariate distribution defined by empirical historical data.

After identifying the boundary region limits, the empirical multivariate distribution of boundary region $f_I(x)$ is begotten from historical data by filtering the records within the identified boundary limits. When $p = 1$ in Eq. (8), we have complete sampling bias towards the boundary region $f_I(x)$. The inter-dependencies between various individual loads are captured in the sampling process that use

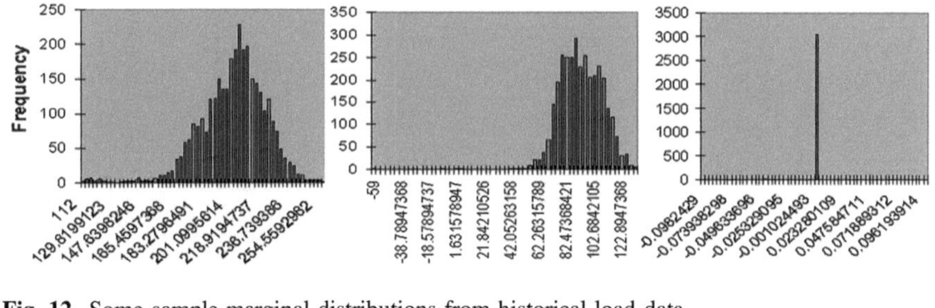

Fig. 12 Some sample marginal distributions from historical load data

copulas, and correlated multivariate random vectors from $f_I(x)$ are generated. The generated samples are for real power values only, and the reactive power at the corresponding individual load buses are obtained by maintaining the power factor constant.

5.2.3 Results and Discussions

The training database generated within the boundary region contains 2,852 operating conditions. The test database includes 1,976 independent instances unseen by training set, covering a wide range of operating conditions. The candidate attributes available for rule formation consists of 46 and 102 node voltages at 400 and 225 kV voltage levels respectively, reactive power flows (*Q flow*) in 16 tie lines, real power reserve (*P res*) in SEO from 10 generator group's, and reactive power reserve (*Q res*) in SEO from 10 generator group's and 2 SVCs.

Table 4 shows the effectiveness of various combinations of attribute sets in terms of classification accuracy and error rates. Accuracy is defined as the percentage of points correctly classified, false alarm rate is defined as the ratio of total misclassified unacceptable instances among all unacceptable classifications, and risk rate is defined as the ratio of total misclassified acceptable instances among all acceptable classifications. The attribute set "400 kV + Q res" proves to be a good set with lowest risk and high classification accuracy. It has to be noted that the accuracy listed in the Table 4 are for trees that are pruned by restricting the minimum number of instances per leaf node.

Effect of Bias Factor "P"

This section sheds light on the quantitative impact of biasing the sampling process towards the boundary region by presenting results for various values of bias factor

Table 4 Attribute set selection

Attribute Set	Accuracy	False alarm	Risk	Tree size
400 kV + Q res	87.9079	0.193	0.073	15
Q res	87.7159	0.183	0.083	15
225 kV	82.8215	0.243	0.124	15
400 kV + 225 kV	82.7255	0.253	0.12	15
400 kV + 225 kV + Q res	82.6296	0.236	0.132	13
All	82.6296	0.236	0.132	13
225 kV + Q res	82.4376	0.231	0.139	13
400 kV	80.8061	0.231	0.166	17
Q flow	75.5278	0.325	0.191	23
P res	73.8004	0.402	0.169	13

"p". Specifically, two aspects are discussed, namely (a) computational requirements and accuracy, and (b) economic benefits.

(a) *Computation, Accuracy and Tree Size*: Fig. 13a–d show the total SEO load probability distribution from sampled operating conditions as the sliding factor p increases from the base value in $f(x)$ to 1 (bias towards *boundary*).

Table 5 shows the results when validated using the test database, which confirms that as the sampling of operating conditions is biased towards the boundary region, the entropy of the database increases (a quantitative indicator of information content) and even with lesser database size higher accuracy for decision tree is obtained. The error rates, namely *false alarms* and *risks* are both simultaneously reduced to a great degree. It was also found that as the sampling is biased more towards the boundary region, the size of the decision tree required for good classification also decreased. This is due to the ability of database to capture high information content (i.e., the variability of performance measure across the security boundary) even with smaller number of instances.

(b) *Economically beneficial rules*: Table 6 presents the influence of efficient sampling in producing economical rules. The table shows that for the various possibilities of the decision tree top node attribute among the most influential attributes, the database generated from within boundary region with p = 1 finds rules with attribute thresholds that are always less conservative than from the database that was generated with $p = 0$, i.e., from entire operational state space.

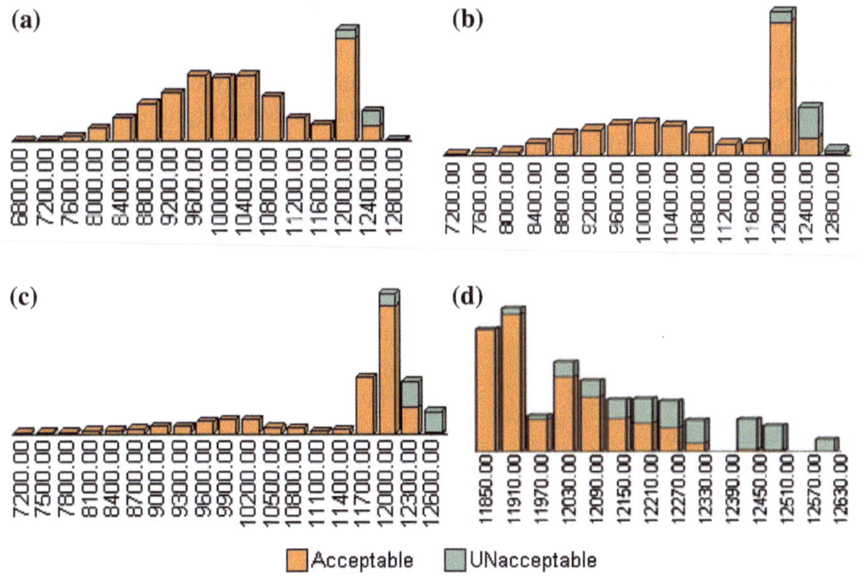

Fig. 13 Effect of p on sampled total SEO load probability distribution. **a** p = 0.25. **b** p = 0.50. **c** p = 0.75. **d** p = 1.0

Table 5 Performance based on sampling bias

p	Size	Entropy	Accuracy	False alarm	Risk
Base	17,748	0.7423	92.51	0.063	0.091
0.25	13,840	0.7716	93.4211	0.064	0.068
0.50	9,932	0.8181	94.9899	0.049	0.051
0.75	6,025	0.9038	96.0526	0.038	0.041
1.0	2,852	0.9993	97.5202	0.021	0.03

Figure 14 shows operational rule formed using two attributes, namely reactive reserves at Chevire unit and Chinon unit respectively. The operating conditions shown in the Fig. 14 are from the entire database. It can be noticed that the rules formed using the database exclusively from the boundary region is providing more operating conditions to be exploited in real time situations, than the rule derived using the database from entire region; because of the increased knowledge and clarity of the boundary limits.

Sampling Strategies Comparison

Table 7 shows the comparison results of two different sampling approaches, namely,

1. Importance sampling (IS) of boundary region, with load distribution modeled with multivariate normal (MVN) distribution (pruned tree).
2. Importance sampling of boundary region, with load distribution modeled with correlated non-parametric multivariate distribution (MVD) (pruned tree).
3. Same as case 2, with un-pruned tree.

Table 6 Economic benefit from efficient sampling

Top Node	$p = 0$	$p = 1$
Cordemais voltage	401.64 kV	399.88 kV
Domloup voltage	397.56 kV	394.51 kV
Louisfert voltage	399.1 kV	396.46 kV
Plaine-Haute voltage	392.26 kV	387.21 kV
Chevire unit reactive reserve	131.38 Mvar	90.76 Mvar
Chinon unit reactive reserve	1,127.54 Mvar	694.62 Mvar
Cordemais unit reactive reserve	70.97 Mvar	16.23 Mvar
Total SEO region reactive reserve	7,395.88 Mvar	6,510.36 Mvar
Plaine-Haute SVC output	11.82 Mvar	13.64 Mvar
Poteau-Rouge SVC output	16.3 Mvar	22.03 Mvar

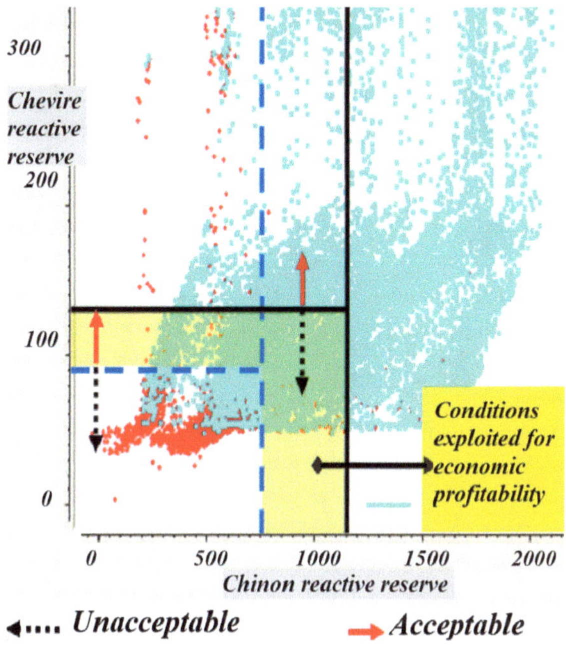

Fig. 14 Economical benefit of operational rules from efficient sampling

Table 7 Comparison between different sampling strategies

Sampling strategy	Size	Accuracy	False alarm	Risk
IS (MVN—pruned)	2,879	80.6142	0.142	0.228
IS (MVD—pruned)	2,852	87.0951	0.094	0.178
IS (MVD)	2,852	97.5202	0.021	0.03

It can be seen from Table 7 that the database produced by importance sampling of correlated-MVD state space definitely shows better performance. When the trees are pruned for operator's convenience of usage the accuracy decreases, which can be improved using the right-hand side loop as shown in the Fig. 3. It also performs better than sampling from MVN load space, which is conventional assumption in many studies due to trivial modeling requirements.

The significance of sampling from correlated-MVD, i.e., capturing the inter-load dependencies, than from MVN is even strongly vindicated by Fig. 15 that shows the top 5 critical attribute locations produced by decision trees from respective databases. The contingency event is shown by a red star. The location of 5 critical monitoring attributes as well as their sequence in the tree matters. Compared to MVN, all the 5 top locations found by correlated-MVD sampling strategy are very interesting ones, with the top node being reactive reserve at a big nuclear plant Chinon, the node in the next level of the tree is closer to the contingency location, the next nodes (3 and 4) in the tree deals with the two SVC locations in Brittany, and finally the attribute of node 5 is right at the contingency location.

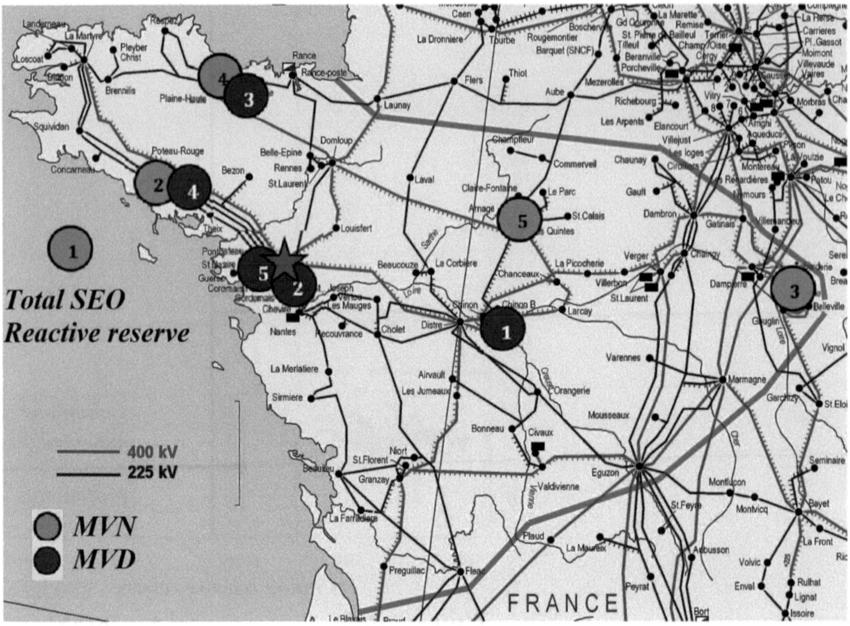

Fig. 15 Critical monitoring locations from decision tree: MVD versus MVN

6 Conclusion

The proposed efficient sampling method based on importance sampling idea is one of the first to be used in power systems for making decision tree based learning methods effective. The thrust of the proposed sampling procedure was to re-orient the sampling process to focus more heavily on points for which post-contingency performance is close to the threshold, i.e., boundary region that contains operating conditions influential for rule formation. The primary goal was to increase the information content in the learning database while reducing the computing requirements, and consequently obtain operational rules that are more accurate for usage in real-time situations.

The developed efficient training database approach was applied for deriving operational rules in a decision tree based voltage stability assessment study on RTE-France's power grid. The results showed that the generated training database enhances rules' accuracy at lesser computation compared to other traditional sampling approaches, when validated on an independent test set. The chapter also emphasized the significance of sampling from non-parametric correlated-multivariate load distribution obtained from historical data, as it is more realistic. Doing so also ensures generating operating rules that provide higher classification accuracy and economics, and selecting interesting monitoring locations that are closer to the contingency event, as corroborated by the results. In order to reduce the computational burden in characterizing multivariate load state space, a linear sensitivity based method supported by Latin hypercube sampling of homothetic stress directions was developed for quickly characterizing the multivariate load state space for various combinations of component unavailabilities. This aided in identifying the boundary region with respect to post-contingency performance measure quickly.

The future directions of research include:

- *Application for other stability problems*: The efficient database generation approach can also be applied to other stability problems such as rotor angle stability, out of step etc. In these problems the performance measure's trajectory sensitivities will have to be used to reduce the computational cost in identifying the boundary region.
- *Optimal placement of Phasor Measurement Units (PMUs)*: The high information content in the training database generated from the proposed efficient sampling method enables finding the most important system attributes for power system's security state monitoring. This concept is highly beneficial in finding the optimal placement of PMUs and extracting relevant knowledge from those PMUs for advancing data-driven power system operation and control.
- *Application in the reliability assessment of Special Protection System (SPS)*: The main difference between deriving operating rules and SPS logic are:
  - The SPS logic is automated.
  - The SPS logic is not only limited to critical operating condition detection with respect to some stability criteria, but also involves automatic corrective action to safeguard the system against impending instability.

Even though many works exist that correspond to SPS "process level" design procedures and failure assessments, there are important questions to be answered about SPS operations from a 'system view-point', such as:

- Are there system operating conditions (topology, loading, flows, dispatch, and voltage levels) that may generate a failure mode for the SPS?
- Are there two or more SPS that may interact to produce a failure mode?

So the objective of this research will be to develop a decision support tool to perform SPS failure mode identification, risk assessment and logic re-design from a 'systems view'. The efficient scenario processing method presented in this chapter has tremendous scope to be used in biasing the sampling process such that SPS failure modes (including multiple SPS interactions) can be identified, risk levels may be estimated, and accordingly the logic may be re-designed using the efficient decision tree process.

Acknowledgments The author acknowledges Professor James D. McCalley at Iowa State University (Ames, Iowa, USA), Sebastien Henry at RTE-France (Versailles, France), and Samir Issad at RTE-France (Versailles, France) for their valuable support during the course of this research project.

References

Abed, A. M. (1999). WSCC voltage stability criteria, under voltage load shedding strategy, and reactive power reserve monitoring methodology. In IEEE Power Engineering Society Summer Meeting (pp. 191–197), July 18–22, 1999, Edmonton, Alta. doi:10.1109/PESS.1999.784345.

Beshir, M. J. (1999). Probabilistic based transmission planning and operation criteria development for the Western Systems Coordinating Council. In IEEE Power Engineering Society Summer Meeting (pp. 134–139), July 18–22, 1999, Edmonton, Alta. doi:10.1109/PESS.1999.784334.

Billinton, R., & Li, W. (1994). *Reliability assessment of electric power systems using Monte Carlo methods*. New York: Plenum Press.

Billinton, R., Salvaderi, L., McCalley, J. D., Chao, H., Seitz, Th, Allan, R. N., et al. (1997). Reliability issues in today's electric power utility environment. *IEEE Transactions Power Systems, 12*(4), 1708–1714.

Cholley, P., Lebrevelec, C., Vitet, S., & De Pasquale, M. (1998). Constructing operating rules to avoid voltage collapse: A statistical approach. In *Proceedings of International Conference on Power System Technology, POWERCON '98* (pp. 1468–1472), August 18–21, 1998, Beijing. doi:10.1109/ICPST.1998.729331.

Chowdhury, A. A., & Koval, D. O. (2006). Probabilistic assessment of transmission system reliability performance. In IEEE Power Engineering Society General Meeting, Montreal, Que. doi:10.1109/PES.2006.1709096.

Cutsem, T. V., Wehenkel, L., Pavella, M., Heilbronn, B., & Goubin, M. (1993). Decision tree approaches to voltage security assessment. In *IEE Proceedings on Generation, Transmission and Distribution* (Vol. 140, No. 3, pp. 189–198).

Devroye, L. (1986). *Non-uniform random variate generation*. New York: Springer.

Dobson, I., & Lu, L. (2002). New methods for computing a closest saddle node bifurcation and worst case load power margin for voltage collapse. *IEEE Transactions Power Systems, 8*(3), 905–913.

Dy-Liacco, T. E. (1997). Enhancing power system security control. *IEEE Computer Applications in Power, 10*(3), 38–41.

Foody, G. M. (1999). The significance of border training patterns in classification by a feed forward neural network using back propagation learning. *International Journal of Remote Sensing, 20*(18), 3549–3562.

Genc, I., Diao, R., Vittal, V., Kolluri, S., & Mandal, S. (2010). Decision tree-based preventive and corrective control applications for dynamic security enhancement in power systems. *IEEE Transactions Power Systems, 25*(3), 1611–1619.

Gentle, J. E. (1998). *Random number generation and Monte Carlo methods*. Newyork: Springer.

Greene, S., Dobson, I., & Alvarado, F. L. (1997). Sensitivity of the loading margin to voltage collapse with respect to arbitrary parameters. *IEEE Transactions on Power Systems, 12*(1), 262–272.

Henry, S., Lebrevelec, C., and Schlumberger, Y. (1999). Defining operating rules against voltage collapse using a statistical approach: The EDF experience. In *International Conference on Electric Power Engineering, PowerTech Budapest*, August 29–September 2, 1999, Budapest, Hungary. doi:10.1109/PTC.1999.826461.

Henry, S., Bréda-Séyès, E., Lefebvre, H., Sermanson, V., & Béna, M. (2006). Probabilistic study of the collapse modes of an area of the French network. In *Proceedings of the 9th International Conference on Probabilistic Methods Applied to Power Systems* (pp. 1–6), June 11–15, 2006, Stockholm. doi:10.1109/PMAPS.2006.360261.

Henry, S., Pompee, J., Bulot, M., and Bell, K. (2004a). Applications of statistical assessment of power system security under uncertainty. In *International Conference on Probabilistic Methods Applied to Power Systems* (pp. 914–919). September 16–16, 2004, Ames, IA.

Henry, S., Pompee, J., Devatine, L., Bulot, M., and Bell, K. (2004b). New trends for the assessment of power system security under uncertainty. In *IEEE PES Power Systems Conference and Exposition* (pp. 1380–1385), Oct 10–13, 2004. doi:10.1109/PSCE.2004.1397731.

Hormann, W., Leydold, J., & Derflinger, G. (2004). *Automatic non-uniform random variate generation*. Newyork: Springer.

Jacquemart, Y., Wehenkel, L., & Pruvot, P. (1996). Practical contribution of a statistical methodology to voltage security criteria determination. In *Proceedings of the 12th Power Systems Computation Conference* (pp. 903–910).

Jiantao, X., Mingyi, H., Yuying, W., & Yan, F. (2003). A fast training algorithm for support vector machine via boundary sample selection. In *Proceedings of the International Conference on Neural Networks and Signal Processing* (pp. 20–22), December 14–17, 2003, Nanjing. doi:10.1109/ICNNSP.2003.1279203.

Krishnan, V., Liu, H., and McCalley, J. D. (2009). Coordinated reactive power planning against power system voltage instability. In *IEEE/PES Power Systems Conference and Exposition* (pp. 1–8), March 15–18, 2009, Seattle, WA. doi:10.1109/PSCE.2009.4839926.

Lebrevelec, C., Schlumberger, Y., & De Pasquale, M. (1999). An application of a risk based methodology for defining security rules against voltage collapse. In IEEE Power Engineering Society Summer Meeting (pp. 185–190), Jul 18–22, 1999, Edmonton, Alta. doi:10.1109/PESS.1999.784344.

Lebrevelec, C., Cholley, P., Quenet, J.F., & Wehenkel, L. (1998). A statistical analysis of the impact on security of a protection scheme on the French power system. In *Proceedings of International Conference on Power System Technology, POWERCON* (pp. 1102–1106), Aug 18–21, 1998, Beijing. doi:10.1109/ICPST.1998.729256.

Li, W., & Choudhury, P. (2007). Probabilistic transmission planning. *IEEE Power and Energy Magazine, 5*(5), 46–53.

Long, B., & Ajjarpau, V. (1999). The sparse formulation of ISPS and its application to voltage stability margin sensitivity and estimation. *IEEE Transactions on Power Systems, 14*(3), 944–951.

Martigne, H., Cholley, P., King, D., & Christon, J. (2001). Statistical method to determine operating rules in the event of generator dropout on EDF French Guyana Grid. In *Proceedings of IEEE Power Tech* (pp. 1–5), Sep 10–13, 2001, Porto. doi:10.1109/PTC.2001.964599.

Niimura, T., Ko, H. S., Xu, H., Moshref, A., and Morison, K. (2004). Machine learning approach to power system dynamic security analysis. *IEEE PES Power Systems Conference and Exposition* (pp. 1084–1088), October 10–13, 2004. doi:10.1109/PSCE.2004.1397549.

Papaefthymiou, G., & Kurowicka, D. (2009). Using copulas for modeling stochastic dependence in power system uncertainty analysis. *IEEE Transactions Power Systems, 24*(1), 40–49.

Paul J., Bell K. (2004). A flexible and comprehensive approach to the assessment of large-scale power system security under uncertainty. *International Journal of Electrical Power & Energy Systems, 26*(4), 265–272.

Pierre, J., Lebrevelec, C., & Wehenkel, L. (1999). Automatic learning methods applied to dynamic security assessment of power systems. In *International Conference on Electric Power Engineering. PowerTech Budapest*, Aug 29–Sept 2, 1999, Budapest, Hungary. doi:10.1109/PTC.1999.826612.

Rencher, A. (1995). *Methods of multivariate analysis*. New York: Wiley.

Ripley, B. D. (1987). *Stochastic Simulation*. New York: Wiley.

Rubinstein, R. Y. (1981). *Simulation and the Monte Carlo method*. New York: Wiley.

Schlumberger, Y., Lebrevelec, C., & De Pasquale, M. (1999). Power systems security analysis-new approaches used at EDF. IEEE Power Engineering Society Summer Meeting (pp. 147–151), Jul 18–22, 1999, Edmonton, Alta. doi:10.1109/PESS.1999.784337.

Schlumberger, Y., Pompee, J., & De Pasquale, M. (2002). Updating operating rules against voltage collapse using new probabilistic techniques. In *IEEE/PES Transmission and Distribution Conference and Exhibition: Asia Pacific* (pp. 1139-1144), October 6–10, 2002. doi:10.1109/TDC.2002.1177638.

Senroy, N., Heydt, G. T., & Vittal, V. (2006). Decision tree assisted controlled islanding. *IEEE Transactions Power Systems, 21*(4), 1790–1797.

Singh, C., & Mitra, J. (1997). Composite system reliability evaluation using state space pruning. *IEEE Transactions Power Systems, 12*(1), 471–479.

Thisted, R. A. (1988). *Elements of statistical computing*. New York: Chapman and Hal Ltd.

Unger, E. A., Harn, L., and Kumar, V. (1990). Entropy as a measure of database information. In *Proceedings of the Sixth Annual in Computer Security Applications Conference* (pp. 80–87), December 3–7, 1990, Tucson, AZ. doi:10.1109/CSAC.1990.143755.

Wan, H., McCalley, J. D., & Vittal, V. (2000). Risk based voltage security assessment. *IEEE Transactions Power Systems, 15*(4), 1247–1254.

WECC. (2003, April). NERC/WECC planning strandards. Available at, http://www.wecc.biz/documents/library/procedures/planning/WECC-NERC_Planning%20Standards_4-10-03.pdf.

Wehenkel, L. (1997). Machine learning approaches to power-system security assessment. *IEEE Expert, IEEE Intelligent Systems and Their Applications, 12*(5), 60–72.

Wehenkel, L. (1998). *Automatic learning techniques in power systems*. Berlin: Kluwer Academic Publishers.

Wehenkel, L., Glavic, M., Geurts, P., & Ernst, D. (2006). Automatic learning of sequential decision strategies for dynamic security assessment and control. In *IEEE Power Engineering Society General Meeting, Montreal, Que*. doi: 10.1109/PES.2006.1708874.

Witten, I. H., & Frank, E. (2000). *Data mining: Practical machine learning tools and techniques with Java implementations*. San Fransisco, CA: Morgan Kaufmann Publishers.

Wyss, W. G., & Jorgensen, K. H. (1998). A user's guide to LHS: Sandia's latin hypercube sampling software. *Sandia National Laboratories Report SAND98-0210*, Albuquerque, NM.

Xiao, F., & McCalley, J. D. (2007). Risk-based security and economy tradeoff analysis for real-time operation. *IEEE Transactions Power Systems, 22*(4), 2287–2288.

Yu, X., & Singh, C. (2004). Expected power loss calculation including protection failures using importance sampling and SOM. In *IEEE Power Engineering Society General Meeting* (pp 206–211), June 6–10, 2004. doi:10.1109/PES.2004.1372787.

Zhou, G., & McCalley, J. D. (1999). Composite security boundary visualization. *IEEE Transactions Power Systems, 14*(2), 725–731.

Zhou, Q., Davidson, J., & Fouad, A. A. (1994). Application of artificial neural networks in power system security and vulnerability assessment. *IEEE Transactions Power Systems, 9*(1), 525–532.

Classification of Normal and Epileptic Seizure EEG Signals Based on Empirical Mode Decomposition

Ram Bilas Pachori, Rajeev Sharma and Shivnarayan Patidar

Abstract Epileptic seizure occurs as a result of abnormal transient disturbance in the electrical activities of the brain. The electrical activities of brain fluctuate frequently and can be analyzed using electroencephalogram (EEG) signals. Therefore, the EEG signals are commonly used signals for obtaining the information related to the pathological states of brain. The EEG recordings of an epileptic patient contain a large amount of EEG data which may require time-consuming manual interpretations. Thus, automatic EEG signal analysis using advanced signal processing techniques plays a significant role to recognize epilepsy in EEG recordings. In this work, the empirical mode decomposition (EMD) has been applied for analysis of normal and epileptic seizure EEG signals. The EMD generates the set of amplitude and frequency modulated components known as intrinsic mode functions (IMFs). Two area measures have been computed, one for the graph obtained as the analytic signal representation of IMFs in complex plane and another for second-order difference plot (SODP) of IMFs of EEG signals. Both of these area measures have been computed for first four IMFs of the normal and epileptic seizure EEG signals. These eight features obtained from both area measures of first four IMFs have been used as input feature set for classification of normal and epileptic seizure EEG signals using least square support vector machine (LS-SVM) classifier. Among all three kernel functions namely, linear, polynomial, and radial basis function (RBF) used for classification, the RBF kernel has provided best classification accuracy in the classification of normal and epileptic seizure EEG signals. The proposed method based on the two area measures of IMFs obtained using EMD process, together with

R.B. Pachori (✉) · R. Sharma · S. Patidar
Discipline of Electrical Engineering, Indian Institute of Technology Indore, Indore 452017, India
e-mail: pachori@iiti.ac.in

R. Sharma
e-mail: phd1301102007@iiti.ac.in

S. Patidar
e-mail: shivnarayan.patidar@iiti.ac.in

Q. Zhu and A.T. Azar (eds.), *Complex System Modelling and Control Through Intelligent Soft Computations*, Studies in Fuzziness and Soft Computing 319,
DOI 10.1007/978-3-319-12883-2_13

LS-SVM classifier has been studied on EEG dataset publicly available by the University of Bonn, Germany. Experimental results have been included to show the effectiveness of the proposed method in comparison to other existing methods.

1 Introduction

Human brain is a highly complex system. The epilepsy is a common neurological disorder of human brain. It affects at least 50 million people of the world (Ngugi et al. 2010). The annual occurrence of epilepsy, 48 per 100,000 populations in developed counteries was reported in Hirtz et al. (2007). The prevalence of epilepsy is higher in low and middle income countries than developed countries (Thurman et al. 2011). At least 50 % of the epileptic cases start developing at childhood or adolescence (World Health Organization 2014). Occurrence of epilepsy can also be noticed in elderly people, which may require special considerations in treatment (Ramsay et al. 2004). If the patient with epilepsy are treated properly, then 70–80 % of them can lead to normal lives (World Health Organization 2014). Therefore, study of epilepsy is an important research area in the field of the biomedical engineering.

The electroencephalogram (EEG) signals are very useful to measure the electrical activity of the human brain. The EEG signals are commonly analyzed by experts in order to assess the states of the brain. The EEG based measures are very helpful for diagnosis of neurological disorders specially epilepsy. Presence of spikes in EEG signals is main indication of epileptic seizure activity in the brain (Ray 1994; Mukhopadhyay and Ray 1998). Automatic detection of epileptic seizure by analyzing EEG signals using advanced signal processing techniques is very useful for diagnosis of epilepsy (Iasemidis et al. 2003).

The EEG recording using electrodes from the scalp is the result of firing of neurons within the brain (Schomer and da Silva 2005). The EEG signals recorded from scalp exhibit oscillations of different range of frequencies, corresponding to the electrical activities of the neurons in the brain. These oscillations of different frequency ranges present in the EEG signals are related with the different states of the functioning of human brain. A typical voltage range of the amplitude of EEG signal for a adult human is about 10–100 μV when measured from the scalp (Aurlien et al. 2004).

Conventional scalp EEG signal recording is performed by placing electrodes over the scalp. The conductive gel or paste is utilized in order to make proper electrical connection between scalp and electrode. The electrode cap is generally used for recording of EEG signals. There are some standards that have been defined for electrode placement over the scalp, out of which international 10–20 system (Andrzejak et al. 2001) is commonly used for EEG signal recording of healthy persons. The intracranial EEG signals obtained using intracranial electrodes which also requires depth electrodes for recording of ictal and seizure-free EEG signals for epilepsy patients can be useful for diagnosis of epilepsy (Andrzejak et al. 2001). The presence of interference or artifacts due to external sources, in the EEG signal recording may create problem in the diagnosis based on these recorded EEG signals. Therefore, filtering is required to remove these artifacts (Senthil et al. 2008).

1.1 Epileptic Seizure EEG Signals

Epilepsy is one of the most common neurological disorders of human brain. As epileptic activity manifests the clear and abnormal transient patterns in a normal EEG signal, therefore EEG signals are widely used in diagnostic application for detection of epilepsy. In epileptic patients, brain exhibits the process known as 'epileptogenesis' (Cross and Cavazos 2007) in which normal neural network abruptly converts into a hyper-excitable network, causing evocation of strange sensations and emotions or sometimes muscle spasms and consciousness loss. In such subjects, the nerve cells in the brain transmit excessive electrical impulses that cause epileptic seizures. Epilepsy is recognized by occurrence of such unprovoked seizures. Evaluation of the epilepsy can be performed by recording and analyzing the epileptic seizure EEG signals from the electrodes which are placed on the affected area on the brain scalp region (Coyle et al. 2010; Ince et al. 2009). The recorded EEG signals are complex, non-linear, and non-stationary in nature (Acharya et al. 2013; Boashash et al. 2003; Pachori and Sircar 2008; Pachori 2008). The epileptic seizures can have severe harmful effect on the brain. Manual process to identify the seizure events, consists of visual inspection and review of the entire recorded EEG signals by trained expert, which is time consuming process and demands considerable skills. Moreover, subjective nature of expert can also affect the judgement of seizure events in EEG records. Therefore, it is appealing to develop computer-aided automatic analysis method that consists of advanced signal processing techniques, for classification between normal and epileptic seizures EEG signals in recorded EEG signals.

1.2 Classification of Epileptic Seizure EEG Signals

Various methods have been developed for automatic classification of the epileptic seizures by extracting parameters from EEG signals. These parameters can be extracted using time-domain, frequency-domain, time-frequency domain and non-linear methods of analysis and serve as the features for classification of EEG signals based on signal processing methods (Acharya et al. 2013).

Many time-domain based techniques have been presented in literature with an objective to detect epileptic seizures from EEG signals. The value of linear prediction error energy has been found to be much higher in seizure EEG signals than that of seizure-free EEG signals and has been used to detect epileptic seizures in EEG signals (Altunay et al. 2010). The fractional linear prediction (FLP) method has been employed to model seizure and seizure-free EEG signals and prediction error energy along with signal energy with support vector machine (SVM) classifier has been used to classify the epileptic seizure EEG signals from the seizure-free EEG signals (Joshi et al. 2014). Epileptic seizures have been detected in EEG signals using principal component analysis in combination with enhanced cosine radial basis function neural network (Ghosh-Dastidar et al. 2008). Artificial neural network (ANN) based methodology has been developed to detect the epileptic seizure using time-domain as well as the frequency-domain features in Srinivasan et al. (2005). Spectral parameters based on the Fourier transformation of EEG signals have been utilized to detect epileptic seizures in EEG signals (Polat and Güneş 2007).

The EEG signals exhibit non-stationary nature (Boashash et al. 2003). In literature, many time-frequency domain based methods have been proposed to detect epileptic seizure EEG signals, these methods include time-frequency distribution (Tzallas et al. 2007, 2009), wavelet transform (Ghosh-Dastidar et al. 2007; Uthayakumar and Easwaramoorthy 2013; Ocak 2009; Subasi 2007; Subasi and Gursoy 2010; Adeli et al. 2007; Acharya et al. 2012), multi-wavelet transform (Guo et al. 2010), and empirical mode decomposition (EMD) (Pachori 2008; Oweis and Abdulhay 2011; Pachori and Patidar 2014; Li et al. 2013; Bajaj and Pachori 2012). The Fourier-Bessel (FB) series expansion of intrinsic mode functions (IMFs) extracted from EMD, has been used to compute mean frequency of IMFs and these mean frequencies have been used as features to discriminate ictal and seizure-free EEG signals (Pachori 2008). In Oweis and Abdulhay (2011), weighted mean frequency of IMFs has been proposed to detect epileptic seizures from EEG signals. Ellipse area of second-order difference plot (SODP) of different IMFs with 95 % confidence limit has been proposed as a feature to classify epileptic seizure and seizure-free EEG signals (Pachori and Patidar 2014). The coefficient of variation and fluctuation index computed from IMFs of EEG signals have been proposed to recognize patterns of ictal EEG signals (Li et al. 2013). The amplitude modulation (AM) and frequency modulation (FM) bandwidths computed from the IMFs together with least square support vector machine (LS-SVM) classifier have been used for classification of seizure and nonseizure EEG signals (Bajaj and Pachori 2012).

Various non-linear parameters have been proposed as features for classification of epileptic seizure EEG signals. The Lyapunov exponent parameter with probabilistic neural network (PNN) in Übeyli (2010) and Güler et al. (2005), correlation integral in Casdagli et al. (1997), fractal dimension parameters in Easwaramoorthy and Uthayakumar (2011) and Accardo et al. (1997), multistage nonlinear pre-processing filter combined with a diagnostic neural network in Nigam and Graupe (2004), entropy based measures with adaptive neuro-fuzzy inference system in Kannathal et al. (2005), and approximate entropy (ApEn) with ANN in Srinivasan et al. (2007) have been proposed for discrimination of epileptic seizure EEG signals.

In this work, we propose a method based on the EMD process for classification of normal and epileptic seizure EEG signals. The area measures namely area of analytic signal representation of IMFs and area of ellipse from SODP of IMFs have been used as an input feature set for LS-SVM classifier.

The rest of the chapter has been organized as follows. In Sect. 2, proposed methodology has been described which includes dataset, EMD method, feature extraction and LS-SVM classifier. Feature extraction section consists of two further subsections: one of which is analytic signal representation and area computation of circular region and other is second-order difference plot and area computation of elliptical region. Results of experimental analysis and comparison with other methods have been discussed in Sect. 3. Finally, conclusion has been provided in Sect. 4.

2 Methodology

2.1 Dataset

In this work, the online publicly available EEG dataset as described in Andrzejak et al. (2001) has been used. Recordings in this dataset include EEG signals which have been acquired for both healthy and epileptic subjects. This dataset contains five subsets denoted as Z, O, N, F, and S, each of which having 100 single-channel EEG signals of duration 23.6 s. The first two subsets Z and O are surface EEG recordings of five healthy volunteers. These subsets contain EEG recordings with eyes open and closed, respectively. The subset F have been recorded in seizure-free intervals from five patients in the epileptogenic zone and the subset N has been acquired from the hippocampal formation of the opposite hemisphere of the brain. The subset S contains seizure activity selected from all recording sites exhibiting ictal activity. The subsets Z and O have been recorded extracranially using standard electrode placement scheme (according to the international 10–20 system (Andrzejak et al. 2001), whereas the subsets N, F, and S have been recorded intracranially using depth electrodes implanted symmetrically into the hippocampal formations. Subsets N and F have EEG signals which were taken from all contacts of the relevant depth electrode (Andrzejak et al. 2001). The strip electrodes were implanted onto the lateral and basal regions (middle and bottom) of the neocortex. The EEG signals of the subset S contains segments taken from contacts of all electrodes (depth and strip). Set N and F contain only activity measured during seizure free intervals, while set S only contains seizure activity. The data were digitized at a sampling rate of 173.61 Hz using 12-bit analog-to-digital (A/D) converter. Bandwidth range of bandpass filter were 0.53–40 Hz. More detail about this dataset can be found in Andrzejak et al. (2001). In this study, we have used subsets Z and S of the dataset to evaluate performance of, proposed method which consists of EMD, feature extraction and classification using LS-SVM classifier.

2.2 Empirical Mode Decomposition

The main idea of empirical mode decomposition (EMD) is based on the assumption that any signal comprises of different simple mode of oscillations (Huang et al. 1998). It is a data dependant signal processing technique that represents any temporal signal into a finite set of amplitude and frequency modulated (AM-FM) oscillating components termed as intrinsic mode functions (IMFs). It is noteworthy that this method of decomposition does not require any prior assumption about the stationarity and linearity of signal. The EMD method decomposes a complicated signal $x(t)$ iteratively into a set of the band-limited IMFs, $I_m(t)$, where $m = 1, 2, \ldots, M$ (Huang et al. 1998). Each of these IMFs satisfies the following two basic conditions:

1. The number of extrema and the number of zero crossings must be either equal or differ at most by one,
2. The mean value of the envelopes defined by the local maxima and that of defined by the local minima must be zero.

The EMD algorithm to extract IMFs from a signal $x(t)$ can be explained in following steps (Huang et al. 1998):

1. Find all the local maxima and local minima in the signal $x(t)$.
2. Connect all the maxima and all the minima separately in order to get the envelopes $E_{\max}(t)$ and $E_{\min}(t)$ respectively.
3. Compute the mean value of the envelopes by using the following formula:

$$m(t) = \frac{E_{\max}(t) + E_{\min}(t)}{2} \tag{1}$$

4. Subtract $m(t)$ from signal $x(t)$ as:

$$g_1(t) = x(t) - m(t) \tag{2}$$

5. Check if the $g_1(t)$ satisfies the conditions for IMF as mentioned above or not.
6. Repeat the steps 2–5 until IMF is obtained.

After obtaining first IMF define $I_1(t) = g_1(t)$ which is smallest temporal scale in $x(t)$. Next IMF can be derived by generating a residue $r_1(t) = x(t) - I_1(t)$ which can be used as the new signal for above algorithm. The process is repeated until the residue obtained becomes a constant or monotonic function from which no more IMF can be generated. The obtained IMFs are a set of narrow-band symmetric waveforms. After the decomposition, the signal $x(t)$ can be represented as follows (Huang et al. 1998):

$$x(t) = \sum_{m=1}^{M} I_m(t) + r_M(t) \tag{3}$$

where, M is the number of IMFs, $I_m(t)$ is the mth IMF and $r_M(t)$ is the final residue.

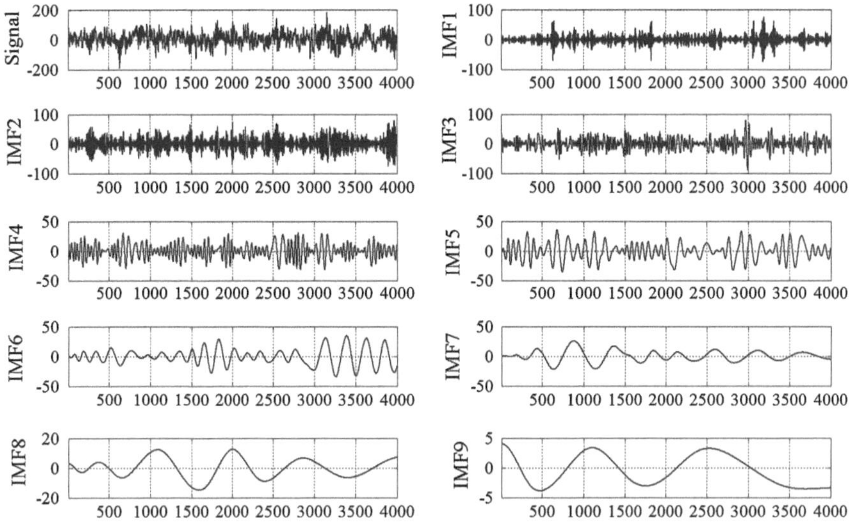

Fig. 1 Normal EEG signal and its first nine IMFs

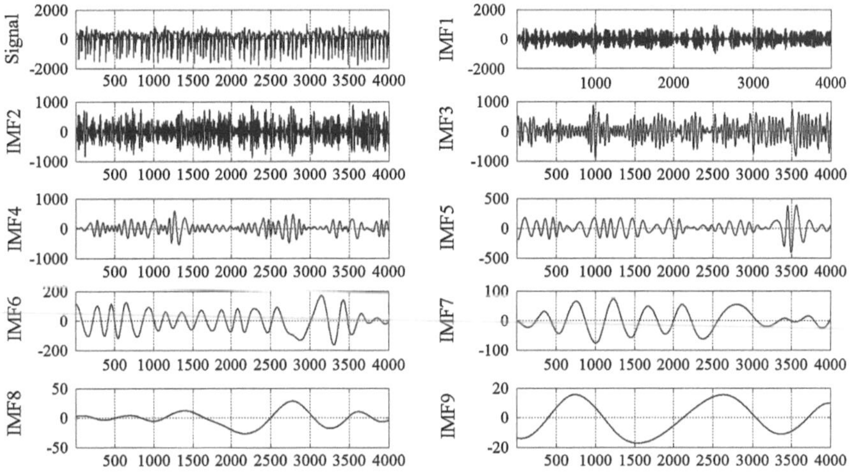

Fig. 2 Seizure EEG signal and its first nine IMFs

The empirical mode decomposition of the normal and epileptic seizure EEG signals are shown in Figs. 1 and 2 respectively. It should be noted that only first nine IMFs of the signals are shown in each figure for EEG signal.

2.3 Feature Extraction

Feature extraction is an important step in pattern recognition and plays a vital role in detection and classification of EEG signals by extracting relevant information. Feature extraction can be understood as finding a set of parameters which effectively represent the information content of an observation while reducing the dimensionality. These parameters explore the property of two classes which has separate range of values for different classes. Two different area measures which are related with the variability of the signal are used here as a feature set. These area measures are computed for first four IMFs to create feature vector space. Final feature set consists of eight features for classification of normal and epileptic seizure EEG signals. The computation of these area measures have been described in detail as follows:

2.3.1 Analytic Signal Representation and Area Computation of Circular Region

The IMFs that have been obtained using EMD method on EEG signals are real signals. These IMFs can be converted to analytic signals by applying the Hilbert transform.

Analytic signal of $x(t)$ can be defined as (Huang et al. 1998; Lai and Ye 2003):

$$z(t) = x(t) + jy(t) \tag{4}$$

where, $y(t)$ represents the Hilbert transform of the real signal $x(t)$, defined as follows:

$$\begin{aligned} y(t) &= x(t) * \frac{1}{\pi t} \\ &= \frac{1}{\pi} \text{p.v.} \int_{-\infty}^{\infty} \frac{x(\tau)}{t - \tau} d\tau \end{aligned} \tag{5}$$

with Fourier transform

$$Y(\omega) = -j \, \text{sgn}\,(\omega) X(\omega) \tag{6}$$

where p.v. indicates the Cauchy principle value, and $X(\omega)$ is Fourier transform of signal $x(t)$.

The signal $z(t)$ can also be expressed as:

$$z(t) = A(t) e^{j\phi(t)} \tag{7}$$

where, $A(t)$ is the amplitude envelope of $z(t)$, defined as:

$$A(t) = \sqrt{x^2(t) + y^2(t)} \tag{8}$$

and $\phi(t)$ is the instantaneous phase of $z(t)$, defined as:

$$\phi(t) = \tan^{-1}\left(\frac{y(t)}{x(t)}\right) \tag{9}$$

The instantaneous frequency of the analytic signal can be obtained by differentiating (9) as:

$$\begin{aligned} \omega(t) &= \frac{d\phi(t)}{dt} \\ &= \frac{x(t)\frac{dy(t)}{dt} - y(t)\frac{dx(t)}{dt}}{A^2(t)}. \end{aligned} \tag{10}$$

The instantaneous frequency $\omega(t)$ of the analytic signal $z(t)$ is a measurement of the rate of rotation in the complex plane. The Hilbert transform can be applied on all IMFs obtained by EMD method. The IMFs are mono-component signals and exhibit property of locally symmetry. Therefore, the instantaneous frequency is well localized in the time-frequency domain and reveals a meaningful feature of the signal (Huang et al. 1998).

The analytic signal can be obtained for all the IMFs using the Hilbert transform. A complex signal can be represented as a sum of proper rotational components using EMD method which makes it possible to compute the area in a complex plane (Amoud et al. 2007). Since each IMF is a proper rotational component and has its own rotation frequency, the plot of the analytic IMF follows circular geometry in complex plane. The complex plane representation can be obtained by tracing the real part against the imaginary part of the analytic signal. The analytic signal representations in complex plane corresponding to the normal and epileptic seizure EEG signals and their first four intrinsic mode functions are depicted in Figs. 3 and 4, respectively. These figures present the traces of entire signals in the complex plane, as well as those of each IMF for both signals. It can be observed that the shape of this trace is similar to a rotating curve. The analytic signal representation of IMFs in complex plane possess a proper structure of rotation with a unique center (Lai and Ye 2003).

Central tendency measure (CTM) provides a rapid way to summarize the visual information related to a graph or plot (Cohen et al. 1996). The modified CTM can be used to measure the degree of variability from analytic signal representation of the signal. CTM can be used to determine the area of the complex plane representation (Pachori and Bajaj 2011). The radius corresponding to 95 % modified CTM can be used to compute the area of analytic signal representation of the IMF in complex plane. The modified CTM provides the ratio of points falling inside the circular region of specified radius to the total number of points in analytic signal

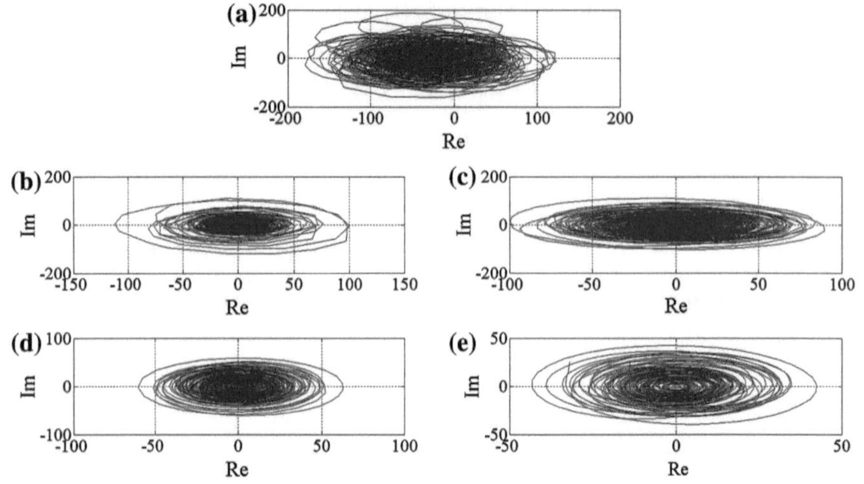

Fig. 3 Analytic signal representation in the complex plane for window size of 4,000 samples: **a** Normal EEG signal, **b** IMF1, **c** IMF2, **d** IMF3 and **e** IMF4

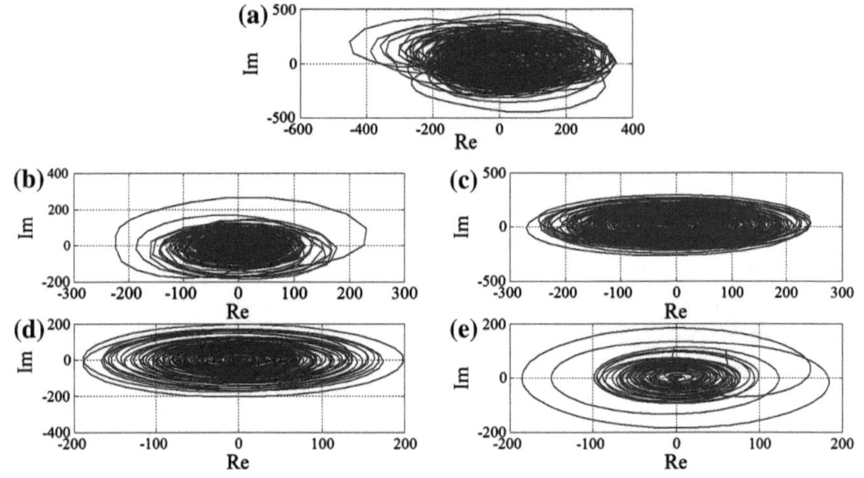

Fig. 4 Analytic signal representation in the complex plane for window size of 4,000 samples: **a** Seizure EEG signal, **b** IMF1, **c** IMF2, **d** IMF3 and **e** IMF4

representation. Let, the total number of points are N and the specified radius of central area is r, then modified CTM can be defined as (Pachori and Bajaj 2011):

$$\mathrm{CTM} = \frac{\sum_{k=1}^{N} \rho(d_k)}{N} \tag{11}$$

$$\rho(d_k) = \begin{cases} 1 & \text{if } ([\Re\{z[n]\}]^2 + [\Im\{z[n]\}]^2)^{0.5} < r \\ 0 & \text{otherwise} \end{cases} \tag{12}$$

where, $1 \leq k \leq N$. If r_{CTM95} is the radius corresponding to 95 % CTM, then area of analytic signal can be defined as:

$$A_{\text{analytic}} = \pi r^2_{\text{CTM95}} \tag{13}$$

2.3.2 Second-Order Difference Plot and Area Computation of Elliptical Region

The second-order difference plot (SODP) provides a graph of successive rates against each other and has been used to measure the variability present in EEG and center of pressure (COP) signals (Thuraisingham et al. 2007; Pachori et al. 2009). Useful diagnostic information can be extracted from SODP of the IMFs of EEG signals. The area of SODP of IMFs of EEG signals can be used as features for classification of normal and epileptic seizure EEG signals. The SODP of signal $x[n]$ can be obtained by plotting $X[n]$ against $Y[n]$ which are defined as (Cohen et al. 1996),

$$X[n] = x[n+1] - x[n] \tag{14}$$

$$Y[n] = x[n+2] - x[n+1] \tag{15}$$

In SODP above mentioned successive rates are plotted against each other, consequently provides rate of variability of data. The 95 % confidence ellipse area can be used to determine the confidence area of SODP of IMFs which covers around 95 % of the points. SODP corresponding to the normal and epileptic seizure EEG signals and their first four intrinsic mode functions are shown in Figs. 5 and 6, respectively. These figures represent trace of two successive rates, $X[n]$ and $Y[n]$ of different IMFs of EEG signals. The SODP of IMFs of EEG signals exhibit elliptical patterns, the area of ellipse in SODP of IMFs has been used as a feature for classification of epileptic seizure and seizure-free EEG signals (Pachori and Patidar 2014). In this work, we have used the area parameter computed from the SODP of IMFs as a feature for classification of normal and epilpetic seizure EEG signals. The procedure to compute the 95 % confidence ellipse area from the SODP can be given as (Prieto et al. 1996; Cavalheiro et al. 2009):

The μ_X and μ_Y are mean values of $X[n]$ and $Y[n]$ as defined in Prieto et al. (1996), Cavalheiro et al. (2009) and μ_{XY} can be defined as,

$$\mu_{XY} = \sqrt{\frac{1}{N}\sum X[n]Y[n]} \tag{16}$$

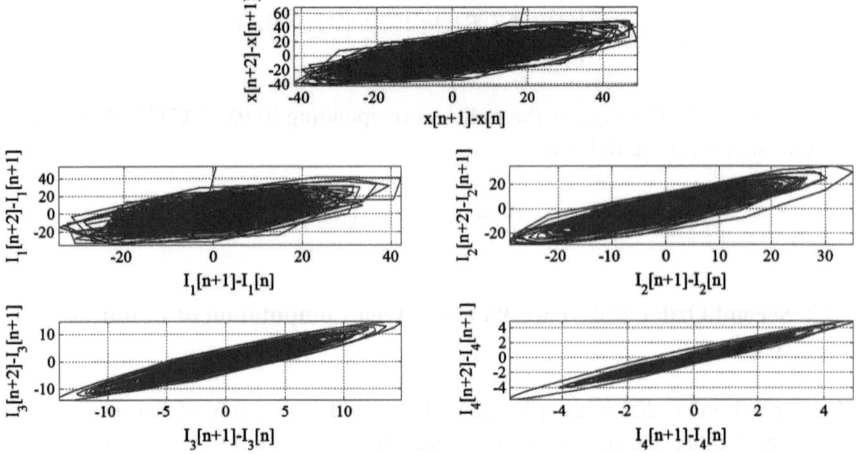

Fig. 5 SODP of the normal EEG signal and its first four IMFs

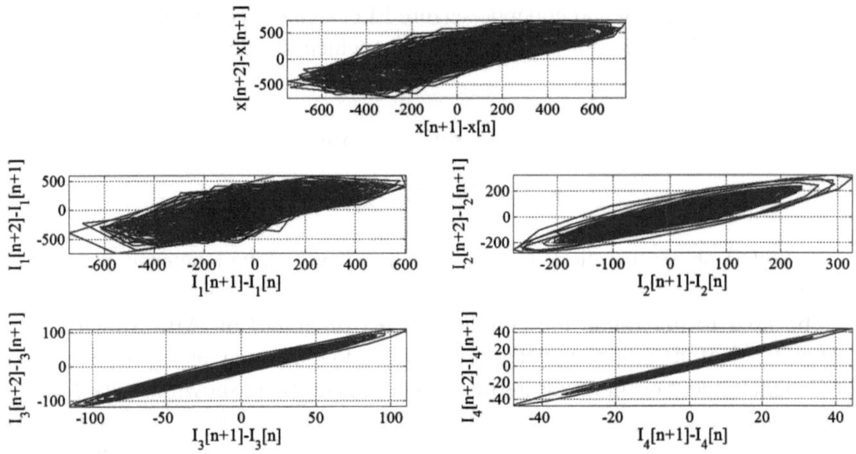

Fig. 6 SODP of the seizure EEG signal and its first four IMFs

The D parameter can be computed as:

$$D = \sqrt{(\mu_X^2 + \mu_Y^2) - 4(\mu_X^2\mu_Y^2 - \mu_{XY}^2)} \tag{17}$$

and,

$$a = 1.7321\sqrt{(\mu_X^2 + \mu_Y^2 + D)} \tag{18}$$

$$b = 1.7321\sqrt{(\mu_X^2 + \mu_Y^2 - D)} \tag{19}$$

The ellipse area can be computed from the parameters a and b as:

$$A_{\text{ellipse}} = \pi ab. \tag{20}$$

2.4 Least Square Support Vector Machine

Classification is a problem of finding out the particular category of data to which the new upcoming observed sample can belong. The decision is made on the basis of the observed samples of data whose category is already known, these sets of observed samples are known as training sets. Support vector machine (SVM) is a machine learning technique used to classify samples belongs to different classes. SVM is a very useful tool for pattern classification problem (Cortes and Vapnik 1995). SVM is trained to search for an optimal separating hyperplane that can provide superior generalization, particularly when dimension of input data is large. Hyper planes are determined to create decision boundaries between two different classes of data in SVM. The effectiveness of the features in classifying normal and epileptic seizure EEG signals has been evaluated using a least square support vector machine (LS-SVM) a least square version of SVM (Suykens and Vandewalle 1999).

Consider a training set of N data points (x_i, y_i), $i = 1, \ldots, N$, where x_i is input data and $y_i = +1$ or -1, class label for two different classes. The SVM approach aims at constructing a discriminant function of the form:

$$f(x) = \text{sign}\left[\omega^T g(x) + b\right] \tag{21}$$

where, ω is the d-dimensional weight vector and b is a bias, and $g(x)$ is a mapping function that maps x into d-dimensional space. The goal of SVM algorithm is to identify optimum separating hyper plane which is able to maximize the distance from either class to the hyperplane. This problem of optimization can be formulated as a quadratic programming problem considering inequality constraints (Suykens and Vandewalle 1999). The LS-SVM is the least square variant of SVM for classification of two class problem. The statement of the problem can be written as in following way:

$$\text{Minimize } J(\omega, b, e) = \frac{1}{2}\omega^T\omega + \frac{\gamma}{2}\sum_{i=1}^{N} e_i^2 \tag{22}$$

subjected to following equality constraints:

$$y_i\left[\omega^T g(x_i) + b\right] = 1 - e_i,\ i = 1, 2, 3, \ldots, N \tag{23}$$

where, $e = (e_1, e_2, \ldots, e_N)^T$. The Lagrangian multiplier α_i can be defined for (22) as:

$$L(\omega, b, e; \alpha) = J(\omega, b, e) - \sum_{i=1}^{N} \alpha_i \{y_i[\omega^T g(x_i) + b] - 1 + e_i\} \tag{24}$$

On solving (24), the LS-SVM classifier can be expressed as:

$$f(x) = \text{sign}\left[\sum_{i=1}^{N} \alpha_i y_i K(x, x_i) + b\right] \tag{25}$$

where, $K(x, x_i)$ is a kernel function. The following kernel functions are used in this work, which have been defined in Khandoker et al. (2007):

1. Linear kernel: The linear kernel can be defined as:

$$K(x, x_i) = x^T x_i \tag{26}$$

2. Polynomial kernel: The polynomial kernel can be defined as:

$$K(x, x_i) = (x^T x_i + 1)^d \tag{27}$$

   where d is the degree of polynomial.
3. Radial basis function (RBF) kernel: The RBF kernel can be defined as:

$$K(x, x_i) = e^{\frac{-\|x - x_i\|^2}{2\sigma^2}} \tag{28}$$

   where, width of RBF kernel can be controlled by varying scaling factor σ. The performance evaluation parameters of the LS-SVM classifier depends on the selection of the kernel parameters. In this work, we have used trial and error method in order to determine the suitable kernel parameters for classification of normal and epilpetic seizure EEG signals.

2.4.1 Performance Evaluation Parameters

The classification performance of the LS-SVM classifier for classification of normal and epileptic seizure EEG signals can be evaluated by computing the sensitivity, specificity, and accuracy. Sensitivity measures the ability of test to identify proportion of actual positives as such. Considering an example where percentage of epileptic seizure signals from test set, correctly falls in the category of epileptic seizure signals after classification. Specificity measures the ability of test to exclude the actual negatives correctly. For example, percentage of normal EEG signals correctly identified as not having seizures. A perfect classification would result in 100 % sensitivity by detecting all epileptic seizure EEG signals correctly. It also exhibits 100 % specificity by not recognizing any normal EEG signal as epileptic seizure signal. Positive predictive value is, the fraction of total positive patterns, which represents the actual positive patterns (Azar and El-Said 2014). Accuracy of classification is proportion of number of patterns which are correctly classified. Similarly, negative predictive value is, the fraction of total identified negative patterns, which represent actual negative patterns. Considering, *TP* and *TN* represent the total number of correctly identified true positive patterns and true negative patterns respectively, along with *FP* and *FN* represents total number of false positive patterns and false negative patterns, respectively. The sensitivity (*SEN*), specificity (*SPF*), accuracy (*ACC*), positive prediction value (*PPV*), negative prediction value (*NPV*) of classifier can be defined as (Azar and El-Said 2014):

$$SEN = \frac{TP}{TP + FN} \times 100\ (\%) \tag{29}$$

$$SPF = \frac{TN}{TN + FP} \times 100\ (\%) \tag{30}$$

$$ACC = \frac{TP + TN}{TP + TN + FP + FN} \times 100\ (\%) \tag{31}$$

$$PPV = \frac{TP}{TP + FP} \times 100\ (\%) \tag{32}$$

$$NPV = \frac{TN}{TN + FN} \times 100\ (\%) \tag{33}$$

Matthews correlation coefficient (*MCC*) is another parameter to measure classification performance, which is the indication of classification accuracy of imbalanced positive and negative patterns in dataset (Azar and El-Said 2014). Higher the value of *MCC* parameter, the better the classifier performance (Yuan et al. 2007). The *MCC* parameter can be defined as follows (Yuan et al. 2007):

$$MCC = \frac{TP \cdot TN - FN \cdot FP}{\sqrt{(TP + FN)(TP + FP)(TN + FN)(TN + FP)}}. \tag{34}$$

3 Experimental Results and Discussion

Main steps of proposed method include applying EMD on EEG signals to obtain IMFs, computation of both area measures for first four IMFs, extraction and formation of feature set, training and testing of LS-SVM classifier. The proposed method has been implemented using Matlab. The Matlab codes for EMD method are available at http://perso.enslyon.fr/patrick.flandrin/emd.html. In this study, the proposed methodology has been validated with one online freely available EEG dataset (Andrzejak et al. 2001). As discussed in Sect. 2.1, this dataset includes EEG signals which have been recorded from both healthy and epileptic subjects. It contains five subsets denoted as Z, O, N, F, and S. Data subsets Z and S have been used to evaluate the performance of the proposed method for classification of normal and epileptic seizure EEG signals. Data subset Z consists of normal EEG recordings taken from 5 healthy volunteers and subset S consists of the EEG recordings of seizure activities. Each of these subsets have 100 single-channel EEG signals of duration 23.6 s.

The decomposition of EEG signals using EMD method results into IMFs that are in decreasing order of frequency, in which first component is associated with highest frequency. As the IMFs can help to compute the area of analytic signal representation of the IMFs in the complex plane and ellipse area parameter obtained from SODP of IMFs, therefore the EMD has been used to decompose the EEG signals into a set of IMFs. These above mentioned two area parameters have been used to create the feature space for classification between normal and epileptic seizure EEG signals.

Recently in Pachori and Bajaj (2011), the ability of the analytic signal representation of IMFs to discriminate EEG signals which contains normal and epileptic seizure EEG signals has been explored. It comes out of this study that the analytic signal representation of IMFs provides a set of proper rotations which facilitates accurate identification of the centers and estimation of surface areas in the complex plane. It has been shown that the area parameter of the analytic IMFs has significant potential to differentiate between epileptic seizure and normal EEG signals. The experimental results of the above mentioned method reveals that the epileptic seizure EEG signals had evidently greater surface area in comparison to that of the normal EEG signals. The increased surface area in the complex plane for IMFs of the epileptic seizure EEG signals could be attributed to large amplitude of EEG signals for seizure subjects. It should be noted that the use of EMD enabled the extraction of individual centers of rotation for each IMF. Furthermore, as discussed in this study, it is evident from experimental analysis, that window size of 2,000 samples has provided better results, therefore the same window size has been used to compute the area parameters in this work. As the analytic signal representation has circular geometry, therefore modified CTM has been measured to compute the area of the analytic signal representation of the IMFs of EEG signals in the complex plane. The radius of the circular region which covers the 95 % of the CTM has been used to determine the area parameter for first four IMFs of EEG signals. In Pachori and Patidar (2014), the efficacy of the ellipse area parameters of SODP of IMFs for classification of seizure-free and ictal EEG signals has been examined. This study has employed the 95 % confidence ellipse area as a feature for discrimination of ictal EEG signals from the seizure-free EEG signals, and the classification performance of the ellipse area parameter have been evaluated for various window sizes (500, 1,000, 2,000, 4,000 samples) of seizure-free and ictal EEG signals.

Table 1 Classification performance of the proposed method for different kernel functions

Performance parameters	Linear	Polynomial ($d = 2$)	RBF ($\sigma = 1$)
SEN (%)	100	100	100
SPF (%)	97.00	90.00	100
ACC (%)	98.50	99.00	100
PPV (%)	97.69	90.91	100
NPV (%)	100	100	100
MCC	0.97	0.98	1

Along with area parameter of analytic IMFs, we have also computed 95 % confidence ellipse area of SODP for first four IMFs of EEG signals which covers around 95 % points in SODP. By considering both area measures for first four IMFs, lead to eight features that forms the final input feature set for LS-SVM based classification of normal and epileptic seizure EEG signals.

SVM is a supervise machine learning approach, suitable for small-sample dataset (Azar and El-Said 2014). LS-SVM is the least square reformulation of the SVM problem (Suykens and Vandewalle 1999) which uses equality constraints, instead of inequality constraint used in standard SVM. Consequently, solution follows from set of linear equations instead of quadratic programming problem. Hence, LS-SVM offers less computational complexity with excellent generalised performance (Suykens and Vandewalle 1999). In this work, the area parameters computed from the IMFs has been used as input feature set for LS-SVM classifier for classification of normal and epileptic seizure EEG signals. In order to evaluate the classification performance, different kernel functions have been utilized and their performance parameter values have been shown in Table 1. Various performance parameters discussed in previous section have been computed for three kernel functions which are linear kernel, polynomial kernel, and radial basis function (RBF) kernel. It can be observed that performance parameter values for RBF kernel are best among three kernel functions. The value of scaling factor associated with RBF kernel has been set empirically as 1. The ten-fold cross validation procedure is suitable for evaluating classification accuracy of a classifier for classification of biomedical signals (Sharma et al. 2014; Pachori and Patidar 2014). In this study, ten-fold cross validation procedure has been employed to evaluate the classification performance of LS-SVM classifier.

The classification accuracy achieved using proposed method with RBF kernel is 100 % which suggests successful identification of all, normal and epileptic seizure EEG signals. The resulting 100 % sensitivity shows the correct identification of all epileptic seizure EEG signals and 100 % specificity shows adequate classification by not recognizing any normal EEG signal as epileptic seizure EEG signal. Moreover, Table 2 shows the results obtained with proposed method and some other existing methods using the same dataset. Different parameters analysed for classification in other compared methods have also been mentioned in Table 2. It should be noted that the performance of the proposed method in terms of classification accuracy is same as that of discussed in Tzallas et al. (2007), in which time-frequency analysis based parameters have been used for classification. The area

Table 2 Comparison of the proposed method for classification of normal and epileptic seizure EEG signals with the existing methods studied on same dataset

Authors	Method	Accuracy (%)
Nigam and Graupe (2004)	Nonlinear pre-processing filter and diagnostic neural network	97.2
Srinivasan et al. (2005)	Time and frequency domain based features and recurrent neural network	99.6
Kannathal et al. (2005)	Entropy based measures and adaptive neuro-fuzzy inference system	about 90
Polat and Güneş (2007)	Fast Fourier transform based features and decision tree	98.72
Subasi (2007)	Discrete wavelet transform based features and mixture of expert model	94.5
Tzallas et al. (2007)	Time-frequency analysis based features and artificial neural network	100
This work	Proposed method	100

measures used in this work are the simple and can be used as indicators for diagnosis of epilepsy. Moreover, these parameters are defined in time domain which can help us to implement the proposed methodology for epileptic seizure detection with low computational complexity. It can be observed that performance of the proposed method in terms of accuracy is better than that of the other compared methods. The experimental analysis of the proposed method shows that features based on area measures are very effective to represent the behavior of epileptic seizure EEG signals giving excellent classification performance.

4 Conclusion

This book chapter has developed a novel approach for classification of the normal and epileptic seizure EEG signals using empirical mode decomposition and computing two area parameters for IMFs. Since the EEG signal is non-linear and non-stationary in nature, the EMD which is data dependent approach and suitable for analysis of nonlinear and non-stationary signals, efficaciously decompose the EEG signals into IMFs which are oscillatory components. In this work, we have explored the capability of two area parameters as the features for classification of normal and epileptic seizure EEG signals. It is noteworthy that the symmetric nature of IMFs, makes it possible to compute these two area measures and justifies the application of EMD before feature extraction from EEG signals. Computation of area measures uses the analytic signal representation of IMFs and SODP of IMFs. The IMFs have single center of rotation with circular geometry in analytic signal representation. Similarly, IMFs also exhibit elliptical patterns in SODP. Consequently, these obtained geometrical patterns help to compute the area of analytic signal representation in complex plane with 95 % CTM and 95 % confidence area of ellipse in SODP. It has been found that these two area parameters have significantly higher values for seizure EEG signals as compared to normal EEG signals. The performance of LS-SVM classifier is best when RBF kernel has been employed to create decision boundary between two classes (normal and seizure) and consequently have provided 100 % classification accuracy. The features of the proposed method are suitable for real time implementation of an expert system for detection of the epileptic seizure in EEG signals. This system can act as an important diagnostic tool for clinician to detect the epilpesy automatically by analysing EEG signals.

In future, performance of the proposed methodology can be evaluated for classification between different classes of EEG signals like normal, inter-ictal and ictal EEG signals. The future direction of research may also include the application of the proposed methodology for identification of different psychological states of brain from EEG signals. Moreover, it would be of interest to study the expert system based on the proposed methodology for classification of other signals like electromyogram (EMG) signals, center of pressure (COP) signals, electrocardiogram (ECG), and speech signals corresponding to normal and abnormal conditions.

References

Accardo, A., Affinito, M., Carrozzi, M., & Bouquet, F. (1997). Use of the fractal dimension for the analysis of electroencephalographic time series. *Biological Cybernetics, 77*, 339–350.

Acharya, U. R., Sree, S. V., Alvin, A. P. C., & Suri, J. S. (2012). Use of principal component analysis for automatic classification of epileptic EEG activities in wavelet framework. *Expert Systems with Applications, 39*(10), 9072–9078.

Acharya, U. R., Sree, S. V., Swapna, G., Martis, R. J., & Suri, J. S. (2013). Automated EEG analysis of epilepsy: A review. *Knowledge-Based Systems, 45*, 147–165.

Adeli, H., Ghosh-Dastidar, S., & Dadmehr, N. (2007). A wavelet-chaos methodology for analysis of EEGs and EEG sub-bands to detect seizure and epilepsy. *IEEE Transactions on Biomedical Engineering, 54*(2), 205–211.

Altunay, S., Telatar, Z., & Erogul, O. (2010). Epileptic EEG detection using the linear prediction error energy. *Expert Systems with Applications, 37*(8), 5661–5665.

Amoud, H., Snoussi, H., Hewson, D. J., and Duchêne, J. (2007). Hilbert-Huang transformation: Application to postural stability analysis. In: *29th Annual International Conference of the IEEE Engineering in Medicine and Biology Society* (pp. 1562–1565), Lyon, France , 29–23 Aug 2007.

Andrzejak, R. G., et al. (2001). Indications of nonlinear deterministics and finite-dimensional structures in time series of brain electrical activity: Dependence on recording region and brain state. *Physical Review E, 64*(6), 061907.

Aurlien, H., et al. (2004). EEG background activity described by a large computerized database. *Clinical Neurophysiology, 115*(3), 665–673.

Azar, A. T., & El-Said, S. A. (2014). Performance analysis of support vector machines classifier in breast cancer mammography recognition. *Neural Computings and Applications. 24*(5), 1163–1177. doi:10.1007/S00521-012-1324-4.

Bajaj, V., & Pachori, R. B. (2012). Classification of seizure and nonseizure EEG signals using empirical mode decomposition. *IEEE Transactions on Information Technology in Biomedicine, 16*(6), 1135–1142.

Boashash, B., Mesbah, M., & Colditz, P. (2003). Time–frequency detection of EEG abnormalities. In B. Boashash (Ed.), *Time-frequency signal analysis and processing: A comprehensive reference* (pp. 663–670). Oxford: Elsevier.

Casdagli, M. C., et al. (1997). Non-linearity in invasive EEG recordings from patients with temporal lobe epilepsy. *Electroencephalography and Clinical Neurophysiology, 102*(2), 98–105.

Cavalheiro, G. L., Almeida, M. F. S., Pereira, A., & Andrade, A. O. (2009). Study of age-related changes in postural control during quiet standing through linear discriminant analysis. *BioMedical Engineering Online, 8*(35), 10–1186.

Cohen, M. E., Hudson, D. L., & Deedwania, P. C. (1996). Applying continuous chaotic modeling to cardic signal analysis. *IEEE Engineering in Medicine and Biology Magazine, 15*(5), 97–102.

Cortes, C., & Vapnik, V. (1995). Support-vector networks. *Machine Learning, 20*(3), 273–297.

Coyle, D., McGinnity, T. M., & Prasad, G. (2010). Improving the separability of multiple EEG features for a BCI by neural-time-series-prediction-preprocessing. *Biomedical Signal Processing and Control, 5*(3), 196–204.

Cross, D. J., & Cavazos, J. E. (2007). The role of sprouting and plasticity in epileptogenesis and behavior. In S. Schachter, G. L. Holmes, & D. G. Trenite (Eds.), *Behavioural Aspects of Epilepsy* (pp. 51–57). New York: Demos Medical Publishing.

Easwaramoorthy, D., & Uthayakumar, R. (2011). Improved generalized fractal dimensions in the discrimination between healthy and epileptic EEG signals. *Journal of Computational Science, 2*(1), 31–38.

Ghosh-Dastidar, S., Adeli, H., & Dadmehr, N. (2007). Mixed-band wavelet-chaos neural network methodology for epilepsy and epileptic seizure detection. *IEEE Transactions on Biomedical Engineering, 54*(9), 1545–1551.

Ghosh-Dastidar, S., Adeli, H., & Dadmehr, N. (2008). Principal component analysis enhanced cosine radial basis function neural network for robust epilepsy and seizure detection. *IEEE Transactions on Biomedical Engineering, 55*(2), 512–518.

Güler, N. F., Übeyli, E. D., & Güler, I. (2005). Recurrent neural networks employing Lyapunov exponents for EEG signals classification. *Expert Systems with Applications, 29*(3), 506–514.

Guo, L., Rivero, D., & Pazos, A. (2010). Epileptic seizure detection using multiwavelet transform based approximate entropy and artificial neural networks. *Journal of Neuroscience Methods, 193*(1), 156–163.

Hirtz, D., Thurman, D. J., Gwinn-Hardy, K., Mohamed, M., Chaudhuri, A. R., & Zalutsky, R. (2007). How common are the "common" neurologic disorders? *Neurology, 68*(5), 326–337.

Huang, N. E., et al. (1998). The empirical mode decomposition and Hilbert spectrum for nonlinear and non-stationary time series analysis. *Proceedings of the Royal Society of London Series A: Mathematical, Physical and Engineering Sciences, 454*(1971), 903–995.

Iasemidis, L. D., et al. (2003). Adaptive epileptic seizure prediction system. *IEEE Transactions on Biomedical Engineering, 50*(5), 616–627.

Ince, N. F., Goksu, F., Tewfik, A. H., & Arica, S. (2009). Adapting subject specific motor imagery EEG patterns in space–time–frequency for a brain computer interface. *Biomedical Signal Processing and Control, 4*(3), 236–246.

Joshi, V., Pachori, R. B., & Vijesh, A. (2014). Classification of ictal and seizure-free EEG signals using fractional linear prediction. *Biomedical Signal Processing and Control, 9*, 1–5.

Kannathal, N., Choo, M. L., Acharya, U. R., & Sadasivan, P. K. (2005). Entropies for detection of epilepsy in EEG. *Computer Methods and Programs in Biomedicine, 80*(3), 187–194.

Khandoker, A. H., Lai, D. T. H., Begg, R. K., & Palaniswami, M. (2007). Wavelet-based feature extraction for support vector machines for screening balance impairments in the elderly. *IEEE Transactions on Neural Systems and Rehabilitation Engineering, 15*(4), 587–597.

Lai, Y. C., & Ye, N. (2003). Recent developments in chaotic time series analysis. *International Journal of Bifurcation and Chaos, 13*(6), 1383–1422.

Li, S., Zhou, W., Yuan, Q., Geng, S., & Cai, D. (2013). Feature extraction & recognition of ictal EEG using EMD and SVM. *Computers in Biology and Medicine, 43*(7), 807–816.

Mukhopadhyay, S., & Ray, G. C. (1998). A new interpretation of nonlinear energy operator and its efficacy in spike detection. *IEEE Transactions on Biomedical Engineering, 45*(2), 180–187.
Ngugi, A. K., Bottomley, C., Kleinschmidt, I., Sander, J. W., & Newton, C. R. (2010). Estimation of the burden of active and life-time epilepsy: A meta-analytic approach. *Epilepsia, 51*, 883–890.
Nigam, V. P., & Graupe, D. (2004). A neural-network-based detection of epilepsy. *Neurological Research, 26*, 55–60.
Ocak, H. (2009). Automatic detection of epileptic seizures in EEG using discrete wavelet transform and approximate entropy. *Expert Systems with Applications, 36*(2), 2017–2036.
Oweis, R. J., & Abdulhay, E. W. (2011). Seizure classification in EEG signals utilizing Hilbert-Huang transform. *BioMedical Engineering Online, 10*, 38.
Pachori, R. B. (2008). Discrimination between ictal and seizure-free EEG signals using empirical mode decomposition. *Research Letters in Signal Processing, 293056*, 5 p.
Pachori, R. B., & Bajaj, V. (2011). Analysis of normal and epileptic seizure EEG signals using empirical mode decomposition. *Computer Methods and Programs in Biomedicine, 104*(3), 373–381.
Pachori, R. B., Hewson, D., Snoussi, H., & Duchêne, J. (2009). Postural time-series analysis using empirical mode decomposition and second-order difference plots. In *IEEE International Conference on Acoustics, Speech and Signal Processing* (pp. 537–540), Taipei, Taiwan, 19–24 Apr 2009.
Pachori, R. B., & Patidar, S. (2014). Epileptic seizure classification in EEG signals using second-order difference plot of intrinsic mode functions. *Computer Methods and Programs in Biomedicine, 113*(2), 494–502.
Pachori, R. B., & Sircar, P. (2008). EEG signal analysis using FB expansion and second-order linear TVAR process. *Signal Processing, 88*(2), 415–420.
Polat, K., & Güneş, S. (2007). Classification of epileptiform EEG using a hybrid system based on decision tree classifier and fast Fourier transform. *Applied Mathematics and Computation, 187* (2), 1017–1026.
Prieto, T. E., Myklebust, J. B., Hoffmann, R. G., Lovett, E. G., & Mykelbust, B. M. (1996). Measures of postural steadiness: Differences between healthy young and elderly adults. *IEEE Transactions on Biomedical Engineering, 43*(9), 956–966.
Ramsay, R. E., Rowan, A. J., & Pryor, F. M. (2004). Special considerations in treating the elderly patient with epilepsy. *Neurology, 62*(5 suppl 2), S24–S29.
Ray, G. C. (1994). An algorithm to separate nonstationary part of a signal using mid-prediction filter. *IEEE Transactions on Signal Processing, 42*(9), 2276–2279.
Schomer, D. L., & da Silva, F. L. (Eds.) (2005). *Electroencephalography: Basic Principles, Clinical Applications, and Related Fields*. Philadelphia: Lippincot Williams & Wilkins.
Senthil, P. K., Arumuganathan, R., Sivakumar, K., & Vimal, C. (2008). Removal of artifacts from EEG signals using adaptive filter through wavelet transform. In *9th IEEE International Conference on Signal Processing*, 2008 (pp. 2138–2141).
Sharma, R., Pachori, R. B., & Gautam, S. (2014). Empirical mode decomposition based classification of focal and non-focal EEG signals. In *IEEE International Conference on Medical Biometrics* (pp. 135–140), Shenzhen, China, 30 May–01 June 2014.
Srinivasan, V., Eswaran, C., & Sriraam, N. (2005). Artificial neural network based epileptic detection using time–domain and frequency–domain features. *Journal of Medical Systems, 29* (6), 647–660.
Srinivasan, V., Eswaran, C., & Sriraam, N. (2007). Approximate entropy-based epileptic EEG detection using artificial neural networks. *IEEE Transactions on Information Technology in Biomedicine, 11*(3), 288–295.
Subasi, A. (2007). EEG signal classification using wavelet feature extraction and a mixture of expert model. *Expert Systems with Applications, 32*(4), 1084–1093.
Subasi, A., & Gursoy, M. I. (2010). EEG signal classification using PCA, ICA, LDA and support vector machine. *Expert Systems with Applications, 37*(12), 8659–8666.

Suykens, J. A. K., & Vandewalle, J. (1999). Least squares support vector machine classifiers. *Neural Processing Letters, 9*(3), 293–300.

Thuraisingham, R. A., Tran, Y., Boord, P., & Craig, A. (2007). Analysis of eyes open, eye closed EEG signals using second-order difference plot. *Medical & Biological Engineering & Computing, 45*(12), 1243–1249.

Thurman, D. J., et al. (2011). Standards for epidemiologic studies and surveillance of epilepsy. *Epilepsia, 52*(s7), 2–26.

Tzallas, A. T., Tsipouras, M. G., & Fotiadis, D. I. (2007). The use of time–frequency distributions for epileptic seizure detection in EEG recordings. In *Proceedings of 29th Annual International Conference of the IEEE on Engineering in Medicine and Biology Society* (pp. 3–6), August 2007.

Tzallas, A. T., Tsipouras, M. G., & Fotiadis, D. I. (2009). Epileptic seizure detection in EEGs using time–frequency analysis. *IEEE Transactions on Information Technology in Biomedicine, 13*(5), 703–710.

Übeyli, E. D. (2010). Lyapunov exponents/probabilistic neural networks for analysis of EEG signals. *Expert Systems with Applications, 37*(2), 985–992.

Uthayakumar, R. & Easwaramoorthy, D. (2013). Epileptic seizure detection in EEG signals using multifractal analysis and wavelet transform. *Fractals*, 21(2).

World Health Organization. (2014). *Neurological disorders, including epilepsy*. Retrieved from http://www.who.int/mental_health/management/neurological/en/. Accessed 8 Apr 2014.

Yuan, Q., Cai, C., Xiao, H., Liu, X., & Wen, Y. (2007). Diagnosis of breast tumours and evaluation of prognostic risk by using machine learning approaches. *Communications in Computer and Information Science, 2*, 1250–1260.

A Rough Set Based Total Quality Management Approach in Higher Education

Ahmad Taher Azar, Renu Vashist and Ashutosh Vashishtha

Abstract Contrary to the popular belief that TQM is a poor fit in higher education sector, this research proposes a Rough Set Theory (RST) based model for grading educational institution using TQM parameters. It is a well established fact that TQM needs major reshaping before it can be effectively applied in higher education sector for quality assessment and improvement. This chapter takes a balance view by employing RST approach in TQM architecture and eliminating the much publicized shortcomings of TQM approach. RST theory has advantage of working on a small size of data containing vague and imprecise information which is widely prevalent in education sector. A carefully drafted questionnaire, containing nine attributes, is used for generating research data from the different stake holders in higher educational institutes of India. Nine modified condition attributes are selected on the basis of literature survey and expert views which are subsequently treated with RST analysis. One decision parameter 'Grade' depends on nine independent condition attributes. The resultant model contains only two significant attributes namely, 'Effective Learning and Teaching' and 'Administrative Setup' which can effectively determine the grading of educational institutions. Results of this study may be utilized to improve the higher education quality through appropriate grading mechanism based on self assessment of quality parameters by the different stakeholders of the education sector. The study confirms that TQM can be useful to enhance both quasi-academic areas such as 'administrative setup' along with core academic area 'effective teaching and learning'.

A.T. Azar (✉)
Faculty of Computers and Informations, Benha University, Benha, Egypt
e-mail: ahmad_t_azar@ieee.org

R. Vashist
Faculty of Computer Science, Shri Mata Vaishno Devi University, Katra, (J & K), India
e-mail: vashist.renu@gmail.com

A. Vashishtha
Faculty of Management, Shri Mata Vaishno Devi University, Katra, (J & K), India
e-mail: ashu.vashishtha@smvdu.ac.in

Q. Zhu and A.T. Azar (eds.), *Complex System Modelling and Control Through Intelligent Soft Computations*, Studies in Fuzziness and Soft Computing 319,
DOI 10.1007/978-3-319-12883-2_14

Keywords Rough set theory · Total quality management · Core · Higher education · Education institutes

1 Introduction

Over the last several decades the term 'quality management' has evolved as an obsession with business and non business entities for achieving the goals of sustainable profits, competitiveness and long term survival. It is equally gaining attention from educational sector companies, universities, colleges and government agencies of education sector. The genesis of quality management has its roots in manufactured product, productive process and can be traced to the work of Shewhart in the 1930s (Shewhart 1931). Many years later after the World War II Deming, Juran, Feigenbaum and others formulated quality based management techniques which inherited the Shewhart's philosophy of quality but extended it into business applications across many organizations. Total Quality Management (TQM) increasingly used as an effective business strategy to capture wider market share and gaining competitive edge (Rehder and Ralston 1984; Fortuna 1990; Fisher 1993; Ruben 1995). TQM is not only a tool, which is ready to use, but there are number of principles and methods, which needs to be applied according to organizational needs. There are some early evidences of adoption of TQM in USA higher education system in the non academic areas such as administration and support functions (Ruben 1995; Koch and Fisher 1998; Yorke 1999) whereas application of TQM in core academic areas remains debatable and very selective (Vazzana et al. 2000). This is largely because TQM is essentially evolved in manufacturing based industries and switching it to extensively human specific higher education sector pauses incompatibility issues (Houston and Studman 2001).

All over the world Higher Education in general and technical education in particular transforming its focus from 'elite oriented' to 'mass oriented' (Weeks 2000) and therefore, developing market orientation like other business organizations. Higher educational institutions with traditional ways of working are finding it increasingly difficult to cope with the pressure of change. This is more so in developing and under developing countries where public funding for higher education is limited and national objective of increasing higher education enrollment may not be achieved without private funding. This brings into focus the utility of TQM into higher education sector with the business like goals of reducing operating costs; increasing fees based revenue, improving student's satisfaction, employability and faculty retention (Zabadi 2013).

Successful Implementation of TQM to improve academic content delivery and overall functioning of higher education institution continues to be a daunting task for the reasons spelled out earlier. This chapter aims to improve TQM utility in professional higher education sector by introducing Rough Set Theory (RST) based approach. RST of (Pawlak 1982, 1991) was developed as an alternative data analysis tool but subsequently made inroads into the areas of Artificial Intelligence,

Cognitive Sciences, Knowledge Discovery, Decision Analysis and Expert Systems. RST has an important advantage that it can handle inexact, uncertain and vague datasets (Maji and Roy 2002). The chapter attempts to fit TQM model into the professional higher education system on the basis of selected ten attributes out of which nine are conditional or independent attributes and one is decision or dependent attribute. The empirical data is collected from the respondents in select higher education institutions of India through a questionnaire.

2 Related Work

Total Quality Management has been successfully implemented in some Higher Education Institutions, and it has improved the quality of higher education in those institutions. In the last decade, TQM has emerged an important topic of research. There is a growing interest of researcher in this topic, which can be tested by the number of publications in this particular area.

The theoretical essence of the Deming approach to TQM concerns the creation of an organizational system that fosters cooperation and learning for facilitating the implementation of process management practices, which, in turn, leads to continuous improvement of processes, products, and services as well as to employee fulfillment, both of which are critical to customer satisfaction, and ultimately, to firm survival (Deming 1982, 1994). It is possible to extract from the total quality management philosophy, a set of traits, values, and behaviors that can lead to positive outcomes for organizations (Anderson et al. 1994).

Quality in higher education means enabling students to achieve learning goals and academic standards in effective educational environment (Venkatraman 2007).

Research proved that faculty has a major impact on students teaching (Hill et al. 2003) and is the main strength in an educational institution (Gary et al. 2005). Quality in teaching and learning can only be enhanced if the faculty members are satisfied and content (Chen et al. 2006). According to Imai (2006), Kaizen theory is all about employing small continuous steps to improve business organizations. Consequently, it is a humanistic approach that involves everyone in the organization from top managers to the employees. The concept is communicated from the top and implemented by the employees.

Extant literature emphasized the importance of employee's job satisfaction and performance in higher education (Ooi et al. 2007). Universities must provide competitive levels of work environment conducive to faculty needs in order to attain faculty commitment. This can only be achieved if universities emphasize continuous improvement and identify mechanisms for quality improvement (Chen et al. 2006). In literature, number of areas for faculty development can be found with reference to TQM, such as teaching and research activities, administration and management support, salary and promotion, professional development, overall working environment, and decision making (Chen et al. 2006). An excess of research can be found regarding student satisfaction in education (Sirvanci 2004).

Employees are internal customers in any organization (Sallis 2002) and quality of that organization cannot be improved without the satisfaction of their employees (Ooi et al. 2007). Becket and Brookes (2008) undertakes a critical evaluation of the different methods used to assess the quality of provision in higher education departments in the UK. Al-Tarawneh (2011) studies the role of management in higher education institutions and for implementing TQM in universities which need the participation of all to ensure survival and continuity of universities.

Manjula et al. (2012) propose a new capability maturity decision making model based on rough computing for extracting key process areas and its relevance for the development of quality education.

Andollo et al. (2013) examined the influence of training and empowerment and effective communication as an aspect of quality management system on service provision.

Acharjya and Bhattacharjee (2013) propose a performance evaluation for educational institutions using rough set on fuzzy approximation spaces with ordering rules and information entropy. In order to measure the performance of educational institutions, they construct an evaluation index system.

The study conducted by Altahayneh (2014) indicated that TQM principles were poorly implemented in Jordanian colleges of physical education and the findings revealed that academic rank, years of experience, and education level did not significantly affect the faculty members' perceptions of TQM implementation.

3 A Systemic View of TQM

A comprehensive examination of TQM literature provides an insight into major quality improvement parameters like, strategic planning, customer focus, leadership, information analysis, process management, supplier management, and human resource management (Sila 2007). This is a macro view on TQM practices in various organizations most of which are manufacturing or production based organizations. Another view on TQM states that it is an integration of all organizational functional areas such as marketing, finance, production, human resource, design, engineering, so that customer need and organizational objectives can be synchronized (Hung 2007). There is another extremely interesting interpretation of TQM which highlights TQM as a system of three inter related and interdependent components, namely, values, methodologies, and tools that are used tighter to enhance the satisfaction of internal as well as external customers (Hellsten and Klefsjo 2000). **This view is important in the light of fact that education institutions have faculty as internal customers and students as external customers**. Deming's 14 TQM principles (see Table 1) provides a general direction to any organization for improving overall quality in most holistic fashion but these principles need to be reengineered when applying them to professional higher education sector.

Organization lacking a democratic culture and participative management style may not be perfect candidate for applying TQM, moreover, if the organization is

Table 1 Deming's 14 principles

S. No.	Deming's principles
1	Create constancy of purpose for improving products and services
2	Adopt the new philosophy
3	Cease dependence on inspection to achieve quality
4	End the practice of awarding business on price alone; instead, minimize total cost by working with a single supplier
5	Improve constantly and forever every process for planning, production and service
6	Institute training on the job
7	Adopt and institute leadership
8	Drive out fear
9	Break down barriers between staff areas
10	Eliminate slogans, exhortations and targets for the workforce
11	Eliminate numerical quotas for the workforce and numerical goals for management
12	Remove barriers that rob people of pride of workmanship, and eliminate the annual rating or merit system
13	Institute a vigorous program of education and self-improvement for everyone
14	Put everybody in the company to work accomplishing the transformation

Source Deming (1982)

too rigid to accept the new reforms than TQM applications will be a futile exercise. Figure 1 shows different organizations that follows some or all the principles of TQM. It is important to understand that Fig. 1 represents organizational alignment with external environment which, in other words, means that organizations learn to change, as the surrounding environment changes in the presence of TQM. Obviously, this realignment is absent in non TQM organizations and present in case of fully TQM compliant organization.

Unlike business organization, educational institutions of highest learning produce social goods in the form of intellectual capital for the economy as a whole and therefore the word 'management' needs to be interpreted differently. The *management* in TQM refers to everyone, starting from the top level to the lower level, who behaves as the manager of his own responsibilities (Sallis 2002). Similarly, in educational institution faculty and students are two focal points around which the entire organizational support system and auxiliary services revolve. Student and faculty have diversified responsibilities and manage these in the same manner as does the business manager and thereby suggesting the possible success of any TQM approach.

4 Rough Set Methodology for TQM in Higher Education

4.1 Empirical Analysis

Self assessment is a critical component of TQM process as it provides the deep insight of the educators or other higher education stake holders (Higher education

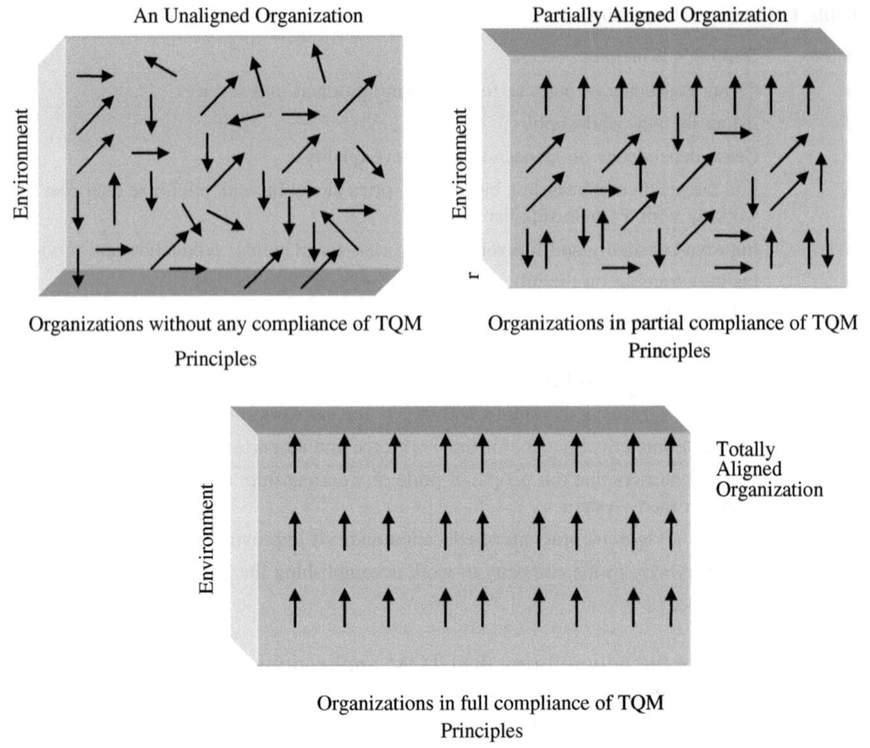

Fig. 1 Organizational alignment

Management boards and National Education Regulatory bodies) to make a considered judgment about institution performance and plugging the gaps to achieve the TQM goals. In order to further the goal of continuous improvement for delivering the quality education it is imperative for the educators to identify the roadblocks and weaker areas. Ten self assessment indicators developed by Sallis (2002) have been globally used and acknowledged in the realm of higher education. With a view to administer the research tool (Questionnaire) to a select pool of professional higher education respondents (Students, faculty, staff) belonging to Engineering and Management stream. Self assessment indicators have reduced to nine, retaining three generic parameters (effective learning and teaching, leadership and students) and including six new parameters (Administrative setup, Research, Faculty with experience and industrial exposure, Industry—Institute Interface, Placement, Infrastructure). However, these six parameters may have some indirect similarities to generic parameters of Sallis in terms of sub parameters but by and large the modified parameters remain different and serve the purpose to generate the data set from the particular pool of survey respondents. It is obvious that higher education is vastly diversified domain and TQM practices are likely to yield better results if these are designed with some tailor made adjustments to address the specific needs of a particular sub set of this domain.

A research instrument in the form of a survey questionnaire containing nine parameters or attributes has been developed (see Table 2). Selection of six modified parameters is done on the basis of literature review, inputs from higher education experts and interviews with various stakeholders in higher education. Prior to finalizing the questionnaire, it was pilot tested on carefully selected small group of respondents. Finalized questionnaire has been administered to faculty, students and administrative staff from Indian higher education institutions. A sample size of 50 respondents is being used for this study. The sample size has been kept small because Rough Set Analysis yields much accurate results on a smaller dataset. Respondents were asked to rate each condition attribute on a scale of 1–4 where 1 refers to "poor" and 4 to "excellent".

In Table 3 the criterion for Grade (Decision) is:

- Grade 1 is awarded for excellent performance of the educational Institution implying that there are majority of strengths and little or no weaknesses.
- Grade 2 is awarded for good performance of the educational Institution which implies that strengths outweigh weaknesses.
- Grade 3 is awarded for average performance of the educational Institution implying that there is a balance of strengths and weaknesses.
- Grade 4 is awarded for below average performance of the educational Institution. This means that weaknesses outweigh strengths.
- Grade 5 is awarded for poor performance of the educational Institution indicating that there are majority of weaknesses and little or no strength.

4.2 Rough Set Analysis

Rough Set Theory (RST) is useful and valid mathematical tool which deals with imprecise, vague and uncertain information. RST treats knowledge as an ability to classify objects relative to classes using indiscernible relation. Rough set analysis basically starts from a table called information table because with every object in this universe some information is associated. Information table contains objects which are represented by values of attributes. Objects containing the same information are indiscernible.

An information system is a pair $S = (U, A), X \subseteq U \text{ and } P \subseteq A$ where U is a nonempty finite set called the universe and A is a nonempty finite set of attributes, i. e., $a: U \rightarrow V_a$ for $a \in A$, where V_a is called domain of a. A decision table is a special case of information system

Table 2 Attribute normalization and classification

S. no.	Attribute	Notation	Classification
1	Effective learning and teaching	a	Poor (1)
	• Appropriateness of learning methods		Good (2)
			V.good (3)
	• Curriculum		Excellent(4)
	• Teaching and evaluation		
2	Leadership	b	Poor (1)
	• Top management composition		Good (2)
	• Departmental/school level supervision		V.good (3)
	• Student leadership		Excellent (4)
3	Administrative setup	c	Poor (1)
	• Timely availability of information		Good (2)
			V.good (3)
	• Implementation of decisions		Excellent (4)
	• Quality of support staff		
4	Research	d	Poor (1)
	• Faculty research publications		Good (2)
	• Sponsored research projects and consultancy		V.good (3)
	• Student research projects		Excellent (4)
5	Faculty	e	Poor (1)
	• Faculty teaching experience		Good (2)
	• Faculty with industrial exposure		V.good (3)
	• Faculty qualifications and communication skills		Excellent (4)
6	Industry institute interface	f	Poor (1)
	• Special lectures by industrial experts		Good (2)
			V.good (3)
	• Industrial tours		Excellent (4)
	• Industrial sponsorship to various events		
7	Placement	g	Poor (1)
	• Placement cell and staff		Good (2)
	• Campus placements		V.good (3)
	• Salary package offered		Excellent (4)
8	Infrastructure	h	Poor (1)
	• Physical environment (class rooms, labs, sports, canteen)		Good (2)
			V.good (3)
	• Academic infrastructure(library, online journals)		Excellent (4)
	• Health facilities		
9	Students	i	Poor (1)
	• Handling of student affairs		Good (2)
	• Monitoring students progress		V.good (3)
	• Student satisfaction		Excellent (4)

Table 3 Sample dataset

Objects	a	b	c	d	e	f	g	h	i	Grade(D)
1	3	3	4	4	4	3	4	3	2	2
2	2	2	2	2	2	1	1	3	3	2
3	3	3	4	2	3	2	3	4	2	2
4	2	1	3	2	2	1	1	4	4	2
5	2	2	2	2	2	2	3	3	2	2
6	3	2	4	3	3	2	2	3	2	2
7	3	4	4	3	3	4	4	3	3	2
8	2	2	3	2	3	2	1	3	2	3
9	3	2	3	2	3	2	1	3	4	3
10	3	2	4	4	4	3	2	3	3	3
11	3	2	4	3	4	2	2	4	3	3
12	3	2	3	3	3	2	1	2	4	3
13	2	2	2	2	3	2	1	2	4	3
14	3	2	3	2	2	1	1	2	3	3
15	3	4	3	3	3	4	2	3	2	3
16	3	3	3	3	4	3	2	3	2	3
17	3	3	3	2	4	2	1	3	4	3
18	2	2	2	3	2	2	2	1	2	4
19	2	3	3	2	3	2	3	2	3	4
20	3	3	2	3	2	3	3	2	3	4
21	3	3	2	3	3	3	3	2	3	4
22	2	2	2	1	3	2	1	1	2	4
23	2	2	1	1	2	1	1	3	3	4
24	2	2	3	2	3	2	1	2	3	4
25	2	1	1	2	2	2	2	3	3	4
26	3	3	3	3	2	3	3	3	2	4
27	3	3	3	3	1	2	2	2	3	4
28	1	1	1	1	2	1	1	2	1	5
29	3	3	3	3	1	3	3	2	2	5
30	1	1	1	1	1	2	1	1	2	5
31	1	1	1	2	1	1	1	2	1	5
32	1	3	3	1	2	1	1	3	2	5
33	1	1	1	2	1	2	1	2	1	5
34	1	1	1	1	1	1	2	2	1	5
35	1	2	1	1	1	2	2	2	1	5
36	1	1	1	1	2	1	1	3	2	5
37	1	2	2	2	1	1	2	3	1	5
38	1	2	1	1	1	4	3	2	1	5
39	2	3	2	2	3	2	1	2	3	5
40	4	3	3	3	2	4	3	2	2	1
41	4	3	3	3	4	4	4	3	3	1

(continued)

Table 3 (continued)

Objects	a	b	c	d	e	f	g	h	i	Grade(D)
42	2	3	3	4	2	3	4	4	1	1
43	4	4	4	3	4	3	4	3	4	1
44	4	4	4	4	4	4	4	4	2	1
45	4	4	3	2	4	4	3	3	3	1
46	4	4	3	3	4	3	3	4	3	1
47	4	3	3	4	3	4	3	4	3	1
48	4	4	3	4	4	3	4	3	4	1
49	1	2	2	2	2	2	3	3	2	1
50	3	3	2	2	2	4	3	4	3	1

$$S = (U, A = CU\{d\})$$

where attribute in C are called condition attributes and d is a designated attribute known as decision attribute.

Now, define two approximations $\underline{P(X)}$ and $\overline{P(X)}$ called the P-lower and the P-upper approximation of X respectively where

$$\underline{P(X)} = \bigcup x \in U\{P(x) : P(x) \subseteq X\} \text{ and}$$
$$\overline{P(X)} = \bigcup x \in U\{P(x) : P(x) \cap X \neq \phi\}.$$

Lower approximation will consist of all the members of the information system which surely belongs to the set and Upper approximation consist of all the members of the information system which possibly belongs to the set.

The boundary region is given by the set difference $\overline{P(X)} - \underline{P(X)}$ consists of those objects that can neither be ruled in nor ruled out as members of the target set X. If the boundary region is empty i.e. $\overline{P(X)} = \underline{P(X)}$ then the set is crisp otherwise the set is rough (inexact).

Rough set is organized in the form of information table or decision table, whose columns are labeled as condition and decision attributes and rows of the table contain the example (Pawlak and Skowron 2007a). Entries in the table represent the attribute values. Rows of Table 3 which is a decision table are called *examples* (objects, entities). Properties of examples are perceived through assigning values to some variables. Condition attributes of decision table are also called independent variable and decision attribute is called the dependent variable. The dependent variable is a function of independent variable and the value of dependent variable is solely depends on the values of independent variable. Analysis of Table 3 is done using Rose 2 S/W of rough set (Predki and Wilk 1999).

The set P of attributes is the *reduct* (or covering) of another set Q of attributes if P is minimal and the Indiscernibility relations, defined by P and Q are same.

A reduct can be thought of as a sufficient set of features sufficient, that is, to represent the category structure and no attribute can be removed from reduct set without changing the equivalence classes. There may be $2^n - 1$ reducts of a decision table and it is not always feasible to find all the reducts of a set (Pawlak and Skowron 2007c). Therefore the reduct of an information system is *not* unique.

Reducts of Table 3 as discover by Rose2 s/w are

$R_1 = \{a, b, c, e, f, h\}$
$R_2 = \{a, c, e, h, i\}$
$R_3 = \{a, c, d, e, i\}$
$R_4 = \{a, b, d, e, g, i\}$
$R_5 = \{a, c, e, f, g, i\}$
$R_6 = \{a, b, c, d, g, h\}$
$R_7 = \{a, b, c, e, g, h\}$
$R_8 = \{a, b, c, f, g, h\}$
$R_9 = \{a, b, d, g, h, i\}$
$R_{10} = \{a, b, d, g, h, i\}$

The set of attributes which is common to all reducts is called the *core*. The core is the set of attributes which is possessed by every legitimate reduct, and therefore core consists of attributes which cannot be removed from the information system without causing collapse of the equivalence class structure. RST considers that the core is the set of necessary attributes and it is the set of most important attributes of the dataset and if any of the core attribute is eliminated from the dataset then it shoddily affect the classification (Pawlak and Skowron 2007b). It is pertinent here to mention that the core set may be empty for some datasets.

$$Core = \cap\ Reduct$$

where *Reduct* is the set of all the reducts.

$$Core = \mathrm{R}_1 \cap \mathrm{R}_2 \cap \mathrm{R}_3 \cap \mathrm{R}_4 \cap \mathrm{R}_5 \cap \mathrm{R}_6 \cap \mathrm{R}_7 \cap \mathrm{R}_8 \cap \mathrm{R}_9 \cap \mathrm{R}_{10}$$

Therefore core of Table 3 is:

$$Core = \{\mathrm{a}\} = \{\text{Effective learning and teaching}\}$$

This is the most significant attribute of Table 3.

The lower and upper approximation of the table is given by the Fig. 2. The accuracy of approximation is given by

$$\alpha_P(X) = \frac{|\underline{P}(X)|}{|\overline{P}(X)|}$$

where |X| denotes the cardinality of $X \neq \varphi$ and Obviously $0 \le \alpha \le 1$.

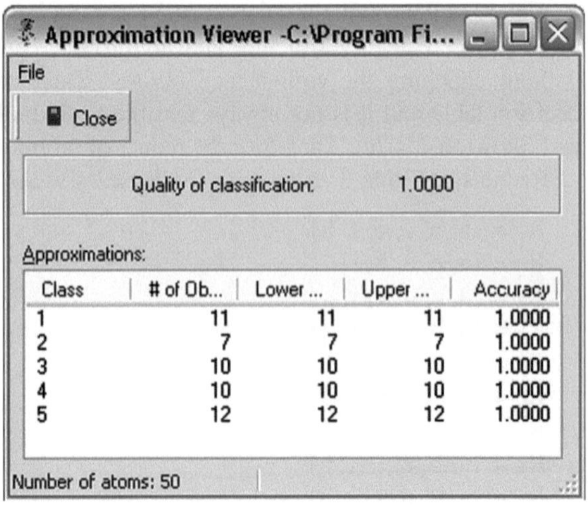

Fig. 2 Lower and upper approximation and accuracy of classification

If

$$\alpha_P(X) = 1,$$

then the set is crisp with respect to *P* and if

$$\alpha_P(X) < 1,$$

which means set is rough with respect to *P*.

The lower and upper approximation and the classification accuracy of decision Table 3 are shown in Fig. 2.

Decision Rules

Extracting decision rules from the decision table is one of the important aspects of RST. Numbers of attribute reduction algorithm are available which can lead to more accurate and simple decision rules. These decision rules can directly determine the performance of information system. Decision rules are generally represented in the form of 'if-then' form. Reduct based rules can also be generated which are lesser in number and yet significant (Vashist and Garg 2011, 2012). The set of decision rules are also called decision algorithm.

Heuristic rules for decision Table 3 as extracted by Rose 2 S/W are represented as follows:

Rule 1 (a = 4) => (D = 1); [8, 8, 72.73 %, 100.00 %] [8, 0, 0, 0, 0] [{40, 41, 43, 44, 45, 46, 47, 48}, {}, {}, {}, {}] or

Rule 1 If (Effective Learning and Teaching = excellent) then(Grade = excellent).

Rule 2 (a = 1) & (c = 1) => (D = 5); [8, 8, 66.67 %, 100.00 %] [0, 0, 0, 0, 8][{}, {}, {}, {}, {28, 30, 31, 33, 34, 35, 36, 38}] or

Rule 2 If (Effective Learning and Teaching = poor) and(Administrative Setup = poor) then (Grade = poor).

5 Discussion

Higher education institutions like universities, colleges, research institutes have extensively relied on qualitative and quantitative tools of quality (such as interviews, focus groups, survey questionnaire and observation studies) to measure the perception of students and other stakeholders regarding the quality of education. This chapter also employed one such tool, namely, survey questionnaire because of its ease and accuracy in comparison to other tools. More importantly, RST as data mining technique can better extract information from a close ended questionnaire rather than other highly qualitative data collection instruments.

Several quality models have been employed and tested in higher education to achieve the perennial goal of quality improvement like TQM, QFD, Six Sigma, ISO 9001, the Malcolm Baldrige National Quality Award, the EFQM Model, SERVQUAL and many more (Houston and Studman 2001; Wiklund et al. 2003; Kanji et al. 1999; Talib 2013). However, our choice of TQM attributes or parameters is largely based on the fact that TQM remains the generic philosophy which continues to influence other quality models in one way or other. Nine condition attributes and one decision attribute are used for RST analysis. The value of decision or dependent attribute depends on the values of nine condition or independent attributes and each condition attribute further has three sub attributes. Decision attribute assume values ranging from grade 1 to grade 5 and condition attributes assume values ranging from 1 to 4. RST analysis (see Table 3) returns 'Effective Learning and Teaching' attribute as the most significant attribute of the dataset because it appears as 'Core'. A brief discussion of this attribute follows:

Core: Effective Learning and Teaching
Condition attribute 'Effective Learning and Teaching' has sub attributes 'Appropriateness of learning methods', 'Curriculum' and 'Teaching and Evaluation'. There are varieties of learning methods (Traditional Class room teaching, Information Technology based teaching aids, case study method, industrial training and projects etc.) and the choice of a particular learning method or combination of those depends on intra institution and inter institution factors. However, their appropriateness certainly influences the learning and teaching potential of the students and faculty. Similarly, curriculum quality and design is institution specific and depends on the competence of students and faculty. Modern day higher education imbibes the practice of flexible curriculum and regular revision of the same. Since, this study deals with professional education belonging to the disciplines of management and engineering therefore industry participation in curriculum development constitute an important element. 'Teaching and evaluation' sub attribute refers to quality of teaching and academic evaluation. This includes student perception of teaching

effectiveness and accuracy of academic evaluation as reflected from number of re-evaluation requests or display of answer sheets to students. 'Effective Learning and Teaching' as Core attribute signifies that this is indispensible for the purpose of quality grading of the higher education institutions and any attempt to eliminate this attribute will result into incorrect grading. The Rough Set analysis also generated two rules from heuristic method. The first rule establish that

If (Effective Learning and Teaching = excellent) then (Grade = excellent).

This rule indicates that if 'Effective Learning and Teaching' is 'excellent' then the resulting grade of the institution will also be 'excellent'. The strength (or accuracy) of this rule is 72.73 %.

Similarly, second rule states that

If (Effective Learning and Teaching = poor) and (Administrative Setup = poor) then (Grade = poor).

Thereby, indicating that if 'Effective Learning and Teaching' is 'poor' and also, 'Administrative Setup' is 'poor' then the resulting grade of the institution is also 'poor'. The strength (or accuracy) of this rule is 66.67 %.

In other words, the interpretation of second rule implies that the condition attribute 'Administrative Setup' must be poor along with 'Effective Learning and Teaching' in order to grade the institution 'poor'. It must be notices that Condition Attribute 'Administrative Setup' is second most significant attribute of the dataset containing nine condition attributes.

Condition Attribute: 'Administrative Setup'

This attribute includes three sub attributes i.e. 'Timely availability of information', 'Implementation of decisions' and 'Quality of support staff'. 'Timely availability of information' refers to quick, accurate and timely distribution of critical information to various stakeholders of the education system. This has different meaning for different persons, for example there is plethora of information such as examination date sheet, results declaration, shortage of attendance, change in class time table, change of guest lecture venue, new eligibility norms for admission, Campus placements, training schedule, sponsored research etc. which has different meaning for different persons. Similarly, speedy implementation of decisions is equally critical in a tightly packed academic schedule. Delayed decisions regarding students, faculty and support staff may tarnish the image of the institution and brings in a typical bureaucratic organizational culture. 'Quality of support staff' refers to the non teaching, administrative and ministerial employees who perform range of service like typing, accounting, finance, purchase, laboratory staff, engineering wing, transportation services and much more. Any compromise on this parameter may derail the education institution and cripple the academic functioning altogether. Moreover, there is tendency in institution to cut down the operating expenses and shift the administrative task to faculty who is primarily for research

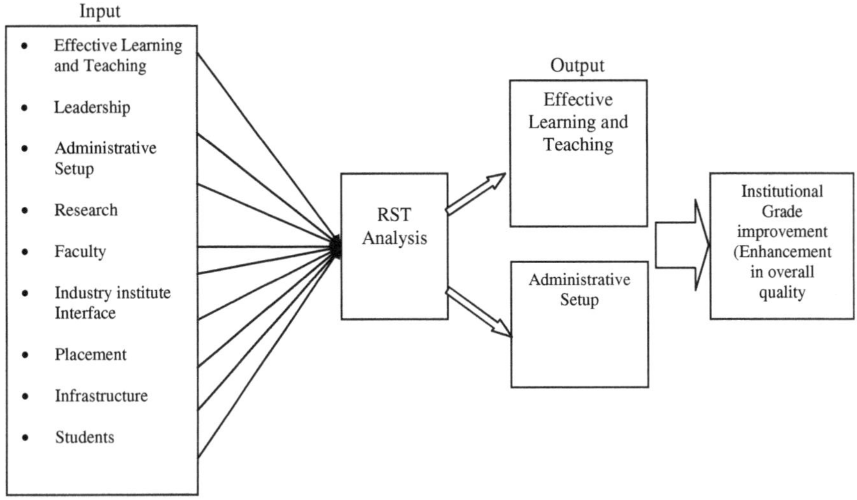

Fig. 3 RST based TQM model for higher education

and teaching. This may result into overburdened and disoriented faculty which ultimately leads to quality degradation.

According to the results of RST analysis the remaining seven condition attributes are lesser significant and therefore does not contribute much to the decision (Grading of the institution).

The proposed RST based TQM model emerges from the results of RST analysis as depicted in Fig. 3. Self assessment of quality parameters in higher education by three different stake holders, namely, faculty, students and staff are treated with RST analysis and two significant condition attributes emerges as the major pillars of this model. Grading of higher education institution may be improved if these two attributes are given maximum attention. However, this should not be interpreted that other attributes are meaningless but the outcome of this model reflects the RST approach of fine tuning the vague and uncertain data.

6 Conclusion

The last 5 years have seen a phenomenal surge in the popularity of global and local rankings of universities along with some complexities and problems (Hazelkorn 2011). The sudden rush for prestigious rankings and grading by different rating agencies has initiated a debate about their validity, accuracy and real worth. This ascertains the growing importance of institutional grading as an important measure of quality assessment in the highly competitive arena of global higher education. This research aims to contribute through the use of the proposed model to enhance the total quality education through a modified grading mechanism which is focused exclusively on two parameters rather than bundle of vague parameters. It has been argued that TQM philosophy is based on customer defined quality concept which is unique to a business organization and poorly fits in the context of higher education which is sensitive to economic and social environment surrounding it. Opponents of TQM in higher education also reason that quality is subjective perception in education sector for students, faculty and other stakeholders whereas it has a definite meaning in business units producing goods and services for a specific customer. We disagree with this view as students employability is a major concern in professional education which in turns depends on skill sets of graduating students in tune with industry's expectation. Any mismatch at this stage (i.e. between education supplier and employers expectations) may defeat the entire purpose of higher education and may result into unemployable graduates who are a poor fit to the industry.

This model is an attempt to highlight the application of TQM approach and RST in improving the quality of professional higher education. Results are based on a country specific respondent's survey from engineering and management segments of higher education. In order to extend the results across the other segments of higher education a cross country survey may be conducted which is likely to give more generalized version of the proposed model. Further, similar research may be carried out with different set of TQM attributes and RST may be replaced by some other data mining tool.

References

Acharjya, D. P., & Bhattacharjee, D. (2013). A rough computing based performance evaluation approach for educational institutions. *International Journal of Software Engineering and Its Applications, 7*(4), 331–347.

Al-Tarawneh, H. (2011). The implementation of Total Quality Management (TQM) on higher educational sector in Jordan. *International Journal of Industrial Marketing, 1*(1), 4–5.

Altahayneh, Z. L. (2014). Implementation of Total Quality Management in colleges of physical education in Jordan. *International Journal of Business and Social Science, 5*(3), 109–117.

Andollo, A. A., Rambo, C. M., & Monari, F. (2013). Influence of quality management systems on service provision in The University Of Nairobi, Kenya. *African Journal of Business and Management, 3*, 98–116.

Anderson, J. C., Rungtusanatham, M., & Schroeder, R. G. (1994). A theory of quality management underlying the Deming management method. *Academy of Management Review, 19*(3), 472–509.

Becket, N., & Brookes, M. (2008). Evaluating qualtiy management in university departments. *Quality Assurance in Education, 14*(2), 123–142. doi:10.1108/09684880610662015.

Chen, S. H., Yang, C. C., Shiau, J. Y., & Wang, H. H. (2006). The development of an employee satisfaction model for higher education. *The TQM Magazine, 18*(5), 484–500.

Deming, W. E. (1982). *Out of the crisis*. MA: The MIT Press.

Deming, W. E. (1994). Report card on TQM. *Management Review, 83*(1), 22–25.

Fisher, J. L. (1993). TQM: A warning for higher education. *Educational Record, 74*(2), 15–19.

Fortuna, R. M. (1990). The quality imperative. In E. C. Huge (Ed.), *Total quality: An executive's guide for the 1990s*. Homewood, IL: Dow Jones-Irwin.

Gary, W., David, S., & Derek, Z. (2005). Academic preparation, effort and success: A comparison of student and faculty Perceptions. *Educational Research Quarterly, 29*(2), 29–36.
Hazelkorn, E. (2011). *Rankings and the reshaping of higher education: The battle for world class excellence*. Basingstoke, UK: Palgrave Macmillan.
Hellsten, U., & Klefsjo, B. (2000). TQM as a management system consisting of values, techniques and tools. *The TQM Magazine, 12*(4), 238–244.
Hill, Y., Lomas, L., & MacGregor, J. (2003). Students' perceptions of quality in higher education. *Quality Assurance in Education, 11*(1), 15–20.
Houston, D. J., & Studman, C. J. (2001). Quality management and a university: A deafening clash of metaphors? *Assessment and Evaluation in Higher Education, 26*(5), 475–487.
Hung, H. M. (2007). Influence of the environment on innovation performance of TQM. *Total Quality Management, 18*(7), 715–730.
Imai, M. (2006). *Kaizen, the key to Japan's competitive success*. United State of America: Mc-Graw-Hill Inc.
Kanji, G. K., Tambi, A. M. B. A., & Wallace, W. (1999). A comparative study of quality practices in higher education institutions in the US and Malaysia. *Total Quality Management. 10*(3), 357–371.
Koch, J. V., & Fisher, J. L. (1998). Higher education and total quality management'. *Total Quality Management, 9*(8), 659–668.
Maji, P. K., & Roy, A. R. (2002). An application of soft sets in a decision making problem. *Computers and Mathematics with Applications, 44*(8–9), 1077–1083.
Manjula, R., Vaideeswaran, J., & Acharjya, D. P. (2012). A capability maturity decision making system for educational quality based on rough computing. *International Journal of Advanced Science and Technology, 45*, 55–72.
Ooi, B. K., Bakar, A. N., Arumugam, V., Vellapan, L., & Loke, Y. K. A. (2007). Does TQM influence employees' job satisfaction? *An empirical case analysis, International Journal of Quality and Reliability Management, 24*(1), 62–77.
Pawlak, Z. (1982). Rough sets. *International Journal of Computer and International Sciences, 11* (3), 341–356.
Pawlak, Z. (1991). *Rough sets, theoretical aspects of reasoning about data*. Dordrecht, Boston, London: Kluwer Academic Publishers.
Pawlak, Z., & Skowron, A. (2007a). Rough sets: Some extensions. *Information Sciences, 177*(1), 28–40.
Pawlak, Z., & Skowron, A. (2007b). Rudiments of rough sets. *Information Science, 177*(1), 3–27.
Pawlak, Z., & Skowron, A. (2007c). Rough sets and Boolean reasoning. *Information Sciences, 177* (1), 41–73.
Prędki, B., & Wilk, S. (1999). Rough set based data exploration using ROSE system. In Z.W. Ras, A. Skowron (Eds.), *Foundations of intelligent, lecture notes in artificial intelligence* (Vol. 1609, pp. 172–180). Berlin: Springer.
Rehder, R., & Ralston, F. (1984). Total quality management: A revolutionary management philosophy. *SAM Advanced Management Journal, 49*, 24–33.
Ruben, B. D. (1995). *Quality in higher education*. New Brunswick: Transaction Publishers.
Sallis, E. (2002). *Total quality management in education* (3rd ed.). UK: Kogan Page LTD.
Sila, I. (2007). Examining the effects of contextual factors on TQM and performance through the lens of organizational theories: An empirical study. *Journal of Operations management, 25*(1), 83–109.
Sirvanci, M. B. (2004). Critical issues for TQM implementation in higher education. *The TQM Magazine, 16*(6), 382–386.
Shewhart, W. A. (1931). *Economic control of quality of manufactured product*. New York: D. Van Nostrand Company, Macmillan.
Talib, F. (2013). An overview of total quality management: Understanding the fundamentals in service organization. *International Journal of Advanced Quality Management, 1*(1), 1–20.
Vashist, R., & Garg, M. L. (2011). Rule generation based on reduct and core: A rough set approach. *International Journal of Computer Application, 29*(9), 1–5.

Vashist, R., & Garg, M. L. (2012). A rough set approach for generation and validation of rules for missing attribute values of a data set. *International Journal of Computer Application, 42*(14), 31–35.

Vazzana, G., Elfrink, J., & Bachman, D. P. (2000). A longitudinal study of total quality management processes in business colleges. *Journal of Education for Business, 76*(2), 69–74.

Venkatraman, S. (2007). A framework for implementing TQM in higher education programs. *Quality Assurance in Education, 15*(1), 92–112.

Weeks, P. (2000). Benchmarking in higher education: An Australian case study. *Innovations in Education and Training International, 37*(1), 59–67.

Wiklund, H., Klefsjö, B., Wiklund, P. S., & Edvardsson, B. (2003). Innovation and TQM in Swedish higher education institutions—possibilities and pitfalls. *The TQM Magazine, 15*(2), 99–107.

Yorke, M. (1999). Assuring quality and standards in globalised higher education. *Quality Assurance in Education, 7*(1), 14–24.

Zabadi, A. M. (2013). Implementing Total Quality Management (TQM) on the higher education institutions: A conceptual model. *Journal of Economics and Finance, 1*(1), 42–60.

Iterative Dual Rational Krylov and Iterative SVD-Dual Rational Krylov Model Reduction for Switched Linear Systems

Kouki Mohamed, Abbes Mehdi and Abdelkader Mami

Abstract Methods reductions of large scale linear time invariant systems are numerous, include among these which are based on the projection onto the Krylov subspace (Arnoldi, Lanczos, Arnoldi rational, Lanczos rational, Adaptive rational Arnoldi, rational Krylov) and methods based on singular values decomposition. Against, the reduction approaches of large scale switched linear systems are very limited (LMI, Arnoldi). In this chapter, two model reductions algorithms for approximation of large-scale linear switched systems are proposed, which are based on the Krylov subspace on the one hand and on the singular value decomposition on the other hand. At first the principle of the Dual rational Krylov based method is presented, based on this method for presenting at first the iterative dual rational Krylov approach that constructs a union of Krylov subspaces to generate two projection matrices. The iterative dual rational Krylov is low in cost, numerically efficient but the stability of reduced linear switched system is not always guaranteed. In the second part, the iterative SVD-Dual Rational Krylov approach is presented. This method is a combining of two sided-projections, one side is generated by the dual Rational Krylov-based model reduction techniques and the other side is generated by the SVD model reduction techniques, while the SVD-side depends on the observability gramian. This method is numerically efficient, minimize the H_∞ error between the original switched system and reduced one and preserve the stability of reduced systems. A simulation two examples are considered in order to take a performance study of these proposed approaches.

K. Mohamed (✉) · A. Mehdi · A. Mami
Laboratoire d'Analyse, de Commande et de Conception des Systemes,
Université de Tunis El Manar, École Nationale d'Ingenieurs de Tunis, LR-11-ES20,
BP 37, LE BELVEDERE 1002 Tunis, Tunisia
e-mail: koukimohammed2014@outlook.com

A. Mehdi
e-mail: mehdi.abbes@enit.rnu.tn

A. Mami
e-mail: abdelkader.mami@fst.rnu.tn

Q. Zhu and A.T. Azar (eds.), *Complex System Modelling and Control Through Intelligent Soft Computations*, Studies in Fuzziness and Soft Computing 319,
DOI 10.1007/978-3-319-12883-2_15

1 Introduction

The technological advance means the increase of the complexity of industrial systems. They operate in different environments with changing conditions and characteristics (quickly and brutally). Such as the industries aerospace, automotive, aggro-food, process engineering, chemical process, electrical circuit, power electronic systems, Thermal fluid systems and Mechanical system.... The modeling of these system types usually leads to the production of non-linear complex models of high order. However the dynamic is influenced by both discrete and continuous event which leads to the hybrid dynamical systems which are divided into two broad classes of hybrid systems, the first class is the multi-model system, which assumes that it is always possible to model a complex system with simple models, often linear models, assigning each model an operating area of the system. The second class is the linear piecewise system or called the switched system. This class of models is widely used for the analysis and control tools for linear systems which are very developed and also because much of the actual process can be represented by models from this class. Recent research on switching systems are mainly focusing on the modeling area, design control law and stability study. However, in the case of working for the development of the control law, the order of the controlled system must be taken into consideration because there are several hybrid systems of high order. These later are difficult to manipulate and the resolution of such models is indeed very demanding in computational resources. However, reduction of switched systems is an important solution for these problems. In this chapter, the iterative dual rational Krylov algorithm and the iterative SVD-dual rational Krylov algorithm for switched linear system are presented.

The model reduction problem, focus in this work, can be stated as follows.

Given a switched linear dynamical system in state space form (Dongmei et al. 2008; Kouki et al. 2013a, b; Zhendong and Shuzhi 2009):

$$\Sigma_q = \begin{cases} E\frac{dx(t)}{dt} = A_q x(t) + B_q u(t) \\ y(t) = C_q x(t) + D_q u(t) \end{cases} \tag{1}$$

In which $E_q \in \mathbb{R}^{n\times n}$, $A_q \in \mathbb{R}^{n\times n}$, $B_q \in \mathbb{R}^{n\times p}$, $C_q \in \mathbb{R}^{p\times n}$, $D_q \in \mathbb{R}^{p\times p}$, $u(t) \in \mathbb{R}^{n\times p}$, $y(t) \in \mathbb{R}^{p\times n}$ and q is a switching signal.

Applying the Laplace transform to the system (1), the following frequency formulation is obtained (Kouki et al. 2014a, b):

$$\Sigma_q = \begin{cases} sEX(s) = A_q X(s) + B_q U(s) \\ Y(s) = C_q X(s) + D_q U(s) \end{cases} \tag{2}$$

where $X(s)$, $Y(s)$ and $U(s)$ are the Laplace transform of $x(t)$, $y(t)$ and $u(t)$ respectively. For simplicity, supposing that the $u(t)$ is an impulse response, then the transfer function of the original switched linear system is given by (Grimme 1997; Kouki et al. 2013c):

$$f_q(s) = C_q(sI_q - A_q)^{-1}B_q + D_q \quad (3)$$

The problem consists in approximating: $E_{r_q} \in \mathbb{R}^{r_q \times r_q}$, $A_{r_q} \in \mathbb{R}^{r_q \times r_q}$, $B_{r_q} \in \mathbb{R}^{r_q \times p}$, $C_{r_q} \in \mathbb{R}^{p \times r_q}$, $D_{r_q} \in \mathbb{R}^{p \times p}$ and $y_r(t) \in \mathbb{R}^{p \times r_q}$, the matrices of the each reduced subsystem of order r_q, where $r_q \ll n$.

The state space representation of reduction switched dynamical linear systems is as follows (Benner et al. 2003; Gaoa et al. 2006; Grimme 1997; Kouki et al. 2013a, b; Tulpule et al. 2011):

$$\hat{\Sigma}_q = \begin{cases} E_{r_q} \frac{dx_{r_q}(t)}{dt} = A_{r_q} x_q(t) + B_{r_q} u(t) \\ y_{r_q}(t) = C_{r_q} x_q(t) + D_{r_q} u(t) \end{cases} \quad (4)$$

The Laplace transform is applied to the system (4), this relation is obtained:

$$\hat{\Sigma}_q = \begin{cases} sE_{r_q} X_{r_q}(s) = A_{r_q} X_{r_q}(s) + B_{r_q} U(s) \\ Y_{r_q}(t) = C_{r_q} X_{r_q}(s) + D_{r_q} U(s) \end{cases} \quad (5)$$

where $X_{r_q}(s)$ and $Y_{r_q}(s)$ represents the Laplace transform of the reduces $x_{r_q}(t)$ and $y_{r_q}(t)$. The transfer function of the reduced linear switched system is as follows:

$$f_{r_q}(s) = C_{r_q}(sI_{r_q} - A_{r_q})^{-1}B_{r_q} + D_{r_q} \quad (6)$$

This chapter is organized as follows. Section 2, briefly presents an overview of the Lyapunov equations and the H_∞ error. In Sect. 3, the Dual Rational Krylov is presented. Section 4, the Iterative Dual Rational Krylov method for switched linear systems, will be presented with application on the numerical examples. In Sect. 5, the Iterative SVD-Dual Rational Krylov method for switched linear systems is detailed and evaluated by the use of the numerical examples. In Sect. 6, a comparative study between the Iterative Dual Rational Krylov method and the Iterative SVD-Dual Rational Krylov method is given. The last section is dedicated to conclude this paper.

2 Preliminaries

2.1 *Lyapunov Equations*

Let a switched linear stable system as in (1). The infinite observability and reachability gramians in the continuous time of this system are obtained by this two relations (Antoulas 2009; Diepold and Eid 2011):

$$Q_q = \int_0^{\infty} e^{A_q^T t} C_q^T C_q e^{A_q t} dt \tag{7}$$

$$P_q = \int_0^{\infty} e^{A_q t} B_q B_q^T e^{A_q^T t} dt \tag{8}$$

The solution of the stable switched linear system in the sense of Lyapunov is obtained by solving the following two equations for each subsystem (Andres et al. 2013; Antoulas 2009; Kouki et al. 2013b; Mehrmann et al. 2012):

$$A_q P_q + P_q A_q^T + B_q B_q^T = 0 \tag{9}$$

$$A_q^T Q_q + Q_q A_q + C_q^T C_q = 0 \tag{10}$$

The solutions of these two equations are P and Q (Antoulas 2009; Gugercin 2008). $P_q \in \mathbb{R}^{n\times n}$ and $Q_q \in \mathbb{R}^{n\times n}$ are called the reachability and the observability gramians matrices, respectively.

The Eq. (9) is proved by the following formulation (Bao et al. 2006):

$$\begin{aligned} A_q P_q + P_q A_q^T &= \int_0^{\infty} [A_q e^{A_q t} B_q B_q^T e^{A_q^T t} + e^{A_q t} B_q B_q^T e^{A_q^T t} A^T] dt \\ &= \int_0^{\infty} d(e^{A_q t} B_q B_q^T e^{A_q^T t}) - B_q B_q^T \end{aligned} \tag{11}$$

The Eq. (10) is proved by the following formulation:

$$\begin{aligned} A_q^T Q_q + Q_q A_q &= \int_0^{\infty} [A_q^T e^{A_q^T t} C_q^T C_q e^{A_q t} + e^{A_q^T t} C_q^T C_q e^{A_q t} A] dt \\ &= \int_0^{\infty} d(e^{A_q^T t} C_q^T C_q e^{A_q t}) C_q^T C_q \end{aligned} \tag{12}$$

The infinite gramians in the frequency time of the system (1) are obtained by these relationships:

$$Q_q = \frac{1}{2\Pi} \int_{-\infty}^{+\infty} (-iw_q I_q - A_q^T)^{-1} C_q^T C_q (iw_q I_q - A_q)^{-1} dw \tag{13}$$

and

$$P_q = \frac{1}{2\Pi} \int_{-\infty}^{+\infty} (iw_q I_q - A_q)^{-1} B_q B_q^T (iw_q I_q - A_q)^{-1} dw \quad (14)$$

Owing to the complexity dot matrix, solving of the Lyaponuv equations is very complicated for switched linear system of high order.

2.2 The Singular Value Decomposition

Given a square matrix $A_q \in \mathbb{R}^{n \times n}$, their singular value decomposition is defined by this relation (Antoulas 2009; Antoulas et al. 2001):

$$A_q = U_q \Sigma_q V_q^T \quad (15)$$

where $\Sigma_q = diag(\sigma_{1_q}, \ldots, \sigma_{n_q}) \in \mathbb{R}^{n \times n}$, $U_q = (u_{1_q}, u_{2_q}, \ldots, u_{n_q}) \in \mathbb{R}^{n \times n}$, $V_q = (v_{1_q}, v_{2_q}, \ldots, v_{n_q}) \in \mathbb{R}^{n \times n}$ and $\sigma_{1_q} \geq \sigma_{1_q} \cdots \geq \sigma_{n_q}$.

Gramians matrices plays an important part in the reduction methods based on singular value decomposition. The relationship between the singular value decomposition and the gramians matrices is as follows (Antoulas 2009):

$$\sigma_{i_q} = \sqrt{\lambda_{i_q}(P_q Q_q)} \quad for \quad i_q = 1, \ldots, n \quad (16)$$

where, σ_{i_q} presents the Hankel singular value of Σ_q and λ_{i_q} presents the eigenvalues of the product $P_q Q_q$.

2.3 H_∞ of Dynamical Switched Systems

In this work, determining the error between the original switched system and reduced one is obtained by using the H_∞ technique knowing that (Antoulas 2009; Gugercin 2008):

$$\|\Sigma_q(jw) - \hat{\Sigma}_q(jw)\|_{H_\infty} \leq 2(\sigma_{(r+1)_q} + \cdots + \sigma_{n_q}) \quad (17)$$

2.4 Krylov Subspace

Given a square matrix A_q and a vector b_q, the spanned by the vectors $\{b_q, A_q b_q, \ldots, A_q^{m-1} b_q\}$ is called a standard Krylov subspace of dimension m denoted $K_{m_q}\{A_q, b_q\}$ for each sub-matrix (Awais et al. 2007; Heyouni and Jbilou 2006):

$$K_{m_q}\{A_q, b_q\} = span\{b_q, A_q b_q, \ldots, A_q^{m-1} b_q\} \tag{18}$$

An effective reduced model in the form (2) by the projection onto the Krylov subspace of the states matrices of system (1) is obtained. But there is another method to generate the Krylov subspace, which is more efficient that called rational Krylov subspace defined as :

$$K_{m_q}\{A_q, b_q, s_q\} = span\{(A_q - s_{1_q} I_q)^{-1} b_q, \ldots, \prod_{j_q}^{m_q} (A_q - s_{j_q} I_q)^{-1} b_q\} \tag{19}$$

where, $s_q = (s_{1_q}, s_{2_q}, \ldots, s_{m_q})$

3 Dual Rational Krylov for Switched Linear System

In this section, the details of the Dual Rational Krylov algorithm for computing of two projection matrices V_{r_q} and Z_{r_q} for each subsystem according to switching signal q are briefly recalled. Dual Rational Krylov is among the best approaches to reduce the large-scale linear switched systems. It is easy to implemented, numerically stable and to avoid the difficulties in the constructing of the two projections matrices. V_{r_q} and Z_{r_q} are constructed column by column during the iteration process using a Gram Schmidt techniques in orthogonalization procedure, such as the condition of biorthogonalithy is satisfied $Z_{r_q}^T V_{r_q} = I_{r_q}$. Take a switched linear system as a form (1) and assume that a sequence of expansion points $\{s_{1_q}, s_{2_q}, \ldots, s_{r_q}\}$ is given, with r is the order of reduced subsystem. These expansion points are interspersed. For each expansion point of each subsystem a two column vectors are generated, i.e in the first iteration uses s_{1_q}, the second iteration uses s_{2_q} until *rth iteration.*

The details of the Dual Rational Krylov algorithm for switched linear system can be found in Table 1 (Antoulas 2009; Druskin and Simoncini 2011; Flagg et al. 2012; Zhanga et al. 2008).

The main steps of this method are:

Step 1: Choose the interpolation points for each subsystem by the use of the eigenvalues criterion (Gugercin 2008).

Table 1 DRK-SLS Algorithm

DRK-SLS Algorithm:(input:$I_q, A_q, B_q, C_q, D_q, S_q$; output:$V_{r_q}, Z_{r_q}$)
Switch q
1/*Choose the Initial Interpolation points*/
s_{i_q} for $i_q = 1$ to r_q
2/*Construction of the matrices V_q and Z_q by the dual rational-Krylov based method*/
for k_q **= 1 to** r_q
if k_q**:=1**
$v0_q = ((A_q - s_q * I_q)^{-1} * B_q$
$v0_q = v0_q/norm(v0_q,' fro')$
$V_q(:,1) = v0_q$
$z0_q = ((A_q - s_q * I_q)^{-T} * C_q^T$
$z0_q = z0_q/norm(z0_q,' fro')$
$Z_q(:,1) = z0_q$
else
$v_q(:,k) = ((A_q - s_q * I_q)^{-1} * B_q$
$v_q(:,k) = v_q(:,k) - V_q(:,k-1) * V_q(:,k-1)^T * v_q(:,k)$
$V_q(:,k) = v_q(:,k)/norm(v_q(:,k),' fro')$
$z_q(:,k) = ((A_q - s_q * I_q)^{-T} * C_q^T$
$z_q(:,k) = z_q(:,k) - Z_q(:,k-1) * Z_q(:,k-1)^T * z_q(:,k)$
$Z_q(:,k) = z_q(:,k)/norm(z_q(:,k),' fro')$
End if
End for
End Switch

Step 2: Compute the V_{r_q} and Z_{r_q} bases with Rational Krylov subspaces, such as the condition of biorthogonalithy is satisfied :

$$((Z_{r_q}^T * V_{r_q})^{-1} Z_{r_q}^T) V_{r_q} = I_{r_q}. \tag{20}$$

The parameters of the reduced system can be obtained by the congruence transformation:

$$A_{r_q} = Z1 A_q V_q,\ B_{r_q} = Z1 B_q,\ C_{r_q} = C_q V_q,\ D_{r_q} = D_q$$
$$\text{where, } Z1 = ((Z_q^T V_q)^{(-1)} * Z_q^T).$$

3.1 Numerical Example

To evaluate this approach, two switched linear stable models (FOM of order 1006 and Clamped Beam of order 348) and a switched signal where q = 1, 2 are taken (Chahlaoui and Dooren 2005; Diepold and Eid 2011; Mignone et al. 2000):

Example 1 Switched FOM model of order 1006

The FOM model is a theoretical SISO (single-input single output) model of order 1006 proposed by Penzl in 1999, it is composed of two subsystems, every one of order 1006. The state-space matrices are given by: $A_1 = diag\,(\gamma_1, \gamma_2, \gamma_3, \gamma_4)$ with,

$$\gamma_1 = \begin{pmatrix} -1 & -100 \\ -100 & -1 \end{pmatrix},$$
$$\gamma_2 = \begin{pmatrix} -1 & -200 \\ -200 & -1 \end{pmatrix},$$
$$\gamma_3 = \begin{pmatrix} -1 & -400 \\ -400 & -1 \end{pmatrix},$$
$$\gamma_4 = diag(-1, \ldots, -1{,}000),$$
$$B_1 = [10 * ones(6, 1);\ ones(1{,}000, 1)],\ C_1 = B_1^T,\ D_1 = 0.$$
$$A_2 = A_1 - 5 * I,\ B_2 = B_1,\ C_2 = C_1,\ D_2 = D_1.$$

The Fig. 1 presents the largest singular value of the frequency response of the original switched system FOM of order 1006.

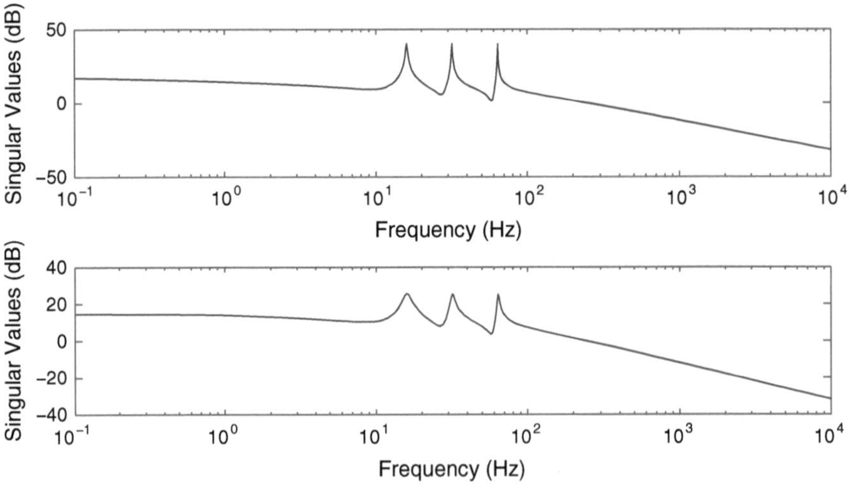

Fig. 1 Largest singular value of the frequency response of the original switched system (FOM of order 1006)

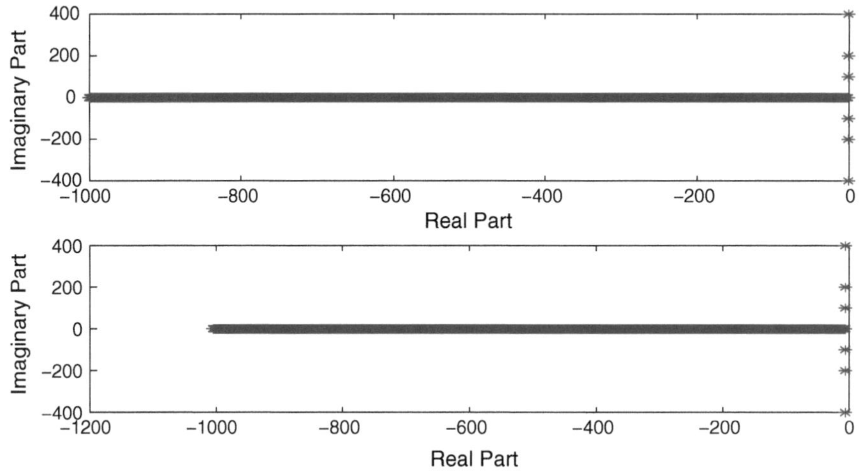

Fig. 2 Poles distribution of original switched system (FOM of order 1006)

The Fig. 2 presents the poles distribution of the original switched system FOM of order 1006, note that all the poles are negative real part, then the original subsystems are stable.

Example 2 Switched Clamped Beam model of order 348

The Clamped Beam is a SISO model, composed of two subsystems, every one having 348 states, it is obtained using the spatial discretization of a suitable partial differential equation. The input model is a force applied to the structure beside the free extremity, and the output is obtained from the resulting displacement The state matrices of the first subsystem is introduced in Chahlaoui and Dooren (2005),

The state matrices of the second subsystem is as follows:

$$A_2 = A_1 - 5 * 10^{1} * I,\ B_2 = B_1,\ C_2 = C_1,\ D_2 = 0.$$

The Fig. 3 shows the largest singular value of the frequency response of the original switched system Clamped Beam of order 348.

The Fig. 4 depicts the poles distribution of the original switched system Clamped Beam of order 348, note that all poles are negative real part, then the original subsystems are stable.

3.1.1 Example 1: Simulation Results

In this part the largest singular value of the frequency response, the distribution poles of the reduced switched system (FOM order 10) and the absolute error between original subsystem (FOM order 1006) and reduced one (FOM order 10) are presented.

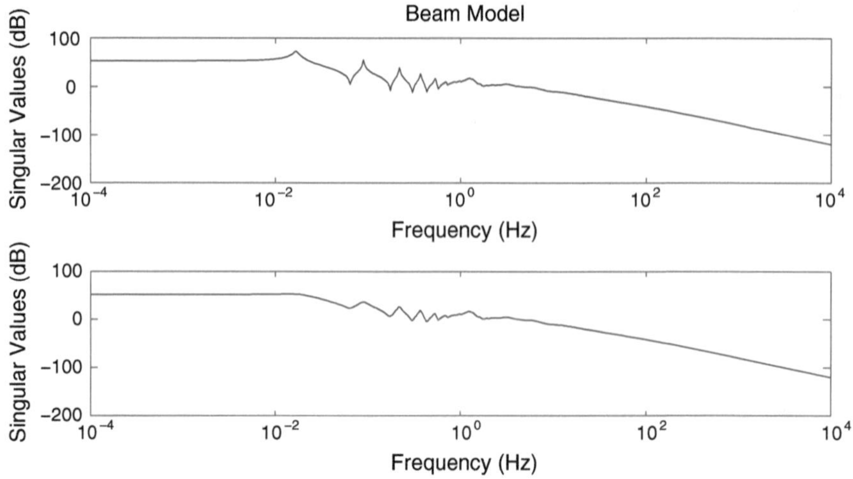

Fig. 3 Largest singular value of the frequency response of the original switched system of order (Clamped Bam of order 348)

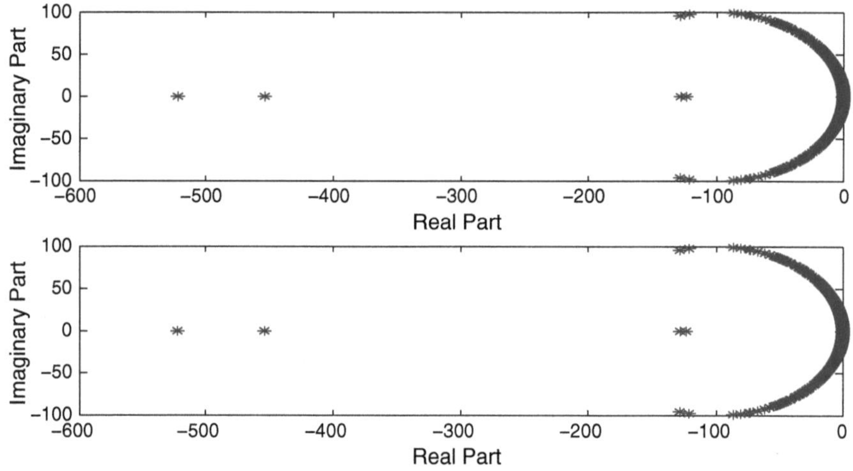

Fig. 4 Poles distribution of original switched system (Clamped beam of order 348)

The Fig. 5 presents the largest singular value of the frequency response of the original switched linear system (FOM order 1006) and reduced one (order 10) to a frequency range by DRK-SLS method. Note that when a correlation over the entire frequency range shape with a low error rate for low frequency. The Fig. 6 shows the variation of the singular value of the absolute error between the original switched linear system and the reduced one, note that that the error is small around the low

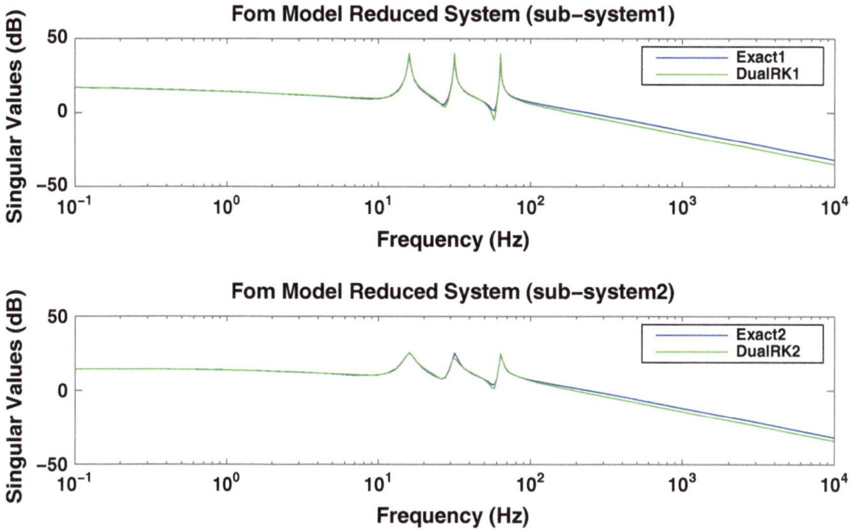

Fig. 5 Largest singular value of the frequency response of the original switched system of order (1006) and reduced one of order (10) to a frequency range with DRK-SLS method

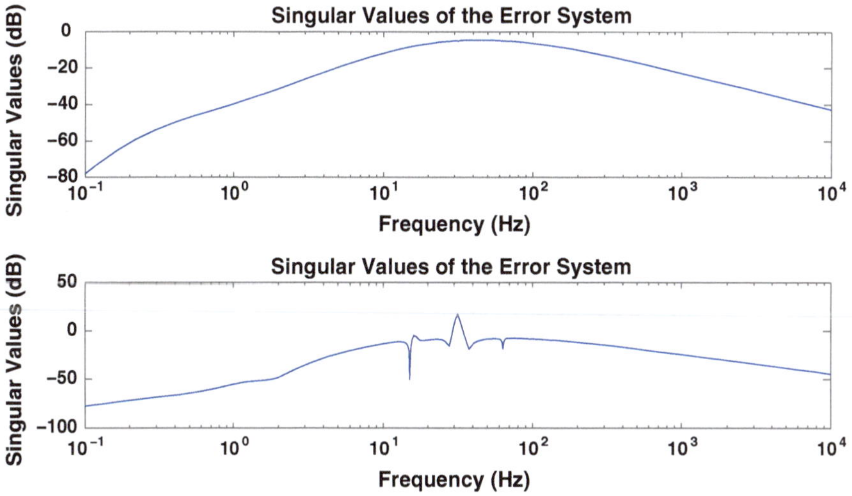

Fig. 6 Absolute error system between original switched system of order (1006) and the reduced one of order (10) with DRK-SLS method

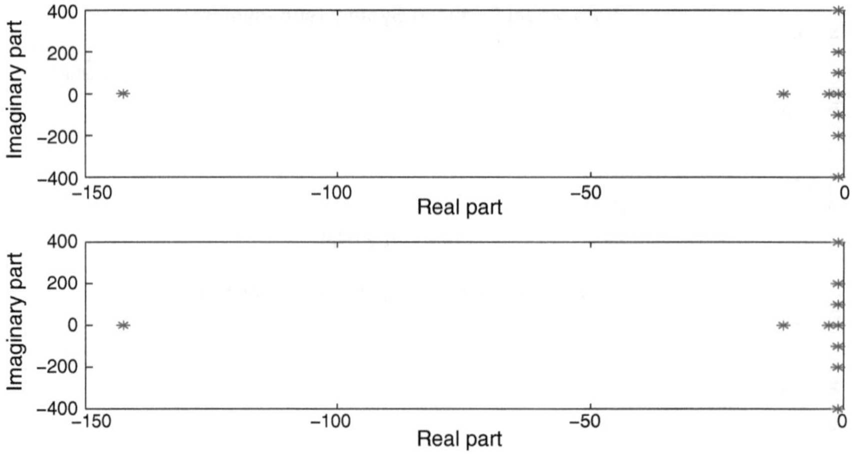

Fig. 7 Poles distribution of FOM reduced switched system (order 10) with DRK-SLS method

frequency. The distribution poles in the complex plane of each subsystem is depicted in Fig. 7, all poles are negative real part, then the reduces switched linear subsystems are stable.

3.1.2 Example 2: Simulation Results

In this part the largest singular value of the frequency response, the distribution poles of the reduced switched system (BEAM order 24) and the absolute error between original subsystem (BEAM order 348) and reduced one (BEAM order 24) are presented.

The Fig. 8 presents the largest singular value of the frequency response of the original switched linear system (Beam order 348) and reduced one (order 24) to a frequency range by DRK-SLS method.

Note that when a correlation over the entire frequency range shape with a low error rate for low frequency for the second subsystem. However for the first subsystem, note that the correlation is not good.

The Fig. 9 shows the variation of the singular value of the absolute error between the original switched linear system and the reduced one, note that the error is small around the low frequency for the second subsystem, which is not the case for the first subsystem. The distribution poles in the complex plane of each subsystem is depicted in Fig. 10, noting that the existence of positive real part poles, then the reduces switched linear subsystems are unstable.

The use of the dual rational Krylov method does not guarantee the stability of the reduced systems, so it does not minimise the error between the original subsystems and the reduced ones over the entire frequency range. For these reasons, the iterative dual rational Krylov method will be presented in the next section.

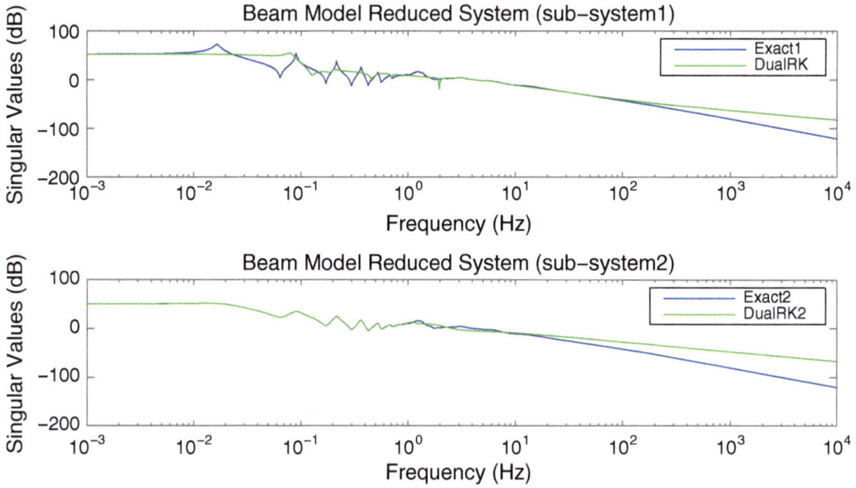

Fig. 8 Largest singular value of the frequency response of the original switched system (beam of order 348) and reduced one of order (24) to a frequency range with DRK-SLS method

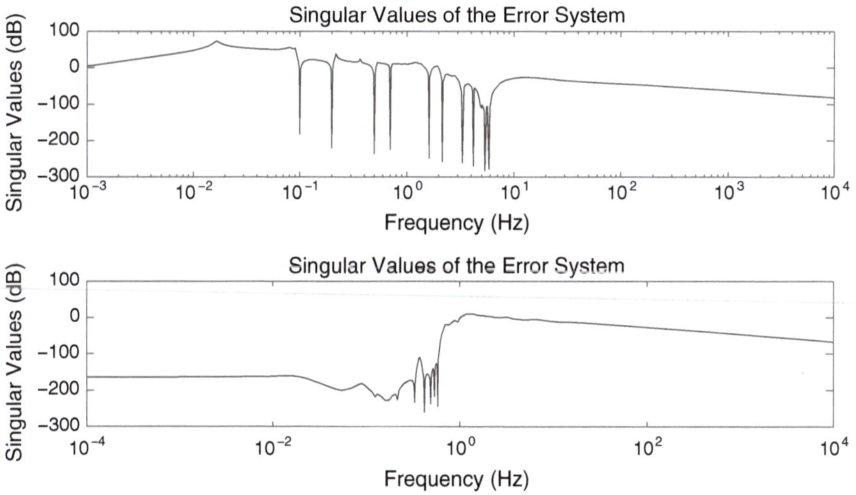

Fig. 9 Absolute error system between original switched system [Beam of order (348)] and the reduced one of order (24) with DRK-SLS method

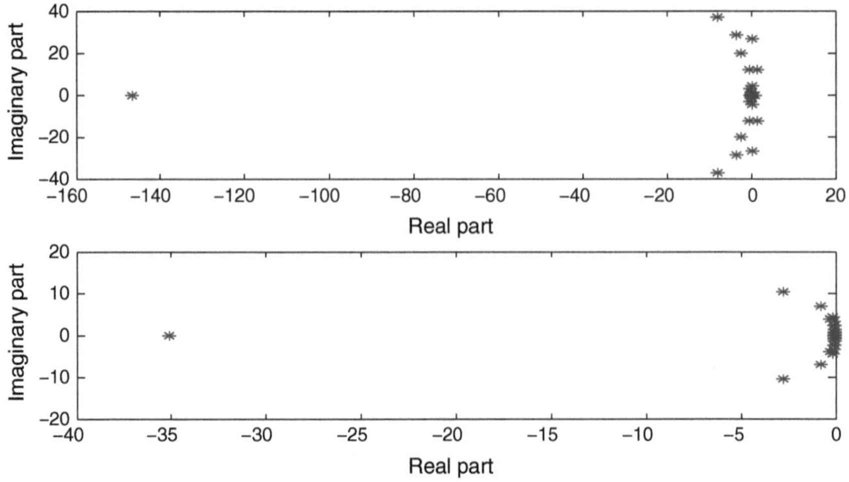

Fig. 10 Poles distribution of beam reduced switched system (order 24) with DRK-SLS method

4 Iterative Dual Rational Krylov for Linear Switched Systems

In this section the proposed method is given, the iterative dual rational Krylov model reduction for switched linear system, is an extended version of the dual rational Krylov method for switched linear system. Iterative dual rational Krylov is a connection between the Krylov-based reduction method and the interpolation of the expansion points. Given a stable switched linear system as the form (1) and using the eigenvalues criterion in the choice of the interpolation points (Flagg et al. 2012; Kouki et al. 2013a, 2014a, b). However, this method generated two Krylov subspaces V_{r_q} and Z_{r_q} for each subsystem, the generation of the two Krylov subspaces is performed iteratively until the satisfaction of the stopping criterion $((s_{(i+1)_q} - s_{i_q})/s_{(i+1)_q})$ and guarantees the biorthogonalithy condition of the two Krylov subspaces for each subsystem (i.e. $Z_{r_q}^T V_{r_q} = I_{r_q}$ where r_q is the order of reduced system) (Druskin and Simoncini 2011; Gallivan et al. 1996; Quarteroni et al. 2007). Theorem 1 summarizes this result:

Theorem 1 Take a switched linear system as a form (1) and the interpolation point $\{s_{i_q}\}$ for $i_q = 1, \ldots, r_q$. Let $V_{r_q} \in \mathbb{R}^{n\times r}$ and $Z_{r_q} \in \mathbb{R}^{n\times r}$ be obtained as follows (Grimme 1997):

$$\begin{cases} V_{r_q} = Span(s_{1_q}I - A_q)^{-1}B_q, \ldots, (s_{r_q}I - A_q)^{-1}B_q \\ Z_{r_q} = Span(s_{1_q}I - A_q)^{-T}C_q^T, \ldots, (s_{r_q}I - A_q)^{-T}C_q^T \end{cases} \quad (21)$$

with $Z_{r_q}^T V_{r_q} = I_{r_q}$.

The transfer function $f_{r_q}(s)$ of the reduced switched system (2) is matched with the transfer function $f_q(s)$ of original switched linear system in (1):

$$f_q(s_{i_q}) = f_{r_q}(s_{i_q}) \quad for \quad i_q = 1, \ldots, r_q \quad and \quad s_{i_q} = -\lambda_{i_q}(A_{r_q}) \tag{22}$$

where, λ_{i_q} is the eigenvalues of A_{r_q}.

The details of the Iterative Dual Rational Krylov algorithm for switched linear system (IDRK-SLS) can be found in Table 2 (Druskin and Simoncini 2011; Gallivan et al. 1996; Quarteroni et al. 2007):

The main steps of this method are:

Table 2 IDRK-SLS Algorithm

IDRK-SLS Algorithm:(input:$I_q, A_q, B_q, C_q, D_q, S_q, tol$; output:$V_{r_q}, Z_{r_q}$)
Switch q
1/*Choose the Initial Interpolation points*/
s_{i_q} for i=1 to r
2/*Construction of the matrices V_q and Z_q by the dual rational-Krylov based method knowing that*/
(a):$V_{r_q} = \text{Span}(s_{1_q}I_q - A_q)^{-1}B_q, \ldots, (s_{r_q}I_q - A_q)^{-1}B_q$
(b):$Z_{r_q} = \text{Span}(s_{1_q}I_q - A_q)^{-T}C_q^T, \ldots, (s_{r_q}I_q - A_q)^{-T}C_q^T$
With $Z_{r_q}^T V_{r_q} = I_{r_q}$, where $Z_{r_q} = (Z_{r_q}^T * V_{r_q})^{-1} Z_{r_q}^T$
3/*While (the relative change in s_i : $((s_{i+1} - s_i)/s_i) \geq tol$
(a):$A_{r_q} = Z_{r_q} A_q V_{r_q}$
(b):$s_{i_q} = -\lambda_{i_q}(A_{r_q})$ for $i = 1 : r_q$
(c):Construction of the matrices V_{r_q} and Z_{r_q} by the rational-Krylov based method knowing that:
(d):$V_{r_q} = Span(s_{1_q}I_q - A_q)^{-1}B_q, \ldots, (s_{r_q}I_q - A_q)^{-1}B_q$
(e):$Z_{r_q} = Span(s_{1_q}I_q - A_q)^{-T}C_q^T, \ldots, (s_{r_q}I_q - A_q)^{-T}C_q^T$
With $Z_{r_q}^T V_{r_q} = I_{r_q}$, where $Z_{r_q} = (Z_{r_q}^T * V_{r_q})^{-1} Z_{r_q}^T$
(4)/*Generate real V_{r_q} and Z_{r_q} for complex interpolation point*/
if there exist any s_{i_q} is not a real number
then $V_{r_q}(real) = real(V_{r_q})$, $V_{r_q}(imaginairy) = imag(V_{r_q})$,
$Z_{r_q}(real) = real(Z_{r_q})$, $Z_{r_q}(imaginairy) = imag(Z_{r_q})$,
$[V_{r_q}, rr_q] = QR[V_{r_q}(real), V_{r_q}(imaginairy)]$,
$[Z_{r_q}, rr_q] = QR[Z_{r_q}(real), Z_{r_q}(imaginairy)]$,
end if
(5)/*parameters of reduced model*/
$A_{r_q} = Z_{r_q} A_q V_{r_q}, \quad B_{r_q} = Z_{r_q} B_q, \quad C_{r_q} = C_q V_q, \quad D_{r_q} = D_q$
End Switch

Step 1: Choose the interpolation points for each sub-system by the use of the eigenvalues criterion (Gugercin 2008).
Step 2: Compute the V_{r_q} and Z_{r_q} bases with Dual Rational Krylov knowing that the orthogonality condition is satisfied $((Z_{r_q}^T * V_{r_q})^{-1} Z_{r_q}^T) V_{r_q} = I_{r_q})$.
Step 3: Calculate the reduced states matrices A_{r_q} and the corresponding eigenvalues. Using the mirror of these eigenvalues as interpolation points and recalculate the new bases V_{r_q} and Z_{r_q} applying again the Dual Rational Krylov method. Repeat these instructions until the satisfaction of a stopping criterion in the expansion frequency.

4.1 Numerical Example

To evaluate this approach, the same models used previously are taken with the same switching signal. fixing the largest singular value of the frequency response of each original subsystem and the reduced one, the variation of absolute error between each original subsystem and the reduced one is presented and the poles distribution of the reduces subsystems are given (Chahlaoui and Dooren 2005; Diepold and Eid 2011).

The Fig. 11 presents the largest singular value of the frequency response of the original switched linear system (order 1006) and reduced one (order 10) to a frequency range by IDRK-SLS method. Note that when a correlation over the entire frequency range shape with a low error rate. The Fig. 12 shows the variation of the singular value of the absolute error between the original switched linear system and

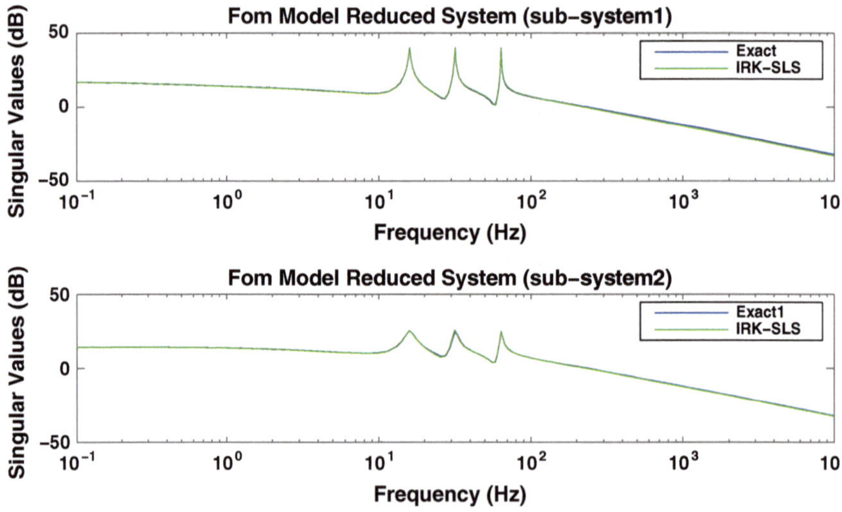

Fig. 11 Largest singular value of the frequency response of the original switched system of order (1006) and reduced one of order (10) to a frequency range with IDRK-SLS method

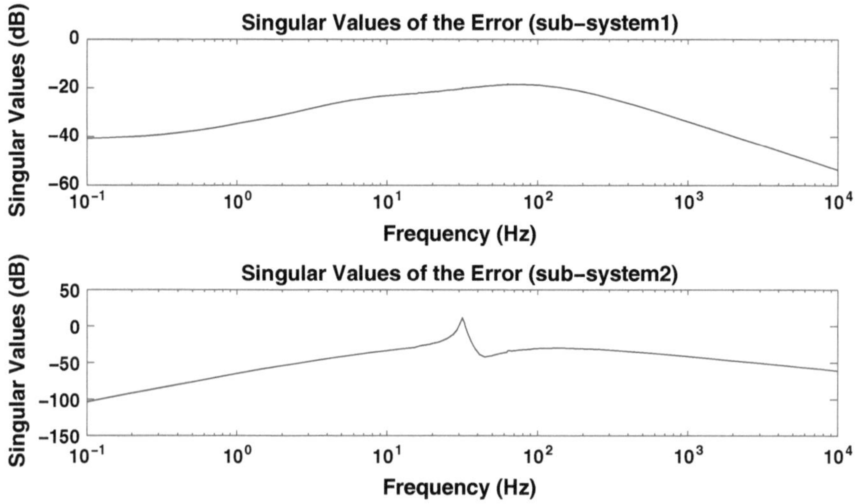

Fig. 12 Absolute error system between original switched system (FOM of order 1006) and the reduced one of order (10) with IDRK-SLS method

the reduced one, note that the error is small over the entire frequency range. The distribution poles in the complex plane of each subsystem is depicts in Fig. 13, all poles are negative real part, then the subsystems are stable.

The Fig. 14 presents the largest singular value of the frequency response of the original switched linear system (order 348) and reduced one (order 24) to a frequency range by IDRK-SLS method. When a good correlation between the original

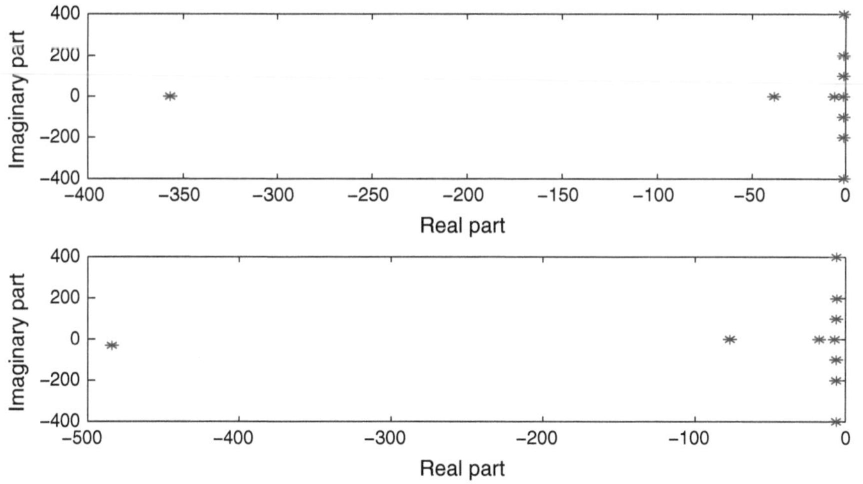

Fig. 13 Poles distribution of FOM reduced switched system (order 10) with IDRK-SLS method

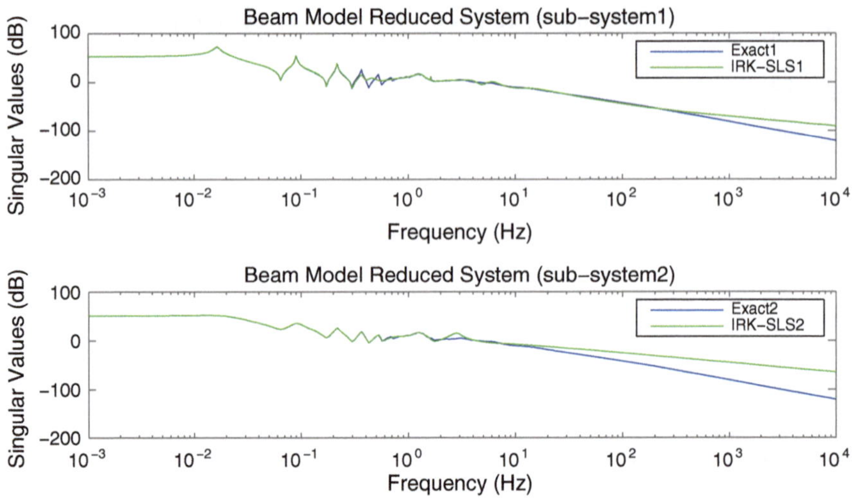

Fig. 14 Largest singular value of the frequency response of the original switched system (beam of order 348) and reduced one of order (24) to a frequency range with DRK-SLS method

switched linear system and reduced one over the entire low frequency range shape with a low error rate.

The Fig. 15 shows the variation of the singular value of the absolute error between the original switched linear system and the reduced one, note that the error is small over the entire low frequency range. The distribution poles in the complex

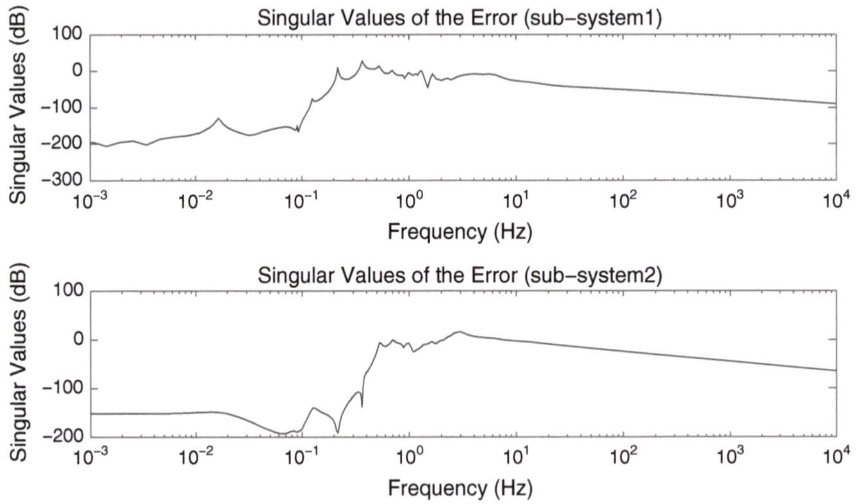

Fig. 15 Absolute error system between original switched system (beam of order 348) and the reduced one of order (24) with DRK-SLS method

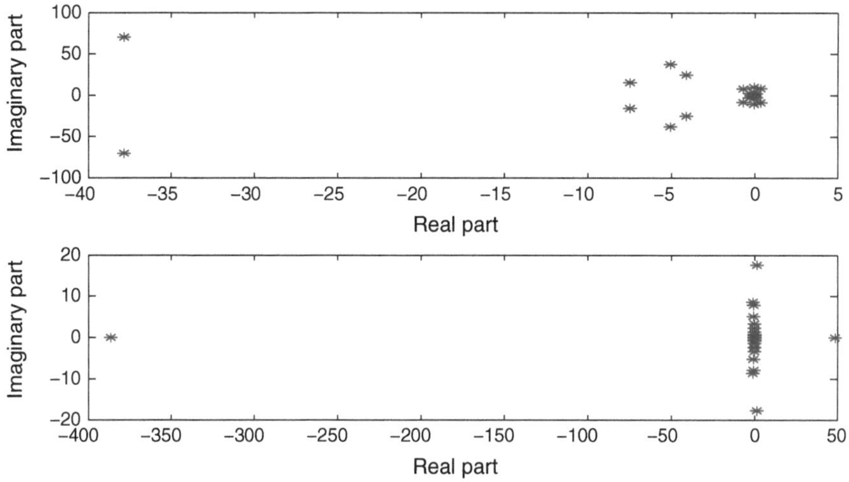

Fig. 16 Poles distribution of beam reduced switched system (order 24) with DRK-SLS method

plane of each subsystem is depicts in Fig. 16, noting the existence of positive real part poles, then the subsystems are unstable.

It is obvious the results obtained by this method is better than that obtained by the previous method, but this method does not guarantee the stability of reduced system. To solve this problem, a new method that minimizes the H_∞ error between the original switched linear system and the reduced one and guarantee the stability of reduced switched system is proposed in the next section.

5 Iterative SVD-Dual Rational Krylov for Switched Linear Systems

While IDRK-SLS algorithm do not always guarantee stability of the each reduced subsystem, Iterative SVD-Dual Rational Krylov algorithm for linear switched system gives a reduced model with guaranteed stability and minimize the error between the original system and reduced one for each sub-system. Hence, Iterative SVD-Dual Rational Krylov algorithm for linear switched system combines the advantages of the dual rational Krylov based method and the singular value decomposition based method, the use of SVD provide stability for reduced system. This method can generate two matrices, one matrix generated by the Dual Rational Krylov method (V_{r_q}) depends on the observability gramian and the other generated by the singular value decomposition (Z_{r_q}). The two matrices Z_{r_q} and V_{r_q} satisfies the following orthogonality relation (Gugercin 2008; Gugercin and Antoulas 2006; Gugercin et al. 2003; Quarteroni et al. 2007):

$$Z_{r_q}^T V_{r_q} = I_r \tag{23}$$

Theorem 2 Take a stable switched linear system Σ_q with the transfer functions $f_q(s)$ as in (1) and fix the the interpolation points s_{i_q}, Let Σ_{r_q} be an *rth* reduced subsystems with transfer functions $f_{r_q}(s)$ having fixed stable reduced poles $\lambda_{1_q}, \ldots, \lambda_{r_q}$. Then the error between each original subsystem and reduced one is minimized if and only if (Grimme 1997):

$$f_q(s) = f_{r_q}(s) \quad for \quad s = -\lambda_{1_q}, \ldots, -\lambda_{r_q} \tag{24}$$

The reduced order model is defined by these relationships:

$$A_{r_q} := Z_{r_q}^T A_q V_{r_q}, \quad B_{r_q} := Z_{r_q}^T B_q, \quad C_{r_q} := C_q V_{r_q}; \quad D_{r_q} := D_q. \tag{25}$$

The different steps of the Iterative SVD-Dual Rational Krylov Algorithm for switched linear system can be found in Table 3 (Gugercin 2008; Lee et al. 2006; Quarteroni et al. 2007):

The main steps of Iterative SVD-Dual Rational Krylov algorithm for switched linear system are:

Step 1: Choose the interpolation points for each subsystem by the use of criterion poles, the number of interpolation points must be equal to the order of reduced subsystem.

Step 2: Use the based method of dual rational Krylov for constructing the V_{r_q} basis knowing that the orthogonality condition is satisfied $(V_{r_q}^T V_{r_q} = I_{r_q})$.

Step 3: Calculate the gramian matrix of observability g_{o_q} for each subsystem.

Step 4: Construct the matrix Z_{r_q} using the observability matrix gramian and the matrix V_{r_q}.

Step 5: Calculate the reduced matrices states A_{r_q} and the corresponding eigenvalues. Determine the eigenvalues of these matrices to re-initialize the interpolation points of each subsystem. Then, recalculate the orthonormal basis V_{r_q} and the matrix Z_{r_q}. Repeat these instructions until the satisfaction of a stoping criterion in the interpolation points.

The stoping criterion is a tolerance which was set at the beginning, that denote the relative change between two successive interpolation points.

Step 6: Generate the real orthogonal matrices V_{r_q} using the reduced *QR* factorization if there exists any complex interpolation points.

Step 7: The reduced order model is defined as:

$$A_{r_q} = Z_{r_q}^T A_q V_{r_q}, \quad B_{r_q} = Z_{r_q}^T B_q, \quad C_{r_q} = C_q V_{r_q}, \quad D_{r_q} = D_q.$$

Table 3 Iterative SVDDRK-SLS Algorithm

Iterative SVDDRK-SLS Algorithm: (input:$I_q, A_q, B_q, C_q, D_q, S_q, tol$; output:$V_{r_q}, Z_{r_q}$)
Switch q
1/*Choose the initial interpolation points*/
s_{i_q} for $i_q = 1$ to r_q
2/*Construction of the matrices V_q by the dual rational-Krylov based method knowing that*/
(a):$V_{r_q} = Span(s_{1_q} I_q - A_q)^{-1} B_q, \ldots, (s_{r_q} I_q - A_q)^{-1} B_q$
With $V_{r_q}^T V_{r_q} = I_{r_q}$
3/*Calculate the garamian matrix of observability for each subsystem*/
(a)$g_{o_q} = \int_0^\infty e^{tA_q^T} C_q^T C_q e^{tA} dt$
4/*Construction of the matrix Z_q by the SVD based method knowing that*/
(b):$Z_{r_q} = Q_q V_{r_q} (V_{r_q}^T)$
5/*While (the relative change in s_i : $((s_{i+1} - s_i)/s_i) \geq tol$
(a):$A_{r_q} = Z_{r_q}^T A_q V_{r_q}$
(b):$s_{i_q} = -\lambda_{i_q}(A_{r_q})$ for $i_q = 1{:}r_q$
(c):Construction of the matrix V_q by the rational-Krylov based method knowing that:
(d):$V_{r_q} = Span(s_{1_q} I_q - A_q)^{-1} B_q, \ldots, (s_{r_q} I_q - A_q)^{-1} B_q$
With $Z_{r_q}^T V_{r_q} = I_{r_q}$
(e):*Construction of the matrix Z_{r_q} by the SVD based method knowing that*/
$Z_{r_q} = Q_q V_{r_q} (V_{r_q}^T)$
6/*Generate real V_{r_q} and Z_{r_q} for complex interpolation point*/
if there exist any s_{i_q} is not a real number
then $V_{r_q}(real) = real(V_{r_q})$, $V_{r_q}(imaginairy) = imag(V_{r_q})$,
$Z_{r_q}(real) = real(Z_{r_q})$, $Z_{r_q}(imaginairy) = imag(Z_{r_q})$,
$[V_{r_q}, rr_q] = QR[V_{r_q}(real), V_{r_q}(imaginairy)]$,
$[Z_{r_q}, rr_q] = QR[Z_{r_q}(real), Z_{r_q}(imaginairy)]$,
end if
67*Parameters of reduced model*/
$A_{r_q} = Z_{r_q}^T A_q V_{r_q}, \quad B_{r_q} = Z_{r_q}^T B_q, \quad C_{r_q} = C_q V_{r_q}, \quad D_{r_q} = D_q$
End Switch

5.1 Numerical Example

To evaluate this approach the same model used previously is taken with the same switching signal. Given the largest singular value of the frequency response of each original subsystem and the reduced one, the variation of absolute error between each original subsystem and the reduced one and the poles distribution of the each reduced subsystem (Chahlaoui and Dooren 2005; Diepold and Eid 2011).

The Fig. 17 presents the largest singular value of the frequency response of the original switched linear system (order 1006) and reduced one (order 10) to a frequency range by Iterative SVDDRK-SLS method.

Note that when a correlation over the entire frequency range shape with a low error rate. The Fig. 18 shows the variation of the singular value of the absolute error between the original switched linear system and the reduced one, note that the error is inconsiderable over the entire frequency range.

The distribution poles in the complex plane of each subsystem is depicted in Fig. 19, all poles are negative real part, then the subsystems are stable.

The Fig. 20 presents the largest singular value of the frequency response of the original switched linear system (order 348) and reduced one (order 24) to a frequency range by Iterative SVDDRK-SLS method.

When a correlation over the entire frequency range shape with a low error rate. The Fig. 21 shows the variation of the singular value of the absolute error between the original switched linear system and the reduced one, note that the error is inconsiderable over the entire frequency range.

The Fig. 22 presents the distribution poles in the complexes plane of each subsystems, all poles are negative real part, then the subsystems are stable.

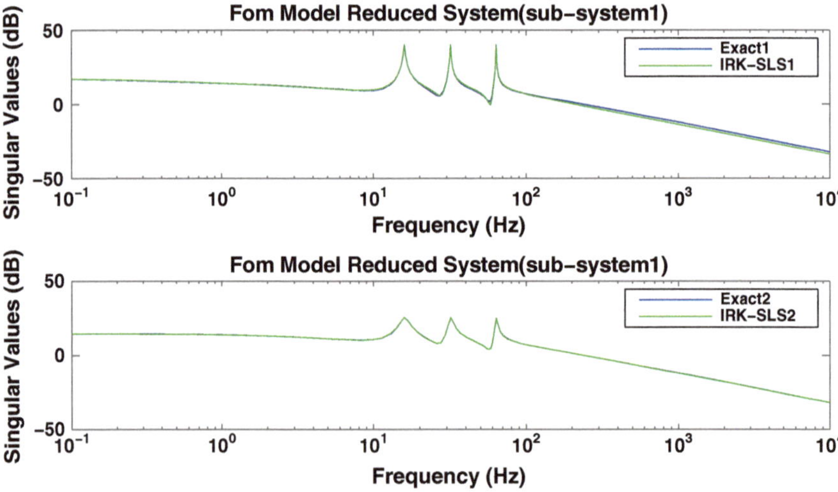

Fig. 17 Largest singular value of the frequency response of the original switched system of order (1006) and reduced one of order (10) to a frequency range with iterative SVDDRK-SLS method

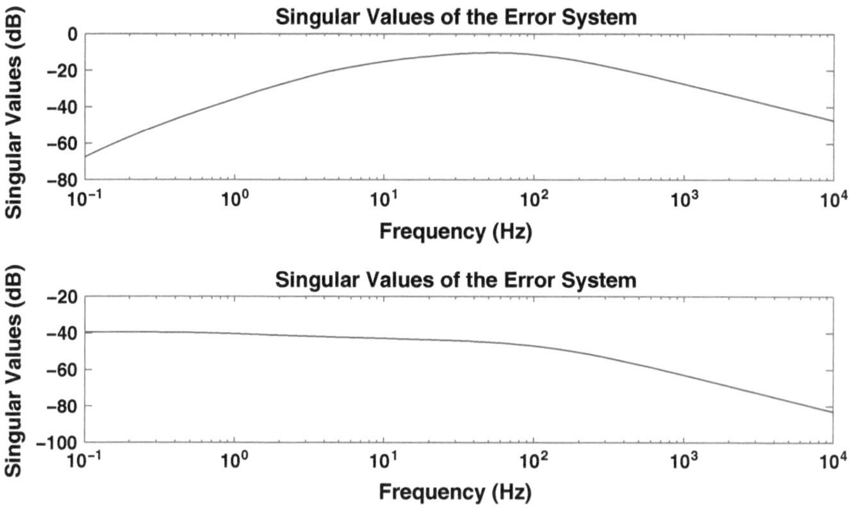

Fig. 18 Absolute error system between original switched system of order (1006) and the reduced one of order (10) with iterative SVDDRK-SLS method

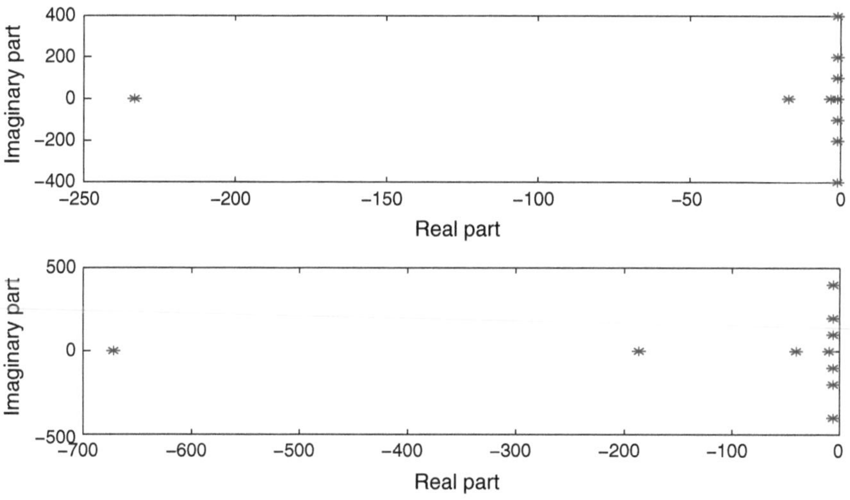

Fig. 19 Poles distribution of FOM reduced switched system (order 10) with iterative SVDDRK-SLS method

In order to prove the efficiency of the proposed iterative SVD-dual Krylov rational method, a comparative study between the tree methods will be presented in the next section.

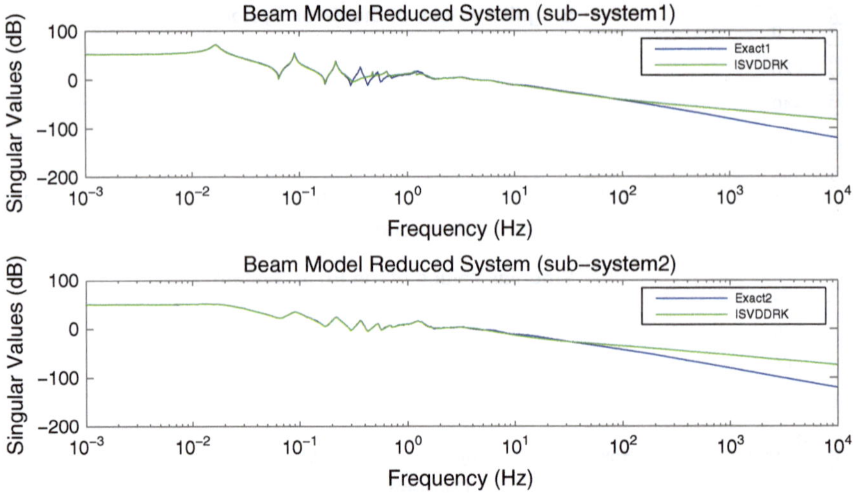

Fig. 20 Largest singular value of the frequency response of the original switched system (beam of order 348) and reduced one of order (24) to a frequency range with DRK-SLS method

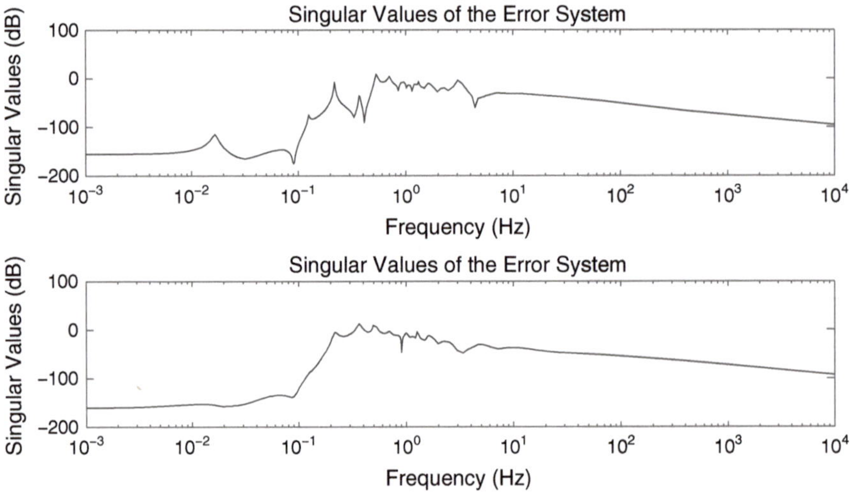

Fig. 21 Absolute error system between original switched system (beam of order 348) and the reduced one of order (24) with DRK-SLS method

6 Comparative Study

In this section, the Iterative Dual Rational Krylov method is compared with Iterative SVD-Dual Rational Krylov method for switched linear system. At first, the largest singular value of the frequency response of the original system and of the

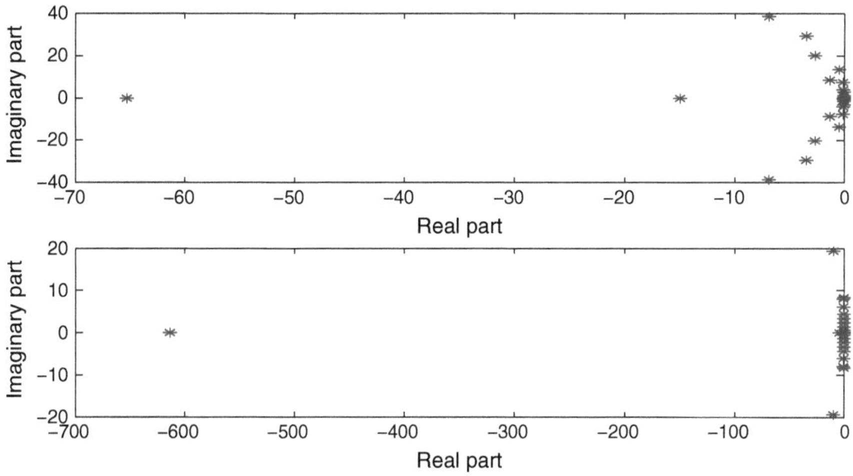

Fig. 22 Poles distribution of beam reduced switched system (order 24) with DRK-SLS method

Table 4 H_∞ error, CPU-time and the tolerance of iterative DRK-SLS and Iterative SVDDRK-SLS

Models	Methods	H_∞ Error	CPU-time (s)	Tol
FOM 1006	Iterative DRK-SLS	6.207×10^{-9}	127	10^{-3}
FOM 1006	Iterative SVDDRK-SLS	1.877×10^{-9}	110	10^{-9}
Beam 348	Iterative DRK-SLS	1.545×10^{-8}	30	5×10^{-3}
Beam 348	Iterative SVDDRK-SLS	1.081×10^{-8}	55	2×10^{-2}

reduces systems using the two proposed methods are presented, also the variation of the absolute error between the original system and reduces systems are given. The Table 4 contains the H_∞ error, the CPU-time and the tolerance of each model. From these figures and this table, note that the iterative SVDDRK-SLS has a better results compare to IDRK-SLS. Figure 23 shows the largest singular values of the frequency response of the original switched linear system of order (1006) and reduced one of order (10) to a frequency range using the IDRK-SLS and ISVDRK-SLS methods, note a good correlation between the responses obtained by the reduces systems and the original system, to understand the efficiency of one relative to the other in the process of comparing the curves of absolute error variation.

The Fig. 24 presents the variation of the absolute error between the original system (FOM of order 1006) and reduces systems (of order 10), note that the best result is obtained by the ISVDDRK-SLS method. Figure 23 presents the largest singular values of the frequency response of the original switched linear system of order (348) and reduced one of order (24) to a frequency range using the IDRK-SLS and ISVDRK-SLS methods. A good correlation over the whole frequency range between the responses is obtained by the reduced system and the original one

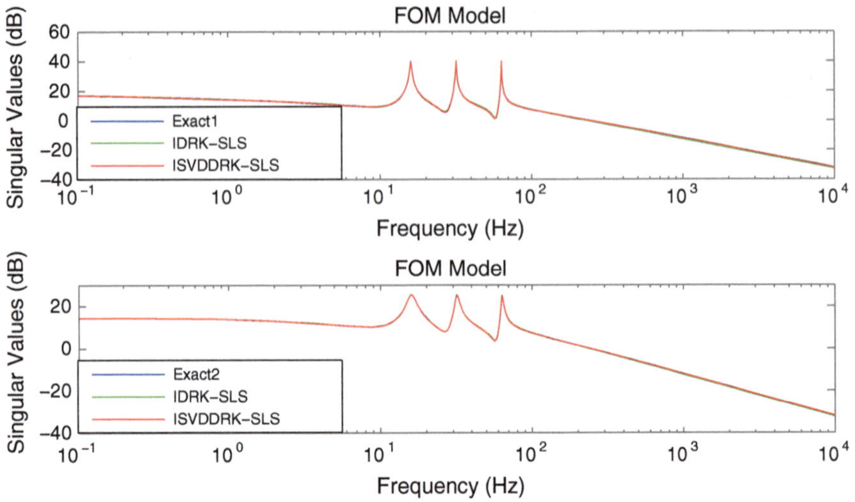

Fig. 23 Largest singular value of the frequency response of the original switched system of order (1006) and reduces systems of order (10) to a frequency range with IDRK-SLS and ISVDDRK-SLS methods

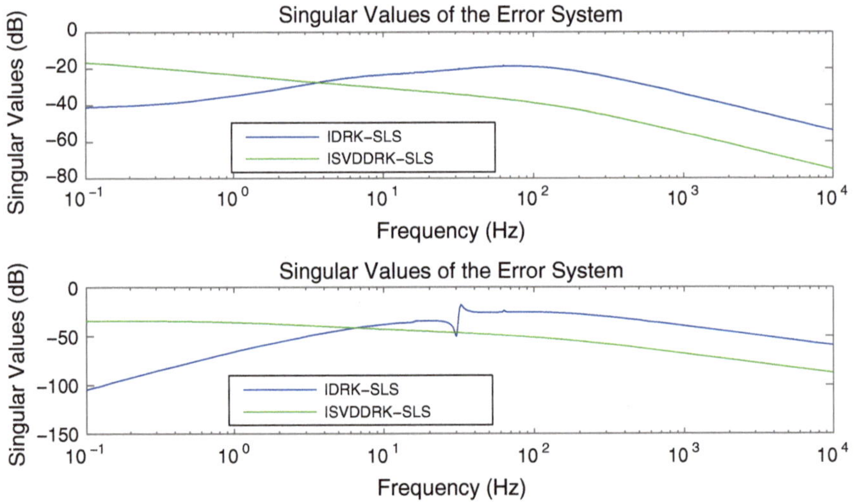

Fig. 24 Absolute error system between original switched system (FOM of order 1006) and the reduces systems of order (10) with IDRK-SLS and ISVDDRK-SLS methods

using the ISVDDRK-SLS method, which is not the case for the high frequency using the IDRK-SLS method, hence the performance of the ISVDDRK-SLS method (Fig. 25).

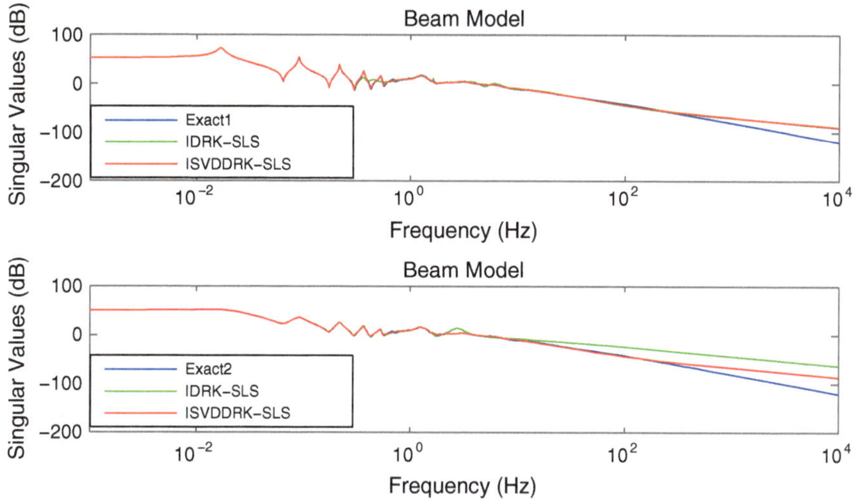

Fig. 25 Largest singular value of the frequency response of the original switched system of order (348) and reduces systems of order (24) to a frequency range with IDRK-SLS and ISVDDRK-SLS methods

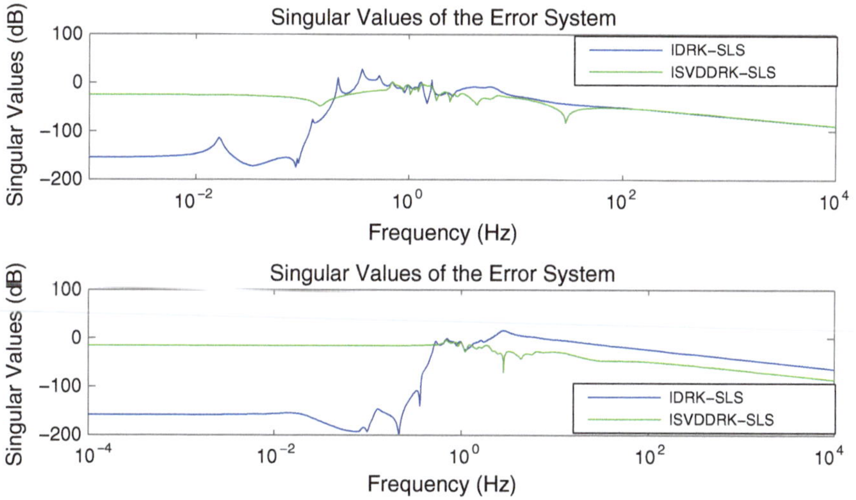

Fig. 26 Absolute error system between original switched system (beam of order 348) and the reduces systems of order (24) with IDRK-SLS and ISVDDRK-SLS methods

The Fig. 26 presents the variation of the absolute error between the original system (BEAM of order 348) and reduces systems (of order 24), note that the best result is obtained by the ISVDDRK-SLS method.

Table 4 contains a comparison between the iterative DRK-SLS and iterative SVDDRK-SLS methods, determining for each method the H_∞ Error, the CPU-Time and the Tolerance(Tol) for both models. Seeing the values of H_∞ error for two models obtained by the iterative SVDDRK-SLS method are better then obtained by the iterative DRK-SLS method, also be seen that the CPU-time and the tolerance differs from one method to another, they are mainly dependent on the order model.

7 Conclusion

In this chapter, two methods for reduction of linear switched systems have been proposed. At first the iterative dual rational Krylov method based on generation of Krylov subspaces is presented. This method have low cost, but the stability of reduced system not always guaranteed.

In the second part, the iterative SVD-Dual rational Krylov based on the SVD and Krylov subspace techniques in generating of the projection matrices V_{r_q} and Z_{r_q} for each subsystem is presented. This method is numerically efficient using the Krylov technique and guaranteed the stability of each reduced subsystems using the observability matrix obtained from the Lyaponuv equation.

To evaluate the accuracy and efficient of these methods, a numerical examples is presented.

As a future works the development and validation of control algorithms for switched linear system based on reduces models obtained by Iterative DRK-SLS and Iterative SVDDRK-SLS methods is recommended. The implementation of the control algorithm determined from the reduced systems obtained by the dual rational Krylov methods on a microcontroller to control the original switched system is proposed. Addaptation of the previous methods to be applied on non-linear systems and on the other hybrid systems is also proposed.

References

Andres, L., Diego, P., RafaelGa, D., & Briel, P. (2013). An equivalent continuous model for switched systems. *Systems and Control Letters, 62*(2), 124–131.

Antoulas, A. C. (2009). *Approximation of large-scale dynamical systems(ed. 1)* Advances in Design and Control Series, Society for Industrial and Applied Mathematics.

Antoulas, A. C., Sorensen, D., & Gugercin, S. (2001). A survey of model reduction methods for large-scale systems. *Contemporary Mathematics, 280*, 193–219.

Awais, M. M., Shamail, S., & Ahmed, N. (2007). Dimensionally reduced Krylov subspace model reduction for large scale systems. *Applied Mathematics and Computation, 191*(1), 21–30.

Bao, L., Lin, Y., & Wei, Y. (2006). Krylov subspace methods for the generalized sylvester equation. *Applied Mathematics and Computation, 175*(1), 557–573.

Benner, P., Mehrmann, & Sorensen, D. C. (2003). *Dimension reduction of large-scale systems.* Proceedings of a workshop held in Oberwolfach, Germany. Number 67–78 in Springer.

Chahlaoui, Y., & Dooren, P. V. (2005). *A collection of benchmark examples for model reduction of linear time invariant dynamical systems.* Technical report, Manchester Institute for Mathematical Sciences School of Mathematics.

Diepold, K. J., & Eid, R. (2011). *Guard-based model order reduction for switched linear systems*. Methoden und Anwendungen der Regelungstechnik. Erlangen-MÃ¼nchener Workshops. Number 67–78 in Shaker-Verlag.

Dongmei, X., Ning, X., & Chen, X. (2008). LMI Approach to H_2 model reduction for switched systems. *The 7th World Congress on Intelligent Control and Automation*, pp. 6381–6386. June 25–27 2008, Chongqing. doi: 10.1109/WCICA.2008.4593893.

Druskin, V., & Simoncini, V. (2011). Adaptive rational Krylov subspaces for large-scale dynamical systems. *Systems and Control Letters, 60*(8), 546–560.

Flagg, G., Beattie, C., & Gugercin, S. (2012). Convergence of the iterative rational Krylov algorithm. *Systems and Control Letters, 61*(6), 688–691.

Gallivan, K., Grimme, E., & Dooren, P. V. (1996). A rational lanczos algorithm for model reduction. *Numerical Algorithms, 12*(1), 33–63.

Gaoa, H., Lamb, J., & Wanga, C. (2006). Model simplification for switched hybrid system. *Systems and Control Letters, 55*(12), 1015–1021.

Grimme, E. J. (1997). Krylov Projection Methods For Model Reduction. (PhD thesis, University of Illinois at Urban Champaign).

Gugercin, S. (2008). An iterative svd-Krylov based method for model reduction of large-scale dynamical systems. *Linear Algebra and its Applications, 428*(8–9), 1964–1986.

Gugercin, S., & Antoulas, A. C. (2006). Model reduction of large scale systems by least squares. *Linear Algebra and its Applications, 415*(2–3), 290–321.

Gugercin, S., Sorensen, D. C., & Antoulas, A. C. (2003). A modified low-rank smith method for large-scale lyapunov equations. *Numerical Algorithms, 32*(1), 27–55.

Heyouni, M., & Jbilou, K. (2006). Matrix Krylov subspace methods for large scale model reduction problems. *Applied Mathematics and Computation, 181*(2), 1215–1228.

Kouki, M., Abbes, M., & Mami, A. (2013a). Arnoldi model reduction for switched linear systems.

Kouki, M., Abbes, M., & Mami, A. (2013b). Lanczos model reduction for switched linear systems.

Kouki, M., Abbes, M., & Mami, A. (2013c). A survey of linear invariant time model reduction. *ICIC Express Letters*, An International Journal of Research and Surveys, 7(3(B)): 909–916.

Kouki, M., Abbes, M., & Mami, A. (2014a). Non symmetric and global lanczos model reduction for switched linear systems. *International Journal of Mathematics and Computers in Simulation, 8*, 67–72.

Kouki, M., Abbes, M., & Mami, A. (2014b). Rational arnoldi & adaptive order rational arnoldi for switched linear systems. *Neural, Parallel, and Scientific Computations, 22*, 75–88.

Lee, H. J., Chu, C. C., & Feng, W. S. (2006). An adaptive-order rational arnoldi method for model-order reductions of linear time-invariant systems. *Linear Algebra and its Applications, 415*(2–3), 235–261.

Mehrmann, V., Schroder, C., & Simoncini, V. (2012). An implicitly-restarted Krylov subspace method for real symmetric/skew-symmetric eigenproblems. *Linear Algebra and its Applications, 436*(10), 4070–4087.

Mignone, D., Ferrari-Trecate, G., & Morari, M. (2000). Stability and stabilization of piecewise affine and hybrid systems. In the 39th IEEE Conference on Decision and Control, (pp. 504–509) Sydney, Australia. doi: 10.1109/CDC.2000.912814.

Quarteroni, A., Sacco, R., & Saleri, F. (2007). *Methodes Numeriques: Algorithmes, analyse et applications* (Vol. 538). Milano: Springer.

Tulpule, P., Yurkovich, S., Wang, J., and Rizzoni, G. (2011). Hybrid large scale system model for a dc microgrid. In *American Control Conferences*, June 29 2011–July 1 2011, San Francisco, CA, pp. 3900–3904.

Zhanga, L., Shi, P., Boukasc, E., & Wanga, C. (2008). H_∞ model reduction for uncertain switched linear discrete-time systems. *Automatica, 44*(11), 2944–2949.

Zhendong, S., & Shuzhi, S. G. (2009). *Switched linear systems: Control and design*. Berlin: Springer.

Household Electrical Consumptions Modeling and Management Through Neural Networks and Fuzzy Logic Approaches

Lucio Ciabattoni, Massimo Grisostomi, Gianluca Ippoliti and Sauro Longhi

Abstract In recent years the European Union and, moreover, Italy has seen a rapid growth in the photovoltaic (PV) sector, following the introduction of the feed in tariff schemes. In this scenario, the design of a new PV plant ensuring savings on electricity bills is strongly related to household electricity consumption patterns. This chapter presents a high-resolution model of domestic electricity use, based on Fuzzy Logic Inference System. The model is built with a "bottom-up" approach and the basic block is the single appliance. Using as inputs patterns of active occupancy and typical domestic habits, the fuzzy model give as output the likelihood to start each appliance within the next minute. In order to validate the model, electricity demand was recorded over the period of one year within 12 dwellings in the central east coast of Italy. A thorough quantitative comparison is made between the synthetic and measured data sets, showing them to have similar statistical characteristics. The focus of the second part of this work is to develop a neural networks based energy management algorithm coupled with the fuzzy model to correctly size a residential photovoltaic plant evaluating the economic benefits of energy management actions in a case study. A cost benefits analysis is presented to quantify its effectiveness in the new Italian scenario and the evaluation of energy management actions.

L. Ciabattoni (✉) · M. Grisostomi · G. Ippoliti · S. Longhi
Dipartimento Di Ingegneria Dell'Informazione, Università Politecnica Delle Marche,
Via Brecce Bianche 12, 60131 Ancona, Italy
e-mail: l.ciabattoni@univpm.it

M. Grisostomi
e-mail: m.grisostomi@univpm.it; massimo.grisostomi@metasistemi.it

G. Ippoliti
e-mail: gianluca.ippoliti@univpm.it

S. Longhi
e-mail: sauro.longhi@univpm.it

Q. Zhu and A.T. Azar (eds.), *Complex System Modelling and Control Through Intelligent Soft Computations*, Studies in Fuzziness and Soft Computing 319,
DOI 10.1007/978-3-319-12883-2_16

1 Introduction

The rapid depletion of conventional energy sources and the ever-increasing demand for more energy coupled with the focus on environmental issues has encouraged intensive research into new sources of energy and clean fuel technologies that utilize the latest technology. Most renewable sources use wind, micro-hydro, tidal, geothermal, biomass and solar energy. This energy is then converted into electrical energy to be delivered either to the utility grid directly or to isolated loads. From ancient times, the human race has harnessed solar energy, radiant light and heat from the sun using a range of different technologies. Some modern solar energy technologies include solar heating, solar photovoltaic, solar thermal and solar architecture. These methods can have the potential to make a significant contribution to resolving the pressing energy problems that the world faces.

Photovoltaic (PV) systems and some other renewable energy systems (such as wind, tidal, waves, geothermal) are excellent choices in remote areas for low to medium power levels due to the easy scaling of the input power source (e.g. the use of solar inverters). The main attraction of the PV systems is that they produce electric power without harming the environment by directly transforming the free inexhaustible solar energy into electricity. Distributed grid-connected photovoltaics (PV) is playing an increasingly significant role as an electric supply resource and as an integral part of the electrical grid, due to the continual decrease in costs and the increase in their efficiency. Although inferior to other technologies in terms of installed capacity, PV is currently the most important Distributed Generation (DG) technology all over the world, due to financial support from the government (Timilsina et al. 2012; Yang 2010).

As is well known, electricity systems can benefit from the integration of small-scale PV-DG. For instance, since distributed generation produces electricity where needed, it helps reducing the electric load on transmission lines and the need for costly new lines associated with new power plants far from towns and cities. However, PV poses notable challenges to grid engineers, planners and operators.

Sometimes and especially when having high penetration of PV in parts of the distribution system dominated by residential end-users, the amount of power generated by the PV may exceed the total demand being served by a given part of the distribution system. In those circumstances, "excess" power can have a dramatic effect on the electric service voltage.

Another effect is known as "back-flow". This entails the current flow from the "low voltage side" of electrical transformers (also known as the transformers' secondary side) to the higher voltage side (also known as the transformers' primary side). This challenge tends to be more common in parts of the distribution system that serve primarily residential end-users, because demand in those parts of the grid tend to be relatively low during the day (residents may be at work or school).

In this scenario the modeling of residential energy use and the planning of energy management actions can play a crucial role (Ciabattoni et al. 2013b). The pattern of electricity use for any individual domestic dwelling is highly dependent upon the activities of the occupants and their associated use of electrical appliances. In this chapter we present a high-resolution model of domestic electricity use, based upon a combination of patterns of active occupancy and daily activity profiles (typical appliances usage frequency and starting time). The model is built using a "bottom-up" approach, according to Richardson et al. (2010). The basic building block is the appliance, i.e. any individual domestic electric load. The model, managing the start of each appliance in the household through a fuzzy logic inference system, gives as output the 1 min resolution electricity usage pattern. All data necessary to build the fuzzy inference system are obtained from 2 weeks of measures through wireless smart plugs installed in the households appliances with an automatic procedure. Fuzzy sets and rules are determined with an automatic procedure analyzing sensors measures. In order to validate the model, electricity demand was recorded over the period of a year within 12 dwellings in the central east coast of Italy. A through quantitative comparison is made between the synthetic and measured data sets, showing them to have similar statistical characteristics.

The problem of household energy management has been discussed and a possible solution presented through neural network based forecasts of consumption and PV production used to inform and influence prosumers on the way they use electricity to increase the amount of self consumed energy.

The fuzzy model has been used for a case study on the proper sizing of a PV plant (Benghanem and Mellit 2010; Jakhrani et al. 2012; Jallouli and Krichen 2012; Kaabeche et al. 2011) in the central east region of Italy and the evaluation of Energy Management potential benefits based on a costs benefits analysis (CBA). The installation in a dwelling of all the devices necessary to actuate proper EM policies has a relatively high cost compared to that of a PV system (Di Giorgio et al. 2012; Sawyer et al. 2009). The focus of this analysis is to set an upper limit for the equipment cost in order to obtain real savings for a specific household through the CBA.

The chapter is organized as follows. An overview of the related works appears in the second section. A brief introduction on the fuzzy inference system modeling is reported in the third section. The structure of the model, a human interaction based classification of the appliances into different categories, a sample of the rule set, the National Instruments Labview software implementation details are reported in the fourth one. Model validation results are given in Section five, where the simulator output is compared with one year data sets recorded from 12 dwellings in the central east coast of Italy. In the sixth section energy management problem and neural network based forecasting algorithms for both photovoltaic production and home consumptions are described. In Section seven is presented the application of the FIS consumption simulator for the PV optimal sizing and energy management benefits evaluation in a case study.

2 Related Work

The analysis and identification of energy consumption pattern nowadays is receiving strong interest together with fault diagnosis of appliances components and there have been a large number of researches in this area (Ferracuti et al. 2013a, b; Ihbal et al. 2011; Zaidi et al. 2010; Zia et al. 2011).

Another related research field regards the forecast and simulation of households' electricity consumption patterns, see, e.g., Azadeh et al. (2008), Barbato et al. (2011), Ciabattoni et al. (2013c), Gruber and Prodanovic (2012), Murata and Onoda (2002), Osman et al. (2009), Subbiah et al. (2013). Most of the existing models and analysis focus on data from specific geographic regions and try to explain the results in a local perspective (Guo et al. 2011; Suh et al. 2012).

Photovoltaic sizing is an important research field in this area but most of the works concern with the optimization of stand alone systems without an analysis of the demand response scenario for grid connected users, see e.g. Benghanem and Mellit (2010), Ciabattoni et al. (2013a), Jakhrani et al. (2012), Jallouli and Krichen (2012), Kaabeche et al. (2011). In this scenario only the knowledge of the typical demand pattern for each household will make possible the proper sizing of a photovoltaic plant, the design of demand response techniques and energy management actions. The pattern of electricity use for any individual domestic dwelling is highly dependent upon the activities of the occupants and their associated use of electrical appliances.

Energy usage models developed in literature e.g. in Bernard et al. (2011), Richardson et al. (2010) are configured using statistics describing mean total annual energy demand and associated power use characteristics of household appliances. Furthermore these modeling approaches (Bernard et al. 2011; Richardson et al. 2010; Widen et al. 2009) concern specific household energetic behavior without an easy customization capability. It is often impossible to add every appliance and predispose a "seasonal behavior" (Bernard et al. 2011), (Richardson et al. 2010) without using the flexibility of a fuzzy inference systems, as proposed in this work.

It is well known that overall cost-saving by distributed generation would only have a marginal impact if the demand pattern does not match with the production one and no actions of energy management are performed. In this scenario only the knowledge of the typical demand pattern and the forecast of the generation pattern for each household will make possible the design of proper demand response techniques and the planning of energy management actions. In this context energy management for residential consumers has become a significant research and development field for both electrical (Ciabattoni et al. 2013d) and thermal side (Giantomassi et al. 2014a, b), as a result of the advances in the electrical power grid technologies and the high penetration of solar, wind and other forms of Distributed Generation (DG) (Ciabattoni et al. 2012, 2013e; Cimini et al. 2013; Kanchev et al. 2011). Less attractive feed-in-tariffs for new installations of renewable energy DGs (solar, wind and geothermal plants) and incentives to promote self-consumption suggest that new operation modes should be explored in order to reach grid parity, which has been predicted to become a reality in the next years in the European Union (Fazeli et al. 2011; Kanchev et al. 2011; Palensky and Dietrich 2011; Zong et al. 2012). By increasing the self consumed local generated energy, the grid parity could be achieved earlier and DG of renewable energies will finally make economic sense becoming cheaper (over the lifetime of the system) than to buy it from utility (Aghaei and Alizadeh 2013; Lewis 2009; Lopez-Polo et al. 2012).

There are increasing numbers of studies on smart homes and the benefits of demand-side management (Di Giorgio et al. 2011; Shahgoshtasbi and Jamshidi 2011; Zeilinger 2011) and control and monitoring techniques to reduce overall energy usage Meyers et al. (2010).

3 Fuzzy Inference System

Fuzzy rule-based systems (FRBS) have been successfully employed for system modeling in many areas Azar (2010b). Existing fuzzy systems in the literature Azar (2010a, 2012) can be classified into three main categories: Mamdani, Takagi-Sugeno (T-S) and Tsukamoto systems based on their implemented fuzzy rule structures. Furthermore, depending on the intended application, the fuzzy modeling research field can be divided into two main approaches.

The first is the linguistic fuzzy modeling (LFM) where good human interpretability of the underlying fuzzy model is paramount for tasks such as knowledge mining and data analysis. This is usually achieved by adopting the Mamdani rule structure for knowledge representation.

The other is the precise fuzzy modeling (PFM) where T-S and Tsukamoto fuzzy rule structures are generally used in the learned fuzzy model to achieve high output accuracies for function approximation and regression-centric applications. Having good fuzzy rule-base interpretability and high modeling accuracy are contradictory requirements and one usually prevails over the other based on the modeling objective and fuzzy rule structure employed.

Generally, Mamdani fuzzy models are more interpretative than T-S fuzzy models from a human perspective and thus can better explain and describe a modeled system's behaviors.

3.1 Fuzzy Modeling

The modeling of the appliance's usage has been performed with a LFM approach to determine if wether or not it is going to be started. Since the aim of this work is to represent the household energetic behavior we choose Mamdani model, in order to give the best interpretability to the rules.

The usage pattern, depending on the appliance's category, can be related to many variables, such as the number of active people in the house, the typical frequency of the appliance, the time of the day, the temperature. For example, when people are not at home, most appliances will not be used (only the so called continuous use appliances).

In daily appliance electricity profile, the occupants use virtually little power (stand by and fridge-freezer) during the night, may wake up and have breakfast, vacate the house during the morning and then return around mid-day for lunch, e.g. starting the microwave. In the evening, the meal is cooked, television is watched, lights are on, showers are taken, etc.

This typical pattern can drastically change during the weekend and holidays (when people can be in the house mostly during daytime) and, moreover, it can change from dwelling to dwelling due to different life styles. The main factors influencing occupancy pattern and appliances usage are: the number of occupants, the time the first person gets up in the morning and last person goes to sleep, the periods house is unoccupied during work days, holidays and weekends. When analyzing the households load profile we need information on the active occupants of the dwelling. To compute the overall occupancy pattern a specified model can be used, for instance that one developed by Richardson et al. (2008).

Starting from basic information in this chapter we build a 1 minute resolution daily active occupants pattern for each day of the week. To compute the number of the busy occupants a counter is used; this counter is increased every time an appliance that requires interaction with a person is switched on, and decreased every time it is switched off. The number of unoccupied people in the dwelling can be computed from the active occupants pattern and the current value of the busy occupants counter. Knowing this value for each time of the day, we can enable or interdict the switching on of the appliance.

A further important feature is to identify the typical frequency of each appliance's starting for each household. This parameter is rarely a crisp value, e.g. "the washing machine starts usually from 2 to 3 times a week", and often related to the time of the day, e.g. "the television starts some hours a day usually at night". In this work all information regarding occupancy, appliances frequency and typical start time are taken with a brief interview. The former are used to build the active occupancy pattern and the latter to build fuzzy rules.

4 Appliances Fuzzy Inference System

The electricity consumption pattern model for any individual domestic dwelling is developed using a "bottom-up" approach, according to those proposed by Richardson et al. (2010). The basic block is the appliance, i.e. any individual domestic electric load. As it is well known, home appliances differ one from each other by size, functions, human interaction level, automation level.

4.1 Appliances Classification

In particular to build the fuzzy rules a load classification based on the human interaction has been used and four different groups found:

- Continuous use appliances, characterized by a 24/7 use, not depending on factors like the time of the day and the number of active occupants of the dwelling (e.g. refrigerator).
- Periodical use appliances without human interaction during the operation (e.g. washing machine, dishwasher, oven).
- Periodical use appliances with human interaction during the operation (e.g. vacuum cleaner, iron).
- Multimedia appliances and lighting, with a strongly intermittent use, directly related to the number of active occupants of the dwelling.

These 5 different categories of appliances have different fuzzy input-output variables. Input variables for the FIS inference are the time $h(t)$ of the day, the percentage $p(t)$ of unoccupied people in the dwelling and $DT/T(t)$ that is the time elapsed since the last appliance start normalized on his period. The outputs of the FIS engine are the probability $P(t)$ to start a certain appliance and the total time $D(t)$ the appliance will be on. In particular Table 1 contains inputs and outputs for each category.

Another classification method considered is based on the automation level, in particular we can find:

- Smart Appliances: loads for which consumption profile is available and it is possible to choose the start time (remotely or locally).
- Controllable Loads: loads which are connected to smart plugs and can be remotely switched on/off without damage and degradation of consumer quality of experience.
- Monitorable Loads: connected to smart plugs to monitor their consumptions; they can not be switched on/off.
- Detectable Loads: the consumption of which can be estimated by performing the difference among the power measures provided by the smart meter and all the smart plugs and appliances, being them not smart appliances and not connected to smart plugs.

Table 1 Fuzzy input output variables for the different appliance's categories

Category	IN	IN	IN	OUT	OUT
Continuous	–	–	–	–	–
Periodic without human	$h(t)$	$DT/T(t)$	–	$P(t)$	–
Periodic with human	$h(t)$	$DT/T(t)$	–	$P(t)$	–
Multimedia	$h(t)$	$DT/T(t)$	$p(t)$	$P(t)$	$D(t)$
Lighting	$h(t)$	$DT/T(t)$	$p(t)$	$P(t)$	–

Due to the lack of smart appliances on the market it has been necessary to configure standard appliances into monitorable loads to extract their consumption profiles. On the same time if a user plans to use energy management actions controllable loads are necessaries. In particular the appliance remote start can be performed only for some of the so called "periodical use without human interaction", due to their features.

4.2 Appliances Fuzzy Rules

The membership functions of the input variables (samples shown in Figs. 1–3) consist of triangular asymmetric and trapezoidal functions. The trapezoidal function is totally represented with four points, known also as fuzzy set: $A = (a_1, a_2, a_3, a_4)$. This representation is interpreted as membership functions:

$$\mu_A(x) = \begin{cases} 0, & x < a_1 \\ \frac{x-a_1}{a_2-a_1}, & a_1 < x < a_2 \\ 1, & a_2 < x < a_3 \\ \frac{a_4-x}{a_4-a_3}, & a_3 < x < a_4 \\ 0, & x > a_4 \end{cases} \tag{1}$$

when $a_2 = a_3$, the triangular function can be considered as a particular case of the trapezoidal one. Table 2 shows the fuzzy sets for the input variables.

Table 2 Considered fuzzy sets for input variables

	Abbreviation	a_1	a_2	a_3	a_4
h(*t*)					
Early morning	EM	0	0	300	450
Morning	M	300	400	750	800
Afternoon	A	650	750	1,000	1,150
Evening	E	1,050	1,100	1,250	1,300
Late evening	LE	1,250	1,300	1,440	1,440
DT/T(*t*)					
Very advance	VA	0	0	0.3	0.6
Advance	A	0.5	0.75	0.75	1
In time	IT	0.9	1	1	1.1
Late	L	1	1.25	1.25	1.5
Very late	VL	1.4	1.8	2	2
p(*t*)					
Very low	VL	0	0	0.2	0.4
Low	L	0.2	0.3	0.4	0.5
Medium	M	0.3	0.5	0.7	0.9
High	H	0.7	0.8	1.0	1.1
Very high	VH	1.0	1.1	*inf*	*inf*

Table 3 Dishwasher FIS sample

	VA	A	IT	L	VL
EM	VL	VL	VL	VL	VL
M	VL	VL	VL	L	L
A	VL	VL	L	L	M
E	VL	L	M	H	VH
LE	VL	VL	VL	VL	VL

Input *DT*/*T*(*t*) is in the first row, while *h*(*t*) is in the first column. Probability *P*(*t*) are the central values of the table

A sample of the fuzzy control rule base for a "Periodical use appliance without human interaction" (e.g. the dishwasher) is shown in Table 3; the Max-Min fuzzy inference algorithm is considered, (Bose 2011). The outputs of the FIS engine are the probability *P*(*t*) to start a certain appliance: (N) None, (VL) Very Low, (L) Low, (M) Medium, (H) High, (VH) Very High and the total time *D*(*t*) the appliance will be on: (VL) Very Low, (L) Low, (M) Medium, (H) High, (VH) Very High. Output membership functions, shown as example in Fig. 4, consist of sigmoid functions with different values for each appliance category.

Concerning the defuzzyfication we use the modified Center of Area defuzzyfication method since the centroid method evaluates the area under the scaled membership functions only within the range of the output linguistic variable and the resulting crisp output values could not span the full range. The fuzzy logic controller uses the following equation to calculate the geometric center of the full area under the scaled membership functions:

$$\mathrm{mCoA} = \frac{\int \mathrm{f(x)} \cdot \mathrm{xdx}}{\int \mathrm{f(x)dx}} \tag{2}$$

where mCoA is the modified center of area. The interval of integration is between the minimum membership function value and the maximum membership function value. Note that this interval might extend beyond the range of the output variable.

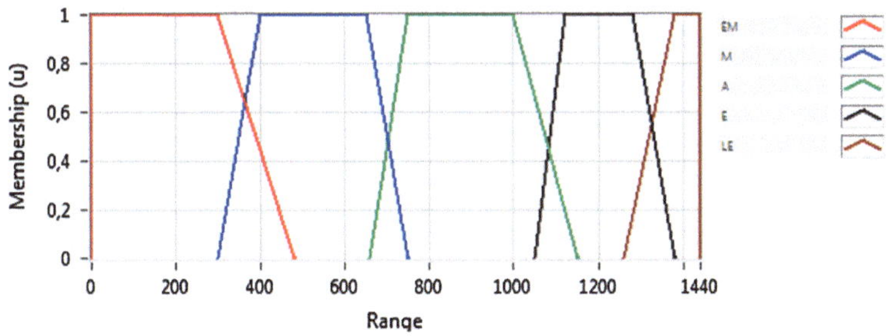

Fig. 1 Membership function of the input variable *h*(*t*). The x-axis is the time of the day in minutes

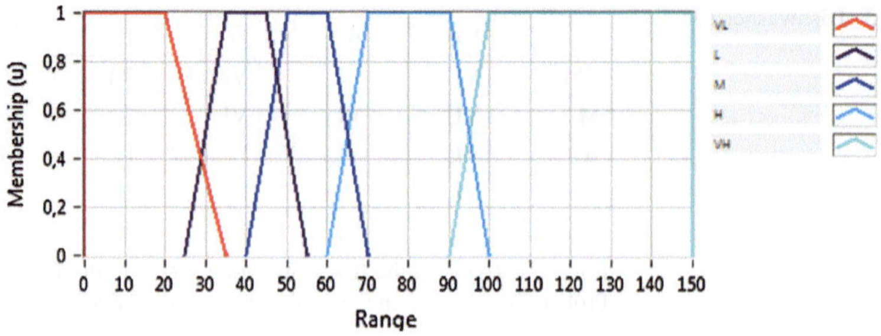

Fig. 2 Membership function of the input variable *p*(*t*). The x-axis is the percentage of occupancy of the dwelling

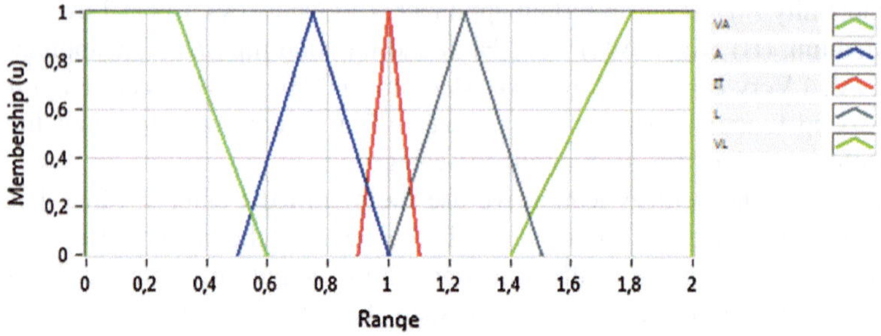

Fig. 3 Membership function of the input variable *DT*/*T*(*t*). The x-axis is the ratio between the time elapsed since the last start and the average starting period

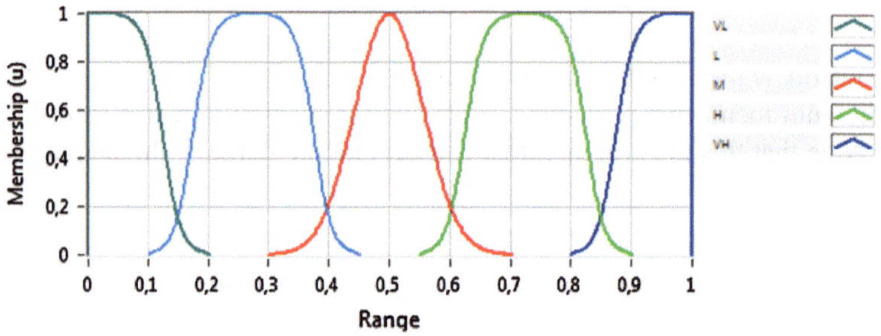

Fig. 4 Membership function of the output variable *P*(*t*). The x-axis is the probability to start an appliance

4.3 Model Implementation

The aim of the simulation tests is to evaluate the potentialities of an energy management technique applied for different households, in order to evaluate the economic benefits users can obtain.

The model has been realized using LabVIEW, the graphical programming environment of National Instruments. In particular the FIS has been realized using the LabVIEW fuzzy toolkit while the input-output membership functions and the rule set with the fuzzy system designer. As the simulator is not time driven when a simulation runs one-min resolution electricity demand data can be generated for a specified time period using two nested FOR loops (the outer for the days of the year and the inner for the minutes of each day) as shown in Fig. 5.

Each single appliance block, implemented as a functional global variable, is in the inner loop and runs in two phases. During the first iteration of the simulation all the configuration parameters are loaded, e.g. the fuzzy rule set of the appliance, the consumption profile, the maximum power, the typical starting frequency, number of people typically interacting with the appliance (all the mentioned parameters are fully editable in text files and fuzzy rules through LabVIEW graphical interface).

After the first iteration the likelihood an appliance will start within the next minute is evaluated with a time resolution of one-minute (except for the so called "Continuous use appliances"). In particular, since the FIS output is a probability value, to manage the start of an appliance this value is multiplied by a calibration

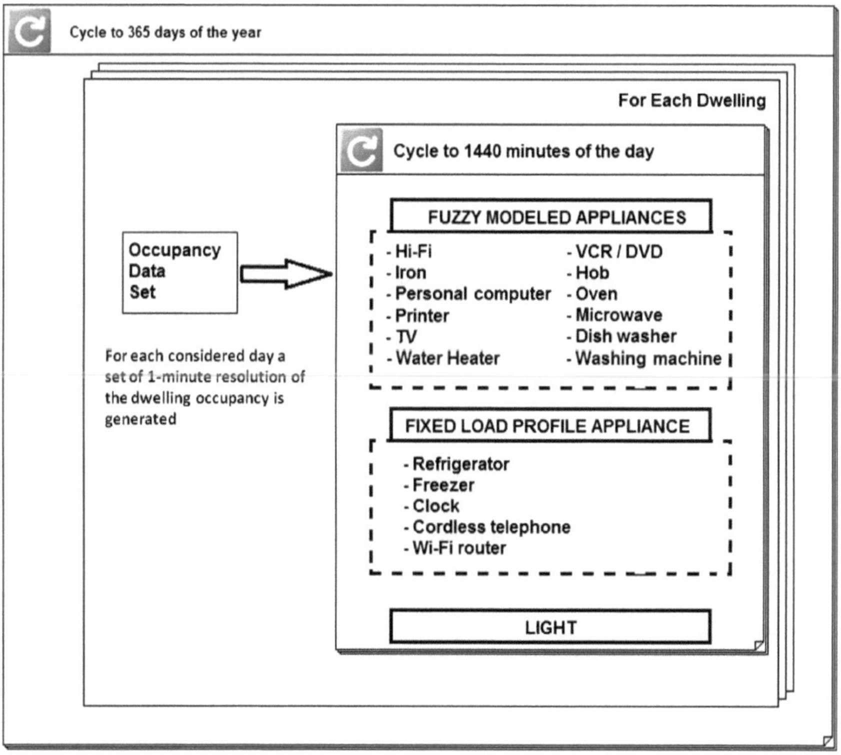

Fig. 5 Structure of the LabVIEW simulator

factor (equal to the difference in hours between the average period of use of the appliance and the time elapsed since the last start), as stated in Richardson et al. (2010). The result is then compared with a random number (within the real interval 0–1). The appliance will start if:

- this number is less than the scaled probability
- there is at least one person in the house
- there are sufficient active people in the house (only for some appliance's categories)
- the sum between the current electrical consumption and the max power of the appliance is less than the power the customer can absorb from the grid.

Table 1 shows the need of taking into account also the number of active people in the dwelling for "Periodical use appliances with human interaction" and "Multimedia Appliances". Starting from the typical pattern of people in the household we decrement this number when an appliance of one of these categories starts and increment this number when the appliance is turned off.

To simulate EM actions, fuzzy rules have been modified to approximate a different user behavior regarding the starting time of the two main shiftable appliances (dishwasher and washing machine). As an example, without any action, fuzzy input sets for "periodical use appliances without human interaction" are:

- the time of the day $h(t)$
- the time elapsed since the last appliance start multiplied his typical start frequency $DT/T(t)$

and a typical rule formulation is:

if *h(t) is afternoon* **and** *DT/T(t) is late,* **then** *the probability to start the appliance is low.*

In the following section we will describe tests performed to validate the model.

5 Model Validation

We validated the model collecting a set of consumption data from 12 volunteer dwellings in and around the town of Ripatransone in the province of Ascoli Piceno, Italy. All people in these households have been briefly interviewed to build occupancy patterns and fuzzy rule sets starting from their typical energy habits. A set of data loggers were installed in the dwellings and configured to record demand at 1 min intervals. An example 24 h demand profile for a single dwelling taken from the measured data set is shown in Fig. 6. In order to create a consumption database we installed in four of these dwellings individual appliance monitors (IAMs from Current Cost company) to extract 6 s resolution consumption data of every household monitorable load (e.g. washing machine, dishwasher, multimedia appliances, iron, oven, microwave). For the remaining 8 dwellings, appliances were not directly monitored, but the profiles were used choosing for

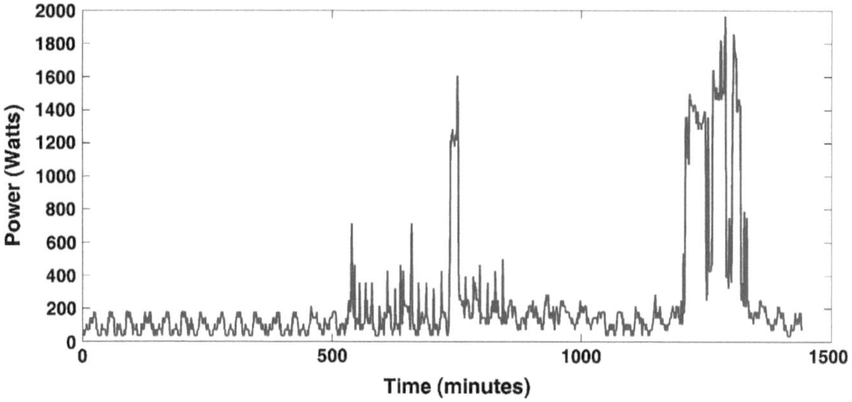

Fig. 6 1 *min* Resolution consumption for one of the considered households in Ripatransone (AP), Italy on a spring day (March 12 2012). One the x-axis are represented the minutes in a day

each appliance the most similar profile in the database (e.g. same brand for the dishwasher, same size for the TV or the laptop battery charger).

It is important to emphasize that the differences between single appliance blocks for the different dwellings are taken into account changing the fuzzy rules, the occupancy profile and using different consumption patterns from the database (according to the different appliances).

The final aim of this simulation tool is the prediction of the human behavior (e.g. the starting of an appliance within 1 h of its real start) especially during daylight periods, in order to give a good method to correctly design a PV plant and evaluate Energy Management actions benefits.

Consequently the purpose of the following validation is to show that the measured and simulated data have similar statistics and differ only for limited quantities. To this end, the model was used to create synthetic data for 12 dwellings covering a full year at 1 min resolution.

Table 4 reports the RMS error, the standard deviation and the RMS% error of measured and simulated values for all 12 dwellings. These values are computed for different time scales, showing a good modeling performance in particular for what regards the daytime period, our main focus to compute the self consumption percentage. Indeed the *RMSE%* calculated from 9 a.m. to 5 p.m. in the whole year for the 12 dwellings is 8.02 %, showing a good capability of the simulator to model the human behavior during the day.

Table 4 Model validation results. percentage mean error, *RMSE*, *SD* and *RMSE%* between the simulated and measured values

Time Scale	Mean error (%)	RMSE (KWh)	SD (Wh)	RMSE (%)
Daytime	0.56	0.514	0.408	8.02
Daily	0.35	1.062	0.890	11.58
Weekly	0.29	6.012	4.360	7.11

Figure 7 shows a 1 min data comparison between the simulator output and the measured values for one day and one dwelling.

Figure 8 shows a comparison between the daily energy simulated and measured during an entire month.

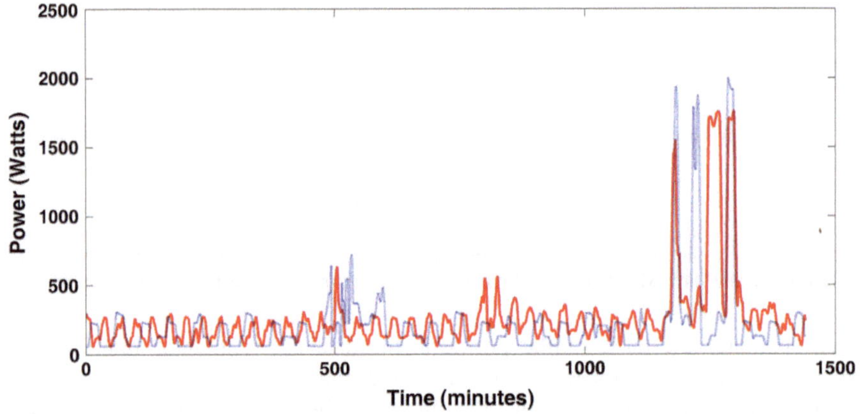

Fig. 7 March 23 2012. 1 min resolution data for one of the considered households in Ripatransone (AP), Italy. The *dotted blue line* is the simulation load profile, the *red continuous line* is the measured one

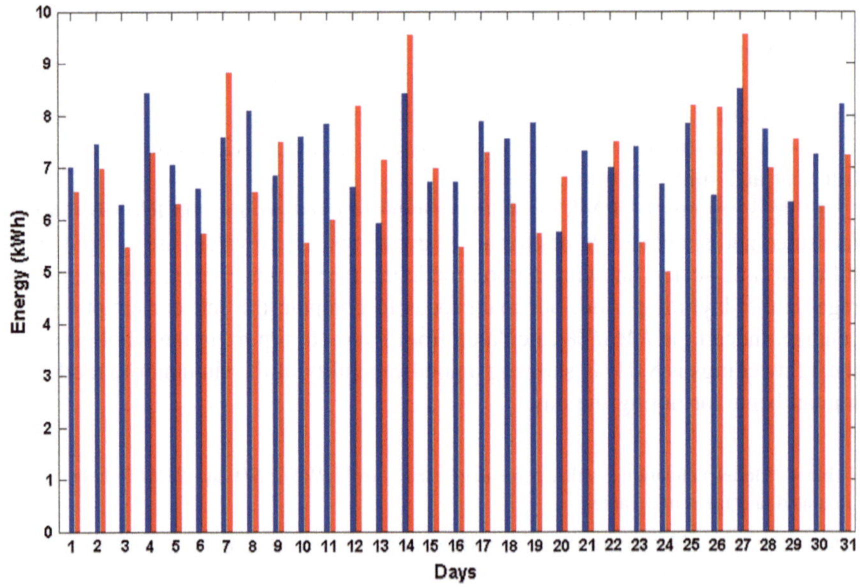

Fig. 8 March 2012. Daily data for one of the considered households in Ripatransone (AP), Italy. *Blue bars* are the simulated values, *red bars* are the measured ones

In the following section a description of the energy management problem and a set of solution will be presented.

6 Energy Management Techniques

The installation of a PV plant can have a great impact on the energy behavior of users. They can use an energy manager, forecasting tools or simply plan to start appliances according to weather forecast. The energy management problem can be expressed as the minimization of a cost function X, given a certain number of electrical tasks N_{TASKS} (e.g. the appliance starts) to arrange in N_{TIME} time samples:

$$\min x = \sum_{k=1}^{N_{TASKS}} \sum_{i=0}^{N_{TIME}} \omega_k(i) \cdot L_k(i) \cdot C_k(i) \tag{3}$$

where $\omega_k(i)$ express if wether or not a task is running at time i, $C_k(i)$ is the energy cost at time i, as computed in 6, $L_k(i)$ is the energy consumed by the task in the time interval i. In particular this minimization problem is subject to the following constrains considering the total power absorbed at each time and the absence of interruptions for each task:

$$\forall \bar{i} \rightarrow \sum_k \omega_k(\bar{i}) \cdot L_k(\bar{i}) \leq P_{\max} \tag{4}$$

$$\forall \bar{k} \rightarrow \omega_{\bar{k}}(i) = 1\,, \quad \forall i : T^{start} < i < T^{end} \tag{5}$$

$$C_{\bar{k}}(\bar{i}) = \begin{cases} R(\bar{i}) & if \sum_{k \neq \bar{k}} \omega_k(\bar{i}) \cdot L_k(\bar{i}) > P_{PV} \\ 0 & if \sum_{k \neq \bar{k}} \omega_k(\bar{i}) \cdot L_k(\bar{i}) \leq P_{PV} \end{cases} \tag{6}$$

In the energy management problem considered in this work we use only two shiftable tasks: the dishwasher and the washing machine. For these tasks $\omega_k(i)$ can be set to 1 according to a forecasting policy.

In particular since in this model we represent the typical user behavior, for what regards the starting of one of these two tasks we need to consider the best time to start the appliance according to user needs (the maximum end cycle time of the appliance has to be fixed by the user) and the optimization of the cost function.

To model the user behavior a new input $DX(t)$ is added in the model, representing the time distance from the best time to start an appliance.

According to this new input, the same rule discussed above will change:

if *h*(*t*) *is afternoon* **and** *DT/T*(*t*) *is late* **and** *DX*(*t*) *is very low,* **then** *the probability to start the appliance is very high.*

6.1 Prediction Algorithms

Valid and reliable forecast information on the expected PV power production and home consumptions play a primary role for the design of an energy management system and to find the best time to start an appliance.

The following approach to implement a Minimal Resource Allocating Network (MRAN) is based on a sequential learning algorithm and an Extended Kalman Filter (EKF) Kadirkamanathan and Niranjan (1993), Platt (1991), Sundararajan et al. (2002). In particular the sequential learning algorithm adds and removes neurons on-line to the network according to a given criterion (Platt 1991), (Sundararajan et al. 2002; Yingwei et al. 1998), and an EKF is used to update the net parameters (Kadirkamanathan and Niranjan 1993).

6.1.1 Radial Basis Function Neural Network

A RBFN with input pattern $\boldsymbol{x} \in \mathbb{R}^m$ and a scalar output $\hat{y} \in \mathbb{R}$ implements a mapping $f : \mathbb{R}^m \to \mathbb{R}$ according to

$$\hat{y} = f(\boldsymbol{x}) = \lambda_0 + \sum_{i=1}^{K} \lambda_i \phi(\|\boldsymbol{x} - \boldsymbol{c}_i\|) \tag{7}$$

where $\phi(\cdot)$ is a given function from $\mathbb{R}^+$ to $\mathbb{R}$, $\|\cdot\|$ denotes the Euclidean norm, λ_i, $i = 0, 1,\ldots, K$ are the weight parameters, $\boldsymbol{c}_i \in \mathbb{R}^m$, $i = 1, 2, \ldots, K$, are the radial basis function centers (called also units or neurons) and K is the number of centers Chen et al. (1991). The terms:

$$o_i = \lambda_i \phi(\|\boldsymbol{x} - \boldsymbol{c}_i\|), \quad i = 1, \ldots, K \tag{8}$$

are called the hidden unit outputs.

In this work the RBFN is used for the prediction of the output of a dynamical system and the system dynamics can be taken into account through the network input pattern **x**, that must be composed of a proper set of system input and output samples acquired in a finite set of past time instants Hunt et al. (1992), i.e. $\boldsymbol{x} \in \mathbb{R}^{n_y + n_u}$ and it is defined as:

$$\boldsymbol{x}(n) = [y(n-1), \ldots, y(n-n_y), u(n-1), \ldots, u(n-n_u)]^T \tag{9}$$

where $n = 1,2,\ldots$ are the time instants, $y(\cdot)$ and $u(\cdot)$ are the system output and inputs (for a detailed description see Sect. 6.2), respectively; n_y, n_u are the lags of the output and input, respectively.

Theoretical investigation and practical results show that the choice of the non-linearity $\phi(\cdot)$, a function of the distance d_i between the current input $\boldsymbol{x}$ and the centre $\boldsymbol{c}_i$, does not significantly influence the performance of the RBFN Chen et al. (1991). Therefore, the following gaussian function is considered:

$$\phi(d_i) = \exp\left(-d_i^2/\beta_i^2\right), \quad i = 1, 2, \ldots, K \tag{10}$$

where $d_i = \|\boldsymbol{x} - \boldsymbol{c}_i\|$ and the real constant β_i is a scaling or "width" parameter (Chen et al. 1991).

Minimal Resource Allocating Network Algorithm

The learning process of MRAN involves allocation of new hidden units and a pruning strategy as well as adaptation of network parameters (Kadirkamanathan and Niranjan 1993; Platt 1991; Sundararajan et al. 2002). The network starts with no hidden units and as input-output data $(\boldsymbol{x}(\cdot), y(\cdot))$ are received, some of them are used to generate new hidden units based on a suitably defined growth criteria. In particular at each time instant n the following three conditions are evaluated to decide if the input $\boldsymbol{x}(n)$ should give rise to a new hidden unit:

$$\|e(n)\| = \|y(n) - f(\boldsymbol{x}(n))\| > E_1 \tag{11}$$

$$e_{rms}(n) = \sqrt{\sum_{j=n-(M-1)}^{n} \frac{e(j)^2}{M}} > E_2 \tag{12}$$

$$d(n) = \|\boldsymbol{x}(n) - \boldsymbol{c}_r(n)\| > E_3 \tag{13}$$

where $\boldsymbol{c}_r(n)$ is the centre of the hidden unit that is nearest to $\boldsymbol{x}(n)$ and M represents the number of past network outputs for calculating the output error $e_{rms}(n)$. The terms E_1, E_2 and E_3 are thresholds to be suitably selected. As stated in Sundararajan et al. (2002), Yingwei et al. (1998) these three conditions evaluate the novelty in the data. If all the criteria of relations (11)–(13) are satisfied, a new hidden unit is added and the following parameters are associated with it:

$$\lambda_{K+1} = e(n) \tag{14}$$

$$\mathbf{c}_{K+1} = \boldsymbol{x}(n) \tag{15}$$

$$\beta_{K+1} = \alpha\|\boldsymbol{x}(n) - \mathbf{c}_r(n)\| \tag{16}$$

where α determines the overlap of the response of a hidden unit in the input space as specified in Kadirkamanathan and Niranjan (1993), Sundararajan et al. (2002). If the observation $(\boldsymbol{x}(n), y(n))$ does not satisfy the criteria of relations (11)–(13), an EKF is used to update the following parameters of the network:

$$\boldsymbol{w} = \left[\lambda_0, \lambda_1, \boldsymbol{c}_1^T, \beta_1, \ldots, \lambda_N, \boldsymbol{c}_N^T, \beta_N\right]^T. \tag{17}$$

The update equation is given by:

$$\boldsymbol{w}(n) = \boldsymbol{w}(n-1) + \boldsymbol{k}(n)e(n) \tag{18}$$

where the gain vector $\boldsymbol{k}(n)$ is expressed by:

$$\boldsymbol{k}(n) = \boldsymbol{P}(n-1)\mathbf{a}(n)\left[r(n) + \mathbf{a}^T(n)\boldsymbol{P}(n-1)\mathbf{a}(n)\right]^{-1} \tag{19}$$

with $\mathbf{a}(n)$ the gradient vector of the function $f(\boldsymbol{x}(n))$ with respect to the parameter vector $\boldsymbol{w}(n-1)$ Kadirkamanathan and Niranjan (1993), Sundararajan et al. (2002), $r(n)$ is the variance of the measurement noise and $\boldsymbol{P}(n-1)$ is the error covariance matrix which is updated by:

$$\boldsymbol{P}(n) = \left[I - \boldsymbol{k}(n)\mathbf{a}^T(n)\right]\boldsymbol{P}(n-1) + \boldsymbol{Q}(n-1) \tag{20}$$

where $\boldsymbol{Q}(n-1)$ is introduced to avoid that the rapid convergence of the EKF algorithm prevents the model from adapting to future data Kadirkamanathan and Niranjan (1993), Sundararajan et al. (2002). The $z \times z$ matrix $\boldsymbol{P}(n)$ is positive definite symmetric and z is the number of parameters to be adjusted. When a new hidden neuron is allocated, the dimension of $\boldsymbol{P}(n)$ increases to:

$$\boldsymbol{P}(n) = \begin{pmatrix} \boldsymbol{P}(n-1) & 0 \\ 0 & p_0 \boldsymbol{I}_{z_1 \times z_1} \end{pmatrix} \tag{21}$$

where p_0 is an estimate of the uncertainty in the initial values assigned to the parameters and z_1 is the number of new parameters introduced by adding the new hidden neuron. As stated in Sundararajan et al. (2002), Yingwei et al. (1998) to keep the RBF network in a minimal size a pruning strategy removes those hidden units that contribute little to the overall network output over a number of consecutive observations. To carry out this pruning strategy, for every observation $(\boldsymbol{x}(n), y(n))$ the hidden unit outputs are computed:

$$o_i(n) = \lambda_i \phi(\|\boldsymbol{x}(n) - \boldsymbol{c}_i\|), \quad i = 1, \ldots, K \tag{22}$$

and normalized with respect to the highest output:

$$\overline{o}_i(n) = \frac{o_i(n)}{\max\{o_i(n)\}}, \quad i = 1, \ldots, K \tag{23}$$

The hidden units for which the normalized output (23) is less than a threshold δ for ξ consecutive observations are removed and the dimensionality of all the related matrices are adjusted to suit the reduced network (Sundararajan et al. 2002; Yingwei et al. (1998).

The EKF has been implemented with the assumption that $\boldsymbol{Q}(n) = I_{n,n}\sigma_\eta^2$ and $r(n) = \sigma_v^2$.

The MRAN prediction algorithm Sundararajan et al. (2002), Yingwei et al. (1998), with the EKF, here called MRANEKF algorithm, is shown in Fig. 9 and it is summarized as follow:

1. For each observation $(\boldsymbol{x}(n), y(n))$ do: compute the overall network output: $\hat{y}(n) = f(\boldsymbol{x}(n)) = \lambda_0 + \sum_{i=1}^{K} \lambda_i \phi(\|\boldsymbol{x}(n) - \boldsymbol{c}_i\|)$ where K is the number of hidden units;
2. Calculate the parameters required by the growth criterion:

- $\|e(n)\| = \|y(n) - f(\boldsymbol{x}(n))\|$
- $e_{rms}(n) = \sqrt{\sum_{j=n-(M-1)}^{n} \frac{e(j)^2}{M}}$
- $d(n) = \|\boldsymbol{x}(n) - \boldsymbol{c}_r(n)\|$

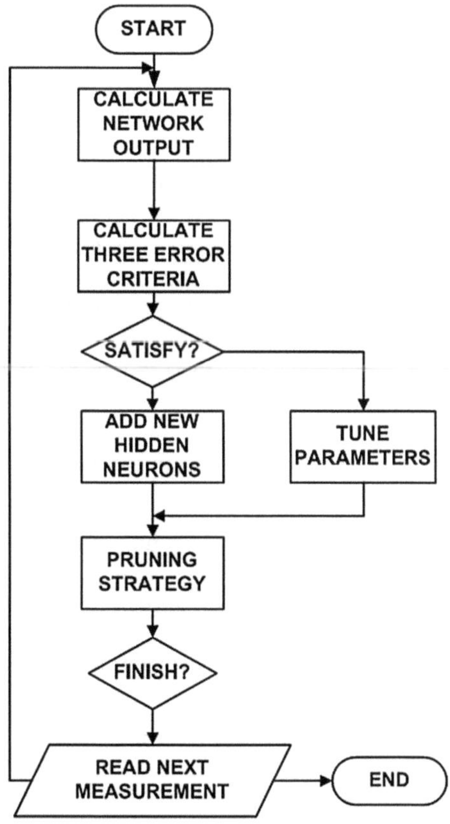

Fig. 9 Flow chart of the MRANEKF algorithm

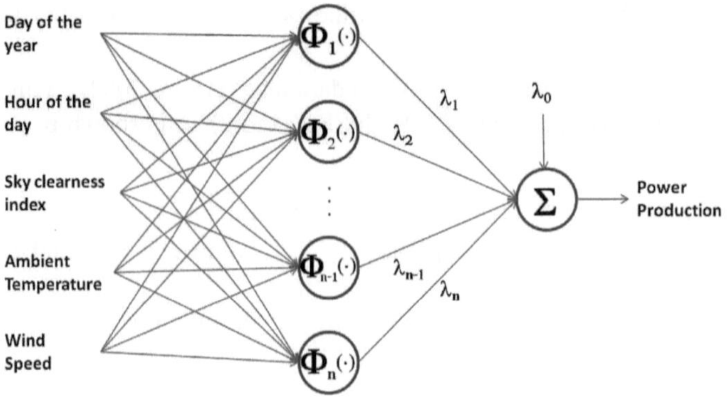

Fig. 10 Input Output structure of the PV production forecast neural network

3. Apply the criterion for adding a new hidden unit:
   if
   $\|e(n)\| > E_1$ **and** $e_{rms}(n) > E_2$ **and** $d(n) > E_3$ allocate a new hidden unit $K + 1$ with:
   - $\lambda_{K+1} = e(n)$
   - $\boldsymbol{c}_{K+1} = \boldsymbol{x}(n)$
   - $\beta_{K+1} = \alpha\|\boldsymbol{x}(n) - \boldsymbol{c}_r(n)\|$

   else
   - tune the network parameters:

   $$\boldsymbol{w}(n) = \boldsymbol{w}(n-1) + \boldsymbol{k}(n)e(n)$$

   - update the error covariance matrix:

   $$\boldsymbol{P}(n) = \left[I - \boldsymbol{k}(n)\mathbf{a}^T(n)\right]\boldsymbol{P}(n-1) + \boldsymbol{Q}(n-1)$$

   end
4. Check the criterion to prune hidden units:
   - compute the hidden unit outputs:

   $$o_i(n) = \lambda_i\phi(\|\boldsymbol{x}(n) - \boldsymbol{c}_i\|), i = 1, \ldots, K$$

   - compute the normalized outputs:

   $$\bar{o}_i(n) = \frac{o_i(n)}{\max\{o_i(n)\}}, i = 1, \ldots, K$$

- **if** $\overline{o}_i(\cdot) < \delta$ for ξ consecutive observations **than** prune the ith hidden unit and reduce the dimensionality of the related matrices

 end

5. $n = n + 1$ and **go** to step 1.

6.2 Prediction Algorithm Results

Forecast algorithms performance have been evaluated through real experimental tests based on data acquired from March 2012 to August 2012. In particular the 3 houses with 3.3 KWp PV plant considered, are located in Ripatransone (AP), Italy. The MRANEKF learning algorithm starts with a pre-trained net based only on few information found on the web, such as power production profile of clear sky days and cloudy days for the specified location Pvg (2011), panel orientation and tilting and typical electrical load profile of a house. This is a common operating condition, when no sensors and measures are available before the forecast begins.

The inputs of the production forecasting network, as shown in Fig. 10, are:

- the day of the year (from 1 to 365)
- the hour of the day (considered from 0 to 24)
- the ambient temperature (in Kelvin)
- the sky clearness index (a coefficient ranging from 0 to 10 mapping the website forecast, e.g. "clear and sunny" is 10 while "clouds and heavy rain" is 0)
- the wind speed (in m/s)

The input pattern of the consumption forecasting net, as shown in Fig. 11, consists of:

- the day of the week (e.g. Monday is day 1, Tuesday is day 2)
- the hour of the day (considered from 0 to 24)

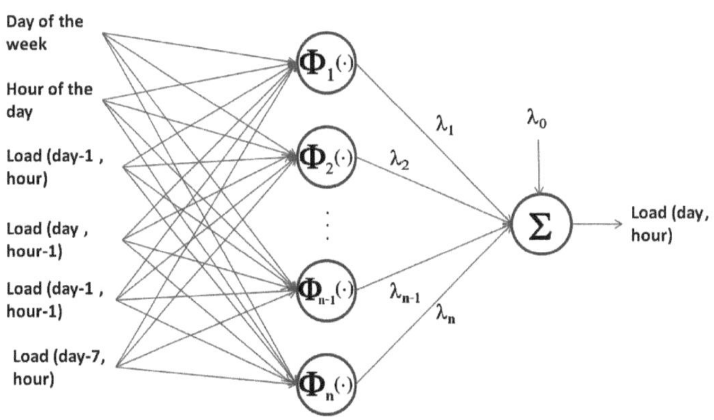

Fig. 11 Input-output structure of the load forecast network

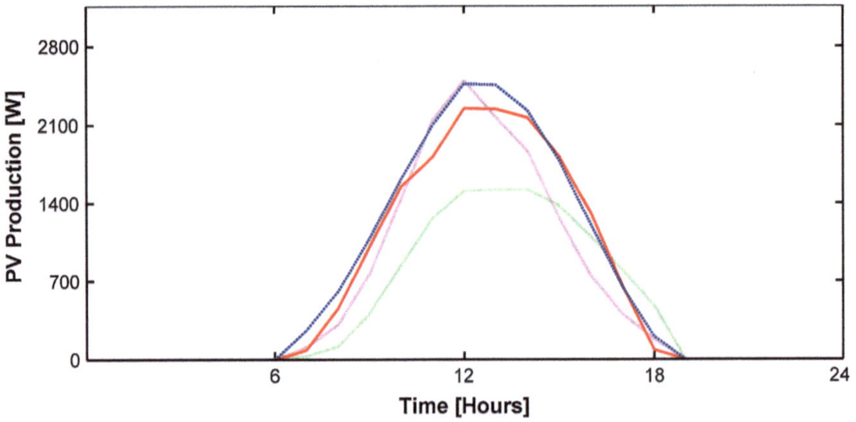

Fig. 13 The *continuous red line* is the measured PV production, the *dashed blue line* is the 3 h ahead MRANEKF network forecast, the *dotted purple line* and *green line* are respectively the 8 and 18 h ahead forecast

a specified end time, customers will reach the major economical benefits starting appliances when $P(k)$ has the minimum value $P_{\min}$ in peak off periods, paying

$$X^* = C_1(k + i - 1) * P_{\min}, i = 1, \ldots, J \tag{26}$$

This will be the reference value that the algorithm has to improve taking into account that the self-consumption reduces the feed in tariff of δ (production and consumptions are in [KWh], while prices and feed in tariffs in [eur/KWh]). When the user plan to start an appliance with end time H, the algorithm used to minimize the price of the appliance cycle and find the best hour to start it is summarized as follows:

min = X^*;
for $k = 1, \ldots, H - J$
$X(k) = 0$;
for $i = 1, \ldots, J$
if $P_{PV}(k + i - 1) - C_0(k + i - 1) - C_1(k + i - 1) < 0$
$X(k) = X(k) + (P_{PV}(k + i - 1) - C_0(k + i - 1) - C_1(k + i - 1)) * P(k) + C_1(k + i - 1) * \delta$
else $X(k) = X(k) + C_1(k + i - 1) * \delta$
end; end;
if $X(k) <$ min
min = $X(k)$; $hour = k * 2$;
end; end

In the following section we will examine a case study on the proper sizing of a PV plant together with an economic evaluation of energy management benefits.

7 Photovoltaic Plant Sizing: A Case Study

Due to the random nature of solar energy, great effort must be made to design PV systems that optimize energy savings, self consumption and costs. In this section we propose a PV sizing case study in the Italian scenario using the consumption pattern of one of the previously considered household with an annual electrical consumption of 2,300 KWh.

The key of the proposed sizing method is the self consumption percentage, computed by the simulation tool. In Italy the government took the decision to cut PV incentives on June 2013, instead of 2016 as previously expected. An example on how PV incentives varied for a building integrated 3 kWp plant since their introduction in 2005 is shown in Fig. 14.

To provide support to PV industry a new net metering scheme has been amended Regulatory Authority for Electricity and Gas (2013) and came into effect on 1st January 2013. Under this decree PV system owners can get credits for the value of the excess of electricity fed into the grid over a time period. Further encouraging self-consumption, the Italian Revenue Agency introduced tax breaks for off-grid PV systems installed on buildings.

A 3 year historical solar irradiance data set is used to calculate the output of a varying size PV plant (1–3.5 kWp) and compared with the consumption pattern computed by the simulator in order to obtain the self consumption percentage for each considered PV plant size. A financial evaluation technique is used to compare the different investments under the revised Italian net metering scheme known as "scambio sul posto" in which GSE pays a contribution E_t to the customer equal to:

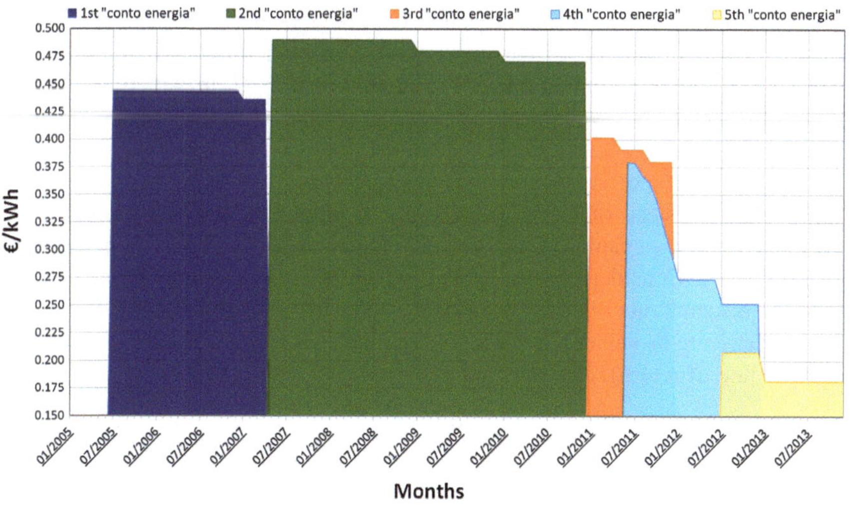

Fig. 14 Year 2005–2013. Evolution of the Italian FITs, according to the Ministerial Decrees, for a 3 kWp building integrated PV plant

$$E_t = C_t \cdot \min(F_t, W_t) \quad (27)$$

where F_t and W_t are respectively the injected and withdrawn electricity in KWh and C_t represents a coefficient comprehensive of the electricity cost and net services cost in eur/KWh. For the global cost of the PV plant, an average of the main solar installer prices in the considered area has been considered.

7.1 Economical Analysis

The cost-benefit analysis (CBA) is a financial valuation technique used to predict the effects of a project, a program or an investment, verifying its benefits. CBA, as an alternative to traditional methods of economic analysis, represents also a method of ex-ante evaluation by external parties that have to decide on the financial viability of an investment or have to choose how to allocate scarce financial resources among different possible investments.

To evaluate the economic convenience of PV systems on the considered building we carried out the CBA of different sizes of PV plants to choose the best one. The discounted cash flows generated from each investment have been calculated for 20 years, equal to the period in which PV module producers guarantee at least 80 % of their initial performance. The net present value (NPV), calculated for each PV plant size, is:

$$NPV = \sum_{t=0}^{K} \frac{C_t}{(1+r)^t} \quad (28)$$

where C_t is the cash flow at time t, r the discount rate (equal to 5 % in our case) and K the considered lifetime of the investment. The cash flow C_t is the difference between the discounted annual cash inflows I_t and outflows O_t. In particular I_t consists of the annual directly saved energy by self consumption (considering a 3 % annual increase of the unitary energy price), the net metering contribution E_t and government contributions (50 % of the plant cost in taxes deduction for the first 10 years).

O_t consists instead of the initial cost of the plant (we consider an investment made only with own capital) and the annual maintenance costs (0.5 % of the initial cost per year). Considering that NPV calculation strongly depends on the used reference discount rate r used (for which the same investment may be convenient or less in relation to its value) it is useful to consider as financial indicator also the IRR (internal rate of return), calculated as the rate r^* for which results:

$$NPV(r^*) = 0 \quad (29)$$

Table 6 reports the unitary costs (Cost), the self consumption percentages of two simulated scenarios (user performing EM actions and user maintaining the same behavior) and CBA results for different PV plant sizes in the analyzed case study.

Table 6 Unitary costs, self consumption percentages (SC) and CBA results (NPV and IRR) for the considered case study with and without energy management actions

		No EM actions			EM actions		
3–5 7–9 Size (KWp)	Cost (€/KWp)	SC (%)	NPV (€)	IRR (%)	SC (%)	NPV (€)	IRR (%)
1.00	3,850	41.1	787	7.91	53.4	1,005	8.64
1.25	3,750	35.3	937	7.85	47.3	1,208	8.60
1.50	3,500	31.3	1,251	8.35	42.9	1,566	9.11
1.75	3,150	27.4	1,711	9.28	38.5	2,067	10.07
2.00	2,950	24.2	2,048	9.71	35.2	2,443	10.51
2.25	2,750	22.5	2,069	9.47	32.9	2,501	10.30
2.50	2,700	20.6	1,730	8.51	30.6	2,198	9.36
2.75	2,500	19.4	1,716	8.42	29.4	2,215	9.32
3.00	2,450	17.5	1,363	7.60	26.9	1,888	8.51
3.25	2,320	16.2	1,310	7.46	25.8	1,879	8.44
3.50	2,260	15.7	1,047	6.89	24.7	1,624	7.85

Figure 15 shows the trend of NPV and IRR depending on the PV plant size when a user do not perform any EM action. The values of NPV, which range between 790 and 2,070 €, IRR, between 6.89 and 9.71 %, show better results for a 2.25 KWp plant. In particular revenues decrease from 2,070 to 1,360 € with a 3 KWp plant and IRR decrease of 2 %, emphasizing the need of the correct sizing of the plant.

We have furthermore analyzed the situation in which the user performs basic EM actions (starting the 2 main shiftable appliances around the peak production hours of each day, according to the results provided by the energy management algorithms presented).

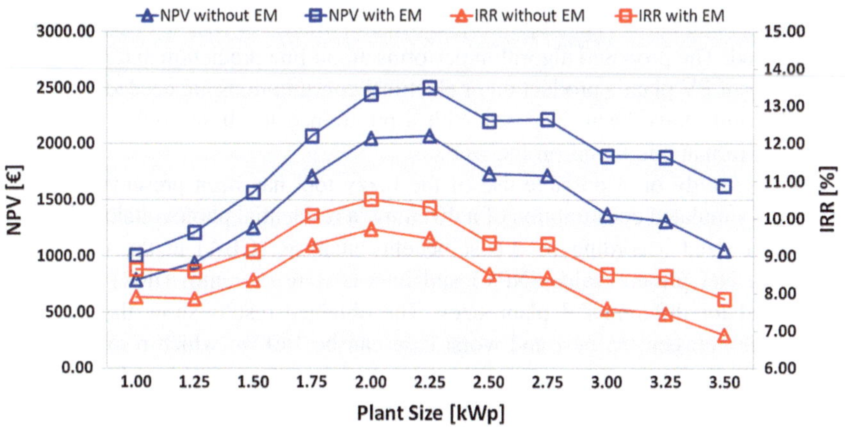

Fig. 15 Results of the cost benefits analysis. NPV (*blue line*) and IRR (*red line*) for the different sizes of PV plants computed when a user performs EM actions (*squared markers*) or maintains the same energy behavior (*triangular markers*)

8 Conclusions

In recent years (2005–2013), the whole Europe has highlighted a rapid growth of the photovoltaic sector, after the introduction of economical incentives by governments.

In this context, due to higher feed in tariffs, the expansion involved mainly PV systems on buildings. This situation poses notable issues since demand tend to be relatively low during peak power production periods. On the same time a new PV installation needs aa accurate sizing to smooth the so called "back-flow" effect and maximize economical benefits to owners.

In this scenario the modeling of residential energy use and the planning of energy management actions can play a crucial role. Indeed the matching of the production and consumption patterns is the only way to achieve satisfying economical benefits.

This chapter deals with the description of a novel Fuzzy approach to model household electrical consumption. The model is built using a "bottom-up" approach and the basic block is the single appliance. Using as inputs patterns of active occupancy (i.e. when people are at home and awake) and typical domestic habits (i. e. start frequency of some appliances), the FIS model give as output the starting probability of each appliance. To validate the model we have recorded electricity demand within 12 dwellings in Ripatransone (AP), in the central east coast of Italy, over the period of 12 months. Simulation performances, in particular for what regards daytime period (the mean error is 0.52 %), make possible its use for self consumption estimation and PV sizing.

Energy management problem has been introduced and a neural network based algorithm to forecasts of both photovoltaic production and home consumptions presented. The considered algorithm, based on the minimal resource allocating networks method, is used to perform long range predictions. In particular the power production and home consumptions presented in the above tests is forecasted up to 24 h ahead. The proposed algorithm performs an on-line prediction and no previous measures of PV plant's production or electrical consumptions are needed. Therefore the algorithm have been proposed with a pre-trained net based only on few historical informations found on the web.

A case study on a possible use of the fuzzy tool has been presented. Starting from the simulated consumption of a dwelling, a residential photovoltaic (PV) plant has been sized according to a cost benefits analysis (CBA) in the new Italian scenario. Net present value (NPV) and internal rate of return (IRR) have been computed for different PV plant sizes. The obtained results show that the NPV difference between the best and worst case can be 140 % (which results in more than 1,200 €). Furthermore a parallel analysis of the economical benefits of energy management actions (shifting of the two main appliances) has been performed. The CBA analysis shows that revenues can further increase from 250 to 600 € (depending on the plant size) thus imposing cost limitation for the EM equipment.

References

Aghaei, J., & Alizadeh, M.-I. (2013). Demand response in smart electricity grids equipped with renewable energy sources: A review. *Renewable and Sustainable Energy Reviews, 18*, 64–72.

Azadeh, A., Seraj, O., & Saberi, M. (2008). A total fuzzy regression algorithm for energy consumption estimation. In *6th IEEE International Conference on Industrial Informatics, (INDIN 2008)*. (pp 1562–1568).

Azar, A. T. (2010a). *Adaptive neuro-fuzzy systems*. Vienna: InTech.

Azar, A. T. (2010b). *Fuzzy systems*. Vienna: InTech.

Azar, A. T. (2012). Overview of type-2 fuzzy logic systems. *International Journal of Fuzzy System Applications (IJFSA), 2*(4), 1–28.

Barbato, A., Capone, A., Rodolfi, M., & Tagliaferri, D. (2011). Forecasting the usage of household appliances through power meter sensors for demand management in the smart grid. In *IEEE International Conference on Smart Grid Communications* (pp. 404–409).

Benghanem, M., & Mellit, A. (2010). Radial basis function network-based prediction of global solar radiation data: Application for sizing of a stand-alone photovoltaic system at al-madinah, saudi arabia. *Energy, 35*(9), 3751–3762.

Bernard, J.-T., Bolduc, D., & Yameogo, N.-D. (2011). A pseudo-panel data model of household electricity demand. *Resource and Energy Economics, 33*(1), 315–325.

Bose, B. (2011). Fuzzy logic and neural networks in power electronics and drives. *IEEE Industry Applications Magazine, 6*(3), 57–63.

Chen, S., Cowan, C. F. N., & Grant, P. M. (1991). Orthogonal least squares learning algorithm for radial basis function networks. *IEEE Transaction on Neural Networks, 2*(2), 302–309.

Ciabattoni, L., Cimini, G., Grisostomi, M., Ippoliti, G., Longhi, S., & Mainardi, E. (2013a). Supervisory control of PV-battery systems by online tuned neural networks. In *IEEE International Conference on Mechatronics (ICM)* (pp. 99–104), Vicenza, Italy.

Ciabattoni, L., Grisostomi, M., Ippoliti, G., & Longhi, S. (2013b). A fuzzy logic tool for household electrical consumption modeling. In *Industrial Electronics Society, IECON 2013—39th Annual Conference of the IEEE* (pp. 8022–8027).

Ciabattoni, L., Grisostomi, M., Ippoliti, G., & Longhi, S. (2013c). *Neural networks based home energy management system in residential pv scenario* (pp. 1721–1726).

Ciabattoni, L., Grisostomi, M., Ippoliti, G., Longhi, S., & Mainardi, E. (2012). Online tuned neural networks for PV plant production forecasting. In *38th IEEE Photovoltaic Specialists Conference (PVSC)* (pp. 2916–2921), Austin, TX.

Ciabattoni, L., Ippoliti, G., Benini, M., Longhi, S., & Pirro, M. (2013d). Design of a home energy management system by online neural networks. In 11th ifac international workshop on adaptation and learning in control and signal processing (pp. 677–682), Caen, France.

Ciabattoni, L., Ippoliti, G., Longhi, S., & Cavalletti, M. (2013e). Online *Tuned Neural Networks for Fuzzy Supervisory Control of PV-Battery Systems*. In *IEEE PES Innovative Smart Grid Technologies Conference (ISGT)*.

Cimini, G., Corradini, M., Ippoliti, G., Malerba, N., & Orlando, G. (2013). Control of variable speed wind energy conversion systems by a discrete-time sliding mode approach. In *IEEE International Conference on Mechatronics (ICM)* (pp. 736–741).

Di Giorgio, A., Pimpinella, L., & Liberati, F. (2012). A model predictive control approach to the load shifting problem in a household equipped with an energy storage unit. In *20th Mediterranean Conference on Control Automation* (pp. 1491–1498).

Di Giorgio, A., Pimpinella, L., Quaresima, A., & Curti, S. (2011). An event driven smart home controller enabling cost effective use of electric energy and automated demand side management. In *Mediterranean Conference on Control Automation* (pp. 358–364).

Fazeli, A., Christopher, E., Johnson, C., Gillott, M., & Sumner, M. (2011). Investigating the effects of dynamic demand side management within intelligent smart energy communities of future decentralized power system. In *IEEE PES International Conference and Exhibition on Innovative Smart Grid Technologies* (pp. 1–8).

Ferracuti, F., Giantomassi, A., Iarlori, S., Ippoliti, G., & Longhi, S. (2013a). Induction motor fault detection and diagnosis using kde and kullback-leibler divergence. In *Industrial Electronics Society, IECON 2013—39th Annual Conference of the IEEE* (pp. 2923–2928).

Ferracuti, F., Giantomassi, A., & Longhi, S. (2013b). MSPCA with KDE thresholding to support QC in electrical motors production line. In *Manufacturing Modelling, Management, and Control* (Vol. 7, pp. 1542–1547).

Giantomassi, A., Ferracuti, F., Iarlori, S., Longhi, S., Fonti, A., & Comodi, G. (2014a). Kernel canonical variate analysis based management system for monitoring and diagnosing smart homes.

Giantomassi, A., Ferracuti, F., Iarlori, S., Puglia, G., Fonti, A., Comodi, G., et al. (2014b). Smart home heating system malfunction and bad behavior diagnosis by multi-scale PCA under indoor temperature feedback control. In *22nd Mediterranean Conference on Control & Automation (MED).*

Gruber, J. & Prodanovic, M. (2012). Residential energy load profile generation using a probabilistic approach. In *European Symposium on Computer Modeling and Simulation* (pp. 317–322).

Guo, R., Ren, Z., & Li, F. (2011). A preliminary analysis on household energy consumption of shanghai. In *International Conference on Bioinformatics and Biomedical Engineering* (pp. 1–4).

Hunt, K. J., Sbarbaro, D., Zbikowski, R., & Gawthrop, P. J. (1992). Neural networks for control systems—a survey. *Automatica, 28*(6), 1083–1112.

Ihbal, A., Rajamani, H., Abd-Alhameed, R., & Jalboub, M. K. (2011). Identifying the nature of domestic load profile from a single household electricity consumption measurements. In *International Multi-Conference on Systems, Signals and Devices* (pp. 1–4).

Jakhrani, A. Q., Othman, A.-K., Rigit, A. R. H., Samo, S. R., & Kamboh, S. A. (2012). A novel analytical model for optimal sizing of standalone photovoltaic systems. *Energy, 46*(1), 675–682.

Jallouli, R., & Krichen, L. (2012). Sizing, techno-economic and generation management analysis of a stand alone photovoltaic power unit including storage devices. *Energy, 40*(1), 196–209.

Kaabeche, A., Belhamel, M., & Ibtiouen, R. (2011). Sizing optimization of grid-independent hybrid photovoltaic/wind power generation system. *Energy, 36*(2), 1214–1222.

Kadirkamanathan, V., & Niranjan, M. (1993). A function estimation approach to sequential learning with neural network. *Neural Computation, 5*, 954–975.

Kanchev, H., Lu, D., Colas, F., Lazarov, V., & Francois, B. (2011). Energy management and operational planning of a microgrid with a PV-based active generator for smart grid applications. *IEEE Transactions on Industrial Electronics, 58*(10), 4583–4592.

Lewis, D. (2009). Solar grid parity—[power solar]. *Engineering Technology, 4*(9), 50–53.

Ljung, L. (1999). *System identification, theory for the use*. Information and System Sciences Series. New Jersey: Prentice Hall PTR.

Lopez-Polo, A., Haas, R., Panzer, C., & Auer, H. (2012). Prospects for grid-parity of photovoltaics due to effective promotion schemes in major countries. In *Asia-Pacific Power and Energy Engineering Conference* (pp. 1–4).

Meyers, R. J., Williams, E. D., & Matthews, H. S. (2010). Scoping the potential of monitoring and control technologies to reduce energy use in homes. *Energy and Buildings, 42*(5), 563–569.

Murata, H., & Onoda, T. (2002). Estimation of power consumption for household electric appliances. In *International Conference on Neural Information Processing* (Vol. 5, pp. 2299–2303).

Osman, Z., Awad, M., & Mahmoud, T. (2009). Neural network based approach for short-term load forecasting. In *IEEE/PES Power Systems Conference. and Exposition* (pp. 1–8).

Palensky, P., & Dietrich, D. (2011). Demand side management: Demand response, intelligent energy systems, and smart loads. *IEEE Transactions on Industrial Informatics, 7*, 381–388.

Platt, J. (1991). A resource allocating network for function interpolation. *Neural Computation, 3*, 213–225.

PVG (2011). Photovoltaic geographical information system website.

Regulatory Authority for Electricity and Gas (2013). net metering scheme regulation. (http://www.autorita.energia.it/allegati/docs/12/322-12.pdf). Last Access 28 October 2013.

Richardson, I., Thomson, M., & Infield, D. (2008). A high-resolution domestic building occupancy model for energy demand simulations. *Energy and Buildings, 40*(8), 1560–1566.

Richardson, I., Thomson, M., Infield, D., & Clifford, C. (2010). Domestic electricity use: A high-resolution energy demand model. *Energy and Buildings, 42*(10), 1878–1887.

Sawyer, R., Anderson, J., Foulks, E., Troxler, J., & Cox, R. (2009). Creating low-cost energy-management systems for homes using non-intrusive energy monitoring devices. In *Energy Conversion Congress and Exposition, ECCE 2009* (pp. 3239–3246). IEEE.

Shahgoshtasbi, D., & Jamshidi, M. (2011). Energy efficiency in a smart house with an intelligent neuro-fuzzy lookup table. In *6th International Conference on System of Systems Engineering (SoSE)* (pp. 288–292).

Subbiah, R., Lum, K., Marathe, A., & Marathe, M. (2013). Activity based energy demand modeling for residential buildings. In *IEEE PES Innovative Smart Grid Technologies* (pp. 1–6).

Suh, D., Yoo, Y.-S., Lee, I.-W., & Chang, S. (2012). An electricity energy and water consumption model for korean style apartment buildings. In *International Conference on Control, Automation. and Systems* (pp. 1113–1117).

Sundararajan, N., Saratchandran, P., & Li, Y. (2002). *Fully tuned radial basis function neural networks for flight control.* London: Kluver Academic.

Timilsina, G. R., Kurdgelashvili, L., & Narbel, P. A. (2012). Solar energy: Markets, economics and policies. *Renewable and Sustainable Energy Reviews, 16*(1), 449–465.

Widen, J., Lundh, M., Vassileva, I., Dahlquist, E., Ellegard, K., & Wackelgard, E. (2009). Constructing load profiles for household electricity and hot water from time-use data—modelling approach and validation. *Energy and Buildings, 41*(7), 753–768.

Yang, C.-J. (2010). Reconsidering solar grid parity. *Energy Policy, 38*(7), 3270–3273.

Yingwei, L., Sundararajan, N., & Saratchandran, P. (1998). Performance evaluation of a sequential minimal radial basis function (RBF) neural network learning algorithm. *IEEE Transaction on Neural Networks, 9*(2), 308–318.

Zaidi, A., Kupzog, F., Zia, T., & Palensky, P. (2010). Load recognition for automated demand response in microgrids. In *IECON 2010—36th Annual Conference on IEEE Industrial Electronics Society* (pp. 2442–2447).

Zeilinger, F. (2011). Simulation of the effect of demand side management to the power consumption of households. In *3rd International Youth Conference on Energetics (IYCE), Proceedings of the 2011* (pp. 1–9).

Zia, T., Bruckner, D., & Zaidi, A. (2011). A hidden markov model based procedure for identifying household electric loads. In *37th Annual Conference on IEEE Industrial Electronics Society* (pp. 3218–3223).

Zong, Y., Kullmann, D., Thavlov, A., Gehrke, O., & Bindner, H. (2012). Application of model predictive control for active load management in a distributed power system with high wind penetration. *IEEE Transactions on Smart Grid, 3*(2), 1055–1062.

Modeling, Identification and Control of Irrigation Station with Sprinkling: Takagi-Sugeno Approach

Wael Chakchouk, Abderrahmen Zaafouri and Anis Sallami

Abstract The spray under pressure is an effective save on water. This task should be automated and controlled in order to limit the water waste and the facilities of damages. For this reason, it's necessary to find a mathematical model describing the irrigation process. In order to facilitate this step the Takagi-Sugeno fuzzy model is the best approaches of nonlinear systems representation. Various techniques are used in the literature of such systems; the clustering technique is one of the best solutions. In this paper, we'll model the irrigation station with the T-S algorithm and use the fuzzy c-means (FCM) algorithm and present the results of simulation and some validation tests and we present the stability of T-S irrigation station model.

1 Introduction

The development of a mathematical model making it possible to represent "as well as possible" the dynamic behavior of a complex real process represents a very important problem in the practical world. In recent years, and with the evolution of technology, a significant effort has been given to modeling, identification and control of such systems. The Takagi-Segeno fuzzy model (Takagi and Sugeno 1985; Grisales 2007; Li et al. 2012; Chakchouk et al. 2014) is one of the best approaches to the representation of such a process, it was widely used in many research areas, since it has an excellent ability to describe the nonlinear system.

W. Chakchouk (✉) · A. Zaafouri · A. Sallami
Higher School of Sciences and Techniques of Tunis, University of Tunisia,
Taha Hussein, B.P. 56, Bab Menara 1008, Tunisia
e-mail: waelchakchouk@hotmail.fr

A. Zaafouri
e-mail: abderrahmen.zaafouri@isetr.rnu.tn

A. Sallami
e-mail: anis.sellami@esstt.rnu.tn

Q. Zhu and A.T. Azar (eds.), *Complex System Modelling and Control Through Intelligent Soft Computations*, Studies in Fuzziness and Soft Computing 319,
DOI 10.1007/978-3-319-12883-2_17

Indeed, the T-S fuzzy model can approximate highly nonlinear system into several locally linear subsystems interconnected. The identification problem in the T-S fuzzy model can be summarized in two steps: structure identification and parameter estimation. Several techniques were developed to conclude the modeling of these systems: we quote primarily the neuro-fuzzy technique (Daneshwar and Noh 2013; Azar 2010a) and clustering technique (Daneshwar and Noh 2013; Azar 2010a) and clustering technique (Troudi et al. 2011; Li et al. 2013; Jang et al. 2007; Pingli et al. 2006; Xu and Zhang 2009; Zahid et al. 2001; Chakchouk et al. 2014). Indeed Several researchers have noticed that a nonlinear system can be approximated by the sum of several linear sub-systems. Method of clustering proves to be an interesting technique for identification and the modelisation of the nonlinear systems. Indeed, this technique consists in approximating the total nonlinear system by a vague model of Takagi-Sugeno type. In this case, each cluster represents one fuzzy rule of Takagi-Sugeno. The number of clusters is fixed by an expert according to the type and the performances of application considered. By consequent to each cluster one correspond homogeneous zone of operation such that is defined in the form of a linear local model. We are interested to model and identify a nonlinear system by the fuzzy logic approach such as Takagi-Sugeno (T-S) approach. The latter, uses modeling containing linguistic rules to obtain the model of system outputs. Initially, we present the fuzzy logic approach design, we gives an outline on the first two models. Then, we detail (T-S) model, uses the method of fuzzy coalescence for the identification of the nonlinear systems by the fuzzy C-means (FCM) algorithm. We will in addition present tests of validation of (T-S) model. Then, we will give the results of identification and modeling of the station of irrigation by sprinkling.

The remainder of this chapter is described such as the following section. In the first section we have describe the station of irrigation by sprinkling, in which we define the practical constraints existing on the outputs pressure and flow and other components of our station, secondly, in this section we detail the operation mode and the flowcharts of the closed loop mode with any controller and how select the operation mode. In the second section we have describe the Fuzzy coalescence algorithms. Thirdly, we spend to detail the FCM algorithm step by step. Finally, we finished by application of FCM algorithm to the irrigation station by sprinkling located in the laboratory shown in the Fig. 1. After identification and modelisation with FCM algorithm it is necessary to validate our simulation results (model mathematic of our pumping station) with Root Mean Square Error test (RMSE) and the Variance accounting for test (VAF) and many other validation tests we have test the stability of our open loop model, after modelisation and identification we control our T-S obtained model by two types of controllers, PI controller of the station of irrigation and Fuzzy logic regulator, and we finish our chapter with a comparative study between these controllers.

Fig. 1 Overview of the irrigation station by sprinkling

2 Description of the Irrigation Station with Sprinkling

The French company LEROY–SOMMER makes available to researchers an irrigation station (Fig. 1) with sprinkling but with practical constraints existing in the real irrigation stations (Mejri et al. 2013), this station is composed of two parts: hydraulic circuit and an electrical cabinet (Sommer 1996).

2.1 Practical Constraints of Irrigation Station

Before going to modeling our irrigation station, we will submit all practical constraints existing in the real irrigation stations, because the desired performances it is necessary that it respects the following constraints:

- Regulation of the flow and water pressure:

$$\begin{aligned} Q(F,t) &\Rightarrow Q_{ref} \\ P(F,t) &\Rightarrow P_{ref} \end{aligned} \quad (1)$$

- Constraints on the control:

$$N_{\min} \leq N(t) \leq N_{\max} \quad (2)$$

The fixed speed pump will be active or not. N the numbers of turns of the variable speed pump.

- Constraints on the state:

$$Q_{\min} \leq Q(x,t) \leq Q_{\max} \tag{3}$$

  x unspecified position of drain.
- Constraints on the output:

$$P_{\min} \leq P(F,t) \leq P_{\max} \tag{4}$$

- Constraints over the computing time:
  Sample time: $Te = 0.2\,\text{s}$
- Energy constraints:
  Concerning the operation of electrical equipment, cost optimization of pumping and turbine.
- The constraints of operation:
  As they may be related to the geometry of the system levels maximum, minimum, and so on. How it should be managed to ensure the functions given to him: instructions, etc.
- The constraints of safety:
  This may result in the need to keep such a volume of safety in reserve, ensuring the supply in case of unforeseen demand or incidents on the network.

2.2 Operation Mode of the Irrigation Station

The general diagram of the hydraulic system is given by the following Figs. 2 and 3:

Our station of irrigation is fed with an electrical network $400\,\text{V}(TRI + N + PE), 50\,\text{Hz}$. $(TRI + N + PE), 50\,\text{Hz}$ (Sommer 1996).

The station of irrigation starting from the cabin, we can select the operating process of the station through a selector with 6 positions

- 0 Stop;
- 1 operation in Automatic mode;
- 2 operation in Semi-automatic mode;
- 3 operation in mode Forced;
- 4 operation in mode API;
- 5 operation in mode Open loop;

Then the selection of the operating process be described in this following. (Fig. 4)

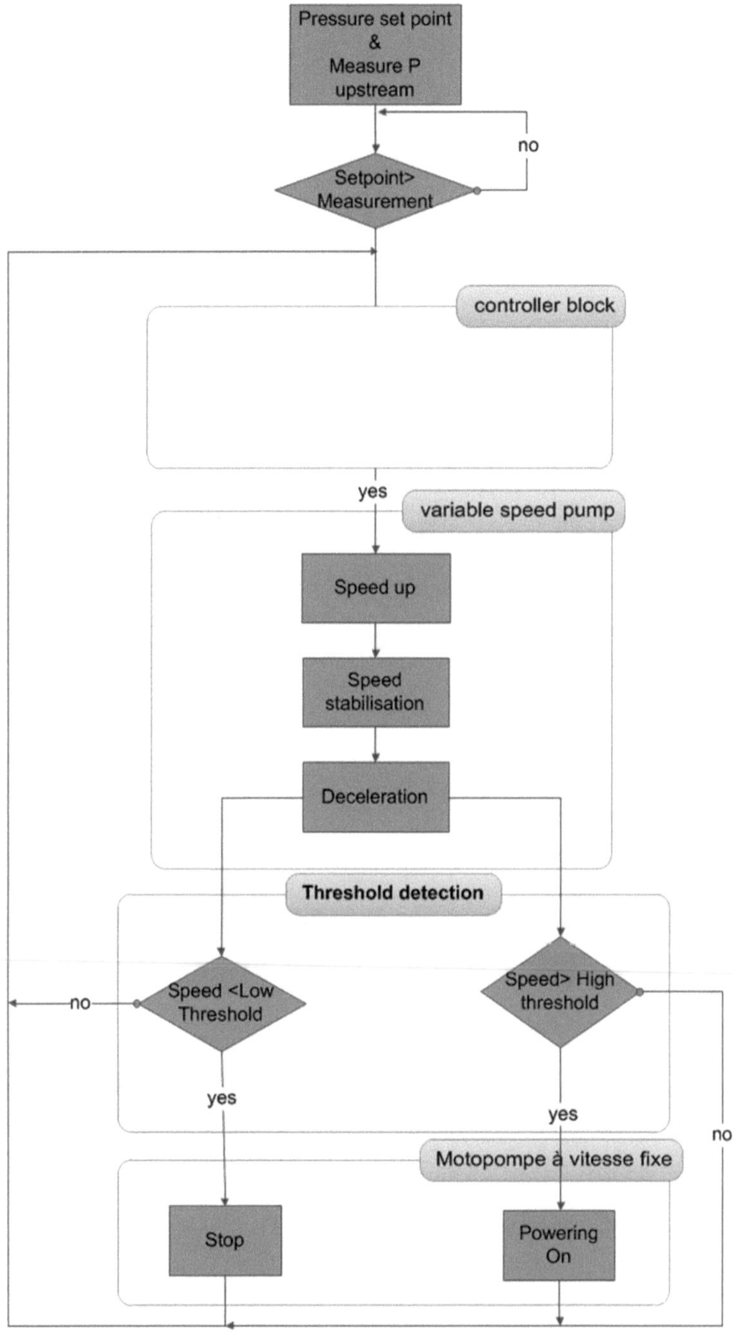

Fig. 2 Closed loop operational flowchart of the irrigation station

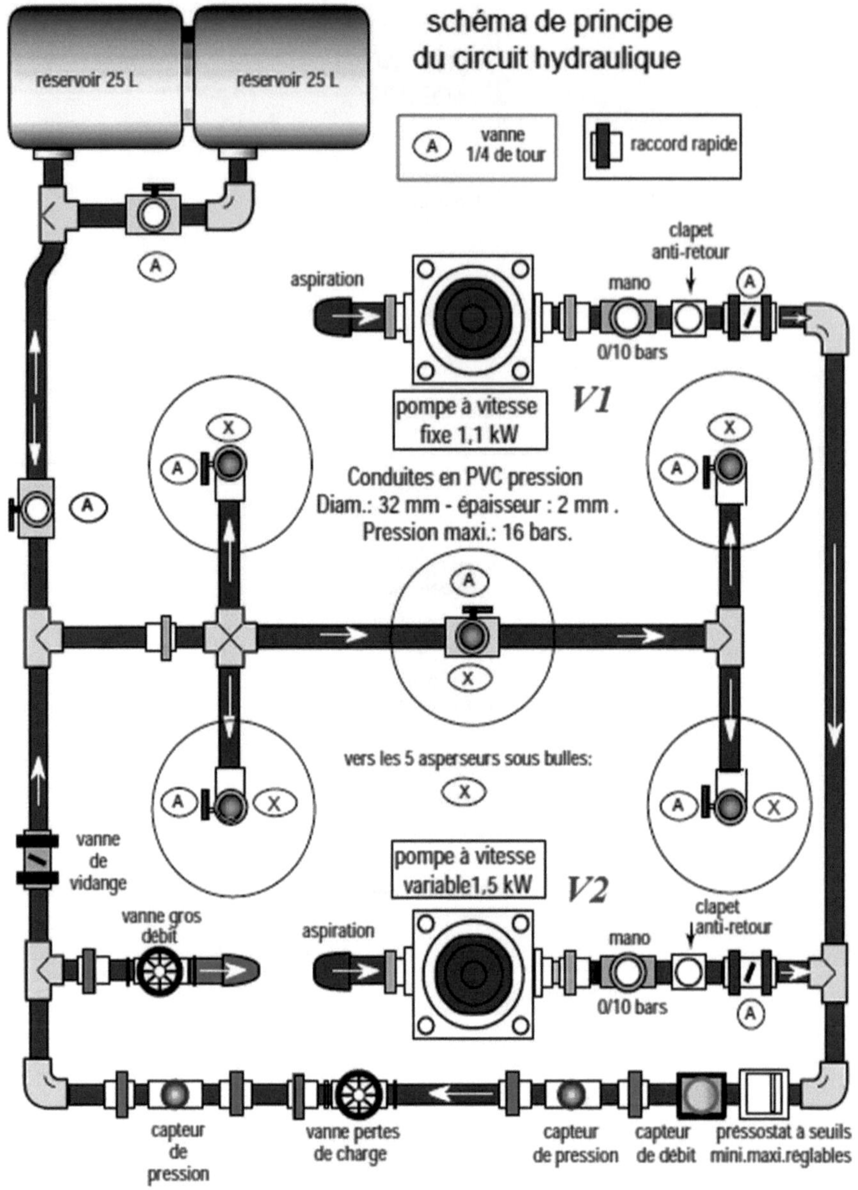

Fig. 3 General diagram of the hydraulic system

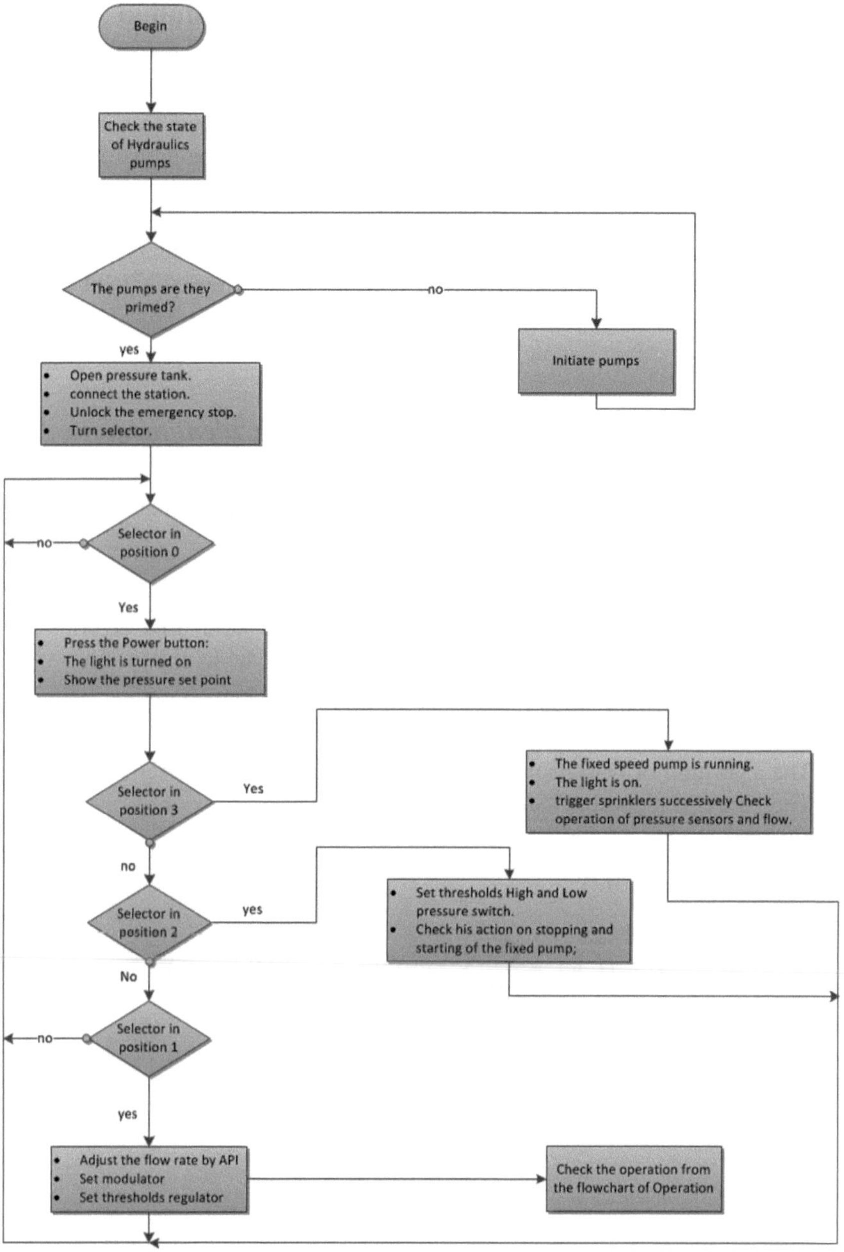

Fig. 4 Function diagram of operating modes

3 Identification and Modeling of the Irrigation Station

The implementation of a mathematical model of a complex real process operating in a stochastic environment draw the attention of many researchers in various disciplines of science and technology. In this context the use of traditional methods of modeling and identification in order to estimate the parameters of such a type of process cannot satisfy the desired performance indices (speed, accuracy and stability). To overcome this problem, other techniques such as fuzzy logic (Azar 2010b, 2012) and more particularly the T-S fuzzy model showed a good result in the identification of these processes types.

3.1 Fuzzy Coalescence Algorithms for System Identification

Let us consider a system described by the following differential equation:

$$y(k) = f_{NL}(x_k) \tag{5}$$

with x_k represent the observation vector, $x_k \in R^n$. The most used algorithms of fuzzy coalescence for the identification parameters of 5 are as follows:

- The algorithm of the fuzzy C-averages, or fuzzy c-Means (FCM) (Bezdek 1981; Chen et al. 1998),
- The algorithm of Gustafson-Kessel (GK) (Gustafson and Kessel 1979),
- The NRFCM algorithm (Soltani et al. 2012).

All these algorithms are based on their minimization of a function objectifies form (Troudi et al. 2012):

$$J(X, U, V) = \sum_{k=1}^{N} \sum_{i=1}^{c} (\mu_{ik})^m (x_k - v_i)^T M (x_k - v_i) \tag{6}$$

where: $X = \{x_k/k = 1, 2, \ldots, N\}$, such that N donate the number of observations; $U = [\mu_{ik} \in [0, 1]^{(c \times N)}]$, the fuzzy partition matrix of data vector X: with

$$\sum_{i=1}^{c} \mu_{ik} = 1 \quad 1 \leq i \leq c \tag{7}$$

V: The prototype clusters vector,

$V = \{v_1, v_2, \ldots, v_c\}$, where c represents the rule number (or of clusters) and $v_i \in R^n$,

m: represent the weighting degree

This parameter influences directly on the form of cluster in data space. Indeed, when m is close to 1, the function of the membership of each cluster becomes almost Boolean i.e., $\mu_{ik} \in \{0, 1\}$. Whereas when m becomes very large, the partition becomes fuzzier and $\mu_{ik} = 1/c$

Generally m is selected between 1.5 and 2.5 but in several applications, it is selected between 2 and 4.

In the following section, we present the fuzzy c-means algorithm.

3.2 Fuzzy c-Means (FCM) Algorithm

This method is based on minimization of the criterion obtained by the addition of the standardization constraint (Troudi et al. 2011).

$$J(X, U, V) = \sum_{k=1}^{N} \sum_{i=1}^{c} (\mu_{ik})^m (x_k - v_i)^T M (x_k - v_i) + \sum_{k=1}^{N} \lambda_k \left[\sum_{i=1}^{c} \mu_{ik} - 1 \right] \quad (8)$$

In this case the minimization of the criterion 8 can be solved by cancelling the derivative of J where the variables are U, V and λ. The solution of this criterion is given by:

$$\begin{aligned} v_i &= \frac{\sum_{k=1}^{N} (\mu_{ik})^m \cdot x_k}{\sum_{k=1}^{N} (\mu_{ik})^m} \\ \mu_{ik} &= \frac{1}{\sum_{j=1}^{c} (d_{ik}/d_{jk})^{\frac{2}{m-1}}} \end{aligned} \quad (9)$$

where d_ik: represent the distance enters X_k and v_i

$$d_{ik} = (x_k - v_i)^T M (x_k - v_i) \quad (10)$$

M: generally selected equal to the identity. The prototype vector of the clusters is given by:

$$d_{ik}^2 = (x_k - v_i)^T (x_k - v_i) \quad i = 1, \ldots, c;\ k = 1, \ldots, N \quad (11)$$

The iteration count of c-means algorithm is selected according to the precise details required by the expert and according to the type of application considered. The criterion of the stop is selected by satisfying the following condition:

$$\left\| U^{(l)} - U^{(l-1)} \right\| < \delta \quad (12)$$

where l is the iteration count.

Fuzzy c-means algorithm (FCM): Being given a whole of data X, FCM algorithm is described by the following stages (Fig. 5):

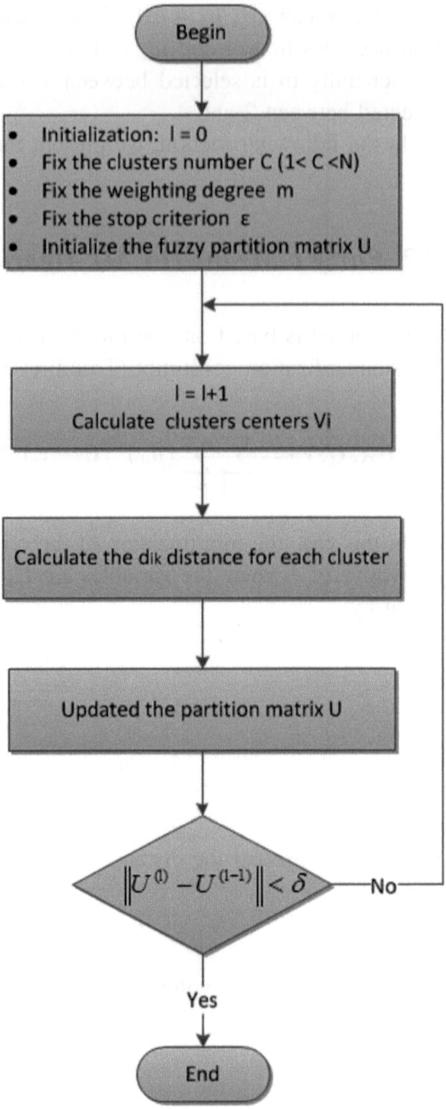

Fig. 5 Fuzzy c-means algorithm (FCM)

The FCM Algorithm converges in general towards a local minimum of the objective function. Its performance depends on several factors such as:

- The cluster number;
- Choice m;
- Choice of stop criterion.

3.3 Determination of Consequent System Parameters

The identification of consequent parameters is necessary to determine the equivalent TS model such system, we find in the literature many identification methods such as the method of ordinary least square (Bertrand and Moonen 2012) (LMS used for linear system) Method of recursive least square (RLS) (Duan et al. 2011), weighted least square (WLS) (Li et al. 2009), recursive least square weighted (RWLS) (Soltani and Chaari 2013) (this method is used for the noisy nonlinear systems).

In our case we used in the identification algorithm method of recursive least square (RLS) (Duan et al. 2011; Chakchouk et al. 2014).

We know the form of T-S model $f_i = a_i^T x + d_i$, then the vector of consequent parameters written as follow:

$$\theta_i = \left[a_i^T, d_i\right]^T \tag{13}$$

the increased regression matrix is defined by:

$$X_e = [X, 1] \tag{14}$$

then we defined the gains matrix with the follow equation:

$$P(N) = \left[X^T(N) \cdot X(N)\right]^{-1} \tag{15}$$

$P(N)$ can be written as follows:

$$P^{-1}(N) = \lambda(N) + \lambda(N) \cdot \mu(N) \cdot x(N) \cdot x^T(N)) \tag{16}$$

If we applied the matrix inverse theorem then:

$$P(N) = \frac{1}{\lambda}\left[P(N-1) - \frac{P(N-1) \cdot x^T(N) \cdot x(N) \cdot P(N-1)}{\frac{1}{\mu(N)} + x^T(N) \cdot P(N-1) \cdot x(N)}\right] \tag{17}$$

Then we defined the gain $G(N)$ with the following equation:

$$G(N) = \left[\frac{P(N-1) \cdot x^T(N)}{\frac{1}{\mu(N)} + x^T(N) \cdot P(N-1) \cdot x(N)}\right] \tag{18}$$

Then the regression matrix and the parameters consistent vector is as follow:

$$\begin{aligned}\theta(N) = {} & \left[I - G(N)x^T(N)\right]\theta(N-1) \\ & + \mu(N)\left[P(N-1)x(N) - G(N)x^T(N)P(N-1)x(N)\right]y(N)\end{aligned} \tag{19}$$

If we factorize the Eq. 15, we have:

$$\theta(N) = \theta(N-1) + G(N)\left[y(N) - x^T(N)\theta(N-1)\right]y(N) \quad (20)$$

3.4 Application of FCM Algorithm on the Station of Irrigation by Sprinkling

Let us consider a system described by the Eq. 6. Firstly, we approximate the nonlinear function Eq. 6 by the model of Takagi-Sugeno (TS):

$$R^i : if\ x_{k1}\ is A_{i1}\ and\ x_{k2}\ is\ A_{i2}\ and \ldots and\ x_{kn}\ is\ A_{in}\ then\ y^i = a_i^T x_k + b_i \quad (21)$$

To represent the rule, we need use observations vector $x_k = \left[x_{k1}, x_{k2}, \ldots, x_{kn}\right]^T$ the units fuzzy $A_{i1}, A_{i2}, \ldots, A_{in}$ to identify the parameters in the model 21, we builds the matrix of regression X and the vector of the output Y starting from measurements resulting from the system such as: $X = \left[x_1^T, x_2^T, \ldots, x_N^T\right]^T$ *and* $Y = \left[y_1, y_2, \ldots, y_N\right]^T$ with $N \gg n$.

The identification of T-S model parameters requires a taking away of the real signals of irrigation station. Using a numerical oscilloscope, we took the real dynamics of pressure and flow of the station of irrigation by sprinkling, then (Figs. 6, 7 and 8):

These results are taken from connectors of the cabinet.

In order to initialize the iteration count l = 0, we fix the weighting degree m = 2.75 what makes it possible to initialize the partial random matrix U. We pass

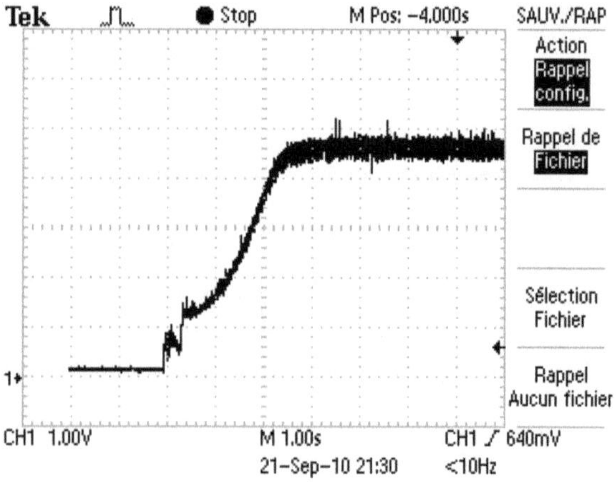

Fig. 6 Real curve of the pressure evolution

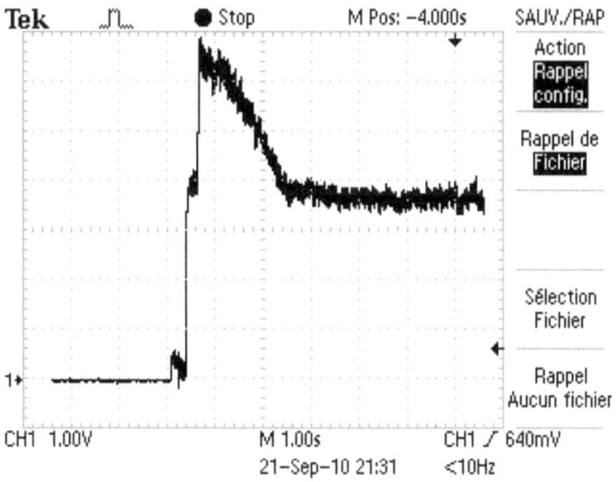

Fig. 7 Real curve of the flow evolution

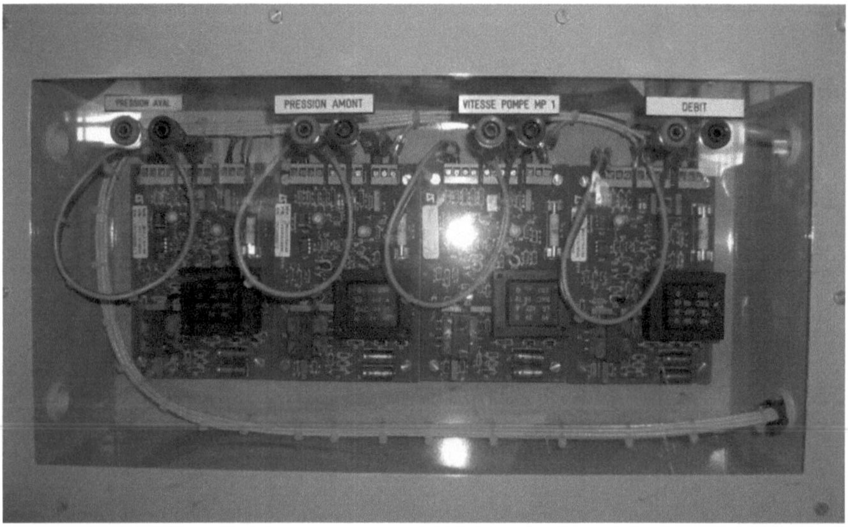

Fig. 8 Connectors of sampling signals (pressure and flow)

then to the choice of the number of clusters. We apply the classification entropy test CE for each outputs pressure (*P*) and flow (*Q*). We noted *CEP CEQ* respectively.

$$\begin{aligned} C_{ec}(c) &= \frac{1}{N}\sum_{k=1}^{N}\sum_{i=1}^{c}\mu_{ik}\log(\mu_{ik}) \\ C_{opt} &= \min[C_{ec}(c)] \end{aligned} \tag{22}$$

Then the optimal number of clusters is equal to 3 as it indicates in the 1 (Figs. 9, 10 and Table 1).

The excitation signal must be rich to run the system in all operating region. In order to reach all steps, the simulation results of the FCM algorithm are given by the following Figs. 11 and 12.

Algorithm FCM is followed the real data input of pressure and flow. It is noticed that the error between the evolution of the real and estimated pressure is almost null even for flow. The station of irrigation by sprinkling made up of two nonlinear systems in the same way input and different output, one of pressure and the other of flow, each one partitioned in 3 subsystems. We obtain the following results:

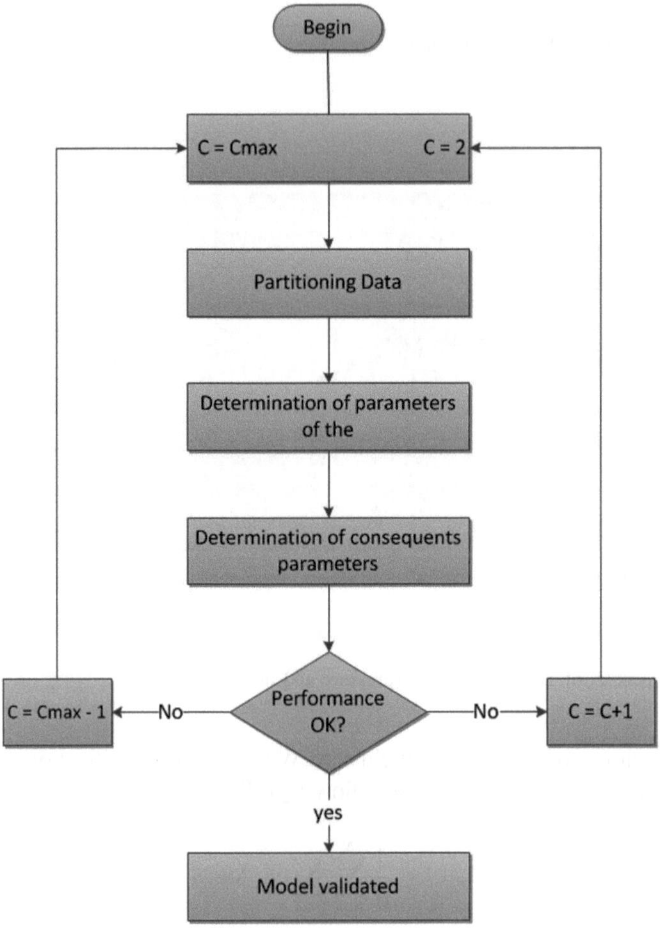

Fig. 9 Diagram of cluster number choice

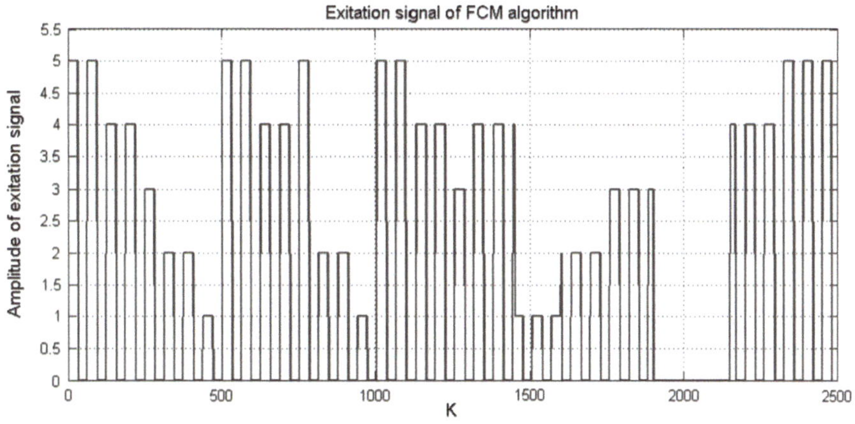

Fig. 10 Excitation signal of FCM algorithm

Table 1 Results of classification entropy test

	C = 2	C = 3	C = 4	C = 5
CEP (10^{-6})	−0.491	−5.21	−0.41	−1.18
CEQ (10^{-6})	−5.4	−10.8	−3.35	−3.58

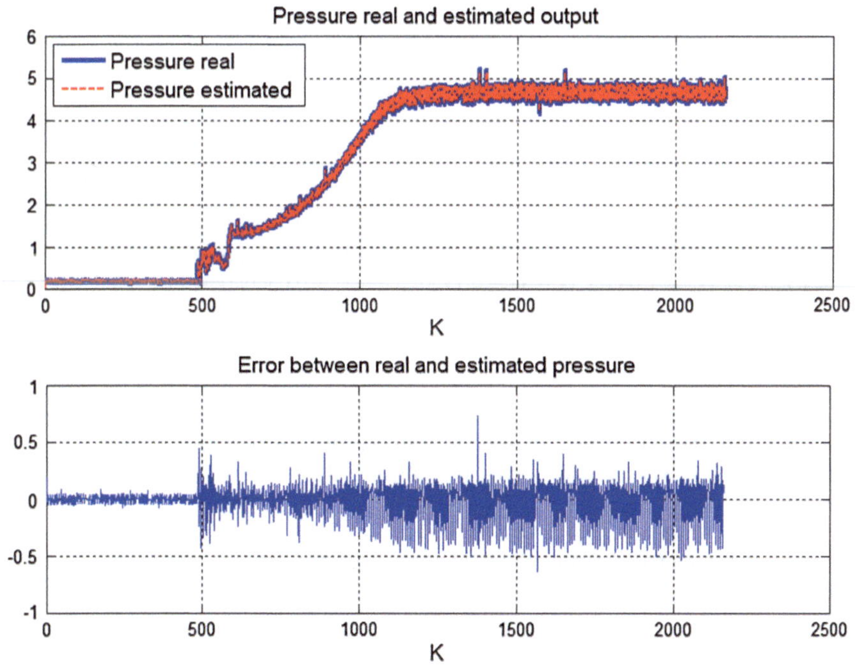

Fig. 11 Simulation results of FCM algorithm for the pressure output

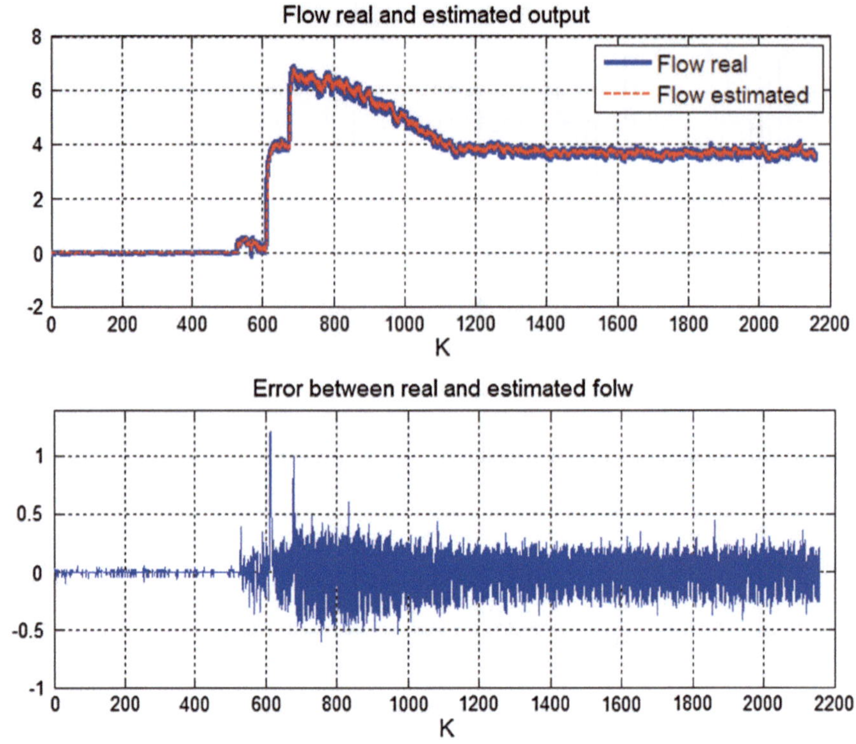

Fig. 12 Simulation results of FCM algorithm for the flow output

- For the pressure sub-systems:

$$\begin{cases} R_{p1} : y_{P1}(k) = 1.0853y_p(k-1) - 0.1744y_p(k-2) + 0.0570u(k-1) + 0.0318u(k-2) \\ R_{p2} : y_{P2}(k) = 1.0851y_p(k-1) - 0.1743y_p(k-2) + 0.0565u(k-1) + 0.0320u(k-2) \\ R_{p2} : y_{P2}(k) = 1.0852y_p(k-1) - 0.1750y_p(k-2) + 0.0560u(k-1) + 0.0315u(k-2) \end{cases}$$

- For the flow sub-systems:

$$\begin{cases} R_{Q1} : y_{Q1}(k) = 1.0853y_Q(k-1) - 0.1744y_Q(k-2) + 1.4118u(k-1) - 1.31u(k-2) \\ R_{Q1} : y_{Q1}(k) = 1.0851y_Q(k-1) - 0.1743y_Q(k-2) + 1.4116u(k-1) - 1.33u(k-2) \\ R_{Q1} : y_{Q1}(k) = 1.0852y_Q(k-1) - 0.1750y_Q(k-2) + 1.4120u(k-1) - 1.31u(k-2) \end{cases}$$

For the total identification of system we can draw a rule for each subsystem (flow and pressure) as being modeling and linearization of the whole system, through intermediary of the Eq. 23:

$$y(k+1) = \frac{\sum_{i=1}^{c} \mu_{ik} \cdot (x(k)) \cdot y_i(k+1)}{\sum_{i=1}^{c} \mu_{ik} \cdot (x(k))} \tag{23}$$

Then the global rule of the pressure output is as follow:

$$\begin{aligned} R_{PG} : y_{PG}(k) =& 1.0851 y_p(k-1) - 0.1745 y_p(k-2) \\ &+ 0.0563 u(k-1) + 0.0317 u(k-2) \end{aligned} \tag{24}$$

and the global rule of flow output is as follow:

$$\begin{aligned} R_{QG} : y_{QG}(k) =& 1.0851 y_Q(k-1) - 0.1745 y_Q(k-2) \\ &+ 1.4116 u(k-1) - 1.32 u(k-2) \end{aligned} \tag{25}$$

thus, the open loop transfer functions are:

$$\begin{cases} H_{BOP} = \dfrac{0.05632z + 0.0317}{z^2 - 1.0851z + 0.1745} \\ H_{BOQ} = \dfrac{1.4116z - 1.32}{z^2 - 1.0851z + 0.1745} \end{cases} \tag{26}$$

The discrete state representation associated with system 26:

$$\begin{cases} \begin{bmatrix} P_{k+1} \\ Q_{k+1} \end{bmatrix} = \begin{bmatrix} 0.1422 & -0.4403 \\ 0.0917 & 0.9428 \end{bmatrix} \begin{bmatrix} P_k \\ Q_k \end{bmatrix} + \begin{bmatrix} 0.0917 \\ 0.0119 \end{bmatrix} u_k \\ y_k = \begin{bmatrix} 0 & 4.7235 \\ 14.7535 & 4.9165 \end{bmatrix} \begin{bmatrix} P_k \\ Q_k \end{bmatrix} + \begin{bmatrix} 0 \\ 0 \end{bmatrix} u_k \end{cases} \tag{27}$$

We introduce the delay $\tau = 5\,\text{s}$ into the model obtained, The system sampling period is chosen $T_e = 0.2$ s then the delay $\tau = 5\,\text{s}$ is calculated at field discrete time by $z^{-\frac{\tau}{T_e}} = z^{-\frac{5}{0.2}} = z^{-25}$. The system 26 becomes:

$$\begin{cases} H_{BOP} = z^{-25} \dfrac{0.05632z + 0.0317}{z^2 - 1.0851z + 0.1745} \\ H_{BOQ} = z^{-25} \dfrac{1.4116z - 1.32}{z^2 - 1.0851z + 0.1745} \end{cases} \tag{28}$$

3.5 Validation Tests of T-S Model

Therefore, to ensure that the model obtained from the estimation it is compatible with other forms of inputs to represent correctly system functioning to identify. In this paragraph, statistical tests to validate a fuzzy model based on Root Mean Square Error test, Variance accounting for, the residues autocorrelation function and on the cross-correlation between residues and other inputs in the system.

- Root Mean Square Error test (RMSE) (Troudi et al. 2011):
  This is an overall measure of the deviation of total points number from the expected value.

$$RMSE = \sqrt{\frac{1}{N}\sum_{k=1}^{N}(y_k - \hat{y}_k)^2} \tag{29}$$

- Variance accounting for test (VAF) (Troudi et al. 2011):
  This criterion evaluates the quality percentage of a model by measuring the normalized variance of the difference between two signals.

$$VAF = 100\,\%\left[1 - \frac{var(y - \hat{y})}{var(y)}\right] \tag{30}$$

- Autocorrelation function of the residues:

$$\hat{r}_{\varepsilon\varepsilon}(\tau) = \frac{\sum_{k=1}^{N-\tau}\left(\varepsilon(k,\hat{\theta}) - \bar{\varepsilon}\right)\left(\varepsilon(k-\tau,\hat{\theta}) - \bar{\varepsilon}\right)}{\sum_{k=1}^{N}\left(\varepsilon(k,\hat{\theta}) - \bar{\varepsilon}\right)^2} \tag{31}$$

- Cross-correlation between residues and inputs previous:

$$\hat{r}_{u\varepsilon}(\tau) = \frac{\sum_{k=1}^{N-\tau}(u(k) - \bar{u})\left(\varepsilon(k-\tau,\hat{\theta}) - \bar{\varepsilon}\right)}{\sqrt{\sum_{k=1}^{N}(u(k) - \bar{u})^2}\sqrt{\sum_{k=1}^{N}\left(\varepsilon(k,\hat{\theta}) - \bar{\varepsilon}\right)^2}} \tag{32}$$

with

$$\begin{aligned}\bar{\varepsilon} &= \frac{1}{N}\sum_{k=1}^{N}\varepsilon(k)\\ \bar{u} &= \frac{1}{N}\sum_{k=1}^{N}u(k)\end{aligned} \tag{33}$$

ε: Is the prediction error and $u(k)$ is the system input. $x(k)$ can take either the value ε or $u(k)$. Ideally, if the model is valid, the result of these correlation tests gave the following results:

$$\hat{r} = \begin{cases} 1, & \tau = 0 \\ 0, & \tau \neq 0 \end{cases} \text{et} \hat{r}_{u\varepsilon}(\tau) = 0, \quad \forall \tau \tag{34}$$

Typically, we verified that the functions $\hat{r}$ are zero for the interval $\tau \in [-20, 20]$ with a confidence interval of 95 %, then:

$$\frac{-1.96}{\sqrt{N}} < \hat{r} < \frac{1.96}{\sqrt{N}}. \tag{35}$$

3.6 Results of Validation Tests

- Root Mean Square Error test (RMSE):

$$\begin{cases} RMSE_{pressure} = 0.1471 \\ RMSE_{flow} = 0.1926 \end{cases} \tag{36}$$

- Variance accounting for test (VAF):

$$\begin{cases} VAF_{pressure} = 99.6090\ \% \\ VAF_{flow} = 99.3272\ \% \end{cases} \tag{37}$$

- Autocorrelation and cross-correlation function results (Figs. 13 and 14):

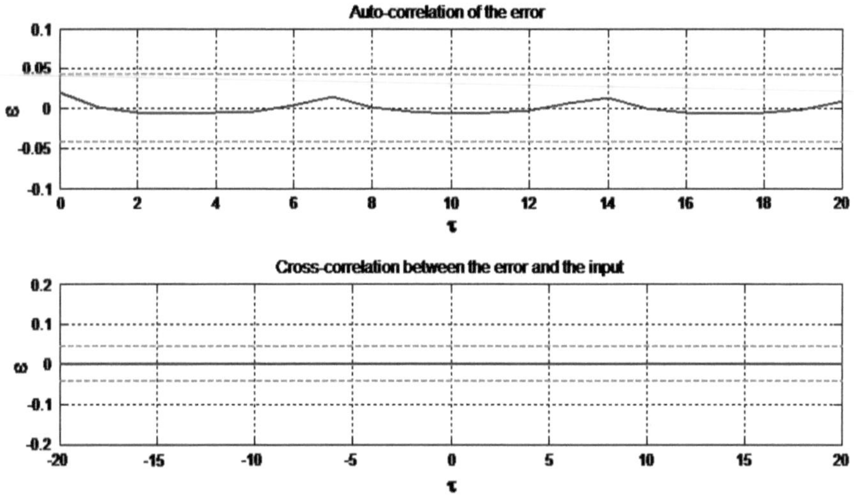

Fig. 13 Validation results autocorrelation and cross-correlation of pressure output

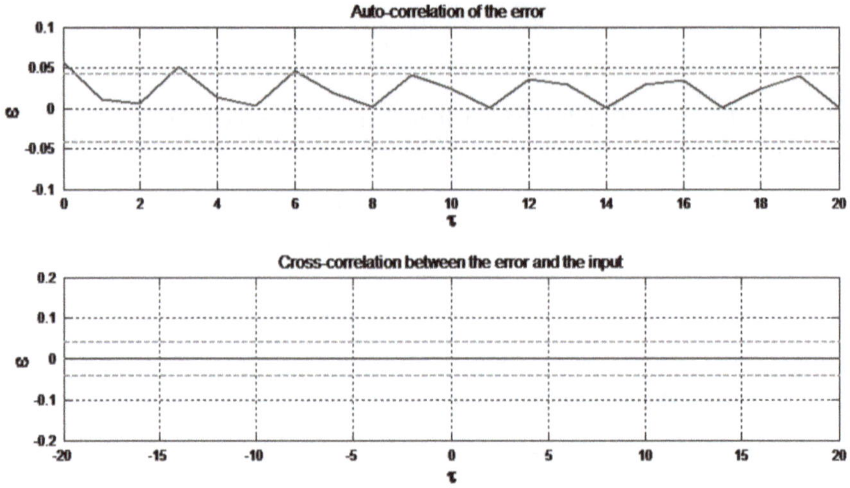

Fig. 14 Validation results autocorrelation and cross-correlation of flow output

Fig. 15 Comparison between measured and estimated pressure

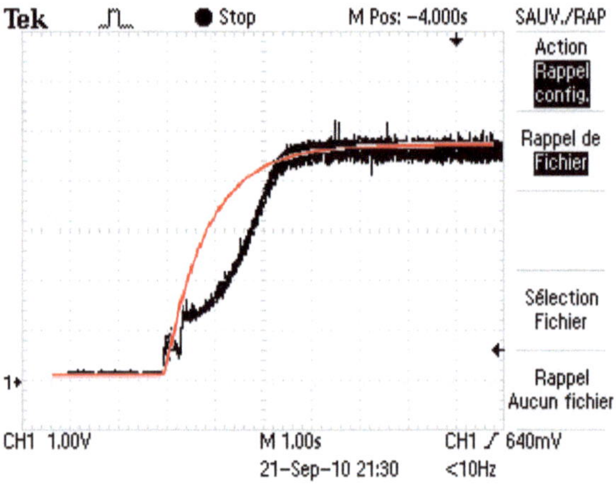

A comparison was made between the estimated outputs and actual outputs collected using a digital oscilloscope (Figs. 15 and 16).

The results simulations of irrigation station model are confused with those of the real taking away.

3.7 Stability Analysis

In this part we interested to study the stability of estimated model, first of all will analyze the behavior of the discrete model obtained.

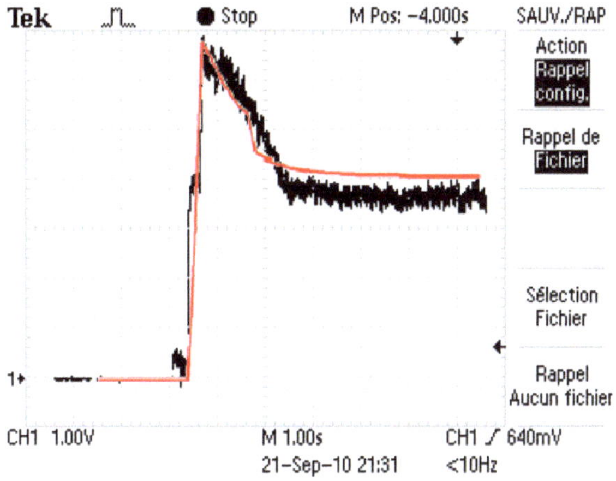

Fig. 16 Comparison between measured and estimated flow

- Lemma 1:
  A linear dynamic system is stable if and only if, isolated from its equilibrium position by an external request, the system returns to this position when the request ceased (Eivd 2005).
- Lemma 2:
  A discrete linear dynamic system is stable, if and only if, all poles of transfer function are located inside the unit disc.

$$|p_i| < 1 \tag{38}$$

Initially, we referring to lemma 1 we will test the stability of irrigation station model by impulse response which gives Fig. 17:

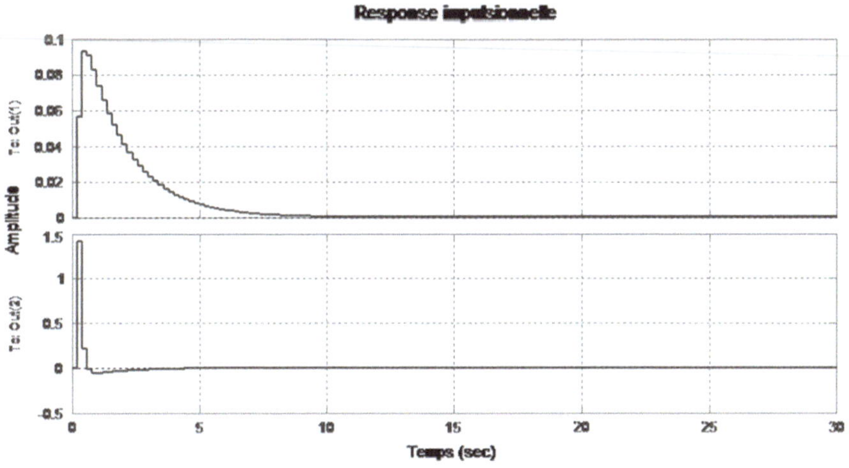

Fig. 17 Impulse response of the system

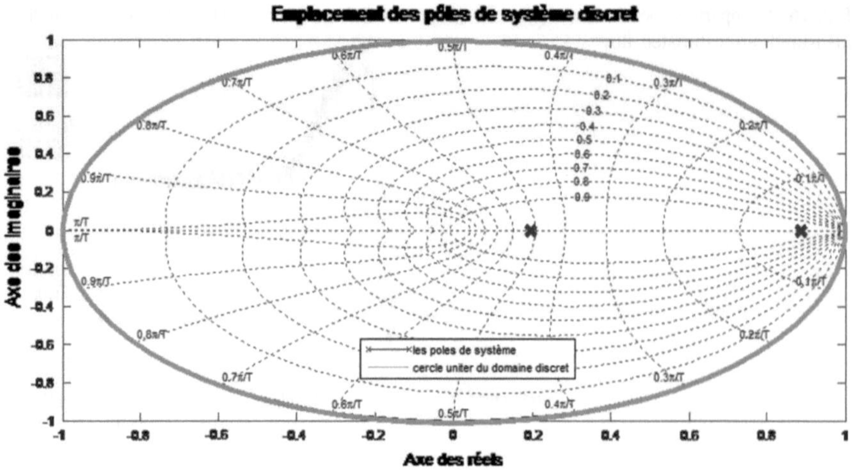

Fig. 18 Location of discrete time system poles

The test of stability by placement of the poles in the discrete place of the poles, gives Fig. 18:

We noticed well that the model obtained is a model associated with a stable system from where the poles modules are strictly lower than one.

4 Control of Irrigation Station with Sprinkling

4.1 Control of Station with PI Regulation

The irrigation station is equipped with a PI controller card which is provided by LEROY-SOMMER, this controller ensures specific control for the pumps. The originators in the LEROY-SOMMER company (Sommer 1996), chooses the parameters of following adjustments $K_p = 0.5 \;\; T_i = 1$ m.

$$\frac{\mathrm{U(s)}}{\varepsilon(s)} = K_p\left(1 + \frac{1}{T_i s}\right) \tag{39}$$

The form of discrete regulator PI is given by (Chakchouk et al. 2014) (Fig. 19):

$$\frac{\mathrm{U(s)}}{\varepsilon(s)} = \frac{K_p\left(1 + \frac{T}{T_i}\right) - K_p z^{-1}}{1 - z^{-1}} = \frac{r_0 + r_1 z^{-1}}{1 - z^{-1}} = \frac{r_0 z + r_1}{z - 1} \tag{40}$$

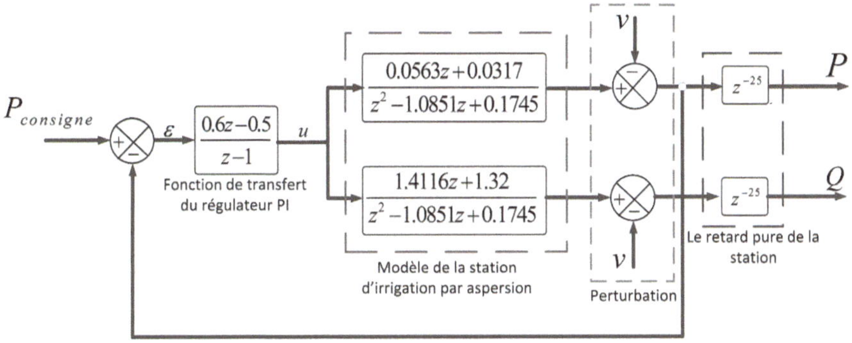

Fig. 19 Functional diagram of the system buckled with PI controller

4.2 *Simulation Results of PI Controller*

The result obtained by PI regulator ensures the control of the pressure because the answer follows the instruction given. In the presence of 20 % disturbance, the robustness of this technique of regulation appears in the compensation of the latter. The major disadvantage of this method of regulation resides primarily at the problem of adaptation of the controller opposite the external variations such as the extension of network of drain, the escapes, etc. (Figs. 20 and 21).

4.3 *Fuzzy Logic Control of the Irrigation Station*

To use the fuzzy controller (Chakchouk et al. 2014), this last must be programmed through the tool FUZZY OF MATLAB. Entries and are chosen of Gaussian form

Fig. 20 Evolution of pressure in presence of PI regulator

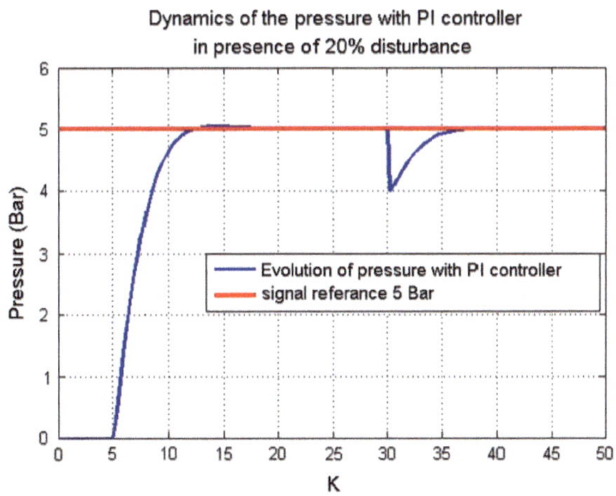

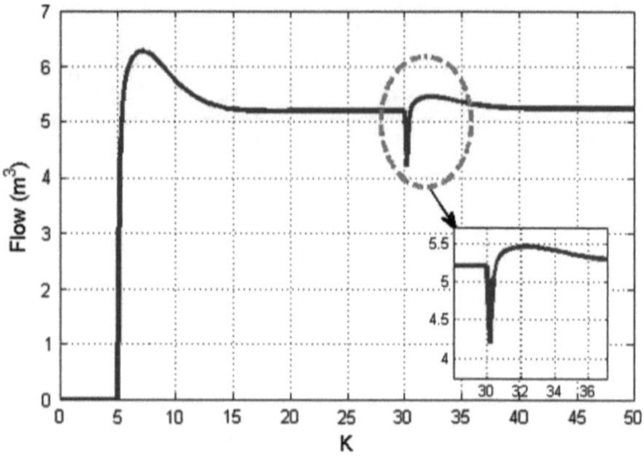

Fig. 21 Evolution of flow in presence of PI regulator

(bell) and one divided the universe of speech of each one into three sets: Z, P, and N. Thus, by using all the possible combinations, nine fuzzy rules were generated for five singletons on the level of the consequence part as it shows in Table 2 (Fig. 22). The rules can be written in the following way:

$$if\,(\varepsilon\, is\, A)\; and\; (\dot{\varepsilon}\, is\, B)\; then\; U_{cf} = S_i(\varepsilon, \dot{\varepsilon}) \tag{41}$$

One uses the method min max like engine of inferences and the centre of gravity for the defuzzification. The exit of the fuzzy controller can be written in the following form:

Table 2 Inference matrix of the fuzzy controller

	N	Z	P
N	PG	PN	NM
Z	PM	Z	NM
P	PM	NM	NG

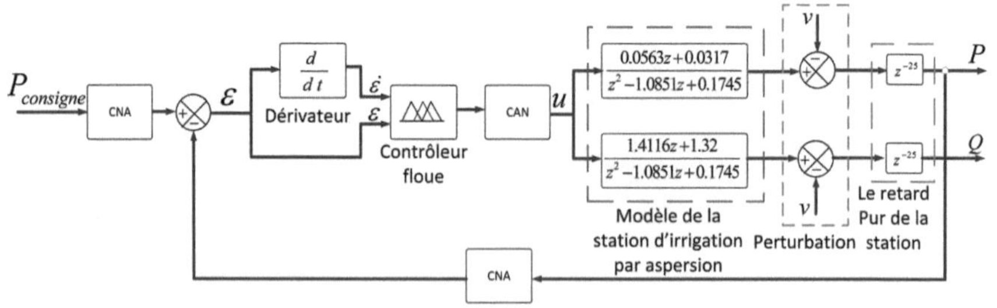

Fig. 22 Functional diagram of system buckled with a fuzzy

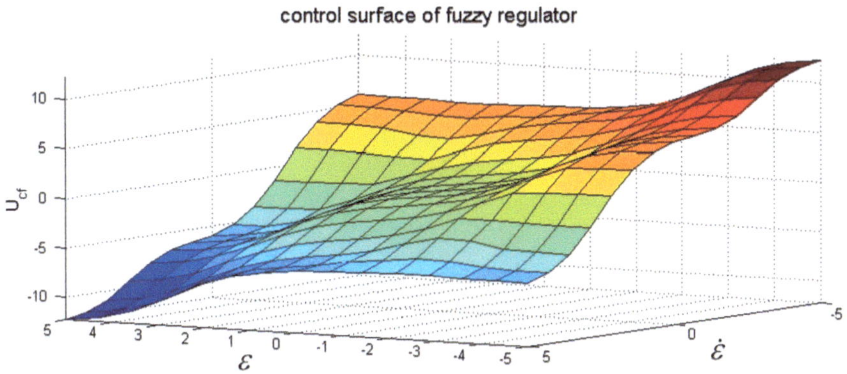

Fig. 23 The decision surface of fuzzy controller

$$S_i(\varepsilon, \dot{\varepsilon}) = \min\left(\mu_A^i(\varepsilon), \mu_B^i(\dot{\varepsilon})\right) \tag{42}$$

$$U_{cfG} = \max(S_i(\varepsilon, \dot{\varepsilon})) \tag{43}$$

The discourse universe of output U_{cf} currency in five fields (Fig. 23)

The decision surface of fuzzy controller reflect a probably smooth law of order what provides us an energy saving on the output of the fuzzy controller (Fig. 24).

4.4 Simulation Results of Fuzzy Controller

The response of the flow if the system is regulated by the fuzzy controller exceeds the maximum flow (8 m^3/h) accepted by the station of irrigation. One thus proposes to add a saturation to compensate for this going beyond (Figs. 25, 26, 27 and, 28).

4.5 Comparative Study

Taking into account the results obtained, we note that for the two examples of regulators, the fuzzy approach suggested makes it possible to obtain the best speed ratio/energy of order however the fuzzy regulation brings a static error to the evolution of the two outputs. The recourse has a profit inserted into the exit makes it possible to reduce this error (Fig. 29, Tables 3 and 4).

The fuzzy controller ensures perfectly the control of the irrigation station by sprinkling, on the other hand regulator PI used in the model appears robust from point of view stabilization in transitory mode (Fig. 30).

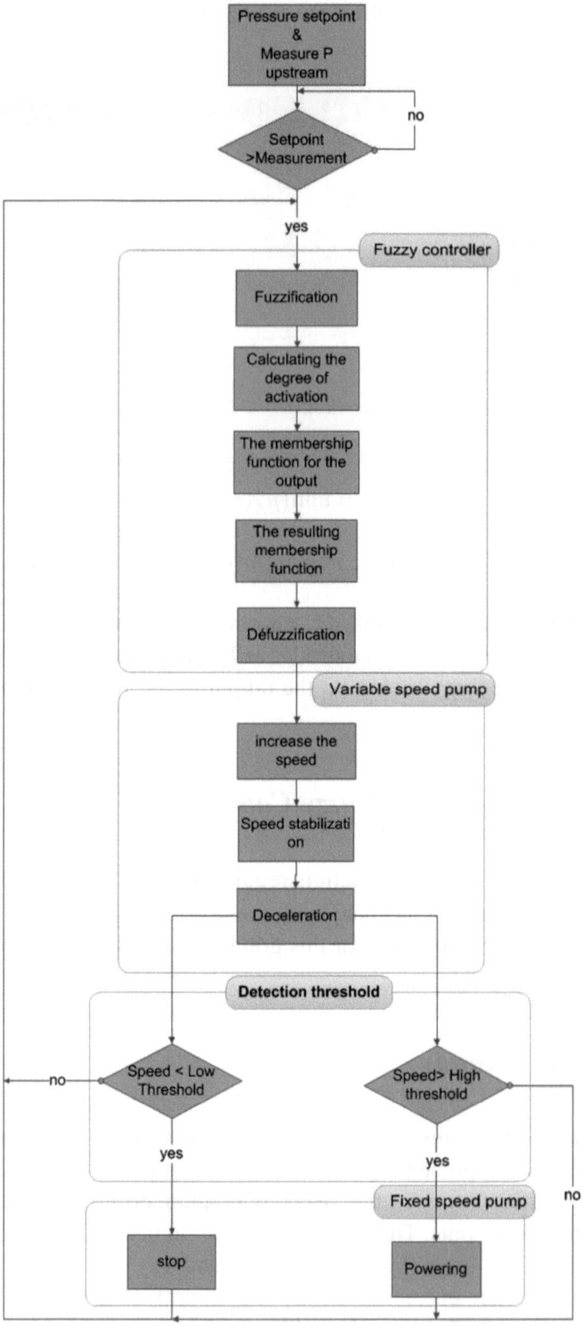

Fig. 24 Flow chart of a regulation cycle with the fuzzy controller

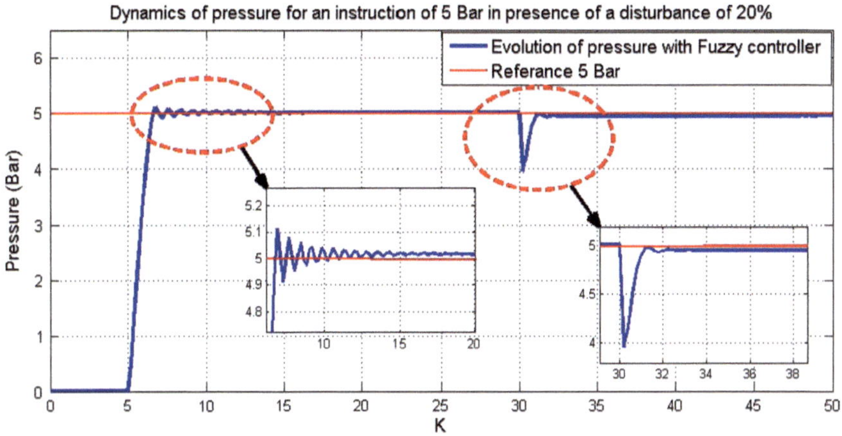

Fig. 25 Evolution of pressure with the Fuzzy controller

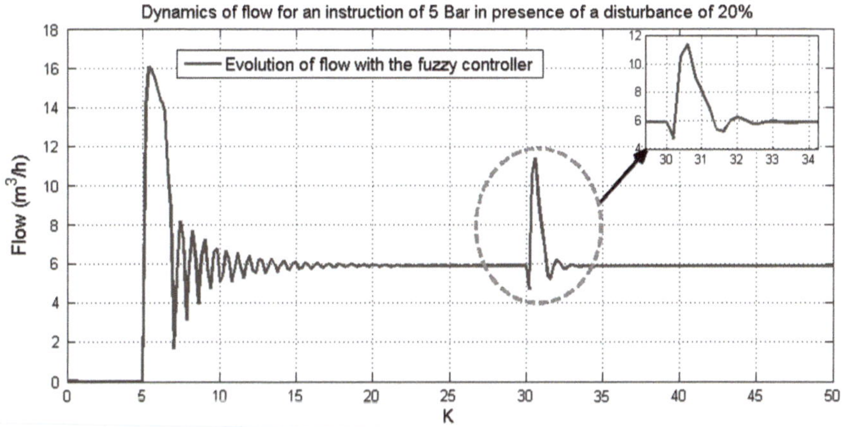

Fig. 26 Evolution of flow with the Fuzzy controller

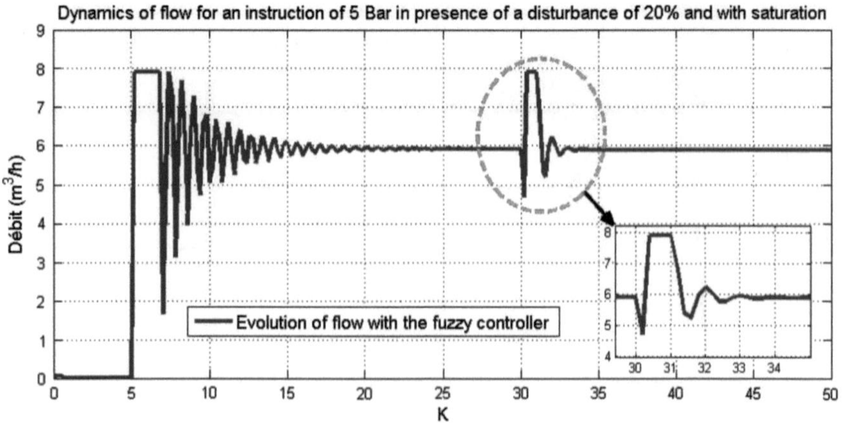

Fig. 27 Evolution of flow with the Fuzzy controller with saturation

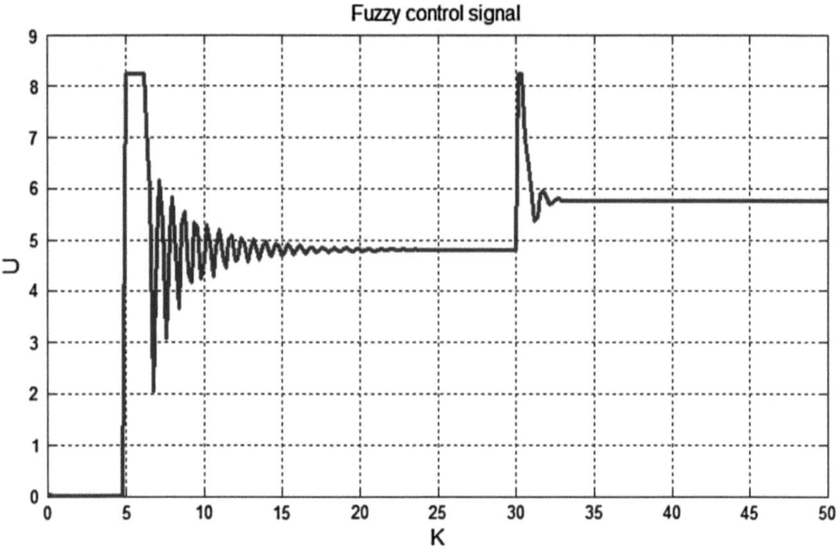

Fig. 28 Fuzzy control signal

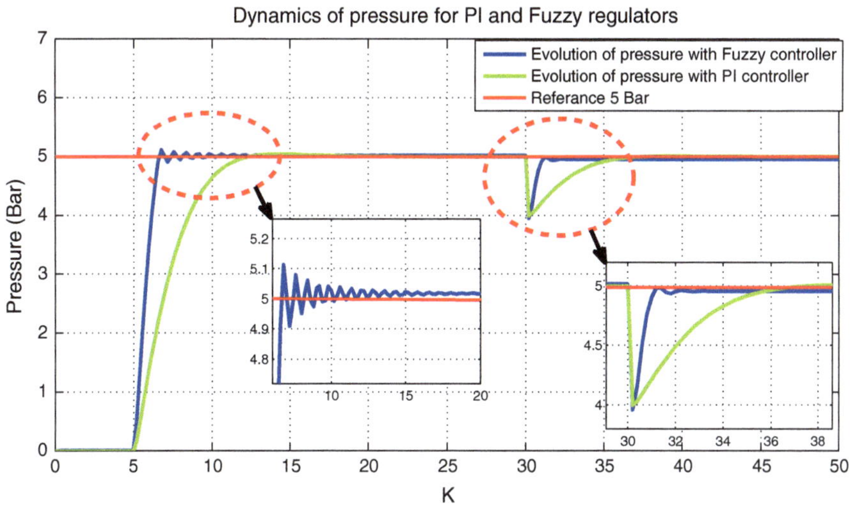

Fig. 29 Evolution of the pressure for the two types of regulators

Table 3 Comparative table enters pi and fuzzy controllers, relating to the pressure

	PI controller	Fuzzy controller
Response time at ±5%	12.6 s	6.6 s
Static error of position (in bar)	0.014	0.01

Table 4 Comparative table enters pi and fuzzy controllers, relating to flow

	PI controller	Fuzzy controller
Response time at ±5%	13.4 s	11.6 s
Static error of position (in bar)	0.8	0.088

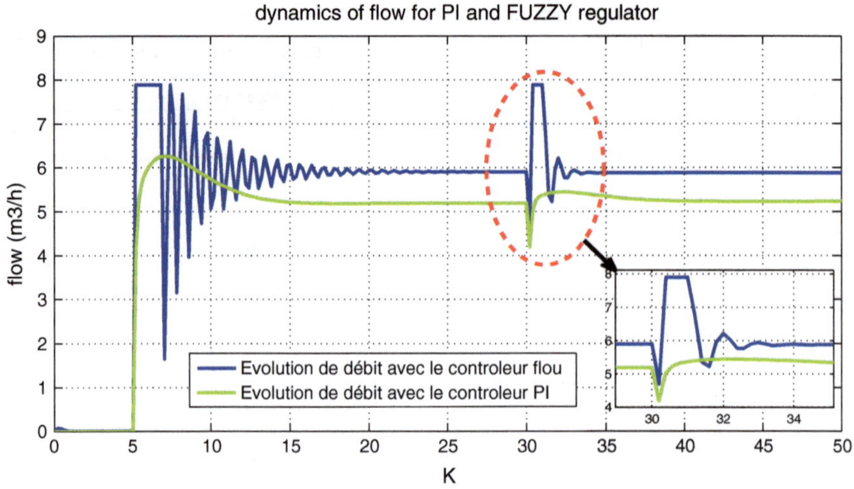

Fig. 30 Evolution of flow for the two types of regulators

5 Conclusion

In this work, we have applied the Takagi-Sugeno algorithm to a station of irrigation with sprinkling (real pumping station) and obtained real values from the station. The system is taken as a black box with outputs pressure and flow. We have modeled and identified the system by the FCM algorithm.

After obtained the T-S model we have validated curves is almost identical to the real ones. The obtained linear model gives a good description of the system behavior in the particle area of nonlinear system, and the importance of the clustering methods.

Even for the control results, the comparison between the results obtained of the two controls types (PI, Fuzzy) enables us to conclude that the fuzzy controller makes it possible to cost reduce of the water pumping.

References

Azar, A. T. (2010a). Adaptive neuro-fuzzy systems. In A.T. Azar (Ed.), *Fuzzy systems. IN-TECH,* Vienna, Austria. ISBN 978-953-7619-92-3.

Azar, A. T. (2010b). *Fuzzy systems*. Vienna, Austria: IN-TECH. ISBN 978-953-7619-92-3.

Azar, A. T. (2012). Overview of type-2 fuzzy logic systems. *International Journal of Fuzzy System Applications (IJFSA), 2*(4), 1–28.

Bertrand, A., & Moonen, M. (2012). Low complexity distributed total least squares estimation in ad hoc sensor networks. *IEEE Transactions on Signal Processing, 60*(8), 4321–4333.

Bezdek, J. (1981). *Pattern recognition with fuzzy objective function algorithms*. New York: Plenum Press.

Chakchouk, W., Zaafouri, A., & Sallami, A. (2014). Control and modelling using Takagi-Sugeno fuzzy logic of irrigation station by sprinkling. *World Applied Sciences Journal, 29*(10), 1251–1260.

Chen, J. Q., Xi, Y. G., & Zhang, Z. J. (1998). A clustering algorithm for fuzzy model identification. *Fuzzy Sets and Systems, 98*(4), 319–329.

Daneshwar, M. & Noh, N. M. (2013). *Adaptive neuro-fuzzy inference system identification model for smart control valves with static friction.* In *IEEE International Conference on Control System, Computing and Engineering (ICCSCE),* pp. 122–126, Mindeb. doi: 10.1109/ICCSCE.2013.6719944.

Duan, H., Jia, J., & Ding, R. (2011). *Two stage recursive least squares parameter estimation algorithm for output error models.* UK: Elsevier.

Eivd, M. (2005). *Digital control course.*

Grisales, V. (2007). Modélisation et commande floue de type Takagi-Sugeno appliquées a un Bioprocédé de traitement des eaux uses. (PhD thesis, Paul Sabaties University Toulouse III and Laos the Andes University, Colombie).

Gustafson, D. & Kessel, W. (1979). Fuzzy clustering with a fuzzy covariance matrix. In *Proceedings of IEEE CDC,* pp. 761–766, San Diego, CA, USA.

Jang, W., Kang, H., Lee, B., Kim, K., Shin, D., & Kim, S. (2007). Optimized fuzzy clustering by predator prey particle swarm optimization. In *IEEE Congress on Evolutionary Computation.*

Li, Z., Liang, H., and Zengjun, B. (2013). T-s fuzzy modeling method based on c-means clustering. In *International Conference on Intelligent Human-Machine Systems and Cybernetics (IHMSC).*

Li, C. S., Zhou, J. Z., Fu, B., Kou, P., & Xiao, J. (2012). T-s fuzzy model identification with gravitational search based hyper-plane clustering algorithm. *IEEE Transaction. Fuzzy System, 20*, 305–317.

Li, C., Zhou, J., Xiang, X., Li, Q., & An, X. (2009). T-s fuzzy model identification based on a novel fuzzy c-regression model clustering algorithm. *Engineering Applications of Artificial Intelligence, 22*(4), 646–653.

Mejri, M. R., Zaafouri, A., & Chaari, A. (2013). Hybrid control of a station of irrigation by sprinkling. *International Journal of Engineering and Innovative Technology (IJEIT), 3*(1), 9–17.

Pingli, L., Yang, Y., & Wenbo, M. (2006). Random sampling fuzzy c-means clustering and recursive least square based fuzzy identification. In *Proceedings of the American control conference.*

Soltani, M., & Chaari, A. (2013). A novel weighted recursive least squares based on euclidean particle swarm optimization. *Kybernetes, 42*(2), 268–281. Emerald Group.

Soltani, M., Chaari, A., & Benhmida, F. (2012). A novel fuzzy c regression model algorithm using new measure of error and based on particle swarm optimization. *International Journal of Applied Mathematics and Computer Science, 22*(3), 617–628.

Sommer, L. (1996). *Technical catalog of irrigation station by sprinkling.* French company LEROY SOMMER.

Takagi, T., & Sugeno, M. (1985). Fuzzy identification of systems and its application to modelling and control. *IEEE Transaction, System, Man Gyber, 15*(1), 116–132.

Troudi, A., Houcine, L., & Chaari, A. (2011). New extended possibilistic c-means algorithm for identification of an electreo-hydraulic system. In *12th, International Conference, STA, ACS.*

Troudi, A., Houcine, L., & Chaari, A. (2012). Nonlinear system identification using clustering algorithm and particle swarm optimization. *Scientific Research and Essays,* 7(13):1415–1431.

Xu, Y. and Zhang, S. (2009). Fuzzy particle swarm clustering of infrared images. In *Second International Conferance on Information and Computing Science.*

Zahid, N., Abouelala, O., Limouri, M., & Essaid, A. (2001). Fuzzy clustering based on K-nearest-neighbours rule. *Fuzzy Sets Systems, 120*:239–247.

Review and Improvement of Several Optimal Intelligent Pitch Controllers and Estimator of WECS via Artificial Intelligent Approaches

Hadi Kasiri, Hamid Reza Momeni and Mohammad Saniee Abadeh

Abstract Wind turbines in megawatt classification ordinarily rotate at variable speed in wind farm. Therefore turbine operation must be managed in order to maximize the conversion efficiency below rated power and reduce loading on the drive-train. In addition, to control the energy captured throughout operation above and below rated wind speed, researchers particularly employ pitch control of the blades. Thus, we could manage the energy captured throughout operation above and below rated wind speed using pitch control of the blades. This chapter suggests six new plans to conquer wind fluctuation problems based on a new Nero Fuzzy and Nero Fuzzy Genetic Controller where the fuzzy knowledge based are tuned automatically by Genetic Algorithm (GA) as known Tuned Fuzzy Genetic System (TFGS). Additionally In this Chapter, a new hybrid control has been trained that Wind Energy Conversion System (WECS) has optimal performance. This method contains a Multi-Layer Perceptron (MLP) Neural Network (NN) (MLPNN) and a Fuzzy Rule extraction from a Trained Artificial Neural Network using Genetic Algorithm (FRENGA). Proposed Hybrid method recognizes disturbance wind with sensors and it generates desired pitch angle control by comparison between FRENGA and MLPNN results. One of them has better signal control is selected to send to pitch blade controller. Consequently Proposed strategies reject wind disturbance in Wind Energy Conversion Systems (WECSs) input with pitch angel control generation. Consequently, proposed approaches have regulated output aerodynamic power and torque in the nominal range. Results indicate that the new proposed Artificial Intelligent (AI) methods extraction system outperform the best and earliest methods in controlling the output during wind fluctuation.

H. Kasiri (✉) · H.R. Momeni · M.S. Abadeh
Faculty of Electrical and Computer Engineering, Tarbiat Modares University,
P.O.Box: 14115-143, Tehran, Iran
e-mail: hadi.kasiri@modares.ac.ir

H.R. Momeni
e-mail: momeni_h@modares.ac.ir

M.S. Abadeh
e-mail: saniee@modares.ac.ir

Q. Zhu and A.T. Azar (eds.), *Complex System Modelling and Control Through Intelligent Soft Computations*, Studies in Fuzziness and Soft Computing 319,
DOI 10.1007/978-3-319-12883-2_18

1 Introduction

Since 1985 wind power has improved essentially in European Union (EU) every year growth at least 20 %. Wind power's natural popular of generating carbon-emission free electricity ascribes increment, Wind Energy Conversion System (WECS), this is the well-known growing method of new electrical generation in the world (Boukhezzar et al. 2007). Wind energy increased very hasty through the previous 25 years and became a significant branch of universal electrical power supply. Progressively requirements for network connection of Wind Turbines and Wind Farms were recognized by system operators in order to develop the grid constancy.

In addition as we know the increase in the average temperature of the Earth's near-surface air and oceans is a major issue all over the world. In accordance with the Intergovernmental Panel on Climate Change (IPCC) the mainstream of the temperature increase is caused by the greenhouse effect (Verdonschot 2009).

The greenhouse effect refers to the change in steady state temperature of a planet by the presence of an atmosphere containing gas that absorbs and emits infrared radiation. More than seventy five percent of the greenhouse gas releases on earth are caused by the combustion of fossil fuels. Wind power, solar power, hydropower and biomass and other renewable energy technologies can decrease the production of greenhouse gases extensively and at the same time reduce the addiction on the oil industry. These two point of view combined make renewable energy a major item all over the world. Wind power is the fastest growing renewable energy source. Wind power is presently responsible for about 1.5 % of the world's electricity use (Kanellos et al. 2000).

Due to this high interest in wind energy, it becomes to a greater extent important to increase the efficiency of wind energy conversion systems (WECS), also called wind turbines. Wind turbines extract kinetic energy from the wind and convert this into mechanical energy. This mechanical energy is then converted to electrical energy by means of a generator. The maximum amount of energy from the wind is extracted using wind turbine when it operates at an optimal rotor speed. Optimal operation of rotor depends on the wind speed. Because the wind speed is variable by nature, the optimal rotor speed also varies. Earlier research has shown that variable speed operation of the rotor results in a higher energy production compared to a system operating at one constant speed. Beside an increase in energy production, variable speed operation enables a reduction in dynamic loads acting on the mechanical components (Kanellos et al. 2000; Carlin et al. 2001; Mangialardi and Mantriota 1996; Cotrell 2004). On the other hand, a problem arises when the speed of the rotor varies while the wind turbine must deliver AC power with a fixed phase and frequency to the electrical grid. To match the grid requirements, current variable speed wind turbines incorporate expensive power electronics to convert the variable frequency power to a constant frequency (Verdonschot 2009).

The power electronics contain a partial efficiency and they can introduce harmonic distortion of the AC current in the electrical grid, reducing the quality of the produced power. Next to these disadvantages, power electronics are one of the main sources of failure in wind turbines. They account for about 25 % of turbine failures and, unlike mechanical failures, they are not predictable and therefore in-crease maintenance costs. Additionally, when looking at the North American market, GE owns a patent on variable speed wind turbines incorporating power electronics. This increases the interest for variable speed systems using a different technology. At the moment, a number of wind turbines using a different technology to obtain the variable speed are available on the market. These technologies incorporate some kind of variable transmission in the drive train to control the rotor speed. For example, Wikov Wind in cooperation with ORBITAL2 Ltd. has launched a 2 MW wind turbine equipped with a variable transmission.

Because of this growth Wind Turbine simulation became admired to verify the particular performance. For researchers modeling and simulation is beneficial with the purpose of diminish expenses for expansion of components, controller design and the whole system of a Wind Turbine.

For testing and estimation parameter of the control, e.g. for the functionality during voltage dip (LVRT), simulations are very useful to reduce cost and time. The test of the Wind Turbine carried out via serious check tools creates capacity consistent with an examination sketch. This test plan includes requirements of the international technical guideline, the IEC 61400-21, limit values and further requirements from grid codes. The next step of this development is the recreation of these examinations, as distinct in the examination sketch. The comparison of the measurement and simulation results and the validation according to defined validation routines are necessary for the assessment and results in any case in a Validation Report. If the validation criteria are fulfilled the Wind Turbine Model is validated, if the differences are bigger than acceptable according to the validation criteria the Wind Turbine Model has to be remodeled (Martin, Hamburg, Germany).

In any action, control has two major objective protection and optimization of operation additionally, when relate to WECS, control turn more important in all particular, as the central qua lit of WECS, is that they have to look with the particularly variable, random and unpredictable nature of the wind, all WECS have some variety of power control. Fixed speed WECS, with either acquiescent or occupied or occupied stall, dominated the wind power energy for long time. Their major catch is their inflexible, as the fixed generator speed does not provide any control suppleness. This pass form sight with the use of fall deposit power converts WECS. Variable-speed performance become possible by include power electronics transform (Brice et al. 2009).

Variable-speed WECS control organization generally comprises three main control subsystems: One is aerodynamic power control, through pitch control. Thirds Variable-speed action and energy apparel end maximization by utilizing generator control. And finally Grid power shift control, throughout the power electronics convertor. One of the research area improvements can be reached professional control, so wind power production the evolvement of efficient manufacture tools has been increased. In addition the research works presented in this chapter is in the famous frame works in the aim to introduce of the mechanics for best growth used from this source energy.

In this chapter the main goal of the controllers is to slow decrement rotor speed swing function and electrical power while minimizing the control activating loads.

The blade pitch angel and the generator torque are available control inputs. Since wind speeds over high, its nominal count, power is fixed to the turbines rated output by altering place the control objective from maximum power catch (Boukhezzar et al. 2007).

Linear controllers have been widely used for power regulation (fixed) thought the control of blade pitch angel (Hand 1999; Ma 1997), suggested PI, PID pitch controllers. LQ, LQG control techniques have also been created in Wermter and Sun (2000).

Output power of wind turbine generator (WTG) is not stable due to wind speed changes. To decrease the unpleasant effects of the power system introducing WTGs, there are a number of available reports on output power control of WTGs detailing various researches based on pitch angle control, variable speed wind turbines, energy storage systems, and so on.

Kaneko et al. (2010) presents an incorporated control technique for a WF to diminish frequency deviations in a small power system. The WF achieves the frequency control with two control schemes: load estimation and short-term ahead wind speed prediction. For load estimation in the small power system, a minimal-order observer is used as disturbance observer. The estimated load is utilized to determine the output power command of the WF. To regulate the output power command of the WF according to wind speed changing, short-term ahead wind speed is predicted by using least-squares method. The predicted wind speed adjusts the output power command of the WF as a multiplying factor with fuzzy reasoning.

A typical Wind Energy Conversion System (WECS) is authoritative (mighty) of changing rotational speed and blade pitch angel is given as a block chart in Fig. 1.

Several purposes of these rule extraction methods are data discover knowledge acquisition for symbolic AI systems and improved autarchy for data mining applications (Pao 1989; Jianjun et al. 2006). Six popular techniques, Genetic Fuzzy System GFS, Fuzzy Rule Extraction from Neural Network using Genetic Algorithm (FREGNA), Hybrid technique and Nero Fuzzy Genetic Controller where the fuzzy knowledge based are tuned automatically by Genetic Algorithm (GA) as known Tuned Fuzzy Genetic System (TFGS), refuses wind disturbance in WECS input with pitch angel control generation and estimation parameters.

This chapter is organized as follows: in Sect. 2 related work has extended. A brief modeling of the wind turbine characteristic has been presented in Sect. 3. A simplified mathematical model is derived. Then multi Layer Percepron (MLP) neural network and Radial Basis Function networks (RBF) have been discussed in Sect. 4 and we will indicate the control objectives of this work. Section 5 starts with a brief description of some Genetic Fuzzy System (GFS) techniques and protocols of rules. Thus Rule extraction and the FRENGA controller in pitch and torque is then presented in Sects. 7 and 8. Section 9 extended Hybrid Optimal control strategy and finally Tuned Fuzzy Genetic System (TFGS) proposed in Sect. 10.

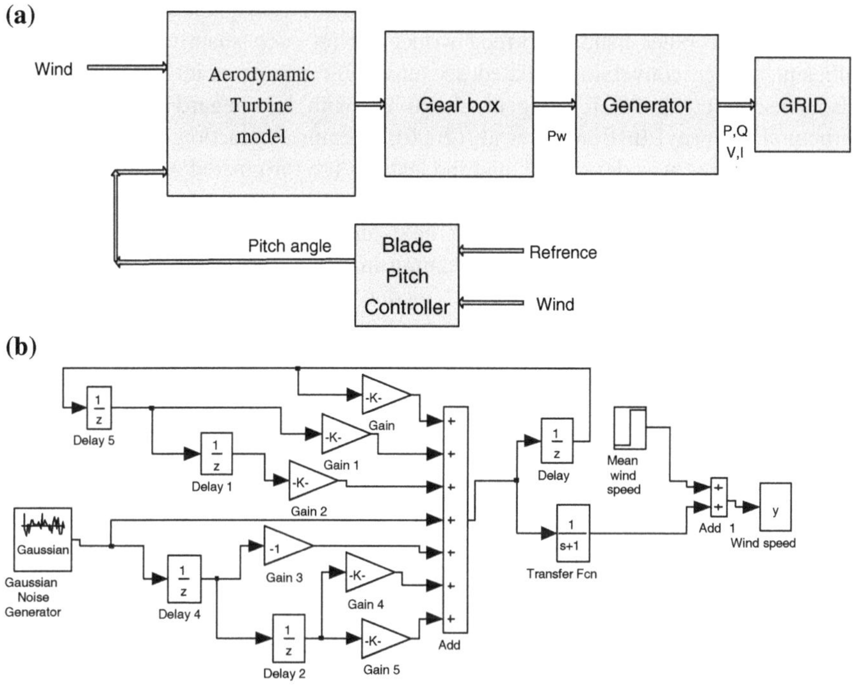

Fig. 1 **a** Block diagram of a WECS and **b** generation of wind speed by the ARMA model

Thus out put power has been fixed in optimal and nominal rang by pitch angel regulation.

Consequent these proposed approaches optimal have been adjusted turbines out power by pitch angel fixed simulation results of six these methods in Sect. 11. Resultant these simulations have verified that in collation to other suggested accedes; intelligent controllers have touched theses demand with higher accuracy in wind turbulence rejection.

2 Related Works

Whenever turbulence wind speeds takes place, the blade pitch control practically decreases the fluctuation of WECSs. There are several ways to determine suitable pitch angle for steady operation (Sakamoto et al. 2005). In some cases this control is accomplished using generalized predictive models, while in other cases PID controllers are employed. In other word the benefits when operating in the immediate surroundings of gusty, nonlinear and adaptive model-based controllers, large wind speeds have been documented (Horiuchi and Kawahito 2001; Bianchi et al. 2008).

Multivariable control describes an approach that afford pitch angle (Laks et al. 2009). On the other hand, informed works exploit such adaptive controllers for efficient power conversion, related to maximum power point (MPP) tracking (Boukhezzar et al. 2007; Xing et al. 2009), with less regard to the WECS's structural integrity. In Bianchi et al. (2006), a feature-extraction algorithm, a frequency analyzer, was developed, and the features are formulated as the inputs of an artificial previous term neural network next term using back propagation. An artificial-previous term neural-network next term-based controller has been presented in Kuo (1995), to realize fast valuing in a power-generation plant.

Progression of hybrid architectures (Yingduo et al. 1997), knowledge acquisition for symbolic AI systems and improved adequacy for data mining applications (Horiuchi and Kawahito 2001; Bianchi et al. 2006, 2008), these show some purposes of rule extraction approaches are data exploration. Genetic algorithms (GAs) have been used in various problems, such as nonlinear optimization, combinatorial optimization and machine learning (Muhando 2008; Gen and Cheng 1997; Goldberg 1989 ; Andrew and Haiyang 2010). Also genetic algorithms are applied for selecting fuzzy if-then rules, modification of nonlinear scaling functions, and for determining hierarchical structures of fuzzy rule-based systems. A cascade GA (Prakash et al. 2011), a micro-GA (Heider and Drabe 1997) is uncommon genetic algorithm that was used for designing fuzzy rule-based systems.

Since the wind speed exceeds its nominal value, power is regulated to the turbine's rated output by shifting the control objective from maximizing power catch (Glorennec 1997). One of the powerful universal predictors that show very good performance solving complex problems is Neural Network (NN). Several purposes of these rule extraction methods are data exploration, progression of hybrid architectures (Boukhezzar et al. 2007), knowledge acquisition for symbolic AI systems and improved adequacy for data mining applications (Wermter and Sun 2000; Mitra et al. 2002; Mitra 1994). Three popular techniques that extract rules from a trained NN are Neurorule, Trepan and Nefclass (Witten and Frank 1999; Baesens et al. 2003).

Rules from each unit in a NN have been extracted by the decompositional algorithms. The so called extracted rules are then aggregated to form the final fuzzy forecasting system. In Craven and Shavlik (1996) an online training fuzzy neural network controls the induction generator via a high performance speed observer. Hong et al. (Nauck 2000) propose a new method of wind power and speed forecasting using a multi-layer feed-forward neural network. They develop a forecasting system for time-scales that might vary from a few minutes to an hour. In Lin and Hong (2010), Lin et al. (2010a, b), a Wilcoxon radial basis function network with hill-climb searching maximum power point tracking strategy is suggested for a permanent magnet synchronous generator with a variable-speed wind turbine. An approach for optimization of power factor and the power produced by wind turbines was presented in Hong et al. (2010). Data-mining algorithms capture the relationships among the power output, power factor, and controllable and non-controllable variables of a 1.5 MW wind turbine. In Lin and Hong (2010), Lin et al. (2010a, b), the modeling and the control of a WECS associated to a super capacitor module as an energy storage system have been presented. Moreover, paper (Abdelkarim et al. 2011) presents a wind-driven induction generator system with a hybrid controller, which combines the advantages of the integral proportional and the sliding mode controllers. The proposed controller in Abdelkarim et al. (2011) is designed to adjust the turbine speed to extract maximum power from the wind.

Sometimes for the below-rated wind speed conditions, controlled torque of generator is known as the indirect control in torque control technique. Characteristic control schemes in power control employ blade pitch angle as the only controller input. Mono variable control keeps the generator torque constant at its nominal value in most controllers (Lin and Hong 2010; Lin et al. 2010a, b; Hand 1999; Bossanyi 2000; Camblong 2004; van der Hooft and van Engelen 2003, 2004).

Large winds farms consist of MW class wind turbine connect directly to transmission networks. Nevertheless, the extensive power system performance and designing in terms of power security, quality, stability, and voltage control have been influenced by raised wind power generation (Boukhezzar et al. 2007; Salman and Teo 2003). Researchers in Litipu and Nagasaka (2004) describe an operational optimization strategy to be adopted at the wind park control level, that enables defining the commitment of wind turbines and their active and reactive power outputs following requests from Wind Park Dispatch Centers, assuming that individual wind turbines short-term wind speed forecasts are known and are expressed as power availability. Connection of wind turbine to the utility grid is changed the system's dynamic characteristics (Moyano and Lopes 2009). The most great significance subjects in wind turbine are Aerodynamic and structural optimization. This optimization requires the definition of an aerodynamic formation, which gratifies definite aims issue to limitations.

Most engineering computation methods have reverse nature. Thus reversed design process is represented in conventional methods for the application scheme. Ordered geometric shape often manually changed in a trial and error manner and accordingly aerodynamic performance is considered by it. Prescribed target distributions of some aerodynamic quantity determine the optimum architect or geometric shape. As result inverse methods are designed by it. Ultimate goal is obtained by a direct control process, where the control space is searched for the optimum control in an intelligent way. The data in real-world applications generally include quantitative value, but most preceding reports concentrated on database with binary values (Gjengedal 2004). Where the membership function (MFs) is recognized in progress, some approach for mining fuzzy organization rules from quantitative data have been proposed in many studies (Yue et al. 2000; Agrawal et al. 1993). When MFs set in the best statue, it may have an important effect on the result. Studies in Hong et al. (2001), Wang and Bridges (2000) have been achieved a learning or training of the MMFs by different methods. Newly a genetic lateral tuning of MFs has been performed by a new proposed linguistic rule representation model (Kaya and Alhajj 2003).

3 WECS Modelling

3.1 Wind Modelling

Wind time model has been created founded on the ARMA series in this section. The wind speed $v_w(t)$ has two parts declared as:

$$v_w(t) = v_m + v_t(t) \tag{1}$$

where v_m is the signify wind speed at hub height and $v_t(t)$ is the instantaneous turbulent part, whose linear model is composed by a first-order filter disturbed by Gaussian noise. The instantaneous turbulence component of wind speed is obtained as Endusa and Aki (2009):

$$v_t(t) = \delta_t \vartheta_t \tag{2}$$

where δ_t is the standard deviation and ϑ_t is the ARMA time series model, which may be expressed as:

$$\vartheta_t = \delta_1 \vartheta_{t-1} + \delta_2 \vartheta_{t-2} + \cdots + \delta_n \vartheta_{t-n} + \eta_t - \theta_1 \eta_{t-1} - \cdots - \theta_m \eta_{t-m} \tag{3}$$

where η_t is the white noise process with zero mean, $\delta_i(i = 1, 2, \ldots, n)$ and $\theta_j(j = 1, 2, \ldots, m)$ are the autoregressive parameters and moving average parameters, respectively. Wind speed by the ARMA model is presented in Simulink environment as Fig. 1b.

3.2 Simulation of Wind Model

The main parameters of wind turbine have been simulated in this section. First the wind speed is modeled by the ARMA series in MATLAB simulation. The mean wind speed is determined based on spectral energy distribution of the wind and a superimposed noise signal. Figure 2 shows the response of the wind turbine to a wind speed profile above and below rated wind.

3.3 Wind Turbine Modeling

Considered wind turbines in this work, operate at variable speed in wind farms. They can be represented by the modeling of rotor, drive train and generation system with generator and power-factor-correction capacitors. The aerodynamic power captured by the rotor is given by the nonlinear expression:

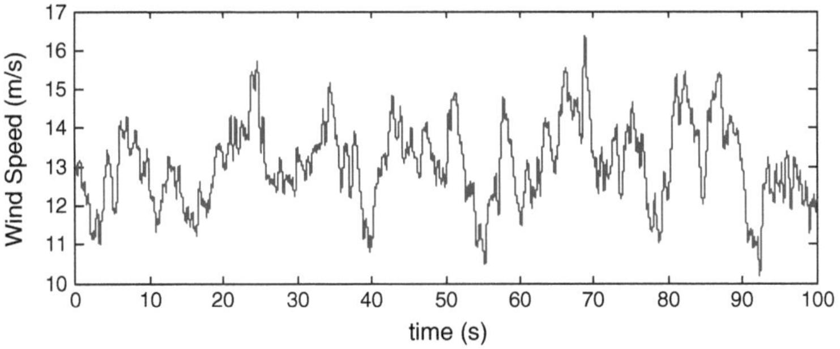

Fig. 2 Simulated wind speed

$$P_a = \frac{1}{2}\rho\pi R^2 v_w^3 C_p(\lambda, \beta) \tag{4}$$

That ρ is the air density, R is rotor radius, v_w wind speed. The aerodynamic performance is the ratio of turbine power to wind power and is known as the turbine's power coefficient, C_p. Power coefficient is related to turbine characteristic among blade pitch angle β and tip speed ratio (TSR) λ. The relationship between performance C_p with λ and β can be presented by Endusa and Aki (2009):

$$C_p(\lambda, \beta) = 0.5176\left(\frac{116}{\lambda_i} - 0.4\beta - 5\right)e^{-21/\lambda_i} + 0.0068\lambda \tag{5}$$

where λ proportion between the blade tip speed and the wind speed and λ_i depends on instantaneous $\{\lambda, \beta\}$, thus we have:

$$\lambda = \frac{\omega_t R}{v_w}, \ \lambda_i = \frac{1}{\lambda + 0.8\beta} - \frac{0.035}{\beta^3 + 1} \tag{6}$$

In a control model the controlled torque Γ_c, which it has been used for the variable speed turbines, is given by Camblong (2004):

$$\Gamma_c = \frac{1}{2}\rho A R^3 \frac{C_{p\max}}{\lambda_*^3}\omega^2 \tag{7}$$

R is the rotor radius, and λ_* is the tip-speed ratio at which the maximum power coefficient $C_{p\max}$ occurs.

4 Artificial Neural Networks

Artificial Neural Networks (ANNs) are mathematical representations inspired by the functioning of the human brain (Bishop 1995; Pao 1989; Jianjun et al. 2006). In some frequent cases the linear approximation is not valid and the accuracy of system modeling decreases significantly. Thus ANNs are capable of modeling very complex functions. In addition they keep in check the curse of dimensionality problem that bedevils efforts to model nonlinear functions with large numbers of variables.

4.1 Multi-layer Perceptron Networks

The multilayer neural network is typically composed of an input layer, one or more hidden layers, and an output layer, each consisting of several neurons. Each neuron processes its input and generates one output value which is transmitted to the neurons in the subsequent layer. All neurons and layers are arranged in a feed forward manner, and no feedback connections are allowed. Training process and computation in layers and neurons happen by the following equation (Kathryn et al. 2006):

$$y_p^{(k)} = sgm_p^{(k)}\left[W_{ip}^{(k-1)}.y_i^{(k-1)} - \beta_i^k\right]; (p = 1, 2, \ldots, N_k; k = 1, 2, \ldots, M) \quad (8)$$

where W_{ip}^k is the connection weight between the ith neuron in the (k−1)th layer and pth neuron in the kth layer, y_p the output of the pth neuron in the kth layer, sgm_p the sigmoid activation function of the pth neuron in the kth layer and β_p^k is the threshold of the pth neuron in the kth layer. Sigmoid activation function is given as:

$$sgm(x) = \frac{1}{1 + \exp(-x)} \quad (9)$$

Training process of the back propagation algorithm runs according to the following steps (Oh 2010; Wang et al. 2010):

1. Initialize all weights at random.
2. Calculate the output vector.
3. Calculate the error propagation terms.
4. Update the weights by using Eq. (10).
5. Calculate the total error "ε" by using Eq. (11).
6. Iterate the calculation by returning to error is less than the

$$W_{ip}^{(k-1)}(t+1) = W_{ip}^{k-1}(t) + \alpha \sum_{n=1}^{I} \delta_{np}^{(k)} y_{ni}^{(k-1)} \tag{10}$$

$$\delta_{np}^{(k)} = sgm_{np}^{k}(.).\left[\sum_{n=1}^{I} \delta_{np}^{(k)} W_{pl}^{(k)}(t)\right] \tag{11}$$

where t is the iteration number and α is the learning rate.

4.2 Radial Basis Function Networks

A radial basis function (RBF) network, therefore, has a hidden layer of radial units, each actually modeling a Gaussian response surface. Since these functions are nonlinear, it is not actually necessary to have more than one hidden layer to model any shape of function: sufficient radial units will always be enough to model any function. RBF networks have a number of advantages over MLPs. First, as previously stated, they can model any nonlinear function using a single hidden layer, which removes some design-decisions about numbers of layers. Second, the simple linear transformation in the output layer can be optimized fully using traditional linear modeling techniques, which are fast and do not suffer from problems such as local minima which plague MLP training techniques.

Radial basis Gaussian transfer function is considered as (12) in this study

$$F(u,c,\sigma) = \exp\left\{-\left(\frac{u-c}{\sigma}\right)^2\right\} \tag{12}$$

where c is the center, σ is the variance and u is the input variable. The output of the ith neuron in the output layer at time n is

$$y_i = \sum_{j=1}^{H} W_{ij} F_j(u,c,\sigma) \tag{13}$$

Training process of the radial basis function neural network runs according to the following steps (Oh 2010; Wang et al. 2010).

Initialize all weights at random.

Calculate the output vector by Eq. (13).

Calculate the error term "ε" of each neuron in the output layer according to (14).

$$\varepsilon_i(n) = y_i(n) - \hat{y}_i(n); \;\; (i = 1, 2, \ldots, L) \tag{14}$$

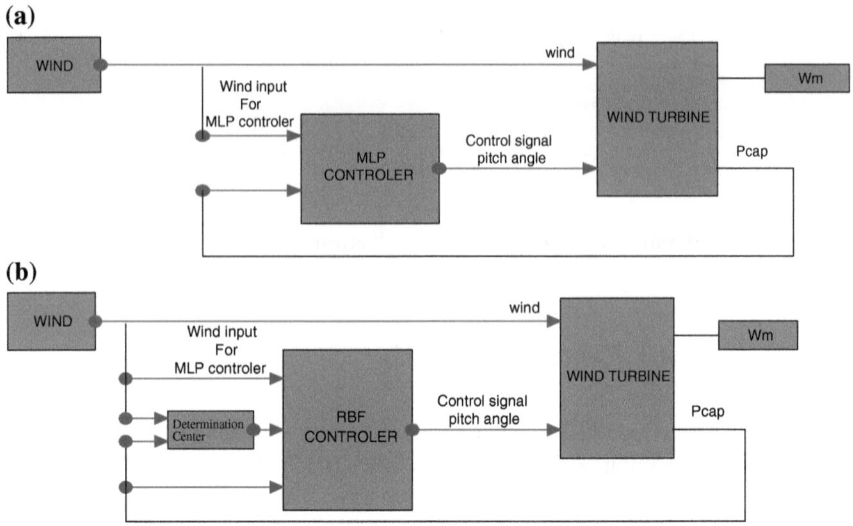

Fig. 3 Block diagram of controllers. **a** MLP controller diagram. **b** RBF controller diagram

where $\hat{y}_i$ is the desired output vector of the ith neuron in the output layer and y_i the calculated output vector of the ith neuron in the output layer by using (13).

Update the weights by using Eq. (15).

Calculate the total error "ε_T" according to (16).

Iterate the calculation by returning to Step 2 until the total error is less than the desired error.

$$W_{ij}(n+1) = W_{ij}(n) + \alpha\varepsilon_i(n)F_j(u, c, \sigma); (i = 1, 2, \ldots, L; j = 1, 2, \ldots, H) \quad (15)$$

where n is the iteration and α the learning rate

$$\varepsilon_T = \sum_{n=1}^{I}\sum_{j=1}^{N_M}\left(y_{nj} - \hat{y}_{nj}\right)^2 \quad (17)$$

According to Fig. 3a the MLP controller in this paper has two inputs (wind speed and output power) and one output (pitch angle value). At first this controller is trained with credible optimal values of wind turbine input-output (Kasiri et al. 2011b, 2012b).

Therefore this approach procures the best weight between input output using desired error produces. The training of NN computes an appropriate pitch angle upon catch wind speed. There are five neurons in the hidden layer. The hidden layer has nonlinear activation functions, but the output layer has a linear one.

The RBF neural network controller has three inputs and one output. Figure 3b shows block diagram of RBF controller (Kasiri et al. 2012b). Similar to previous controller at first this controller is trained with full credible rang optimal value of wind turbine input-output. Thus RBF controller results the best weight between input output using desired error produces, and produces an appropriate pitch angle upon catch wind speed. There are 10 neurons in the hidden layer.

The hidden layer has nonlinear activation functions, but the output layer has a linear one.

4.3 Simulation Results of Artificial Neural Networks Controller

To facilitate perform all of the simulation results this chapter, a 2 MW wind turbine is chosen with the important parameters given in Table 1 (W2000 2 MW Wind Turbine 2007). Simulation results brightly confirm the truth of the proposed control methods.

In this section controller output and output power of wind turbine have been simulated. According to Fig. 4a the angular speed of wind has been controlled actually. As plotted nearby rated wind speed, at large wind speeds, a dynamic variation of the generator speed have been permitted by controller.

Because of it absorbs rapid alterations in power through wind gusts thus escaping mechanical pressures. Afterward Fig. 4b presented output power of wind turbine in Pu, which has been managed by MLP controller perfectly.

As plotted power reduces in some spots that it has been resulting of intense and sudden wind speed reduction, but MLP algorithm with changes in pitch angle almost control it. Similarly, in RBF algorithm output simulation and analysis Fig. 5a reveals angular speed of wind turbine blades that has been controlled by RBF neural network. Lastly, Fig. 5b presented output power of wind turbine in Pu, which has been regulated by RBF controller. Similar to MLP simulation output the power curve in RBF distinguishes power drops in some spots that it's result of turbulence wind speed.

Table 1 Specifications of the wind power generating facility

Wind turbine and rotor	
Blade radius, R	37.5 m
Number of blades	3
Cut-in/cut-out wind speed	4/25 m/s
Rated capacity, P_r	2 MW
Pitch controller	
Max/min pitch angle	30/−2°
Max/min pitch rate	8/−8°/s
Wind field	
Rated wind speed	12 m/s
Air density	1.225 kg/m^3
Turbulence intensity	16 %

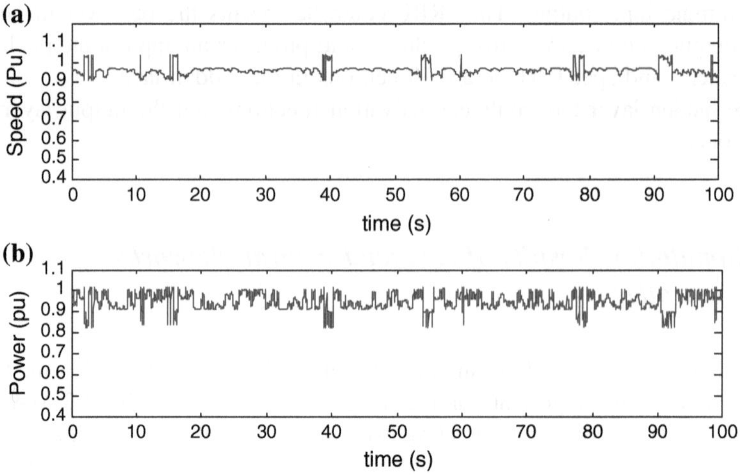

Fig. 4 Evolution of control and power alteration parameters in MLP neural network. **a** Angular speed of wind turbine blades. **b** Output power of wind turbine in Pu

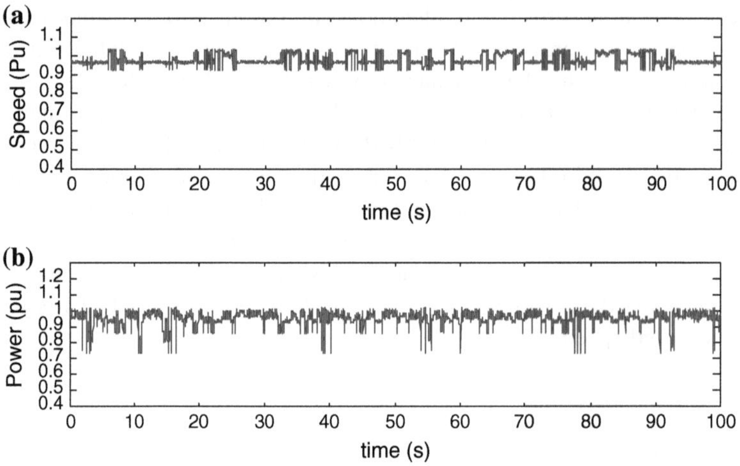

Fig. 5 Evolution of control and power alteration parameters in RBF neural network. **a** Angular speed of wind turbine blades. **b** Output power of wind turbine in Pu

However, RBF algorithm has controlled it with change pitch angle. In addition, we observe respond of MLP NN shows a better result than RBF NN because the desired values in training have not cluster mode and variation of wind speed have the relatively large scatter.

5 Genetic Fuzzy System (GFS) Strategy

In these days multifarious problem, like nonlinear optimization and machine learning have been resolved by Genetic Algorithms (GAs) (Wermter and Sun 2000). A lot of Genetic Algorithm-based methods have been planed of inbreeding fuzzy if-then states (Kasabov 1998).

These systems can be classified into two approaches in the same manner as the graduation non-fuzzy genetic-based machine leaning methods has 2 types.

5.1 The Michigan Approach

In the Michigan Approach (MA), only unique fuzzy rule-based system is store the perforation of Genetic Algorithm, where as a number of fuzzy rule-based system are saved in the Pittsburgh Approach (Hisao et al. 1999; Azar 2012). This required for actuates to the short computation time required for federation update by genetic steer age consequently one profit Michigan Approach is, Its teeny retention necessity. But one disservice of Michigan Approach is the shortcoming of a proximate adhesion between the transaction of Genetic Algorithm and the optimization of fuzzy rule-based systems. Because the espasuration of each fuzzy rule-based prees, hence algorithm distention for proper fuzzy if-then rules to perform its optimization sideway.

5.2 The Pittsburg Approach

In the Pittsburg Approach a set of if-then laws is symbolic as a strand to witch genetic inflection like crossover, mutation, selection are exerted (Hisao et al. 1999).

One profit of this approach is that performance of each fuzzy rule-based system can be straightly used as a Fitness Function (FF). Because of population consists of a number of fuzzy rule-based systems PA is its huge retention necessity thus it one disservice for this approach. The Pa postulates much more retention storage than the MA, in which a unique fuzzy rule-based system corresponds to the complete hesitancy.

6 GFS Strategy

In subsumes systems to model categorization problems, a fuzzy if-then law or suit consist of coefficient (SCCs) is exert as an exclusive in GA, thus it can show the subsumed system as a kind of Michigan Approach (Kasiri et al. 2012b).

6.1 Classifier System

For classifier first, genetic algorithm generates an inceptive resident of fuzzy if-then rules or SCC. Each of them is leaded arbitrary. The perquisite class and the assuredly grad of every single fuzzy if-then mandate or SCC are manifest by Ishibuchi (Hisao et al. 1999; Azar 2010b). In second step it estimates each fuzzy if-then edict or SCC in the progression residents. Next, genetic algorithm results in new fuzzy if-then laws or SCC by genetic operation like: crossover, mutation, selection.

In semifinal it restores a part of the progression public with newly generated fuzzy if then rules or SCC. Finally if a pre-identified stopping share isn't content, comeback to second step.

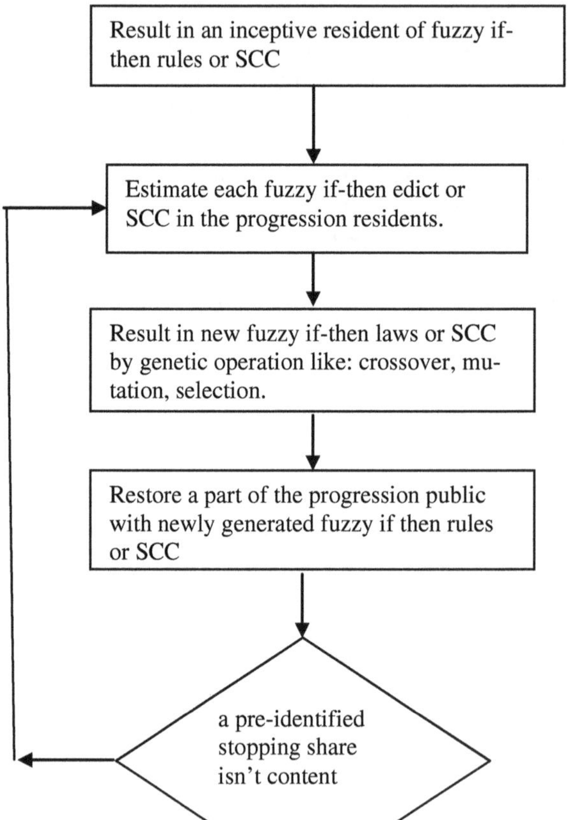

6.2 Protocol of Fuzzy Rule

In proposed method fuzzy if-then rules are coded as a numeral string and are implied by its corresponding random fuzzy sets. Consequent class and the certainty grade can be absolutely detailed by the heuristic fitness function.

The following symbols are used for indicating the seven used linguistic values: (1) very small, (2) small, (3) medium small, (4) medium, (5) medium large, (6) large, (7) very large.

For example, the following fuzzy if-then rule is coded as "172": If Xl is very small then Y1 is very large and Y2 is small.

6.3 Interpretation of Fitness Function

In fuzzy exclusive systems a population of fuzzy if-then laws equate to a fuzzy rule-based classification methodology. Fitness function is GFS encompass 2 section, one of them analogized generated laws with optimal upscale, thus a rule that conceal almost all the optimal values could be a genuine law. In the second section, symbol alternative of laws have computation on wind turbine power formula, correspondingly a law that has the best control on power and adjust it well, could be a goal rule. Finally each piece of these parts has been distributed weight. As an outcome the leading law will be nominated from inceptive random laws (Kasiri et al. 2012b).

6.4 Genetic Operations (GO)

A GA usually has the three genetic operators that act on the chromosomes of each generation of the genetic algorithm. These operators include (Hisao et al. 1999):

1. Selection: In GA, pair of parents is selected from the current population according to Darwin's survivor of the fittest principle. Methods in this works have employed the well-known roulette wheel selection method.
2. Crossover: After couples are formed an n-point crossover is performed. The position of the n crossover points is determined randomly and according to gene boundaries. Each couple will produce two off-springs.
3. Mutation: Mutation is a probabilistic choice of *k* chromosomes of the pool and performing a random alternation of the genes at *l* points. The values of *n, k* and *l* are among the dynamics of the GA and are fixed by the user before the execution of GA.

Finally, the algorithm replaces the worst fuzzy if-then rules with the smallest fitness values with the newly generated fuzzy if-then rules with the utmost fitness

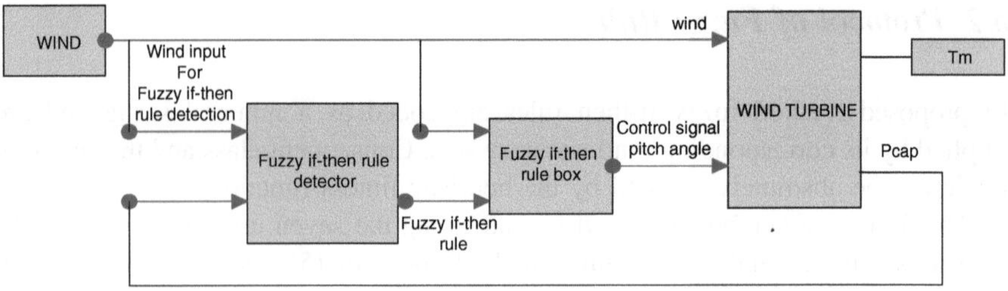

Fig. 6 Block diagram of controllers GFS controller diagram

values. The number of removed fuzzy if-then rules is usually the same as that of added rules in classic genetic algorithm. Figure 6 shows block diagram of the proposed controller.

According to block diagrams of GFS controller in Fig. 6 controller has two inputs and one output. This controller provides a suitable pitch angle upon catch wind speed.

6.5 Simulation Results of GFS Controller

Figure 7a depicts the established generator torque for GFS controller. The output mechanical power of GFS is shown in Fig. 7b.

Figure 7d shows the blade pitch angle control signal for GFS controller. In below rated wind speed, optimal power is attained by regulating (optimal performance coefficient); thus, the pitch angle is kept at a mechanical minimum and rotor speed is controlled in such a way that (optimal tip speed ratio) is always acquired, akin to MPP tracking.

7 Rule Extraction

The idea of rule extraction from a neural network involves definite activity. For example, parameters have read from a network, which is not permitted by a traditional connectionist structure that these neural networks are established on Gallant (1998). Gallant presented a routine for extracting propositional rules from a simple network, in order to explain the inference process in the system at the end of 1980s Gallant (1998). This could be surveyed as the origin of the examination on rule extraction from neural networks. In addition some algorithms have been developed to extract deterministic finite-state automata (DFA) from recurrent neural networks (Giles and Omlin 1993; Giles et al. 1992).

Rules from each unit in a neural network have been extracted by the decomposition algorithms and these methods aggregate them. The pedagogical algorithms

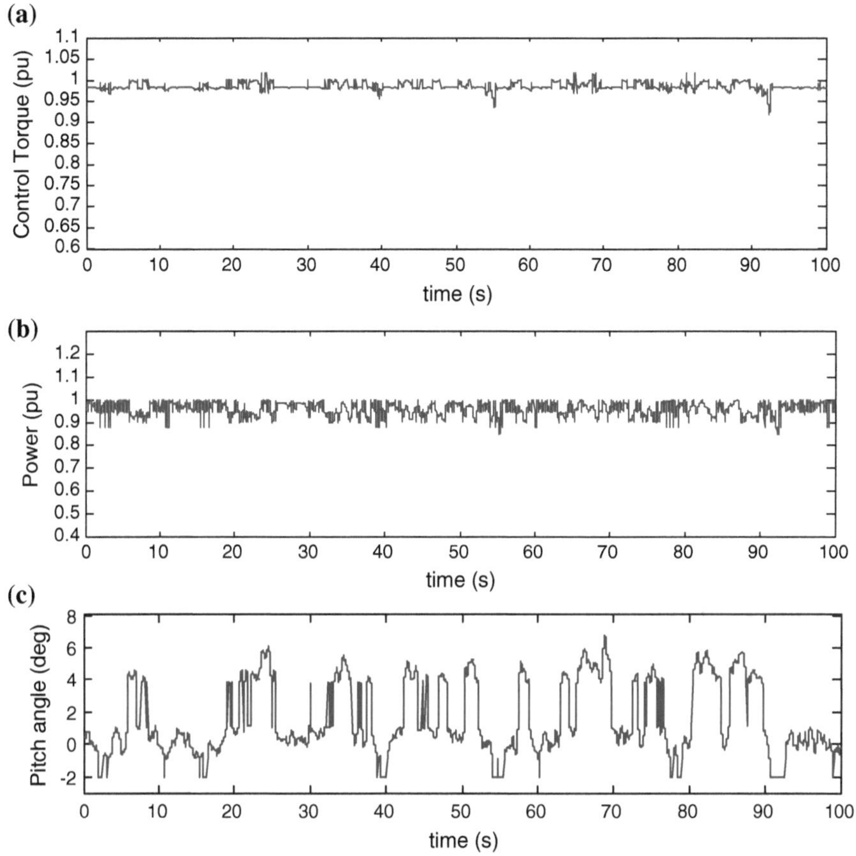

Fig. 7 Evolution of control and power alteration parameters in GFS and output power for two compared approaches. **a** Controlled torque. **b** Output aerodynamic power of wind turbine in Pu. **c** Pitch angle signal for GFS

regard the trained neural network as an opaque and aim to extract rules that map inputs directly into outputs.

The calculative complexity of extracting rules from trained neural networks and the complexity of extracting the rules straight from the data are both NPhard (Golea 1996), thus many of these cases contain a salient theoretical discovery in this area. Roy intelligently made known the difference between the idea of rule extraction and traditional connectionism (Roy 2000).

There are two theory intentions in Rule Extraction from trained neural network. The ANNs or weights are evaluated to extract the whole rule set witch should illustrate he complete wisdom.

These concepts are as follows:

1. Black box, we can handle an ANN as a black box and requiring considering its building and weight values. Input vectors are mapped by the ANN to quality vectors and algorithm extracts rules by analyzing the link between input and output.
2. Dissolution, some algorithms try to collect around corresponding weights to simplify the extracted rules. these kinds of algorithms take a look at the weights and search for study patch through the net.

In one of the express activities of RE from neural net, variables have read from a network. This diversely is not qualify by the network traditional connections structure (Luo and Unbehaben 1998). Also some algorithms have been developed to extract deterministic finite-state automata DFA from frequent neural network (Omlin et al. 1992; Giles et al. 1992; Giles and Omlin 1993).

A little while back some researchers even have aim to generate suppression rules from neural regressors (Saito and Nakano 2002; Setiono 2002).

8 Rule Extraction from AI

Neural network rule extract approach attempt to open the NN black box generate representative rules NN. One of the important catch in some application of NN is the annoyance with take in system. Therefore extracting comprehension the method accordingly extracting knowledge from NN in an extensive way has born originated (Darbari 2000; Mitra and Hayashi 2000; Santos et al. 2000; Zhou et al. 2000). Commonly, it has the form of hypothesis rules many rules extraction approached that advanced in last few years. A new rule extraction method supported on MLP NN and GFS optimization has been conferred in this part. It caused called FRENGA.

The meta-heuristic explore method extracts various rule from MLP network using GA, for searching optimal solution in enormous space of possible solutions (Kasiri et al. 2011a, 2012a).

In this part o NN characterization in classification questions by a new genetic fuzzy algorithm has been presented by proposed method FRENGA completely uses from investigational data in verifiable article. NN has been trained by this speculative data. That being so this method uses the NN results in definitional of Fitness Function (FF). In conclusion GFS has been trained with this FF to extract GFS fuzzy set rule. Reactions of these processes are given in Table 2. Table 2 encompasses extracted five rules from NN using GFS. These rules set pitch angle in the best setting to optimally control wind turbine. It can be simply noticed that the set of if-then rules be in need of cycle, which is composed of five steps as follows (Hisao et al. 1999).

Table 2 If then rules

IF X1 IS MEDIUM SMALL THEN Y1 IS VERY SMALL AND Y2 IS MEDIUM LARGE
IF X1 IS MEDIUM THEN Y1 IS SMALL AND Y2 IS MEDIUM
IF X1 IS MEDIUM LARGE THEN Y1 IS MEDIUM AND Y2 IS MEDIUM SMALL
IF X1 IS LARGE THEN Y1 IS LARGE AND Y2 IS SMALL
IF X1 IS VERY LARGE THEN Y1 IS VERY LARGE AND Y2 IS VERY SMALL

8.1 Coding of Fuzzy Rule

In this proposed method fuzzy if-then rules are coded as GFS method with membership function according to Fig. 8 (Kasiri et al. 2011a, 2012a). This innovative membership function is suited for the determination of better rules.

8.2 Definition of Fitness Function

A particular type of objective function that prescribes the optimality of a solution (that is, a chromosome or string) in a genetic algorithm is fitness function. Hence that particular string may be ranked against all the other strings. Optimal strings, or at least strings which are more optimal, are allowed to create and compound their datasets by any of several techniques, producing a new generation that will hopefully be even better.

Fitness function in FRENGA includes two parts; one of them compares generated rules with optimal values, thus a rule that covers most of the best values could be a desired rule. In another part; numeral equivalent of rules are being calculated on wind turbine power formula, thus these rules calculated on MLP NN either, consequently a rule that least slide from trained neural network value could be a good rule. Finally, for each piece of these parts has been allocated a weight. As a result the best rules have selected from initial random rules. Figure 9 shows block diagram of the FRENGA controller (Kasiri et al. 2011a, 2012a).

8.3 Simulation Results of FRENGA

Figure 10a reveals Tip-Speed Ratio of wind turbine. Figure 10b shows the pitch angle response. It is corroborated that the pitch actuator does not suffer from excessive activity despite the strong turbulence.

Figure 10c shows the evolution of the controlled torque of wind turbine in Pu that is controlled by FRENGA. The turbine is required to operate at its rated power.

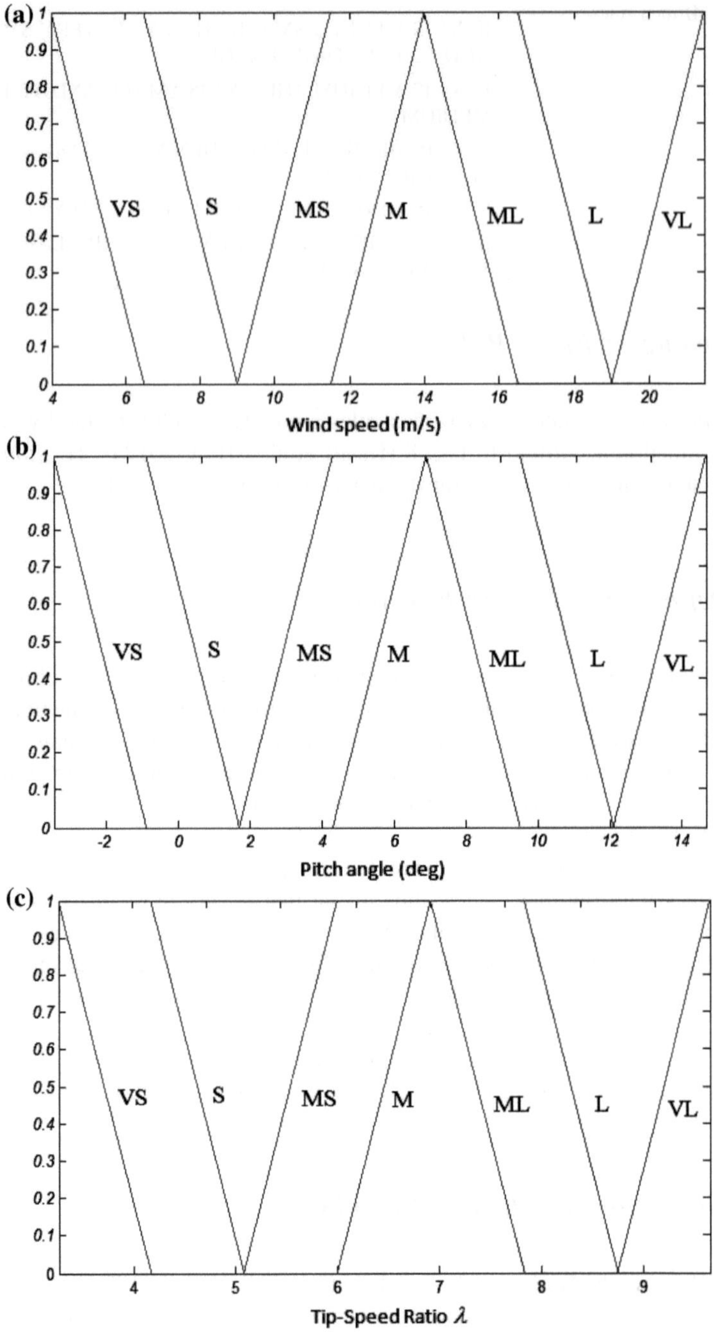

Fig. 8 Initial membership functions. **a** For wind speed. **b** For pitch angle. **c** For tip-speed ratio. *VS* very small, *S* small, *MS* medium small, *M* medium, *ML* medium large, *L* large, *VL* very large

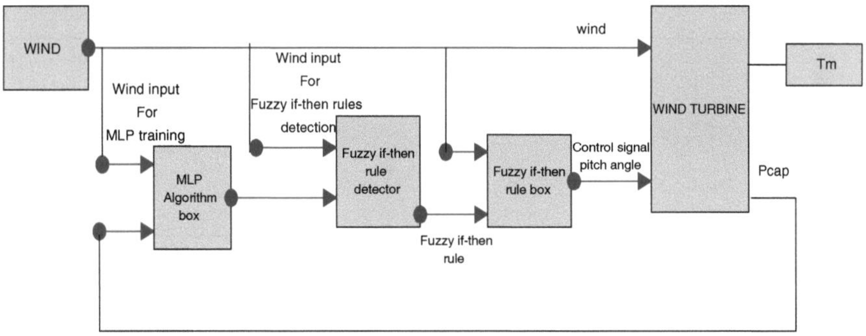

Fig. 9 Block diagram of FRENGA controllers

Figure 10d shows the evolution of the output power of wind turbine in Pu that is controlled by FRENGA.

By analyzing the simulation results, it can be concluded that the present models allow an accurate approximation of the dynamic response of the wind turbine operating with different winds, although the wind turbines generate the maximum reactive power.

9 Hybrid Optimal Control Strategy

The annoyance with understanding the method is the basic obstacle in some application of neural networks (NNs). Thus extracting knowledge from NN in the comprehensible way has been developed (Santos et al. 2000; Mitra and Hayashi 2000). Generally, it has the form of propositional rules. Many rule extraction approaches were advanced in the last few years. Framework of these methods determines the expressive power of extracted rules.

In this section, first MLP neural network has been trained with data extracted from authoritative articles next we extracted Fuzzy If-Then rules from Multi-Layer Perceptron NN using FGS optimization, which called FRENGA. In the case of rule extraction, this work is interested in receiving the set of Fuzzy If-Then rules and satisfying different criteria. Two approaches proposed a signal control to manage output power and torque.

At the end online form both methods were tested on a turbine and one of them that product better respond is selected. Figure 11 shows block diagram of the proposed Hybrid controller (Kasiri 2011b). Figure 12a reveals controlled torque of wind turbine blades that has been structured by Hybrid controller very well. Figure 12b presented output power of wind turbine in Pu that is controlled by Hybrid controller.

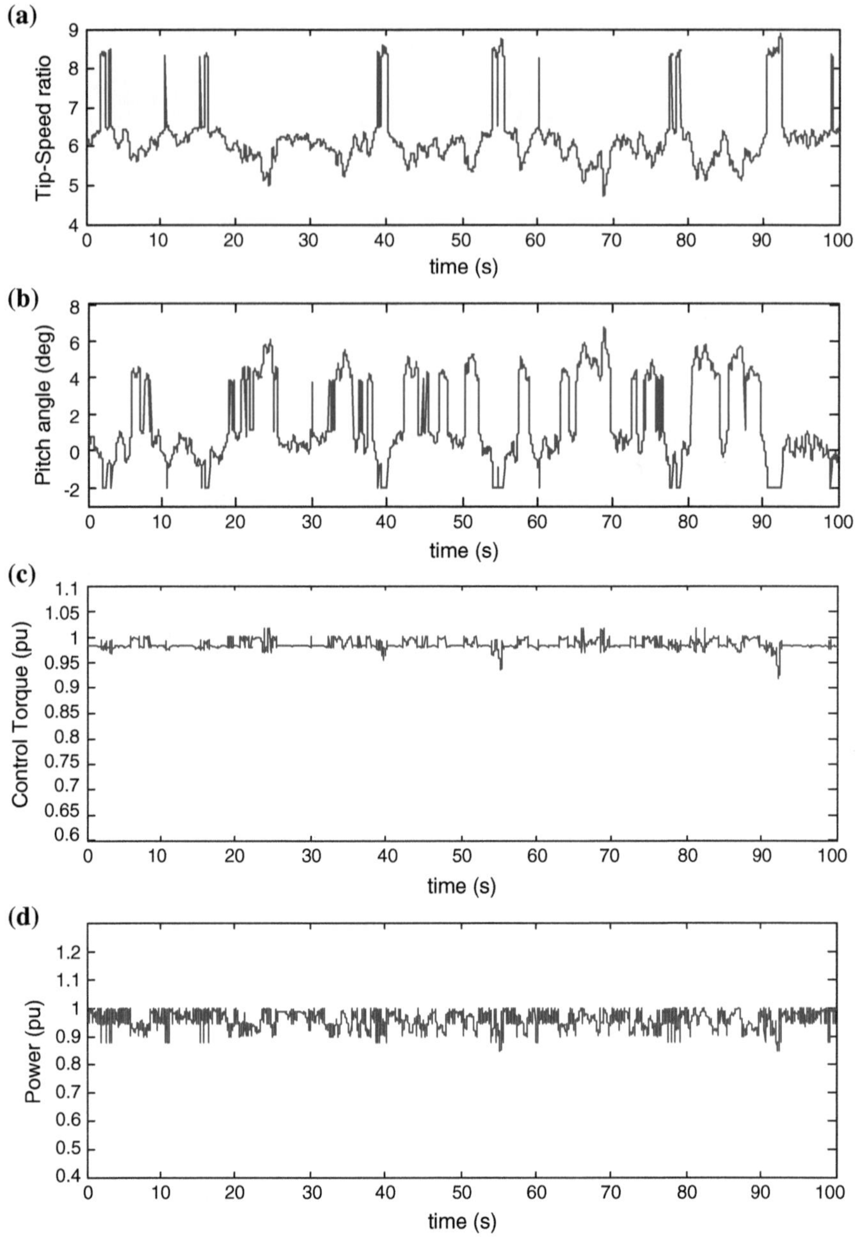

Fig. 10 Evolution of control and power alteration parameters in FRENGA. **a** Tipspeed ratio. **b** Pitch angle. **c** Controlled torque. **d** Output aerodynamic power of wind turbine in Pu

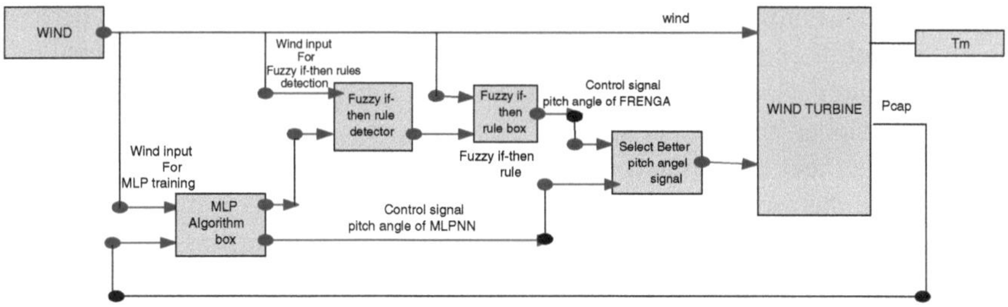

Fig. 11 Block diagram of proposed Hybrid controller

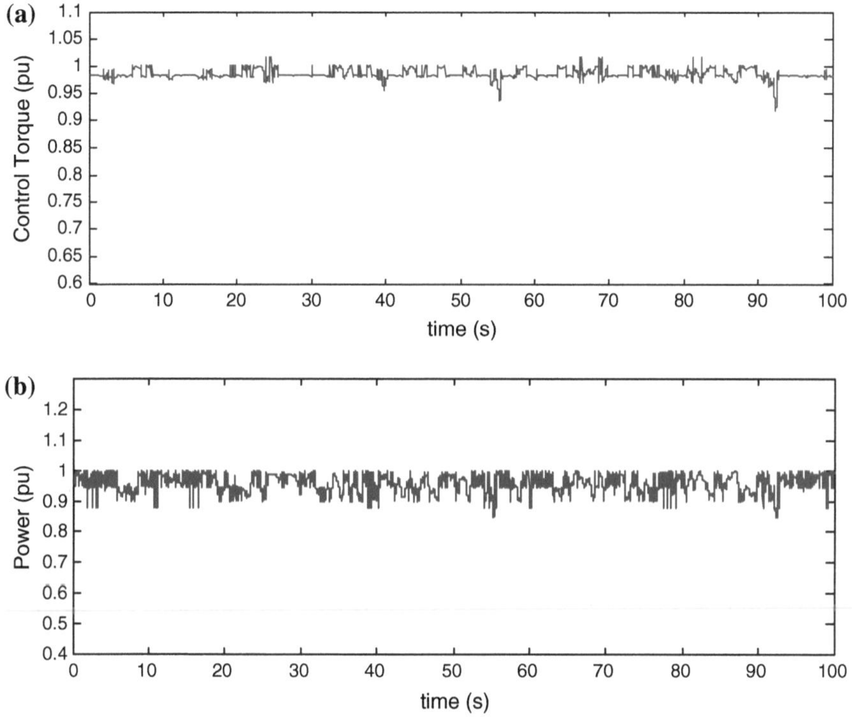

Fig. 12 Evolution of control and power alteration parameters in hybrid controller. **a** Controlled torque. **b** Output aerodynamic power of wind turbine in Pu

Application of electricity from wind energy conventionally has been growing. Therefore, directly connection of wind turbines to public transmission networks is very important. Thus the output power should be stable to allow it to be connected to national network. Because of instability creates many problems for electrical devices. In this paper has been tried to control turbine in wind turbulence.

Consequently, output power has least variations. Based on Based on a perform ability model, a control strategy is devised for maximizing energy conversion in low to medium winds, and maintaining rated output in above rated winds while keeping tensional torque fluctuations.

Controlling and estimating of a parameter in variable speed wind turbines was performed in this work using a new hybrid controller. MLPNN learns experimental and optimal data and FRENGA can be easily used for extracting a set of fuzzy if-then rules directly from NN black box. However, because of the good ability of NN to extract noise its operating is advised. As plotted from simulation results, is realized with change of turbine blades pitch angle, system can obtain most suitable performance coefficient (Cp) and tip speed ratio (TSR). Proposed controller by changing Pitch angle optimally Controls system. Experimental study has shown that it works efficiently for both continuous and enumerates attributes.

Hence we use NN to control wind turbine and produce a training example for the rule extraction method. Consequently output power has been regulated successfully. Finally we have been compared between proposed approaches and two different methods an MLPNN and FRENGA alone. We see our output power almost is most proper than them.

10 Tuning of Membership Function

As we know, the knowledge Base (BS) is formulate by membership rule MR Membership Function MF and fuzzy rule base RB. Thus ditty or intellect membership methods, or fuzzy mandate base or both of them are some resource to originate genetic fuzzy system (Magdalena 2001).

When fuzzy statue base is violated in advance, for regulate membership function an independent population denotes parameters of the membership province appearance. Against for adapt fuzzy rule base, all of fuzzy laws possibility that membership tasks is divert before have been personify the population Fig. 13 shows these syllabify. In this section of work initial shapes for the membership function have been shown in Fig. 8 and GFS will tune this shapes.

10.1 Proposed Tuned Fuzzy Genetic System (TFGS)

Thus this method is absorption in acquire the set of fuzzy If-Then rules that assuage different criteria.

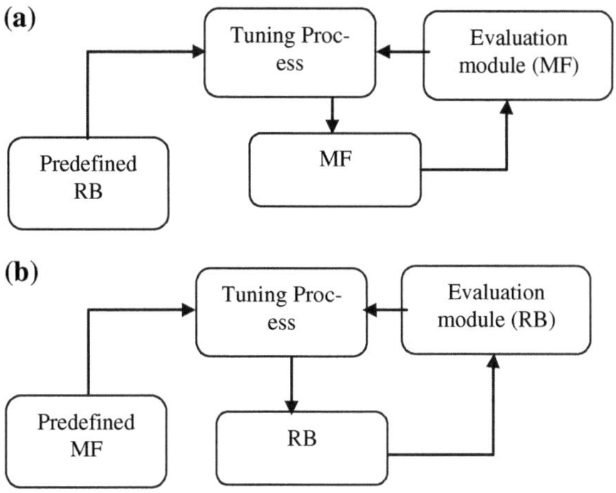

Fig. 13 Design of genetic fuzzy system in tuning **a** tuning membership function **b** tuning fuzzy rules

In this part knowledge base tuning in classification problems by a new genetic fuzzy system has been presented it can be simply awareness that eh set of if-then rules involved a cycle, which is constitute of five stairway follows (Hisao et al. 1999).

In proposed method fuzzy if-then rules are coded as a numeral string and are implied by its corresponding random fuzzy sets. Consequent class and the certainty grade can be absolutely detailed by the heuristic fitness function.

The following symbols are used for indicating the seven used linguistic values: (1) very small, (2) small, (3) medium small, (4) medium, (5) medium large, (6) large, (7) very large. For example, the following fuzzy if-then rule is coded as "172": If Xl is very small then Y1 is very large and Y2 is small.

10.2 Membership Function Tuning

The composition of the chromosome is deferent in this domination. Chromosome is nominated for an express and mystic by numerals.

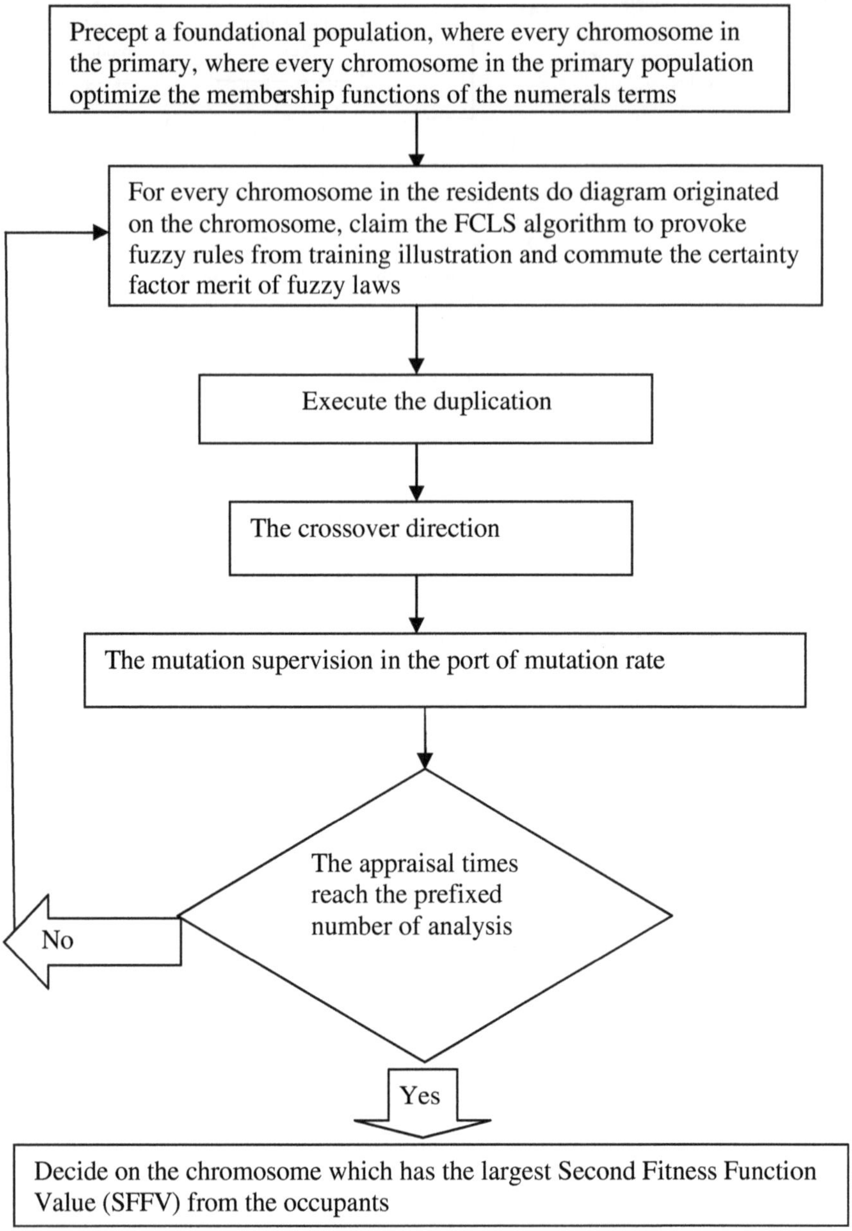

Therefore at the whole lot the chromosome is nominated for a express meantime. The goal of this work is done with the best the best option for any act to regulate the spin angel for the turbine that is conscious in Fitness Function.

The aspire algorithm for tuning the membership tasks of the linguistic terms of the property for approximation null values in relational database organization is ready as follows above algorithm (Chen and Yeh 1997).

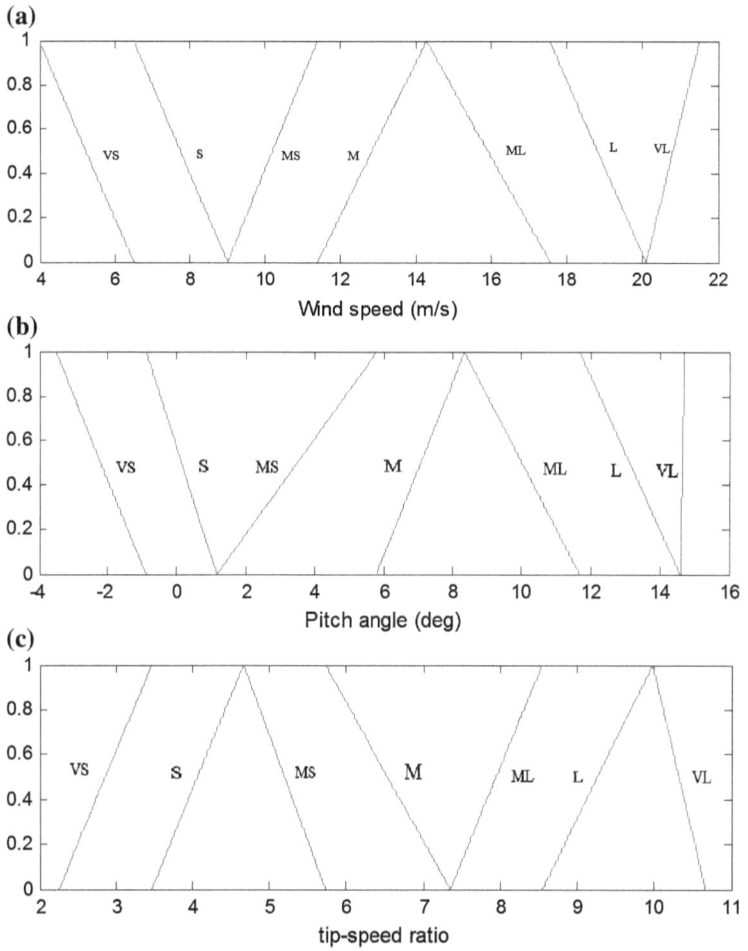

Fig. 14 Tuned Membership functions. **a** For wind speed. **b** For pitch angle. **c** For tip-speed ratio. *VS* very small, *S* small, *MS* medium small, *M*, medium, *ML* medium large, *L* large, *VL* very large

This chromosome issued to illustrate the membership function of the numeral's terms for approximate revoked in relational database arrangement Fig. 14 visible tuned membership function.

10.3 Explanation of Fitness Function

Aspect kind of objective function that recommends the optimality of an explanation (that is a chromosome or string) in a genetic algorithm is Fitness Function. Accordingly that definite string way be rated aggressive all the stings. Ideal string, or not less than strings which are more optimal, are sanction to fabricate and combination.

Their databases by any of various techniques, assemble new generation that will informative be even better.

In Sect. 1, FFFV comprises two portions: one of them differentiates generated rules with optimal worth, thus a rule that shields most of the best values could be inclination rule.

In a further part; integer similar of rules are being computed on wind turbine power formula thus these rules calculated on case (1–4) either, accordingly a rule that minutest slide from nominal value could be good rule.

At the end, for each segment for these parts has been distributed a weight. As a reaction the best rules have nominated from primary random rules.

In Sect. 2, SFFV committed the greatest specific meantime for any rule to compute the twist angel for the turbine.

10.4 Simulation Result of TFGS

This part includes Simulation result of TFGS. Figure 15a reveals Tip-Speed Ratio of wind turbine. Figure 15b shows the pitch angle response. It is corroborated that the pitch actuator does not suffer from excessive activity despite the strong turbulence. Figure 15c shows the evolution of the controlled torque of wind turbine in Pu that is controlled by TFGS. The turbine is required to operate at its rated power. Figure 15d shows the evolution of the output power of wind turbine in Pu that is controlled by TFGS.

11 Discussion and Comparison on the Presented Methods

By considering the simulation results, it can be concluded that the present models allow an accurate approximation of the dynamic response of the wind turbine operating with different winds, although the wind turbines generate the maximum reactive power. In this chapter it has been tried to control turbine in wind turbulence. Consequently, output power in any methods has least variations with differences that have been compared them in this section. In addition proposal methods are compared with other related works. Classical methods based on PI(D) algorithms are a good starting point for many aspects of closed-loop controller design for variable-speed turbines. Control design is a task that demands rigorous test data and extensive engineering judgment. The presented approach in Endusa and Aki (2009) has been employed to model the WECS subsystems as a basis for multi objective controller design. Model validation is carried out to authenticate the results. Produced power and generator torque from this strategy is shown in the Fig. 16b. A PID controller based on the BP neural network has been offered and used in the variable pitch angle wind turbine control system in Xing et al. (2009). Figure 16c, indicates output power of this strategy. The System response has sudden change, and hence is sensitive to

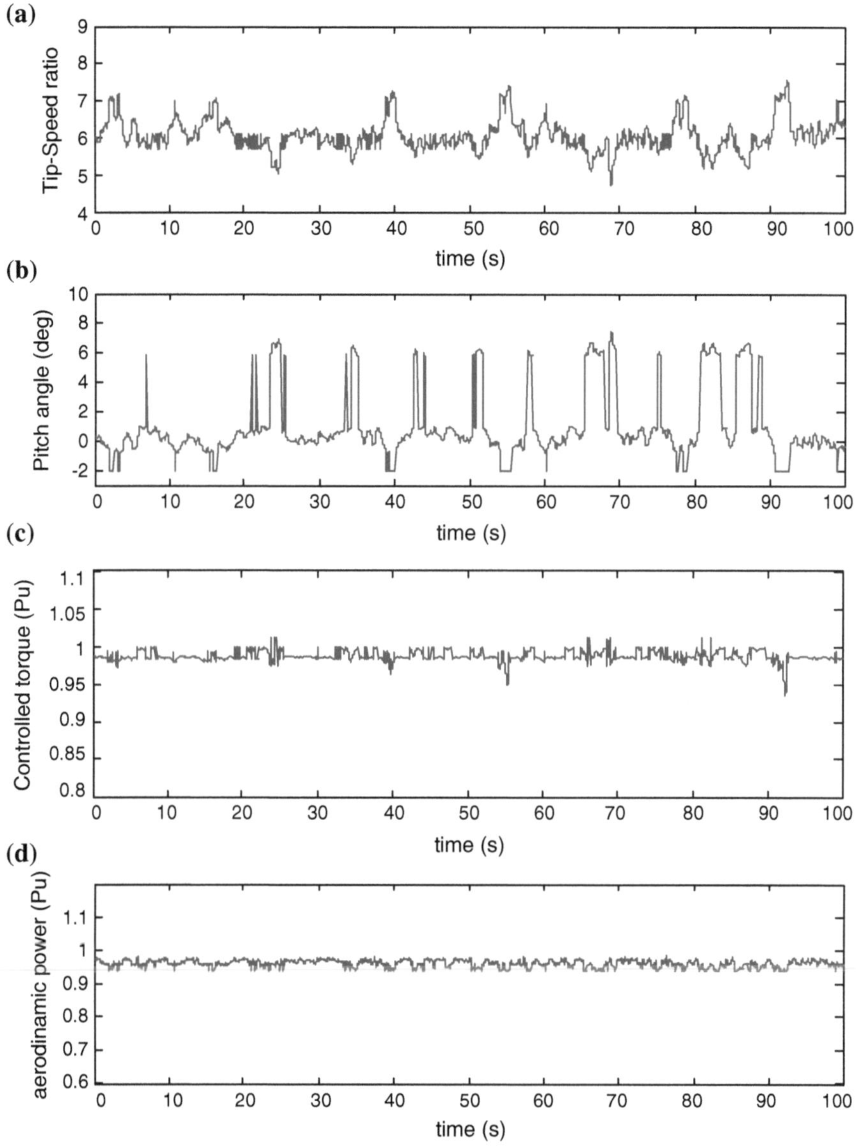

Fig. 15 Evolution of control and power alteration parameters in TFGS. **a** Tipspeed ratio. **b** Pitch angle. **c** Controlled torque. **d** TFGS output aerodynamic power of wind turbine in Pu

wind turbulence extremely. Simulation results of this method indicate that output power intensity had downfall and rise dramatically.

An output power leveling control strategy of wind farm based on both average wind farm output power and standard deviation of wind farm output power has been offered fuzzy logic neural network and used in the variable pitch angle wind

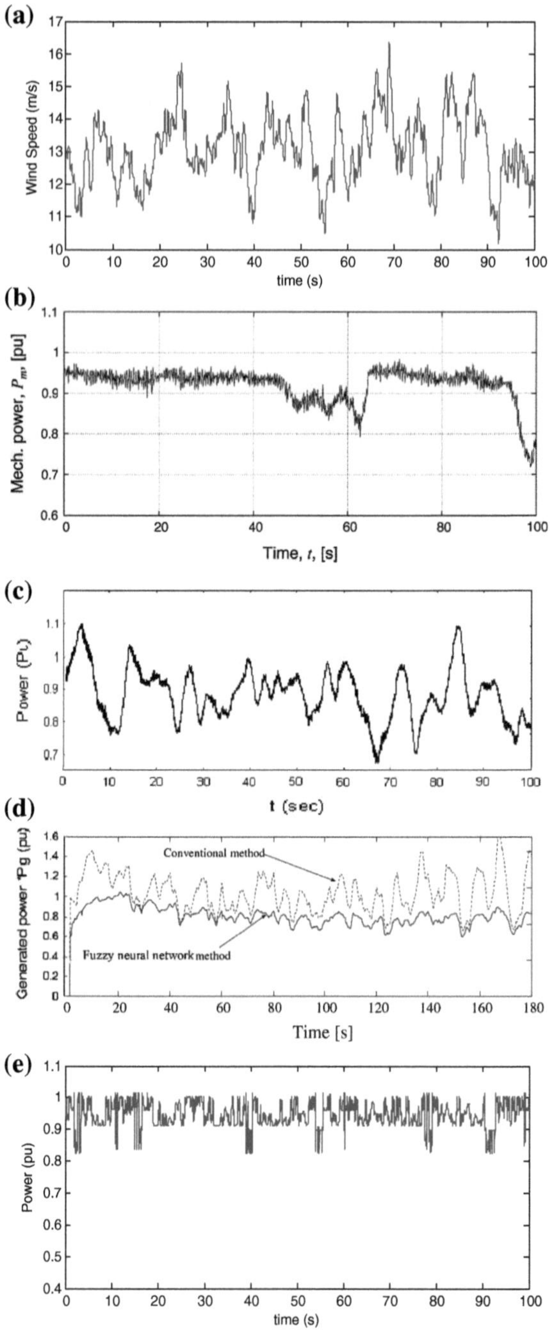

Fig. 16 Evolution of power alteration. **a** Simulated wind speed. **b** Output power of LQG algorithm in Pu. **c** Output power for method based on the BP neural network. **d** Output power for method based on the Fuzzy Logic neural network in Pu. **e** Output power of MLPNN in Pu. **f** Output power of FRENGA in Pu. **g** Output power of GFS controlled wind turbine in Pu. **h** output power of Hybrid controller. **i** TFGS output aerodynamic power of wind turbine in Pu

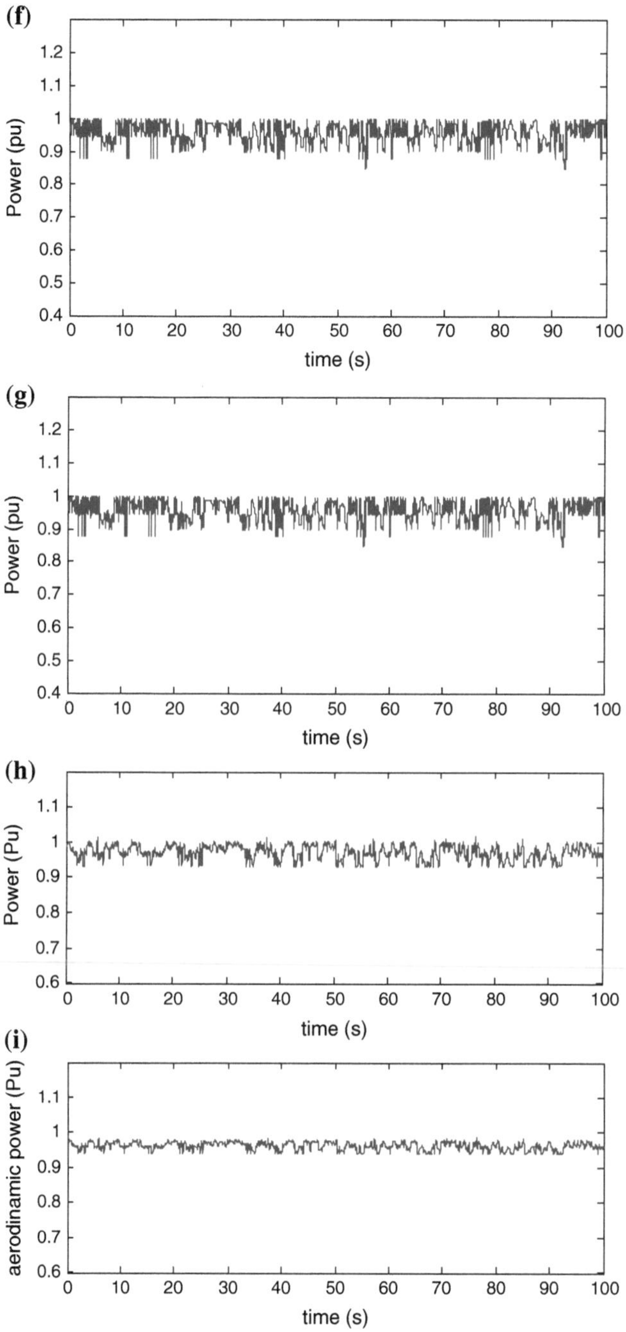

Fig. 16 (continued)

Table 3 Comparision of two rule extraction algorithms

Algorithms / Operation	FRENGA	FGS	TFGS
Iteration	50	100	65
Mutation probability	0.7	0.6	0.7
Cross over probability	0.4	0.7	0.6
Population size	65	60	65
Membership size	7	7	7
Time required to Rule extraction	302.646775	501.212821	**Rule extraction** time = 302.646775
			Tuning time = 459.895189
			Total = 762.541964
Lowest error in the Rule extraction	0.1226	0.1445	0.0615

turbine control system in Senjyu et al. (2006). Figure 16d, indicates output power of this strategy in 180 s. The System response has sudden change, and hence is sensitive to wind turbulence extremely. This method has saved power in same places, but it has severe change when power reduces in long time that is an effect of the wind downfall. Figure 16e shows that proposed MLP algorithm has produced controlled power with fewer drops than output power and torque of LQG algorithm (Endusa and Aki 2009), for example, between 45 and 70 s. In those situations that wind speed reduces intensely, output power downfall for a while. Note that this event is unavoidable.

Figure 16f, g indicates that FRENGA and GFS have better effect on output power of wind turbine than MLP algorithm.

Figure 16h shows hybrid controller output power that proposed in this chapter. According to this figure, proposed hybrid algorithms have produced power with fewer drops than output power of Fig. 16b–d, for example, between 45 and 70 s. In addition hybrid method has better respond than FRENGA and MLPNN alone.

As a result in Fig. 16i TFGS has improved responses better of them. Thus TFGS has best response and controls turbine very well.

In the power curve, power reduction in some spots is distinguished that it's result of wind sudden reduction, but proposed algorithms with shift pitch angle almost control it. Also FRENGA is compared by Fuzzy Genetic system without Neural Network and TFGS in Table 3. Despite the same and rules, TFGS extracts rules with great accuracy and speed.

12 Conclusions

Most important success factor of neural network structure is the accessibility of valuable learning algorithms. Planned approaches optimally control Wind Energy Conversion Systems with changing Pitch angle and estimates parameters.

In this work, several new methods estimate and predict pitch angle value for a wind turbine. Introduced proposed controllers are trained using multi-layer perceptron and radial basis function neural networks.

As plotted from simulation results, controllers change pitch angle of turbine blades to achieve most suitable output power. Since the FRENGA was tested with the Multi-Layer Perceptron NN, it is independent from the NN construction phase. These methods used NN to produce a training example for the rule extraction method. FRENGA can be easily employed for extracting a set of fuzzy if-then rules directly from data. In Hybrid controller MLP and FRENGA proposed a signal control to manage output power and torque. Finally online form both methods were tested on a turbine and one of them that product better respond is selected.

Experimental results admitted that FRENGA, GFS and TFGS work efficiently for both continuous and enumerates attributes. As plotted, wind turbine could obtain the most suitable performance coefficient (Cp) and tip speed ratio (TSR) with change of turbine blades pitch angle. Consequently, simulation results realize which output power has been regulated successfully.

As results indicated, the new proposed genetic fuzzy rule extraction system with tuning membership function (TFGS) outperformed one of the best and earliest approaches in controlling the production through wind fluctuation.

In future work we will try adjust and improvement our result with another intelligent methods.

References

Abdelkarim, M., Achraf, A., & Lotfi, K. (2011). Electric power generation based on variable speed wind turbine under load disturbance. *Energy, 36*, 5016–5026.

Agrawal, R., Imielinski, T., & Swami, A. (1993). *Mining association rules between sets of items in large databases* (pp. 207–216). Washington D.C: SIGMOD.

Andrew, K., & Haiyang, Z. (2010). Optimization of wind turbine energy and power factor with an evolutionary computation algorithm. *Energy, 35*, 1324–1332.

Azar, A. T. (2010b). Adaptive neuro-fuzzy system. In A. T Azar (Ed.) Fuzzy systems. Vienna: IN-TECH. ISBN 978-953-7619-92-3.

Azar, A. T. (2012). Overview of Type-2 fuzzy logic Systems. *International Journal of Fuzzy System Applications (IJFSA), 2*(4), 1–28.

Baesens, B., Setiono, R., Mues, C., & Vanthienen, J. (2003). Using neural network rule extraction and decision tables for credit-risk evaluation. *Management Science, 49*(3), 313–329.

Bianchi, F. D., Battista, H. D., & Mantz, R. J. (2006). *Wind turbine control systems: Principles, modeling and gain scheduling*. New York: Springer.

Bianchi, F. D., Battista, H. D., & Mantz, R. J. (2008). Optimal gain-scheduled control of fixed-speed active stall wind turbines. *IET Renewable Power Generation*, 14–29. doi:10.1049/iet-rpg: 20070106.

Bishop, C. M. (1995). *Neural networks for pattern recognition*. U.K: Oxford University Press.

Bossanyi, E. A. (2000). The design of closed loop controllers for wind turbines. *Wind Energy, 3*, 149–163.

Boukhezzar, B., Lupu, L., Siguerdidjane, H., & Hand, M. (2007). Multivariable control strategy for variable speed, variable pitch wind turbines. *Renewable Energy, 32*, 1273–1287.

Brice, B., Tarek, A. A., & Mohamed, E. H. B. (2009). High-order sliding-mode control of variable-speed wind turbines. *IEEE Transactions on Industrial Electronics, 56*(9), 361–376.

Camblong, H. (2004). Minimisation de l'impact des perturbations d'origine éoliennes dans la génération d'électricité par des aérogénérateurs à vitesse variable.

Carlin, P. W., Laxson, A. S., & Muljadi, E. B. (2001). The history and state of the art variable-speed wind turbine technology. NREL. Technical Report.

Chen, S. M., & Yeh, M. S. (1997). Generating fuzzy rules from relational database systems for estimating null values. *Cybernetics and Systems: An International Journal, 28*(8), 695–723.

Cotrell, J. (2004). Motion technologies CRADA CRD-03-130: Assessing the potential of amechnical continuously variable transmission. NREL, Technical Report.

Craven, M. W., & Shavlik, J. W. (1996). Extracting tree-structured representations of trained networks. In *Advances in Neural Information Processing Systems 8*, 24–30.

Darbari, A. (2000). Rule extraction from trained ANN: A survey. Technical report Institute of Artificial intelligence, Department of Computer Science, TU Dresden.

Endusa, B. M., & Aki, U. (2009). LQG design for megawatt-class WECS with DFIG based on functional models' fidelity prerequisites. *IEEE Transactions on Energy Conversion, 24*(4), 321–340.

Gallant, S. I. (1998). Connectionist expert systems. *Communications of the ACM, 31*(2), 152–169.

Gen, M., & Cheng, R. (1997). *Genetic algorithms and engineering design* (pp. 32–55). New York: Wiley.

Giles, C. L., Miller, C. B., Chen, D., Chen, H., Sun, G. Z., & Lee, Y. C. (1992). Learning and extracting finite state automata with second-order recurrent neural networks. *Neural Computation, 4*(3), 393–405.

Giles, C. L., & Omlin, C. W. (1993). Extraction, insertion, and refinement of symbolic rules in dynamically driven recurrent networks. *Connection Science, 5*(3–4), 307–328.

Gjengedal, T. (2004). Large scale wind power farms as power plants. In *Proceedings Nordic Wind Power Conference* (pp. 48–55).

Glorennec, P. Y. (1997). Coordination between autonomous robots. *International Journal of Approximate Reasoning, 17*(4), 433–446.

Goldberg, D. E. (1989). *Genetic algorithms in search, optimization, and machine learning*. Reading, MA: Addison Wesley.

Golea, M. (1996). On the complexity of rule extraction from neural networks and network querying. In *Proceedings of the AISB'96 Workshop on Rule Extraction from Trained Neural Networks* (pp. 51–59), Brighton, UK.

Hand, M. M. (1999). Variable-speed wind turbine controller systematic design methodology: A comparaison of nonlinear and linear model-based designs. NREL report TP-500-25540, National Renewable Energy Laboratory, Golden, CO, July.

Heider, H., & Drabe, T. (1997). A cascaded genetic algorithm for improving fuzzy-system design. *International Journal of Approximate Reasoning, 17*(4), 351–368.

Hisao, I., Tomoharu, N., & Tadahiko, M. (1999). Techniques and applications of genetic algorithm-based methods for designing compact fuzzy classification systems. *(Chap. 40) Fuzzy Theory Systems: Techniques and Applications, 3*, 35–45.

Hong, Y. Y., Chang, H. L., & Chiu, C. S. (2010). Hour-ahead wind power and speed forecasting using simultaneous perturbation stochastic approximation (SPSA) algorithm and neural network with fuzzy inputs. *Energy, 3*, 3870–3876.

Hong, T. P., Kuo, C. S., & Chi, S. C. (2001). Trade-off between time complexity and number of rules for fuzzy mining from quantitative data. *International Journal Uncertain Fuzziness Knowledge-Based Systems, 9*(5), 587–604.

Horiuchi, N., & Kawahito, T. (2001). Torque and power limitations of variable speed wind turbines using pitch control and generator power control. *IEEE Power Engineering Society Summer Meeting, 2*(1), 638–643.

Jianjun, L., Shozo, T., & Yoshikazu, I. (2006). Explanatory rule extraction based on the trained neural network and the genetic programming. *Journal of the Operations Research Society of Japan, 49*(1), 66–82.

Kaneko, T., Uehara, A., Senjyu, T., Yona, A., & Urasaki, N. (2010). An integrated control method for a wind farm to reduce frequency deviations in a small power system. *Applied Energy, 88*, 1049–1058.

Kanellos, F. D., Papathanassiou, S. A., & Hatziargyriou, N. D. (2000). Dynamic analysis of a variable speed wind turbine equipped with a voltage source ac/dc/ac converter interface and a reactive current control loop. In *10th Mediterranean Electrotechnical Conference*, IEEE, *3*, pp. 986–989.

Kasabov, N. K. (1998). *Foundation of neural networks, fuzzy systems and knowledge engineering* (2nd ed.), Cambridge: A Bradford Book/the MIT Press.

Kasiri, H., Momeni, H. R., Azimi, M., & Motavalian, A. R. (2011b). A New hybrid optimal control for WECS using MLP neural network and genetic neuro fuzzy. In 2nd *IEEE International Conference on Control, Instrumentation and Automation (ICCIA)* (pp. 361–366), 978-1-4673-1690-3 IEEE.

Kasiri, H., Momeni, H. R., & Kasiri, A. (2012a). Optimal intelligent control for wind turbulence rejection in WECS using ANNs and genetic fuzzy approach. *International Journal of Soft Computing and Soft Engineering, Jscse, 2*(9), 16–34. doi:10.7321/jscse.v2.n9.2.

Kasiri, H., Sanee Abadeh, M., & Momeni, H. R. (2012b). Optimal estimation and control of WECS via a genetic neuro fuzzy approach. *Energy, 40*, 438–444.

Kasiri, H., Sanee Abadeh, M., Momeni, H. R., & Motavalian, A. R. (2011a). Fuzzy rule extraction from a trained artificial neural network using genetic algorithm for WECS control and parameter estimation. In *Proceedings of the 8th IEEE international Conference on Fuzzy Systems and Knowledge Discovery (FSKD11)* (pp. 635–639), Shanghai, China. doi:978-1-61284-181-6/11.

Kathryn, E. J., Lucy, Y. P., Mark, J. B., & Lee, J. F. (2006). Standard and adaptive techniques for maximizing energy capture. *IEEE Control Systems Magazine (June), 3*(2), 232–240.

Kaya, M., & Alhajj, R. (2003). A clustering algorithm with genetically optimized membership functions for fuzzy association rules mining. In *IEEE International Conference on Fuzzy Systems* (pp. 881–886), St. Louis, Missouri.

Kuo, R. J. (1995). Intelligent diagnosis for turbine blade faults using artificial neural networks and fuzzy logic. *Engineering Applications of Artificial Intelligence, 8*(1), 25–34. doi:10.1016/0952-1976(94)00082-X.

Laks, J. H., Pao, L. Y., & Wright, A. D. (2009). Control of wind turbines: Past, present, and future. In *American Control Conference Hyatt Regency Riverfront* (pp. 10–12).

Lin, W. M., & Hong, C. M. (2010). Intelligent approach to maximum power point tracking control strategy for variable-speed wind turbine generation system. *Energy, 35*, 2440–2447.

Lin, W. M., Hong, C. M., & Fu, S. C. (2010a). Fuzzy neural network output maximization control for sensorless wind energy conversion system. *Energy, 35*, 592–601.

Lin, W. M., Hong, C. M., & Fu, S. C. (2010b). On-line designed hybrid controller with adaptive observer for variable-speed wind generation system. *Energy, 35*, 3022–3030.

Litipu, Z., & Nagasaka, K. (2004). Improve the reliability and environment of power system based on optimal allocation of WPG. IEEE Power System Conference Exposition Proceedings (Vol. 1, pp. 524–532).

Luo, F. L., & Unbehaben, R. (1998). *Applied neural networks for signal processing*. Cambridge: Cambridge University Press.

Ma, X. (1997). Adaptive extremum control and wind turbine control PhD thesis, Danemark.

Magdalena, L. (2001). *Genetic fuzzy systems—Evolutionary tuning and learning of fuzzy knowledge bases*. Singapore: World Scientific.

Mangialardi, L., & Mantriota, G. (1996). Dynamic behaviour of wind power systems equipped with automatically regulated continously variable transmission. *Renewable Energy, 7*(2), 185–203.

Martin, F., Purellku, I., & Gehlhaar, T. (2014). Modelling of and simulation with grid code validated wind turbine models. Germanischer Lloyd Industrial Services GmbH, Competence Centre Renewables Certification, Steinhöft 9, 20459 Hamburg, Germany.

Mitra, S. (1994). Fuzzy MLP based expert system for medical diagnosis. *Fuzzy Sets and Systems, 65*(2–3), 285–296.

Mitra, S., & Hayashi, Y. (2000). Neuro-fuzzy rule generation: Survey in soft computing framework. *IEEE Transaction on Neural Networks, 11*(3), 748–768.

Mitra, S., Pal, S. K., & Mitra, P. (2002). Data mining in soft computing framework: A survey. *IEEE Transactions on Neural Networks, 13*(1), 3–14.

Moyano, C. F., & Lopes, J. A. (2009). An optimization approach for wind turbine commitment and dispatch in a wind park. *Electric Power Systems Research, 79*, 71–79.

Muhando, E. B. (2008). Modeling-based design of intelligent control paradigms for modern wind generating systems. (Doctoral dissertation, University of the Ryukyus, Nishihara, Japan).

Nauck, N. (2000). Data analysis with neuro-fuzzy methods. (Habilitation Thesis University of Magdeburg).

Oh, S. H. (2010). Error back-propagation algorithm for classification of imbalanced data. *Neurocomputing, 5*(3), 23–35. doi:10.1016/j.neucom.2010.11.024.

Omlin, C. W., Giles, C. L., & Miller, C. B. (1992). Heuristics for the extraction of rules from discrete time recurrent neural networks. In *Proceedings of the International Joint Conference on Neural Networks* (Vol. 1, pp. 33–38), Baltimore, MD.

Pao, Y. H. (1989). Adaptive *pattern recognition and neural networks*. Reading, MA: Addison-Wesley Publishing Co., Inc.

Prakash, A., Chan Felix, T. S., & Deshmukh, S. G. (2011). FMS scheduling with knowledge based genetic algorithm approach. *Expert Systems with Applications, 38*, 3161–3171.

Roy, A. (2000). On connectionism, rule extraction, and brain-like learning. *IEEE Transactions on Fuzzy Systems, 8*(2), 222–227.

Saito, K., & Nakano, R. (2002). Extracting regression rules from neural networks. *Neural Networks, 15*(10), 1279–1288.

Sakamoto, R., Senjyu, T., Kinjo T., Urasaki, N., Funabashi, T., & Fujita, H. (2005). Output power leveling of wind turbine generator for all operation regions by pitch angle control. In IEEE *Power Engineering Society General Meeting* (pp. 2274–2281).

Salman, K. S., & Teo, A. L. J. (2003). Windmill modeling consideration and factors influencing the stability of a grid-connected wind power-based embedded generator. *IEEE Transactions on Power Systems, 18*, 793–802.

Santos, R., Nievola, J., & Freitas, A. (2000). Extracting comprehensible rules from neural networks via genetic algorithm. In *Proceedings of the IEEE Symposium on Combination of Evolutionary Algorithm and Neural Network* (pp. 130–139), S. Antonio, RX, USA.

Senjyu, T., Sakamoto, R., Urasaki, N., Funabashi, T., & Sekine, H. (2006). Output power leveling of wind farm using pitch angle control with fuzzy neural network. 1-4244-0493-2 © IEEE.

Setiono, R., Leow, W. K., & Zurada, J. M. (2002). Extraction of rules from artificial neural networks for nonlinear regression. *IEEE Transactions on Neural Networks, 13*(3), 564–577.

van der Hooft, E. L, & van Engelen, T. G. (2003). Feed forward control of estimated wind speed. Technical report ECN-C-03-137, ECN Windenergie.

van der Hooft, E. L, & van Engelen, T. G. (2004). Estimated wind speed feed forward control for wind turbine operation optimisation. In *European Wind Energy Conference Proceedings* (pp. 35–42), London.

Verdonschot, M. J. (2009). Modeling and control of wind turbines using a continuously variable transmission. *Master's thesis*, Eindhoven University of Technology Department Mechanical Engineering Dynamics and Control Technology Group, April.

W2000 2 MW Wind Turbine, Wikov Wind in partnership with WINDTEC ORBITAL2, September 2007, Technical brochure.
Wang, W., & Bridges, S.M. (2000). Genetic algorithm optimization of membership functions for mining fuzzy association rules. In *International Joint Conference on Information Systems, Fuzzy Theory and Technology Conference* (pp. 1–4), Atlantic City.
Wang, X. Z., Zhang, T., & He, L. (2010). Application of fuzzy adaptive back-propagation neural network in thermal conductivity gas analyzer. *Neurocomputing, 73*, 679–683.
Wermter, S., & Sun, R. (2000). *Hybrid neural systems*. Berlin: Springer.
Witten, I. H., & Frank, E. (1999). *Data mining. practical machine learning tools and techniques with java implementations*. San Diego: Academic Press.
Xing, Z., Li, Q., Su, X., & Guo, H. (2009). Application of BP neural network for wind turbines. In *Second International Conference on Intelligent Computation Technology and Automation* (pp. 10–18). doi:10.2119.
Yingduo, H., Zonghong, W., Qi, C., & Shaohua, T. (1997). Artificial-neural-network-based fast valving control in a power-generation system. *Engineering Applications of Artificial Intelligence, 10*(2), 139–155. doi:10.1016/S0952-1976(96)00071-1.
Yue, S., Tsang, E., Yeung, D., & Shi D. (2000). Mining fuzzy association rules with weighted items. In *IEEE International Conference on Systems, Man and Cybernetics* (pp. 1906–1911), Nashville, Tennessee.
Zhou, Z. H., Chen, S. F., & Chen, Z. Q. (2000). A statistics based approach for extracting priority rules from trained neural networks. In *Proceedings of the IEEE-INNS-ENNS International Joint Conference on Neural Networks* (Vol. 3, pp. 401–406), Como, Italy.

Secondary and Tertiary Structure Prediction of Proteins: A Bioinformatic Approach

Minu Kesheri, Swarna Kanchan, Shibasish Chowdhury and Rajeshwar Prasad Sinha

Abstract Correct prediction of secondary and tertiary structure of proteins is one of the major challenges in bioinformatics/computational biological research. Predicting the correct secondary structure is the key to predict a good/satisfactory tertiary structure of the protein which not only helps in prediction of protein function but also in prediction of sub-cellular localization. This chapter aims to explain the different algorithms and methodologies, which are used in secondary structure prediction. Similarly, tertiary structure prediction has also emerged as one of developing areas of bioinformatics/computational biological research owing to the large gap between the available number of protein sequences and the known experimentally solved structures. Because of time and cost intensive experimental methods, experimentally determined structures are not available for vast majority of the available protein sequences present in public domain databases. The primary aim of this chapter is to offer a detailed conceptual insight to the algorithms used for protein secondary and tertiary structure prediction. This chapter systematically illustrates flowchart for selecting the most accurate prediction algorithm among different categories for the target sequence against three categories of tertiary structure prediction methods. Out of the three methods, homology modeling which is considered as most reliable method is discussed in detail followed by strengths and limitations for each of these categories. This chapter also explains different practical and conceptual problems, obstructing the high accuracy of the protein structure in each of the steps for all the three methods of tertiary structure prediction. The popular hybrid methodologies which further club together a number of features such as structural alignments, solvent accessibility and secondary structure information are also discussed. Moreover, this chapter elucidates about the Meta-servers that generate consensus result from many servers to build a protein

M. Kesheri · R.P. Sinha
Laboratory of Photobiology and Molecular Microbiology, Centre of Advanced Study in Botany, Banaras Hindu University, Varanasi 221005, India

S. Kanchan (✉) · S. Chowdhury
Department of Biological Sciences, Birla Institute of Technology and Science, Pilani, Rajasthan 333031, India
e-mail: swarnabioinfo@gmail.com

Q. Zhu and A.T. Azar (eds.), *Complex System Modelling and Control Through Intelligent Soft Computations*, Studies in Fuzziness and Soft Computing 319,
DOI 10.1007/978-3-319-12883-2_19

Table 1 Current PDB holdings (as on April 22nd, 2014)

Experimental methods	Molecule types				
	Proteins	Nucleic acids	Protein/NA complexes	Other	Total
X ray	82,406	1,516	4,287	4	88,213
NMR	9,129	1,078	206	7	10,420
Electron microscopy	523	52	173	0	748
Hybrid	59	3	2	1	65
Other	155	4	6	13	178
Total	92,272	2,653	4,674	25	99,624

highly needed to fill the gap between the protein sequences available in public domain databases and their experimentally solved structures.

Historically, protein structure prediction was classified into three categories: (i) Ab initio modeling (Liwo et al. 1999; Zhang et al. 2003; Bradley et al. 2005; Klepeis et al. 2005; Klepeis and Floudas 2003) (ii) Threading or Fold recognition (Bowie et al. 1991; Jones et al. 1992; Xu and Xu 2000; Zhou and Zhou 2005; Skolnick et al. 2004) and (iii) Homology or Comparative modeling (Šali and Blundell 1993; Fiser et al. 2000). Threading and comparative modeling build protein models by aligning query sequences onto solved template structures by X-ray crystallography or NMR. When close templates are identified, high-resolution models could be built by the template-based methods. If templates are absent from the PDB, the models need to be built from scratch, i.e. ab initio modeling.

Nowadays, these prediction categories are clubbed into two major groups: free modeling (FM) involving Ab initio folding and template-based modeling (TBM), which includes comparative/homology modeling and threading. These predicted models must be checked for protein structure quality validation by various programmes available.

This chapter is broadly divided under 9 sections which are further divided into sub-headings wherever required. Section 2.1 describes about amino acid propensity based secondary structure prediction method. Section 2.2 discusses about template based secondary structure predictions and the accuracy obtained by these methods. Section 2.3 explains the secondary structure prediction methods based on machine learning approaches. Ab initio folding/modeling and its limitations are is described in Sect. 3.1. Threading and Homology modeling methods with their strengths and their weakness are explained in Sects. 3.2 and 3.3 respectively. Hybrid and Meta-Servers which aid in accuracy of protein models are described in Sects. 4 and 5. Section 6 describes about the protein structure prediction community, Critical Assessment of protein Structure Prediction (CASP). Section 7 describes about the various application of protein models generated by the three major prediction methods. Future prospects of protein secondary and tertiary structure prediction methodologies or algorithms as well as key steps which need to be improved are discussed in Sect. 8. Finally, Sect. 9 provides a comprehensive conclusion for the entire chapter.

2 Secondary Structure Prediction

2.1 Amino Acid Propensity Based Prediction

Early prediction methods as proposed by Chou and Fasman (1974) and the Garnier-Osguthorpe-Robson (GOR) (Garnier et al. 1978) rely on the propensity of amino acids that belong to a given secondary structure. These are simple and direct methods, devoid of complex computer calculations, that utilize empirical rules for predicting the initiation and termination of helical regions in proteins. The relative frequencies of each amino acid in each secondary structure of known protein structures are used to extract the propensity of the appearance of each amino acid in each secondary structure type. Propensities are then used to predict the probability that amino acids from the protein sequence would form a helix, a beta strand, or a turn in a protein. These methods have introduced the conditional probability of immediate neighbor residues for computation. The web-servers based on Chou and Fasman (1974) and GOR showed prediction accuracy between 60–65 %. However the updated, GOR V algorithm which is available as web-server at http://gor.bb.iastate.edu/ combines information theory, bayesian statistics and evolutionary information and has reached an accuracy of prediction to 73.5 % (Sen et al. 2005).

2.2 Template Based Prediction

This method uses the information from database of proteins with known secondary structures to predict the secondary structure of a query protein by aligning the database sequence with the query sequence and finally assigning the secondary structures to the query sequence. The nearest-neighbor method belongs to this category. This category is reliable if both sequences have good identical or homologous regions as compared to a threshold value. The two most successful template-based methods are Nearest-neighbor Secondary Structure Prediction (NNSSP) (Yi and Lander 1993) and PREDATOR (Frishman and Argos 1997). The accuracy of these methods lies in the range 63–68 % (Runthala and Chowdhury 2013).

2.3 Sequence Profile Based Method

This method uses the machine learning algorithms to predict the secondary structure of the query protein. Artificial Neural Networks (ANNs), Support Vector Machines (SVMs) and Hidden Markov Models (HMMs) are the most widely used machine learning algorithms that come under this category (Jones 1999a; Karplus et al. 1998; Kim and Park 2003; Chandonia and Karplus 1995). Currently, most effective PSS prediction methods are based on machine learning algorithms, such as PSIPRED (McGuffin et al. 2000), SVMpsi (Kim and Park 2003), PHD (Rost et al. 1994), PHDpsi (Przybylski and Rost 2002), Porter (Pollastri and McLysaght 2005), JPRED3 (Cole et al. 2008), STRIDE (Heinig and Frishman 2004), SPARROW (Bettella et al. 2012) and SOPMA (Geourjon and Deléage 1995) and which employ Artificial Neural Network (ANN) or Support Vector Machines (SVM) learning models. In addition to protein secondary structure, these servers also make predictions on Solvent Accessibility and Coiled-coil regions etc. These programmes or web-servers are listed in Table 2. These methods have an accuracy ranging 72–80 %, depending on the method, the training and the test datasets.

Two types of errors are most prevalent in secondary structure prediction of proteins. One of these errors is called local errors which occur when a residue is wrongly predicted. Second type of error is called structural error, which occur when the structure is altered globally. Sometimes, errors that alter the function of a protein should be avoided whenever possible. Q3 is the most commonly used measures of local errors, whereas the Segment Overlap (SOV) Score (Zemla et al. 1999) is the most well known measure for structural errors. These measures have been adopted by various communities in these research areas e.g. CASP (Moult et al. 1995) and EVA (Eyrich et al. 2001). Good secondary structures lay the foundation for better prediction of tertiary structures of proteins. The following section provides an insight into the methods for predicting the tertiary structures of proteins.

3 Tertiary Structure Prediction

As discussed in introduction, tertiary structure prediction methods are categorized into three major methods to model a target protein sequence. Flowchart for selecting the most accurate prediction algorithm/method among these three categories for the target sequence is schematically represented in Fig. 1.

Table 2 List of sequence profile-based web servers and programmes for secondary structure prediction along with the webpage URL and the programme description

S. no.	Name of the web server/group (URL)	Description of the web server/group
1	PSIPRED (McGuffin et al. 2000) [http://bioinf.cs.ucl.ac.uk/psipred/]	A simple and accurate secondary structure prediction server, incorporating two feed-forward neural networks which perform an analysis on output obtained from PSI-BLAST
2	PORTER (Pollastri and McLysaght 2005) [http://distill.ucd.ie/porter/]	A server which relies on bidirectional recurrent neural networks with shortcut connections, accurate coding of input profiles obtained from multiple sequence alignments, second stage filtering by recurrent neural networks
3	PHD (Rost et al. 1994) [http://npsapbil.ibcp.fr/cg-bin/npsa_automat.pl?page=/NPSA/npsa_phd.html]	An automated server which uses evolutionary information from multiple sequence alignment to predict the secondary structure prediction of proteins
4	JPRED3 (Cole et al. 2008) [http://www.compbio.dundee.ac.uk/www-jpred/]	Jpred incorporates the Jnet algorithm in order to make more accurate predictions. In addition to protein secondary structure Jpred also makes predictions on solvent accessibility and coiled-coil regions
5	STRIDE (Heinig and Frishman 2004) [http://webclu.bio.wzw.tum.de/stride/]	This server implements a knowledge-based algorithm that makes combined use of hydrogen bond energy and statistically derived backbone torsional angle information
6	SPARROW (Bettella et al. 2012) [http://agknapp.chemie.fu-berlin.de/sparrow/]	This server uses a hierarchical scheme of scoring functions and a neural network to predict the secondary structure
7	SOPMA (Geourjon and Deléage 1995) [http://npsa-pbil.ibcp.fr/cgi-bin/npsa_automat.pl?page=npsa_sopma.html]	A web-server which improved their prediction accuracy when combined with PHD secondary structure prediction method

3.1 Ab Initio Folding/Modeling

This method is simply based on elementary fundamentals of energy and geometry (Moult and Melamud 2000). Ab initio structure prediction seeks to predict the native conformation of a protein from the amino acid sequence alone. Ab initio prediction of protein structures makes no use of information available in databases mainly PDB (Nanias et al. 2005). The goal of this method is to predict the structure of a protein based entirely on the laws of physics and chemistry. It is assumed that the actual native state of a protein sequence has the lowest free energy. It means that the protein native state conformation is basically a model at the global minima of

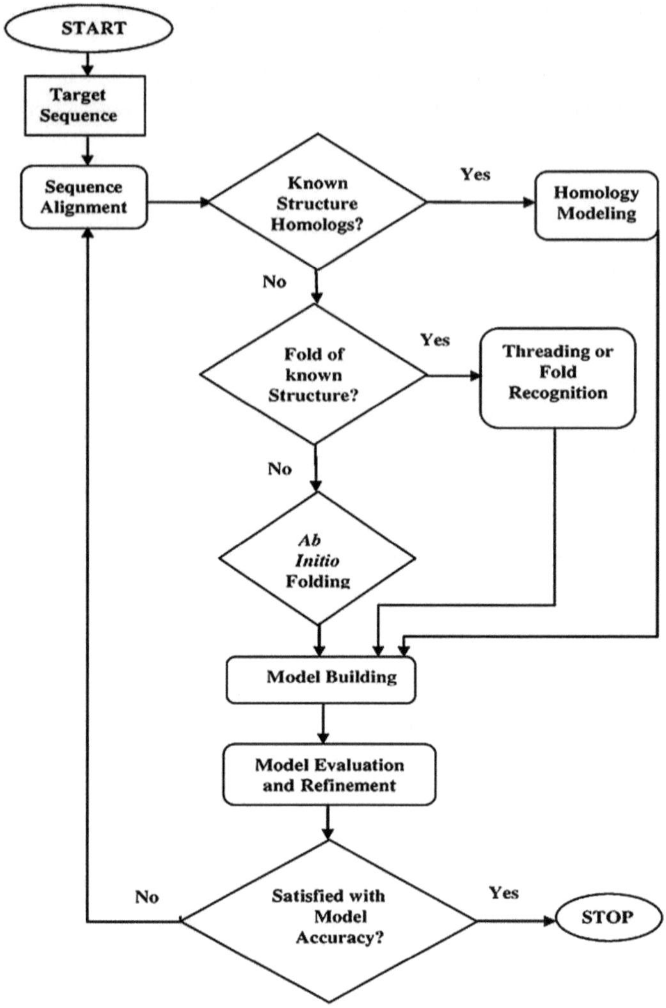

Fig. 1 Flowchart for selecting the most accurate algorithm for prediction of the target sequence against three categories of tertiary structure prediction

the energy landscape. Hence, ab initio algorithm actually searches the entire possible conformational space of a target sequence, in order to find the native state among all conformations.

For example, if we consider only three allowed conformations per residue, then a protein of 200 residues can have 3^{200} different conformations (Runthala and Chowdhury 2013). Hence, searching this huge conformational space will be extremely challenging task. This is the most difficult category of protein structure prediction among all the three different methods of structure prediction which

Table 3 List of web servers for modeling protein structure by ab initio folding method along with the webpage URL and the programme description

S. no.	Name of the web server/group (URL)	Description of the web server/group
1	ROBETTA (Kim et al. 2004; Bradley et al. 2005) [http://robetta.bakerlab.org]	This web-server provides ab initio and comparative models of protein domains. Domains having no sequence similarity with PDB sequences are modeled by Rosetta de novo protocol
2	QUARK (Xu and Zhang 2012) [http://zhanglab.ccmb.med.umich.edu/QUARK/]	De novo protein structure prediction web server aims to construct the correct protein 3D model from amino acid sequence by replica-exchange Monte Carlo simulation under the guide of an atomic-level knowledge-based force field
3	PROTINFO (Hung et al. 2005) [http://protinfo.compbio.washington.edu]	De novo protein structure prediction web server utilizes simulated annealing for 3D structure generation and different scoring functions for selection of final five conformers
4	SCRATCH (Cheng et al. 2005) [http://www.igb.uci.edu/servers/psss.html]	This server utilizes recursive neural networks, evolutionary information, fragment libraries and energy to build protein 3D model
5	BHAGEERATH (Jayaram et al. 2006) [http://www.scfbio-iitd.res.in/bhageerath]	Energy based methodology for narrowing down the search space and thus helps in building a good protein 3D model

completely predicts a new fold (Skolnick and Kolinski 2002; Floudas et al. 2006). With increasing protein size, the conformational space to be searched increases sharply, this makes the ab initio modeling of larger proteins extremely difficult (Zhang and Skolnick 2004).

Currently, the accuracy of ab initio modeling is limited to small proteins having length less than 50 amino acid residues. Ab initio structure prediction requires an efficient potential function to find the conformation of the modeled protein near to native state protein structure with lowest free energy. Ab initio structure prediction is challenging because current potential functions have limited accuracy. Few popular web servers for modeling of the protein structure by ab initio folding/modeling method are listed in Table 3.

3.2 *Fold Recognition (FR) or Threading*

Fold recognition or threading method aims to fit a target sequence to a known structure in a library of folds and the model built is evaluated using residue based contact potentials (Floudas 2007). Although fold recognition will not yield equivalent results as those from X-ray crystallography or NMR yet, it is a comparatively fast and inexpensive way to build a close approximation of a structure from a sequence without involving the time and costs of experimental procedures. Fold Recognition (FR) was reserved for methods which did not rely on sequence searching and where the sequence identity between target and template was below the so-called "twilight zone" spanning between 25–30 %. The rationale behind the threading method is that total number of experimentally solved 3D structure deposited in PDB database doesn't have a new fold. The nature has limited number of basic folds which form the framework of most of the protein structures available in PDB. Generally, similar sequence implies similar structure but the reverse is not true. Similar structures are often found for proteins for which no sequence similarity to any known structure can be detected (Floudas et al. 2006). Using fold recognition or threading, we are able to identify proteins with known structures that share common folds with the target sequences. Fold recognition methods work by comparing each target sequence against a library of potential fold templates using energy potentials and/or other similarity scoring methods. For such comparison, we first need to define a library of potential folds. Once the library is defined, the target sequence will be fitted into each library entry and an energy function is used to evaluate the fit between the target sequence and the library entries to determine the best possible templates. The template with the lowest energy score is then assumed to best fit the fold of the target protein.

Fold recognition methods also includes various properties of structural environment of the amino acid residue. Structural environments are more conserved than the actual type of residue, therefore in the absence of homology, a fold could be predicted by measuring the compatibility of a sequence with template folds in terms of amino acid preferences for certain structural environments. These amino acid preferences for structural environment provide sufficient information to choose among the folds. The amino acid preferences for three main types of structural environment comprise of the solvent accessibility, the contact with polar atoms and the secondary structure. The main limitation of this method is high computational cost, since each entry in the whole library of thousands of possible folds needs to be aligned in all possible ways to select the fold(s). Another major bottleneck is the energy function used for the evaluation of alignment. It is not reasonable to expect to find the correct folds in all cases with a single form of energy function. Few popular web servers for modeling the protein structure by threading method are listed in Table 4.

3.3 Homology Modeling or Comparative Modeling

Comparative or homology protein structure modeling builds a three-dimensional model for a protein of unknown structure (the target) based on one or more related proteins of known structure. The necessary conditions for getting a useful model are

Table 4 List of web servers for modeling the protein structure by threading or fold recognition method along with the webpage URL and the description of programme

S. no.	Name of the web server/group [URL]	Description of the web server/group
1	I-TASSER (Zhang et al. 2005) [http://zhanglab.ccmb.med.umich.edu/I-TASSER/]	3D models are built based on multiple-threading alignments by LOMETS and iterative template fragment assembly
2	SPARKS^X (Yang et al. 2011) [http://sparks-lab.org/yueyang/server/SPARKS-X/]	This server employs significantly improved secondary structure prediction, real value torsion angle prediction and solvent accessibility prediction to model a more accurate protein structure
3	LOOPP (Teodorescu et al. 2004) [http://cbsuapps.tc.cornell.edu/loopp.aspx]	A fold recognition program based on the collection of numerous signals to build the target structure
4	PROSPECT (Xu and Xu 2000) [http://compbio.ornl.gov/structure/prospect]	PROSPECT is based on scoring function, which consists of four additive terms: (i) a mutation term, (ii) a singleton fitness term, (iii) a pairwise-contact potential term, and (iv) alignment gap penalties
5	MUSTER (Wu and Zhang 2008) [http://zhanglab.ccmb.med.umich.edu/MUSTER/]	Muster generates sequence-template alignments by combining sequence profile-profile alignment with multiple structural information
6	PHYRE2 (Kelley and Sternberg 2009) [http://www.sbg.bio.ic.ac.uk/~phyre2/html/page.cgi?id=index]	A server which uses profile–profile matching algorithms to build the protein model

(a) Detectable similarity (Greater than or equal to 30 %) between the target sequence and the template structures and
(b) Availability of a correct alignment between them.

Homology or Comparative modeling is a multistep process that can be summarized in following six steps:

3.3.1 Template Search, Selection and Alignment

Template search is generally done by comparing the target sequence with the sequence of each of the structures in the PDB database. The performance depends on the sensitivity of the comparison of target and template sequences by various programmes e.g. FASTA which is available at http://www.ebi.ac.uk/Tools/sss/fasta/ while, BLAST and PSI-BLAST (Altschul et al. 1997) are available at http://blast.ncbi.nlm.nih.gov/Blast.cgi. The simplest template selection rule is to select the structure with the highest sequence similarity with the target sequence. The quality of a template increases with its overall sequence similarity with the target and decreases with the number and length of gaps in the alignment. Multiple sequence alignment by various freely available programmes e.g. ClustalW (Larkin et al. 2007) Mufft (Katoh et al. 2002), Kalign (Lassmann and Sonnhammer 2005), Probcons (Do et al. 2005) etc. and a development of phylogenetic tree by freely available programmes e.g. MEGA (Tamura et al. 2013) and PHYLIP etc. can help in selecting the template from the subfamily that is closest to the target sequence. HHpred (http://toolkit.tuebingen.mpg.de/hhpred) is one of the best servers which the can even detect very distant relationships between the target sequence and the solved PDB structures significantly. This is the first server that is based on the pairwise comparison of profile Hidden Markov Models (HMMs) (Söding et al. 2005).

The similarity between the 'environment' of the template and the environment in which the target needs to be modeled should also be considered. The quality of the experimentally determined structure is another important factor in template selection whereby high resolution X-ray crystal structure is more preferred for template selection than that of low resolution crystal structure. Multiple templates rather than selecting a single template, generally increases the model accuracy. A good protein structure model depends on alignment between the target and template.

3.3.2 Alignment Correction in Core Regions

An accurate alignment can be calculated automatically using standard sequence-sequence alignment methods, for example, Blast2seq (http://blast.ncbi.nlm.nih.gov/Blast.cgi) and dynamic programming based Needle global sequence alignment (https://www.ebi.ac.uk/Tools/psa/emboss_needle/). In low sequence identity cases, the alignment accuracy is the most important factor which affects the quality of the predicted model. Alignments can be improved by including structural information from the template protein structure. Gaps should be avoided in core regions mainly in secondary structure elements (which are found to be conserved in most cases), buried regions and between two residues that are far in space. It is important to inspect and edit the alignment manually by many tools e.g. Bioedit (www.mbio.ncsu.edu/bioedit/bioedit.html) etc., especially if the target-template sequence identity is low.

3.3.3 Backbone, Loop and Side-Chain Modeling

Creating the backbone is essential for modeled protein structure. For backbone, we simply copy the coordinates of those template residues that show up in the alignment with the model sequence. If two aligned residues differ, only the backbone coordinates (N, Cα, C and O) can be copied. If they are the same, we can also include the coordinates of side chain amino acid residues.

In comparative modeling, target sequences often have few inserted residues as compared to the template structures. Thus, no structural information about these inserted regions could be obtained from the template structures. These regions are called surface loops. Loops often play an important role in defining the functional specificity of a given protein structure, forming the active and binding sites for drug molecules. The accuracy of loop modeling is a major issue for comparative models for applications such as protein-ligand docking i.e. structure based drug design. There are two main classes of loop modeling methods:

(a) Database search approaches, where a small loop of 3–10 amino acid residues are searched in a database of known protein structures and if such loops fit the criteria of lowest energy, such loops are selected and added to the model structure. All major molecular modeling programs and servers support this approach e.g. Modeller (Šali and Blundell 1993), Swiss-Model (Guex and Peitsch 1997).
(b) The conformational search approaches mainly depend on an efficient energy function to choose the loop with lowest energy. If required, energy of the selected loop is minimized using Monte Carlo or molecular dynamics simulations by AMBER, and GROMACS techniques in order to arrive at the best loop conformation with lowest energy.

Side chain modeling is also one of the essential components in structure prediction of proteins. When we compare the side-chain conformations (rotamers) of residues that are conserved in structurally similar proteins, we copy coordinates of conserved amino acid residues entirely from the template to the model. But when we have different residues, side chains are added to each amino acid and their all possible rotamers are searched to find the most stable (having least energy) rotamer from rotamer library.

3.3.4 Model Refinement

One of the major limitations of computational protein structure prediction is the deviation of predicted models from their experimentally derived true, native structures. Refinement of the protein model is required, if there is problem in structural packing of side chains, loops, and secondary structural elements in the target model. For any error in backbone or side chain packing, energy minimization is done which requires an enormous precision in the energy function. At every minimization step, a few big errors (like bumps, i.e., too short atomic distances) are removed while many small errors might be introduced which lead to another distortion in the structure. In energy minimization, force fields must be fast to handle these large molecules efficiently. Refinement of the low resolution predicted models to high resolution structures are close to the native state, however, it has proven to be extremely challenging. There are various programmes e.g. GROMACS (http://

www.gromacs.org/), AMBER (www.amber.scripps.edu), and CHARMM (http://www.charmm.org/) which are freely as well commercially available for protein model refinement by correcting the overall protein structural geometry. One of the recently developed refinement methods called 3Drefine is computationally inexpensive and consumes only few minutes of CPU time to refine a protein of typical length of 300 amino acid residues (Bhattacharya and Cheng 2013).

3.3.5 Model Evaluation or Validation

The predicted model must be checked for

(a) Errors or distortion in side chain packing of the modeled structure.
(b) Distortions or shifts in correctly aligned region of target with the template structures.
(c) Distortions or shifts of a region that does not align with any of the template structures.
(d) Distortions or shifts of a region that is aligned incorrectly with the template structures.

Structural model accuracy is mainly based on global distance test (GDT), which is an average percentage of model Cα atoms within a specified distance threshold to actual native conformation (Jauch et al. 2007).

$$\mathrm{GDT} = \frac{1}{4}(\max{}^{C}_{1A}0 + \max{}^{C}_{2A}0 + \max{}^{C}_{4A}0 + \max{}^{C}_{8A}0) \tag{1}$$

Equation 1 GDT score where $\mathrm{C}^{0}_{\mathrm{nA}}$ is the number of atom pairs closer than distance of n = 1, 2, 4 and $8\mathrm{A}^{0}$.

TM score is another method for validating the model accuracy to score the topological similarity of target and template structures, where the score near to 1.00 is the best predicted near-native model against the actual experimental structure for a target (Xu and Zhang 2010). MaxSub is another new and independently developed method which aims at identifying the largest subset of C(alpha) atoms of a model that superimpose 'well' over the experimental structure, and produces a single normalized score that represents the quality of the model (Siew et al. 2000).

Various programmes and web-servers are available for checking the quality of the model. One of these is Procheck (Laskowski et al. 1993) that generates the Ramachandran Plot, which illustrates the stereo chemical quality of the protein model. Few popular web servers for protein structure quality validation and their description are listed in Table 5.

Few popular web servers for modeling the protein structure by homology or comparative modeling method along with the webpage URL and description are listed in Table 6.

Table 5 List of web servers for protein structure quality validation along with the webpage URL and the programme description

S. no.	Name of the web server/group [URL]	Description of the web server/group
1	QMEAN (Benkert et al. 2009) [http://swissmodel.expasy.org/qmean/cgi/index.cgi]	Quality estimate is based on geometrical analysis of single model, and the clustering-based scoring function
2	PROSA-WEB (Wiederstein and Sippl 2007) [https://prosa.services.came.sbg.ac.at/prosa.php]	Quality is checked by generation of Z-scores and energy plots that highlight potential problems spotted in protein structures
3	PROCHECK (Laskowski et al. 1993) [http://services.mbi.ucla.edu/SAVES/]	Stereo chemical quality of a protein structure is checked by analyzing residue-by-residue geometry and overall structural geometry
4	VERIFY-3D (Bowie et al. 1991; Luthy et al. 1992) [http://services.mbi.ucla.edu/SAVES/]	Determines the compatibility of an atomic model (3D) with its own amino acid sequence (1D) by assigning a structural class based on its location and environment
5	ERRAT (Colovos and Yeates 1993) [http://services.mbi.ucla.edu/SAVES/]	This server analyzes the statistics of non-bonded interactions between different atom types and plots the value of the error function

3.3.6 Homology Models Repositories

However, there are many repositories available, which contain protein homology models generated using various automated methods that provide models which serve as starting points for biologists/experimentalists. SWISS-MODEL repository (http://swissmodel.expasy.org/repository/) is one of the databases of annotated three-dimensional comparative protein structure models generated by the fully automated homology-modelling pipeline SWISS-MODEL. Protein Model Portal (http://proteinmodelportal.org) is another repository aimed at storing manually built 3D models of proteins (Arnold et al. 2009). The most recent database is Modbase (http://modbase.compbio.ucsf.edu) which contains the datasets of comparative protein structure models, calculated by modeling pipeline ModPipe (Pieper et al. 2011).

Several additional features when clubbed to the methods for tertiary structure prediction generate hybrid methods which are used to produce more accurate protein tertiary structures. Following section discusses about these hybrid methods for the protein tertiary structure prediction.

Table 6 List of web servers for modeling the protein structure by homology modeling or comparative modeling method along with the webpage URL and the programmes description

S. no.	Name of the web server/group [URL]	Description of the web server/group
1	GENO3D (Combet et al. 2002) [http://geno3d-pbil.ibcp.fr/]	A web server which builds the model based on distance geometry, simulated annealing and energy minimization algorithms to build the protein 3D model
2	M4T (Fernandez-Fuentes et al. 2007) [http://manaslu.aecom.yu.edu/M4T/]	A fully automated comparative protein structure modeling server with two major modules, Multiple Templates (MT) and Multiple Mapping Method (MMM)
3	CPHMODELS 3.2 (Nielsen et al. 2010) [http://www.cbs.dtu.dk/services/CPHmodels/]	Protein modeling is based on profile-profile alignment guided by secondary structure and exposure predictions
4	3DJIGSAW (Bates et al. 2001) [http://www.bmm.icnet.uk/servers/3djigsaw/]	An automated server to build three-dimensional models for proteins based on homologues of known structure
5	PUDGE (Norel et al. 2010) https://bhapp.c2b2.columbia.edu/pudge/cgi-bin/pipe_int.cgi	A server that includes secondary structure predictions, domains predictions and disorder prediction to predict the high quality homology model
6	SWISS-MODEL (Guex and Peitsch 1997) [http://swissmodel.expasy.org/SWISSMODEL.html]	A fully automated protein structure homology-modeling server
7	ESYPRED3D (Lambert et al. 2002) [http://www.fundp.ac.be/sciences/biologie/urbm/bioinfo/esypred/]	This server results in good protein model by using several multiple alignment programs by combining, weighing and screening

4 Hybrid Methods for Protein Tertiary Structure Prediction

Nowadays, a number of fully automated hybrid methods are designed in order to perform rapid, completely automated fold recognition on a proteome wide scale. These hybrid methods club together a number of features such as structural alignments, solvent accessibility and secondary structure information in order to produce a protein model with high accuracy. Some such methods are discussed below.

GenTHREADER (Jones 1999b) is a fully automated hybrid method for fold recognition which uses a traditional sequence alignment algorithm to generate alignments. These generated alignments are thereafter evaluated by a method derived from threading techniques. The algorithm for GenTHREADER is divided into three stages: alignment of sequences, calculation of pair potential as well as solvation terms and finally, evaluation of the alignment using a neural network (Jones 1999b). GenTHREADER is advantageous as apart from being very fast, it requires no human intervention in the prediction process.

FUGUE (Shi et al. 2001) is another example of hybrid server for recognizing distant homologues by sequence-structure comparison. FUGUE utilizes environment-specific substitution tables and structure-dependent gap penalties. Here scores for amino acid matching and insertions/deletions are evaluated based on the local structural environment of each amino acid residue in a known structure. Local structural environment defined in terms of secondary structure, solvent accessibility, and hydrogen bonding status, are used by FUGUE to produce a high quality 3D protein model. FUGUE also encompasses scanning database of structural profiles, calculation of the sequence-structure compatibility scores and prediction of alignment of multiple sequences against multiple structures in order to enrich the conservation/variation information (Shi et al. 2001).

123D+ (http://pole-modelisation.univ-bpclermont.fr/prive/fiches_HTML/123D+.html) is another hybrid server which combines sequence profiles, secondary structure prediction and contact capacity potential to thread a protein sequence through asset of structures.

RaptorX (http://raptorx.uchicago.edu/) is a protein structure prediction hybrid server that excels in predicting 3D structures for protein sequences without close homologs in the PDB (Källberg et al. 2012). It predicts secondary and tertiary structures, contacts, solvent accessibility, disordered regions and binding sites for a given input sequence. Raptor X, first of all uses profile-entropy scoring method to assess the quality of information content in sequence profiles (Peng and Xu 2010). Thereafter it uses conditional random fields to integrate a variety of biological signals in a nonlinear threading score. Finally, multiple-template threading procedure (Peng and Xu 2009), which enables the use of multiple templates to model a single target sequence is used to produce a high quality protein 3D model.

MULTICOM toolbox (http://sysbio.rnet.missouri.edu/multicom_toolbox/) is another programme consisting of a set of protein structure and structural feature prediction tools. Secondary structure prediction, solvent accessibility prediction, disorder region prediction, domain boundary prediction, contact map prediction, disulfide bond prediction, beta-sheet topology prediction, fold recognition, multiple template combination and alignment, template-based tertiary structure modeling, protein model quality assessment, and mutation stability prediction are some of the functions facilitated by MULTICOM toolbox (Cheng et al. 2012).

Hybrid methods use various aspects for predicting an accurate protein tertiary structure. However Meta-servers discussed in the following section deals with generation of a consensus prediction of protein tertiary structure assembled from different servers.

5 Meta-Servers for Protein Tertiary Structure Prediction

Several meta-servers not only integrate protein structure predictions performed by various methods but also assemble and interpret the results to come up with a consensus prediction. This section deals with a comprehensive discussion of such meta-servers.

Pcons.net meta-server (Wallner et al. 2007) retrieves results from several publically available servers which are then analyzed and assessed for structural correctness using Pcons as well as ProQ, thus presenting the users a ranked list of possible models (Lundström et al. 2001). In combination of several publically available servers, Pcons.net meta-server also uses Reversed Position Specific BLAST (RPS-BLAST) to parse the sequence into structural domains by analyzing the significance and span of the best RPS-BLAST alignment.

3D-Jury (Ginalski et al. 2003) are the meta-servers which focus on the selection of high quality obtained from different servers. 3D-Jury, takes groups of models generated by a set of servers as input which are then compared with each other and a similarity score is assigned to each pair by MaxSub tool (Siew et al. 2000) followed by providing ranking to the models.

3D-SHOTGUN (Fischer 2003) meta server does not just select the best model but also refines initial models for building the protein structure model with high accuracy. 3D-SHOTGUN meta-predictor consists of three steps: (i) assembly of hybrid models, (ii) confidence assignment, and (iii) selection. 3D-SHOTGUN first assembles hybrid models from the initial models and then assigns scores to each of the assembled models by using the original models scores and the structural similarities between them. Thereby resulting a highly sensitive and ensuring a significantly higher specificity of the models than that of individual servers (Fischer 2003).

GeneSilico (Kurowski and Bujnicki 2003) is another meta-server which combines the useful features of other meta-servers available, but with much greater flexibility of the input in terms of user-defined multiple sequence alignments. However, there are several drawbacks reported in the current meta-servers including 3D-Jury (Ginalski et al. 2003) and GeneSilico (Kurowski and Bujnicki 2003). They take the initial threading inputs from remote computer which are occasionally shut down or are not available. Secondly, the instability of the algorithms of the remote servers is another drawback of these meta-servers (Wu and Zhang 2007).

LOMETS (Wu and Zhang 2007), overcomes the above drawbacks. It is one of the good performing meta-servers in which all nine individual threading servers are installed locally, which facilitates controlling and tuning of Meta-server algorithms in a consistent manner making the users able to obtain quick final consensus. It facilitates quick generation of initial threading alignments owing to the nine state of art threading programs that are installed and run in a local computer cluster, thus ensure faster results as compared to the traditional remote-server-based meta-servers. Based on TM-score, the consensus models generated from the top

Table 7 List of meta-servers for protein tertiary structure prediction along with the webpage URL and the description of the programs

S. no.	Name of the web server/group [URL]	Description of the web server/group
1	LOMETS (Wu and Zhang 2007) [http://zhanglab.ccmb.med.umich.edu/LOMETS/]	Meta server that includes locally installed threading programs FUGUE, HHpred, SPARKS. LOMETS generates the final models using a consensus approach
2	3D-Jury (Ginalski et al. 2003) [http://BioInfo.PL/Meta/]	The meta server provides access and results assessment from various remote predictors including, 3DPSSM, ESyPred3D, FUGUE, HHpred, mGenTHREADER etc.
3	GeneSilico (Kurowski and Bujnicki 2003) [https://genesilico.pl/meta2/]	The meta server provides access to various remote and local predictors including 3DPSSM, FUGUE, HHpred, mGenTHREADER, Pcons, Phyre, etc.
4	Pcons.net (Lundström et al. 2001), (Wallner et al. 2007) [http://pcons.net/]	The Pcons protocol analyzes the set of protein models and looks for recurring three-dimensional structural patterns and assigns a score
5.	3D-SHOTGUN (Fischer 2003) [http://bioinfo.pl/meta]	This meta-predictor consists of three steps: (i) assembly of hybrid models, (ii) confidence assignment, and (iii) selection

predictions by LOMETS were at least 7 % more accurate than the best individual servers. In addition to the 3D structure prediction by threading, LOMETS also provides highly accurate contact and distance predictions for the query sequences. The performance of LOMETS can be analyzed by the fact that average CPU time for a medium size protein (~200 residues) is less than 20 min when the programs are run in parallel on nine nodes of the cluster.

A List of Meta-servers for protein tertiary structure prediction along with the webpage URL and the description of the programs in Table 7.

The need for critical evaluation of various methods and developments in the field of protein structure prediction is successfully fulfilled by CASP meetings. The following section gives an overview of several agenda of CASP.

6 CASP

Protein structure prediction algorithms are constantly being developed and redefined to reach the experimental accuracy. Therefore, protein structure prediction strategies and methodologies are tested every 2 years in the Critically Assessment of techniques for protein Structure Prediction (CASP) meeting, which started since 1994. Since then, ten successful CASP meetings are over by 2012 and CASP11 is due in 2014. The participation by various research groups in the CASP are increasing by each successive CASP meetings. The main goal of CASP is to obtain an in-depth and objective assessment of the current abilities and inabilities in the area of protein structure prediction. It critically evaluates the various protein structure programmes and servers besides assigning ranks for the same. CASP also tests the prediction accuracy of those protein sequences, whose solved experimental structures are kept undisclosed until the end of summit. Predictors/participants in CASP, fall into two categories. The first category comprises of teams of human participants who devote considerable time, usually a period of several weeks in order to model each target, to complete their work. The second category involves automatic servers with a target time period of 48 h for the completion of the assigned task (Moult 2005). Participant registration, target management, prediction collection and numerical analysis are all handled by the Protein Structure Prediction Center (http://predictioncenter.org/). The later also provides access to details of all experiments and results apart from providing a discussion forum for the CASP community. CASP also monitors progress in identification of disordered regions in proteins, and the ability to predict three-dimensional (3D) contacts which can be used as restraints during tertiary structure prediction of proteins (Moult et al. 2014). Ab initio modeling methods have also improved substantially and now we have topologically accurate models for small residues (<100 residues) having single domain non-template proteins due to regular CASP experiments (Kryshtafovych et al. 2014). Homology models vary greatly in accuracy depending on a number of factors, and for that reason CASP has encouraged the development of methods that can estimate overall accuracy of a model and accuracy at the individual amino acid level. The accuracy of homology models monitored by CASP, has improved dramatically, through a combination of improved methods. In CASP10, a new "contact-assisted" category has been introduced apart from the already existing previous categories. The idea in the CASP contact-assisted category is to investigate the extent to which experimental information is needed in order to deliver a given level of model accuracy besides encouraging the development of new methods for the same (Moult et al. 2014).

In CASP10 experiment, 114 protein sequences were released as modeling targets. Among these, 53 were designated "all groups" (human and server) targets. Finally 96 experimental structures were available for evaluation and assessment after cancellation of 18 targets (Moult et al. 2014). In CAS10, 217 groups registered, from several relevant communities. Finally, 41,740 predicted models submitted by 150 predictor groups were assessed as template-based modeling predictions where Zhang-Server, QUARK, PMS, Leecon and Zhang groups provided the most accurate models for the assessment units targets (Huang et al. 2014). Thus, CASP meeting is the best way to keep updated with the advancement in protein structure prediction strategies and methodologies.

Any development in the field of science is considered important if it has applications which are of significance to biological systems. The following section deals with various applications of the above discussed methods for protein structure prediction.

7 Applications of Protein Structure Prediction

Homology/Comparative modeling plays an essential role in structure based drug design. For example representative structures produced by in silico screening forms the basis of generation of three-dimensional structures of the remaining proteins encoded in the various genomes that can be predicted by homology modeling (Takeda-Shitaka et al. 2004). Comparative modeled proteins may be used for predicting the binding modes and affinities of different drug compounds as they interact with protein binding sites in structure-based drug design. Computational approach to this problem is usually termed as molecular docking. The goal of ligand-protein docking is to predict the predominant binding mode(s) of a ligand with a protein of known three-dimensional structure. Docking can be used to perform virtual screening on large libraries of compounds, rank the results, and propose structural hypotheses of how the ligands inhibit the target (Morris and Lim-Wilby 2008). However, it is widely accepted that docking with comparative models is more challenging and less successful than docking with crystallographic structures. Comparative models are not only useful in protein-ligand, but also useful in protein-protein docking (Vakser 1997).

Comparative models can also be used for testing and improving sequence structure alignment (Wolf et al. 1998). Based on the alignment of known structures, alignments can be well defined even for a new target sequence. Apart from the presence of functional motifs or the signature sequences, calculated electrostatic potential around the protein structure may help in predicting the protein function (Drew et al. 2011).

Protein models by comparative method can be also used to decipher important residues for biological activity as well as function of the protein. These models can be helpful in designing mutants to test hypotheses about protein functions (Boissel et al. 1993). On the basis of its primary sequence and the location of its disulfide bonds, erythropoietic hormone erythropoietin was modeled by homology modeling which predicts a four alpha-helical bundle motif, in common with other cytokines. Deletions of 5–8 residues from erythropoietin hormone erythropoietin protein within predicted alpha-helices resulted in the failure of export of the mutant protein from the cell (Boissel et al. 1993).

Comparative models can also be used to explore the substrate specificity in several enzymes. After the crystallization of the bacterial leucine transporter protein LeuT, development of 3-D computational models were used for structure-function studies on the plasmalemmal monoamine transporters (MATs). LeuT-based MAT models were used to guide elucidation of substrate and inhibitor binding pockets. Moreover, molecular dynamics simulations using these models provided insight into the conformations involved in the substrate translocation cycle (Manepalli et al. 2012).

Comparative models have been used in conjunction with virtual screening to successfully identify novel inhibitors over the past few years. Novel inhibitors of dihydrofolate reductase in *Typnosoma. cruzi* (the parasite that causes Chagas disease) was discovered by docking into a comparative model to dihydrofolate reductase in *L. major*, a related parasite (Zuccotto et al. 2001). Since the crystal/NMR structure of various drug targets are not available so far, comparative models of drug targets could also be used for computational screening of new inhibitors for *Mycobacterium tuberculosis* drug target proteins (Gahoi et al. 2013).

Comparative modeled structure of cell receptors responsible for binding of foreign particles and thus causing diseases may also be used to study these interactions and may facilitate in investigating the mechanism. Comparative models can be also used to predict the antigenic epitopes. Mouse mast cell protease (mMCP) 1, mMCP-2, mMCP-4, and mMCP-5 models were used to predict immunogenic epitopes and surface regions that are likely to interact with proteoglycans (Sali et al. 1993).

Native PAGE results illustrated the presence of variations in number of isoforms of superoxide dismutase antioxidative enzymes in different cyanobacterial samples (Kesheri et al. 2011). Comparative modeling may be used to generate antioxidative enzymes models that may further help in studying the binding of metal cofactors with the isoforms. Comparative modeling may also be used to study the drug resistance in many vectors.

Garg et al. (2009) constructed the comparative model of dihydropteroate synthase protein which illustrated that novel point mutations at two positions may lead to sulphadoxine drug resistance in *Plasmodium falciparum*. Compararative models facilitates molecular replacement in X-ray structure/NMR models which allows refinement of a determined structure through the knowledge of already known structures.The computational prediction of protein structure also serves as an alternative to produce raw informations that may be validated by wet lab experiments. Following section produces an overview of further developments that may be made in the field of protein structure prediction.

8 Future Prospects

Homology modeling and protein threading are becoming more powerful and important for structure prediction along with the PDB growth and the improvement of prediction protocols. The error of a template-based model comes from template selection and sequence-template alignment. So, the identification of the best template is still a challenging task in protein structure prediction. However, HMM based template search algorithms like HHpred has solved this issue to some extent. Now, another big dilemma is of generation and choosing the correct alignment between target sequence and template sequence. Still, there is no set benchmark available for selection of the best alignment between the target and template sequence.

Model building is also one of the challenging task in structure prediction, in which a number of times it has been seen that side chains are not added properly in their proper conformations which mostly need structure refinement. Model

Refinement algorithms mostly don't fold a target structure to its possible native state. Model refinement is still obstructed with incorrect energy function, integrated with an additional complication of erroneous conformational search programs.

Model selection among hundreds of models generated by Modeller is still a challenging task. However, these issues have been solved to some extent by evaluating these models by various scores e.g. GDT-TS and TM Score etc. Improvement in the current algorithms is needed for the selection of the best model since till date there is no set benchmark for selection of the best model, even by top ranked servers as per CASP.

9 Conclusion

Correct prediction of secondary structure is the key to predict a good or satisfactory tertiary structure of the protein. Secondary structure not only helps in predicting the tertiary structure but also helps in predicting the function as well as sub-cellular localization of proteins. Staring from the amino acid propensity based secondary structure prediction methods, machine learning approaches has revolutionized the prediction accuracy of secondary structure from 60 to 80 %.

Tertiary structure prediction by bioinformatics or computational biology tools is always a challenging task for scientists. Ab initio folding and threading are computationally expensive methods for tertiary structure prediction which, also results in protein structural models having low accuracy. Tertiary structure prediction by ab initio folding/modelling still has a limitation due to searching a large number of conformations generated as well as absence of suitable potential functions as the number of amino acid increases. Another method is fold recognition where, the prediction accuracy is better than ab initio folding/modeling. Homology modeling, the third prediction method, has emerged as the sole method which can build the model close to X-ray crystal/NMR structure. Therefore, among the three methods, comparative or homology modeling is considered as the best method for protein structure predication with high accuracy in such cases where the sequence identity between the target and template sequence is more than 30 %. These comparative models may be used for structure based drug designing as well as virtual screening to identify novel inhibitors. Selecting the best model in homology modelling is one of the major challenging tasks to look into. In homology modeling, the major chances of error may be in loop modeling if long loop is present in the target protein molecule. Side chain modeling is another challenging area where prediction accuracy should be increased. Now a day, hybrid methods became popular because they club together a number of features such as structural alignments, solvent accessibility and secondary structure information in order to produce a protein model with high accuracy. Along with hybrid methods, several meta-servers are also available which integrate protein structure predictions performed by various methods that assemble and interpret the results to come up with a consensus model prediction. Nevertheless, we have not reached the pinnacle of that modelling accuracy till date. However, it is interesting to discuss that, all our predictions may take a long time, while a cell takes only a few micro-seconds to fold a primary sequence into fully functional global native minima structure. Hence, further research to improve the algorithms is still needed to make the prediction close to native state or in other words close to fold adopted by the nature.

Acknowledgments Minu Kesheri is thankful to University Grant Commission, Govt. of India, New Delhi, for providing financial assistance in the form of research fellowship. Swarna Kanchan is thankful to University Grant Commission, Govt. of India, New Delhi for providing the financial support in the form of the Basic Science Research Fellowship under University Grant Commission (New Delhi) Special Assistance Programme to Department of Biological Sciences, Birla Institute of Technology and Science, Pilani, India.

References

Altschul, S. F., Madden, T. L., Schäffer, A. A., Zhang, J., Zhang, Z., Miller, W., et al. (1997). Gapped BLAST and PSI-BLAST: A new generation of protein database search programs. *Nucleic Acids Research, 25*(17), 3389–3402.

Arnold, K., Kiefer, F., Kopp, J., Battey, J. N., Podvinec, M., Westbrook, J. D., et al. (2009). The protein model portal. *Journal of Structural and Functional Genomics, 10*(1), 1–8.

Baker, D., & Sali, A. (2001). Protein structure prediction and structural genomics. *Science, 294* (5540), 93–96.

Bates, P. A., Kelley, L. A., MacCallum, R. M., & Sternberg, M. J. E. (2001). Enhancement of protein modeling by human intervention in applying the automatic programs 3D-JIGSAW and 3D-PSSM. *Proteins: Structure, Function, and Bioinformatics, 45*(5), 39–46.

Benkert, P., Künzli, M., & Schwede, T. (2009). QMEAN server for protein model quality estimation. *Nucleic Acids Research, 37*(Web Server issue), W510–W514.

Berman, H. M., Westbrook, J., Feng, Z., Gilliland, G., Bhat, T. N., Weissig, H., et al. (2000). The protein data bank. *Nucleic Acids Research, 28*(1), 235–242.

Bettella, F., Rasinski, D., & Knapp, E. W. (2012). Protein secondary structure prediction with SPARROW. *Journal of Chemical Information and Modeling, 52*(2), 45–56.

Bhattacharya, D., & Cheng, J. (2013). 3Drefine: Consistent protein structure refinement by optimizing hydrogen bonding network and atomic-level energy minimization. *Proteins: Structure, Function, and Bioinformatics, 81*(1), 119–131.

Boissel, J. P., Lee, W. R., Presnell, S. R., Cohen, F. E., & Bunn, H. F. (1993). Erythropoietin structure-function relationships. Mutant proteins that test a model of tertiary structure. *Journal of Biological Chemistry, 268*(21), 15983–15993.

Bowie, J., Luthy, R., & Eisenberg, D. (1991). A method to identify protein sequences that fold into a known three-dimensional structure. *Science, 253*(5016), 164–170.

Bradley, P., Misura, K. M. S., & Baker, D. (2005). Toward high-resolution de novo structure prediction for small proteins. *Science, 309*(5742), 1868–1871.

Chandonia, J.- M., & Karplus, M. (1995). Neural networks for secondary structure and structural class predictions. *Protein Science, 4*(2), 275–285.

Cheng, J., Li, J., Wang, Z., Eickholt, J., & Deng, X. (2012). The MULTICOM toolbox for protein structure prediction. *BMC Bioinformatics, 13*, 65.

Cheng, J., Randall, A. Z., Sweredoski, M. J., & Baldi, P. (2005). SCRATCH: A protein structure and structural feature prediction server. *Nucleic Acids Research, 33*(Web Server issue), W72–W76.

Chou, P. Y., & Fasman, G. D. (1974). Prediction of protein conformation. *Biochemistry, 13*(2), 222–245.

Cole, C., Barber, J. D., & Barton, G. J. (2008). The Jpred3 secondary structure prediction server. *Nucleic Acids Research, 36*(Web Server issue), W197–W201.

Colovos, C., & Yeates, T. O. (1993). Verification of protein structures: Patterns of non-bonded atomic interactions. *Protein Science, 2*(9), 1511–1519.

Combet, C., Jambon, M., Deléage, G., & Geourjon, C. (2002). Geno3D: Automatic comparative molecular modelling of protein. *Bioinformatics, 18*(1), 213–214.

Do, C. B., Mahabhashyam, M. S. P., Brudno, M., & Batzoglou, S. (2005). ProbCons: Probabilistic consistency-based multiple sequence alignment. *Genome Research, 15*(2), 330–340.

Drew, K., Winters, P., Butterfoss, G. L., Berstis, V., Uplinger, K., Armstrong, J., et al. (2011). The Proteome folding project: Proteome-scale prediction of structure and function. *Genome Research, 21*(11), 1981–1994.

Eyrich, V. A., Martí-Renom, M. A., Przybylski, D., Madhusudhan, M. S., Fiser, A., Pazos, F., Valencia, A., Sali, A., & Rost, B. (2001). EVA: Continuous automatic evaluation of protein structure prediction servers. *Bioinformatics, 17*(12), 1242–1243.

Fernandez-Fuentes, N., Madrid-Aliste, C. J., Rai, B. K., Fajardo, J. E., & Fiser, A. (2007). M4T: A comparative protein structure modeling server. *Nucleic Acids Research, 35*(Web Server issue), W363–W368.

Fischer, D. (2003). 3D-SHOTGUN: A novel, cooperative, fold-recognition meta-predictor. *Proteins: Structure, Function, and Bioinformatics, 51*(3), 434–441.

Fiser, A., Do, R. K. G., & Šali, A. (2000). Modeling of loops in protein structures. *Protein Science, 9*(9), 1753–1773.

Floudas, C. A. (2007). Computational methods in protein structure prediction. *Biotechnology and Bioengineering, 97*(2), 207–213.

Floudas, C. A., Fung, H. K., McAllister, S. R., Mönnigmann, M., & Rajgaria, R. (2006). Advances in protein structure prediction and de novo protein design: A review. *Chemical Engineering Science, 61*(3), 966–988.

Frishman, D., & Argos, P. (1997). Seventy-five percent accuracy in protein secondary structure prediction. *Proteins: Structure, Function, and Bioinformatics, 27*(3), 329–335.

Gahoi, S., Mandal, R. S., Ivanisenko, N., Shrivastava, P., Jain, S., Singh, A. K., et al. (2013). Computational screening for new inhibitors of M. tuberculosis mycolyltransferases antigen 85 group of proteins as potential drug targets. *Journal of Biomolecular Structure and Dynamics, 31*(1), 30–43.

Garg, S., Saxena, V., Kanchan, S., Sharma, P., Mahajan, S., Kochar, D., et al. (2009). Novel point mutations in sulfadoxine resistance genes of Plasmodium falciparum from India. *Acta Tropica, 110*(1), 75–79.

Garnier, J., Osguthorpe, D. J., & Robson, B. (1978). Analysis of the accuracy and implications of simple methods for predicting the secondary structure of globular proteins. *Journal of Molecular Biology, 120*(1), 97–120.

Geourjon, C., & Deléage, G. (1995). SOPMA: Significant improvements in protein secondary structure prediction by consensus prediction from multiple alignments. *Computer applications in the biosciences: CABIOS, 11*(6), 681–684.

Ginalski, K., Elofsson, A., Fischer, D., & Rychlewski, L. (2003). 3D-Jury: A simple approach to improve protein structure predictions. *Bioinformatics, 19*(8), 1015–1018.

Gong, H., & Rose, G. D. (2005). Does secondary structure determine tertiary structure in proteins? *Proteins: Structure, Function, and Bioinformatics, 61*(2), 338–343.

Guex, N., & Peitsch, M. C. (1997). SWISS-MODEL and the Swiss-Pdb viewer: An environment for comparative protein modeling. *Electrophoresis, 18*(15), 2714–2723.

Heinig, M., & Frishman, D. (2004). STRIDE: A web server for secondary structure assignment from known atomic coordinates of proteins. *Nucleic Acids Research, 32*(Web Server issue), W500–W502.

Huang, Y. J., Mao, B., Aramini, J. M., & Montelione, G. T. (2014). Assessment of template-based protein structure predictions in CASP10. *Proteins: Structure, Function, and Bioinformatics, 82*(2), 43–56.

Hung, L.- H., Ngan, S.- C., Liu, T., & Samudrala, R. (2005). PROTINFO: New algorithms for enhanced protein structure predictions. *Nucleic Acids Research, 33*(Web Server issue), W77–W80.

Jauch, R., Yeo, H. C., Kolatkar, P. R., & Clarke, N. D. (2007). Assessment of CASP7 structure predictions for template free targets. *Proteins: Structure, Function, and Bioinformatics, 69*(8), 57–67.

Jayaram, B., Bhushan, K., Shenoy, S. R., Narang, P., Bose, S., Agrawal, P., et al. (2006). Bhageerath: An energy based web enabled computer software suite for limiting the search space of tertiary structures of small globular proteins. *Nucleic Acids Research, 34*(21), 6195–6204.

Jones, D. T. (1999a). Protein secondary structure prediction based on position-specific scoring matrices. *Journal of Molecular Biology, 292*(2), 195–202.

Jones, D. T. (1999b). GenTHREADER: An efficient and reliable protein fold recognition method for genomic sequences. *Journal of Molecular Biology, 287*(4), 797–815.

Jones, D. T., Taylort, W. R., & Thornton, J. M. (1992). A new approach to protein fold recognition. *Nature, 358*, 86–89.

Källberg, M., Wang, H., Wang, S., Peng, J., Wang, Z., Lu, H., et al. (2012). Template-based protein structure modeling using the RaptorX web server. *Nature Protocols, 7*(8), 1511–1522.

Karplus, K., Barrett, C., & Hughey, R. (1998). Hidden Markov models for detecting remote protein homologies. *Bioinformatics, 14*(10), 846–856.

Katoh, K., Misawa, K., Kuma, K., & Miyata, T. (2002). MAFFT: A novel method for rapid multiple sequence alignment based on fast Fourier transform. *Nucleic Acids Research, 30*(14), 3059–3066.

Kelley, L. A., & Sternberg, M. J. E. (2009). Protein structure prediction on the Web: A case study using the Phyre server. *Nature Protocols, 4*(3), 363–371.

Kesheri, M., Richa, & Sinha, R. P. (2011). Antioxidants as natural arsenal against multiple stresses in cyanobacteria. *International Journal of Pharma and Bio Sciences, 2*(2), B168–B187.

Kim, D. E., Chivian, D., & Baker, D. (2004). Protein structure prediction and analysis using the Robetta server. *Nucleic Acids Research, 32*(Web Server issue), W526–W531.

Kim, H., & Park, H. (2003). Protein secondary structure prediction based on an improved support vector machines approach. *Protein Engineering, 16*(8), 553–560.

Klepeis, J. L., & Floudas, C. A. (2003). ASTRO-FOLD: A combinatorial and global optimization framework for ab initio prediction of three-dimensional structures of proteins from the amino acid sequence. *Biophysical Journal, 85*(4), 2119–2146.

Klepeis, J. L., Wei, Y., Hecht M. H., & Floudas, C. A. (2005). Ab initio prediction of the three-dimensional structure of a de novo designed protein: A double-blind case study. *Proteins: Structure, Function, and Bioinformatics*, *58*(3), 560–570.

Kryshtafovych, A., Fidelis, K., & Moult, J. (2014). CASP10 results compared to those of previous CASP experiments. *Proteins: Structure, Function, and Bioinformatics, 82*(2), 164–174.

Kurowski, M. A., & Bujnicki, J. M. (2003). GeneSilico protein structure prediction meta-server. *Nucleic Acids Research, 31*(13), 3305–3307.

Lambert, C., Léonard, N., De, B. X., & Depiereux, E. (2002). ESyPred3D: Prediction of proteins 3D structures. *Bioinformatics, 18*(9), 1250–1256.

Larkin, M. A., Blackshields, G., Brown, N. P., Chenna, R., McGettigan, P. A., McWilliam, H., et al. (2007). Clustal W and Clustal X version 2.0. *Bioinformatics*, *23*(21), 2947–2948.

Laskowski, R. A., Macarthur, M. W., Moss, D. S., & Thornton, J. M. (1993). PROCHECK: A program to check the stereochemical quality of protein structures. *Journal of Applied Crystalography, 26*, 283–291.

Lassmann, T., & Sonnhammer, E. (2005). Kalign—An accurate and fast multiple sequence alignment algorithm. *BMC Bioinformatics, 6*(1), 298.

Lisewski, A. M., & Lichtarge, O. (2006). Rapid detection of similarity in protein structure and function through contact metric distances. *Nucleic Acids Research, 34*(22), e152.

Liwo, A., Lee, J., Ripoll, D. R., Pillardy, J., & Scheraga, H. A. (1999). Protein structure prediction by global optimization of a potential energy function. *Proceedings of the National Academy of Sciences, USA, 96*(10), 5482–5485.

Lundström, J., Rychlewski, L., Bujnicki, J., & Elofsson, A. (2001). Pcons: A neural-network–based consensus predictor that improves fold recognition. *Protein Science, 10*(11), 2354–2362.

Luthy, R., Bowie, J. U., & Eisenberg, D. (1992). Assessment of protein models with three-dimensional profiles. *Nature, 356*, 83–85.

Manepalli, S., Surratt, C., Madura, J., & Nolan, T. (2012). Monoamine transporter structure, function, dynamics, and drug discovery: A computational perspective. *American Association of Pharmaceutical Scientists Journal, 14*(4), 820–831.

McGuffin, L. J., Bryson, K., & Jones, D. T. (2000). The PSIPRED protein structure prediction server. *Bioinformatics, 16*(4), 404–405.

Morris, G. M., & Lim-Wilby, M. (2008). Molecular docking. *Methods in Molecular Biology*, *443*, 365–382.

Moult, J. (2005). A decade of CASP: Progress, bottlenecks and prognosis in protein structure prediction. *Current Opinion in Structural Biology, 15*(3), 285–289.

Moult, J., Fidelis, K., Kryshtafovych, A., Schwede, T., & Tramontano, A. (2014). Critical assessment of methods of protein structure prediction (CASP)-round x. *Proteins: Structure, Function, and Bioinformatics, 82*(2), 1–6.

Moult, J., & Melamud, E. (2000). From fold to function. *Current Opinion in Structural Biology, 10*(3), 384–389.

Moult, J., Pedersen, J. T., Judson, R., & Fidelis, K. (1995). A large-scale experiment to assess protein structure prediction methods. *Proteins: Structure, Function, and Bioinformatics*, *23*(3), ii–iv.

Nair, R., & Rost, B. (2003). Better prediction of sub-cellular localization by combining evolutionary and structural information. *Proteins: Structure, Function, and Bioinformatics, 53* (4), 917–930.

Nair, R., & Rost, B. (2005). Mimicking cellular sorting improves prediction of subcellular localization. *Journal of Molecular Biology, 348*(1), 85–100.

Nanias, M., Chinchio, M., Ołdziej, S., Czaplewski, C., & Scheraga, H. A. (2005). Protein structure prediction with the UNRES force-field using replica-exchange Monte Carlo-with-minimization; comparison with MCM, CSA, and CFMC. *Journal of Computational Chemistry, 26*(14), 1472–1486.

Nielsen, M., Lundegaard, C., Lund, O., & Petersen, T. N. (2010). CPHmodels-3.0—Remote homology modeling using structure-guided sequence profiles. *Nucleic Acids Research, 38*(Web Server issue), W576–W581.

Norel, R., Petrey, D., & Honig, B. (2010). PUDGE: A flexible, interactive server for protein structure prediction. *Nucleic Acids Research, 38*(Web Server issue), W550–W554.

Peng, J., & Xu, J. (2009). *Boosting protein threading accuracy* (Vol. 5541, pp. 31–45). Lecture Notes in Computer Science.

Peng, J., & Xu, J. (2010). Low-homology protein threading. *Bioinformatics, 26*(12), i294–i300.

Pieper, U., Webb, B. M., Barkan, D. T., Schneidman-Duhovny, D., Schlessinger, A., Braberg, H., et al. (2011). MODBASE, a database of annotated comparative protein structure models and associated resources. *Nucleic Acids Research, 39*(Database issue), D465–D474.

Pollastri, G., & McLysaght, A. (2005). Porter: A new, accurate server for protein secondary structure prediction. *Bioinformatics, 21*(8), 1719–1720.

Przybylski, D., & Rost, B. (2002). Alignments grow, secondary structure prediction improves. *Proteins: Structure, Function, and Bioinformatics, 46*(2), 197–205.

Przytycka, T., Aurora, R., & Rose, G. D. (1999). A protein taxonomy based on secondary structure. *Nature Structural & Molecular Biology, 6*(7), 672–682.

Rost, B., Sander, C., & Schneider, R. (1994). PHD-an automatic mail server for protein secondary structure prediction. *Computer Applications in the Biosciences: CABIOS, 10*(1), 53–60.

Runthala, A., & Chowdhury, S. (2013). Protein structure prediction: Are we there yet?. In D. P. Tuan, & L. C. Jain (Eds.), *Knowledge-based systems in biomedicine and computational life science* (Vol. 450, pp. 9–115). Berlin, Heidelberg: Springer.

Šali, A., & Blundell, T. L. (1993). Comparative Protein modelling by satisfaction of spatial restraints. *Journal of Molecular Biology, 234*(3), 779–815.

Sali, A., Matsumoto, R., McNeil, H. P., Karplus, M., & Stevens, R. L. (1993). Three-dimensional models of four mouse mast cell chymases. Identification of proteoglycan binding regions and protease-specific antigenic epitopes. *Journal of Biological Chemistry, 268*(12), 9023–9034.

Sen, T. Z., Jernigan, R. L., Garnier, J., & Kloczkowski, A. (2005). GOR V server for protein secondary structure prediction. *Bioinformatics, 21*(11), 2787–2788.

Shi, J., Blundell, T. L., & Mizuguchi, K. (2001). FUGUE: Sequence-structure homology recognition using environment-specific substitution tables and structure-dependent gap penalties. *Journal of Molecular Biology, 310*(1), 243–257.

Siew, N., Elofsson, A., Rychlewski, L., & Fischer, D. (2000). MaxSub: An automated measure for the assessment of protein structure prediction quality. *Bioinformatics, 16*(9), 776–785.

Skolnick, J., Fetrow, J. S., & Kolinski, A. (2000). Structural genomics and its importance for gene function analysis. *Nature Biotechnology, 18*(3), 283–287.

Skolnick, J., Kihara, D., & Zhang, Y. (2004). Development and large scale benchmark testing of the PROSPECTOR_3 threading algorithm. *Proteins: Structure, Function, and Bioinformatics, 56*(3), 502–518.

Skolnick, J., & Kolinski, A. (2002). A unified approach to the prediction of protein structure and function. In R. Friesner (Ed.), *A Computational Methods for Protein Folding* (Vol. 120, pp. 131-192). USA: Wiley.

Söding, J., Biegert, A., & Lupas, A. N. (2005). The HHpred interactive server for protein homology detection and structure prediction. *Nucleic Acids Research, 33*(Web Server issue), W244–W248.

Su, E., Chiu, H.- S., Lo, A., Hwang, J.- K., Sung, T.- Y., & Hsu, W.- L. (2007). Protein subcellular localization prediction based on compartment-specific features and structure conservation. *BMC Bioinformatics, 8*(1), 330.

Takeda-Shitaka, M., Takaya, D., Chiba, C., Tanaka, H., & Umeyama, H. (2004). Protein structure prediction in structure based drug design. *Current Medicinal Chemistry, 11*(5), 551–558.

Tamura, K., Stecher, G., Peterson, D., Filipski, A., & Kumar, S. (2013). MEGA6: Molecular evolutionary genetics analysis version 6.0. *Molecular Biology and Evolution, 30*(12), 2725–2729.

Teodorescu, O., Galor, T., Pillardy, J., & Elber, R. (2004). Enriching the sequence substitution matrix by structural information. *Proteins: Structure, Function, and Bioinformatics, 54*(1), 41–48.

Vakser, I. A. (1997). Evaluation of GRAMM low-resolution docking methodology on the hemagglutinin-antibody complex. *Proteins: Structure, Function, and Bioinformatics, 29*(1), 226–230.

Wallner, B., Larsson, P., & Elofsson, A. (2007). Pcons.net: Protein structure prediction meta server. *Nucleic Acids Research, 35*(Web Server issue), W369–W374.

Wiederstein, M., & Sippl, M. J. (2007). ProSA-web: Interactive web service for the recognition of errors in three-dimensional structures of proteins. *Nucleic Acids Research, 35*(Web Server issue), W407–W410.

Wolf, E., Vassilev, A., Makino, Y., Sali, A., Nakatani, Y., & Burley, S. K. (1998). Crystal structure of a GCN5-related N-acetyltransferase: Serratia marcescens aminoglycoside 3-N-acetyltransferase. *Cell, 94*(4), 439–449.

Wu, S., & Zhang, Y. (2007). LOMETS: A local meta-threading-server for protein structure prediction. *Nucleic Acids Research, 35*(10), 3375–3382.

Wu, S., & Zhang, Y. (2008). MUSTER: Improving protein sequence profile–profile alignments by using multiple sources of structure information. *Proteins: Structure, Function, and Bioinformatics, 72*(2), 547–556.

Xu, Y., & Xu, D. (2000). Protein threading using PROSPECT: Design and evaluation. *Proteins: Structure, Function, and Bioinformatics, 40*(3), 343–354.

Xu, J., & Zhang, Y. (2010). How significant is a protein structure similarity with TM-score = 0.5? *Bioinformatics, 26*(7), 889–895.

Xu, D., & Zhang, Y. (2012). Ab initio protein structure assembly using continuous structure fragments and optimized knowledge-based force field. *Proteins: Structure, Function, and Bioinformatics, 80*(7), 1715–1735.

Yang, Y., Faraggi, E., Zhao, H., & Zhou, Y. (2011). Improving protein fold recognition and template-based modeling by employing probabilistic-based matching between predicted one-dimensional structural properties of query and corresponding native properties of templates. *Bioinformatics, 27*(15), 2076–2082.

Yi, T.-M., & Lander, E. S. (1993). Protein secondary structure prediction using nearest-neighbor methods. *Journal of Molecular Biology, 232*(4), 1117–1129.

Zemla, A., Venclovas, Č., Fidelis, K., & Rost, B. (1999). A modified definition of Sov, a segment-based measure for protein secondary structure prediction assessment. *Proteins: Structure, Function, and Bioinformatics, 34*(2), 220–223.

Zhang, Y., Arakaki, A. K., & Skolnick, J. (2005). TASSER: An automated method for the prediction of protein tertiary structures in CASP6. *Proteins: Structure, Function, and Bioinformatics, 61*(7), 91–98.

Zhang, Y., Kolinski, A., & Skolnick, J. (2003). TOUCHSTONE II: A new approach to Ab initio protein structure prediction. *Biophysical Journal, 85*(2), 1145–1164.

Zhang, Y., & Skolnick, J. (2004). Tertiary structure predictions on a comprehensive benchmark of medium to large size proteins. *Biophysical Journal, 87*(4), 2647–2655.

Zhou, H., & Zhou, Y. (2005). Fold recognition by combining sequence profiles derived from evolution and from depth-dependent structural alignment of fragments. *Proteins: Structure, Function, and Bioinformatics, 58*(2), 321–328.

Zuccotto, F. Z. M., Brun, R., Chowdhury, S. F., Di, L. R., Leal, I., Maes, L., et al. (2001). Novel inhibitors of Trypanosoma cruzi dihydrofolate reductase. *European Journal of Medicinal Chemistry, 36*(5), 395–405.

Approximation of Optimized Fuzzy Logic Controller for Shunt Active Power Filter

Asheesh K. Singh, Rambir Singh and Rakesh K. Arya

Abstract The exponential growth of power electronic controlled equipment and non-linear loads have given rise to a type of voltage and current waveform distortion, termed as 'harmonics', adversely affecting the power quality (PQ). Moreover, the sensitivity of these equipments to PQ disturbances has motivated the researchers to develop dynamic and adjustable solutions for harmonic mitigation. Active power filters (APFs) address almost each attribute of PQ depending upon the topology used. The current controlled voltage source inverter (VSI) based shunt active power filter (SAPF) emerges out to be an undisputed alternative for current harmonic mitigation. SAPF having pulse width modulation (PWM) controlled voltage source inverter (VSI) topology is extensively used in distribution power systems, which conventionally utilizes the PI controller for reference voltage tracking. In recent times, Fuzzy logic controllers (FLCs) have been established as viable alternatives of conventional PI controllers in highly non-linear control applications, with varying operating conditions. The improved performance of conventionally used large rule FLC is achieved at the cost of increased complexity, leading to large computational time, and memory requirement. Conventionally triangular membership functions (MFs) are used to represent input and output variables of an FLC. In this chapter other less explored MFs such as generalized bell (Gbell), Gaussian and difference sigmoid (Dsig) are also investigated to find optimal membership function. Gaussian MFs based FLC evolves as the optimized FLC in terms of providing effective harmonic compensation along with efficient dynamic response under randomly varying loading conditions. The chapter focuses

A.K. Singh
Electrical Engineering Department, Motilal Nehru National Institute of Technology Allahabad, Allahabad, India

R. Singh (✉)
Electrical and Electronics Engineering Department, Inderprastha Engineering College, Ghaziabad, India
e-mail: rambir29@gmail.com

R.K. Arya
Remote Sensing Application Centre, M. P. Council of Science and Technology, Bhopal, India

Q. Zhu and A.T. Azar (eds.), *Complex System Modelling and Control Through Intelligent Soft Computations*, Studies in Fuzziness and Soft Computing 319,
DOI 10.1007/978-3-319-12883-2_20

on three main areas, i.e., PQ problem of current harmonics and its mitigation, selection of optimized FLC for shunt APF and complexity reduction of optimized FLC using an approximation technique. The proposed approximation is based on minimizing the sum of square errors, between the outputs of large rule FLC and simplest 4-rule FLC. This approximation of large rules optimized FLC results in reduced computational and functional complexity and less memory requirement without compromising the control performances of FLC in terms of dynamic response and harmonic compensation capabilities. Proposed approximation technique considerably improves the harmonic compensation performance of shunt APF, due to effective approximation and smoother transition of output in the entire UOD.

1 Introduction

Electrical power is one of the most dominant factors in our society. Reliability and quality are the two most important facets of any power delivery system. Power generation, transmission, distribution and usage are undergoing significant changes that affect the electrical quality. The recent deregulation of electric power industry has also contributed to the need for high quality of power. The load side is also experiencing some significant changes such as increased use of power electronics based converters and devices. These devices are main contributors and simultaneously most sensitive to PQ problems.

Technological advancement has lead to the spread of electronic equipments in residential, commercial and industrial sectors owing to their numerous advantages. Due to the inherent non-linear characteristics of these devices, their increased application cause serious side effects on the distribution system, resulting in various PQ problems (Akagi et al. 1984; Arrillaga et al. 1985; Subjak and McQuilkin 1990). Conventionally, the passive filters were used to provide solution to these problems, but their various demerits such as bulky size, detuning with ageing effect, resonance issues, etc., have encouraged the research community to explore more effective alternatives. Consequently, shunt active power filter (APF) has emerged as a potential alternative to conventional passive filters for providing effective current harmonic mitigation and reactive power compensation (Akagi et al. 1984; Singh et al. 1999, 2007; Jain et al. 2002).

The conventionally used control techniques for shunt APF are not always sufficient to deal with the more demanding requirements of the system to deliver with high precision and improved efficiency. The advancement of artificial intelligence (AI) based techniques has open up a new horizon for the control engineers to investigate the control methodologies derived from human behaviour and experience based fuzzy computation, mathematical models based on human nervous system analogy and nature inspired optimization algorithms (Dixon et al. 1999; Jain et al. 2005; Singh et al. 2011a).

In Fuzzy logic controllers (FLCs), the rule base size plays a vital role to decide the control action. With the increase in number of rules, the partitions of universe of discourse (UOD) become finer, resulting in a better control action. As the number of rules increases, the amount of information to be stored in knowledge-base also increases. On the other hand, a smaller rule base is not able to map the non-linear control action with same accuracy as the large rule FLC. Although, a large rule FLC provides precise control action, but at the cost of increased complexity, large memory requirement to store the knowledge base information and more execution time to process a specific control action (Singh et al. 2011b). In this chapter, an effort is made to overcome these limitations by proposing an approximated fuzzy logic controller consisting of a simplest 4-rule FLC with a compensating polynomial.

In the recent past some research on reduction of rule base size has been reported. Some issues on design and rule base size reduction for the fuzzy control of robot manipulators are addressed (Bezine et al. 2002), whereas resizing of rule base by removing inconsistent and redundant rules for the application of vacuum cleaner is discussed and implemented (Ciliz 2005). However, these two studies were application specific. Hampel and Chaker (1998) provided some conclusions for minimization of number of variable parameters for optimization of fuzzy controller. Moser and Navara (2002) proposed a fuzzy controller with conditional firing rules, where only the rule firing conditions, not the number of rules, are reduced. Zeng and Singh (1994, 1995) presented a mathematical description of approximation theory of fuzzy systems for single input single output (SISO), and multi input multi output (MIMO) systems. These works are about the approximation capabilities of the fuzzy systems for approximating a mathematical polynomial rather than on the rule reduction. As the reported works are either application specific or not focused on rule base size reduction, provide the motivation to explore the possibility of a process independent, less complex, approximation scheme capable of providing an equally comparable control action as provided by a large rule FLC (Singh et al. 2011b, 2013; Singh and Singh 2012).

An approximation methodology is proposed in this chapter and its effectiveness is validated with shunt APF providing harmonic and reactive power compensation in an electrical distribution system supplying highly non-linear and randomly varying loads. The proposed approximation technique is process independent, and derived by minimizing the deviation between the responses of a large rule and a reduced rule (4-rule) FLC in the UOD of $[-1, 1]$.

The objectives of this chapter are manifold, some of them are listed below:

- To understand the causes and mitigation of one of the most encountered PQ problem, i.e., harmonics.
- To discuss the working principle and control scheme of shunt APF, a mitigation device used for current harmonics compensation.
- To provide a prime introduction of fuzzy logic controller (FLC).

- To explore the suitability of optimized membership function for a conventional large rule FLC based on various performances indices in terms of dynamic response and harmonic compensation capability of shunt APF.
- Complexity reduction of the optimized FLC by an approximated FLC, without compromising the control performance.
- Performance analysis of approximated optimized FLC under randomly varying load conditions.

The rest of this chapter is organized as following: Basic compensation principle and control scheme of shunt APF is discussed in Sect. 2. Section 3 deals with the introduction to FLC, need of approximation of large rule FLC and technique used for approximation. Simulation results of 49-rule FLC with different membership functions (MFs) are presented in Sect. 4, where, optimized MF is selected, based on the performance comparison. Also, numerical results, their analysis, and discussions obtained using FLC with proposed approximation technique; and optimized MF, are presented. Finally, the major contributions and conclusions of the work are summarized in Sect. 5.

2 Compensation Principle and Control Scheme of Shunt Active Power Filter (APF)

The basic topology of shunt APF, providing harmonic and reactive power compensation is shown in Fig. 1. The compensation is based on the principle of injecting equal and opposite distorted current in the supply line (Akagi et al. 1984).

The shunt APF acts as a current source producing harmonic compensating current component. The harmonic current components in the source currents are cancelled, making source currents sinusoidal and in phase with the supply voltage, thereby providing harmonic and reactive power compensation.

Bose (1994) and Ibrahim and Morcos (2002) have explored the possibilities of applications of artificial intelligence (AI), expert system, fuzzy logic and neural network in power electronics, motion control and PQ related areas. This work provides a new space of opportunities for control engineers.

The inherent nature of FLC is explored in this chapter to develop a flexible control strategy. The schematic diagram of control scheme of shunt APF using FLC is shown in Fig. 2. The entire control task comprises of two loops, i.e., outer voltage control loop and inner current control loop, as shown in Figs. 3 and 4, respectively. In voltage control loop, the DC link voltage of shunt APF is compared with a set reference value. The error (e) and change in error (ce) between actual and reference values of DC link capacitor voltages are used as the input variables to the FLC.

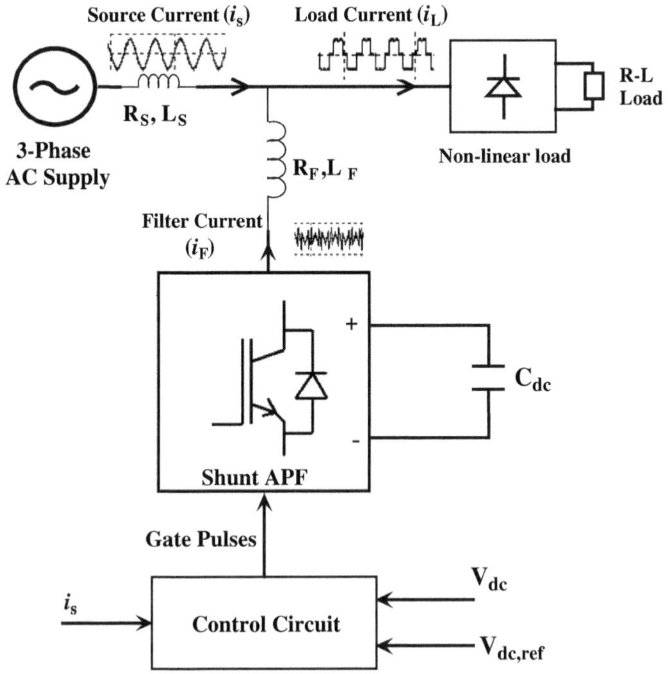

Fig. 1 Single line diagram of shunt active power filter

The change in error (ce) at any sampling instant ‘k’ can be calculated as:

$$\mathrm{ce}(k) = \mathrm{e}(k) - \mathrm{e}(k-1) \tag{1}$$

In Eq. (1), e(k) and e(k − 1) are error values at kth and (k − 1)th sampling instants, respectively. The output of controller is the incremental change in peak value of reference source current (δI_{max}). This incremental change (δI_{max}) is added with the peak value of current at previous sampling instant to obtain the peak reference source current (I_{max}), as given in (2).

$$I_{\max}(k) = \delta I_{\max}(k) + I_{\max}(k-1) \tag{2}$$

In voltage control loop, the peak value of reference currents is estimated by regulating the dc-link capacitor voltage. The actual capacitor voltage is compared with a set reference value, and error signal is then processed in a FLC, to estimate the peak value of reference source current (I_{max}).

In current control loop, the peak value of the current (I_{max}) so obtained, is multiplied by the unit sinusoidal vectors, in phase with the respective source voltages, to obtain the instantaneous reference compensating currents. Using indirect current control technique, these estimated reference currents (i^*_{sa}, i^*_{sb}, i^*_{sc})

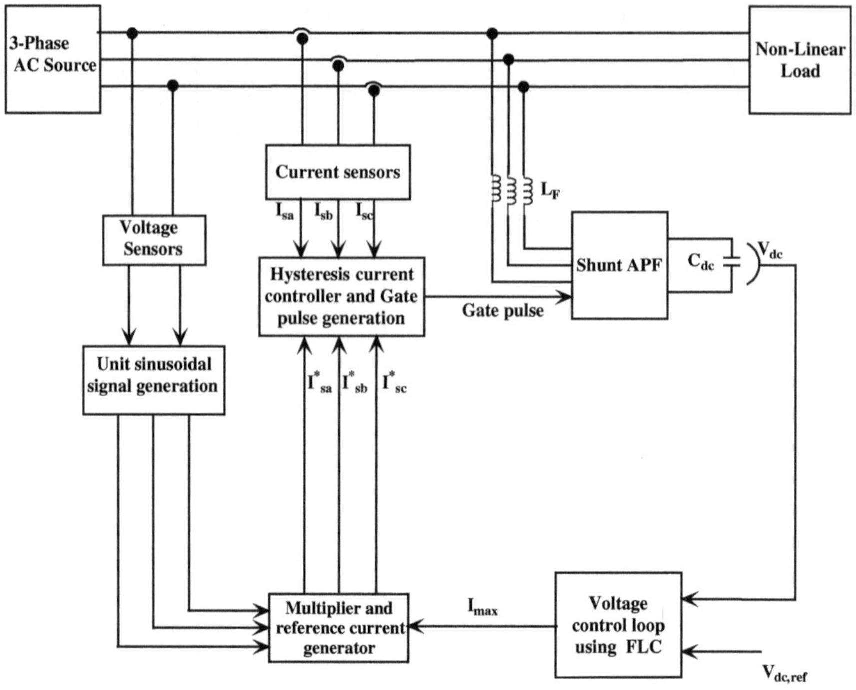

Fig. 2 Control circuit of fuzzy logic controlled shunt APF

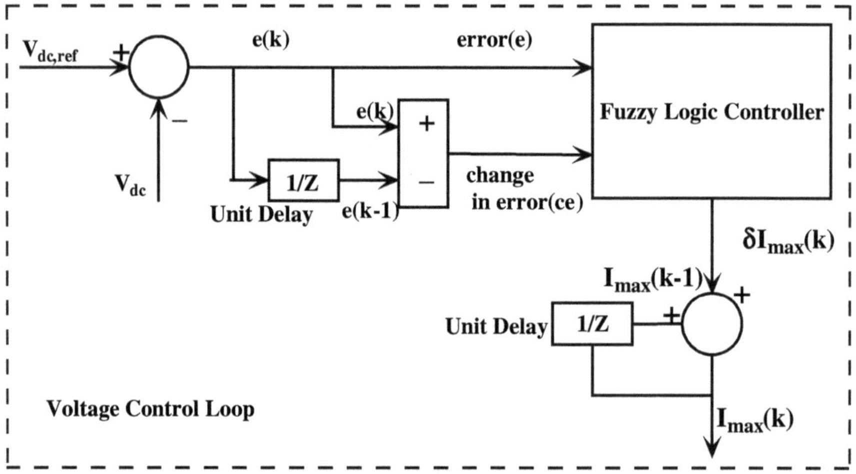

Fig. 3 Block diagram of voltage control loop, of FLC based shunt APF

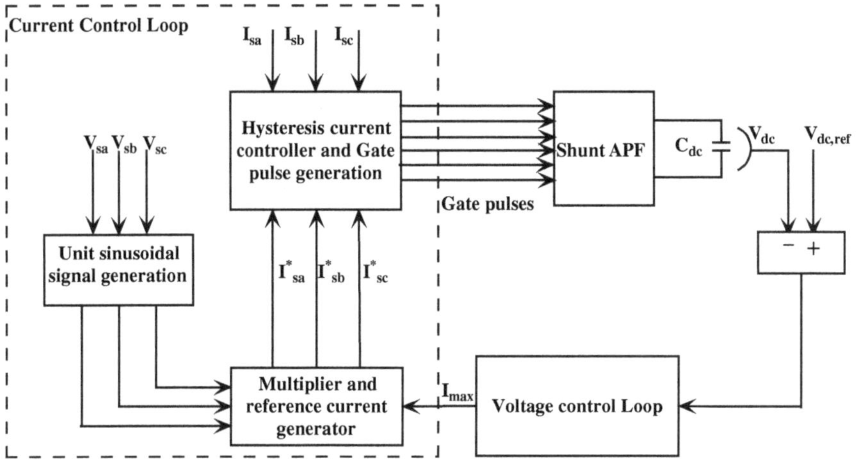

Fig. 4 Block diagram of current control loop of FLC based shunt APF

and sensed actual currents (i_{sa}, i_{sb}, i_{sc}) are compared in hysteresis current controller to generate the switching signals for converter devices.

The switching signals of any phase (say phase 'a') are obtained using following methodology:

If $i_{sa} > \left(i_{sa}^{*} + hb\right)$, the upper switch of first arm of converter is on and lower switch is off and if $i_{sa} < \left(i_{sa}^{*} + hb\right)$, the upper switch of first arm of converter is off and lower switch is on, where 'hb' is the hysteresis band on either side around the reference current. Similarly, switching patterns for other two phases are also obtained. In this way, source currents and DC link voltage are regulated to follow their corresponding reference signals. The source currents become sinusoidal and in phase with respective phase voltages to achieve the objective of efficient harmonic and reactive power compensation.

Using hysteresis current control, the source currents are forced to follow the sinusoidal reference current of corresponding phase, within a fixed hysteresis band. The width of hysteresis window determines the source current pattern, its harmonic spectrum and switching frequency of the devices. The reference source currents for three different phase a, b, and c can be given as:

$$\begin{aligned} i_{sa}^{*} &= i_{s\max} \sin \omega t \\ i_{sb}^{*} &= i_{s\max} \sin(\omega t - \frac{2\pi}{3}) \\ i_{sc}^{*} &= i_{s\max} \sin(\omega t + \frac{2\pi}{3}) \end{aligned} \quad (3)$$

As this chapter basically deals with control technique using FLC, next section provides an insight into basic structure of FLC, design of rule-base, different membership functions, etc. This section also discuss the proposed approximation technique to approximate the control actions of a 49-rule FLC by a simplest 4-rule FLC. The proposed approximation results in a less complex fuzzy structure, with tremendously reduced memory requirement and computational time.

3 Fuzzy Logic Controller

Fuzzy logic is a branch of artificial intelligence that deals with the reasoning algorithms used to emulate human thinking and decision making in machines. The fuzzy logic controller (FLC) based on fuzzy logic provides a method of converting a linguistic control strategy using expert knowledge into an automatic control strategy. The primary thrust of this novel control paradigm is to utilize the human operator's knowledge and experience to develop controllers intuitively. In this chapter, an FLC is designed to improve the dynamic response of shunt APF. The conventionally designed FLCs use large number of rules to provide precise control, but with an increased complexity, memory requirement and computation time.

3.1 *Basic Structure of Fuzzy Logic Controller*

The basic structure of fuzzy logic controller (FLC) is shown in Fig. 5. In general, a fuzzy logic system maps crisp input into crisp output and in such case contains three major modules:

1. Fuzzification,
2. Knowledge Base and Inference engine, and
3. Defuzzification

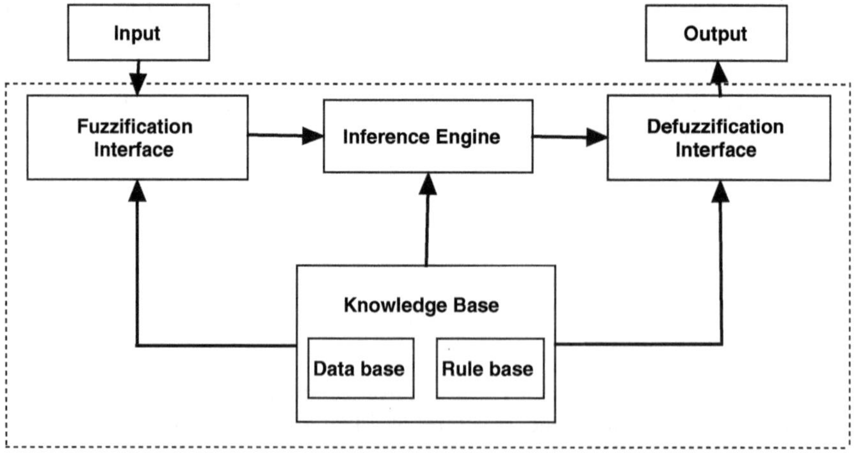

Fig. 5 Basic configuration of fuzzy logic controller

Fuzzification converts input data into suitable linguistic values which may be viewed as labels of fuzzy sets. Data base and rule base are the two components of knowledge base of an FLC. Database provides necessary definitions used for linguistic control rules and fuzzy data. If-then rule statements are used to formulate the conditional statements that comprise fuzzy logic. A rule base is the collection of all these statements. The rules are a set of linguistic statements based on expert knowledge, including experience and heuristics, instead of detailed mathematical model. In the fuzzy inference engine, fuzzy logic principles are used to combine fuzzy rules into a mapping from fuzzy input sets to fuzzy output sets. The process of fuzzy inference involves: membership functions, fuzzy logic operators, and if-then rules. All the membership degrees associated to a fuzzy set generate a shape called membership functions (MFs). An MF is a curve that defines how each point in the input space is mapped to a membership value (or degree of membership) between 0 and 1. The distribution of membership does not always vary linearly with universe of discourse. Hence, we have many types of membership distribution generally referred as membership functions. Various types of MFs are triangular, trapezoidal, sigmoid, exponential, and bell shaped MF etc. Mathematically, fuzzy rule-based inference can be viewed as an interpolation scheme because it enables the fusion of multiple fuzzy rules when their conditions are all satisfied to a degree. Defuzzification produces a crisp output from the fuzzy set that is the output of the inference engine.

The rule base plays a key role in representing expert control, modeling knowledge and experience, and linking the input variables of the controller to the output variable (Ying 2000). Rule base design methodology based on dynamic behaviour and process state is discussed in Lee (1990), Lee and Gatland (1995). This approach helps in understanding the nature of required control action (i.e., its sign). However, its level or degree depends on the instantaneous values of error and change in error. Based on the combinational effect of error and change in error, the rule base as given in Table 1, is used for 49-rule FLC. Large number of rules provides fine control action.

The performance of FLC controlled shunt APF is analyzed and compared using four different membership functions (MFs), viz., triangular, Generalized bell shaped (Gbell), Gaussian, and difference sigmoid (Dsig) MFs, as depicted in Fig. 6.

Table 1 Rule base for 49-rule FLC

ce	e						
	NB	NM	NS	ZE	PS	PM	PB
NB	NB	NB	NB	NB	NM	NS	ZE
NM	NB	NB	NB	NM	NS	ZE	PS
NS	NB	NB	NM	NS	ZE	PS	PM
ZE	NB	NM	NS	ZE	PS	PM	PB
PS	NM	NS	ZE	PS	PM	PB	PB
PM	NS	ZE	PS	PM	PB	PB	PB
PB	ZE	PS	PM	PB	PB	PB	PB

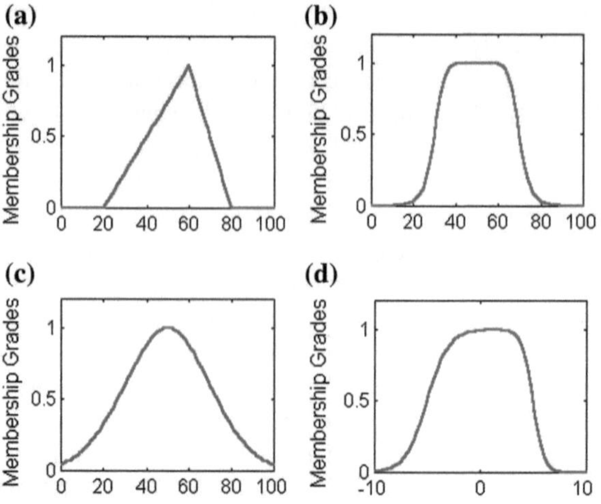

Fig. 6 Different membership functions used for performance comparison and selection of optimized membership function

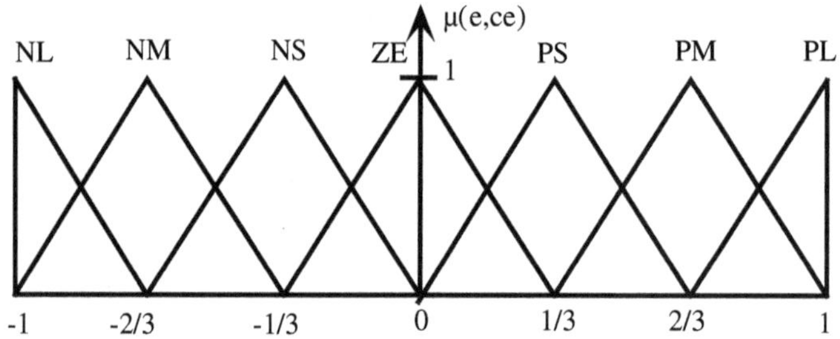

Fig. 7 Triangular membership functions for input variables error (*e*) and change in error (*ce*) of 49-rule FLC

In 49-rule FLC, seven MFs are used for both input and output variables to cover entire UOD as shown in Figs. 7 and 8 (for triangular MFs), similar uniform distribution of UOD is used for other MFs.

The linguistic variables used to represent seven membership functions are negative Big (NB), negative medium (NM), negative small (NS), zero (ZE), positive small (PS), positive medium (PM) and positive big (PB). The range of UOD is taken [−1, 1], for input as well as output variables.

Based on the performance of shunt APF in terms of dynamic and harmonic compensation indices the optimized large rule FLC is selected for approximation. The approximation technique is discussed in next section.

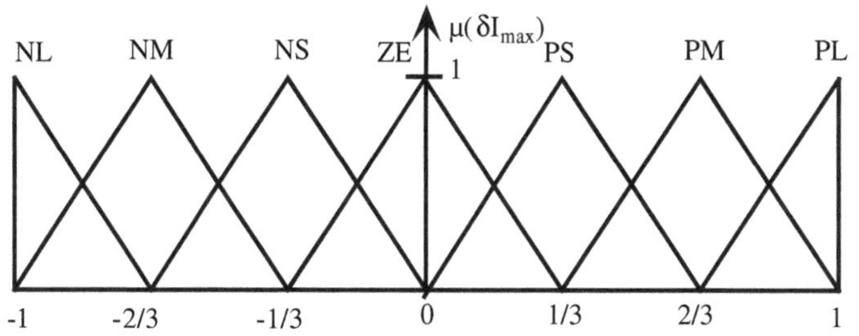

Fig. 8 Triangular membership functions for output variable δI_{max} of 49-rule FLC

3.2 Approximation Technique

A 4-rule simplest FLC is proposed to approximate the control functionality of its 49-rule counterpart. The outputs of a 49-rule FLC and a 4-rule FLC are compared at regular intervals in the entire range of input variables to find the deviation in the responses of two controllers. The approximation is based on evaluating a compensating polynomial such that its series (cascade) combination with simplest FLC, approximately maps the output of the 49-rule FLC.

The concept of simplest FLC was introduced by Ying (2000). The term simplest refers to a minimal possible configuration in terms of number of input variables, fuzzy sets and fuzzy rules for any properly functional FLC. To realize a simplest FLC, two membership functions for each input variable, i.e., error (e) and change in error (ce) are used in the UOD of [−L, L] as shown in Fig. 9a, b, respectively for triangular membership functions. Three triangular membership functions are used for output variable δI_{max} in the UOD of [−H, H] as shown in Fig. 9c.

Similarly, simplest FLC can be realized for other membership functions also. The UOD for input and output variables is taken [−1, 1]. The centre of gravity defuzzification method is used to obtain the output as crisp value.

Let $u(k)$ and $u_1(k)$ are the outputs of a 49-rule FLC and 4-rule FLC at kth sampling instant, respectively. Then the deviation or error can be given as:

$$e(k) = u(k) - u_1(k) \tag{4}$$

The sum of square error (SSE) can be represented as:

$$SSE = \sum_{k=1}^{N} e^2(k) \tag{5}$$

This SSE will be used as a cost function, to be minimized for achieving the accurate approximation. To understand the least square fitting process, let us consider N data points to measure the error in responses. To implement the approximation scheme, an nth order polynomial is used in cascade with the 4-rule

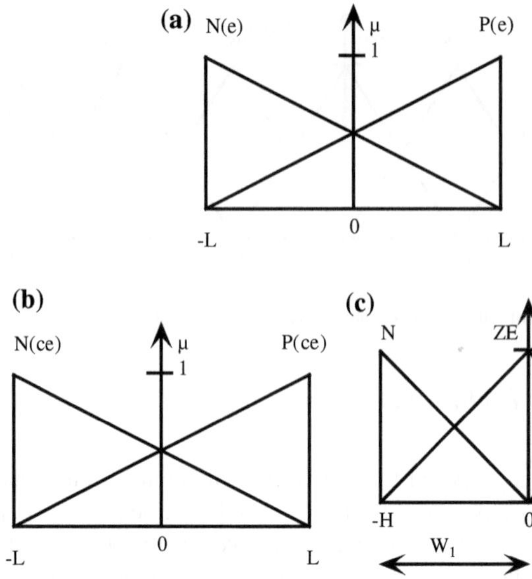

Fig. 9 Triangular membership functions of simplest FLC for **a** error, **b** change in error and **c** output δI_{max}

FLC, in a way that this cascaded combination (i.e., approximated FLC) maps the output of a 49-rule FLC, with least square error. The output of approximated FLC in terms of nth order polynomial of $u_1(k)$ is given as:

$$u_2(k) = a_n u_1^n(k) + a_{n-1} u_1^{n-1}(k) + \cdots + a_1 u_1(k) + a_0 \tag{6}$$

where, $u_1(k)$ and $u_2(k)$ are the outputs of 4-rule FLC and proposed approximated FLC, respectively. a_0, a_1, a_2,…a_n are the coefficients of nth order polynomial. The block diagram of approximation scheme is shown in Fig. 10.

Sum of square errors (SSE), at N data points is represented as:

$$SSE = \sum_{t=1}^{N} \{u(k) - u_2(k)\}^2 \tag{7}$$

$$SSE = \sum_{t=1}^{N} [u(k) - \{a_n u_1^n(k) + a_{n-1} u_1^{n-1}(k) + \cdots + a_1 u_1(k) + a_0\}]^2 \tag{8}$$

To minimize SSE, its partial derivatives with respect to each unknown coefficients are equated to zero to get as many equations as the number of unknown coefficients. The solution of these equations gives the values of these unknown coefficients. The flow chart, to find the unknown coefficients a_n, a_{n-1}, a_{n-2}, …a_1 and a_0, of compensating polynomial, is shown in Fig. 11.

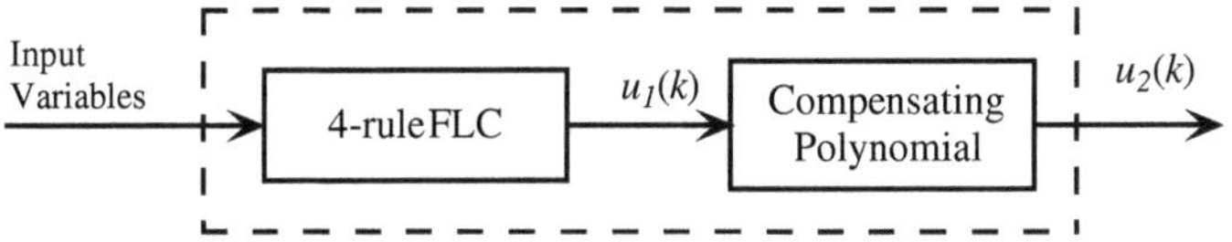

Fig. 10 Block diagram of approximation scheme

Fig. 11 Flow chart showing design steps of approximated FLC

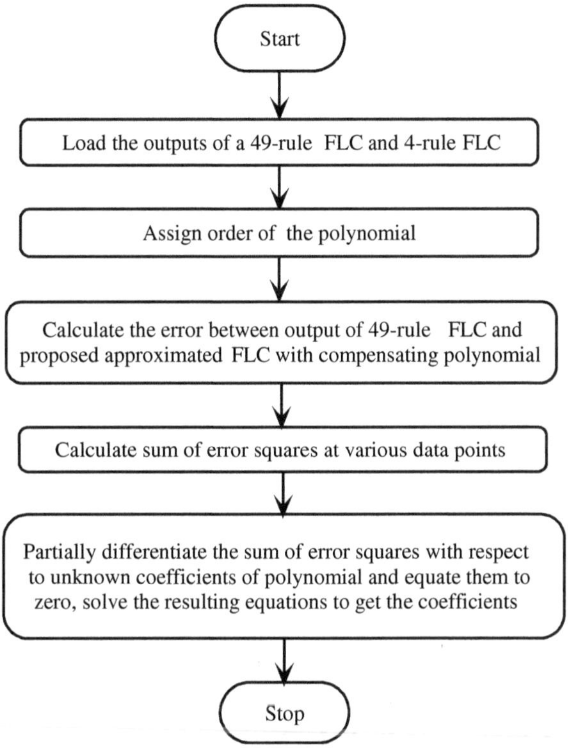

The order of compensating polynomial plays a critical role in approximation. A large order polynomial is avoided due to the following reasons:

(a) A large order polynomial needs more computational time and memory, defeating the basic objective of designing a reduced rule approximated FLC.
(b) Higher order polynomial can be highly oscillatory and an order larger than the exact fit case may lead to multiple solutions, resulting in a confusing state for designer to select one solution.

On the contrary, a lower order polynomial may not provide the sufficient approximation. This situation leads towards maintaining a tradeoff between the order of polynomial and degree of fitness for adequate approximation. A 7th order polynomial is derived in using the proposed approximation technique.

Table 2 System parameters

System parameter	Value
Source voltage (V_s)	230 V (rms/phase)
System frequency (f)	50 Hz
Source impedance (R_s, L_s)	0.1 Ω, 0.5 mH
Filter impedance (R_F, L_F)	0.4 Ω, 3.35 mH
Reference DC link voltage ($V_{dc, ref}$)	680 V
DC link capacitance (C_{dc})	2,000 μF

Performance analysis of shunt APF using 49-rule FLCs with different MFs and approximated FLC using optimized MFs is presented in next section.

4 Simulation Results

Using the system parameters given in Table 2, simulation results of shunt APF with 49-rule FLC using different MFs are compared to find the optimized membership function. Simulations are performed on MATLAB/Simulink, using same normalization and de-normalization factors for all the MFs.

The performance of shunt APF is analyzed for the three randomly varying loading conditions. Initially, the filter is switched on at 0.05 s, to compensate the current harmonics injected by a non-linear load. The load is varied in three steps, each of 10 cycles (i.e., 0.2 s), discussed as three different cases, i.e., *Case*-1, *Case*-2, and *Case*-3, in following subsections. The waveforms during switch-on as well as during load perturbation are shown in Figs. 12, 13, 14 and 15, for triangular, Gbell, Gaussian, and Dsig MFs, respectively.

4.1 Case-1: Switch-On Response

During switch-on at 0.05 s, the rectifier fed load consists of a resistance and inductance of 30 Ω and 20 mH, respectively. The comparison of dynamic performance reveals that the peak overshoot in DC link voltage is 1.32 % with triangular MFs as compared to 0.51 % with Gaussian and 0.05 % with Dsig MFs, respectively. The response of Gbell MFs is not found to maintain the dc link voltage at reference level as depicted in Fig. 13. The switch on response with Dsig MFs with minimum overshoot suffers from steady state error in regulating the dc link voltage. The settling time of dc link voltage within ±1 % of reference value is least with Gaussian MFs, as represented in Table 3.

The THD profile for each cycle, after switch-on, is presented in Fig. 16, All the controllers are found capable to bring the THD of source current well within the limits (5 %) imposed by IEEE-519 (1993), just in the very next cycle of switching

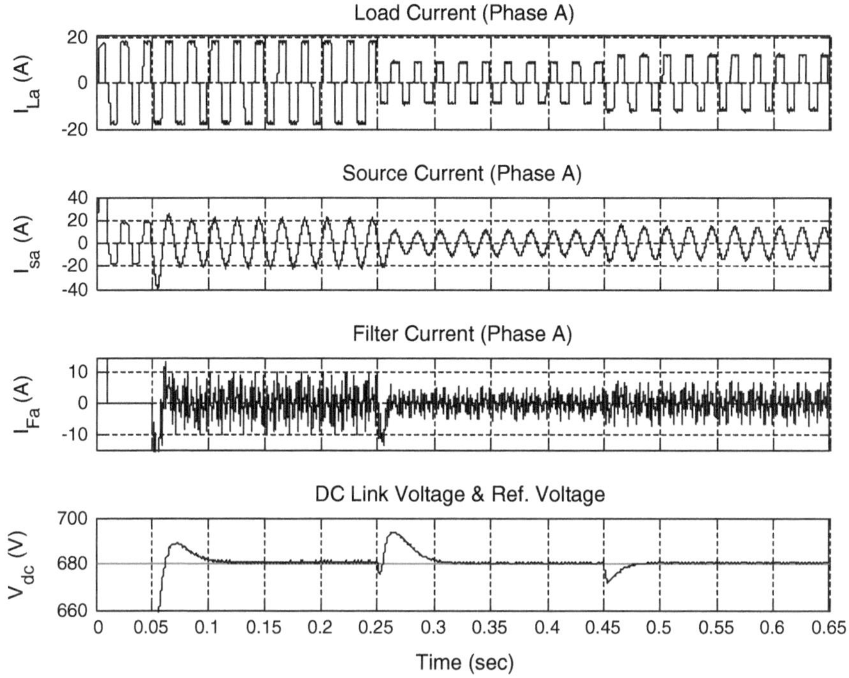

Fig. 12 Dynamic response of shunt APF with 49-rule FLC using triangular MFs

transient However, the THD profile obtained using triangular MFs is slightly inferior to other MFs, which show better immediacy with each other.

In *Case*-1, the transient and steady-state response, as well as harmonic compensation provided by FLC with Gaussian MFs is superior to other MFs.

4.2 Case-2: Load Perturbation Response Followed by Sudden Reduction in Load

In this *Case*-2, at 0.25 s, the load current is suddenly reduced from 13.83 to 6.95 A (rms), as evident from the waveforms shown in Figs. 12, 13, 14 and 15. The comparison of performances of various controllers regulating the DC link capacitor voltage is presented in Table 3.

Here, the peak overshoot in capacitor voltage due to load change is least with triangular MFs, followed by Dsig, Gaussian and Gbell. But due to inferior switch-on response of Gbell and Dsig MFs based FLCs, Gaussian MFs and Triangular MFs based FLC are the true contenders. The settling times of both of them are less than two cycle of fundamental frequency as shown in Table 3. These observations

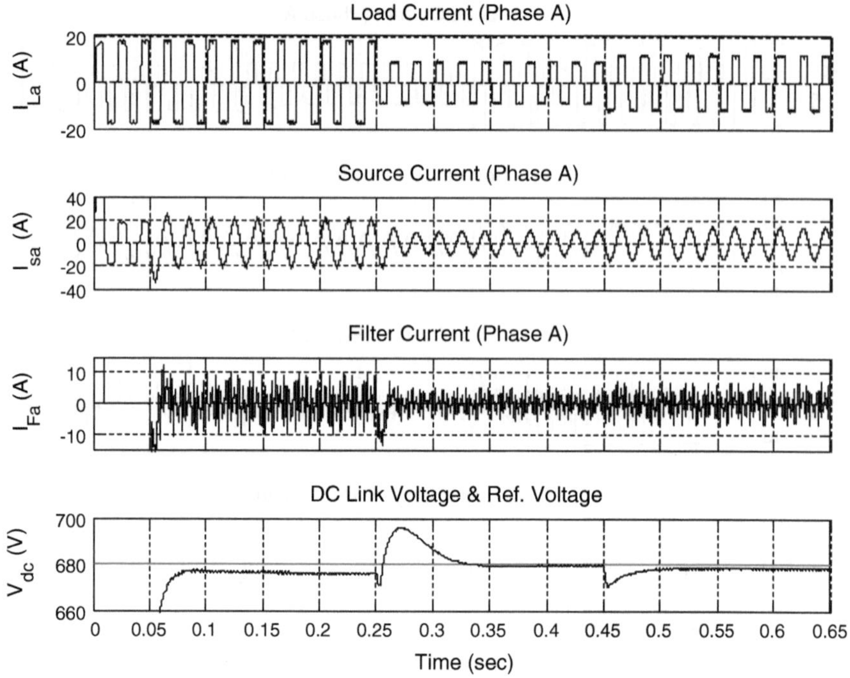

Fig. 13 Dynamic response of shunt APF with 49-rule FLC using Gbell MFs

establish that in this case triangular MFs based FLC has a slightly upper edge over Gaussian MFs based FLC.

However, the Gaussian MFs based FLC exhibits better harmonic compensation performances than triangular MFs based FLC during transient as well as steady-state operation. The harmonic compensation performance of other two MFs, i.e., Gbell and Dsig is also comparable with that of Gaussian MFs but suffers on account of inferior dynamic response in terms of settling time.

In *Case*-2, both, the triangular MFs based FLC performs better in terms of dynamic response, while Gaussian MFs based FLC exhibits better harmonic compensation capabilities.

4.3 Case-3: Load Perturbation Response Followed by Sudden Increase in Load

At t = 0.45 s, a change in loading condition makes the load current to increase from 6.95 to 9.26 A (rms). The effect of this load change on the responses of various controllers regulating the DC link capacitor voltage is shown in Figs. 12, 13, 14 and 15.

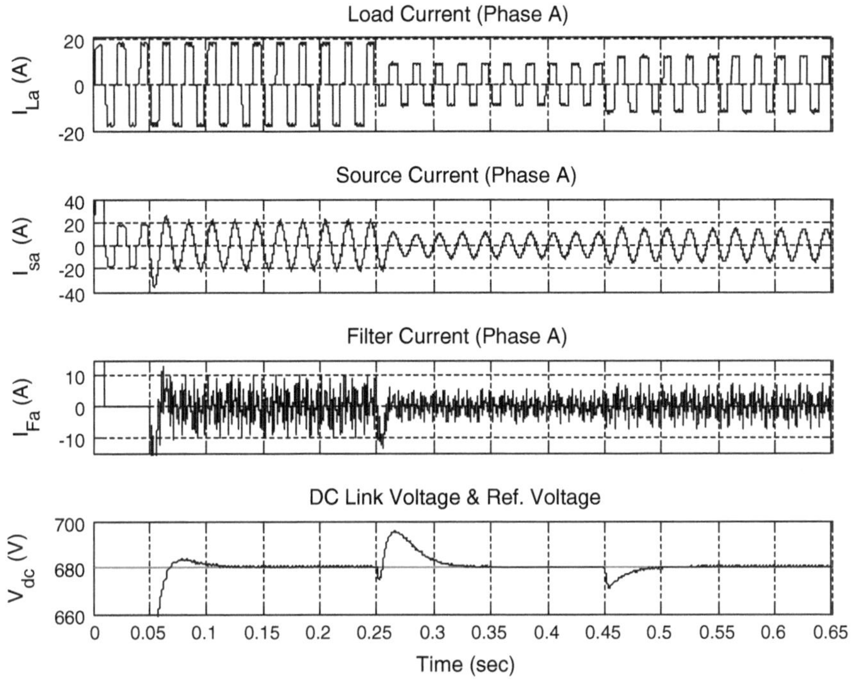

Fig. 14 Dynamic response of shunt APF with 49-rule FLC using Gaussian MFs

Table 3 clearly presents a better value of under-shoot for triangular MFs based FLC, while minimum settling time for Gaussian MFs.

However, the THD profile of source current is superior for Gaussian MFs than triangular, as shown in Fig. 16.

The dynamic response of Gaussian membership function outperforms others due to its constantly varying curved surface incorporating the nonlinear effect as well as smooth transition profile.

Based on these three cases discussed above, the dynamic response of Gaussian MFs based FLC is superior in *Case*-1, while triangular MFs dominates during *Case*-2, and in *Case*-3, the performance with both the MFs is at par. However Gaussian MFs outperforms the triangular MFs in terms of the better harmonic compensation capabilities throughout the three cases. Hence, the Gaussian MFs emerges out as the optimum MFs among the considered ones.

4.4 Comparison in Terms of Performance Indices

Automatic control emphasizes on the mathematical formulation and measurement of system performance for analyzing the controller behaviour using some

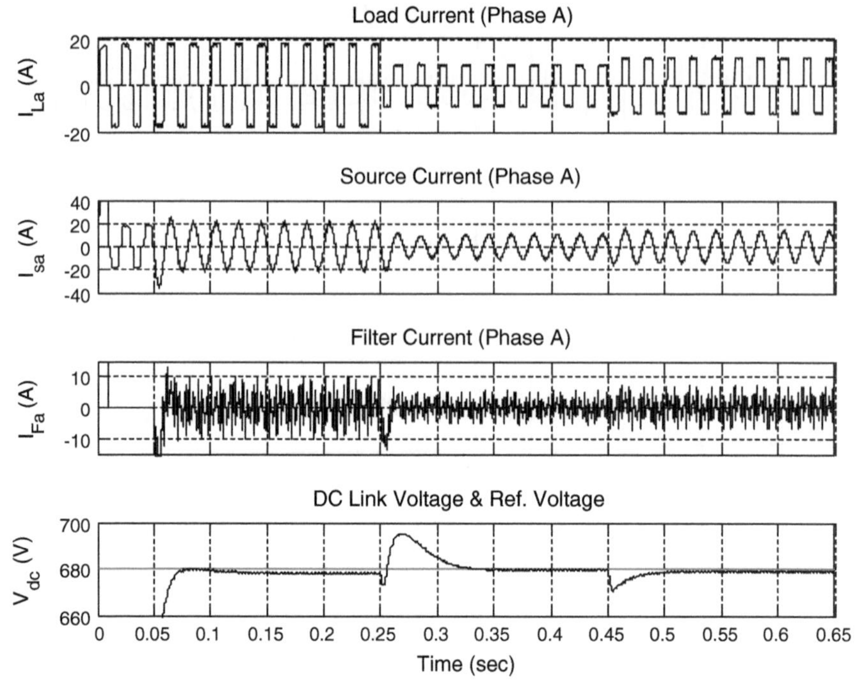

Fig. 15 Dynamic response of shunt APF with 49-rule FLC using Dsig MFs

Table 3 Dynamic response of shunt APF controlled by 49-rule FLC with different membership functions

Load	% peak overshoot/under-shoot				Settling time (cycles)			
	Triangular MFs	Gbell MFs	Gaussian MFs	Dsig MF	Triangular MFs	Gbell MFs	Gaussian MFs	Dsig MF
Case-1	1.32	–	0.51	0.05	1.81	2.0	0.92	1.45
Case-2	2.05	2.35	2.29	2.28	1.80	2.49	1.86	2.19
Case-3	1.17	1.44	1.32	1.40	0.47	0.59	0.42	0.47

quantitative indices. These indices are termed as performance indices and are used for parameter optimization of a control system resulting in the design of optimum control system. In optimum control system, the systems parameters are adjusted such that the error between the reference and actual output is minimized and performance index reaches an extreme, commonly a minimum value. The commonly used performance indices (Nagrath and Gopal 2005) are listed below and their comparison for different controllers is presented in Table 4.

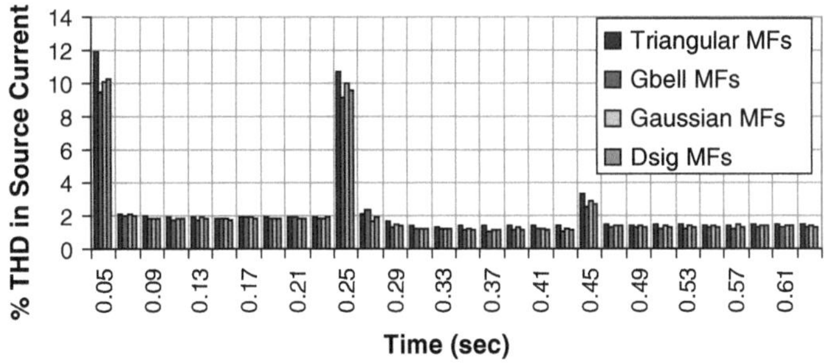

Fig. 16 Comparison of THD profile of source current with 49-rule FLC using various MFs

Table 4 Comparison of controllers on the basis of performance indices

Performance index	MFs			
	Triangular	Gbell	Gaussian	Dsig
ITAE	0.36	0.66	0.35	0.51
IAE	7.62	8.56	7.46	7.91
ISE	1,970.76	1,977.91	1,970.45	1,972.73
ITSE	11.88	13.65	12.05	12.65

(a) Integral square error (ISE),
(b) Integral absolute error (IAE),
(c) Integral time absolute error (ITAE), and
(d) Integral time square error (ITSE).

4.5 Selection of Optimal Membership Function and Performance Analysis of Approximated Optimized FLC

The Gaussian MFs based FLC dominates in three out of four performance indices, and therefore the 49-rule FLC using Gaussian MFs is selected for approximation. The approximated optimized FLC is a cascade combination of a 4-rule FLC using optimized MFs (Gaussian, here) and a compensating polynomial derived based on minimization of SSE. The outputs of a 49-rule FLC and a 4-rule FLC using Gaussian MFs are compared at regular intervals in the entire range of input variables as shown in Fig. 17.

Using the methodology of approximation discussed in Sect. 4, a seventh order polynomial, as given in (8), is derived, whose cascaded combination with a 4-rule FLC approximates the control behaviour of a 49-rule FLC.

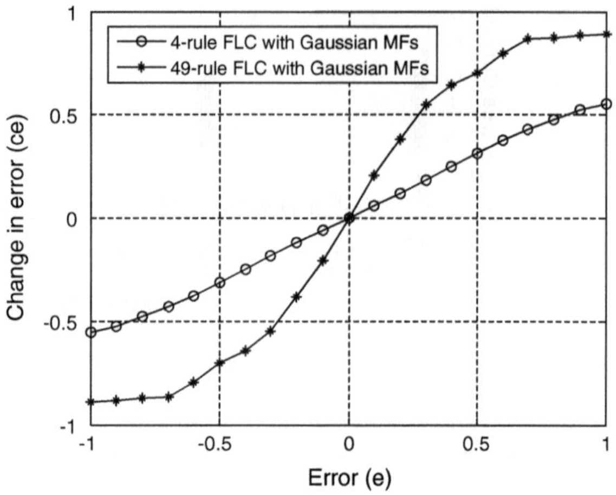

Fig. 17 Comparison of control actions of 4-rule FLC and 49-rule FLC with Gaussian MFs

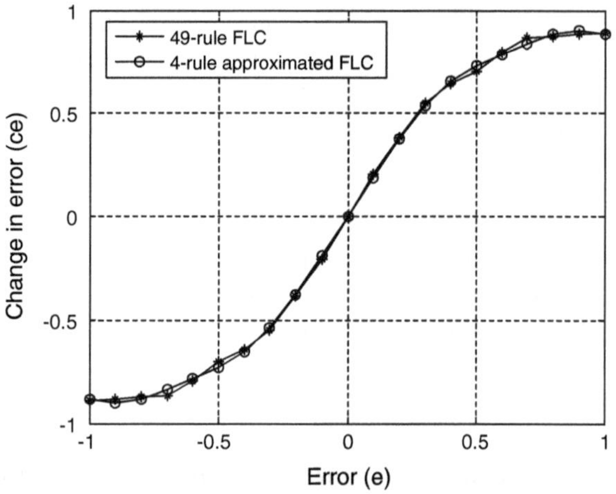

Fig. 18 Comparison of control actions of 4-rule approximated FLC and 49-rule FLC with Gaussian MFs

$$u_2(t) = -61.2748u_1^7(t) + 60.8078u_1^5(t) - 21.4164u_1^3(t) + 4.2446u_1(t) \quad (8)$$

The comparison of control actions of 49-rule FLC and approximated 4-rule FLC using Gaussian MFs is depicted in Fig. 18.

The control performance of approximated 4-rule FLC is almost overlapping with the control performance of 49-rule FLC, thereby minimizing the deviation and leading to effective approximation.

The dynamic response with the proposed approximated 4-rule FLC, for the three different cases considered here, is presented in Fig. 19, and its quantified analysis is shown in Table 5. The results, in terms of peak overshoot /undershoot and settling

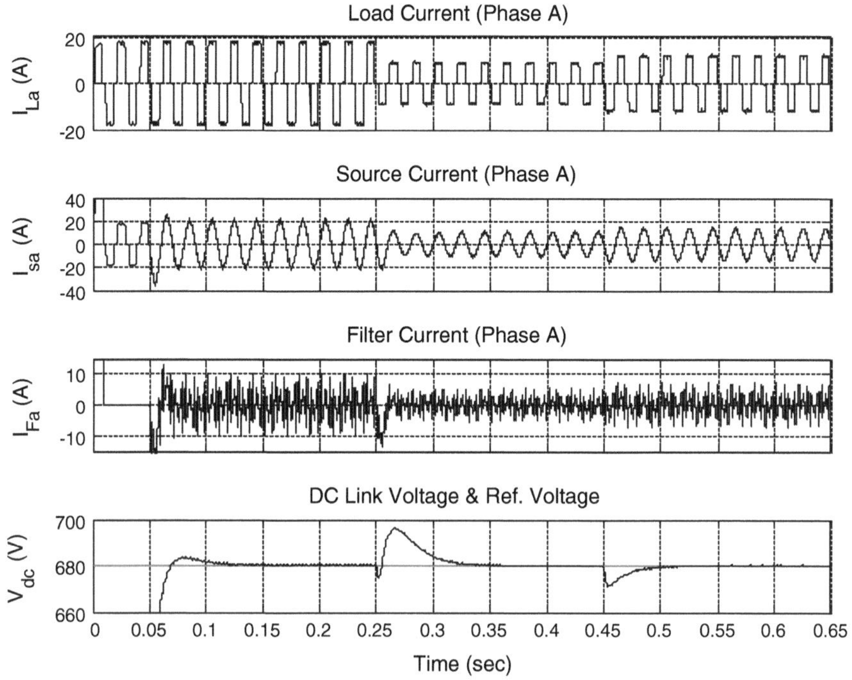

Fig. 19 Dynamic response of shunt APF with 4-rule approximated FLC using Gaussian MFs

Table 5 Dynamic response of shunt APF controlled by 49-rule FLC and approximated FLC using Gaussian MFs

Load	% peak overshoot/under-shoot		Settling time (cycles)	
	49-rule FLC	Approximated 4-rule FLC	49-rule FLC	Approximated 4-rule FLC
Case-1	0.51	0.53	0.92	0.95
Case-2	2.29	2.33	1.86	1.90
Case-3	1.32	1.34	0.42	0.44

time, are comparable with that of a 49-rule FLC and hence confirm the effectiveness of proposed approximation methodology.

The comparison of THD profile of source current with proposed approximated FLC and 49-rule FLC using Gaussian MFs is shown in Fig. 20. The harmonic compensation performance of proposed scheme is at par with 49-rule FLC and even better at the instant of load change.

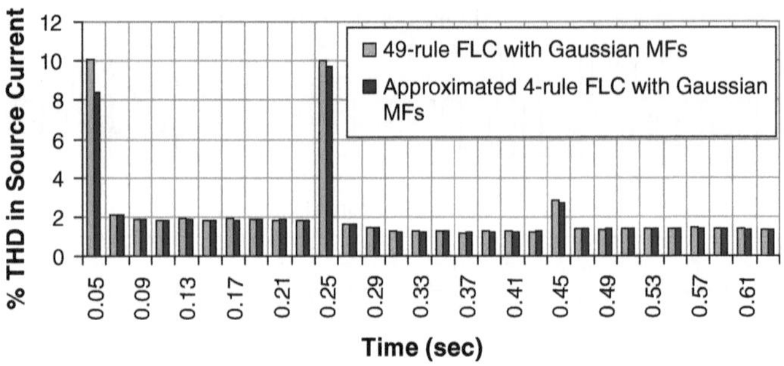

Fig. 20 Comparison of THD profile of source current with 49-rule FLC and approximated 4-rule FLC using Gaussian MFs

Table 6 Comparison of computational memory requirement

Sl. no.	Module	Parameter	49-Rule FLC	4-Rule approximated FLC
1.	Fuzzification	No. of input variables	2	2
		Number of MF for each input	7	2
		Memory units required	14 * m	4 * m
2.	Fuzzy inference and knowledge base	Number of rules	49	4
		Number of antecedent	7	4
		Number of consequent	7	4
		Memory units required	686	32
3.	Defuzzification	Number of output variable	1	1
		MF for output variable δI_{max}	7	3
		Memory units required	7 * m	3 * m
4.	Approximation	Number of comparators	–	–
		Number of multipliers	–	9
		Number of adders	–	3
		Number of terms with non zero coefficients	–	4
		Memory units required	–	16
Total memory requirement			21 * m + 686	7 * m + 48

where 'm' represents unit memory required for each membership function

4.6 Comparative Analysis of Computational Memory

A comparison of memory requirement of 49-rule and proposed approximated FLC is presented in Table 6. For comparative analysis of computational memory requirement, an FLC is functionally divided into following three sections:

(a) Fuzzification,
(b) Fuzzy inference and knowledge base, and
(c) Defuzzification.

Although, a 4-rule FLC require minimum computational memory but suffers with large deviation in control action from 49-rule FLC, as shown in Fig. 17. The proposed approximated FLC overcomes this drawback as shown in Fig. 18. In proposed approximated FLC, some additional memory is required than a 4-rule FLC, to perform approximation using compensating polynomial. Even after this additional requirement, total memory requirement is much lesser than a 49-rule FLC.

5 Conclusions

The aim of this chapter is to investigate and analyze the performance of an FLC with different MFs. Out of the four membership functions used for analysis, the Gaussian MFs based FLC outperforms others in terms of better dynamic response in tracking the reference voltage, and effective harmonic compensation in source current. The harmonic compensation is maintained well within the limits imposed by IEEE-519 standards.

Hence, Gaussian MFs based FLC is recommended as an optimized FLC confirming better performance throughout transient and steady-state conditions and also justifying its robustness in terms of performance indices depicting minimum error in regulating the dc-link voltage.

Then, a stepwise design procedure of approximation techniques focused on rule base size reduction, without compromising the control performance, is discussed. The Main features of design approach used in approximation technique are:

1. Use of simplest possible FLC
2. Reduced rule base size
3. Process independent design
4. Less memory requirement
5. Less computational efforts
6. Reduced computational time, and
7. Faster response

The performance of designed approximation techniques is validated through simulation results. The dynamic performance of performance of approximated FLC is found comparable with the 49-rule FLC and harmonic compensation is even better at the instants of load perturbation due to smooth transition in entire range of UOD.

As research is a never ending process the future work may include

- Study the operation under distorted and/or unbalanced supply voltage conditions.
- Extension of control scheme to multilevel converters.
- Extension of on-line scheme to design adaptive membership functions.

References

Akagi, H., Kanazawa, Y., & Nabae, A. (1984). Instantaneous reactive power compensators comprising switching devices without energy storage components. *IEEE Transactions on Industrial Applications, IA 20*(3), 625–630.

Arrillaga, J., Bradley, D. A., & Bodger, P. S. (1985). *Power system harmonics*. London: Wiley.

Bezine, H., Derbel, N., & Alimi, A. M. (2002). Fuzzy control of robotic manipulator: Some issues on design and rule base size reduction. *Engineering Applications of Artificial Intelligence, 15*, 401–416.

Bose, B. K. (1994). Expert system, fuzzy logic, and neural network applications in power electronics and motion control. *Proceedings of the IEEE, 82*(8), 1303–1323. doi:10.1109/5.301690.

Ciliz, M. K. (2005). Rule base reduction for knowledge based fuzzy controller with application to vacuum cleaner. *Expert System with Applications, 28*, 175–184.

Dixon, J. W., Contardo, J. M., & Moran, L. A. (1999). A fuzzy controlled active front end rectifier with current harmonics filtering characteristics and minimum sensing variables. *IEEE Transactions on Power Electronics, 14*(4), 724–729.

Hampel, R., & Chaker, N. (1998). Minimizing the variable parameters for optimizing the fuzzy controller. *Fuzzy Sets and Systems, 100*, 131–142.

Ibrahim, W. R. A., & Morcos, M. M. (2002). Artificial intelligence and advanced mathematical tools for power quality applications: A survey. *IEEE Transactions on Power Delivery, 17*(2), 668–673.

IEEE Recommended Practices and Requirements for Harmonic Control in Electrical Power Systems (1993). *IEEE Standard 519-1992*, New York.

Jain, S., Agarwal, P., & Gupta, H. O. (2002). Fuzzy logic controlled shunt active power filter for power quality improvement. *Proceedings Electric Power Applications, 149*(5), 317–328. doi:10.1049/ip-epa:20020511.

Jain, S., Agarwal, P., Gupta, H.O., & Agnihotri, G. (2005). Modeling of frequency domain control of shunt active power filter using MATLAB simulink and power system blockset. In *Proceedings of 8th International Conference on Electrical Machine and Systems (ICEMS)* (vol 2 pp. 1124–1129), Sept 27–29 2005, Nanjing, Beijing, China. doi 10.1109/ICEMS.2005.202721.

Lee, C. C. (1990). Fuzzy logic in control systems: Fuzzy logic controller-part I. *IEEE Transactions on System, Man, and Cybernetics, 20*(2), 404–435.

Lee, H. X., & Gatland, H. B. (1995). A New methodology for designing a fuzzy logic controller. *IEEE Transactions on System, Man, Cybernetics, 25*(3), 505–512.

Moser, B., & Navara, M. (2002). Fuzzy controllers with conditionally firing rules. *IEEE Transactions on Fuzzy Systems, 10*(3), 340–349.

Nagrath, I. J., & Gopal, M. (2005). *Control System Engineering*. New Delhi: New Age International Publishers.

Singh, B., Al-Haddad, K., & Chandra, A. (1999). A review of active filters for power quality improvement. *IEEE Transactions on Industrial Electronics, 46*(5), 960–971.

Singh, R., & Singh, A. K. (2012). Design and analysis of an improved approximated fuzzy logic controller for shunt active power filter. *International Journal of Fuzzy System Applications (IJFSA), 2*(3), 69–89.

Singh, R., Singh, A. K., & Arya, R. K. (2011b). Approximated simplest fuzzy logic controlled shunt active power filter for current harmonic mitigation. *International Journal of Fuzzy System Applications (IJFSA), 1*(4), 18–36.

Singh, R., Singh, A. K., & Arya, R. K. (2013). Approximated fuzzy logic controlled shunt active power filter for improved power quality. *Expert Systems, 30*, 152–161.

Singh, R., Singh, A. K., & Kumar, P. (2011a). Comparison of three evolutionary algorithms for harmonic mitigation using SAPF. In *6th IEEE International Conference on Industrial and Information Systems (ICIIS)* (pp. 392–397), Aug 16–19, 2011, Kandy, Sri Lanka. doi 10.1109/ICIINFS.2011.6038100.

Singh, G. K., Singh, A. K., & Mitra, R. (2007). A simple fuzzy logic based robust active power filter for harmonic minimization under random load variation. *Electric Power System Research, 77*, 1101–1111.

Subjak, J. S, Jr, & McQuilkin, J. S. (1990). Harmonics-causes, effects, measurements, and analysis: An update. *IEEE Transactions on Industry Applications, 26*(6), 1034–1042.

Ying, H. (2000). *Fuzzy control and modeling: Analytical foundations and applications*. New York: IEEE Press.

Zeng, X., & Singh, M. G. (1994). Approximation theory of fuzzy systems-SISO case. *IEEE Transactions on Fuzzy Systems, 2*(2), 162–194.

Zeng, X., & Singh, M. G. (1995). Approximation theory of fuzzy systems-MIMO case. *IEEE Transactions on Fuzzy Systems, 3*(2), 219–235.

Soft Computing Techniques for Optimal Capacitor Placement

Pradeep Kumar and Asheesh K. Singh

Abstract Distribution system transfers electric energy from the transmission system to electric loads. Majority of losses in power system, i.e., nearly 10 %, occur in distribution system. Rigid distribution system infrastructure and rising load demand lead to increase in losses, thus, degrading the voltage profile. Utilities utilize the capabilities of the shunt capacitors to provide reactive power, for reducing the power losses and improve the voltage profile. The extent of distribution losses reduction and voltage profile improvement depends upon the location of these capacitors in the system. Thus, optimal capacitor placement (OCP) becomes a problem of significance. The problem of OCP is bifurcated into two sub-problems, (i) selection of candidate buses for capacitor placement, and (ii) sizing of the capacitors at the candidate buses. To select candidate buses for OCP, analytical techniques are used. But, soft computing techniques are utilized for sizing the capacitors. As the problem being, both, continuous and discrete in nature, i.e., mixed-integer type, its solution using classical optimization methods becomes impractical, as they are prone to be trapped in local minima. Therefore, soft computing techniques, like genetic algorithms (GA), particle swarm optimization (PSO), Nelder-Mead particle swarm optimization (NM-PSO), etc., capable of providing the global optimum solution, are utilized to obtain a better solution to the OCP problem. Further, a discussion of the previously used analytical techniques and the numerical techniques along with their disadvantages over the soft computing techniques is presented. This chapter is intended to discuss the application issues related to the solution of OCP using soft computing techniques. Further, special emphasis is given to the modeling of the distribution system and capacitor placement problem (CPP), with the relevance of OCP in distributed generation.

P. Kumar (✉)
Electrical Engineering Department, National Institute of Technology Kurukshetra, Kurukshetra 136119, Haryana, India
e-mail: pradeepkumar0802@gmail.com

A.K. Singh
Electrical Engineering Department, Motilal Nehru National Institute of Technology Allahabad, Allahabad 211004, India
e-mail: asheesh@mnnit.ac.in

Q. Zhu and A.T. Azar (eds.), *Complex System Modelling and Control Through Intelligent Soft Computations*, Studies in Fuzziness and Soft Computing 319,
DOI 10.1007/978-3-319-12883-2_21

Keywords Optimal capacitor placement · Mixed-integer type problems · Genetic algorithms (GA) · Particle swarm optimization (PSO) · Nelder-Mead PSO

1 Introduction

Electric power system is one of the largest man-made systems in the world. Since, the first electric power station, started in 1882, it has seen rapid growth across the length and breadth of the countries. The present social infrastructure would not be at all possible without it. No other energy form has proved to surpass its outstanding properties in the form of flexibility, cleanness, and compactness, etc. Its increasing consumption reflects the growing standards of the society. Its optimum utilization by the people can be ensured by effective distribution system (Pabla 2004; Kresting 2002). Thus, distribution system planning becomes vital. It assures that the future demands of the electricity can be adequately satisfied technically and economically, both.

For capacitor placement, it is important to have complete information about the structure of the distribution system, load growth pattern, distribution of the load across the geographic location, etc. The power, from the transmission system to the loads, is transmitted via distribution lines to plan the power distribution effectively. Unlike the transmission system, the distribution system is un-transposed, i.e., the lines in the distribution system are inductively coupled. Thus, the active power losses are very high. Also, these high losses may be attributed to the presence of the highly inductive loads, low distribution voltage levels, and uneven spread of the distribution lines or the single-phasing of loads, etc., thus affecting the flow of power in the distribution system and making it the most inefficient component of the whole power system (Ng et al. 2000). Since, the single-phasing of the distribution system is inherently present in the structure of the distribution system; it is difficult to remove it completely. But, the high reactive power demand of loads can be compensated by supplying reactive power to the loads, locally.

The shunt capacitors are shunt connected to the loads, for compensating the reactive power demand. This reduction in the reactive power demand of the loads, due to placement of capacitors (Gonen 1986), leads to,

- Released generation capacity.
- Released transmission capacity.
- Released distribution substation capacity.
- Reduced energy losses.
- Reduced voltage drop and consequently improved voltage regulation.
- Released capacity of feeder and associated apparatus.
- Postponement or elimination of capital expenditure due to system improvement and/or expansions.
- Revenue increases due to voltage improvements.

To achieve these targets, and supplying the reactive power locally, it is necessary to place the capacitor optimally in the distribution system. Several methods have been developed, to place the capacitors optimally. These methods guide the distribution companies to obtain the (i) candidate buses for OCP, and (ii) size of the capacitors. The objective of the OCP involves continuous (power and energy loss) and discontinuous (Capacitor sizes), objectives. Such types of problems are termed as mixed integer type problems. Several techniques have been proposed to solve the capacitor placement problem. However, the solutions obtained using the soft computing techniques are optimal as compared to that obtained using other techniques.

This chapter aims to show the application of the soft computing techniques to solve the optimal capacitor placement problem (OCP). In Sect. 2, the structure of the electric power distribution system is discussed along with the role of capacitors in them. Section 3 presents the modelling of capacitor in distribution system and the approach to solve the problem. A discussion on the classical approaches and soft computing techniques used to solve the problem are illustrated in Sect. 4. The results for the application of the soft computing techniques are discussed in Sect. 5. Finally, the conclusions are drawn in Sect. 6.

2 Electric Power Distribution System

The electric power system is divided into three parts, viz., generation, transmission, and distribution, as shown in Fig. 1 Typically, the distribution system starts with distribution substation, fed by one or more sub-transmission lines. However, in some cases, the distribution substation is fed directly from the high-voltage transmission line. These substations serve as one or more primary feeders. Largely, these feeders are radial in nature, i.e., there is only one path for the power to flow from the distribution substation to the user as shown in Fig. 2 (Kresting 2002; Gonen 1986). Further, the distribution system is divided into primary and secondary distribution system. The high capacity substations located at outskirts of the populous area are connected via primary distribution system, whereas small substations to supply the consumers are connected through secondary distribution.

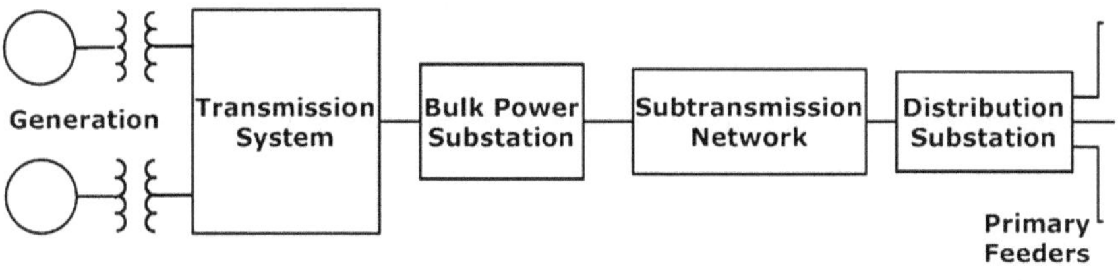

Fig. 1 Power system components (Kresting 2002)

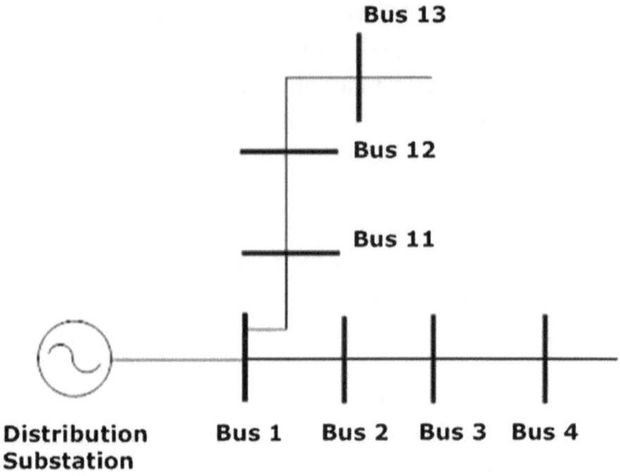

Fig. 2 An example of radial distribution system

Due to the challenges posed by the design and operation of the electric power generation and transmission, they have been the prime area of interest for the researchers. However, during this development, distribution system remained untouched. The differences between the developed generation and transmission system and under-developed distribution system were not able to satisfy the customer demands. The major problems occurring in the distribution system (Haghifam and Malik 2007) were,

- High var demand,
- High voltage drop,
- Reduced system capacity,
- High amount of losses in the system,
- Inherently unbalanced distribution system, and
- Harmonics due to non-linear loads.

The modern, high-efficiency apparatuses at the load side are very sensitive in nature. Their malfunction may result in loss of the manufacturing process and the associated cost. Thus, the distribution system problems may have significant economic impact. Also, in deregulated and competitive schemes, providing reliable and high-quality power is essential to attract and keep customers. To overcome these problems a number of methods are developed, such as,

- Re-conductoring in primary and secondary feeders,
- Feeder reconfiguration,
- Using distributed generation,
- Load balancing between three-phases and feeders,
- Load factor improvement with demand side management strategies,
- Voltage upgrading, and
- Reactive power control by placement and installation capacitors.

In the next section, role of the capacitors in distribution system and their effects are discussed.

2.1 Role of Capacitors

In distribution system, the fundamental function of the capacitors is to control the flow of reactive power. This can be achieved by connecting the capacitor either, in

1. Series Connections, or
2. Shunt Connections.

In series connection, the capacitors balance the inductive reactance of the circuit, whereas, in shunt connection it supplies the reactive power or current to counteract the out-of-phase component of current as required by an inductive load. They modify the characteristics of the inductive load by supplying a leading current; this counteracts the lagging currents of the loads. In series connection, the change in inductance of the distribution lines leads to phenomena, such as, ferro-resonance in transformer, sub-synchronous resonance during motor starting, shunting of motors during normal operation, etc. (Gonen 1986; Kresting 2002). Hence, much application of series capacitors is not observed in distribution system, yet, shunt connection is extensively used. Table 1 shows the connection of the power capacitor and the phasor diagram for both the capacitor connections. It is important to note that the compensation provided to regulate the voltage and reactive power flow by the capacitor is only applicable at the point of installation.

Table 1 Series connections and shunt connections of the power capacitors (Kresting 2002)

Series Connection	Shunt Connection

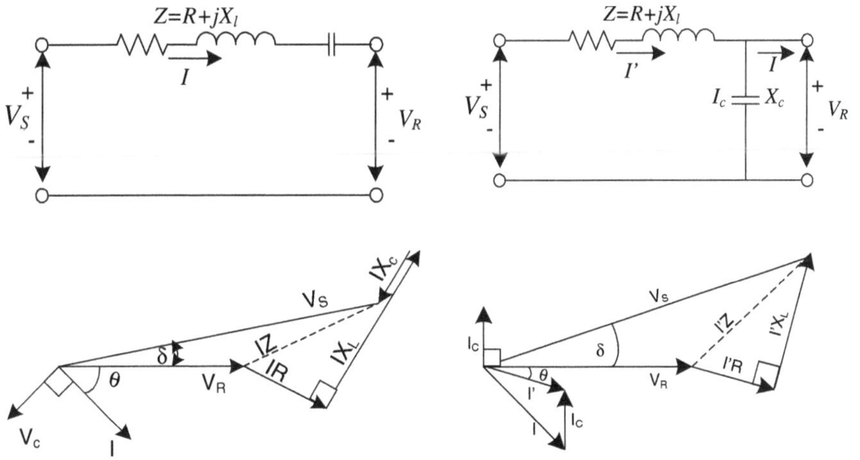

$VD = IR\cos\theta + I(X_L - X_C)\sin\theta$ $\quad$ $VD = I_R R_R + (I_L - I_C)X_L$

In the literature, number of capacitor classifications can be observed, depending upon installation type. However, for the purpose of capacitor installation the classification can be given as

1. Fixed capacitors, and
2. Switched capacitors.

In practice, both fixed and switched capacitors are used. A fixed capacitor has the same kvar values at all the levels. Switched capacitors are connected as per the requirement at the buses. But, the higher cost of the switched capacitors restricts utilities to use them; this also simplifies the nature of the problem. Therefore, the fixed capacitors are considered here.
Only the smallest standard size of capacitors and its multiples are allowed at the buses, to have a more realistic optimal solution that can be implemented later with no difficulties. Capacitor banks are generally supplied in multiples of 150 kvar.

2.2 *Benefits of Capacitor Placement*

The shunt capacitors, connected at the feeder supplying load at lagging power factor, offer several advantages, some of which are,

(a) **Reduced Power System Losses**. The reactive power compensation provided by capacitor, substantially, reduces the network losses in between the substation and point of capacitor installation. In order to maximize the benefits, installation of the capacitor should be implemented as close to the customer load as possible.
(b) **Improved Voltage Profile**. Feeders in the distribution system, with high utilization and reactive power demands, offer large voltage variation and poor voltage profiles. The desirable voltage regulation is within a narrow range, i.e., ±5 % of the nominal value, with balanced loads. However, load fluctuation may lead to voltage deviation beyond their permissible limits. With the use of the capacitors, the use of expensive voltage regulators can be minimized.
(c) **Released Power System Capacity**. The reactive current, supplied by the capacitors connected, furnishes the magnetizing currents for the electromagnetic devices such as, motors, transformers, etc. This reduction, in reactive current, helps in reduction of overloading and permits the addition of additional load to the existing system. Thus, more customers can be connected to the existing system.

In order to place the capacitor, the mathematical modeling of the capacitor placement problem is performed. In modeling, its various components, namely, load flow analysis, modeling at fundamental and harmonic frequencies, load variations, different planning methods and constraints for the problem are discussed.

3 Modeling of Capacitor Placement Problem

For capacitor placement in the distribution system, information about the load flows, capacitor sizes, and the method to solve, is very important. Here, the problem formulation for the capacitor placement is discussed.

The problem of capacitor placement can be solved as a constrained optimization problem

$$\min f(x, u) \tag{1}$$

subjected to

$$F(x, u) = 0 \tag{2}$$

$$G(x, u) \leq 0 \tag{3}$$

where $f(x,u)$ is objective function. The state variable x represents the state of system and capacitor placement scheme is represented by the variable u. $F(x,u)$, represents the set of equality constraints, and $G(x,u)$ represents the set of the inequality constraint of the problem. Following sub-sections provides the necessary methods to formulate the problem.

Capacitors are installed in distribution system to reduce the losses and improve the voltage profile. As the size of capacitor plays a vital role in reactive power flow control. The capacitors are placed and sized by modelling the problem as an optimization problem. The capacitor placement problem is modelled as an objective function, as shown in sections below. The cost of capacitor composed of the capacitor cost and the installation cost, as shown in Fig. 3, is taken into discrete steps of the size of capacitor. Here, the capacitor placement problem is reformulated

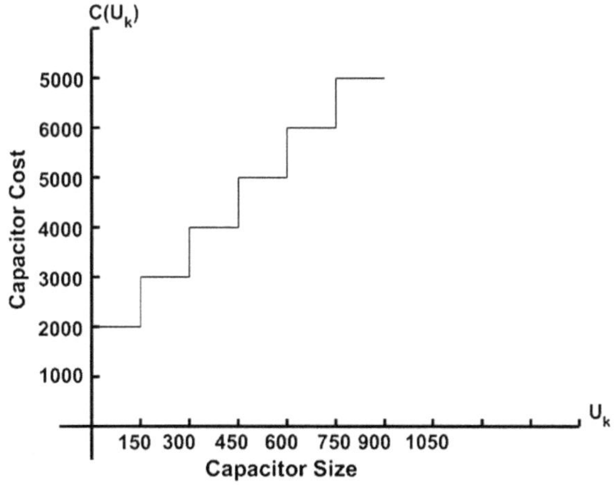

Fig. 3 Capacitor investment cost function

comprehensively. The formulation has given due consideration to the manufacturing cost and economic factors viz., market inflation and creditor interest rates on them. This provides a more realistic approach to the problem.

3.1 Load Flow

In this section, the modelling of the distribution system at fundamental and harmonic frequencies is discussed. Generally, the capacitor placement problem is solved at the fundamental frequency. However, the power distribution system being non-linear in nature, harmonics are present in the voltages and currents. Therefore, the placement problem can be solved at both, fundamental, as well as, harmonic frequencies.

3.1.1 Modelling at Fundamental Frequency

The distribution system is highly coupled system. Therefore, the commonly used load flow models, utilizing the single-phase nature of the distribution system are not utilized to solve it. For solving the radial distribution system, such as given in Fig. 4, with line section l between nodes i and j having shunt admittances and loads attached to each node, three-phase backward-forward load flow (Cheng and Shirmohammadi 1995) is used. Using it, the distribution system can be solved in three steps, as.

1. *Nodal current calculation*

$$I_{i,k} = \left(\frac{y_{c,l}}{2} V_{j,k}\right) + \left[\frac{S_{L,n}}{V_{i,k}}\right]^* \tag{4}$$

where, $y_{c,l}$ is the shunt admittance in section-l, $S_{L,i}$ and $S_{L,j}$ are the apparent power of loads on buses i and j, $V_{i,k}$ is voltage of bus at iteration k.

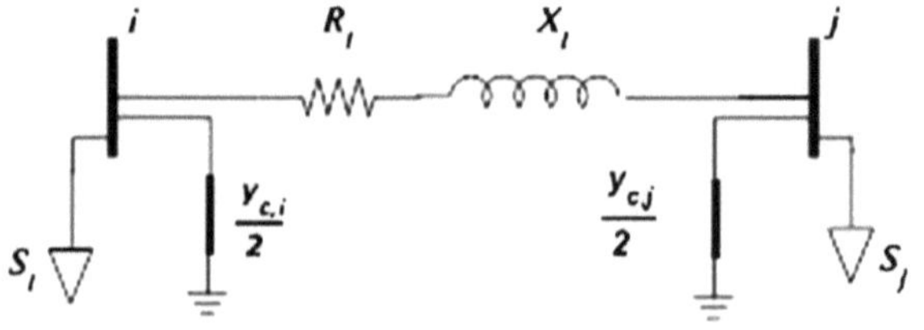

Fig. 4 Line section of distribution system connecting two buses

2. *Section current calculation*

$$J_{l,k} = I_{i,k} + \sum_{b=1}^{N_b} J_{b,k} \tag{5}$$

where, $J_{l,k}$ is the current in section-l at iteration k.

3. *Voltage node calculation*

$$V_{i,k+1} = V_{j,k+1} - (R_l + jX_l)J_{l,k} \tag{6}$$

where, $V_{i,k}$ is the voltage of bus i at iteration k.

To apply the load flow initially the bus voltages are selected as 1 ∠0 per unit. Node currents are then calculated using (4). From the node currents, section currents and bus voltages are calculated using (5) and (6), respectively. The process is continued until the convergence limit is achieved.

3.1.2 Modelling at Harmonic Frequency

At harmonic frequencies, the modelling of line parameter is modified. These harmonics in the distribution system arise due to the non-linear loads in the system (Baghzouz and Ertem 1990; Baghzouz 1991). At any harmonic frequency, of *n*th order, the admittance is given as

$$y_c^n = ny_c^1 \tag{7}$$

where

y_c^1 Line admittance at fundamental frequency, and

y_c^n Line admittance at nth harmonic frequency.

The admittance between buses "*i*" and "*i* + 1" can also be modelled as

$$y_{i,i+1}^n = \frac{1}{R_{i,i+1} + n \cdot X_{i,i+1}} \tag{8}$$

where

$R_{i,i+1}$ Resistance of line between bus i and $i + 1$

$X_{i,i+1}$ Reactance of line between bus i and $i + 1$

$y_{i,i+1}^n$ Admittance of line between bus i and $i + 1$

At harmonic frequencies, the load modelling is also modified. As suggested by Baghzouz and Ertem (1990), Baghzouz (1991) a generalized load model has been used which is composed of resistance in parallel with an inductance selected to account for the respective active and reactive power at fundamental frequency. Load admittance at kth load level is given as

$$y_{ik}^{\ln} = \frac{1 - w_i}{\left|V_{ik}^{1}\right|^2}\left(P_{ik} - j\frac{Q_{ik}}{n}\right) \tag{9}$$

where w_i is the percentage of load at bus 'i'.

The overall voltage at any bus i, at harmonic frequency, is evaluated as

$$|V_{ik}| = \sqrt{\sum_{n=1}^{N} \left|V_{ik}^{n}\right|^2} \tag{10}$$

where,

V_{ik}^{n} nth harmonic voltage at bus i at load level k

N Maximum harmonic order under consideration.

The amount of distortion in the system is calculated based on Total Harmonic Distortions (THD), given as,

$$THD_{ik}(\%) = \frac{\sqrt{\sum_{n \neq 1}^{V} \left|V_{ik}^{n}\right|^2}}{\left|V_{ik}^{1}\right|} \times 100 \tag{11}$$

3.2 Load Variation

To take into account the varying load conditions and to calculate the energy loss at these load conditions, load levels are considered for time duration 'T_i'. These load levels are assumed discrete, as shown in Fig. 5. The load levels are assumed to be a piece-wise linear function of time period, T, which is again divided into intervals during which the load level remains constant. The load levels are divided into three categories, peak, medium, and low.

3.3 Planning Methods

To solve the capacitor placement problem using soft computing techniques the problem is modelled as objective function termed as planning methods. Similar objective functions are applicable for the placement using the analytical, numerical, soft computing techniques. The planning methods used for the capacitor placement are discussed below.

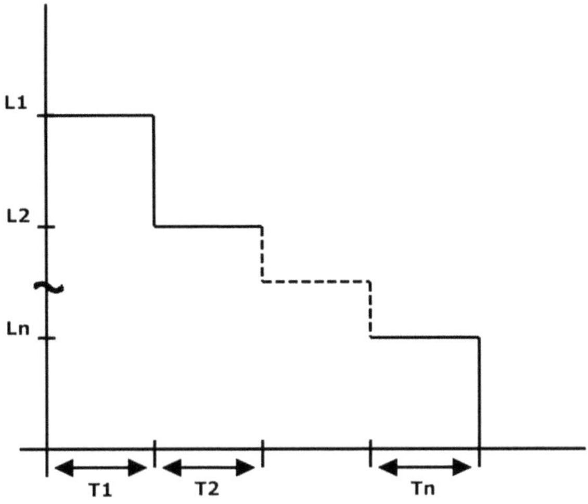

Fig. 5 Variation of load (x-axis) with time (y-axis)

(a) *Static Cost Based Planning Method*
In distribution system, the shunt capacitor increases the system efficiency. The objective of optimal capacitor placement is to reduce the power loss, energy loss, and to minimize the cost of capacitor banks, while maintaining bus voltages within prescribed limits (Kumar and Singh 2011). Based on these objectives in conventional method of planning, the objective function is formulated as

$$\min f(u^i, u^0) = k_e \sum_{i=1}^{nl} T_i P_i + k_p P_l + \sum_{j=1}^{nc} C_j\left(u_j^0\right) \tag{12}$$

where, T_i is time duration, nl is load level, $C_j(u_j^0)$ is cost function of capacitor $C_j(u_j^0) = k_{inst} + k_{cj} u_j^0$, k_{cj} is the capacitor cost, k_{inst} is the capacitor installation cost, nc is number of capacitors, Y is planning years, P_i power loss at load level i, k_e is cost of energy, Pl is power loss at peak load, k_p is cost of peak power loss.

(b) *Variable Cost Based Planning Method*
The static cost based planning method assumes the cost to remain constant over the planning period, which is not the case. In the variable cost based planning method, the objective function of static cost based planning method is reformulated to include the variation in the cost. The variation in the cost is incorporated by inclusion of the cost of maintenance, and the variation in the cost due to certain economic factors. These costs need to be included while minimizing the total cost of the system for achieving the targets (Kumar et al. 2014). The reformulated objective function can be expressed as

$$\min f(u^i, u^0) = k_e \sum_{i=1}^{nl} T_i P_i + k_p P_l + \sum_{j=1}^{nc} C_j\left(u_j^0\right) + \sum_{Y=1}^{L_C} \left(M_{tc}(k_{ef})^{Y-1}\right) \quad (13)$$

where, M_{tc} is Maintenance cost, k_{ef} is economic factor given as $k_{ef} = \frac{1+m}{1+f}$, m is inflation rate, f is interest rate.

The reformulated objective function consists of four terms, i.e., the energy cost, peak power loss, capacitor cost, and maintenance along with the variation in cost due to market inflation, interest rates etc., respectively.

Both the objective functions are non-differentiable, making it unsolvable by conventional methods. The cost function described in this way, being a non-linear mixed integer, optimization problem can be solved efficiently by soft computing techniques. The constraints applied to these objective functions are discussed below.

3.4 Constraints

(a) *Load constraints*
Load constraints are implied by the load connected at the buses. These constraints are imposed on the power flow equation by the loads. A three-phase load flow method has been used in this study for the problem.

(b) *Operational constraints*
Voltages are required to follow a certain limit to remain within upper and lower limits. These constraints need to be followed before and after the capacitor placement.

In the next section, problem of optimal capacitor placement, the approaches and algorithm for its solution are discussed.

4 Optimal Capacitor Placement Problem

The reactive power provided by the power capacitor in the distribution system depends upon the location of the capacitor and the size of the capacitor bank installed. Thus, the problem for the capacitor placement can be sub-divided into two problems, namely, (1) location of the capacitor and the (2) sizing of the capacitors. For both of these problems separate techniques exist. The methods used for determination of location and size are discussed below. The location of the capacitor can be determined using analytical methods based on reasoning, whereas the size of capacitors can be determined by solving the above-mentioned planning methods using soft computing techniques. Sizing of *the capacitor, in itself, is a mixed*-integer programming method, which includes the continuous (power loss and energy loss) and discrete (capacitor sizes) variables.

4.1 Determination of Capacitor Location

The location for the placement of the capacitor can be determined in a number of ways. The simplest method can be to determine the nodes with highest reactive power demand and weak voltage profile. Also, the placement can be done randomly, through simulation, by placing the capacitor and observing its effect on the nearby buses. Moreover, sensitivity based methods can be used to determine the location of the power capacitors.

Sensitivity analysis refers to the determination of how "sensitive" a parameter is to the changes in the value of the other parameters of a model, or to the changes in the structure of the model. Thus, it is defined as "ratio $\Delta x/\Delta y$ relating small changes Δx of some dependent variable to small changes Δy of some independent or controllable variable y" (Peachon et al. 1968). Both Loss Sensitivity (Prakash and Sydulu 2007) and Bus Sensitivity (Silva et al. 2008), can be calculated for any particular bus in power system. Here, these two methods are employed to select the candidate buses for optimal capacitor placement. Both sensitivity methods are discussed in the following paragraphs (Kumar et al. 2011).

4.1.1 Loss Sensitivity

The loss sensitivity analysis is a systematic procedure computing the maximum impact on the real power losses of the system with respect to the nodal reactive power. The relationship for computing the loss sensitivity for any bus can be derived as follows.

Consider a distribution line connected between buses i and j as shown in Fig. 6. Where, r_{ij}, x_{ij} are resistance and reactance of lines between buses i and j, and $S_j = P_j + Q_j$ is the load connected at bus j, P_j is the active power and Q_j is the reactive power.

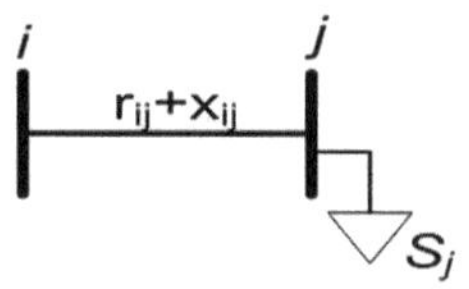

Fig. 6 Section of distribution line

The active power losses (P_{Loss}) and reactive power loss (Q_{Loss}) in the distribution line are given as:

$$P_{Loss} = \frac{r_{ij}\left(P_j^2 + Q_j^2\right)}{V_j^2}, \quad Q_{Loss} = \frac{x_{ij}\left(P_j^2 + Q_j^2\right)}{V_j^2} \tag{17}$$

Loss sensitivity S_L, for any bus j can be given as

$$S_{Lj} = \frac{\partial P_{Loss}}{\partial Q_i} = \frac{2 \times Q_j \times r_{ij}}{V_j^2} \tag{18}$$

Based on S_{Li} the buses are ranked in descending order of its values. The bus having greatest value is ranked top in the priority list and is considered first for capacitor placement. The buses having the highest S, along with voltage ratio V/0.9 > 1.1, is selected as candidate buses for capacitor placement.

4.1.2 Bus Sensitivity

Bus sensitivity method (Silva et al. 2008) is the sensitivity of buses based on reactive power and voltage at the buses. It suggest to provide reactive compensation mainly on those buses where reactive power demand is high and bus voltages are below nominal value. For ith bus the bus sensitivity index (S_{Bi}) is defined as

$$S_{Bi} = \lambda_{Qi} \frac{Ql_i}{V_i} \tag{19}$$

where, λ_{Qi}, Ql_i, and V_i are Lagrange multiplier, load reactive power, and voltage at bus i, respectively.

Lagrange multiplier is calculated as implicit first order derivative of the cost function with respect to the right side with parameter of constraint providing information about the sensitivity. For the capacitor placement problem, λ reflects the sensitivity of objective function to changes in the reactive power injection in ith bus. S_{Bi} gives cost required to increase the voltage at bus i, having unit \$/V.

4.2 Determination of Capacitor Size

To determine the size of the capacitor, for the installation at the locations determined using the methods discussed above, several methods are available. The methods developed can be broadly classified as:

(a) Analytical Methods,
(b) Numerical Programming Methods, and
(c) Soft Computing Based Methods.

4.2.1 Analytical Method

Initially, when the computing resources were limited, the methods were based on the analytical techniques. For capacitor placement, the analytical approaches used (Levitin et al. 2000; Fogel et al. 2000) were based on the maximum economic savings, as given by (12). These initial methods paved the path for the two-third rule. It advocates installation of the capacitor of rating two-third of the peak reactive load, at a position two-third of the distance along the total feeder length. However, it is still used by many utilities, as being based on the assumptions like,

1. Constant feeder conductor sizes,
2. Uniform current loading, and
3. All variables are assumed continuous.

Later, modifications in the techniques using sectionalized normalized equivalent feeders to overcome the assumptions (Lee and Grainger 1981) led to more accurate results.

4.2.2 Numerical Programming Methods

With advancement in computation facilities, numerical programming methods were developed. These iterative procedures maximize or minimize the objective function, satisfying some set of constraints, viz. voltage, capacitor sizes, etc. These approaches allow the inclusion of the mixed variables, i.e., continuous voltage and line loading, and discrete sizes of capacitors. The objective problem for the capacitor takes the same form as (12). Several techniques such as dynamic programming, mixed-integer programming, and integer programming (Aman et al. 2014; Ng et al. 2000) have been developed to solve these problems.

4.2.3 Soft Computing Methods

Soft computing methods utilize the idea and inspiration from the experiences, natural evolution, and adaptation to solve the real word computational problems in an efficient and robust manner (Fogel et al. 2000). The problems, to be dealt with, are non-linear in nature, which are either unsolvable or inaccurately solved using conventional techniques. These soft computing methods provide fast and practical strategies that utilize the exhaustive search space to provide a near optimal solution.

Generally, in soft computing techniques, the terminologies utilized are based on the terminology of the inspiration to reflect their connections, such as in genetic algorithms, we have genotypes, phenotypes, species, etc. (Fogel et al. 2000).

The naïve approach to solve any problem can be given as,

1. *Formation of the search space*: list all the feasible solutions of the given problem,
2. *Evaluate*: evaluate their objective functions,
3. *Best solution*: choose the best solution amongst the various solutions.

These techniques are more flexible and capable of coping with more realistic objective functions and constraints. They are used as a contrast to complete enumeration methods which guarantee to find the global optimum.

The various soft computing techniques used for optimum allocation of the capacitors in the distribution system are discussed below.

Although, in principle, it is possible to solve any problem in this way, in practice it is not due to the vast number of possible solutions to any real-world problem, such as general optimal capacitor placement problem in distribution systems, of a reasonable size.

Genetic Algorithms

Genetic Algorithms (GAs) are the soft computing tools, utilizing the concept of evolution, to determine the optimal solution. It utilizes the survival of the fittest concept to promote the growth of the healthier solutions in the search space, than that of the unhealthier solutions.

Being a powerful tool for the optimization, several variants of the GA have been proposed, viz., binary GA (BGA), real GA, niche GA, etc. These different forms of GAs differ from one another in terms of how the solution space is processed, e.g., in case of BGA the search space is decoded into binary form, whereas RGA processes real form of data. In comparison to the traditional optimization methods, where the solution moves from point to the other in the solution hyperspace, solutions in GA search the solution hyperspace randomly (Aziz et al. 2013).

The various parameters that needs to be defined in a GA are,

1. *Population size*: It represents the random number of solutions in problem hyperspace.
2. *Crossover rate*: It affects the rate at of crossover between the chromosomes. A high probability represents introduction of new strings at a higher rate.
3. *Mutation rate*: It represents the probability of change in bit position of each string in a new population after the selection process.

Figure 7 shows the flowchart for the optimal placement and sizing of capacitor banks in distribution system using GA (Masoum et al. 2004).

In spite of being a popular technique for optimization, GA lacks the flexibility to maintain a balance between global and local exploration of search space. Large number of parameters needs to be decided to begin the solution to the problem, and sometimes the convergence of the algorithm may fail (Boeringer and Werner 2004).

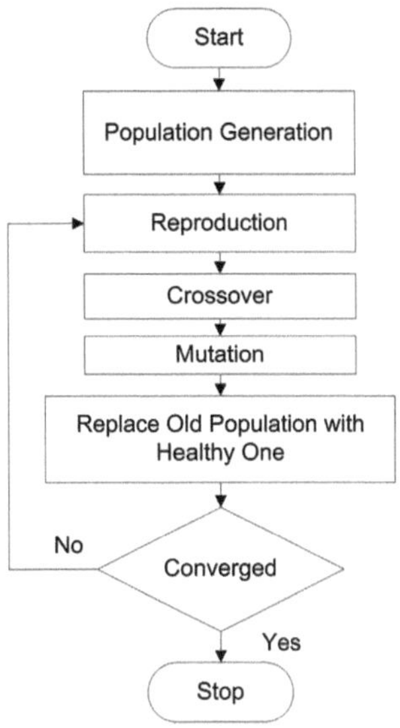

Fig. 7 Flowchart for the optimal placement and sizing of capacitor banks in distribution system using GA

Particle Swarm Optimization (PSO)

PSO, developed by Eberhart and Kennedy (Valle et al. 2008), is a soft computing technique, inspired by the social behaviour of bird flocking and fish schooling In PSO, where swarm imitates the irregular movement of particles in the problem space. It refers to the solutions as mass-less and volume-less particle, which are subjected to velocities and accelerations towards a better mode of behaviour as the possible solutions to the given problem. The population of particle move through the hyperspace through a given velocity. At each iteration, the velocity of the individual particles is stochastically adjusted according to the previous best position for the particle itself, and neighbourhood's best position. Both particles best and neighbourhood best are derived based on the user defined fitness function (Valle et al. 2008). In addition, PSO utilizes the swarm intelligence concept which is the property of the system, whereby collective behaviour of particles interacting locally with their environment creates coherent global function patterns.

PSO is not largely affected by the size and non-linearity of the problem, and can converge to the optimal solution in many problems, where most analytical methods fail to converge. PSO has some advantages over other similar optimization techniques such as (Valle et al. 2008),

1. It is easier to implement and has lesser parameters to adjust.
2. It has certain memory capability as; every particle remembers its own previous best values as well as the neighbourhood best.
3. Since, all the particles use the information related to the most successful particles in order to improve them. Thus, it is more efficient in maintaining the diversity of the swarm.

(a) *Algorithm*

The position of the particle is continuously updated, based upon the velocity of the particle. The velocity of the particle is calculated based on

$$\begin{aligned}\vec{V}_i(t) = \vec{V}_i(t-1) + \varphi_1 \cdot rand_1 \cdot \left(\vec{p}_i - \vec{x}_i(t-1)\right) \\ + \varphi_2 \cdot rand_2 \cdot \left(\vec{p}_g - \vec{x}(t-1)\right)\end{aligned} \tag{20}$$

where,

φ_1, φ_2	are two positive numbers
$rand_1, rand_2$	two random numbers with uniform distribution in the range [0,1]
$V_i(t)$	velocity of particle i at any instant t
$x_i(t)$	position of particle i at any instant t
p_i	local best of the particle
p_g	global best of the particle

Based on the velocity calculated, the updated position is given by

$$\vec{x}_i(t) = \vec{x}_i(t-1) + \vec{V}_i(t) \tag{21}$$

From (20) and (21), it can be clearly seen that the information available for each particle is based on its own experience, and the knowledge of the performance of other individuals is available in its neighbourhood.

The flow chart of the complete algorithm can be given in Fig. 8.

Nelder-Mead Particle Swarm Optimization (NM-PSO)

The hybrid NM-PSO algorithm, as proposed by Fan and Zahara (Zahara and Kao 2009), is based on Nelder-Mead (NM) simplex search method and PSO for optimization. Nelder and Mead proposed a simple local direct search technique, which does not require any derivative for finding solution of any function (Nelder and Mead 1965). The PSO as proposed is a global search technique, but it is limited by high computational cost of the slow convergence rate.

The slow convergence rate of PSO as compared to local search techniques like NM is due to improper utilization of local information to determine a most promising search direction. To overcome this slow convergence, the PSO is combined with local simplex search technique, in a way that both the algorithms enjoy merits of each other. In NM-PSO, the PSO prevents the hybrid approach from being trapped in local optima, whereas the convergence rate is increased by NM algorithm. In other words, the PSO focuses on "exploration" and the simplex method, i.e., NM algorithm, focuses on "exploitation." Out of two components of NM-PSO, viz., NM and PSO, PSO is described in the previous sub-section, while the NM is discussed below.

Fig. 8 Flowchart for the particle swarm optimization (PSO) algorithm

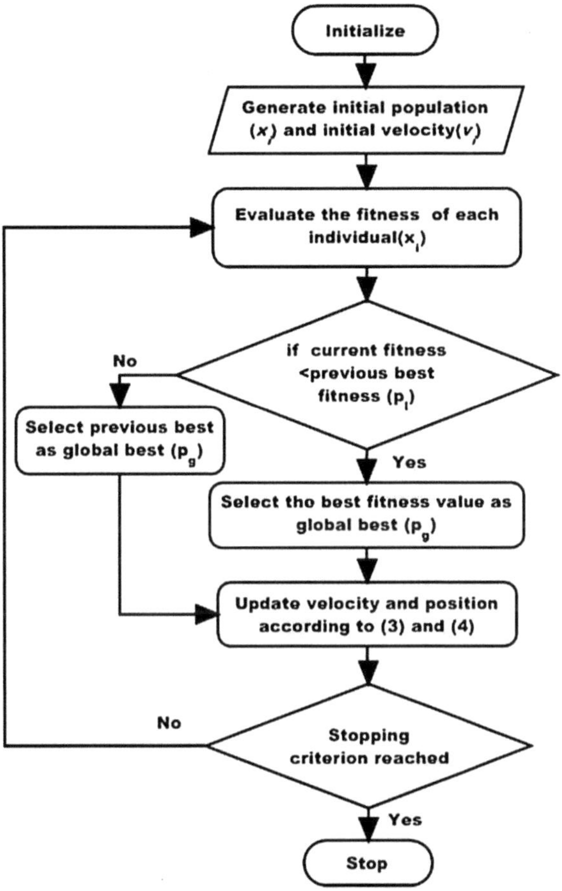

In 1962, an optimum operating tracking method was devised, which was based on set of point by forming a simplex in the factor-space and continually forming new simplexes by reflecting one point in the hyper-plane of the remaining points. It is a local search method designed for unconstrained optimization without using gradient information. Following four basic procedures are used by this method to rescale the simplex based on the local behaviour of the function: reflection, expansion, contraction, and shrinkage. Through these procedures, the simplex can successfully improve itself and get closer to the optimum. The algorithm for NM simplex is outlined below and the steps are illustrated in Fig. 8 through a two-dimensional case (N = 2).

1. *Initialization*: For the minimization of a function of N variables, create N + 1 vertex points to form an initial N-dimensional simplex. Evaluate the functional value at each vertex point of the simplex. See a two-dimensional simplex exhibited in Fig. 9a. For the maximization case, it is convenient to transform the problem into the minimization case by pre-multiplying the objective function by −1.

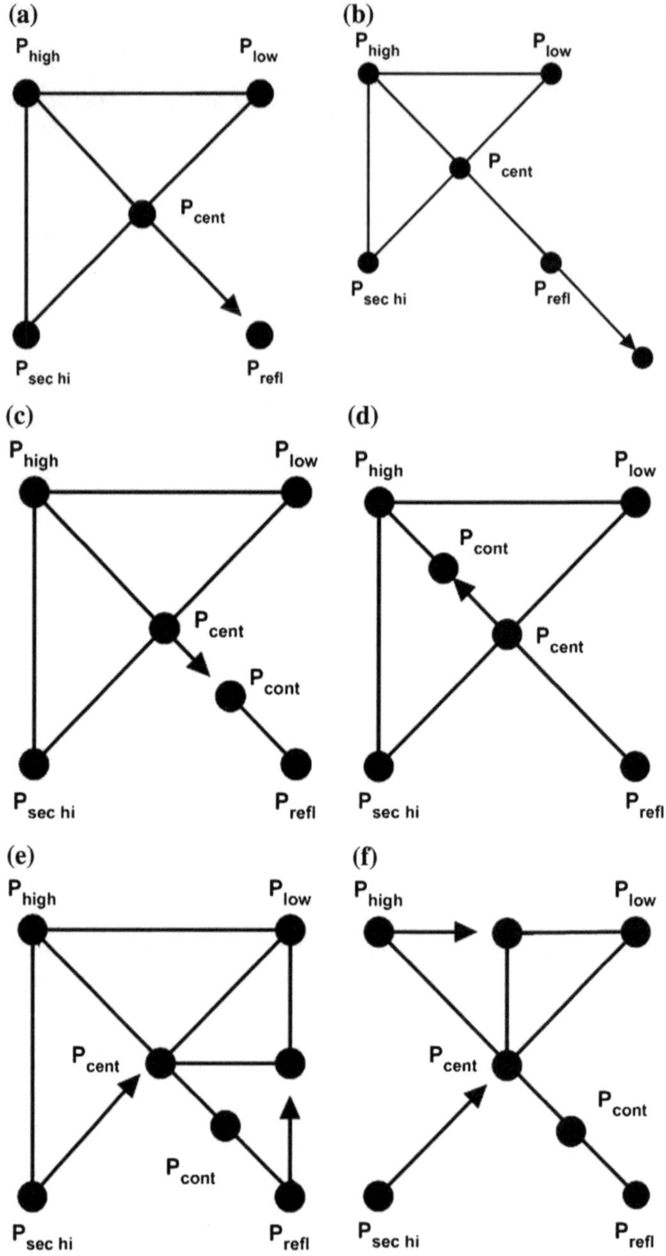

Fig. 9 Nelder-Mead pivot operations. **a** Reflection. **b** Expansion. **c** Contraction when P_{refl} is better than P_{high}. **d** Contraction when P_{high} is better than P_{refl}. **e** Shrink after failed contraction for the case where P_{refl} is better than P_{high}. **f** Shrink after failed contraction for the case where P_{high} is better than P_{refl}

2. *Reflection*: In each iteration step, determine P_{high}, $P_{sec\ hi}$, P_{low} vertices, indicating vertex points that have the highest, the second highest, and the lowest function values, respectively. Let f_{high}, $f_{sec\ hi,}$ represent the corresponding observed function values. Find P_{cent}, the center of the simplex excluding P_{high} in the minimization case. Generate a new vertex P_{refl} by reflecting the worst point according to the following equation (see Fig. 9a):

$$P_{refl} = (1 + \alpha)P_{cent} - \alpha P_{high} \tag{21}$$

where P_{refl} is the reflection coefficient ($\alpha > 0$). Nelder and Mead suggested the use of $\alpha = 1$. If $f_{low} \leq f_{refl} \leq f_{sec\ hi}$, accept the reflection by replacing P_{high} with P_{refl}, and step 2 is entered again for a new iteration.
3. *Expansion*: Should reflection produce a function value smaller than f_{low} (i.e., $f_{refl} < f_{low}$), the reflection is expanded in order to extend the search space in the same direction and the expansion point is calculated by the following equation (see Fig. 9b)

$$P_{\exp} = \gamma \cdot P_{refl} + (1 - \gamma)P_{cent} \tag{22}$$

where γ is the expansion coefficient ($\gamma > 1$). Nelder and Mead suggested $\gamma = 2$. If $f_{exp} < f_{low}$, the expansion is accepted by replacing P_{high} with P_{exp}; otherwise, P_{refl} replaces P_{high}. The algorithm continues with a new iteration in step 2.
4. *Contraction*: When $f_{refl} > f_{sec\ hi}$ and $f_{refl} \leq f_{high}$, then P_{refl} replaces P_{high} and contraction is tried (see Fig. 9c). If $f_{refl} > f_{high}$, then direct contraction without the replacement of P_{high} by P_{refl} is performed (see Fig. 9d). The contraction vertex is calculated by the following equation:

$$P_{cont} = \beta P_{high} + (1 - \beta)P_{cent} \tag{23}$$

where β is the contraction coefficient ($0 < \beta < 1$). Nelder and Mead suggested $\beta = 0.5$. If $f_{cont} \leq f_{high}$, the contraction is accepted by replacing P_{high} with P_{cont} and then a new iteration begins with step 2.
5. *Shrink*: If $f_{cont} > f_{high}$ in step 4, contraction has failed and shrinkage will be the next attempt. This is done by shrinking the entire simplex (except P_{low}) by (see Fig. 9e, f):

$$P_i \leftarrow \delta P_i + (1 - \delta)P_{low} \tag{24}$$

where δ is the shrinkage coefficient ($0 < \delta < 1$). Nelder and Mead suggested $\delta = 0.5$. The algorithm then evaluates function values at each vertex (except P_{low}) and returns to step 2 to start a new iteration.

For an *N*-dimensional problem, the flowchart of the algorithm is given in Fig. 9. Initially, a population of size 3(N + 1) is generated and evaluated. Particles are ranked based on their fitness values. The top N elite values are saved and NM method is applied to the top N + 1 particles. PSO operator is then applied to the remaining 2 N particles with worst fitness. Then, the stopping criterion is checked for all the 3(N + 1). If the criterion is achieved, then the process stops and best solution is achieved, otherwise, the fitness values are again repeated and the complete process repeats.

5 Solution Methodology

Based on this load flow and proposed objective function, the capacitors are placed at respective buses using the algorithm presented in Fig. 10. According to it, the system parameters are entered in the program. Based on the system parameters and sensitivity the candidate buses are selected. On the selected buses, the capacitors are placed using the optimization algorithm and objective function evaluation. After the capacitor placement, the constraints are checked.

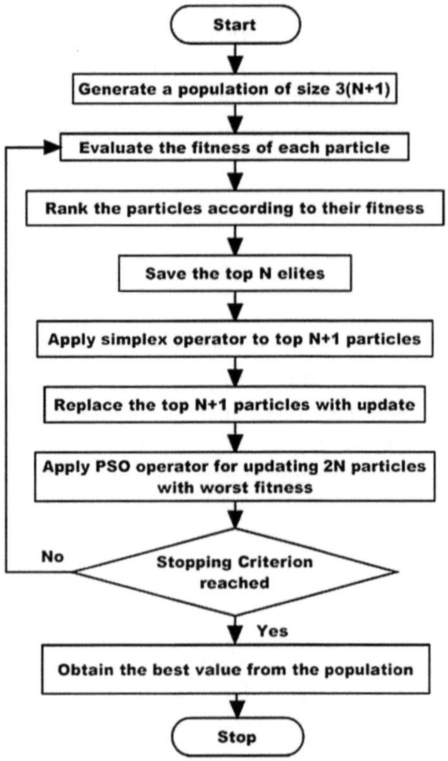

Fig. 10 Flowchart for Nelder-Mead particle swarm optimization (NM-PSO) algorithm

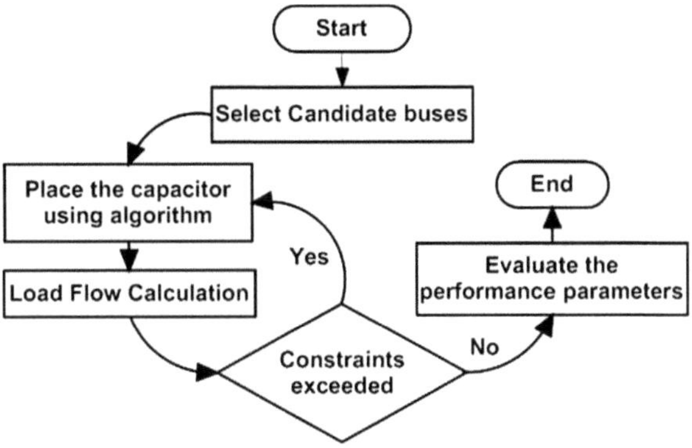

Fig. 11 Flowchart of solution methodology

If they are disobeyed then the whole process is repeated again, otherwise, next step is processed. After the constraints have been followed the system parameters based on which the performance of the system is evaluated, viz., Voltage profile and losses are evaluated. These results are then recorded and compared with the results of the other algorithms implemented. The generalized approach for placing the capacitor is shown in Fig. 11.

6 Results and Discussions

Based on the discussions made above, for the solution methodologies for OCP, the capacitor placement has been performed here for the IEEE 69-bus radial distribution system, as shown in Fig. 12. To show the applicability of soft computing techniques for OCP and the effect of changes in planning method, the study is performed for two different cases,

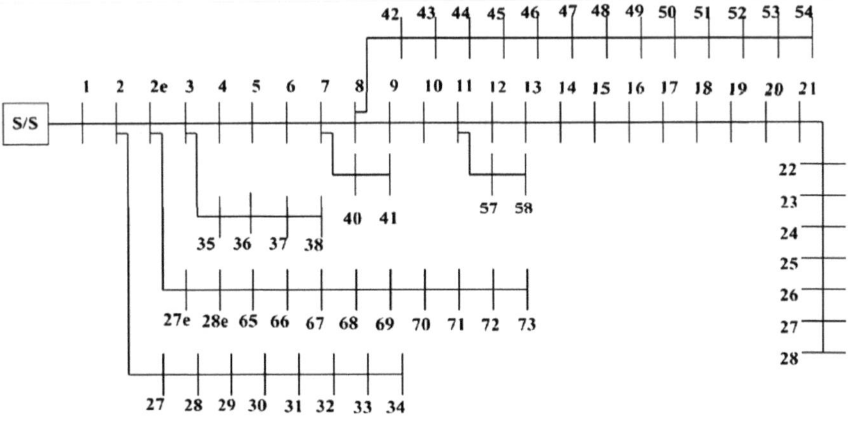

Fig. 12 Single line diagram of IEEE 69-bus radial distribution system

Table 2 Load duration data

	Low	Medium	Peak
Load level	0.5	1	1.4
Time interval (hrs)	1,000	6,760	1,000

Table 3 Parameters and their values

Parameter	Value
Cost of energy (US$), k_e	0.06 US$/kWh
Cost of peak power loss (US$), k_p	168 US$/kW per year
Inflation rate (%), m	5 % per year
Creditor's interest rate (%), f	5 % per year
Maintenance cost in US$, M	2 % of the total cost
Capacitor installation cost (US$), k_{inst}	US$1,000
Capacitor cost (US$), k_c (in multiple of 150 kvar)	3 US$/kvar

1. Case I: Comparison of GA and PSO with P-1 (Static Cost Based Planning Method), and
2. Case II: Comparison of PSO and NM-PSO with P-2 (Variable Cost Based Planning Method).

In Case-I, the best method obtained is used for comparison with the P-2 and NM-PSO combination (Kumar and Singh 2011). For these cases, the load duration data is shown in Table 2. For the simulation, performed on the MATLAB® platform, the different system parameters considered are shown in Table 3.

The bus voltage constraints are taken as $V_{min} = 0.90$ pu, and $V_{max} = 1.10$ pu.

6.1 Candidate Bus Selection

For the selection of the candidate buses the bus sensitivity (BS_i) method has been found superior than the loss sensitivity based method (Kumar et al. 2011). The candidate buses are selected using the numerical value of BS_i in their descending order, for IEEE 69-bus test system, as depicted in Fig. 13. Amongst 69 buses, 9 buses, viz., 11, 12, 21, 48, 49, 50, 59, 61, and 64 are selected for the capacitor placement.

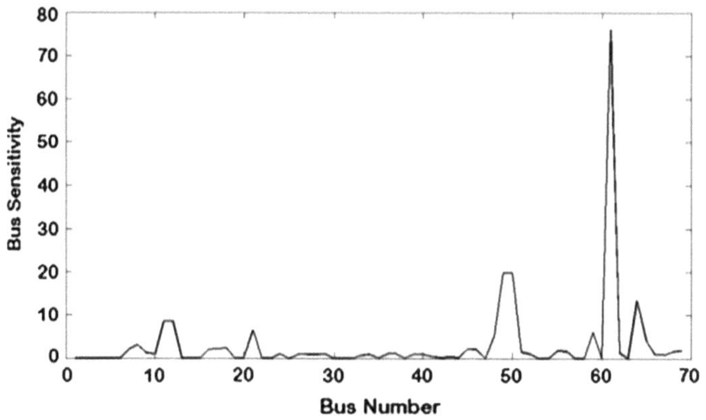

Fig. 13 Bus sensitivity of IEEE 69-bus radial distribution system

6.2 Optimal Capacitor Placement Using Soft-Computing Techniques

6.2.1 Case-I

Here the capacitor placement has been performed using PSO and GA for the IEEE 69-bus radial distribution system. The candidate buses for the placement are selected using the bus sensitivity method as discussed in Sect. 6.1. For PSO algorithm the parameters c_1, c_2, and $V_{(t)}$ are set to 1.5, 2, and difference between maximum and minimum capacitor values divided by number of iteration, respectively. For GA, the population size is selected to be 100, crossover = 0.5, and mutation = 0.01. The experiments are performed up to 100 iterations, for both the algorithms.

Using the algorithms and planning method P-1 (Static Cost Based Planning Method), the total kvar placed are shown in Table 4. It shows, that for the same candidate buses, the capacitors placed are higher, i.e., 9,200 kvar, than that obtained using PSO, 8,400 kvar.

For these kvar, the performance of the network is depicted in Table 5. It depicts that the higher losses (879.07 kvar) are obtained with GA, whereas reduced losses of 830.02 kvar are obtained with PSO. This higher amount of kvar placed in the

Table 4 Capacitors (in kvar) placed using GA and PSO with P1 for IEEE-69 bus radial distribution system

	Bus number									Total (kvar)
Algorithm	11	12	21	48	49	50	59	61	64	
GA	900	600	600	900	600	1,800	600	2,400	600	9,200
PSO	900	600	600	900	300	1,500	600	2,400	600	8,400

Table 5 Performance of IEEE-69 bus radial distribution with capacitors (in kvar) placed using GA and PSO

Algorithm	P_l (kW)	Power loss cost ($/year)	Total cost ($/year)	Benefits ($/year)
GA	879.07	147,683.76	542,800.46	10,243.76
PSO	830.01	139,441.70	532,556.7	7,874.83

system, leads to higher power and energy losses. Thus, total cost obtained with GA is US$542,800.46, while it is US$532,556.7 with PSO. Thus, higher economic benefits are obtained with PSO, in comparison to GA.

Thus, the results show better performance of the network with PSO, than that with GA. Both of these algorithms are stochastic in nature, where the performance of the algorithm may vary with the conditions. The observation of these algorithms can be summarized by running the program many times. In power system problem, evaluating the final design solution is dominating part of the overall computational burden. As a result, to find the correct optimal solution high computational efficiency, less number of design candidate are evaluated. The overall computational efficiency depends upon the population size, and number of iterations or generations. The two components of PSO velocity update (20), provide the global and local search capabilities to the algorithm, which enhances the searching capabilities of the PSO over GA. Further, in OCP discrete nature of the variable (i.e., capacitor sizes) allows the movement in discrete steps in the search space. Thus, the GA fails to provide to better solution, which may be due to entrapment of the GA solutions in the local minima, whereas PSO is able to provide better result.

6.2.2 Case-II

Since, the discussions made above show that better results are obtained with PSO. Here, a comparison is made between the PSO and NM-PSO algorithm, using another planning method P-2, as the objective function. For NM-PSO algorithm, the parameters are selected experimentally, as $\alpha = 1$, $\beta = 2$, $\gamma = 0.5$, $\delta = 0.5$, and N is the number of capacitors to be placed. The experiments are performed up to 100 iterations, for both the algorithms. The capacitors are placed at candidate buses identified using BS_i. The sizing of capacitors at these buses is performed using PSO and NM-PSO separately using the P-2 (Kumar and Singh 2014).

Table 6 Capacitors placed (kvar) at the candidate IEEE 69 bus radial distribution bus system

Algorithm	Bus number									Total (kvar)
	11	12	21	48	49	50	59	61	64	
1	300	600	900	300	300	1,500	600	2,400	900	7,800
2	300	600	600	300	1,500	300	600	2,400	1,200	7,800

Table 7 Results for capacitor placement IEEE 69 bus radial distribution bus system

	Original configuration	PSO	NM-PSO
Peak power loss (kW)	1,149	817.14	812.07
Total power loss (kW)	1,847.06	1,359.98	1,384.04
% Peak power loss reduction	–	28.89	29.32
% Loss reduction	–	25.05	25.07
Cost of peak power loss (US$)	193,032	137,279.5	136,416
Cost of energy loss (US$)	507,918	364,002.4	362,858.4
Benefits in energy loss cost (US$)	–	143,915.6	145,059.6
Capacitor cost (US$)	–	23,400	23,400
Total cost (US$)	700,950	524,681.9	522,674.4
Benefits (US$)	–	176,268.10	178,275.60
% Benefits	–	25.15	25.43

The results for the IEEE-69 bus radial distribution system are summarized in Tables 6 and 7. Table 6 shows that, though, the capacitors (7,800 kvar) placed in the bus system using both the algorithms are same, their internal distribution is different. Performance of the network shown in Table 7, depicts that with solutions obtained using NM-PSO, a reduction of 29.32 % in power loss, and US$145,059.60 in cost of energy loss, respectively, is obtained in comparison to original (i.e. without capacitor placement) system. This subsequently reduces the annual cost of the system by 25.43 % (US$178,275.60) with NM-PSO, in comparison to 25.15 % (US$176,268.10) with PSO. The corresponding improvement in the voltage profile of the bus system is shown in Fig. 14.

Thus, it can be seen that, in spite of the fact that both the algorithms give the same size of the capacitors, NM-PSO provides better results in terms of power loss, energy loss, and total benefits. This is due to better local searching capability of NM-PSO, outreaching to the global optimum solution to the problem. Based on the

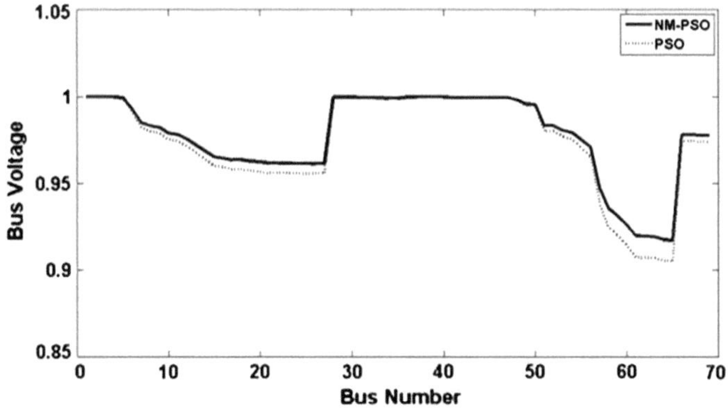

Fig. 14 Bus voltage comparison between the two algorithms

results, it can be concluded that with NM-PSO, better solutions are obtained for the OCP resulting into improved system performance as well as the economic benefits. Also, it depicts capability of the NM-PSO algorithm for handling more complex, nonlinear problems such as CPP, with superior performance.

7 Conclusion

With advancement in the computation power of the computing machines, several new algorithms have evolved for efficient and quick calculations. Soft-computing techniques are algorithms, which allow the user to solve the real-life non-linear problems. These algorithms, inspired from natural experiences, have been found to perform well in comparison to other analytical methods, for solving the problems. Capacitor placement is one such problem which has been a topic of research for a very long time. Being a mixed-integer type problem, i.e., involving both continuous and discrete variables the previously developed analytical and numerical algorithms were not able to provide an optimum solution to the problem. But, the soft computing techniques such as GA, PSO, NM-PSO, etc. are able to provide an optimum solution to the OCP problem.

Here, a study performed on IEEE-69 bus radial distribution system using the Static Cost Based Planning Method, shows that PSO provides better solution than GA for OCP. For the same operating conditions, with solutions using PSO, results obtained using P-2 are found better than that obtained using P-1. Thus, better formulation of objective function leads to superior results. Also, with P-2, both, PSO and NM-PSO algorithms give the same size of the capacitors, in spite of the fact that the NM-PSO provides improved results. This is due to better local searching capability of NM-PSO, outreaching to the global optimum solution to the problem. Thus, for better results better modeling and choice of algorithms is important. To achieve this, it is important to include factors such as, space availability, existing infrastructure, and dynamic network models, etc.

References

Aman, M. M., Jasmon, G. B., Bakar, A. H. A., Mokhlis, H., & Karimi, M. (2014). Optimum shunt capacitor placement in distribution system—A review and comparative study. *Renewable and Sustainable Energy Reviews, 30*, 429–439.

Aziz, A. S. A., Azar, A. T., Salama, M. A., Hassanien, A. E., & Hanafy, S. E. O. (2013). Genetic algorithm with different feature selection techniques for anomaly detectors generation. In *2013 Federated Conference on Computer Science and Information Systems (FedCSIS)* (pp. 769–774), September 8–11 2013, Krakow.

Baghzouz, Y. (1991). Effects of non-linear loads on optimal capacitor placement in radial feeders. *IEEE Transactions on Power Delivery, 6*(1), 245–251.

Baghzouz, Y., & Ertem, S. (1990). Shunt capacitor sizing for radial distribution feeders with distorted substation voltages. *IEEE Transactions on Power Delivery, 5*(2), 650–655.

Boeringer, D., & Werner, D. (2004). Particle swarm optimization versus genetic algorithms for phased array synthesis. *IEEE Transactions Antennas Propagation, 52*(3), 771–779.
Cheng, C. S., & Shirmohammadi, D. (1995). A three-phase power flow method for real-time distribution system analysis. *IEEE Transaction Power System, 10*(2), 671–679.
de Valle, Y., Venayagamoorthy, G. K., Mohagheghi, S., Hernandez, J. C., & Harley, R. G. (2008). Particle swarm optimization: basic concepts, variants, and application in power system. *IEEE Transactions on Evolutionary Computations, 12*(2), 171–195.
Fogel, D. B., Baeck, T., & Michalewicz, Z. (2000). *Evolutionary computation 1: Basic algorithms and operators*. Boca Raton: CRC Press.
Gonen, T. (1986). *Electric power distribution system engineering*. Noida: Tata McGraw-Hill.
Haghifam, M. R., & Malik, O. P. (2007). Genetic algorithm-based approach for fixed and switchable capacitors placement in distribution systems with uncertainty and time varying loads. *IET Generation, Transmission and Distribution, 1*(2), 244–252.
Kresting, W. H. (2002). *Distribution system modeling and analysis*. Boca Raton: CRC Press.
Kumar, P., & Singh, A. K. (2011). Nelder-Mead PSO based approach to optimal capacitor placement in radial distribution system. In *Swarm, evolutionary, and memetic computing*. Lecture Notes in Computer Science (Vol. 7076, pp. 143–150). Heidelberg: Springer.
Kumar, P., Singh, A. K., & Singh, N. (2011). Sensitivity based capacitor placement: A comparative study. In *6th IEEE International Conference on Industrial and Information Systems (ICIIS)* (pp. 381–385), August 16–19 2011, Kandy. doi: 10.1109/ICIINFS.2011.6038098.
Kumar, P., Singh, A. K., & Srivastava, A. (2014). A novel optimal capacitor placement algorithm using Nelder-Mead PSO. *International Journal of Bio inspired Computing, 6*(4), 290–302.
Lee, S. H., & Grainger, J. J. (1981). Optimum placement of fixed and switched capacitors on primary distribution feeders. *IEEE Transactions on Power Apparatus and Systems, 100*(1), 345–352.
Levitin, G., Kalyuzhny, A., Shenkman, A., & Chertkov, M. (2000). Optimal capacitor allocation in distribution systems using a genetic algorithm and a fast energy loss computation technique. *IEEE Transactions on Power Delivery, 15*(2), 623–628.
Masoum, M. A. S., Ladjevardi, M., Jafarian, A., & Fuchs, E. F. (2004). Optimal placement, replacement and sizing of capacitor banks in distorted distribution networks by genetic algorithms. *IEEE Transactions on Power Delivery, 19*(4), 1794–1801.
Nelder, J. A., & Mead, R. (1965). A simplex method for function minimization. *Computer Journal, 7*, 308–313.
Ng, H. N., Salama, M. M. A., & Chikhani, A. Y. (2000). Classification of capacitor allocation techniques. *IEEE Transactions on Power Delivery, 15*(1), 387–392.
Pabla, A. S. (2004). *Electric power distribution*. Noida: Tata McGraw-Hill.
Peachon, J., Piercy, D. S., William, F. T., & Odd, J. T. (1968). Sensitivity in power systems. *IEEE Transactions on Power Apparatus and System, 87*(8), 1687–1697.
Prakash, K., & Sydulu, M. (2007). Particle swarm optimization based capacitor placement on radial distribution systems. In *Proceedings 2007 Power Engineering Society and General Meeting* (pp. 1–5), June 24–28 2007, Tampa, Florida. doi: 10.1109/PES.2007.386149.
Silva, I, Jr, Carneiro, S, Jr, Oliveria, J. E., Costa, J. S., Pereira, J. L. R., & Garcia, P. A. N. (2008). A Heuristic constructive algorithm for capacitor placement on distribution systems. *IEEE Transactions on Power Systems, 23*(4), 1619–1626.
Zahara, E., & Kao, Y. (2009). Hybrid Nelder-Mead simplex search and particle swarm optimization for constrained engineering design problems. *Expert Systems with Applications, 36*(2), 3880–3886.

Advanced Metaheuristics-Based Approach for Fuzzy Control Systems Tuning

Soufiene Bouallègue, Fatma Toumi, Joseph Haggège and Patrick Siarry

Abstract In this study, a new advanced metaheuristics-based optimization approach is proposed and successfully applied to design and tuning of a PID-type Fuzzy Logic Controller (FLC). The scaling factors tuning problem of the FLC structure is formulated and systematically resolved, using various constrained metaheuristics such as the Differential Search Algorithm (DSA), Gravitational Search Algorithm (GSA), Artificial Bee Colony (ABC) and Particle Swarm Optimization (PSO). In order to specify more time-domain performance control objectives of the proposed metaheuristics-tuned PID-type FLC, different optimization criteria such as Integral of Square Error (ISE) and Maximum Overshoot (MO) are considered and compared The classical Genetic Algorithm Optimization (GAO) method is also used as a reference tool to measure the statistical performances of the proposed methods. All these algorithms are implemented and analyzed in order to show the superiority and the effectiveness of the proposed fuzzy control tuning approach. Simulation and real-time experimental results, for an electrical DC drive benchmark, show the advantages of the proposed metaheuristics-tuned PID-type fuzzy control structure in terms of performance and robustness.

S. Bouallègue · F. Toumi (✉) · J. Haggège
Research Laboratory in Automatic Control LA.R.A, National Engineering School of Tunis (ENIT), BP 37, Le Belvdre, 1002 Tunis, Tunisia
e-mail: fatima.toumi@enit.rnu.tn

S. Bouallègue
e-mail: soufiene.bouallegue@issig.rnu.tn

J. Haggège
e-mail: joseph.haggege@enit.rnu.tn

F. Toumi · P. Siarry
Signals, Images and Intelligent Systems Laboratory, LiSSi-EA-3956, University Paris-Est Créteil Val de Marne, 61 Avenue du Général de Gaulle, 94010 Créteil, France
e-mail: siarry@univ-paris12.fr

Q. Zhu and A.T. Azar (eds.), *Complex System Modelling and Control Through Intelligent Soft Computations*, Studies in Fuzziness and Soft Computing 319,
DOI 10.1007/978-3-319-12883-2_22

1 Introduction

Fuzzy logic control approach has been widely used in many successful industrial applications. This control strategy, with the Mamdani fuzzy type inference, demonstrated high robustness and effectiveness properties (Azar 2010a, b, 2012; Lee 1998a, b; Passino and Yurkovich 1998). The known PID-type FLC structure, firstly proposed in Qiao and Mizumoto (1996), is especially established and improved within the practical framework (Eker and Torun 2006; Guzelkaya et al. 2003; Woo et al. 2000). This particular fuzzy controller retains the characteristics similar to the conventional PID controller and can be decomposed into the equivalent proportional, integral and derivative control components (Eker and Torun 2006; Qiao and Mizumoto 1996). In this design case, the dynamic behaviour depends on the adequate choice of the fuzzy controller scaling factors. The tuning procedure depends on the control experience and knowledge of the human operator, and it is generally achieved based on a classical trials-errors procedure. There is not up to now a systematic method to guide such a choice. This tuning problem becomes more delicate and hard as the complexity of the control plant increases.

In order to improve further the performance of the transient and steady state responses of the PID-type fuzzy structure, various strategies and methods are proposed to tune their parameters. In Qiao and Mizumoto (1996) , proposed a peak observer mechanism-based method to adjust the PID-type FLC parameters. This self-tuning mechanism decreases the equivalent integral control component of the fuzzy controller gradually with the system response process time. On the other hand, Woo et al. (2000) developed a method based on two empirical functions evolved with the system's error information. In Guzelkaya et al. (2003), the authors proposed a technique that adjusts the scaling factors, corresponding to the derivative and integral components, using a fuzzy inference mechanism. However, the major drawback of all these PID-type FLC tuning methods is the difficult choice of their scaling factors and self-tuning mechanisms. The time-domain dynamics of the fuzzy controller depends strongly on this hard choice. The tuning procedure depends on the control experience and knowledge of the human operator, and it is generally achieved based on a classical trials-errors procedure. Hence, having a systematic approach to tune these scaling factors is interesting and the optimization theory may present a promising solution.

In solving this kind of optimization problems, the classical exact optimization algorithms, such as gradient and descent methods, do not provide a suitable solution and are not practical. The relative objective functions are non linear, non analytical and non convex (Bouallègue et al. 2012a, b). Over the last decades, there has been a growing interest in advanced metaheuristic algorithms inspired by the behaviours of natural phenomena (Boussaid et al. 2013; Dréo et al. 2006; Rao and Savsani 2012; Siarry and Michalewicz 2008). It is shown by many researchers that these algorithms are well suited to solve complex computational problems in wide and various ranges of engineering applications summarized around domains of robotics, image and signal processing, electronic circuits design, communication networks, but more especially the domain of process control design (Bouallègue et al. 2011,2012a, b; David et al. 2013; Goswami and Chakraborty 2014; Madiouni et al. 2013; Toumi et al. 2014).

Various metaheuristics have been adopted by researchers. The Differential Search Algorithm (DSA) (Civicioglu 2012), Gravitational Search Algorithm (GSA) (Rashedi et al. 2009), Artificial Bee Colony (ABC) (Karaboga 2005) and Particle Swarm Optimization (PSO) (Eberhart and Kennedy 1995; Kennedy and Eberhart 1995) algorithms are the most recent proposed techniques in the literature. They will be adapted and improved for the considered fuzzy control design problem. Without any regularity on the cost function to be optimized, the recourse to these stochastic and global optimization techniques is justified by the empirical evidence of their superiority in solving a variety of non-linear, non-convex and non-smooth problems. In comparison with the conventional optimization algorithms, these optimization techniques are a simple concept, easy to implement, and computationally efficient algorithms. Their stochastic behaviour allows overcoming the local minima problem.

In this study, a new approach based on the use of advanced metaheuristics, such as DSA, GSA, ABC and PSO is proposed for systematically tuning the scaling factors of the particular PID-type FLC structure. The well known classical GAO algorithm is used in order to compare the obtained optimization results (Goldberg 1989; MathWorks 2009). This work can be considered as a contribution on the results given in Bouallègue et al. (2012a, b), Toumi et al. (2014). The synthesis and tuning of the fuzzy controller are formulated as a constrained optimization problem which is efficiently solved based on the proposed metaheuristics. In order to specify more robustness and performance control objectives of the proposed metaheuristics-tuned PID-type FLC, different optimization criteria such as ISE and MO are considered and compared.

The remainder of this chapter is organized as follows. In Sect. 2, the studied PID-type FLC structure is presented and formulated as a constrained optimization problem. An external static penalty technique is investigated to handle the problem constraints. The advanced DSA, GSA, ABC and PSO metaheuristic algorithms, used in solving the formulated problem, are described in Sect. 3. Section 4 is dedicated to apply the proposed fuzzy control approach on an electrical DC drive benchmark. All obtained simulation results are compared with each other and analysed. Experimental setup and results are presented within a real-time framework.

2 PID-Type FLC Tuning Problem Formulation

In this section, the PID-type fuzzy controller synthesis problem is formulated as a constrained optimization problem which will be resolved through the suggested metaheuristics algorithms.

2.1 A Review of PID-Type Fuzzy Control Structure

The particular PID-type fuzzy controller structure, originally proposed by Qiao and Mizumoto within the continuous-time formalism (Qiao and Mizumoto 1996), retains the characteristics similar to the conventional PID controller. This result remains valid while using a particular structure of FLC with triangular uniformly distributed membership functions for the fuzzy inputs and a crisp output, the product-sum inference and the center of gravity defuzzification methods (Bouallègue et al. 2012a, b; Eker and Torun 2006; Guzelkaya et al. 2003; Haggège et al. 2010; Toumi et al. 2014; Woo et al. 2000).

Under these conditions, the equivalent proportional, integral and derivative control components of such a PID-type FLC are given by $\alpha K_e \Pi + \beta K_d \Delta$, $\beta K_e \Pi$ and $\alpha K_d \Delta$, respectively, as shown in Qiao and Mizumoto (1996). In these expressions, Π and Δ represent relative coefficients, K_e, K_d, α and β denote the scaling factors associated to the inputs and output of the fuzzy controller. When approximating the integral and derivative terms within the discrete-time framework (Bouallègue et al. 2012a, b; Haggège et al. 2010; Toumi et al. 2014), we can consider the closed-loop control structure for a digital PID-type FLC, as shown in Fig. 1. The dynamic behaviour of this PID-type FLC structure is strongly depending on the scaling factors, difficult and delicate to tune.

As shown in Fig. 1, this particular structure of Mamdani fuzzy controller uses two inputs: the error e_k and the variation of error Δe_k, to provide the output u_k that describes the discrete fuzzy control law.

2.2 Optimization-Based Problem Formulation

The choice of the adequate values for the scaling factors of the described PID-type FLC structure is often done by a trials-errors hard procedure. This tuning problem becomes difficult and delicate without a systematic design method. To deal with these difficulties, the optimization of these control parameters is proposed like a

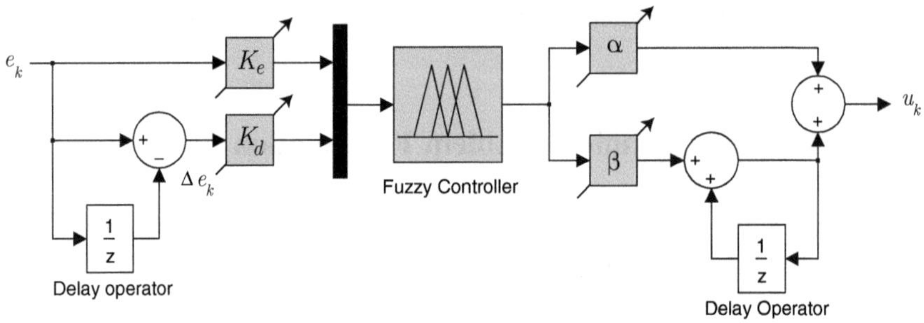

Fig. 1 Proposed discrete-time PID-type FLC structure

promising procedure. This tuning can be formulated as the following constrained optimization problem:

$$\begin{cases} \underset{x=(K_e,K_d,\alpha,\beta)^T\in\mathcal{S}\subset\mathbb{R}_+^4}{minimize} \quad f(x) \\ subject\ \ to: \\ g_1(x) = \delta - \delta^{\max} \leq 0 \\ g_2(x) = t_s - t_s^{\max} \leq 0 \\ g_3(x) = E_{ss} - E_{ss}^{\max} \leq 0 \end{cases} \tag{1}$$

where f: $\mathbb{R}^4 \rightarrow \mathbb{R}$ the cost function, $\mathcal{S} = \left\{x \in \mathbb{R}_+^4, x_{low} \leq x \leq x_{up}\right\}$ the initial search space, which is supposed containing the desired design parameters, and $g_l : \mathbb{R}^4 \rightarrow \mathbb{R}$ the nonlinear problem's constraints.

The optimization-based tuning problem (1) consists in finding the optimal decision variables, representing the scaling factors of a given PID-type FLC structure, which minimizes the defined cost function, chosen as the Maximum Overshoot (MO) and the Integral of Square Error (ISE) performance criteria. These cost functions are minimized, using the proposed particular constrained metaheuristics, under various time-domain control constraints such as overshoot δ, steady state error E_{ss}, rise time t_r and settling time t_s of the system's step response, as shown in Eq. (1). Their specified maximum values constrain the step response of the tuned PID-type fuzzy controlled system, and can define some time-domain templates.

2.3 Proposed Constraints Handling Method

The considered metaheuristics in this study are originally formulated as an unconstrained optimizer. Several techniques have been proposed to deal with constraints. One useful approach is by augmenting the cost function of problem (1) with penalties proportional to the degree of constraint infeasibility. This approach leads to convert the constrained optimization problem into the unconstrained optimization problem. In this paper, the following external static penalty technique is used:

$$\varphi(x) = f(x) + \sum_{l=1}^{n_{con}} \lambda_l \max\left[0, g_l(x)^2\right] \tag{2}$$

where λ_l is a prescribed scaling penalty parameter, and n_{con} is the number of problem constraints $g_l(x)$.

3 Solving Optimization Problem Using Advanced Algorithms

In this section, the basic concepts as well as the algorithm steps of each proposed advanced metaheuristic are described for solving the formulated PID-type FLC tuning problem.

3.1 Differential Search Algorithm

The Differential Search Algorithm (DSA) is a recent population-based metaheuristic optimization algorithm invented in 2012 by Civicioglu (2012). This global and stochastic algorithm simulates the Brownian-like random-walk movement used by an organism to migrate (Civicioglu 2012; Goswami and Chakraborty 2014; Waghole and Tiwari 2014).

Migration behavior allows the living beings to move from a habitat where capacity and diversity of natural sources are limited to a more efficient habitat. In the migration movement, the migrating species of living beings constitute a superorganism containing large number of individuals. Then it starts to change its position by moving toward more fruitful areas using a Bownian-like random-walk model. The population made up of random solutions of the respective problem corresponds to an artificial-superorganism migrating. The artificial superorganism migrates to global minimum value of the problem. During this migration, the artificial-superorganism tests whether some randomly selected positions are suitable temporarily. If such a position tested is suitable to stop over for a temporary time during the migration, the members of the artificial-superorganism that made such discovery immediately settle at the discovered position and continue their migration from this position (Civicioglu 2012).

In DSA metaheuristic, a superorganism X_k^i of N artificial-organisms making up, at every generation $k = 1, 2, \ldots, k_{\max}$, an artificial-organism with members as much as the size of the problem, defined as follows:

$$X_k^i = \left(x_k^{i,1}, x_k^{i,2}, \ldots, x_k^{i,d}, \ldots, x_k^{i,D}\right) \tag{3}$$

A member of an artificial-organism, in initial position, is randomly defined by using Eq. (4):

$$x_0^{i,d} = x_{low}^{i,d} + rand(0,1)\left(x_{up}^{i,d} - x_{low}^{i,d}\right) \tag{4}$$

In DSA, the mechanism of finding a so called Stopover Site, which presents the solution of optimization problem, at the areas remaining between the artificial-organisms, is described by a Brownian-like random walk model. The principle is based on the move of randomly selected individuals toward the targets of a Donor artificial-organism, denoted as (Civicioglu 2012):

$$X_k^{Donor} = X_k^{random\_shuffling(i)} \tag{5}$$

where the index i of artificial-organisms is produced by the Shuffling-random function.

The size of the change occurred in the positions of the members of the artificial-organisms is controlled by the Scale factor given as follows:

$$s_k^i = randG\{2rand_1\}(rand_2 - rand_3) \tag{6}$$

where $rand_1$, $rand_2$ and $rand_3$ are uniformly distributed random numbers in the interval [0, 1], $randG$ is a Gamma-random number.

The Stopover Site positions, which are very important for a successful migration, are produced by using Eq. (7):

$$Y_k^i = X_k^i + s_k^i(X_k^{Donor} - X_k^i) \tag{7}$$

So, the individuals of the artificial-organisms of the superorganism to participate in the search process of Stopover Site are determined by a random process based on the manipulation of two control parameters p_1 and p_2. The algorithm is not much sensitive to these control parameters and the values in the interval [0, 0.3] usually provide best solutions for a given problem (Civicioglu 2012).

Finally, the steps of the original version of DSA, as described by the pseudo code in Civicioglu (2012), can be summarized as follows:

1. Search space characterization: size of superorganism, dimension of problem, random numbers p_1 and p_2, …
2. Randomized generation of the initial population.
3. Fitness evaluation of artificial-organisms.
4. Calculation of the Stopover Site positions in different directions.
5. Randomly select individuals to participate in the search process of Stopover Site.
6. Update the Stopover Site positions and evaluate the new population.
7. Update the superorganism by the new Stopover site positions.
8. Repeat steps 3–7 until the stop criteria are reached.

3.2 *Gravitational Search Algorithm*

The Gravitational Search Algorithm (GSA) is population-based metaheuristic optimization algorithm introduced in 2009 by Rashedi et al. (2009). This algorithm is based on the law of gravity and mass interactions as described in Nobahari et al. (2011),

Precup et al. (2011), Rashedi et al. (2009). The search agents are a set of masses which interact with each other based on the Newtonian gravity and the law of motion.

Several applications of this algorithm in various areas of engineering are investigated (Nobahari et al. 2011; Precup et al. 2011; Rao and Savsani 2012). In GSA, the particles, called also agents, are considered as bodies and their performance is measured by their masses. All these bodies attract each other by the gravity force that causes a global movement of all objects towards the objects with heavier masses. These agents correspond to the optimum solutions in the search space (Rashedi et al. 2009). Indeed, each agent presents a solution of optimization problem and is characterised by its position, inertial mass, active and passive gravitational masses. The GSA is navigated by properly adjusting the gravitational and inertia masses leading masses to be attracted by the heaviest object.

The position of the mass corresponds to a solution of the problem, and its gravitational and inertial masses are determined using a cost function. The exploitation capability of this algorithm is guaranteed by the movement of the heavy masses, more slowly than the lighter ones.

Let us consider a population with N agents. The position of the ith agent at iteration time k is defined as:

$$X_k^i = \left(x_k^{i,1}, x_k^{i,2}, \ldots, x_k^{i,d}, \ldots, x_k^{i,D}\right) \tag{8}$$

where $x_k^{i,d}$ presents the position of the ith particle in the dth dimension of search space of size D.

At a specific time "t", denoted by the actual iteration "k", the force acting on mass "i" from mass "j" is given as follows:

$$F_k^{ij,d} = G_k \frac{M_k^{pi} \times M_k^{aj}}{R_k^{ij} + \varepsilon}\left(x_k^{j,d} - x_k^{i,d}\right) \tag{9}$$

where M_k^{aj} is the active gravitational mass related to agent j, M_k^{pi} is the passive gravitational mass related to agent i, G_k is the gravitational constant at time k, ε is a small constant, and R_k^{ij} is the Euclidian distance between two agents i and j, defined as:

$$R_k^{ij} = \left\|X_k^i, X_k^j\right\|_2 \tag{10}$$

To give a stochastic characteristic to this algorithm, authors of GSA suppose that the total force that acts on agent i is a randomly weighted sum of jth components of the forces exerted from other bodies, given as follows (Rashedi et al. 2009):

$$F_k^{i,d} = \sum_{j=1, j \neq i}^{N} rand^j F_k^{ij,d} \tag{11}$$

where $rand^j$ is a random number in the interval [0, 1].

By the law of motion, the acceleration of the agent i at time k, and in dth direction, is given as follows:

$$a_k^{i,d} = \frac{F_k^{i,d}}{M_k^{ii}} \tag{12}$$

where M_k^{ii} is the inertial mass of ith agent in the search space with dimension d.

Hence, the position and the velocity of an agent are updated respectively by the mean of equations of movement given as follows:

$$x_{k+1}^{i,d} = x_k^{i,d} + v_{k+1}^{i,d} \tag{13}$$

$$v_{k+1}^{i,d} = rand^i v_k^{i,d} + a_k^{i,d} \tag{14}$$

where $rand^i$ is a uniform random number in the interval [0, 1], used to give a randomized characteristic to the search.

To control the search accuracy, the gravitational constant G_k, is initialized at the beginning and will be reduced with time. In this study, we use an exponentially decreasing of this algorithm parameter, as follows:

$$G_k = G_0 e^{-\eta \frac{k}{k_{\max}}} \tag{15}$$

where G_0 is the initial value of G_k, η is a control parameter to set, and $k_{\max}$ is the total number of iterations.

In GSA, gravitational and inertia masses are calculated by the fitness evaluation. A heavier mass means a more efficient agent. Better agents have higher attractions and walk more slowly.

As given in Rashedi et al. (2009), the values of masses are calculated using the fitness function and gravitational and inertial masses are updated by the following equations:

$$M_k^{ai} = M_k^{pi} = M_k^{ii} = M_k^i \tag{16}$$

$$M_k^i = \frac{m_k^i}{\sum_{j=1}^{N} m_k^j} \tag{17}$$

$$m_k^i = \frac{fit_k^i - worst_k}{best_k - worst_k} \tag{18}$$

where fit_k^i represents the fitness value of the agent i at iteration k, and, $worst_k$ and $best_k$ are defined, for a minimization problem, as follows:

$$best_k = \min_{1 \le j \le N} fit_k^j \quad (19)$$

$$worst_k = \max_{1 \le j \le N} fit_k^j \quad (20)$$

To perform a good compromise between exploration and exploitation, authors of GSA choose to reduce the number of agents with lapse of iterations in Eq. (11), which will be modified as:

$$F_k^{i,d} = \sum_{j \in Kbest, j \neq i} rand^j F_k^{ij,d} \quad (21)$$

where *Kbest* is the set of the first *K* agents with best fitness and biggest mass that will attract the others.

The algorithm parameter *Kbest* is a function of iterations with the initial value K_0, usually set to the total size of population *N* at the beginning, and linearly decreasing with time. At the end of search, there will be just one agent applying force to the others.

Finally, the steps of the original version of GSA, as described in Rashedi et al. (2009), can be summarized as follows:

1. Search space characterization: number of agents, dimension of problem, control parameters G_0, K_0, …
2. Randomized generation of the initial population.
3. Fitness evaluation of agents.
4. Update the algorithm parameters G_k, $best_k$, $worst_k$ and M_k^i for each agent and at each iteration.
5. Calculation of the total force in different directions.
6. Calculation of acceleration and velocity.
7. Updating agents' position.
8. Repeat steps 3–7 until the stop criteria are reached.

3.3 Artificial Bee Colony

The Artificial Bee Colony (ABC) is a population-based metaheuristic optimization algorithm introduced in 2005 by Karaboga (2005). The principle of such an algorithm is based on the intelligent foraging behavior of honey bee swarm (Basturk and Karaboga 2006; Karaboga 2005; Karaboga and Akay 2009; Karaboga and Basturk 2007, 2008). The ABC algorithm has been enormously successful in various industrial domains and a wide range of engineering applications as summarized in Karaboga et al. (2012).

In this formalism, the population of the artificial bees' colony is constituted of three groups: employed bees, onlookers and scouts. Employed bees search the destination where food is available, translated by the amount of their nectar. They collect the food and return back to their origin, where they perform waggle dance depending on the amount of nectar's food available at the destination. The onlooker bee watches the dance and follows the employed bee depending on the probability of the available food.

In ABC algorithm, the population of bees is divided into two parts consisting of employed bees and onlooker bees. The sizes of each part are usually taken equal to. Employed bee, representing a potential solution in the search space with dimension, updates its new position by using the movement Eq. (22) and follows greedy selection to find the best solution. The objective function associated with the solution is measured by the amount of food.

Let us consider a population with $N/2$ individuals in the search space. The position of the ith employer at iteration time k is defined as:

$$X_k^i = \left(x_k^{i,1}, x_k^{i,2}, \ldots, x_k^{i,d}, \ldots, x_k^{i,D}\right) \tag{22}$$

where D is the number of decision variables, i is the index on $N/2$ employers.

In the dth dimension of search space, the new position of the ith employer, as well as of the ith onlooker, is updated by means of the movement equation given as follows:

$$x_{k+1}^{i,d} = x_k^{i,d} + r_k^i\left(x_k^{i,d} - x_k^{m,d}\right) \tag{23}$$

where r_k^i is a uniformly distributed random number in the interval $[-1, 1]$. It can be also chosen as a normally distributed random number with mean equal to zero and variance equal to one as given in Karaboga (2005). The employer's index $m \neq i$ is a randomly number in the interval [1, N/2].

Besides, an onlooker bee chooses a food source depending on the probability value of each solution associated with that food source, calculated as follows:

$$p_k^{i,d} = \frac{f_k^{i,d}}{\sum_n^{N/2} f_k^{n,d}} \tag{24}$$

where $f_k^{i,d}$ is the fitness value of the ith solution at iteration k.

When the food source of an employed bee cannot be improved for some predetermined number of cycles, called "Limit for abandonment" and denoted by L, the source food becomes abandoned and the employer behaves as a scout bee and it searches for the new food source using the following equation:

$$x_k^{i,d} = x_{low}^{i,d} + rand(0,1)\left(x_{up}^{i,d} - x_{low}^{i,d}\right) \tag{25}$$

where $x_{low}^{i,d}$ and $x_{up}^{i,d}$ are the lower and upper ranges, respectively, for decision variables in the dimension.

This behaviour of the artificial bee colony reflects a powerful mechanism to escape the problem of trapping in local optima. The value of the "Limit for abandonment" control parameter of ABC algorithm is calculated as follows:

$$L = \frac{N}{2} \times D \tag{26}$$

Finally, the steps of the original version of ABC algorithm, as described in Basturk and Karaboga (2006), Karaboga (2005), Karaboga and Basturk (2007, 2008), can be summarized as follows:

1. Initialize the ABC algorithm parameters: population size N, limit of abandonment L, dimension of the search space D, …
2. Generate a random population equal to the specified number of employed bees, where each of them contains the value of all the design variables.
3. Obtain the values of the objective function, defined as the amount of nectar for the food source, for all the population members.
4. Update the position of employed bees using Eq. (23), obtain the value of objective function and select the best solutions to replace the existing ones.
5. Run the onlooker bee phase: onlookers proportionally choose the employed bees depending on the amount of nectar found by the employed bees, Eq. (24).
6. Update the value of onlooker bees using Eq. (23) and replace the existing solution with the best new one.
7. Identify the abundant solutions using the limit value. If such solutions exist then these are transformed into the scout bees and the solution is updated using Eq. (25).
8. Repeat the steps 3–7 until the termination criterion is reached, usually chosen as the specified number of generations.

3.4 Particle Swarm Optimization

The PSO technique is an evolutionary computation method developed in 1995 by Kennedy and Eberhart (1995), Eberhart and Kennedy (1995). This recent metaheuristic technique is inspired by the swarming or collaborative behaviour of biological populations. The cooperation and the exchange of information between population individuals allow solving various complex optimization problems. The convergence and parameters selection of the PSO algorithm are proved using several advanced theoretical analysis (Bouallègue et al. 2011, 2012a, b; Madiouni et al. 2013).

PSO has been enormously successful in several and various industrial domains and engineering fields (Bouallègue et al. 2012a; Dréo et al. 2006; Rao and Savsani 2012; Siarry and Michalewicz 2008).

The basic PSO algorithm uses a swarm consisting of N particles N_k^i, randomly distributed in the considered initial search space, to find an optimal solution $x^* = \arg\min f(x) \in \mathbb{R}^D$ of a generic optimization problem. Each particle, that represents a potential solution, is characterised by its position and its velocity $x_k^{i,d}$ and $v_k^{i,d}$, respectively.

At each iteration of the algorithm, and in the dth direction, the ith particle position evolves based on the following update rules:

$$x_{k+1}^{i,d} = x_k^{i,d} + v_{k+1}^{i,d} \tag{27}$$

$$v_{k+1}^{i,d} = w_{k+1} v_k^{i,d} + c_1 r_{1,k}^i \left(p_k^{i,d} - x_k^{i,d}\right) + c_2 r_{2,k}^i \left(p_k^{g,d} - x_k^{i,d}\right) \tag{28}$$

where w_{k+1} the inertia factor, c_1 and c_2 the cognitive and the social scaling factors respectively, $r_{1,k}^i$ and $r_{2,k}^i$ the random numbers uniformly distributed in the interval [0,1], $p_k^{i,d}$ the best previously obtained position of the ith particle and $p_k^{g,d}$ the best obtained position in the entire swarm at the current iteration k.

In order to improve the exploration and exploitation capacities of the proposed PSO algorithm, we choose for the inertia factor a linear evolution with respect to the algorithm iteration (Bouallègue et al. 2011, 2012a, b; Madiouni et al. 2013):

$$w_{k+1} = w_{\max} - \left(\frac{w_{\max} - w_{\min}}{k_{\max}}\right) k \tag{29}$$

where $w_{\max} = 0.9$ and $w_{\min} = 0.4$ represent the maximum and minimum inertia factor values, respectively.

Finally, the steps of the original version of PSO algorithm, as described in Eberhart and Kennedy (1995), Kennedy and Eberhart (1995), can be summarized as follows:

1. Define all PSO algorithm parameters such as swarm size N, maximum and minimum inertia factor values, cognitive and social coefficients, …
2. Initialize the particles with random positions and velocities. Evaluate the initial population and determine $p_0^{i,d}$ and $p_0^{g,d}$.
3. For each particle apply the update Eqs. (27)–(29).
4. Evaluate the corresponding fitness values and select the best solutions.
5. Repeat the steps 3–4 until the termination criterion is reached.

4 Case Study: PID-Type FLC Tuning for a DC Drive

This section is dedicated to apply the proposed metaheuristics-tuned PID-type FLC for the variable speed control of a DC drive. All the obtained simulations results are presented and discussed.

4.1 *Plant Model Description*

The considered benchmark is a 250 W electrical DC drive shown in Fig. 2. The machine's speed rotation is 3,000 rpm at 180 V DC armature voltage.

The motor is supplied by an AC-DC power converter. The considered electrical DC drive can be described by the following model (Haggège et al. 2009):

$$G(s) = \frac{\Lambda}{(1 + \tau_e s)(1 + \tau_m s)} \tag{30}$$

The model's parameters are obtained by an experimental identification procedure and they are summarized in Table 1 with their associated uncertainty bounds. This model is sampled with 10 ms sampling time for simulation and experimental setups.

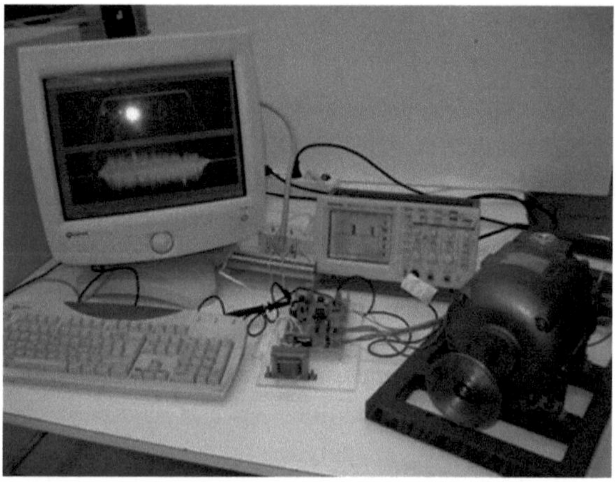

Fig. 2 Electrical DC drive benchmark

Table 1 Identified DC Drive model parameters

Parameters	Nominal values	Uncertainty bounds (%)
Λ	0.05	50
τ_m	300 ms	50
τ_e	14 ms	50

4.2 Simulation Results

For this study case, product-sum inference and center of gravity defuzzification methods are adopted. Uniformly distributed and symmetrical membership functions, are assigned for the fuzzy input and output variables, as shown in Fig. 3.

The linguistic levels assigned to the input variables e_k and Δe_k, and the output variable Δu_k are given as follows: N (Negative), Z (Zero), P (Positive), NB (Negative Big) and PB (Positive Big). The associated fuzzy rule-base is given in Table 2. The view of this rule-base is illustrated in Fig. 4.

For our design, the initial search domain of PID-type FLC parameters is chosen in the limit range of $x_{low} = (1, 5, 2, 25)$ and $x_{up} = (5, 10, 10, 50)$. For all proposed metaheuristics, we use a population size equal to $N = 30$ and run all used algorithms under $k_{\max} = 100$ iterations. The size of optimization problem is equal to $D = 4$. The decision variables are the scaling factors of the studied particular PID-type FLC structure, i.e., α, β, K_e and K_d.

In this study, the control problem constraints are defined by the maximum values of the performance criteria: overshoot ($\delta^{\max} = 20\,\%$), settling time ($t_s^{\max} = 0.9\,\mathrm{s}$) and steady state error ($E_{ss}^{\max} = 0.0001$). The scaling penalty parameter is chosen as constant equal to $\lambda_l = 10^4$. The algorithm stops when the number of generations reaches the specified value for the maximum number of generations.

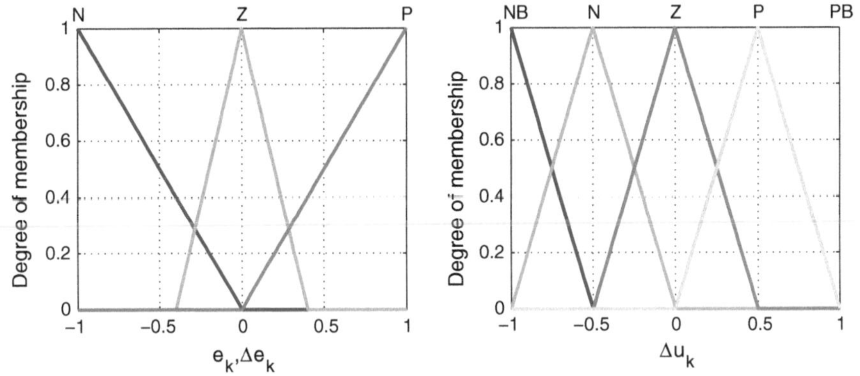

Fig. 3 Membership functions for fuzzy inputs and output variables

Table 2 Fuzzy rule-base for the standard FLC

e_k, Δe_k	N	Z	P
N	NB	N	Z
Z	N	Z	P
P	Z	P	PB

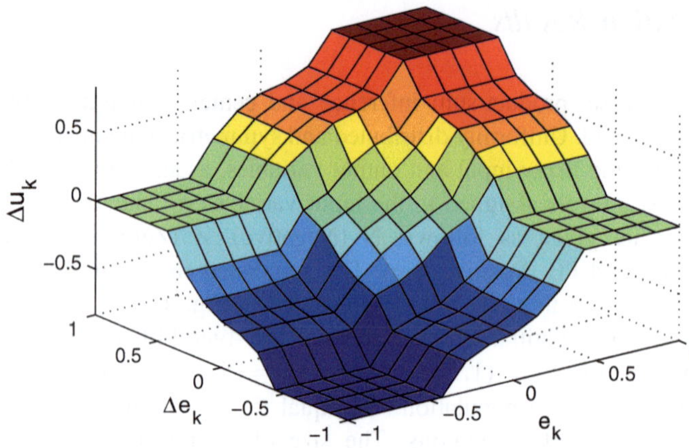

Fig. 4 View of the fuzzy rule-base for the standard FLC

For the software implementation of the proposed metaheuristics, the control parameters of each algorithm are set as follows:

- DSA: random numbers Stopover site research $p_1 = p_2 = 0.3rand\,(0, 1)$;
- GSA: initial value of gravitational constant $G_0 = 75$, parameter $\eta = 20$, initial value of the *Kbest* agents $K_0 = N = 30$ which is decreased linearly to 1;
- ABC: Limit of abandonment $L = 60$;
- PSO: cognitive and social coefficients equal to $c_1 = c_2 = 2$, inertia factor decreasing linearly from 0.9 to 0.4;
- GAO: Stochastic Uniform selection and Gaussian mutation methods, Elite Count equal to 2 and Crossover Fraction equal to 0.8.

In order to get statistical data on the quality of results and so to validate the proposed approaches, we run all implemented algorithms 20 times. Feasible solutions are usually found within an acceptable CPU computation time. The obtained optimization results are summarized in Tables 3 and 4.

4.3 Results Analysis and Discussion

According to the statistical analysis of Tables 3 and 4, as well as the numerical simulations in Figs. 5, 6, 7 and 8, we observe that the proposed approaches produce near results in comparison with each other and with the standard GAO-based method. Globally, the algorithms convergences always take place in the same region of the design space whatever is the initial population. This result indicates that the algorithms succeed in finding a region of the interesting research space to explore.

Table 3 Optimization results from 20 trials of problem (1.1): ISE criterion

Algorithm	Best	Mean	Worst	ST deviation
DSA	0.1621	0.1691	0.1760	0.0045
GSA	0.1556	0.1710	0.1800	0.0193
ABC	0.1928	0.2274	0.2322	0.0134
PSO	0.1600	0.1715	0.1802	0.0140
GAO	0.1643	0.1722	0.1799	0.0086

Table 4 Optimization results from 20 trials of problem (1.1): MO criterion

Algorithm	Best	Mean	Worst	ST deviation
DSA	0.0365	0.0722	0.1307	0.0277
GSA	0.0307	0.0624	0.0096	0.0315
ABC	0.1305	0.1550	0.1972	0.0412
PSO	0.0422	0.0936	0.1420	0.0511
GAO	0.0411	0.0913	0.1300	0.0373

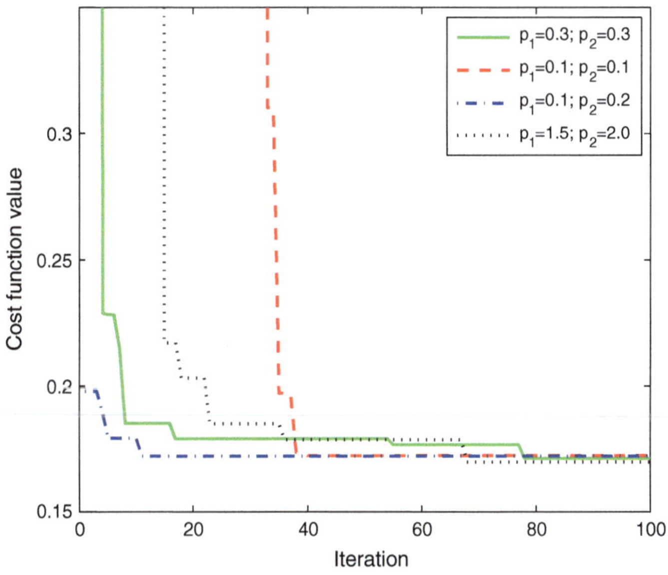

Fig. 5 Robustness convergence under control parameters variation of the DSA-based approach: ISE criterion case

In this case study, we tested the proposed algorithms with different values of the population size in the range of [20, 50]. Globally, all the results found are close to each other. The best values of this control parameter are usually obtained while using a population size equal to 30.

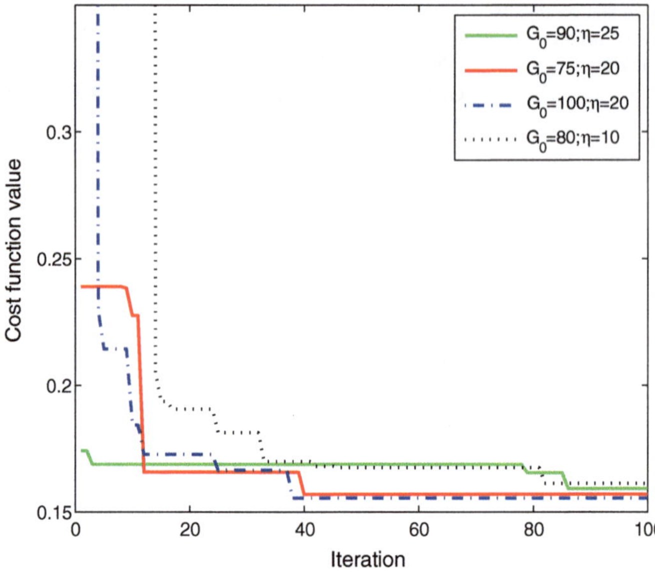

Fig. 6 Robustness convergence under control parameters variation of the GSA-based approach: ISE criterion case

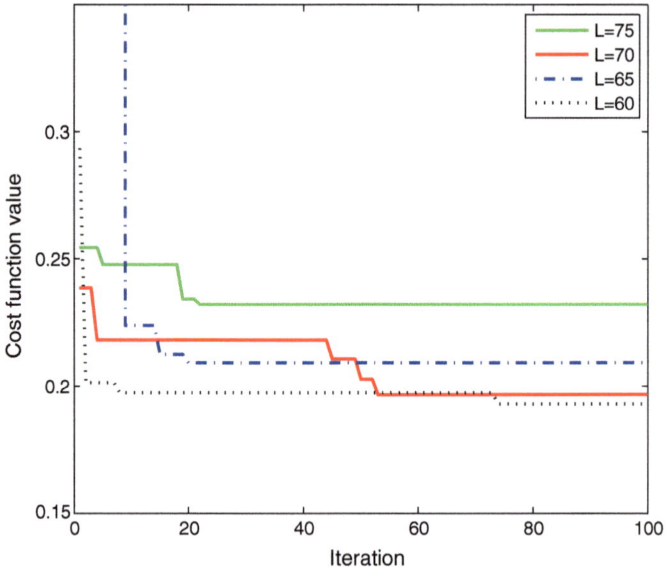

Fig. 7 Robustness convergence under control parameters variation of the ABC-based approach: ISE criterion case

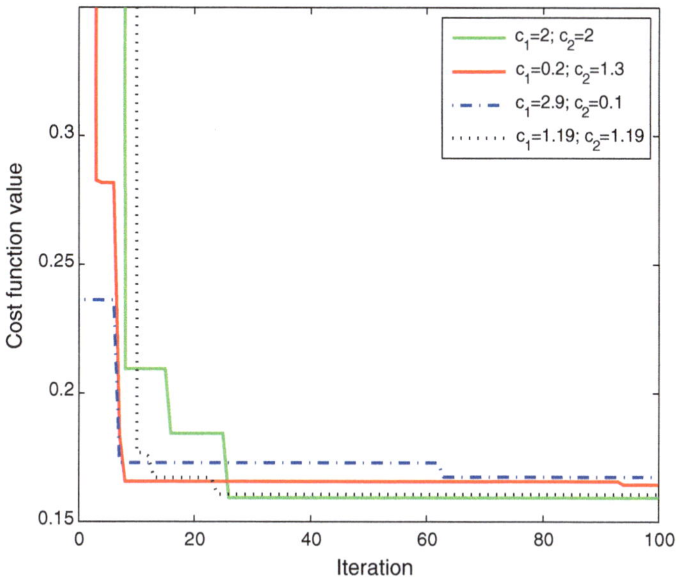

Fig. 8 Robustness convergence under control parameters variation of the PSO-based approach: ISE criterion case

Both for the MO and ISE criteria, the robustness on convergence of the proposed algorithms is guaranteed under their main control parameters variation. The qualities of the obtained solution, the fast convergence as well as the simple software implementation are comparable with the standard GAO-based approach. According to the convergence plots of the implemented metaheuristics, i.e., results of Figs. 5, 6, 7 and 8, the exploitation and exploration capabilities of these algorithms are ever guaranteed.

In this study, only simulation results from the ISE criterion case are illustrated. The main difference between performances of the implemented metaheuristics is their relative quickness or slowness in terms of CPU computation time. For this particular optimization problem, the quickness of DSA and PSO is specially marked in comparison with other techniques. Indeed, while using a Pentium IV, 1.73 GHz and MATLAB 7.7.0, the CPU computation times for the PSO algorithm are about 328 and 360 s in the MO and ISE criterion, respectively. For the DSA algorithm, these are about 296 and 310 s, respectively. For example and in the case of GSA metaheuristic, we obtain about 540 and 521 s for the above criterion respectively.

For the ISE criterion case, all optimization results are close to each other in terms of solutions quality, except those obtained by the ABC-based method. The relative numerical simulation shows the sensitivity of this algorithm under the "Limit for abandonment" parameter variation. The best optimization result, with fitness value equal to 0.1928, is obtained with $L = 60$.

On the other hand, the scaling parameters λ_l, given in Eq. 2, will be linearly increased at each iteration step so constraints are gradually enforced. In a generic and typical optimization problem, the quality of the solution will directly depend on the value of this algorithm control parameter. In this chapter and in order to make the proposed approach simple, great and constant scaling penalty parameters, equal to 10^4, are used for numerical simulations. Indeed, simulation results show that with great values of λ_l, the control system performances are weakly degraded and the effects on the tuning parameters are less meaningful. The proposed constrained and improved algorithms convergence is faster than the case with linearly variable scaling parameters.

The time-domain performances of the proposed metaheuristics-tuned PID-type FLC structure are illustrated in Figs. 9 and 10. Only simulations from the DSA and PSO techniques implementation are presented. All results, for various obtained decision variables, are acceptable and show the effectiveness of the proposed fuzzy controllers tuning method. The robustness, in terms of external disturbances rejection, and tracking performances are guaranteed with degradations for some considered methods. The considered time-domain constraints for the PID-type FC tuning problems, such as the maximum values of overshoot $\delta^{\max} = 20\ \%$, steady state $E_{ss}^{\max} = 0.0001$ and settling time $t_s^{\max} = 0.9\,\mathrm{s}$, are usually respected.

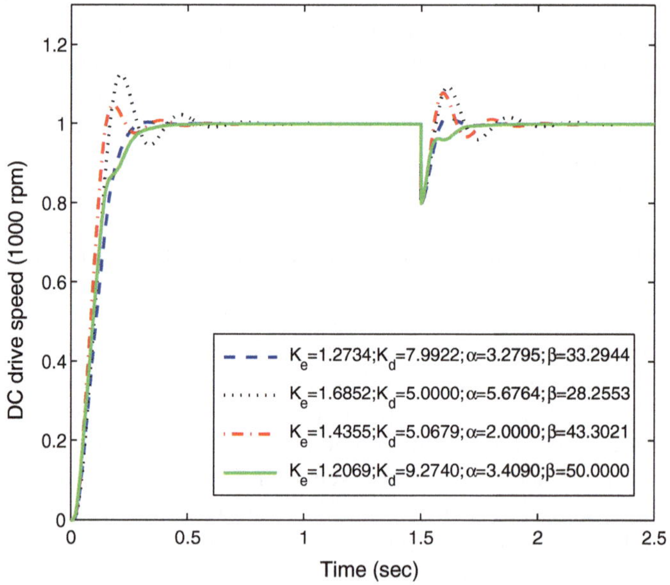

Fig. 9 Step responses of the DSA-tuned PID-type fuzzy controlled system: ISE criterion case

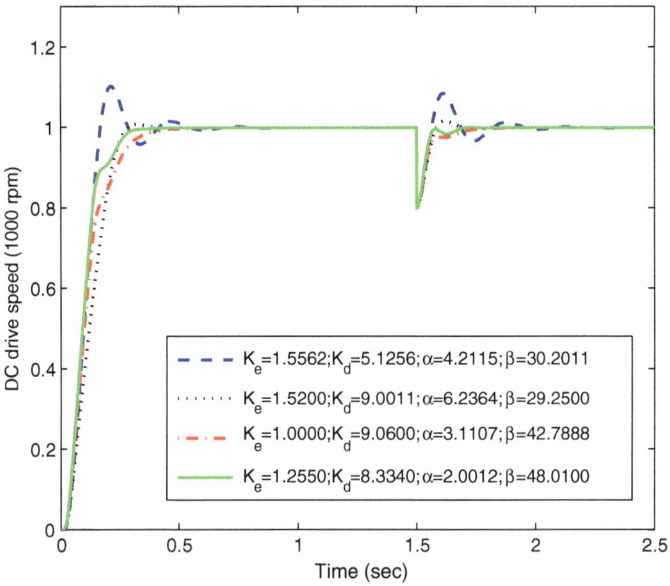

Fig. 10 Step responses of the PSO-tuned PID-type fuzzy controlled system: ISE criterion case

4.4 Experimental Results

In order to illustrate the efficiency of the proposed metaheuristics-tuned PID-type fuzzy control structure, we try to implement the controller within a real-time framework. The developed real-time application acquires input data, speed of the DC drive, and generates control signal for thyristors of AC-DC power converter as a PWM signal (Haggège et al. 2009). This is achieved using a digital control system based on a PC computer and a PCI-1710 multi-functions data acquisition board which is compatible with MATLAB/Simulink as described in Fig. 11.

The power part of the controlled process is constituted of the single-phase bridge rectifier converter. Figure 11 shows the considered half-controlled bridge rectifier, constituted by two thyristors and two diodes. The presence of thyristors makes the average output voltage controllable. A thyristor can be triggered by the application of a positive gate voltage and hence a gate current supplied from a gate drive circuit. The control voltage is generated with the help of a gate drive circuit, which is called a firing or triggering circuit. The used bridge thyristors are switched ON by a train of high-frequency impulses.

In order to obtain an impulse train, beginning with a fixed delay after the AC supply source zero-crossing, it is necessary to generate a sawtooth signal, synchronized with this zero-crossing. This is achieved using a capacitor charged with a constant current during the 10 ms half period of the AC source, and abruptly discharged at every zero-crossing instant. The constant current is obtained using a BC547 bipolar transistor whose base voltage is maintained constant due to a

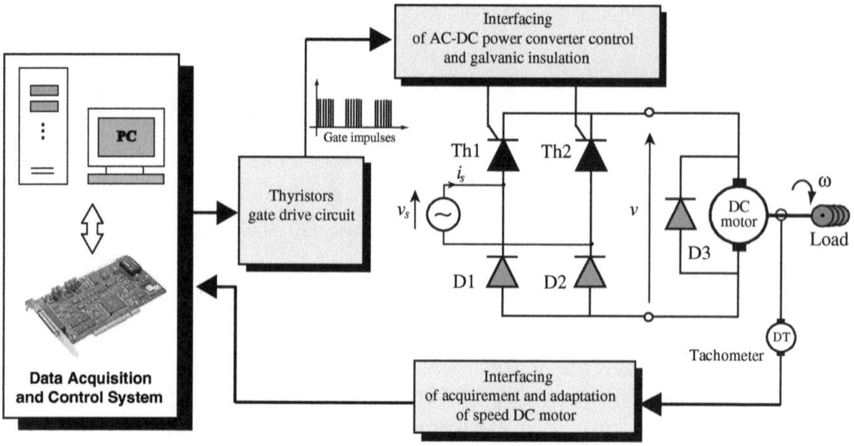

Fig. 11 The proposed experimental setup schematic for DC drive control

polarization bridge constituted by a resistor and two 1N4148 diodes in series. This transistor acts as a current sink, whose value can be determined by an adjustable emitter resistor, to make the capacitor be fully charged after exactly 10 ms. The obtained synchronous saw-tooth signal is compared with a variable DC voltage, using an LM393 comparator, in order to generate a PWM signal which drives a NE555 timer, used as an astable multi-vibrator, producing the impulse train needed to control the thyristors firing. This impulse train is applied to the base of a 2N1711 bipolar transistor which drives an impulse transformer that ensuring the galvanic isolation between the control circuit and the power circuit.

The nominal model of the studied plant and the controller model obtained in synthesis development phase were used to implement the real-time controller. The model of the plant was removed from the simulation model, and instead of it, input device drivers (sensor) and output device driver (actuator) were introduced as shown in Fig. 12. These device drivers close the feedback loop when moving from simulations to experiments. According to this concept, the Fig. 13 illustrates the

Fig. 12 Synoptic of the PCI-1710 based real-time controller implementation

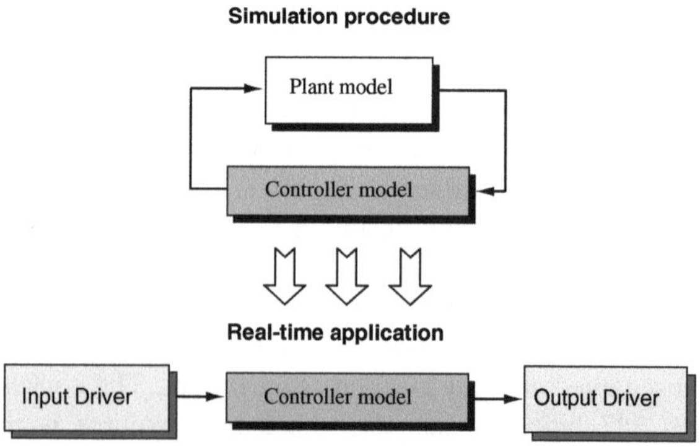

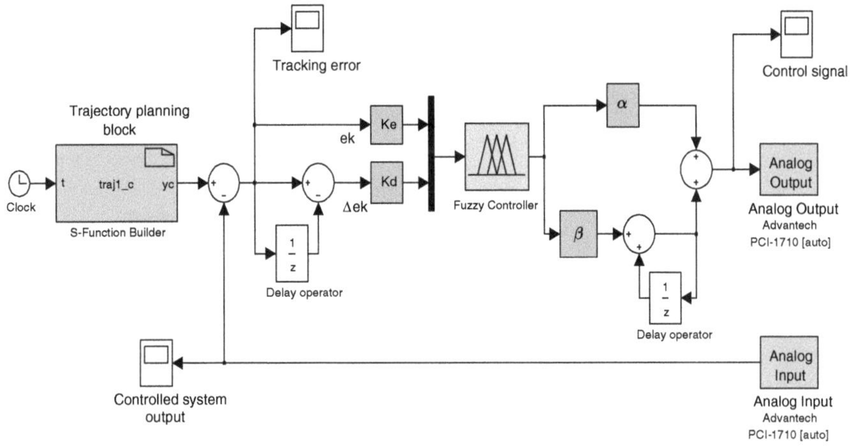

Fig. 13 PCI-1710 board based implementation of the proposed FLC structure

principle of the implementation based on the Real-Time Windows Target tool of MATLAB/Simulink.

The real-time fuzzy controller is developed through the compilation and linking stage, in a form of a Dynamic Link Library (DLL), which is then loaded in memory and started-up. The used environment of real-time controller has some capabilities such as automatic code generation in C language, automatic compilation, start-up of a real-time program and external mode start-up of the simulation phase model allowing for real-time set monitoring and on-line adjustment of its parameters.

The real-time implementation of the proposed metaheuristics-tuned PID-type FLC leads to the experimental results of Figs. 14, 15, 16 and 17.

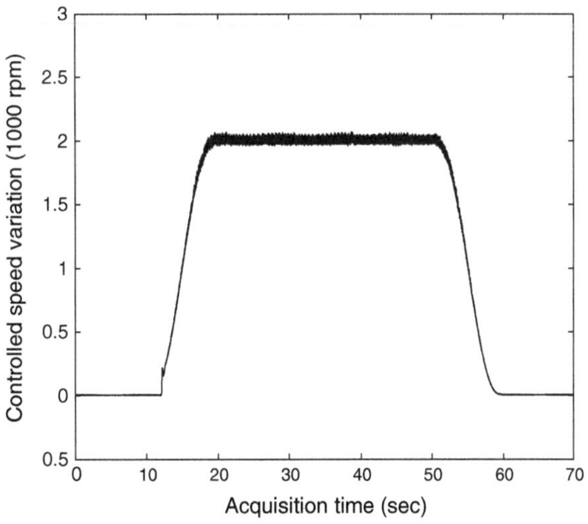

Fig. 14 Experimental results of the PID-type FLC implementation: controlled speed variation

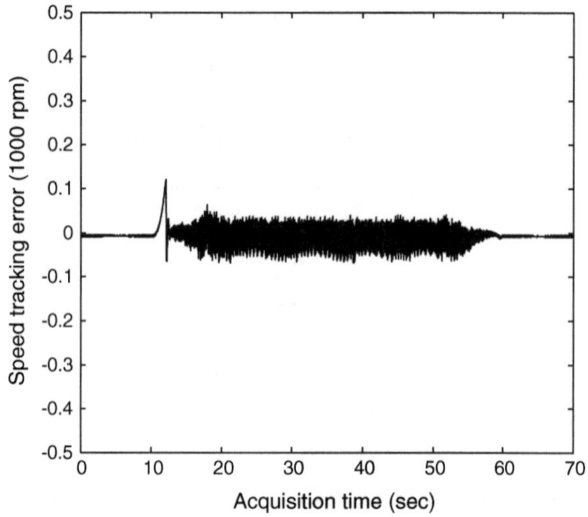

Fig. 15 Experimental results of the PID-type FLC implementation: speed tracking error

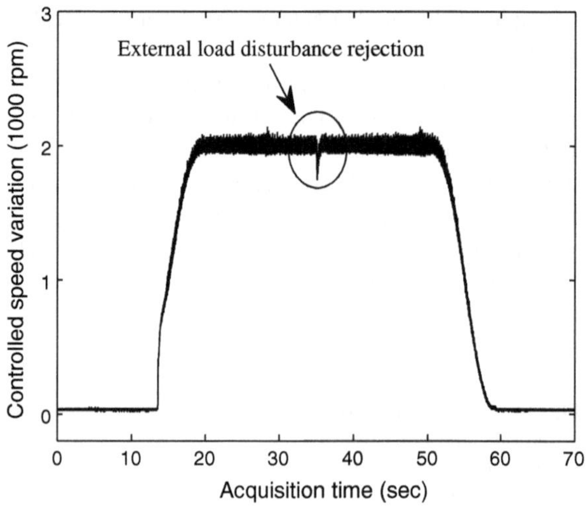

Fig. 16 Robustness of the external disturbance rejection: controlled speed variation

In comparison with the results in Haggège et al. (2009) for a such plant, obtained by the use of a full order $\mathcal{H}_\infty$ controller, as well as those obtained by PID-type FLC with trials-errors tuning method in Haggège et al. (2010), the experimental results of this study are satisfactory for a simple, non conventional and systematic meta-heuristics-based control approach. They point out the controller's viability and performance. As shown in Figs. 14 and 15, the measured speed tracking error is small (less than 10 % of set point) showing the high performances of the proposed control, especially in terms of tracking. The robustness, in terms of external load disturbances of the proposed PID-type FLC approach, is shown in Figs. 16 and 17. The proposed fuzzy controller leads to reject the additive disturbances on the controlled system output with a fast and more damped dynamic.

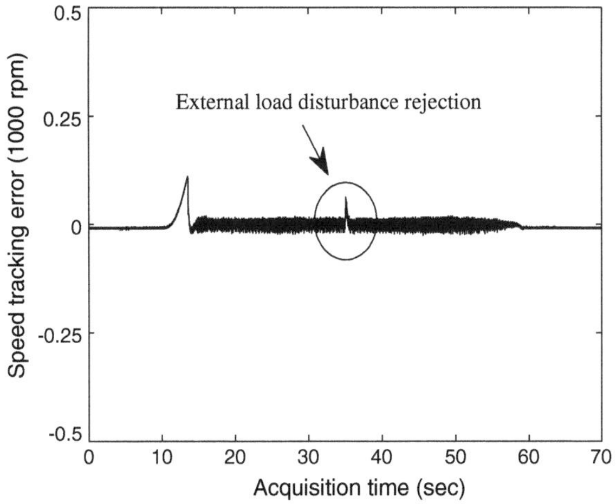

Fig. 17 Robustness of the external disturbance rejection: speed tracking error

Globally, the obtained simulation and experimental results for the considered ISE and MO criteria are satisfactory. Others performance criteria, such as gain and margin specifications (Azar and Serrano 2014), can be used in order to improve the robustness and efficiency of the proposed fuzzy control approach.

5 Conclusion

A new method for tuning the scaling factors of Mamdani fuzzy controllers, based on advanced metaheuristics, is proposed and successfully applied to an electrical DC drive speed control. This efficient metaheuristics-based tool leads to a robust and systematic PID-type fuzzy control design approach. The comparative study shows the efficiency of the proposed techniques in terms of convergence speed and quality of the obtained solutions. This hybrid PID-type fuzzy design methodology is systematic, practical and simple without need to exact analytical plant model description. The obtained simulation and experimental results show the efficiency in terms of performance and robustness. All used DSA, GSA, ABC and PSO techniques produce near results in comparison with each others. Small degradations are always marked by going from one technique to another. The application of the proposed control approach, for more complex and non linear systems, constitutes our future works. The tuning of other fuzzy control structures, such as those described by Takagi-Sugeno inference mechanism, will be investigated.

References

Azar, A. T. (Ed.) (2010a). *Fuzzy systems*. Vienna, Austria: INTECH. ISBN 978-953-7619-92-3.

Azar, A. T. (2010b). Adaptive neuro-fuzzy systems. In *Fuzzy systems*. Vienna, Austria: INTECH. ISBN 978-953-7619-92-3.

Azar, A. T. (2012). Overview of type-2 fuzzy logic systems. *International Journal of Fuzzy System Applications*, *2*(4), 1–28.

Azar, A. T., & Serrano, F. E. (2014). Robust IMC-PID tuning for cascade control systems with gain and phase margin specifications. *Neural Computing and Applications*,. doi:10.1007/s00521-014-1560-x.

Basturk, B. & Karaboga, D. (2006). An artificial bee colony (ABC) algorithm for numeric function optimization. In *Proceedings of IEEE Swarm Intelligence Symposium*, May 12–14, Indianapolis, USA.

Bouallègue, S., Haggège, J., Ayadi, M., & Benrejeb, M. (2012a). PID-type fuzzy logic controller tuning based on particle swarm optimization. *Engineering Applications of Artificial Intelligence, 25*(3), 484–493.

Bouallègue, S., Haggège, J., & Benrejeb, M. (2011). Particle swarm optimization-based fixed-structure $\mathcal{H}_\infty$control design. *International Journal of Control, Automation and Systems, 9*(2), 258–266.

Bouallègue, S., Haggège, J., & Benrejeb, M. (2012b). A new method for tuning PID-type fuzzy controllers using particle swarm optimization. In *Fuzzy Controllers: Recent Advances in Theory and Applications* (pp. 139–162). Vienna, Austria: INTECH. ISBN 978-953-51-0759-0.

Boussaid, I., Lepagnot, J., & Siarry, P. (2013). A survey on optimization metaheuristics. *Information Sciences, 237*(1), 82–117.

Civicioglu, P. (2012). Transforming geocentric Cartesian coordinates to geodetic coordinates by using differential search algorithm. *Computers and Geosciences, 46*(1), 229–247.

David, R. C., Precup, R. E., Petriu, E. M., Radac, M. B., & Preitl, S. (2013). Gravitational search algorithm-based design of fuzzy control systems with a reduced parametric sensitivity. *Information Sciences, 247*(1), 154–173.

Dréo, J., Siarry, P., Pétrowski, A., & Taillard, E. (2006). *Metaheuristics for Hard Optimization Methods and Case Studies*. Heidelberg: Springer.

Eberhart, R. & Kennedy, J. (1995). A new optimizer using particle swarm theory. In *Proceedings of the 6th International Symposium on Micro Machine and Human Science* (pp. 39–43), October 4–6, Nagoya, Japan.

Eker, I., & Torun, Y. (2006). Fuzzy logic control to be conventional method. *Energy Conversion and Management, 47*(4), 377–394.

Goldberg, D. E. (1989). Genetic algorithms in search, optimization and machine learning. Boston: Addison-Wesley Publishing Company.

Goswami, D., & Chakraborty, S. (2014). Differential search algorithm-based parametric optimization of electrochemical micromachining processes. *International Journal of Industrial Engineering Computations, 5*(1), 41–54.

Guzelkaya, M., Eksin, I., & Yesil, E. (2003). Self-tuning of PID type fuzzy logic controller coefficients via relative rate observer. *Engineering Applications of Artificial Intelligence, 16*(3), 227–236.

Haggège, J., Ayadi, M., Bouallègue, S., & Benrejeb, M. (2010). Design of Fuzzy Flatness-based Controller for a DC Drive. *Control and Intelligent Systems, 38*(3), 164–172.

Haggège, J., Bouallègue, S., & Benrejeb, M. (2009). Robust $\mathcal{H}_\infty$Design for a DC Drive. *International Review of Automatic Control, 2*(4), 415–422.

Karaboga, D. (2005). An idea based on honey bee swarm for numerical optimization. Technical report TR06, Erciyes University, Engineering Faculty, Computer Engineering Department, Turkey.

Karaboga, D., & Akay, B. (2009). A comparative study of Artificial Bee Colony algorithm. *Applied Mathematics and Computation, 214*(1), 108–132.

Karaboga, D., & Basturk, B. (2007). A powerful and efficient algorithm for numerical function optimization: artificial bee colony (ABC) algorithm. *Journal of Global Optimization, 39*(3), 459–471.

Karaboga, D., & Basturk, B. (2008). On the performance of artificial bee colony (ABC) algorithm. *Applied Soft Computing, 8*(1), 687–697.

Karaboga, D., Gorkemli, B., Ozturk, C., & Karaboga, N. (2012). A comprehensive survey: Artificial bee colony (ABC) algorithm and applications. *Artificial Intelligent Review, 42*, 21–57. doi:10.1007/s10462-012-9328-0.

Kennedy, J. and Eberhart, R. (1995). Particle swarm optimization. In *Proceedings. of IEEE International Joint Conference on Neural Networks* (pp. 1942–1948), November 27–December 01, Perth, Australia.

Lee, C. C. (1998a). Fuzzy logic in control systems: Fuzzy logic controller-part I. *IEEE Transactions on Systems, Man, and Cybernetics, 20*(2), 404–418.

Lee, C. C. (1998b). Fuzzy logic in control systems: Fuzzy logic controller-part II. *IEEE Transactions on Systems, Man, and Cybernetics, 20*(2), 419–435.

Madiouni, R., Bouallègue, S., Haggège, J., & Siarry, P. (2013). Particle swarm optimization-based design of polynomial RST controllers. In *Proceedings. of the 10th IEEE International Multi-Conference on Systems, Signals and Devices* (pp. 1–7), Hammamet, Tunisia.

MathWorks. (2009). Genetic algorithm and direct search toolbox user's guide.

Nobahari, H., Nikusokhan, M., & Siarry, P. (2011). Non-dominated sorting gravitational search algorithm. In *Proceedings. of the International conference on swarm intelligence* (pp. 1–10), June 14–15, Cergy, France.

Passino, K. M. & Yurkovich, S. (1998). *Fuzzy control.* Boston, Addison Wesley Longman.

Precup, R. E., David, R. C., Petriu, E. M., Preitl, S., & Radac, M. B. (2011). Gravitational search algorithms in fuzzy control systems tuning. In *Proceedings. of the 18th IFAC World Congress* (pp. 13624–13629), August 28–September 02, Milano, Italy.

Qiao, W. Z., & Mizumoto, M. (1996). PID type fuzzy controller and parameters adaptive method. *Fuzzy Sets and Systems, 78*(1), 23–35.

Rao, R. V. and Savsani, V. J. (2012). *Mechanical design optimization using advanced optimization techniques*. Heidelberg: Springer.

Rashedi, E., Nezamabadi-pour, H., & Saryazdi, S. (2009). GSA: A gravitational search algorithm. *Information Sciences, 179*(13), 2232–2248.

Siarry, P., & Michalewicz, Z. (2008). *Advances in metaheuristics for hard optimization.* New York: Springer.

Toumi, F., Bouallègue, S., Haggège, J., & Siarry, P. (2014). Differential search algorithm-based approach for PID-type fuzzy controller tuning. In *Proceedings. of the International Conference on Control, Engineering & Information Technology*, March 22–25, Sousse, Tunisia.

Waghole, V., & Tiwari, R. (2014). Optimization of needle roller bearing design using novel hybrid methods. *Mechanism and Machine Theory, 72*(1), 71–85.

Woo, Z. W., Chung, H. Y., & Lin, J. J. (2000). A PID type fuzzy controller with self-tuning scaling factors. *Fuzzy Sets and Systems, 115*(2), 321–326.

Robust Estimation Design for Unknown Inputs Fuzzy Bilinear Models: Application to Faults Diagnosis

Dhikra Saoudi, Mohammed Chadli and Naceur Benhadj Braeik

Abstract This present chapter addresses the robust estimation problem for a class of nonlinear systems with unknown inputs and bilinear terms. The considered nonlinear system is represented by Takagi-Sugeno (T-S) Fuzzy Bilinear Model (FBM). Two cases are considered: the first one deals with the study of FBM with measurable decision variables and the second one assumes that these decision variables are unmeasurable. Then, the proposed Fuzzy Bilinear Observer (FBO) design for fuzzy bilinear models subject to unknown inputs is developed to ensure the asymptotic convergence of the error dynamic using the Lyapunov method. Stability analysis and gain matrices determination are performed by resolving a set of Linear Matrices Inequalities (LMIs) for both cases. The design conditions lead to the resolution of linear constraints easy to solve with existing numerical tools. The given observer is then applied for fault detection. This chapter studies also the problem of robust fault diagnosis based on a fuzzy bilinear observer. Sufficient conditions are established in order to guarantee the convergence of the state estimation error. Thus a residual generator is determined on the basis of LMI conditions such that the estimation error is sensitive to fault vector and insensitive to the unknown inputs. These results are provided for measurable and unmeasurable decision variables cases. The performances of the proposed estimation and fault diagnosis method is successfully applied to academic examples.

D. Saoudi (✉) · N.B. Braeik
LSA—Laboratoire des Systèmes Avancés, Polytechnic High School of Tunisia, University of Carthage, BP 743, 2078 La Marsa, Tunisia
e-mail: dhikra.saoudi@hotmail.com

N.B. Braeik
e-mail: naceur.benhaj@ept.rnu.tn

M. Chadli
UPJV-MIS—University of Picardie Jules Verne, 33, Rue Saint-Leu, 80039 Amiens, France
e-mail: mchadli@u-picardie.fr

Q. Zhu and A.T. Azar (eds.), *Complex System Modelling and Control Through Intelligent Soft Computations*, Studies in Fuzziness and Soft Computing 319,
DOI 10.1007/978-3-319-12883-2_23

1 Introduction

In the recent past decades, the design of unknown input observer (UIO) plays an essential role in robust model-based fault detection. Fruitful results on the case of unknown input linear system can be found in survey papers (Darouach et al. 1994; Guan and Saif 1991; Hou and Muller 1992; Yang and Wilde 1988) and many types of full order and reduced order unknown input observers are now available.

However, many physical systems are nonlinear in nature. For such system, the use of the well known linear techniques may reduce in bad performance and even instability. Generally, analysis for nonlinear systems is a quite involved procedure. These last decades, a T-S fuzzy approach to represent or approximate a large class of nonlinear systems is developed. It is well known that T-S fuzzy model is an effective tool in the analysis and synthesis for nonlinear control systems (Azar 2010a; Chadli and Borne 2012, 2013; Takagi and Sugeno 1985; Taniguchi et al. 2000). Indeed, the nonlinear model is approximated by a set of linear local models connected by *if-then* rules and the resulting T-S fuzzy model can universally approximate or exactly describe general nonlinear systems (Takagi and Sugeno 1985; Tanaka et al. 1998; Tanaka and Sugeno 1992; Tanaka and Wang 2000). In the last two decade, numerous results have been devoted to observers design of fuzzy T-S systems (Azar 2010b, 2012; Bergsten et al. 2002; Chadli 2010; Chadli and Guerra 2012; Ma 2002; Ma and Sun 2001; Tong and Tang 2000; Yoneyama et al. 2000). These results use different techniques such as linear matrix-inequality approach, sliding mode techniques, adaptive methods, etc. Moreover, it is noted that all of the aforementioned fuzzy systems are based on the T-S fuzzy model with linear rule consequence.

As a special extension, fuzzy bilinear system based on the T-S fuzzy model with bilinear rule consequence has attracted the interest of researchers (Li and Tsai 2007; Li et al. 2008; Saoudi et al. 2010, 2012a, c, 2013b; Tsai and Li 2007). For example robust stabilization for the T-S fuzzy bilinear models has studied in (Li and Tsai 2007) and extension to the T-S fuzzy bilinear models with time-delay is given in Tsai and Li (2007). The problem of robust stabilization for discrete-time fuzzy bilinear models was considered in (Li et al. 2008). The synthesis of fault diagnosis and fault tolerant control, have been proposed for this class of systems in Saoudi et al. (2013a, 2012c). Moreover, several works are devoted to the state estimation by the use of T-S models with measurable decision variables which are especially represented by the input variables or the outputs of the system (Chadli and Coppier 2013; Chadli and Karimi 2012; Gao et al. 2008; Lendek et al. 2010; Liu and Zhang 2003). For example, (Tanaka et al. 1998) proposed a study of stability and stabilization by multiple controllers. Patton et al. (1998) proposed an observer based on the Luenberger observer structure, which was then used for the diagnosis. Ichalal et al. (2009) developed an observer-based approach for robust residual generator and diagnosis in nonlinear systems described by T-S fuzzy models. Gao et al. (2008) presented a fuzzy state observer design for T-S fuzzy systems with application to sensor fault estimation. For T-S fuzzy bilinear models, Saoudi et al. (2010) proposed an observer design method using iterative procedure and extension of this approach in discrete-time case was developed in Saoudi et al. (2012b). By against, in Saoudi et al. (2012a) the proposed design is given in LMI formulation solved simultaneously.

Despite numerous works available, none of them seem able to define an LMI formulation for the problem of state estimation for T-S fuzzy models with unmeasurable premise variables. Few works are dedicated to the use of these models for state estimation (Ichalal et al. 2009; Saoudi et al. 2014) and for fault estimation (Marx et al. 2007). In Bergsten et al. (2002), the authors proposed the Thau-Luenberger observer which is an extension of the classical Luenberger observer. In Yoneyama (2009), the authors proposed a filter estimating the state and minimizing the effect of disturbances. Recently, other approaches for observer design, fault diagnosis and fault tolerant control, have been proposed for this class of systems in Ichalal et al. (2012). But, these results were only obtained for ordinary nonlinear systems, this chapter addresses the state estimation for fuzzy bilinear models with unmeasurable decision variables.

The chapter deals with fuzzy bilinear observers design for a class of nonlinear system in the case of measurable decision variables and in the case of unmeasurable decision variables. The nonlinear system is modeled as a fuzzy bilinear model. This kind of T-S fuzzy model is especially suitable for a nonlinear system with a bilinear term. The considered bilinear observer is obtained by a convex interpolation of unknown input bilinear observers. Based on Lyapunov theory, the synthesis conditions of the given fuzzy observer are expressed in LMI terms for the two cases. The design conditions lead to the resolution of linear constraints easy to solve with existing numerical tools. The given observer is then applied for fault diagnosis. So, this chapter brings some results for the state estimation and fault detection dedicated to fuzzy bilinear models with measurable and unmeasurable decision variables.

The remainder of the chapter is organized as follows: Sect. 2 presents the general structure of a fuzzy bilinear model with unknown input for continuous-time. In Sect. 3, the proposed structure of fuzzy bilinear observer with measurable and unmeasurable decision variables and design conditions are developed. Section 4 is devoted to the problem of fault diagnosis by using unknown input fuzzy bilinear observer for fuzzy bilinear models. A predator-prey model is provided in Sect. 5 to show the effectiveness of the proposed approach. A conclusion finishes the chapter.

Notations. Throughout the chapter, the following useful notations are used: $\Re$ denotes the set of real numbers, I denotes the identity matrix of the appropriate dimension, X^T denotes the transpose of the matrix X, $X > 0$ denotes symmetric positive definite matrix, X^{-1} denotes the Moore-Penrose inverse of X, X^+ denotes the pseudo inverse of X such that $XX^+X = X$, and $\begin{pmatrix} A & * \\ B & C \end{pmatrix}$ denotes symmetric matrix where $(*) = B^T$.

2 General Structure of Fuzzy Bilinear Model

Fuzzy bilinear models based on the T-S fuzzy model with bilinear rule consequence are defined by extending the T-S fuzzy ordinary model. It is proved that often nonlinear behaviors can be approximated by T-S fuzzy bilinear description. This technique is based on the bilinearization of the nonlinear system around some operating points and using adequate weighting functions. This kind of T-S fuzzy model is especially suitable for a nonlinear system with a bilinear term. Moreover, the fuzzy model is described by *if-then* rules and used to present a fuzzy bilinear system. The ith rule of the fuzzy bilinear model for nonlinear systems is represented by the following form:

$$R^i\text{: if } \xi_1(t) \text{ is } F_{i1} \text{ and} \ldots \text{and } \xi_g(t) \text{ is } F_{ig}$$

then

$$\begin{cases} \dot{x}(t) = A_i x(t) + B_i u(t) + N_i x(t)u(t) + F_i d(t) \\ y(t) = Cx(t) \end{cases} \tag{1}$$

where R^i denotes the ith fuzzy rule $\forall i = \{1, \ldots, r\}$, r is the number of *if-then* rules, $\xi_i(t)$ are the premise variables assumed to be measurable and $F_{ij}(\xi_j(t))$ is the membership degree of $\xi_j(t)$ in the fuzzy set F_{ij}, $x(t) \in \Re^n$ is the state vector, $u(t) \in \Re$ is the input vector, $d(t) \in \Re^q$ is the unknown input vector and $y(t) \in \Re^p$ is the system output. The matrices $A_i \in \Re^{n\times n}$, $B_i \in \Re^{n\times 1}$, $N_i \in \Re^{n\times n}$, $C \in \Re^{p\times n}$ are known matrices and define the ith local bilinear model and $F_i \in \Re^{n\times q}$ represent the influence matrix of the unknown input.

Then, the overall fuzzy bilinear model can be described as follows:

$$\begin{cases} \dot{x}(t) = \sum_{i=1}^{r} h_i(\xi(t))(A_i x(t) + B_i u(t) + N_i x(t)u(t) + F_i d(t)) \\ y(t) = Cx(t) \end{cases} \tag{2}$$

where $h_i(.)$ verify the following properties

$$\begin{cases} \sum_{i=1}^{r} h_i(\xi(t)) = 1 \\ \qquad\qquad\qquad\qquad \forall i \in \{1, 2, \ldots, r\} \\ 0 \le h_i(\xi(t)) \le 1 \end{cases} \tag{3}$$

The activation functions $h_i(.)$ depend on the decision vector $\xi(t)$ assumed to depend on measurable variables. It can depend on the measurable state variables, be a function of the measurable outputs of the system and possibly of the known inputs (Murray-Smith and Johansen 1997; Takagi and Sugeno 1985).

Remark 1 Matrices A_i, B_i, N_i, and C can be obtained by using the polytopic transformation (Tanaka et al. 1998). The advantage of this method is in one hand to lead to a bilinear transformation of the nonlinear model without any approximation error, and in another hand to reduce the number of local models compared to other methods (Li and Tsai 2007).

In the following section, a method is proposed to design a fuzzy bilinear observer for fuzzy bilinear models subjects to unknown inputs.

3 Robust Estimation in Fuzzy Bilinear Models

In real life, all the states of the system are not always observed bilinear. Hence, we need to estimate the states by an observer. Here we define a fuzzy observer and make an analysis of the error system, from which we provides a design method of a fuzzy observer for the fuzzy bilinear model (2).

3.1 Problem Formulation

In this subsection, the design of a robust observer for fuzzy bilinear models is developed. In order to estimate the state of the unknown input fuzzy bilinear model (2), the considered unknown input observer structure has the following form:

$$R^{i}\text{:if } \xi_1(t) \text{ is } F_{i1} \text{ and } \ldots \text{ and } \xi_g(t) \text{ is } F_{ig}$$

then

$$\begin{cases} \dot{z}(t) = H_i z(t) + L_i y(t) + J_i u(t) + M_i y(t) u(t) \\ \hat{x}(t) = z(t) - E y(t) \end{cases} \tag{4}$$

The overall FBO can be represented by:

$$\begin{cases} \dot{z}(t) = \sum_{i=1}^{r} h_i(\xi(t))(H_i z(t) + L_i y(t) + J_i u(t) + M_i y(t) u(t)) \\ \hat{x}(t) = z(t) - E y(t) \end{cases} \tag{5}$$

where $z(t) \in \Re^n$ is the observer state and $\hat{x}(t) \in \Re^n$ is the estimated state vector. H_i, M_i, L_i, J_i and E are constant matrices with appropriate dimensions. Our objective is to determine the gains of the observer (5) such that the state estimation error $e(t)$ converges towards zero when $t \to \infty$. Let define

$$e(t) = x(t) - \hat{x}(t) \tag{6}$$

Using the expression of $\hat{x}(t)$ given by the observer (5) and fuzzy bilinear model (2), the state estimation error (6) becomes

$$e(t) = (I_n + EC)x(t) - z(t) \tag{7}$$

According to (2), (5) and (7), the dynamics of the state estimation error is given by

$$\dot{e}(t) = \sum_{i=1}^{r} h_i(\xi(t)) \begin{pmatrix} H_i e(t) + (TA_i - H_i T - L_i C)x(t) \\ +(TN_i - M_i C)x(t)u(t) + TF_i d(t) \\ +(TB_i - J_i)u(t) \end{pmatrix} \tag{8}$$

with

$$T = I_n + EC \tag{9}$$

If the following conditions hold true $\forall i = \{1, \ldots, r\}$,

$$TA_i - H_i T - L_i C = 0 \tag{10}$$

$$TN_i - M_i C = 0 \tag{11}$$

$$TB_i - J_i = 0 \tag{12}$$

$$TF_i = 0 \tag{13}$$

Then the equation of the observing error becomes

$$\dot{e}(t) = \sum_{i=1}^{r} h_i(\xi(t)) H_i e(t) \tag{14}$$

The problem of designing the fuzzy bilinear observer for the fuzzy bilinear model with unknown input is reduced to find the observer gains such that the equation of the dynamic estimation error (7) is stable. This aspect will be subject the next subsection.

3.2 Design and Stability Analysis

This subsection proposes sufficient linear design conditions to guarantee the global asymptotic convergence of state estimation error. Therefore, the stability problem is studied for two cases:

1. when the decision variable of the weighting function depends of a measured variable.
2. when this variable depends of an unmeasured variable.

3.2.1 Observers Design with Measurable Decision Variables

The design of the fuzzy bilinear observer (5) is reduced to satisfy the constraints (10)–(13) by taking into the stability of the observing error (14). In order to establish the gains matrices of the FBO, the substitution of (9) into (10) yields to:

$$H_i = TA_i - K_iC \tag{15}$$

with

$$K_i = H_iE + L_i \tag{16}$$

From the dynamic state estimation error (14) and using (15), it becomes

$$\dot{e}(t) = \sum_{i=1}^{r} h_i(\xi(t))((TA_i - K_iC)e(t)) \tag{17}$$

The following result gives the sufficient linear conditions and the gains determination of the unknown input fuzzy bilinear observer (2) with measurable decision variables.

Theorem 1 *If there exist a symmetric definite positive matrix P, and matrices W_i, V_i, S, R_i such that the following linear conditions hold $\forall i = 1\ldots r$*

$$((P+SC)A_i - W_iC)^T + (P+SC)A_i - W_iC < 0 \tag{18}$$

$$R_i = (P+SC)B_i \tag{19}$$

$$V_iC = (P+SC)N_i \tag{20}$$

$$(P+SC)F_i = 0 \tag{21}$$

then the state estimation of the fuzzy bilinear observer (5) converges globally and asymptotically to the state of the fuzzy bilinear model (2). The observer gains are determined by:

$$E = P^{-1}S \tag{22}$$

$$J_i = P^{-1}R_i \tag{23}$$

$$M_i = P^{-1} V_i \tag{24}$$

$$H_i = (I_n + EC)A_i - P^{-1} W_i C \tag{25}$$

$$L_i = P^{-1} W_i - H_i E \tag{26}$$

Proof Let us consider the following Lyapunov function

$$V(e(t)) = e^T(t) P e(t), \quad P = P^T > 0 \tag{27}$$

Using (17), the derivative of the Lyapunov function (27) is given by

$$\dot{V}(e(t)) = \sum_{i=1}^{r} h_i(\xi(t)) \left(e^T(t)\left((TA_i - K_i C)^T P + P(TA_i - K_i C)\right)e(t)\right) \tag{28}$$

Stability condition for the estimation error yields to that the time derivative of the Lyapunov function should be negative define over (3). Then, one has:

$$(TA_i - K_i C)^T P + P(TA_i - K_i C) < 0 \tag{29}$$

Taking into account (9) and considering the variables change:

$$S = PE \tag{30}$$

$$W_i = PK_i \tag{31}$$

we get the LMI (18). Taking into account (9) and (30), equality (21) is derived from (13).

Similarly, using the following variable change

$$R_i = PJ_i \tag{32}$$

$$V_i = PM_i \tag{33}$$

we get equalities (19) and (20) from (12) and (11) respectively. Which ends the proof.

3.2.2 Observers Design with Unmeasurable Decision Variables

In this section, the design of a robust estimation for fuzzy bilinear models with unmeasurable premise variables is proposed. This case, reputed very difficult, is important in diagnosis method based on observer banks to detect and isolate actuator and/or sensor faults.

Let us denote $\hat{\xi}(t)$ the estimate of the decision variables dependent estimated state variables $\hat{x}(t)$. The system (2) can be rewritten as:

$$\begin{cases} \dot{x}(t) = \sum_{i=1}^{r} h_i(\hat{x}(t)) \begin{pmatrix} A_i x(t) + B_i u(t) + F_i d(t) \\ + N_i x(t) u(t) + \Delta(t) \end{pmatrix} \\ y(t) = Cx(t) \end{cases} \tag{34}$$

where $\Delta(t)$ acts like a disturbance on the dynamic of the fuzzy bilinear model and is defined by:

$$\Delta(t) = \sum_{i=1}^{r} (h_i(x(t)) - h_i(\hat{x}(t))) \begin{pmatrix} A_i x(t) + B_i u(t) \\ + N_i x(t) u(t) + F_i d(t) \end{pmatrix} \tag{35}$$

The dynamic of the state estimation error becomes:

$$\dot{e}(t) = \sum_{i=1}^{r} h_i(\hat{x}(t)) (H_i e(t) + T\Delta(t)) \tag{36}$$

Assumption 1 $\|\Delta(t)\| \leq \gamma \|e(t)\|$, i.e. $\Delta(t)$ is Lipschitz in $e(t)$ where γ is a positive scalar.

This assumption, used in previous works (Bergsten et al. 2002; Khalil 1996), will be useful to derive design conditions. The following lemma will be also used (Boyd et al. 1994).

Lemma 1 *For any matrices X and Y with appropriate dimensions, the following property holds for any positive scalar* ε:

$$X^T Y + Y^T X \leq \varepsilon X^T X + \varepsilon^{-1} Y^T Y \tag{37}$$

The following theorem gives sufficient design conditions for fuzzy bilinear models subjects to unknown inputs (2) with the decision variable is unmeasurable.

Theorem 2 *For a given* $\gamma > 0$, *the fuzzy bilinear observer* (5) *converges asymptotically to the state of the fuzzy bilinear model* (2), *if there exist a symmetric definite positive matrix P, matrices* W_i, V_i, S, R_i *and scalar* ε *such that the following linear conditions hold* $\forall i = 1 \ldots r$:

$$\begin{bmatrix} \Pi + \Pi^T + \varepsilon\gamma^2 I & P + SC \\ * & -\varepsilon I \end{bmatrix} < 0 \tag{38}$$

$$R_i = (P + SC)B_i \tag{39}$$

$$V_i C = (P + SC)N_i \tag{40}$$

$$(P + SC)F_i = 0 \tag{41}$$

where $\Pi = (P + SC)A_i - W_i C$.

The observer gains are determined by:

$$E = P^{-1} S \tag{42}$$

$$J_i = P^{-1} R_i \tag{43}$$

$$M_i = P^{-1} V_i \tag{44}$$

$$H_i = (I_n + EC)A_i - P^{-1} W_i C \tag{45}$$

$$L_i = P^{-1} W_i - H_i E \tag{46}$$

Proof Let us consider the Lyapunov function (27). Using (36), the derivative of the Lyapunov function (27) is given by

$$\dot{V}(e(t)) = \sum_{i=1}^{r} h_i(\hat{x}(t)) \left(e^T \left(H_i^T P + PH_i \right) e + e^T PT\Delta + \Delta^T T^T Pe \right) \tag{47}$$

Using the Lemma 1 and Assumption 1, we get

$$\dot{V}(e(t)) \leq \sum_{i=1}^{r} h_i(\hat{x}(t)) \left(e^T \left(H_i^T P + PH_i + \varepsilon^{-1} PTT^T P + \varepsilon \gamma^2 I \right) e \right) \tag{48}$$

While replacing H_i by the expression (15) and T by the expression (9), the last inequality (48) can be written such that $\forall i \in \{1, 2, \ldots, r\}$:

$$\dot{V}(e(t)) \leq \sum_{i=1}^{r} h_i(\hat{x}(t)) e^T(t) \begin{pmatrix} \Psi^T P + P\Psi + \varepsilon \gamma^2 I \\ + \varepsilon^{-1} P (I_n + EC)(I_n + EC)^T P \end{pmatrix} e(t) \tag{49}$$

where $\Psi = (I_n + EC)A_i - K_i C$.

However the inequality (49) are nonlinear on variables P, E and K_i. In order to linearize, the change of variables (30) and (31) is used. Then integrating (30) and (31), we get

$$\dot{V}(e(t)) \leq \sum_{i=1}^{r} h_i(\hat{x}(t)) e^T(t) \begin{pmatrix} \Pi^T + \Pi + \varepsilon \gamma^2 I \\ + \varepsilon^{-1} (P + SC)(P + SC)^T \end{pmatrix} e(t) \tag{50}$$

where $\Pi = (P + SC)A_i - W_iC$.

Then, the derivative of the Lyapunov function is negative if

$$\Pi^T + \Pi + \varepsilon\gamma^2 I + \varepsilon^{-1}(P + SC)(P + SC)^T < 0 \tag{51}$$

By using the Schur complement on inequality (51), we get (38).

Taking into account (9) and (30), equality (41) is derived from (13).

Similarly, using the following variable change

$$R_i = PJ_i \tag{52}$$

$$V_i = PM_i \tag{53}$$

we get equalities (39) and (40) from (12) and (11) respectively. Which ends the proof.

3.3 Illustrative Example

In this subsection, we apply the proposed method to design a fuzzy bilinear observer for a continuous fuzzy bilinear models subjects to unknown inputs and unmeasurable premise variables. Consider FBM (2) with the following data:

$$A_1 = \begin{bmatrix} -6 & -1 & 0 \\ 5 & -1 & 0 \\ 0 & 1 & -2 \end{bmatrix}, \quad A_2 = \begin{bmatrix} -4 & 0 & -1 \\ 5 & -2 & 0 \\ 0 & 0 & -1 \end{bmatrix}$$

$$B_1 = \begin{bmatrix} 1 \\ 0 \\ 0 \end{bmatrix}, \quad B_2 \begin{bmatrix} 0.6 \\ 0 \\ 1 \end{bmatrix}$$

$$N_1 = N_2 = \begin{bmatrix} -0.5 & 0 & 0 \\ 0 & 0 & 0 \\ 0 & 0 & -0.5 \end{bmatrix}$$

$$F_1 = F_2 = \begin{bmatrix} 0 \\ 0.5 \\ 0 \end{bmatrix}$$

$$C = \begin{bmatrix} 0.2 & -1 & 0 \\ 0 & -0.1 & 1 \end{bmatrix}$$

The case study is envisaged for fuzzy bilinear models subjects to unknown inputs where the decision variable is unmeasurable. Let consider the weighting functions depend on the first component of the estimated state vector $\hat{x}_1(t)$. Then, the weighting functions is described as follows:

$$h_i(\hat{x}_1(t)) = \frac{\mu_i(\hat{x}_1(t))}{\sum_{j=1}^{2} \mu_j(\hat{x}_1(t))}$$

where $\mu_i(\hat{x}_1(t))$ are defined by

$$\mu_1(\hat{x}_1(t)) = \exp(-1/2(\frac{\hat{x}_1 + 5}{2})^2)$$
$$\mu_2(\hat{x}_1(t)) = \exp(-1/2(\frac{\hat{x}_1 - 5}{2})^2)$$

Thus, the resolution of the conditions of Theorem 2 leads to the following matrix:

$$P = 10^4 * \begin{bmatrix} 0.061 & -0.299 & -0.030 \\ -0.299 & 1.495 & -0.006 \\ -0.030 & -0.006 & 1.554 \end{bmatrix}$$

Therefore, the observer gains are computed from (42) to (46) as follows:

$$E = \begin{bmatrix} 1.337 & -3.566 \\ 1.268 & -0.713 \\ 0.127 & -1.071 \end{bmatrix}$$

$$L_1 = \begin{bmatrix} 20.702 & -39.709 \\ 4.140 & -7.942 \\ 0.414 & -0.794 \end{bmatrix}, \quad L_2 = \begin{bmatrix} 10.556 & -31.701 \\ 2.111 & -6.340 \\ 0.211 & -0.634 \end{bmatrix}$$

$$J_1 = \begin{bmatrix} 1.268 \\ 0.254 \\ 0.026 \end{bmatrix}, \quad J_2 = \begin{bmatrix} -2.806 \\ -0.561 \\ -0.056 \end{bmatrix}$$

$$M_1 = M_2 = \begin{bmatrix} -0.718 & 2.273 \\ -0.144 & 0.455 \\ -0.0143 & 0.046 \end{bmatrix}$$

$$H_1 = \begin{bmatrix} -14.445 & 5.607 & 9.443 \\ -2.795 & 0.620 & 2.191 \\ -0.221 & -0.179 & -0.310 \end{bmatrix}, \quad H_2 = \begin{bmatrix} -10.546 & 4.309 & 7.513 \\ -2.009 & 0.362 & 1.503 \\ -0.201 & 0.086 & -0.350 \end{bmatrix}$$

These parameters define completely the given observer:

$$\begin{cases} \dot{z}(t) = \sum_{i=1}^{2} h_i(\hat{x}(t))(H_i z(t) + L_i y(t) + J_i u(t) + M_i y(t) u(t)) \\ \hat{x}(t) = z(t) - Ey(t) \end{cases}$$

To show the effectiveness of the proposed observer, simulation results are presented in Figs. 1, 2 and 3 for the input signal $u(t) = 0.5\sin(0.5\pi t)$ and for the unknown inputs $d(t)$ is a rectangular signal of amplitude 0.5 applied for $1.5 \leq t \leq 2.5$. The evolution of the state estimation error with unmeasurable decision variables are given in these Figs. 1, 2 and 3.

It can be deduced from Figs. 1, 2 and 3 that the state estimation error converges asymptotically tends to zero in spite of the presence of the unknown input and unmeasurable premise variables.

Then, the fuzzy bilinear state and their estimation are given in the following Figs. 4, 5 and 6.

Figures 4, 5 and 6 show respectively the evolution of the state variables x_1, x_2 and x_3 of the considered system and their corresponding observer estimation $\hat{x}_1$, $\hat{x}_2$ and $\hat{x}_3$ with the initial conditions $x_0 = \begin{bmatrix} 0.5 & 0.5 & 0.5 \end{bmatrix}^T$ and $\hat{x}_0 = 0$. Based on Figs. 4, 5 and 6 it can seen that the estimated state converges globally asymptotically to the real state of nonlinear system.

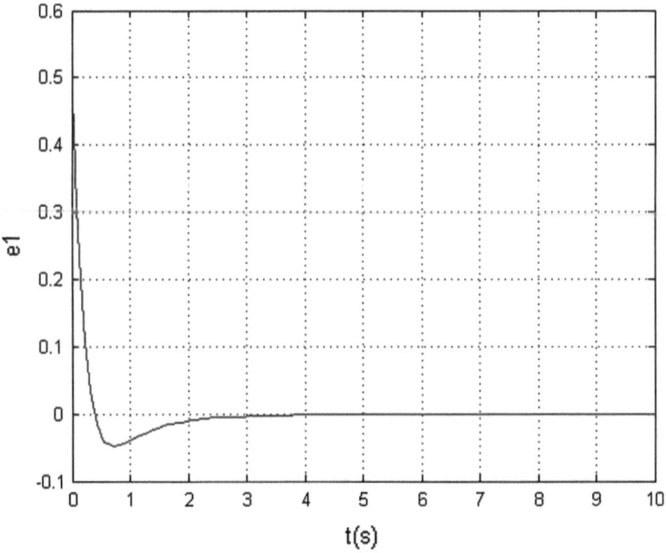

Fig. 1 Trajectories of state estimation error between x_1 and its estimate with unmeasurable decision variables

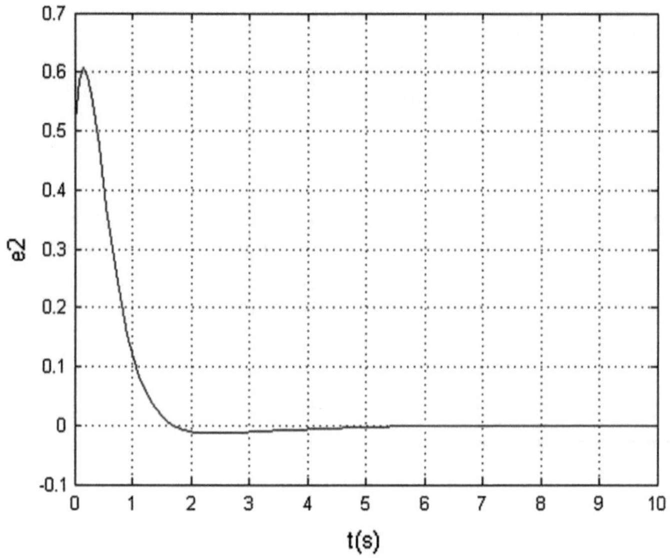

Fig. 2 Trajectories of state estimation error between x_2 and its estimate with unmeasurable decision variables

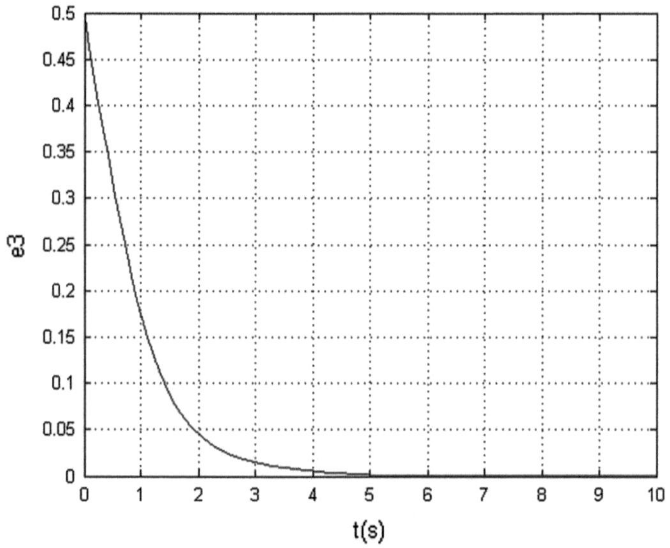

Fig. 3 Trajectories of state estimation error between x_3 and its estimate with unmeasurable decision variables

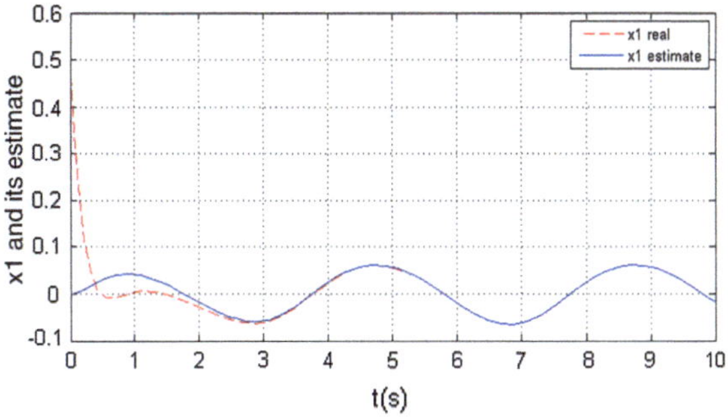

Fig. 4 The state x_1 and its estimate

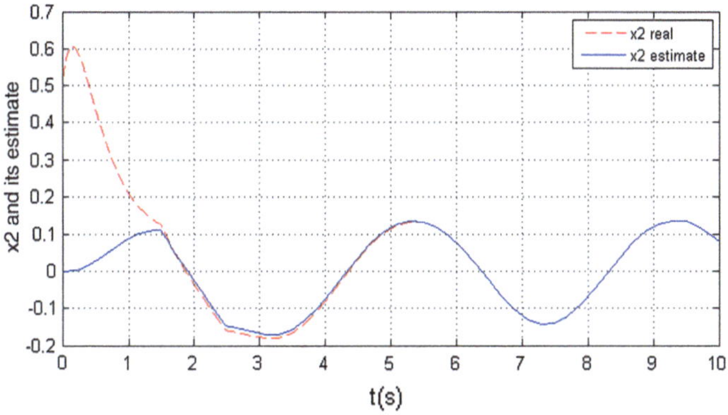

Fig. 5 The state x_2 and its estimate

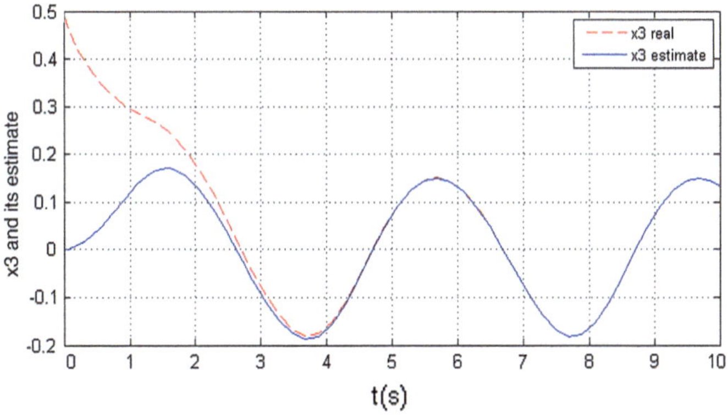

Fig. 6 The state x_3 and its estimate

4 Robust Fault Detection in Fuzzy Bilinear Models

In this section, the design of a residual generator based on unknown input observer for fuzzy bilinear models is developed. A more general situation is analyzed since both unknown input and faults are envisaged. Thus, we consider a fuzzy bilinear system affected by an actuator fault vector $f(t) \in \Re^{nf}$. A residual generator is then synthesized such that it is sensitive to fault vector $f(t)$ and insensitive to the unknown inputs $d(t)$.

4.1 Problem Formulation

The considered fuzzy bilinear model (2) subject to unknown inputs $d(t)$ and affected by a fault vector $f(t)$ is described by the following equation:

$$\begin{cases} \dot{x}(t) = \sum_{i=1}^{r} h_i(\xi(t)) \begin{pmatrix} A_i x(t) + B_i u(t) + N_i x(t) u(t) \\ + F_i d(t) + G_i f(t) \end{pmatrix} \\ y(t) = Cx(t) \end{cases} \tag{54}$$

where $f(t)$ represents the vector of faults and the G_i represents matrix with appropriate dimensions.

The global residual generator is defined by:

$$\begin{cases} \dot{z}(t) = \sum_{i=1}^{r} h_i(\xi(t))(H_i z + L_i y(t) + J_i u(t) + M_i y(t) u(t)) \\ \hat{x}(t) = z(t) - Ey(t) \\ r(t) = \Gamma_1 y(t) - \Gamma_2 z(t) \end{cases} \tag{55}$$

where $z(t)$ represents the estimated vector, and $r(t)$ being the output signal called the residual.

The residual generator design is reduced to determine the gain matrices $H_i, M_i, L_i, J_i, E, \Gamma_1$ and Γ_2 such that the state estimate $\hat{x}(t)$ converges asymptotically to system state $x(t)$. Then, to analyze the convergence of the residual generator, let consider the estimation error from (55) to (54) such that

$$e(t) = x(t) - \hat{x}(t) = \bar{T}x(t) - z(t) \tag{56}$$

where $\bar{T} = I_n + EC$.

The dynamic estimation error is then described by:

$$\dot{e}(t) = \bar{T}\dot{x}(t) - \dot{z}(t) \tag{57}$$

which is equivalent to

$$\dot{e}(t) = \sum_{i=1}^{r} h_i(\xi(t)) \begin{pmatrix} H_i e(t) + (\bar{T}A_i - H_i\bar{T} - L_iC)x(t) \\ +(\bar{T}N_i - M_iC)x(t)u(t) + (\bar{T}B_i - J_i)u(t) \\ +\bar{T}F_i d(t) + \bar{T}G_i f(t) \end{pmatrix} \tag{58}$$

From (55), and by using the estimation error (56), the general expression of the residual vector can be written as:

$$r(t) = \Gamma_2 e(t) + (\Gamma_1 C - \Gamma_2 \bar{T})x(t) \tag{59}$$

If the following conditions hold true $\forall i = 1\ldots r$:

$$\bar{T}A_i - H_i\bar{T} - L_iC = 0 \tag{60}$$

$$\bar{T}N_i - M_iC = 0 \tag{61}$$

$$\bar{T}B_i - J_i = 0 \tag{62}$$

$$\bar{T}F_i = 0 \tag{63}$$

$$\Gamma_1 C - \Gamma_2 \bar{T} = 0 \tag{64}$$

Then the equation of the observing error $e(t)$ and residual $r(t)$ becomes

$$\dot{e}(t) = \sum_{i=1}^{r} h_i(\xi(t))(H_i e(t) + \bar{T}G_i f(t)) \tag{65}$$

$$r(t) = \Gamma_2 e(t) \tag{66}$$

Multiplying (64) by F_i, we get:

$$\Gamma_1 C F_i - \Gamma_2 \bar{T} F_i = 0 \tag{67}$$

Taking into account the constraint (63), Eq. (67) yields:

$$\Gamma_1 C F_i = 0, \quad i = 1, \ldots, r \tag{68}$$

or equivalently

$$\Gamma_1 CF = 0, \quad F = [F_1, F_2, \ldots, F_r] \tag{69}$$

If the condition (70) is satisfied,

$$rank(CF) = rank(F) \tag{70}$$

Then, we get

$$\Gamma_1 = \Omega\left(I_p - CF(CF)^+\right) \tag{71}$$

where $(CF)^+$ is the pseudo inverse of CF, and Ω is an arbitrary matrix.
Substituting Eq. (71) in (64) leads to

$$\Omega C\left(I_n - F(CF)^+ C\right) - \Gamma_2 \bar{T} = 0 \tag{72}$$

A suitable choice of E_1 and $\bar{T}$ satisfying the relation (72) is

$$\Gamma_2 = \Omega C \tag{73}$$

and

$$\bar{T} = I_n - F(CF)^+ C \tag{74}$$

4.2 Design and Stability Analysis

This subsection proposes sufficient linear design conditions to guarantee the global asymptotic convergence of state estimation error. Therefore, the design problem is studied for the two cases; when the decision variables are available or unavailable.

4.2.1 Fault Detection Observer Design with Measurable Decision Variables

The stability of the dynamic estimation error (65) is given by the following theorem.

Theorem 3 *The residual generator* (55) *converges asymptotically to the state of the fuzzy bilinear model* (54)*, if the fault* $f(t)$ *satisfies* $\|f(t)\| \leq \mu$, $\mu > 0$ *and if there exist a symmetric definite positive matrix P, matrices* Z_i, V_i, U_i *and positive scalar* α *such that the following linear conditions hold* $\forall i = 1 \ldots r$:

$$\begin{bmatrix} Z_i + Z_i^T + \alpha I & \mu P\bar{T}G_i \\ * & -\alpha I \end{bmatrix} < 0 \tag{75}$$

$$Z_i\bar{T} + U_iC - P\bar{T}A_i = 0 \tag{76}$$

$$V_iC - P\bar{T}N_i = 0 \tag{77}$$

The observer gains are determined by:

$$H_i = P^{-1}Z_i \tag{78}$$

$$L_i = P^{-1}U_i \tag{79}$$

$$M_i = P^{-1}V_i \tag{80}$$

$$J_i = \bar{T}B_i \tag{81}$$

where $\bar{T}$, Γ_1 and Γ_2 are given in (74), (71), (73), respectively.

Proof In order to establish the stability of the estimation error $e(t)$, let us consider the following Lyapunov function (27). Using (65), the derivative of the Lyapunov function (27) is given by

$$\dot{V}(e(t)) = \sum_{i=1}^{r} h_i(\xi(t))\left(e^T(t)\left(H_i^T P + PH_i\right)e(t) + 2e^T(t)P\bar{T}G_i f(t)\right) \tag{82}$$

If $\|f(t)\| \leq \mu$, then the derivative of $V(e(t))$ becomes:

$$\dot{V}(e(t)) \leq \sum_{i=1}^{r} h_i(\xi(t))[e^T(t)(H_i^T P + PH_i)e(t) + 2\mu\|e^T(t)P\bar{T}G_i\|] \tag{83}$$

For any positive scalar α, we have the following inequality:

$$2\mu\|e^T(t)P\bar{T}G_i\| \leq \alpha^{-1}\mu^2\|e^T(t)P\bar{T}G_i\|^2 + \alpha \tag{84}$$

Hence, one obtains

$$\dot{V}(e(t)) \leq \sum_{i=1}^{r} h_i(\xi(t))[e^T(t)(H_i^T P + PH_i)e(t) + \alpha^{-1}\mu^2\|e^T(t)P\bar{T}G_i\|^2 + \alpha] \tag{85}$$

$$\dot{V}(e(t)) \leq \sum_{i=1}^{r} h_i(\xi(t))[e^T(t)(H_i^T P + PH_i + \alpha^{-1}\mu^2 P\bar{T}G_iG_i^T\bar{T}^T P)e(t) + \alpha] \tag{86}$$

The stability condition $\dot{V}(e(t)) < 0$ is verified if:

$$H_i^T P + PH_i + \alpha^{-1}\mu^2 P\bar{T}G_iG_i^T\bar{T}^T P + \alpha I < 0 \tag{87}$$

Then apply the Schur complement to the condition (87) with change of variables $Z_i = PH_i$, we get the linear matrix inequality (75) and with change of variables $U_i = PL_i$, $V_i = PM_i$, we get the equalities (76) and (77). Thus, the proof is completed.

4.2.2 Fault Detection Observer Design with Unmeasurable Decision Variables

For the case where the decision variable of the weighting function depends of an unmeasured state variable, the equation of the observing error $e(t)$ and residual $r(t)$ becomes

$$\dot{e}(t) = \sum_{i=1}^{r} h_i(\hat{x}(t))(H_i e(t) + \bar{T}G_i f(t) + \bar{T}\Delta(t)) \tag{88}$$

$$r(t) = \Gamma_2 e(t) \tag{89}$$

The following theorem gives linear conditions to design fault detection observer design with unmeasurable decision variables.

Theorem 4 *For a given $\gamma > 0$, the residual generator* (55) *converges asymptotically to the state of the fuzzy bilinear system* (54), *if the fault $f(t)$ satisfies $\|f(t)\| \le \mu$, $\mu > 0$ and if there exist a symmetric definite positive matrix P, matrices Z_i, V_i, U_i and real parameters ε and α such that the following linear conditions hold $\forall i = 1\ldots r$:*

$$\begin{bmatrix} Z_i + Z_i^T + (\varepsilon\gamma^2 + \alpha)I & P\bar{T} & \mu P\bar{T}G_i \\ * & -\varepsilon I & 0 \\ * & * & -\alpha I \end{bmatrix} < 0 \tag{90}$$

$$Z_i\bar{T} + U_iC - P\bar{T}A_i = 0 \tag{91}$$

$$V_iC - P\bar{T}N_i = 0 \tag{92}$$

The observer gains are determined by:

$$H_i = P^{-1}Z_i \tag{93}$$

$$L_i = P^{-1}U_i \tag{94}$$

$$M_i = P^{-1}V_i \tag{95}$$

$$J_i = \bar{T}B_i \tag{96}$$

where $\bar{T}$, Γ_1 and Γ_2 are given in (74), (71), (73) respectively.

Proof To prove the convergence of the state estimation error to zero, we consider the quadratic Lyapunov function (27), differentiating it along (88) and using (48), it becomes:

$$\begin{aligned}\dot{V}(e(t)) \leq \sum_{i=1}^{r} h_i(\hat{x}(t))[e^T(t)\{H_i^T P + PH_i + \varepsilon\gamma^2 I \\ + \varepsilon^{-1} P\bar{T}\bar{T}^T P\}e(t) + 2e^T(t)P\bar{T}G_i f(t)]\end{aligned} \tag{97}$$

If $\|f(t)\| \leq \mu$, then the derivative of $V(e(t))$ becomes:

$$\begin{aligned}\dot{V}(e(t)) \leq \sum_{i=1}^{r} h_i(\hat{x}(t))[e^T(t)\{H_i^T P + PH_i + \varepsilon\gamma^2 I \\ + \varepsilon^{-1} P\bar{T}\bar{T}^T P\}e(t) + 2\mu\|e^T(t)P\bar{T}G_i\|]\end{aligned} \tag{98}$$

Using (84), the last inequality (98) becomes $\forall i = 1\ldots r$

$$\begin{aligned}\dot{V}(e(t)) \leq \sum_{i=1}^{r} h_i(\hat{x}(t))[e^T(t)\{H_i^T P + PH_i + \varepsilon\gamma^2 I \\ + \varepsilon^{-1} P\bar{T}\bar{T}^T P\}e(t) + \alpha^{-1}\mu^2\|e^T(t)P\bar{T}G_i\|^2 + \alpha]\end{aligned} \tag{99}$$

Hence, one has $\forall i = 1\ldots r$

$$\begin{aligned}\dot{V}(e(t)) \leq \sum_{i=1}^{r} h_i(\hat{x}(t))[e^T(t)\{H_i^T P + PH_i + \varepsilon\gamma^2 I \\ + \varepsilon^{-1} P\bar{T}\bar{T}^T P + \alpha^{-1}\mu^2 P\bar{T}G_i G_i^T \bar{T}^T P\}e(t) + \alpha]\end{aligned} \tag{100}$$

The stability condition $\dot{V}(e(t)) < 0$ is verified if $\forall i = 1\ldots r$

$$\begin{aligned}H_i^T P + PH_i + \varepsilon\gamma^2 I + \varepsilon^{-1} P\bar{T}\bar{T}^T P \\ + \alpha^{-1}\mu^2 P\bar{T}G_i G_i^T \bar{T}^T P + \alpha I < 0\end{aligned} \tag{101}$$

Then apply the Schur complement to the condition (101) with change of variables $Z_i = PH_i$, we get the linear matrix inequality (90) and with change of variables $U_i = PL_i$, $V_i = PM_i$, we get the equalities (91) and (92). Thus, the proof is completed.

5 Application: Predator-Prey Model

In this section, an example is given to illustrate the method given in this chapter for the design of the FBO for fuzzy bilinear systems with unmeasurable decision variables. This example concerns the predator-prey model worked out in Keller (1987).

The dynamics of predator-prey model can be described by the following nonlinear second order system:

$$\begin{cases} \dot{x}_1 = ax_1 - bx_1x_2 \\ \dot{x}_2 = cx_1x_2 - dx_2 - fx_2u \\ y = 0.4x_1 + x_2 \end{cases} \tag{102}$$

where the state $x_1(t)$ and $x_2(t)$ represent the prey and predator population, respectively. The predator population may be decimated by men via the input variable $u(t)$. The coefficients a, b, c, d are constant birth and death rates, and f is the extermination rate.

The constants values used in simulation are given by: $a = 1.5$, $b = 1$, $c = 0.3$, $d = 1$ and $f = 0.5$.

Assume that the predator-prey model can be affected by one unknown input. Then the system equation with the term modeling the unknown input is:

$$\begin{cases} \dot{x}_1 = ax_1 - bx_1x_2 + 0.1d \\ \dot{x}_2 = cx_1x_2 - dx_2 - fx_2u + 0.3d \\ y = 0.4x_1 + x_2 \end{cases} \tag{103}$$

5.1 Fuzzy Bilinear Models Representation

The previous nonlinear model (103) can be written as:

$$\begin{cases} \dot{x}(t) = A(x(t))x(t) + Bu(t) + Nx(t)u(t) + Fd(t) \\ y(t) = Cx(t) \end{cases} \tag{104}$$

where the matrices $A(.)$, B, N, F and C are respectively given by:

$$A(x(t)) = \begin{bmatrix} a - bx_2 & 0 \\ cx_2 & -d \end{bmatrix}, \quad B = \begin{bmatrix} 0 \\ 0 \end{bmatrix}$$
$$N = \begin{bmatrix} 0 & 0 \\ 0 & -f \end{bmatrix}, \quad F = \begin{bmatrix} 0.1 \\ 0.3 \end{bmatrix}$$
$$C = \begin{bmatrix} 0.4 & 1 \end{bmatrix}$$

The matrix $A(.)$ contains two nonlinear continuous terms:

$$\xi_1(x(t)) = a - bx_2 \tag{105}$$

$$\xi_2(x(t)) = cx_2 \tag{106}$$

where ξ_j, $(j = 1, 2)$ are two premise variables depend on state $x_2(t)$. Each premise variable is bounded in a compact state space

$$\xi_1(x(t)) \in [\xi_{\min 1}, \xi_{\max 1}] \tag{107}$$

$$\xi_2(x(t)) \in [\xi_{\min 2}, \xi_{\max 2}] \tag{108}$$

Using the polytopic transformation (Chadli and Borne 2012; Tanaka et al. 1998), the nonlinear continuous terms can be written as:

$$\xi_1(x(t)) = F_1^1(x(t)) \cdot \xi_{\max 1} + F_1^2(x(t)) \cdot \xi_{\min 1} \tag{109}$$

$$\xi_2(x(t)) = F_2^1(x(t)) \cdot \xi_{\max 2} + F_2^2(x(t)) \cdot \xi_{\min 2} \tag{110}$$

where the functions F_1^1, F_1^2, F_2^1, and F_2^2 are respectively given by:

$$F_1^1(x(t)) = \frac{\xi_1(x(t)) - \xi_{\min 1}}{\xi_{\max 1} - \xi_{\min 1}}$$
$$F_1^2(x(t)) = \frac{\xi_{\max 1} - \xi_1(x(t))}{\xi_{\max 1} - \xi_{\min 1}}$$
$$F_2^1(x(t)) = \frac{\xi_2(x(t)) - \xi_{\min 2}}{\xi_{\max 2} - \xi_{\min 2}}$$
$$F_2^2(x(t)) = \frac{\xi_{\max 2} - \xi_2(x(t))}{\xi_{\max 2} - \xi_{\min 2}}$$

The decomposition which will be carried out with the combination of nonlinear terms bounds leads to four local models $(r = 2^2 = 4)$. Then, the fuzzy bilinear model is obtained by an interpolation of these local models with four nonlinear activation functions. Let consider the studied case where the decision variable is unmeasurable, the weighting functions depend on the estimated state $\hat{x}_2(t)$. Hence, the fuzzy bilinear representation of the predator-prey system studied subject to unknown input $d(t)$ and unmeasurable decision variable $\hat{x}_2(t)$ can be written as:

$$\begin{cases} \dot{x}(t) = \sum_{i=1}^{4} h_i(\hat{x}_2(t))(A_i x(t) + B_i u(t) + N_i x(t) u(t) + F_i d(t) + \delta(t)) \\ y(t) = Cx(t) \end{cases} \tag{111}$$

where:

$$A_1 = \begin{bmatrix} \xi_{\max 1} & 0 \\ \xi_{\max 2} & -d \end{bmatrix}, B_1 = \begin{bmatrix} 0 \\ 0 \end{bmatrix}, N_1 = \begin{bmatrix} 0 & 0 \\ 0 & -f \end{bmatrix}, F_1 = \begin{bmatrix} 0.1 \\ 0.3 \end{bmatrix}$$
$$A_2 = \begin{bmatrix} \xi_{\max 1} & 0 \\ \xi_{\min 2} & -d \end{bmatrix}, B_2 = \begin{bmatrix} 0 \\ 0 \end{bmatrix}, N_2 = \begin{bmatrix} 0 & 0 \\ 0 & -f \end{bmatrix}, F_2 = \begin{bmatrix} 0.1 \\ 0.3 \end{bmatrix}$$
$$A_3 = \begin{bmatrix} \xi_{\min 1} & 0 \\ \xi_{\max 2} & -d \end{bmatrix}, B_3 = \begin{bmatrix} 0 \\ 0 \end{bmatrix}, N_3 = \begin{bmatrix} 0 & 0 \\ 0 & -f \end{bmatrix}, F_3 = \begin{bmatrix} 0.1 \\ 0.3 \end{bmatrix}$$
$$A_4 = \begin{bmatrix} \xi_{\min 1} & 0 \\ \xi_{\min 2} & -d \end{bmatrix}, B_4 = \begin{bmatrix} 0 \\ 0 \end{bmatrix}, N_4 = \begin{bmatrix} 0 & 0 \\ 0 & -f \end{bmatrix}, F_4 = \begin{bmatrix} 0.1 \\ 0.3 \end{bmatrix}$$
$$C = \begin{bmatrix} 0.4 & 1 \end{bmatrix}$$

and the weighting functions depend on the estimated state $\hat{x}_2(t)$ and are defined by:

$$h_1(\hat{x}_2(t)) = F_1^1(\hat{x}_2(t)).F_2^1(\hat{x}_2(t))$$
$$h_2(\hat{x}_2(t)) = F_1^1(\hat{x}_2(t)).F_2^2(\hat{x}_2(t))$$
$$h_3(\hat{x}_2(t)) = F_1^2(\hat{x}_2(t)).F_2^1(\hat{x}_2(t))$$
$$h_4(\hat{x}_2(t)) = F_1^2(\hat{x}_2(t)).F_2^2(\hat{x}_2(t))$$

- *The numerical values*:

We assume bounded the premise variables by:

$$\xi_1(x(t)) \in [-1.5, -0.5]$$
$$\xi_2(x(t)) \in [0.6, 0.9]$$

The numerical values of the matrices of FBM (111) are given by:

$$A_1 = \begin{bmatrix} -0.5 & 0 \\ 0.9 & -1 \end{bmatrix}, B_1 = \begin{bmatrix} 0 \\ 0 \end{bmatrix}, N_1 = \begin{bmatrix} 0 & 0 \\ 0 & -0.5 \end{bmatrix}, F_1 = \begin{bmatrix} 0.1 \\ 0.3 \end{bmatrix}$$
$$A_2 = \begin{bmatrix} -0.5 & 0 \\ 0.6 & -1 \end{bmatrix}, B_2 = \begin{bmatrix} 0 \\ 0 \end{bmatrix}, N_2 = \begin{bmatrix} 0 & 0 \\ 0 & -0.5 \end{bmatrix}, F_2 = \begin{bmatrix} 0.1 \\ 0.3 \end{bmatrix}$$
$$A_3 = \begin{bmatrix} -1.5 & 0 \\ 0.9 & -1 \end{bmatrix}, B_3 = \begin{bmatrix} 0 \\ 0 \end{bmatrix}, N_3 = \begin{bmatrix} 0 & 0 \\ 0 & -0.5 \end{bmatrix}, F_3 = \begin{bmatrix} 0.1 \\ 0.3 \end{bmatrix}$$
$$A_4 = \begin{bmatrix} -1.5 & 0 \\ 0.6 & -1 \end{bmatrix}, B_4 = \begin{bmatrix} 0 \\ 0 \end{bmatrix}, N_4 = \begin{bmatrix} 0 & 0 \\ 0 & -0.5 \end{bmatrix}, F_4 = \begin{bmatrix} 0.1 \\ 0.3 \end{bmatrix}$$
$$C = \begin{bmatrix} 0.4 & 1 \end{bmatrix}$$

The input signal is defined as follows:

$$u(t) = \begin{cases} 1 & for \quad x_2 > 1 \\ 0 & for \quad x_2 \leq 1 \end{cases}$$

and the unknown input is defined as a sine wave signal of amplitude 1.2 and frequency $0.25 \cdot \Pi$ rad/s.

In order to show the effectiveness of the used modeling method, the nonlinear model and its approximation by the T-S fuzzy model are given in the following Figs. 7 and 8.

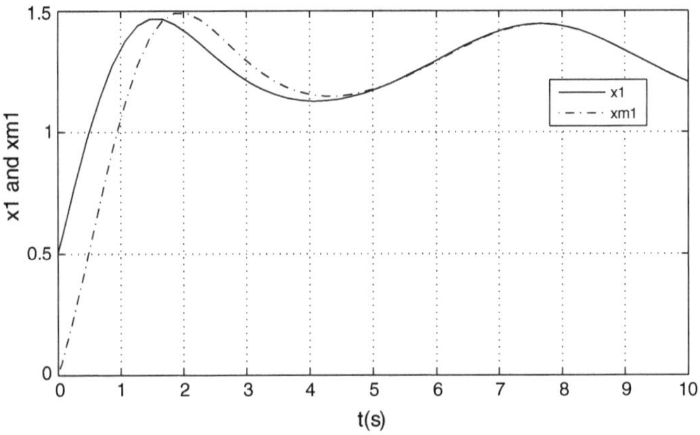

Fig. 7 x_1 of prey population and x_{m1} of the fuzzy bilinear model

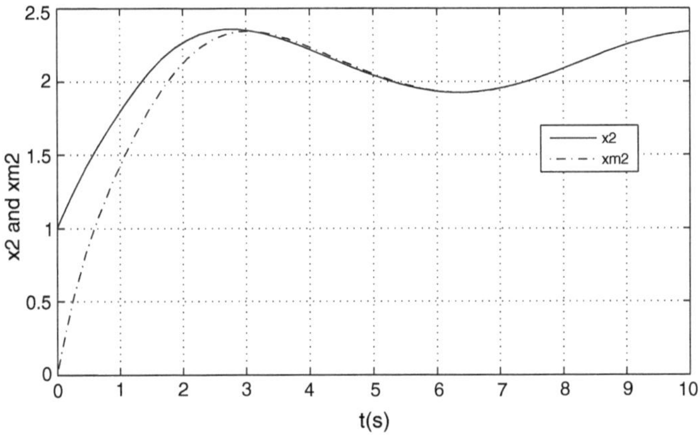

Fig. 8 x_2 of prey population and x_{m2} of the fuzzy bilinear model

These figures consider the fuzzy bilinear models and the nonlinear system. They illustrate the superposition of the nonlinear states with those coming from the T-S bilinear models representation. We can see that the fuzzy bilinear models well approximate the nonlinear dynamic behavior.

5.2 State Estimation Design

The observer gains are obtained by solving the LMIs (38) under constraints (39), (40) and (41). By choosing the scalar $\gamma = 0.895$, their obtained observer gains are:

$$H_1 = \begin{bmatrix} -1.160 & -0.474 \\ 0.264 & -0.311 \end{bmatrix}, L_1 = \begin{bmatrix} 0.013 \\ -0.005 \end{bmatrix},$$

$$J_1 = \begin{bmatrix} 0 \\ 0 \end{bmatrix}, M_1 = \begin{bmatrix} 0.177 \\ -0.071 \end{bmatrix}$$

$$H_2 = \begin{bmatrix} -31.057 & -75.54 \\ 12.223 & 29.716 \end{bmatrix}, L_2 = \begin{bmatrix} 0.072 \\ -0.029 \end{bmatrix},$$

$$J_2 = \begin{bmatrix} 0 \\ 0 \end{bmatrix}, M_2 = \begin{bmatrix} 0.177 \\ -0.071 \end{bmatrix}$$

$$H_3 = \begin{bmatrix} -3.033 & -3.083 \\ 1.013 & 0.733 \end{bmatrix}, L_3 = \begin{bmatrix} -0.342 \\ 0.137 \end{bmatrix},$$

$$J_3 = \begin{bmatrix} 0 \\ 0 \end{bmatrix}, M_3 = \begin{bmatrix} 0.177 \\ -0.071 \end{bmatrix}$$

$$H_4 = \begin{bmatrix} -1.315 & 0.890 \\ 0.326 & -0.856 \end{bmatrix}, L_4 = \begin{bmatrix} -0.287 \\ 0.115 \end{bmatrix},$$

$$J_4 = \begin{bmatrix} 0 \\ 0 \end{bmatrix}, M_4 = \begin{bmatrix} 0.177 \\ -0.071 \end{bmatrix},$$

$$E = \begin{bmatrix} -0.428 \\ -0.829 \end{bmatrix},$$

Then, Figs. 9 and 10 show respectively the evolution of the actual system states and their corresponding observer ones for initial conditions given by $x_0 = [0.5 \quad 1]^T$ and $\hat{x}_0 = 0$.

It can be deduced from these Figs. 9 and 10 that the proposed observer succeeds to track the system trajectories in spite of the presence of the unknown input and with unmeasurable decision variables.

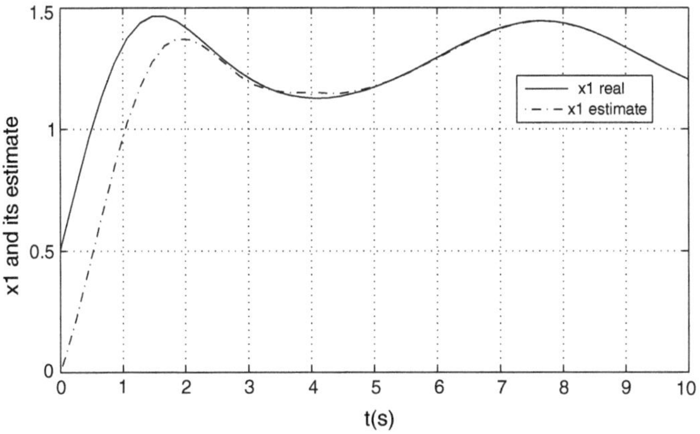

Fig. 9 Evolution of the predator population x_1 and its estimate with unmeasurable decision variables

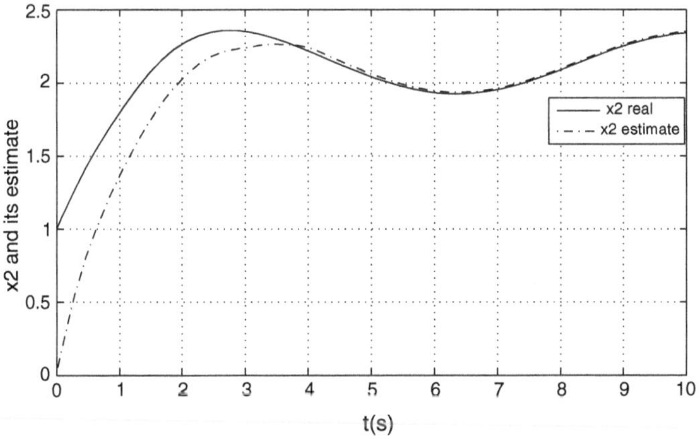

Fig. 10 Evolution of the predator population x_2 and its estimate with unmeasurable decision variables

5.3 Fault Detection

In this subsection, we will consider the same system of predator-prey model subject to fault:

$$\begin{cases} \dot{x}_1(t) = ax_1(t) - bx_1(t)x_2(t) + 0.1d(t) + 0.5f(t) \\ \dot{x}_2(t) = cx_1(t)x_2(t) - dx_2(t) - fx_2(t)u(t) + 0.3d(t) \\ y = 0.4x_1(t) + x_2(t) \end{cases} \tag{112}$$

The considered system can be described using the T-S fuzzy bilinear model of predator-prey model as follows:

$$\begin{cases} \dot{x}(t) = \sum_{i=1}^{4} h_i(\hat{x}(t)) \begin{pmatrix} A_i x(t) + B_i u(t) + N_i x(t)u(t) \\ +F_i d(t) + G_i f(t) + \delta(t) \end{pmatrix} \\ y(t) = Cx(t) \end{cases}$$

with A_i, B_i, N_i, F_i, C are the same previous matrices and

$$G_1 = G_2 = G_3 = G_4 = \begin{bmatrix} 0.5 \\ 0 \end{bmatrix}$$

The system is subjected of fault $f(t)$ of the following form:

$$f(t) = \begin{cases} 0.1 \sin 0.314t & for\ t \in [6\,8] \\ 0 & elsewhere \end{cases}$$

We can check that the condition (70) is verified. Indeed:

$$rank(CF) = rank(F) = 1$$

Then an unknown input fuzzy bilinear fault diagnosis observer given by (55) exists, and it is given by:

$$\begin{cases} \dot{z}(t) = \sum_{i=1}^{4} h_i(\xi(t))(H_i z + L_i y(t) + J_i u(t) + M_i y(t)u(t)) \\ \hat{x}(t) = z(t) - Ey(t) \\ r(t) = \Gamma_1 y(t) - \Gamma_2 z(t) \end{cases}$$

We may choose the arbitrary matrix $\Omega = 1$ to obtain the matrices Γ_1 and Γ_2 using (73) and (71) and the LMIs (90) can be efficiently solved under constraints (91) and (92) via numerical approach within the LMI toolbox in order to compute the gains matrices. Therefore, these inequalities are satisfied with

$$P = 10^4 * \begin{bmatrix} 0.372 & 0.929 \\ 0.929 & 2.321 \end{bmatrix}$$

The remaining matrices H_i, L_i, M_i, J_i can be determined from (93), (94), (95), (96) respectively.

The robust residual signal response is shown in the following figure.

Figure 11 displays the convergence of the residual corresponding to the fault signal. One can see that the residual is almost zero throughout the time simulation run despite the presence of unknown inputs except at time $t = 6\,\mathrm{s}$ where it appears

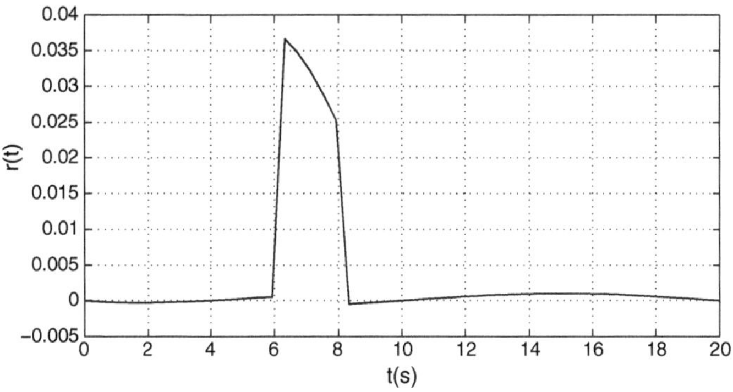

Fig. 11 Residual signal

the fault, and disappears at $t = 8\,\mathrm{s}$. Figure 11 shows that the residual $r(t)$ is sensitive to $f(t)$ and insensitive to $d(t)$. So the designed unknown input fuzzy bilinear fault diagnosis observer can be efficiently used to detect faults.

6 Conclusion

In this chapter, we have presented the design of unknown input observers for nonlinear systems and their application to fault diagnosis. The considered systems are modeled with a T-S fuzzy bilinear structure, particularly suitable for a nonlinear system with a bilinear term. The proposed results are developed for two cases: the first one when the decision variables are measurable, and the second is dedicated to the case when the decision variables are unmeasurable. Convergence conditions are established in order to guarantee the convergence of the state estimation error. The convergence conditions of the given observer are derived using a quadratic Lyapunov candidate function using LMI formulation. The synthesis conditions lead to the resolution of linear constraints easy to solve with existing numerical tools. Then, the proposed unknown input bilinear observer is applied for fault detection. Indeed, a residual generator is considered in order to be sensitive to fault vector and insensitive to the disturbances. These results have been successfully applied to numerical and experimental examples.

References

Azar, A. T. (2010a). A*daptive neuro-fuzzy systems*. In: A.T Azar (ed.), Fuzzy Systems. IN-TECH, Vienna, Austria.

Azar, A. T. (2010b). *Fuzzy systems*. Vienna: IN-TECH.

Azar, A. T. (2012). Overview of Type-2 fuzzy logic systems. *International Journal of Fuzzy System Applications IJFSA, 2*(4), 1–28.

Bergsten, P., Palm, R., & Driankov, D. (2002). Observers for Takagi-Sugeno fuzzy systems. *IEEE Transactions on System, Man, Cybernetics B, Cybernetics, 32*(1), 114–121.

Boyd, S., El Ghaoui, L., Feron, E., & Balakrishnan, V. (1994). *Linear matrix inequalities in system and control theory. Studies in Applied Mathematics (volume 15)*. Philadelphia, PA: SIAM.

Chadli, M. (2010). An LMI Approach to design observer for unknown inputs Takagi-Sugeno fuzzy models. *Asian Journal of Control, 12*(4), 524–530.

Chadli, M. & Borne, P. (2012). *Multimodèles en Automatique: Outils Avancés d'Analyses et de Synthèse*. Hermès-Lavoisier.

Chadli, M. & Borne, P. (2013). *Multiple models approach in automation: Takagi-Sugeno fuzzy systems*. Hardcover.

Chadli, M. & Coppier, H. (2013). Command-control for real-time systems. *Hardcover*, page 368.

Chadli, M., & Guerra, T.-M. (2012). LMI solution for robust static output feedback control of Takagi-Sugeno fuzzy models. *IEEE Transaction on Fuzzy Systems, 20*(6), 1160–1165.

Chadli, M., & Karimi, H. R. (2012). Robust observer design for unknown inputs Takagi-Sugeno models. *IEEE Transaction on Fuzzy Systems, 21*(1), 158–164.

Darouach, M., Zasadzinski, M., & Xu, S. (1994). Full order observer for linear systems with unknown inputs. *IEEE Transaction on Automatic Control, 39*(3), 606–609.

Gao, Z., Shi, X., & Ding, S. X. (2008). Fuzzy state/disturbance observer design for T-S fuzzy systems with application to sensor fault estimation. *IEEE Transaction on System, Man, Cybernetics, Part B., 38*(3), 875–880.

Guan, Y., & Saif, M. (1991). A novel approach to the design of unknown input observers. *IEEE Transaction on Automatic Control, 36*(5), 632–635.

Hou, M., & Muller, P. (1992). Design of observers for linear systems with unknown inputs. *IEEE Transaction on Automatic Control, 37*(6), 871–874.

Ichalal, D., Marx, B., Maquin, D., & Ragot, J. (2012). *Observer design and fault tolerant control of Takagi-Sugeno nonlinear systems with unmeasurable premise variables*. Fault Diagnosis in Robotic and Industrial Systems, Gerasimos Rigatos ed., iConceptPress.

Ichalal, D., Marx, B., Ragot, J., & Maquin, D. (2009). Fault diagnosis for Takagi-Sugeno nonlinear systems. In *7th IFAC Symposium on Fault Detection, Supervision and Safety of Technical Processes*, pages 504–509.

Keller, H. (1987). Non-linear observer design by transformation into a generalized observer canonical form. *International Journal of Control, 46*(6), 1915–1930.

Khalil, H. K. (1996). *Nonlinear systems* (2nd ed.). Englewood Cliffs, NJ: Prentice-Hall.

Lendek, Z., Lauber, J., Guerra, T. M., Babuska, R., & De Schutter, B. (2010). Adaptive observers for TS fuzzy systems with unknown polynomial inputs. *Fuzzy Sets and Systems, 161*(15), 2043–2065.

Li, T. H. S., & Tsai, S. H. (2007). T-S Fuzzy bilinear model and fuzzy controller design for a class of nonlinear systems. *IEEE Transaction on Fuzzy Systems, 15*(3), 494–506.

Li, T. H. S., Tsai, S. H., Lee, J. Z., Hsiao, M. Y., & Chao, C. H. (2008). Robust H infinity fuzzy control for a class of uncertain discrete fuzzy bilinear systems. *IEEE Transaction on System, Man, and Cybernetics, 38*(2), 510–527.

Liu, X., & Zhang, Q. (2003). New approaches to H_∞ controller designs based on fuzzy observers for T-S fuzzy systems via LMI. *Automatica, 39*(9), 1571–1582.

Ma, K. M. (2002). Observer design for a class of fuzzy systems. In *Proceedings of the First International Conference on Machine Learning and Cybernetics*, Vol. 1, pages 46–49.

Ma, X. J., & Sun, Z. Q. (2001). Analysis and design of fuzzy reduced-dimensional observer and fuzzy functional observer. *Fuzzy Sets and Systems, 120*(1), 35–63.

Marx, B., Koenig, D., & Ragot, J. (2007). Design of observers for Takagi-Sugeno descriptor systems with unknown inputs and application to fault diagnosis. *Control Theory and Applications, IET, 1*(5), 1487–1495.

Murray-Smith, R. & Johansen, T. (1997). *Multiple model approaches to modelling and control*. Taylor and Francis.

Patton, R. J., Chen, J., & Lopez-Toribio, C. J. (1998). Fuzzy observers for nonlinear dynamic systems fault diagnosis. In *37th IEEE Conference on Decision and Control*, Vol. 1, pages 84–89.

Saoudi, D., Chadli, M., & Braeik, N. B. (2013a). Design of an active fault tolerant control based on the fuzzy bilinear observer for nonlinear systems. In *10th International Multi-Conference on Systems, Signals and Devices SSD'13, IEEE*, pages 1–6.

Saoudi, D., Chadli, M., & Braeik, N. B. (2013b). State estimation for unknown input fuzzy bilinear systems: application to fault diagnosis. In *European Control Conference ECC'13*, pages 2465–2470.

Saoudi, D., Chadli, M., & Braeik, N. B. (2014). Robust estimation design for fuzzy bilinear systems with unmeasurable premise variables. In *International Conference on Control, Engineering and Information Technology CEIT'14*, pages 1–6.

Saoudi, D., Chadli, M., Mechmeche, C., & Braeik, N. B. (2010). T-S fuzzy bilinear observer for a class of nonlinear systems. In *18th Medit. Conf. Contr. Aut.*, pages 1395–1400.

Saoudi, D., Chadli, M., Mechmeche, C., & Braeik, N. B. (2012a). unknown input observer design for fuzzy bilinear systems: an LMI approach. *Journal of Mathematical Problems in Engineering, MPE'12*, 2012(Special section p1):1–21.

Saoudi, D., Mechmeche, C., Chadli, M., & Braeik, N. B. (2012b). design of multimodel bilinear observers for Takagi-Sugeno discrete models. In *International Symposium on Security and Safety of Complex Systems 2SCS'12*.

Saoudi, D., Mechmeche, C., Chadli, M., & Braeik, N. B. (2012c). Robust residual generator design for Takagi-Sugeno fuzzy bilinear systems subject to unknown inputs. In *8th IFAC Symposium on Fault Detection, Supervision and Safety of Technical Processes, Safeprocess'12*, pages 1023–1028.

Takagi, T., & Sugeno, M. (1985). Fuzzy identification of systems and its applications to modeling and control. *IEEE Transaction on System, Man, Cybernetics, SMC, 15*(1), 116–132.

Tanaka, K., Ikeda, T., & Wang, H. O. (1998). Fuzzy regulators and fuzzy observers: relaxed stability conditions and LMI-based designs. *IEEE Transaction on Fuzzy System, 6*(2), 250–265.

Tanaka, K., & Sugeno, M. (1992). Stability analysis and design of fuzzy control systems. *Fuzzy Sets and Systems, 45*(2), 135–156.

Tanaka, K. & Wang, H. O. (2000). *Fuzzy control systems design and analysis: a linear matrix inequality approach*. Wiley Inter-Science.

Taniguchi, T., Tanaka, K., & Wang, H. O. (2000). Fuzzy descriptor systems and nonlinear model following control. *IEEE Transaction on Fuzzy Systems, 8*(4), 442–452.

Tong, S. & Tang, Y. (2000). Analysis and design of fuzzy robust observer for uncertain nonlinear systems. In *Proceedings of 9th IEEE International Conference on Fuzzy System*, Vol. 2, pages 993–996.

Tsai, S. H., & Li, T. H. S. (2007). robust fuzzy control of a class of fuzzy bilinear systems with time-delay. *Chaos, Solitons and Fractals, 39*(5), 2028–2040.

Yang, F., & Wilde, R. W. (1988). Observers for linear systems with unknown inputs. *IEEE Transaction on Automatic Control, 33*(7), 677–681.

Yoneyama, J. (2009). H_∞ filtering for fuzzy systems with immeasurable premise variables: an uncertain system approach. *Fuzzy Sets and Systems, 160*(12), 1738–1748.

Yoneyama, J., Nishikawa, M., Katayama, H., & Ichikawa, A. (2000). Output stabilization of Takagi-Sugeno fuzzy systems. *Fuzzy Sets and Systems, 111*(2), 253–266.

Unit Commitment Optimization Using Gradient-Genetic Algorithm and Fuzzy Logic Approaches

Sahbi Marrouchi and Souad Chebbi

Abstract The development of the industry and the gradual increase of the population are the main factors for which the consumption of electricity increases. In order to establish a good exploitation of the electrical grid, it is necessary to solve technical and economic problems. This can only be done through the resolution of unit commitment problem (UCP). The decisions are which units to commit at each time period and at what level to generate power meeting the electricity demand. Therefore, in a robust unit commitment problem, first stage commitment decisions are made to anticipate the worst case realization of demand uncertainty and minimize operation cost under such scenarios. Unit Commitment Problem allows optimizing the combination of the production units' states and determining their production planning in order to satisfy the expected consumption with minimal cost during a specified period which varies usually from 24 h to 1 week. However, each production unit has some constraints that make this problem complex, combinatorial and nonlinear. In this chapter, we have proposed two strategies applied to an IEEE electrical network 14 buses to solve the UCP in general and in particular to find the optimized combination scheduling of the produced power for each unit production. The First strategy is based on a hybrid optimization method, Gradient-Genetic algorithm, and the second one relies on a Fuzzy logic approach. Throughout these two strategies, we arrived to develop an optimized scheduling plan of the generated power allowing a better exploitation of the production cost in order to bring the total operating cost to possible minimum when it's subjected to a series of constraints. A comparison was made to test the performances of the proposed strategies and to prove their effectiveness in solving Unit Commitment problems.

S. Marrouchi (✉) · S. Chebbi
Laboratory of Technologies of Information and Comunication and Electrical Engineering (LaTICE), High Engineering School of Tunis, ENSIT, University of Tunis, TUNISIA, 5 Avenue Taha Hussein-Montfleury, BP 56 Bab Mnara, 1008 Tunis, Tunisia
e-mail: sahbimarrouchi@yahoo.fr

S. Chebbi
e-mail: souadchebbi@yahoo.com

Q. Zhu and A.T. Azar (eds.), *Complex System Modelling and Control Through Intelligent Soft Computations*, Studies in Fuzziness and Soft Computing 319,
DOI 10.1007/978-3-319-12883-2_24

Keywords Generation scheduling · Unit commitment · Fuzzy-logic · Optimization · Production cost

1 Introduction

In recent decades, the demand for electricity has undergone an excessive increase with the growth of industrialization. The power system is an interconnection of generating units to load centers through high voltage electric transmission lines and in general is mechanically controlled. It can be divided into three subsystems: generation, transmission and distribution subsystems. Until recently all three subsystems were under supervision of one body within a certain geographical area providing power at regulated rates (Mantawy et al. 1998). Considering the recent electrical network evolutions, the guarantee of stability and to ensure the service continuity became currently the most interested subjects. In addition, these electrical networks must be stable for all small variations as well as for several disturbances cases which able to lead the system to a total voltage collapse (Abbassi et al. 2012; Moez et al. 2011). Thus, different electrical network stability procedures have been procured in order to found desired and acceptable voltage level at any electrical network bus.

In order to provide cheaper electricity the deregulation of power system, which will produce separate generation, transmission and distribution companies, is already being performed. At the same time electric power demand continues to grow and also building of the new generating units and transmission circuits is becoming more difficult because of economic and environmental reasons. Therefore, power utilities are forced to relay on utilization of existing generating units and to load existing transmission lines close to their thermal limits. However, stability has to be maintained at all times. Hence, in order to operate power system effectively, without reduction in the system security and quality of supply, even in the case of contingency conditions such as loss of transmission lines and/or generating units, which occur frequently, and will most probably occur at a higher frequency under deregulation, a new control strategies need to be implemented.

As electrical systems have an important role in modern society, energy managers are trying to ensure the proper functioning of generators while guaranteeing minimal cost. In this context, reliable power production is critical to the profitability of electricity utilities. Power generators (units) need to be scheduled efficiently to meet electricity demand. This dissertation develops a solution method to schedule units for producing electricity while determining the estimated amount of surplus power each unit should produce taking into consideration the stochasticity of the load and its correlation structure (Attaviriyanupap et al. 2002; Cheng et al. 2002). This scheduling problem is known as the unit commitment problem in the power industry. Thus, solving Unit Commitment (UC) problem remains a challenge in optimizing operational planning systems devoted for power production due to its combinatorial nature. In addition, basing on the fact that the total load varies throughout the day and reaches a peak value different from 1 day to another, each utility company must decide in advance which generators must start and when they should be connected to the electrical grid as well as the sequence in which production units should be turned off and for how long. The solution method developed to solve this problem can handle the presence of wind power plants, which creates additional uncertainty. In this problem it is assumed that the system under consideration is an isolated one such that it does not have access to an electricity market. In such a system, the utility needs to specify the probability level that the system should operate under. This is taken into consideration by solving a chance constrained program (Wu et al. 2000; Mantawy et al. 1998). Instead of using a set level of energy reserve, the chance constrained model determines the level probabilistically which is superior to using an arbitrary approximation. Under such probability of generator operating, various number of optimization method are usually used to reduce the production cost during a specified horizon time. This time horizon vary from 24 h to 1 week allowing the determination of a production set units that should be connected to the electrical grid to respond to the request among a minimum production cost (Mantawy et al. 1998; Saber et al. 2007).

We pose the problem of finding the profit-maximizing commitment policy of a generating plant that has elected to self-commit in response to exogenous but uncertain energy and reserve price forecasts. Typically, one generator's output does not physically constrain the output of a different generator 1, so this policy can be applied to each generator in the merchant's portfolio separately and independently. Therefore, for ease of exposition, we assume the case of a single generator. Generators characteristics such as start-up and shutdown costs, minimum and maximum up and down times, ramping rates, etc., of this generator are assumed known. The variation of prices for energy and reserves in future time frames is known only statistically. In particular, the prices follow a stochastic rather than deterministic process. We model the process using a Markov chain. The method is applicable to multiple markets (e.g., day-ahead, hour-ahead) and multiple products (energy, reserves), (Victoire and Jeyakumar 2005; Attaviriyanupap et al. 2002; Juste et al. 1999). Recently, some methods based on artificial intelligence, such as meta-heuristics have been applied to overcome this problem. The introduction of artificial intelligence techniques in control software and decision-making is an essential element in research and in the development of tomorrow's networks (Victoire and Jeyakumar 2005). Thus, to have a good result in operational planning of production units and ensure a minimum production cost, we proposed two strategies for solving the Unit Commitment problem, the first one is based on the combination of two calculations methods, the genetic algorithm and the gradient method and the second one based on the fuzzy logic approach.

The chapter is organized as follows; Sect. 2 describes the reviews for existing works related to the use of the optimization method to solve the Unit Commitment Problem. Section 3 is reserved to formulate the Unit Commitment Problem. Next, in Sect. 4, Methodology of resolution through fuzzy logic and gradient genetic algorithm methods is presented. Section 5 deals with the discussion of simulation results and the main improvements of adopted strategies are highlighted. Finally, Sect. 6 resumes the main conclusions followed by references.

2 Related Work

Since improvement of optimization techniques means decreasing costs of electric power utilities and ensuring continuity of service and a better quality of energy, great effort has been spent on finding a better optimization solution. The complexity of the UC problem and the benefits of its solution improvement keep the research continuously attractive. Therefore, a study of the literature on methods of solving the Unit Commitment Problem (UCP) shows that various numerical optimization techniques have been used to solve this problem. Indeed, dynamic programming (Snyder et al. 1987; Guan et al. 1992; Ouyang and Shahidehpour 1991) is simple but it requires enough computation time to converge to the optimal solution. Dynamic programming searches the solution space for the optimal unit status. The search can proceed in a forward or backward direction. The time periods of the study horizon are known as the stages of the dynamic programming problem. Each stage represents 1 h of operation. The combinations of units within the time period are known as the states, and controls at the stage are again possible combinations of units to be committed. The local cost function at a stage is the production cost in the corresponding hour. Forward dynamic programming consists of two phases, a forward recursion phase in which optimal paths to all reachable states at all stages are established, and a backward recursion phase in which the optimal solution is recovered starting from the feasible terminal state with the List cumulative cost.

As Merlin and Zhuang (Merlin and Sandrin 1983; Zhuang and Galiana 1988), they adopted the method of Lagrangian relaxation because it was more effective than the dynamic programming method due to its better quality of solution and computation time rapidity. Therefore, Lagrangian Relaxation is the combination of a dual optimization techniques and feasibility search procedures. The original mathematical problem is known as the Primal Problem (PP). Corresponding to the primal problem, the Lagrangian dual can be constructed. The Dual Problem (DP) usually has lower dimensions than the PP and is easier to solve.

The Lagrangian dual of the unit commitment problem has a continuous and convex objective function and is subject only to simple bounding constraints on the solution variables. The Lagrangian Relaxation technique uses a different kind of decomposition which generates lower dimension subproblems. Each subproblem consists of determining the commitment schedule for a single unit over the planning horizon. Each subsystem is solved independently, and the only link between these subsystems is Lagrange multipliers, or so called pricing mechanism, which is adjusted to ensure that the system constraints are met.

However, studies (Dekrajangpetch et al. 1999; Guan et al. 1992) have shown that digital convergence and the quality of the solutions are not satisfactory whenever the

Unit Commitment problem is applied to identical units. Furthermore, a recent work (Grey and Sekar 2008) presented a unified solution of the security constrained unit commitment (SCUC) using linear programming (LP) as the optimization tool and an extended DC network model in order to account for the security and the contingency concerns and to calculate the economic dispatch (ED).

By contrast, the occurrence of meta-heuristics methods, genetic algorithm (Cheng et al. 2002; Yingvivatanapong 2006; Damousis et al. 2004), Tabu search (Sudhakaran et al. 2010), simulated annealing (Lin et al. 1993; Rajan et al. 2002) has improved the quality of the optimal solutions. However, these methods require a considerable computation time especially for complex problems. In this context, Maifeld and Sheble (1996) have presented a new strategy for solving the UC problem. The proposed strategy relies on using genetic algorithm (GA) based on a new mutation technology. The results showed that the proposed algorithm have found a good list of planning for production units during a fairly reasonable computation time. However, Genetic algorithms are time-consuming since it requires binary encoding and decoding to represent each unit operation state and to compute the fitness function, respectively, throughout genetic algorithm procedures. This causes huge computation burdens, making it difficult to apply to large-scale systems. GA (Padhy 2001; Wu et al. 2000; Hong and Li 2002) is a general-purpose stochastic and parallel search method based on the mechanics of natural selection and natural genetics. It is a search method, which has the potential of obtaining near-global minimum, and the capability to obtain the accurate results within short time and the constraints are included easily. The ANN (Sasaki et al. 1992; Kohonen 1998; Wood and Woolenberg 1996) has the advantages of giving good solution quality and rapid convergence, and this method can accommodate more complicated unit-wise constraints and is claimed for numerical convergence and solution quality problems. The solution processing in each method is very unique.

Regarding to Zhao et al. (2006), they have applied a hybrid optimization method for solving UC problem: This method is based on the combination of Particle Swarm Optimization (PSO) method, the technique of sequential quadratic programming (SQP), and tabu search (TS) method. The combinatorial part of the UC problem was solved using the TS method. Nevertheless, the nonlinear part of the economic dispatch problem (EDP) was solved using a hybrid technique of PSO and SQP methods. The effectiveness of the hybrid optimization technique has been tested on a network with 7 production units. In the same context, Kazarlis et al. (1996) have developed a genetic algorithm strategy based on different evaluation functions to solve the problem of unit commitment. In order to evaluate the algorithm performances, 100 production units have been tested and the results were compared to those found by the dynamic programming and the Lagrangian method. Using the approach based on fuzzy logic has undergone major progress in effectiveness due to its resolution of the nonlinear difficult problems. Indeed, fuzzy logic follows an approximation of reasoning while enabling effective decision making.

In studies (Kurban and Filik 2009; Dieu and Ongsakul 2007), authors have adopted a fuzzy dynamic programming algorithm to determine the optimal time schedule of a power system interconnected with WECS which considers the wind produced power, the system demand, the reserve requirements and the operational cost as fuzzy quantities. Hence, Saber et al. (2007), have used a twofold simulated annealing method for the optimization of fuzzy-based Unit Commitment model. The adopted strategy has served to offer a robust solution for Unit Commitment problem but it deserves more computing time to converge.

The above said methods are very efficient for solving highly nonlinear and combinatorial optimization problems. When the size of the problem increases, these evolutionary methods (Marrouchi and Chebbi 2013; Rajan and Mohan 2004) will locate the high performance region of the solution space at quick execution time but they face difficulty in locating the exact optimal solution.

3 Problem Formulation

The objective of the Unit Commitment Problem (UCP) is the minimization of total production costs while determining the on/off states of each unit U_{ih} over a period of time H. In the Unit Commitment Production under consideration, an interesting solution would be minimizing the total operating cost of the generating units with several constraints being satisfied.

The total production cost consists of the running, start-up and banking costs.

The first term in the total production cost is associated with the unit in the generation mode and is called the running cost. The running cost of a thermal unit is a function of the power output. It is obtained by adding the fuel cost function and the operational and maintenance costs. The running cost for unit i can be approximated by a quadratic form as follows:

$$\text{Min } C\,(U,P),\ \text{as } C\,(X,P) = \sum_{i=1}^{N_g} \sum_{h=1}^{H} \left[\phi_i(P_{ih}) + ST_i(1 - U_{i(h-1)})\right] U_{ih} \tag{1}$$

where:

$\phi_i(P_{ih})$: Represents the polynomial function defined by:

$$\phi_i(P_{ih}) = a_i P_{ih}^2 + b_i P_{ih} + c_i \tag{2}$$

When a generation unit transits from a shut-down mode to banking or generation mode, certain costs are incurred to bring a boiler to its working temperature and to carry out the start-up procedure until the unit is ready to be synchronized to the power system. These expenses are associated with the second term in the total production cost, called start-up costs.

The start-up process of large steam generators may take several hours and the start-up component in the total production cost for these units is very significant. On the other hand, small gas turbine units can be quickly transferred to the generation mode and the start-up cost for such units is small. Hence, the start-up cost ST_i can be modeled by the following function system:

$$ST_i = \begin{cases} HSC_i \ si \ MDT_i \leq \tau_i^{OFF} \leq MDT_i + SC_i \\ CSC_i \ si \ \tau_i^{OFF} \succ MDT_i + SC_i \end{cases} \quad (3)$$

with:

a_i, b_i and c_i	Coefficients of the production cost,
P_{ih}	Active power generated by the ith unit hth hour, $i = 1, 2, 3, \ldots, N_g$ and $h = 1, 2, 3, \ldots., H$
U_{ih}	On/off status of the ith production unit at the hth hour, $U_{ih} = 0$ for the Off status of one generating unit and $U_{ih} = 1$ for the operating status of one generating unit,
HSC_i	Hot start-up cost of the ith unit,
CSC_i	Cold start-up cost of the ith unit,
MDT_i	Minimum down-time of the unit i,
τ_i^{OFF}	Continuously off-time of unit i,
SC_i	Cold start time of unit i,
N_g	Number of generating units,
H	Time horizon for UC (h).

Unit Commitment is a highly constrained optimization problem. Different power systems have a different set of imposed constraints. The most common can be divided into two categories. The first, called unit constraints, represents the constraints that are applied to the single units; the second type, system constraints, contain those that are applied to the whole power system.

- *System Constraints*
  - Power balance constraints:

  At any time over the planning horizon the total real power generation of the system must be equal to the total demand.

$$\sum_{i=1}^{N_g} P_{ih} U_{ih} - P_{dh} - P_{Lh} = 0 \quad (4)$$

  - Spinning reserve constraints

$$P_{dh} + P_{rh} + P_{Lh} \leq \sum_{i=1}^{N_g} U_i^h P_i^h \quad (5)$$

- *Unit Constraints*

  - Generation limits:

These constraints require that the unit generation be within the minimal, $P_i^{\min}$, and maximal, $P_i^{\max}$, generation levels.

$$P_i^{\min} \leq P_{ih} \leq P_i^{\max} \tag{6}$$

  - Minimum up-time constraint:

The minimum-up time constraint determines the shortest duration a unit must stay in the generation mode, MUP_i after its transit to this mode.

$$U_{ih} = 1 \quad for \quad \sum_{t=h-up_i}^{h-1} U_{ih} \leq MUT_i \tag{7}$$

  - Minimum down-time constraint:

The minimum-down time constraint specifies the shortest duration a unit must stay in the shut-down mode, MDT_i, after it is shut down.

$$U_{ih} = 0 \quad for \quad \sum_{t=h-down_i}^{h-1} U_{ih} \leq MDT_i \tag{8}$$

These constraints are imposed to prevent wear-and-tear of the apparatus due to too frequent transits from one mode to another.

Knowing that:

P_{rh}	System spinning reserve at the *h*th hour,
P_{dh}	Amount of the consumed power at the *h*th hour,
P_{Lh}	Total active losses at the *h*th hour,
$P_i^{\min}$, $P_i^{\max}$	Minimum and maximum power produced by a generator,
MUT_i	Continuously on-time of unit *i*

This chapter attempts to find the solution to the dynamic unit commitment problem by dividing the entire planning horizon into *t* intervals and solving each optimization problem separately. This is equivalent to solving *t* static unit commitment problems. Some heuristics are applied then to combine the unit commitment for each time interval so that they also satisfy minimum-up and minimum-down requirements. In this formulation, the problem becomes much easier to solve and its dimension reduces substantially. We propose to test the hypothesis that the error caused by these assumptions will not be significantly higher compared to the error from dynamic programming. Here in order to transform the complex nonlinear constrained problem into a linear unconstrained problem, we consider the following Lagrangian function:

$$L = \sum_{i=1}^{N_g}\sum_{h=1}^{H}[\phi_i(P_{ih}) + ST_i(1 - U_{i(h-1)})]U_{ih} + \eta \cdot (P_{dh} - P_{Lh} - \sum_{i=1}^{N_g} P_i U_{ih}) \quad (9)$$

Herein, η is the Lagrange coefficient.

The hypothesis being tested in this chapter is that the dynamic unit commitment can combines such solutions to obtain the final unit commitment over the entire planning period. Since the final unit commitment must not only satisfy the demand and reserve constraints, but also minimum-up and minimum-down time constraints, such combinations of independent solutions may not always be found. Therefore, heuristic rules must be applied to the solution of the start-up cost at each consecutive time interval in order to satisfy the time-coupling constraints, since an incorrect choice of the committed units at time t may affect the solutions for the rest of the optimization period.

4 Methodology of Resolution

We have adopted two strategies for solving the Unit Commitment problem, the first one is based on the combination of two calculations methods, the genetic algorithm and the gradient method and the second one based on the fuzzy logic approach. The resolution of the of the Unit Commitment problem through Gradient genetic algorithm method is provided by a specific adjustment of the Lagrangian multipliers λ_i of the Lagrangian function. The combined choice of these two methods is due to inquire about the rapidity of the genetic algorithm in the search for global minimum in first step, and to operate the benefits the gradient method in a second step, since it is effective in terms of the quality of the obtained optimal solutions. Besides, the use of the fuzzy logic approach to resolve this problem is depicted to the effectiveness of this optimization method in solving nonlinear difficult problems.

4.1 Fuzzy Logic

Fuzzy logic provides not only a meaningful and powerful representation for measurement of uncertainties but also a meaningful representation of blurred concept expressed in normal language. Fuzzy logic is a mathematical theory, which encompasses the idea of vagueness when defining a concept or a meaning. For example, there is uncertainty or fuzziness in expressions like ‘low’ or ‘high’, since these expressions are imprecise and relative. Thus, the variables considered are termed ‘fuzzy’ as opposed to ‘crisp’. Fuzziness is simply one means of describing uncertainty (Azar 2012). Such ideas are readily applicable to the unit commitment problem. The application of fuzzy logic allows a qualitative description of the behavior of a certain system, the characteristics of the system, and the response of that system without the need for exact mathematical formulation (Azar 2010a, 2012)

To establish our strategy, we have considered the partial derivatives of the Lagrange function (Eq. 9) with respect to each of the controllable variables equal to zero.

$$\frac{\partial L}{\partial P_{ih}} = \frac{\partial[\phi_i(P_{ih})]}{\partial P_{ih}} - \eta\left(\frac{\partial P_{Lh}}{\partial P_{ih}} - U_{ih}\right) = 0 \tag{10}$$

$$\frac{\partial L}{\partial \eta} = P_{dh} - P_{Lh} - \sum_{i=1}^{N_g} P_i U_{ih} = 0 \tag{11}$$

Equations (10) and (11) represent the optimality conditions necessary to solve equation systems Eqs. (1) and (4) without using inequality constraints (Eqs. 5 and 6). Equation (10) can be written as follows:

$$\eta = \frac{\dfrac{\partial[\phi_i(P_{ih})]}{\partial P_{ih}}}{\dfrac{\partial P_{Lh}}{\partial P_{ih}} - U_{ih}}; \quad i = 1, \ldots, N_G; h = 1, \ldots, H \tag{12}$$

The term $\frac{\partial[\phi_i(P_{ih})]}{\partial P_{ih}}$ represents the incremental cost (IC) of each unit i and $\frac{\partial P_{Lh}}{\partial P_{ih}}$ represents the incremental losses (IL). These terms occur as fuzzy variables associated to our strategy in order to solve the Unit Commitment problem. It should be noted that the strategy is based on the integration of a fuzzy controller to optimize the cost of the production unit while ensuring proper planning of the production units. In the current formulation, the fuzzy input variables associated to the Unit Commitment problem are the load capacity of the generator (LCG), the incremental cost (IC) and the incremental losses (IL). The output variable is the cost of production (C_P). The fuzzy sets related to these variables are selected and normalized between 0 and 1. This normalized value can be multiplied by a scaling factor chosen to accommodate any desired variable.

In the following, a brief description and explanation of the main choice of the mentioned fuzzy variables:

- Load capacity of generator LCG is considered to be fuzzy, as it is based upon the load to be served.
- Incremental losses IL is taken to be fuzzy, because the losses can lead to changes in the total production cost and because losses varies over the holy network architecture.
- Incremental cost IC is taken to be fuzzy, because the cost of fuel may change over the period of time, and because the cost of fuel for each unit may be different.

- Production cost C_P of the system is treated as a fuzzy variable since it is directly proportional to the hourly load.

Fuzzy sets defining the input of the load capacity generator LCG are:
LCG = {Low, Below Average, Average, Above Average, High}
The incremental cost IC is indicated by the following fuzzy sets:
IC = {Zero, Small, Large}
Fuzzy sets representing the incremental losses IL are as follows:
IL = {Low, Medium, High}
The production cost, taken as objective function is:
C_P = *{Low, Below Average, Average, Above Average, High}*

Based on the cited fuzzy sets, membership functions are selected for each fuzzy input/output variables as shown in the following figure (Fig. 1):

In this paper, we have chosen a triangular form to illustrate the membership functions while choosing the If-Then rules to link the input/output fuzzy variables as shown in the following table (Table 1):

The fuzzy decision-making of the fuzzy logic approach is based on three data processing runs of the variables to control. A first data processing run consists of a fuzzification (Azar 2010b), in one second phase the corrector deduces the fuzzy inferences according to imposed conditions' and in a third phase of calculation; each corrector applies a method of defuzzification to deduce a non fuzzy vector of command. The method used in order to evaluate this vector, consists in determining the X-coordinate of the centre of gravity of the surface swept by the fuzzy deductions (Eq. 13).

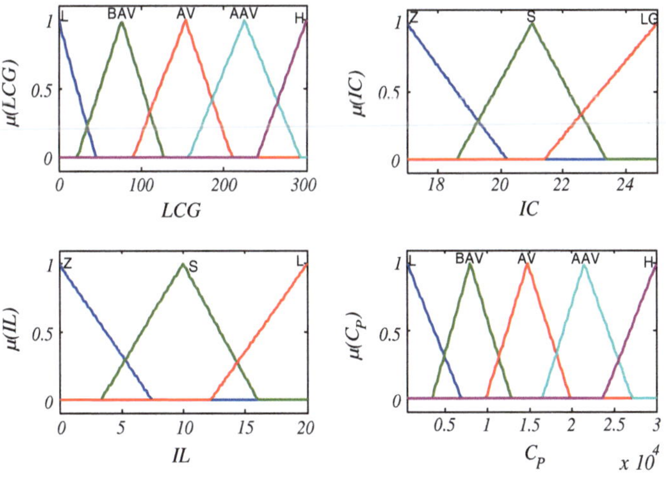

Fig. 1 Membership function of input/output variables

Table 1 Fuzzy rules relating input/output fuzzy variables

LCG	IC	IL	PRC	LCG	IC	IL	PRC	LCG	IC	IL	PRC
L	L	Z	L	BAV	LG	Z	BAV	AAV	M	Z	AAV
L	L	S	L	BAV	LG	S	BAV	AAV	M	S	AAV
L	L	LG	L	BAV	LG	LG	BAV	AAV	M	LG	AAV
L	M	Z	L	AV	L	Z	AV	AAV	LG	Z	AAV
L	M	S	L	AV	L	S	AV	AAV	LG	S	AAV
L	M	LG	L	AV	L	LG	AV	AAV	LG	LG	AAV
L	LG	Z	L	AV	M	Z	AV	H	L	Z	H
L	LG	S	L	AV	M	S	AV	H	L	S	H
L	LG	LG	L	AV	M	LG	AV	H	L	LG	H
BAV	L	Z	BAV	AV	LG	Z	AV	H	M	Z	H
BAV	L	S	BAV	AV	LG	S	AV	H	M	S	H
BAV	L	LG	BAV	AV	LG	LG	AV	H	M	LG	H
BAV	M	Z	BAV	AAV	L	Z	AAV	H	LG	Z	H
BAV	M	S	BAV	AAV	L	S	AAV	H	LG	S	H
BAV	M	LG	BAV	AAV	L	LG	AAV	H	LG	LG	H

$$production\ Cost = \frac{\int_{-1}^{1} C_P \cdot \mu(C_P) \cdot dC_P}{\int_{-1}^{1} \mu(C_P) \cdot dC_P} \tag{13}$$

with:

$\mu(C_P)$ Membership degree of the production cost vector.

Based on the aforementioned fuzzy sets, membership functions are selected for each fuzzy input and the fuzzy output variables. For our case study, a triangular shape is used to illustrate the considered membership functions. Once the membership functions are set, the input variables are then linked to the output variable by IF-THEN rules as shown in the following scheme (Fig. 2).

The Unit Commitment problem can be considered as two linked optimization sub-problems, namely the unit-scheduling problem and the economic load dispatch problem. The second proposed optimization method integrates genetic algorithm with the gradient optimization method to solve the UC problem.

4.2 Gradient-Genetic Algorithm Method

The purpose of this strategy is to validate an approach to apprehend the whole problem by combining an economic model with a model having operational constraints. To achieve this purpose, the approach is to combine a classical gradient method with a meta-heuristic method, genetic algorithm, well suited to take into account new constraints. The fundamental principle of a genetic algorithm is to represent the natural evolution of organisms (individuals).

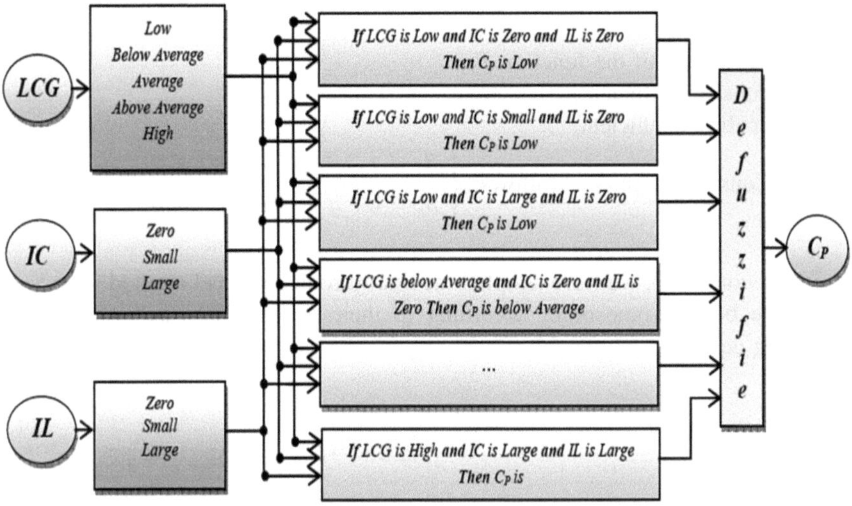

Fig. 2 Configuration of the fuzzy logic system

The solution in the unit commitment problem is represented by a binary matrix U of dimension ($l \times N_g$). The proposed method for coding is a mix of binary and decimal numbers. Each column vector in the solution matrix (which is the operation schedule of one unit) of length l is converted to its equivalent decimal number. The solution matrix is then converted into one row vector (chromosome) of N_g decimal numbers (U_l, $U_2 \ldots U_n$); each variable represents the schedule of one unit. The numbers U_l, $U_2 \ldots U_n$ are integers ranging from 0 to ($2^{N_g} - 1$). Accordingly, a population of size (*POP*) is randomly generated in a matrix $\left(N_{POP} \times N_g\right)$.

In one individual's population, only the strongest, or in other words the best suited to the natural environment, survive and can give offspring. In each evolution stage, the genetic operators (selection, crossover and mutation) operate basing on the data structures in order to allow each individual to sweep the solutions horizon and to distinguish the global optimum among the local optimum (Damousis et al. 2004; Sudhakaran et al. 2010; Marrouchi and Chebbi 2013). At first, from an initial population of individuals, the evaluation function satisfies the following relation:

$$\begin{cases} F(U,P) = \dfrac{1}{1 + K(\frac{F_{\max}}{Fr} - 1)} \\ with \quad F_r = \dfrac{1}{\sum\limits_{i=1}^{N_g} \sum\limits_{h=1}^{H} [\phi_i(P_{ih}) + ST_i(1 - U_{i(h-1)})]U_{ih} + \lambda_i \cdot (P_d - \sum\limits_{i=1}^{N_g} P_i U_{ih}) + \sum\limits_{h=1}^{H} \beta_h \cdot L} \end{cases} \quad (14)$$

With:

$F_{\max}$ Maximum of the function F_r,
L Penalty coefficient,
K Scaling coefficient,
β_h Constant defined as follows: $\begin{cases} \beta_h = 1 & if \quad C(P_{ih}, U) \neq 0 \\ \beta_h = 0 & if \quad C(P_{ih}, U) = 0 \end{cases}$

In a second step, we have adopted the biased roulette wheel method in order to select the best chromosomes according to their performances according to the following equation:

$$perf(c_i) = \frac{f(c_i)}{\sum_{i=1}^{l} f(c_i)} \tag{15}$$

where, l is the length of a binary string.

Subsequently, we cross these chromosomes in order to obtain a population of children. Hence, we can randomly mutate the genes according to (Zhao et al. 2006) (Fig. 3).

In passing from one generation to another, the old population should be replaced by the descendant's population newly created to maintain the search for better

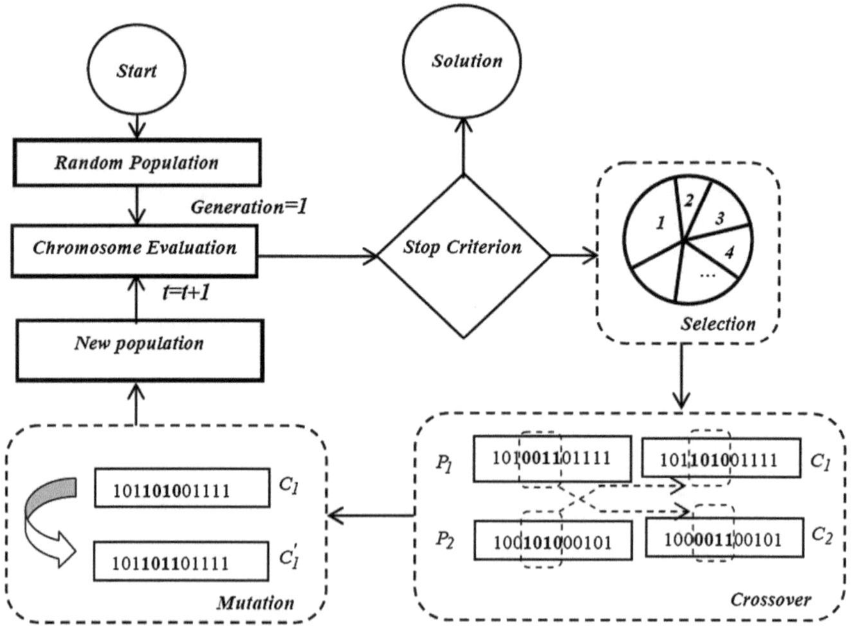

Fig. 3 Operations of the genetic algorithm

solutions. This step is important because it determines the degree of exploitation and advancement of the optimal solution search. This research is based on the save of the best solution until the optimization progresses.

Minimize the objective function, (Eq. 1), is equivalent to minimize the Lagrangian function, (Eq. 9). The process is carried out through the research of the descent direction of the greatest slope corresponding to the minimum production cost. Indeed, we have:

$$\gamma_{k+1} = \gamma_k + d_k \cdot \xi_k \tag{16}$$

The vectors γ_{k+1}, d_k, ξ_k are defined by the following equations system:

$$\begin{cases} \gamma_{k+1} = \begin{bmatrix} P_{ih} \\ X_{ih} \\ \lambda_i \end{bmatrix} \\ \xi_k = \begin{bmatrix} \frac{\partial L}{\partial P_{ih}} \\ \frac{\partial L}{\partial X_{ih}} \\ \frac{\partial L}{\partial \lambda_i} \end{bmatrix} \\ d_k = \dfrac{\xi_k^t \cdot \xi_k}{\xi_k \cdot (A \cdot \xi_k)} \end{cases} \tag{17}$$

where, the Hessian matrix A is defined by:

$$A = \begin{bmatrix}
\frac{\partial^2 L}{\partial P_{1h}\partial P_{1h}} & \frac{\partial^2 L}{\partial P_{1h}\partial P_{2h}} & \cdots & \frac{\partial^2 L}{\partial P_{1h}\partial P_{N_g h}} & \frac{\partial^2 L}{\partial P_{1h}\partial X_{1h}} & \frac{\partial^2 L}{\partial P_{1h}\partial X_{2h}} & \cdots & \frac{\partial^2 L}{\partial P_{1h}\partial X_{N_g h}} & \frac{\partial^2 L}{\partial P_{1h}\partial \lambda_1} & \cdots & \frac{\partial^2 L}{\partial P_{1h}\partial \lambda_{N_g}} \\
\frac{\partial^2 L}{\partial P_{2h}\partial P_{1h}} & \frac{\partial^2 L}{\partial P_{2h}\partial P_{2h}} & \cdots & \frac{\partial^2 L}{\partial P_{2h}\partial P_{N_g h}} & \frac{\partial^2 L}{\partial P_{2h}\partial X_{1h}} & \frac{\partial^2 L}{\partial P_{2h}\partial X_{2h}} & \cdots & \frac{\partial^2 L}{\partial P_{2h}\partial X_{N_g h}} & \frac{\partial^2 L}{\partial P_{2h}\partial \lambda_1} & \cdots & \frac{\partial^2 L}{\partial P_{2h}\partial \lambda_{N_g}} \\
\cdots & \cdots & \cdots & \cdots & \cdots & \cdots & \cdots & \cdots & \cdots & \cdots & \cdots \\
\frac{\partial^2 L}{\partial P_{N_g h}\partial P_{1h}} & \frac{\partial^2 L}{\partial P_{N_g h}\partial P_{2h}} & \cdots & \frac{\partial^2 L}{\partial P_{N_g h}\partial P_{N_g h}} & \frac{\partial^2 L}{\partial P_{N_g h}\partial X_{1h}} & \frac{\partial^2 L}{\partial P_{N_g h}\partial X_{2h}} & \cdots & \frac{\partial^2 L}{\partial P_{N_g h}\partial X_{N_g h}} & \frac{\partial^2 L}{\partial P_{N_g h}\partial \lambda_1} & \cdots & \frac{\partial^2 L}{\partial P_{N_g h}\partial \lambda_{N_g}} \\
\frac{\partial^2 L}{\partial X_{1h}\partial P_{1h}} & \frac{\partial^2 L}{\partial X_{1h}\partial P_{2h}} & \cdots & \frac{\partial^2 L}{\partial X_{1h}\partial P_{N_g h}} & \frac{\partial^2 L}{\partial X_{1h}\partial X_{1h}} & \frac{\partial^2 L}{\partial X_{1h}\partial X_{2h}} & \cdots & \frac{\partial^2 L}{\partial X_{1h}\partial X_{N_g h}} & \frac{\partial^2 L}{\partial X_{1h}\partial \lambda_1} & \cdots & \frac{\partial^2 L}{\partial X_{1h}\partial \lambda_{N_g}} \\
\frac{\partial^2 L}{\partial X_{2h}\partial P_{1h}} & \frac{\partial^2 L}{\partial X_{2h}\partial P_{2h}} & \cdots & \frac{\partial^2 L}{\partial X_{2h}\partial P_{N_g h}} & \frac{\partial^2 L}{\partial X_{2h}\partial X_{1h}} & \frac{\partial^2 L}{\partial X_{2h}\partial X_{2h}} & \cdots & \frac{\partial^2 L}{\partial X_{2h}\partial X_{N_g h}} & \frac{\partial^2 L}{\partial X_{2h}\partial \lambda_1} & \cdots & \frac{\partial^2 L}{\partial X_{2h}\partial \lambda_{N_g}} \\
\cdots & \cdots & \cdots & \cdots & \cdots & \cdots & \cdots & \cdots & \cdots & \cdots & \cdots \\
\frac{\partial^2 L}{\partial X_{N_g h}\partial P_{1h}} & \frac{\partial^2 L}{\partial X_{N_g h}\partial P_{2h}} & \cdots & \frac{\partial^2 L}{\partial X_{N_g h}\partial P_{N_g h}} & \frac{\partial^2 L}{\partial X_{N_g h}\partial X_{1h}} & \frac{\partial^2 L}{\partial X_{N_g h}\partial X_{2h}} & \cdots & \frac{\partial^2 L}{\partial X_{N_g h}\partial X_{N_g h}} & \frac{\partial^2 L}{\partial X_{N_g h}\partial \lambda_1} & \cdots & \frac{\partial^2 L}{\partial X_{N_g h}\partial \lambda_{N_g}} \\
\frac{\partial^2 L}{\partial \lambda_1\partial P_{1h}} & \frac{\partial^2 L}{\partial \lambda_1\partial P_{2h}} & \cdots & \frac{\partial^2 L}{\partial \lambda_1\partial P_{N_g h}} & \frac{\partial^2 L}{\partial \lambda_1\partial X_{1h}} & \frac{\partial^2 L}{\partial \lambda_1\partial X_{2h}} & \cdots & \frac{\partial^2 L}{\partial \lambda_1\partial X_{N_g h}} & \frac{\partial^2 L}{\partial \lambda_1\partial \lambda_1} & \cdots & \frac{\partial^2 L}{\partial \lambda_1\partial \lambda_{N_g}} \\
\cdots & \cdots & \cdots & \cdots & \cdots & \cdots & \cdots & \cdots & \cdots & \cdots & \cdots \\
\frac{\partial^2 L}{\partial \lambda_{N_g}\partial P_{1h}} & \frac{\partial^2 L}{\partial \lambda_{N_g}\partial P_{2h}} & \cdots & \frac{\partial^2 L}{\partial \lambda_{N_g}\partial P_{N_g h}} & \frac{\partial^2 L}{\partial \lambda_{N_g}\partial X_{1h}} & \frac{\partial^2 L}{\partial \lambda_{N_g}\partial X_{2h}} & \cdots & \frac{\partial^2 L}{\partial \lambda_{N_g}\partial X_{N_g h}} & \frac{\partial^2 L}{\partial \lambda_{N_g}\partial \lambda_1} & \cdots & \frac{\partial^2 L}{\partial \lambda_{N_g}\partial \lambda_{N_g}}
\end{bmatrix} \tag{18}$$

ξ_k presents the gradient vector indicating the descent direction to the global minimum, d_k presents the calculation step and A presents the Hessian matrix defined by the partial derivatives of the production function relative to the generated powers and to the various on/off states of each production unit.

The proposed strategy not only helps to search effective solutions corresponding to a minimum production cost, but also to proceed through an acceleration evoked by the second derivatives of the Hessian matrix so as to reach the optimal solution

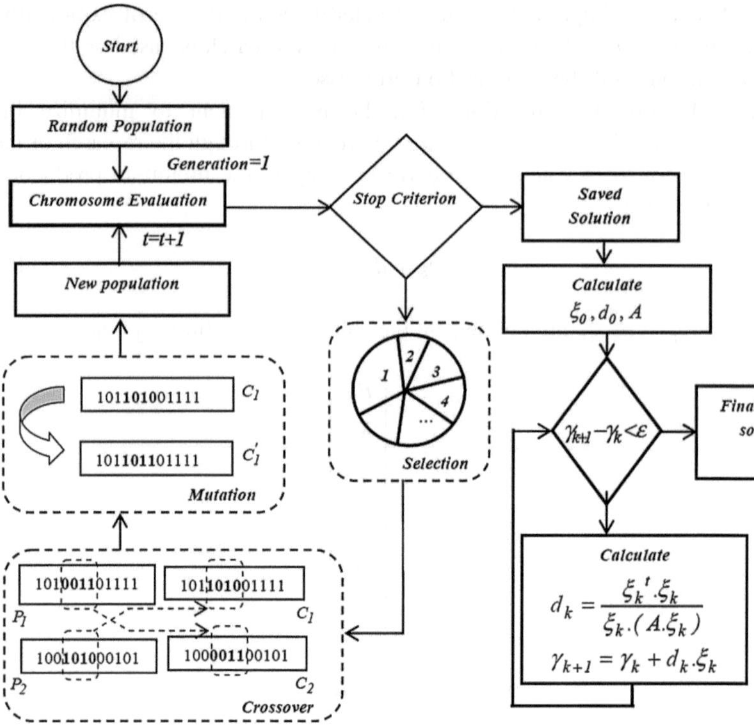

Fig. 4 Flowchart of solving the unit commitment problem via gradient-genetic algorithm method

as quickly as possible. The process of solving the unit commitment problem by gradient-genetic algorithm method is performed according to the following flowchart (Fig. 4):

5 Simulations, Results and Comparative Study

In order to test the performance of the optimization proposed method; the strategy has been applied to an IEEE electrical network 14 buses, having 5 generators, over a period of 24 h (Fig. 5). The strategies are occurring at t = 40 s and the scheduling of the on/off states and the amount of generated power by each production unit is performed for each 3 h.

The characteristics of the different production units are given in Table 2.

We have took as population size = 40, crossover probability = 0.6, mutation probability = 0.02 and the maximum number of generations = 300.

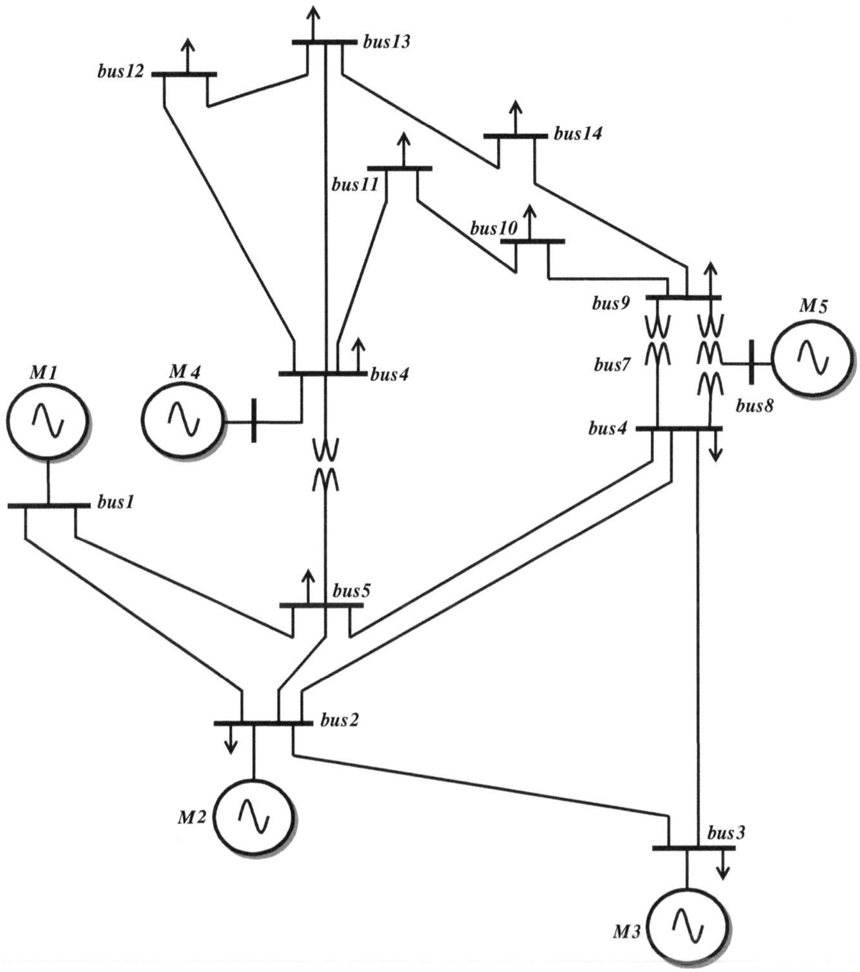

Fig. 5 Studied network (Moez et al. 2011; Abbassi et al. 2012)

Table 2 Characteristics of production units

U	Pmax (MW)	Pmin (MW)	a	b	c	MUT	MDT	Hot start-up cost ($)	Cold start-up cost ($)	Sci (h)
1	582	110	379.2	30.36	0.0756	8	8	4,500	9,000	6
2	55	15	606.6	27.3	0.2274	3	3	170	340	2
3	53	10	454.8	22.74	0.2274	3	3	170	340	2
4	23	8	151.8	22.5	0.1518	1	1	30	60	0
5	23	8	303.6	22.74	0.1518	1	1	30	60	0

Table 3 Amount of load required

Hour	3	6	9	12	15	18	21	24
Demand (MW)	259	200	300	450	527	610	480	320

In this paper, we considered 8 successive periods in order to establish the temporal evolution of the power demand. Each period lasts 3 h, hence, the total period is about 24 h (Table 3).

The IEEE 4th order state model has been adopted for all the five machines as written in (Eq. 19), (Abbassi and Chebbi 2012).

$$\begin{cases} \frac{dE'_q}{dt} = \frac{1}{T'_{do}} \cdot \left[E_{fd} - E'_q + (X_d - X'_d) \cdot I_d\right] \\ \frac{dE'_d}{dt} = \frac{1}{T'_{qo}} \cdot \left[-E'_d + (X_q - X'_q) \cdot I_q\right] \\ \frac{d\omega}{dt} = \frac{1}{M} \cdot \left[P_m - P_e - D \cdot (\omega - 1)\right] \\ \frac{d\delta}{dt} = \omega - 1 \end{cases} \tag{19}$$

where

E'_d, E'_q d and q axis transient *emf*
X_d, X_q d and q axis reactance's,
X'_d, X'_q d and q axis transient reactance's,
w Mechanical speed,
M Moment of inertia,
P_m Maximum available power extracted by the turbine,
P_g, Q_g Active and reactive powers supplied by each machine,
D Friction coefficient,
δ Load angle,
V_r Voltages at bus *i*.

After calculating of the instantaneous (d, q) components of voltages of the generator's terminals and currents as detailed in (Abbassi et al. 2012), the active and reactive powers supplied by each generator are chosen to be:

$$\begin{cases} P_g = \frac{E'_q \cdot V_r \cdot \sin(\delta_r)}{X'_d} + \frac{1}{2} \cdot \left(\frac{1}{X'_q} - \frac{1}{X'_d}\right) \cdot V_r^2 \cdot \sin(2\delta_r) \\ Q_g = \frac{E'_q \cdot V_r \cdot \cos(\delta_r)}{X'_d} + \frac{1}{2} \cdot \left(\frac{1}{X'_q} - \frac{1}{X'_d}\right) \cdot V_r^2 \cdot \cos(2\delta_r) - \frac{1}{2} \cdot \left(\frac{1}{X'_d} - \frac{1}{X'_q}\right) \cdot V_r^2 \end{cases} \tag{20}$$

Figure 6 illustrates the total production cost of various optimization methods for solving the unit commitment problem. Compared to the algorithms of Wei and Cai (Wei and Li 1999; Cai and Cai 1997), we find that these optimization methods present high performance since they improved to win in the production cost.

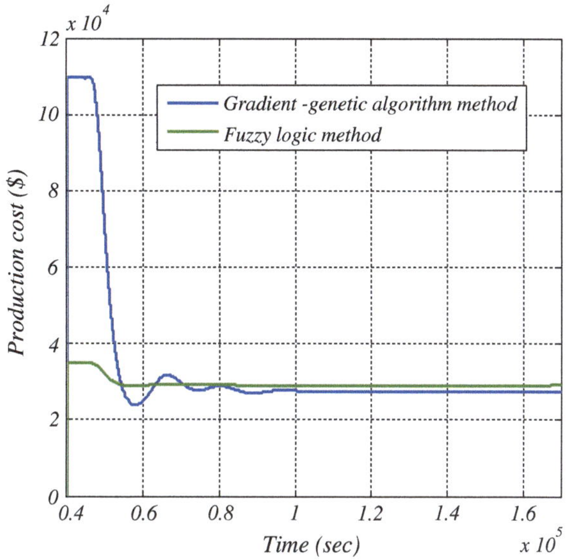

Fig. 6 Production cost of the gradient-genetic algorithm and the fuzzy logic methods

Table 4 Comparison between the optimization methods

	Fuzzy logic	Gradient-genetic algorithm
Production cost ($)	29,210	27,750
Execution time (s)	7.34	12.57

It is clear that through the comparison of production costs using the fuzzy logic by that one obtained using the genetic algorithm method, Table 4, the fuzzy logic approach was reliable and enabled to get a gain of 1 % of the total cost. However, the strategy based on the use of gradient-genetic algorithm method was the most effective and presented high performances not only in the production cost but also in the ability of convergence to the global optimum.

We note the gradient-genetic algorithm method did not presented an efficient resolution time, since it requires enough time to reach the optimal solution depending essentially on the choice of the initial population. Indeed, based on the above table, it is noted that the strategy based on the use of fuzzy logic method is more efficient than the hybrid method in terms of execution time and efficiency of convergence.

Table 5 shows the organization of the on/off states of the production units of the various optimization strategies. Thanks to the hybrid optimization method, we were able to organize the On/Off statements of the various production units through an estimation of the amount of load required by the electrical grid, taking into account the allowable constraints; optimal scheduling can profit of the production cost.

Table 5 Optimal binary combination of units operation

Unit	Units operation scheduling using fuzzy logic	Units operation scheduling using gradient-genetic algorithm
Unit 1	11111111	11111111
Unit 2	00111111	00111110
Unit 3	00011110	00111110
Unit 4	00001110	00001110
Unit 5	00001100	00001000

The superiority of the gradient-genetic algorithm method is obvious. This method operates better than the individual algorithms in terms of On/Off unit commitment states scheduling and in term of optimizing the total production cost.

In fact, based on the probability equation of such a combination planning:

$$P_{Combination} = (2^n - 1)^m \tag{21}$$

where, n is the number of units and m is the discretized duration. For our case study, the combining probability $P_{Combination}$ is about $\boldsymbol{6.2^{35}}$ combinations. This number suggests the ability of the hybrid method to choose a perfect planning, allowing to guarantee the supply/demand balance and a minimal production cost.

We confirm that the two approaches have allowed selecting concisely the production units that should be available to respond to the demand of the electrical network over a future period (Fig. 7).

Moreover, thanks to the best selection of the fuzzy variables, we arrived to develop an optimized scheduling plan of the generated power allowing a better exploitation of the production cost in order to bring the total operating cost to possible minimum in presence of the various constraints. Consequently, basing on the forecasted load curve, this method put generators in guard state for intervening in cases where there is an additional power demand.

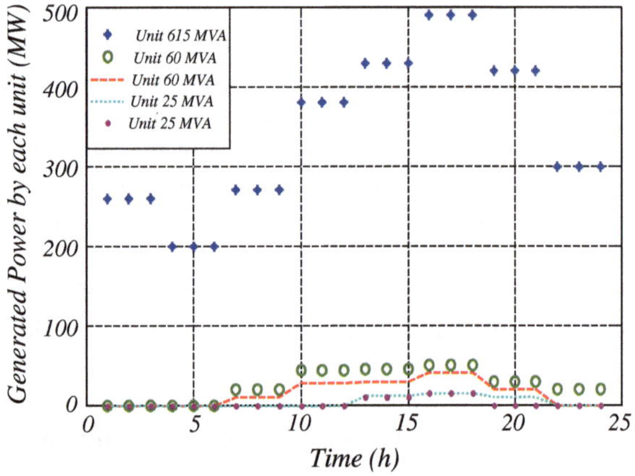

Fig. 7 Unit commitment scheduling over a period of 24 h

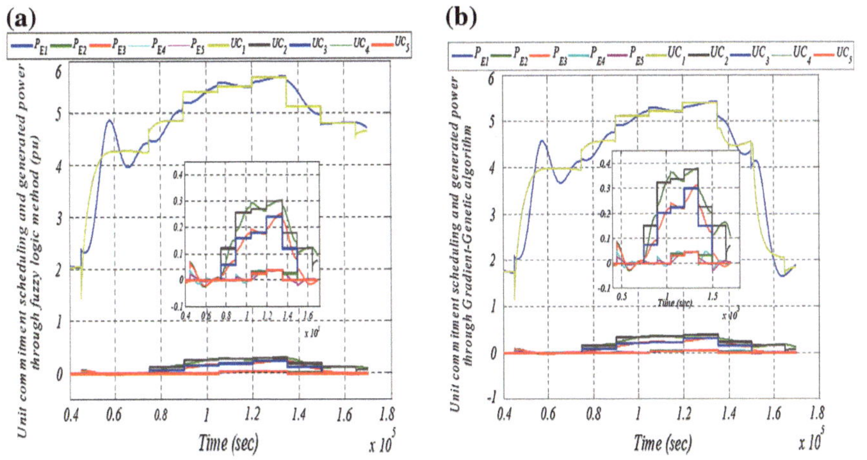

Fig. 8 Unit commitment scheduling and generated power through fuzzy logic (**a**) and gradient-genetic algorithm (**b**) methods

The obtained results through Fig. 7 and Table 5 prove the effectiveness of the Gradient-Genetic algorithm and the fuzzy logic methods in solving the Unit Commitment problem and in choosing the best plan of On/Off production units.

Figure 8a, b show the production scheduling of 5 units for a variable power demand during a discrete time margin (horizon time about 24 h). Indeed, taking into account the technical constraints related to each generator (limited power, minimum down-time before restart, minimum operating time before off state), strategies enabled to get the best On/Off scheduling states of the various units while optimizing the power produced by each unit within the allowable margins. Furthermore, solving the UCP by these optimization methods are considered as reliable and have presented high performances especially for a problem involving identical production units, which is not the case for the application of dynamic programming method to the UCP, established in the work (Dekrajangpetch et al. 1999), which can not in any way applied for the case of identical production units. However, we find that the unit commitment scheduling based on the fuzzy logic theory (Fig. 8a) is effective and this can be observed through the temporal evolution of the power produced by the most powerful generator (615 MVA); which suggests the effectiveness of resolution through the fuzzy logic approach especially in presence of systems that are difficult to model. Nevertheless, the strategy based on the use of gradient-genetic-algorithm method (Fig. 8b) remains the most promising and could be applied to solve the UCP for systems having complicated architecture and for any number of production units. Knowing that the minimization of the production cost equation is closely related to the optimization of the generated power P_{ih}, the efficiency of resolution through Gradient-genetic algorithm approach is guaranteed with great consideration in the limitation of the produced active power by each generator per hour in one side and in the allowable voltage levels margins for each electrical grid in the other side.

The improvement of the production cost for the model based on fuzzy approach depends on the number of fuzzy rules taken in the resolution. However, increasing this number leads to increase the horizon of solutions research which implies the increase of the execution time. Furthermore, the optimization of the production cost through genetic algorithm requires a proper selection of the GA parameters which vary from one system to another. Thus, it is difficult to reduce for both the execution time and the production cost for the mentioned methods. As regards to production cost, the proposed strategy based on Gradient-genetic algorithm method is more promising; Indeed, it leads to a better combination of the production units operating states leading to an optimal production cost While as regards to convergence speed and execution time, the approach based on fuzzy logic has presented high performances. The founded results show the advantage of the proposed strategies. In addition, the adopted approach was promising both in terms of convergence to get the best optimal solutions to minimize the cost production and for an efficient unit commitment scheduling for the different units production.

6 Conclusion

For the present chapter, we analyzed the resolution of the Unit Commitment problem through the combination of genetic algorithm and gradient method in one side and through a fuzzy logic approach in the other side. The simulations done under the Matlab environment on an electrical network having five production units, have proved the effectiveness of these methods as well execution time as production cost. Besides, throughout a comparative study between these two strategies, results showed that in terms of execution time and convergence effectiveness, the resolution through fuzzy logic approach is reliable despite the production cost is relatively minimal but didn't present the best production cost. Yet, the Gradient-genetic algorithm method has presented high performances in optimizing the production cost and capability of convergence to a global optimum. In addition, it is noted that the adopted strategies lead to significant reduction in the number of decision variables and therefore a reduction of the optimization problem size. In addition, they have the potential to reach a global solution of UC problem since they have ensured an optimized unit commitment scheduling of the On/Off unit states which proves their potential to solve problems related to high power electric networks system.

Due to flexibility in Gradient-Genetic Algorithm and Fuzzy logic several other practical constraints can also be easily considered. For future work, the above problem can be solved with artificial intelligence technique like evolutionary programming and artificial neural network. The Unit Commitment problem could be solved if the system complexity increases either by increasing the number of units or adding other constraints.

References

Abbassi, R., & Chebbi, S. (2012). Energy management strategy for a grid-connected wind-solar hybrid system with battery storage: Policy for optimizing conventional energy generation. *International Review of Electrical Engineering, 7*(2), 3979–3990.

Abbassi, R., Marrouchi, S., Moez, B. H., Chebbi, S., & Houda, J. (2012). Voltage control strategy of an electrical network by the integration of the UPFC compensator. *International Review on Modelling and Simulations (I.RE.MO.S), 5*(1), 380–384.

Attaviriyanupap, P., Kita, H., Tanaka, E., & Hasegawa, J. (2002). A new profit-based unit commitment considering power and reserve generating. In *The 2002 IEEE-PES Winter Meeting* (pp. 6–11). New York. January 27–31 2002. doi: 10.1109/PESW.2002.985227.

Azar, A. T. (2012). Overview of type-2 fuzzy logic systems. *International Journal of Fuzzy System Applications (IJFSA), 2*(4), 1–28.

Azar, A. T. (2010a). *Fuzzy systems*. Vienna, Austria: IN-TECH. ISBN 978-953-7619-92-3. 3.

Azar, A. T. (2010b). Adaptive neuro-fuzzy systems. In A. T. Azar (Ed.), *Fuzzy systems*. Vienna, Austria: IN-TECH. ISBN 978-953-7619-92-3.

Cai, C. H., & Cai, Y. Y. (1997). Optimization of unit commitment by genetic algorithm. *Power System Technology, 21*(1), 44–47.

Cheng, C. P., Liu, C. W., & Liu, C. C. (2002). Unit commitment by annealing-genetic algorithm. *Electrical Power and Energy Systems, 24*(2), 149–158.

Damousis, I. G., Bakirtzis, A. G., & Dokopoulos, P. S. (2004). A solution to the unit-commitment problem using integer-coded genetic algorithm. *IEEE Transactions on Power systems, 19*(2), 1165–1172.

Dieu, V. N., & Ongsakul, W. (2007). Improved merit order and augmented lagrange hopfield network for unit commitment. *IET Generation, Transmission and Distribution, 1*(4), 548–556.

Dekrajangpetch, S., Sheble, G. B., & Conejo, A. J. (1999). Auction implementation problems using lagrangian relaxation. *IEEE Transactions on Power Systems, 14*(1), 82–88.

Guan, X., Luh, P. B., Yan, H., & Amalfi, J. A. (1992). An optimization-based method for unit commitment. *Electric power and energy systems, 14*(1), 9–17.

Grey, A., & Sekar, A. (2008). Unified solution of security-constrained unit commitment problem using a linear programming methodology. *IET Generation, Transmission and Distribution, 2* (6), 856–867.

Hong, Y. Y., & Li, C. (2002). Genetic algorithm based economic dispatch for cogeneration units considering multiplant multibuyer wheeling. *IEEE Transactions on Power Systems, 17*(1), 134–140.

Juste, K. A., Kita, H., Tanaka, E., & Hasegawa, J. (1999). An evolutionary programming solution to the unit commitment problem. *IEEE Transactions on Power Systems, 14*(4), 1452–1459.

Kazarlis, S. A., Bakirtzis, A. G., & Petridis, V. (1996). A genetic algorithm solution to the unit commitment problem. *IEEE Transactions on Power Systems, 11*(1), 83–92.

Kohonen, T. (1998). An introduction to neural computing. *Neural Network Journal, 1*(1), 3–16.

Kurban, M., & Filik, U. B. (2009). A comparative study of three different mathematical methods for solving the unit commitment problem. *Mathematical Problems in Engineering, 2009*(1), 1–13, (368024, Hindawi publishing corporation).

Lin, F. T., Kao, C. Y., & Hsu, C. C. (1993). Applying the genetic approach to simulated annealing in solving some NP-hard problems. *IEEE Transactions on Power Systems, Man, and Cybernetics, 23*(6), 1752–1767.

Maifeld, T. T., & Sheble, G. B. (1996). Genetic-based unit commitment algorithm. *IEEE Transactions on Power Systems, 11*(3), 1359–1370.

Mantawy, A. H., AbdelMagid, Y. L., & Selim, S. Z. (1998). A simulated annealing algorithm for unit commitment. *IEEE Transactions on Power System, 13*(1), 197–204.

Marrouchi, S., & Chebbi, S. (2013). Combined use of genetic algorithms and gradient optimization methods for unit commitment problem resolution. *Wulfenia Journal, 20*(8), 357–369.

Merlin, A., & Sandrin, P. (1983). A new method for unit commitment at Electricite De France. *IEEE Transactions on Power Apparatus and Systems, 102*(5), 1218–1225.

Moez, B. H., Sahbi, M., Souad, C., Houda, J., & Rabeh, A. (2011). Preventive and curative strategies based on fuzzy logic for voltage stabilization of an electrical network. *International Review on Modeling and Simulation (I.RE.MO.S), 4*(6), 3201–3207.

Ouyang, Z., & Shahidehpour, S. M. (1991). An intelligent dynamic programming for unit commitment application. *IEEE Transactions on Power Systems, 6*(3), 1203–1209.

Padhy, N. P. (2001). Unit commitment using hybrid models: A comparative study for dynamic programming, expert systems, fuzzy system and genetic algorithms. *International Journal of Electrical Power and Energy Systems, 23*(8), 827–836.

Rajan, C. C. A., & Mohan, M. R. (2004). An evolutionary programming-based Tabu search method for solving the unit commitment problem. *IEEE Transactions on Power Systems, 19* (1), 577–585.

Rajan, C. C. A., Mohan, M. R., & Manivannan, K. (2002). Refined simulated annealing method for solving unit commitment problem, the 2002 neural networks, 2002. In IJCNN '02. *Proceedings of the 2002 International Joint Conference on May 12-17 2002* (pp. 333–338). Honolulu, HI. doi: 10.1109 /IJCNN.2002.1005493.

Saber, A. Y., Senjyu, T., Yona, A., Urasaki, N., & Funabashi, T. (2007). Fuzzy unit commitment solution-A novel twofold simulated annealing approach. *Electric Power Systems Research, 77* (12), 1699–1712.

Sasaki, H., Watanabe, M., Kubokawa, J., Yorino, N., & Yokoyama, R. (1992). A solution method of unit commitment by artificial neural networks. *IEEE Transactions on Power Systems, 7*(3), 974–981.

Snyder, W. L., Powell, H. D., & Rayburn, J. C. (1987). Dynamic programming approach to unit commitment. *IEEE Transactions on Power Systems, 2*(2), 339–350.

Sudhakaran, M., Ajay, D., & Vimal-Raj, P. (2010). Integrating genetic algorithms and tabu search for unit commitment problem. *International Journal of Engineering, Science and Technology, 2*(1), 57–69.

Victoire, T. A. A., & Jeyakumar, A. E. (2005). Unit commitment by a tabu-search-based hybrid-optimization technique. *IEEE Proceedings Generation Transmission and Distribution, 152*(4), 563–574.

Wei, P., & Li, N. H. (1999). Daily generation scheduling based on genetic algorithm. *Automation of Electric Power Systems, 23*(3), 23–27.

Wood, A. J., & Woolenberg, B. F. (1996). *Power generation operation and control* (2nd ed.). New York: Wiley.

Wu, Y. G., Ho, C., & Wang, D. Y. (2000). A diploid genetic approach to short-term scheduling of hydro-thermal system. *IEEE Transactions on Power System, 15*(4), 1268–1274.

Yingvivatanapong, C. (2006, May). Multi-area unit commitment and economic dispatch with market operation components, PhD discussion, University of Texas, Arlington.

Zhao, B., Guo, C. X., Bai, B. R., & Cao, Y. J. (2006). An improved particle swarm optimization algorithm for unit commitment. *International Journal of Electrical Power and Energy Systems, 28*(7), 482–490.

Zhuang, F., & Galiana, F. D. (1988). Towards a more rigorous and practical unit commitment by Lagrangian relaxation. *IEEE Transactions on Power Systems, 3*(2), 763–772.

Impact of Hardware/Software Partitioning and MicroBlaze FPGA Configurations on the Embedded Systems Performances

Imène Mhadhbi, Nabil Litayem, Slim Ben Othman and Slim Ben Saoud

Abstract Due to their flexible architecture, lower-cost and faster processing, Field Programmable Gate Array (FPGA) presents one of the stimulating choices for implementing modern embedded systems. This is due to their intrinsic parallelism, fast processing speed, rising integration scale and lower-cost solution. This kind of platforms can be considered as a futuristic implementation platform. The growing configurable logic capacity of FPGA has enabled designers to incorporate one or more processors in FPGA platform. In contrast to the traditional hard cores, the soft cores processors present an interesting solutions for implementing embedded applications. They give designers the ability to adapt many configurations to their specific application; including memory subsystems, interrupt handling, ISA features, etc. Faced to the various problems related to the selection of an efficient soft-core FPGA embedded processor with appropriate configuration, co-design methodology presents a good deal for embedded designers. The most crucial step in the design of embedded systems is the hardware/software partitioning. This step consists of deciding which component is suitable for hardware implementation and which one is more appropriate for software implementation. This research field is especially active (always on the move) and several approaches are proposed. In this chapter, we will present our contribution on the hardware/software partitioning co-design approach, and discuss their involvement on design acceleration and architecture performances. The first part of this chapter describes the effect of the MicroBlaze Xilinx configuration on the embedded system performance. The second part

I. Mhadhbi (✉) · S.B. Othman · S.B. Saoud
LSA Laboratory, INSAT-EPT, University of Carthage, Tunis, Tunisia
e-mail: imene.mhadhbi@gmail.com

S.B. Othman
e-mail: boslim@yahoo.fr

S.B. Saoud
e-mail: slim.bensaoud@gmail.com

N. Litayem
Department of Computer Science, College of Arts & Science, Salman Bin Abdelaziz University KSA, Al-Kharj, Saudi Arabia
e-mail: nabil.litayem@gmail.com

Q. Zhu and A.T. Azar (eds.), *Complex System Modelling and Control Through Intelligent Soft Computations*, Studies in Fuzziness and Soft Computing 319,
DOI 10.1007/978-3-319-12883-2_25

introduces our new hardware/software partitioning approach on a complex secure lightweight cryptographic algorithm. This work can contribute to enforce the security of SCADA (Supervision Control and Data Acquisition) systems and the DSS (Digital Signal Standard) without compromising the cost and the performance of the final system.

1 Introduction

Embedded systems are now present in practically all domestic and industrial systems (appliances and applications) such as cellular telephones, personal digital assistants (PDAs), digital cameras, Global Positioning System (GPS) receivers, defense systems and security applications. The increased complexities of embedded systems and their real-time operation's constraints allow semiconductor markets to build other solutions for processing. Traditionally, embedded systems were designed and implemented using Microprocessors (MP), Microcontrollers (MCUs), Digital Signal Processors (DSPs), Application-Specific Integrated Circuits (ASICs) and FPGAs. Due to their advantages, FPGAs have substituted DSPs in different applications such as motor controllers (Arulmozhiyal 2012; Xiaoyin and Dong 2007) which are widely used in industrial applications, image processing (Kikuchi and Morioka 2012), wireless (Jing-Jie and Rui 2011; Nasreddine et al. 2010), automotive and aerospace systems. Continuing increases in FPGA performance, capability and architectural features are enabling more embedded systems designs to be implemented using FPGAs. Additionally, FPGAs costs are decreasing, for less than $12, allowing designers to incorporate FPGAs circuits with one million equivalent gates. This made the implementation of Programmable System-On-Chip (SoPCs) possible what also allowed this implementation their pipeline ability, intrinsic parallelism and flexible architecture (Jianzhuang et al. 2008). FPGAs offer a faster processing speed, a lower-cost solution and more functionalities to support more innovative characteristics.

Nevertheless, the increasing complexity of algorithms and the rising integration scale on FPGAs triggered designers into drastically improving design methodologies. In addition, the effort to design complex applications on FPGA is generally much more complicated than impelementing them on programmable processors.

The real challenge, as far as the embedded systems designers are concerned, is how to increase performances (execution time, area and energy consumption) of complex systems and reduce their complexity, and refinement time.

Many interesting design methodologies are presented. Some designers have based their methodologies on reducing development time to implement complex embedded systems. Among the many approaches that have been adapted there is first the automatic transformation of the behavioral system description into structural netlist system components using high level input language such as SpecC

(Fujita and Nakamura 2001) and Bluespec (Dave et al. 2005; Gruian and Westmijze 2008; Talpin et al. 2003). Second, there is Hardware In the Loop (HIL) technique which increase the tractability and earlier testability of the design product (Washington and Dolman 2010). The automation of the hardware/software partitioning step on the co-design methodologies using low-level specification presents the third approach (Stitt et al. 2003).

Other designers have based their methodologies on minimizing the design complexities. One approach is the use of Intellectual Proprieties (IPs) blocs and cores (Mcloone and Mccanny 2003) provided by vendors or designers (Lach et al. 1999). An other is the automating of the hardware code generation HDL (Hardware Description Language) from a high level specification (Samarawickrama et al. 2010; Ku and De Mitcheli 1992). This specification can be defined as language (C, SystemC, etc.) or models (Matlab, Sycos etc.) or UML (Unified Modeling Language) diagrams, called HLS (High Level Synthesis) approach (Lingbo et al. 2006; Wakabayashi and Okamoto 2000).

During last decades, early designers' works have been focused on new contributions of the existing design methodologies, which allow both the high level specification and the automation of the design process to decrease the systems complexities, reduce the development time to enlarge the time reserved to the optimization and increase their performance. None of these approaches deals with the impact of the best configuration selection of soft-cores processors performance in terms of computation acceleration.

The goal of this chapter is to actively contribute to the existing co-design approaches including hardware/software partitioning step using high level specification. The chapter also aims at adding a step to the selection f the best soft-cores processors configuration. These contributions permit the increase of embedded systems performance (soft-cores computation) and the reduction of systems complexities. The remaining parts of this paper are organized as follows: Sect. 2 illustrates the related works and background of design methodologies. Section 3 presents the MicroBlaze soft-core processor. Section 4 depicts our co-design approach. Section 4 presents the performance evaluation techniques and the lightweight cryptographic algorithm. Section 5 determines results of our co-design approach. In Sect. 6, we will discuss results. Finally, Sect. 7 summarizes our study, and gives our perspectives.

2 Related Works and Background

In this section, the different steps of design methodologies will be presented in reverse. We will begin by defining the different architecture of implementation. We will proceed with the design methodologies approaches specifically HLS approach and hardware/software partitioning approach. Finally, system level specification of embedded systems will be dealt with.

2.1 Design Methodologies, Challenges

Embedded applications require increasingly sophisticated functionalities and severe constraints. They incorporate many application areas such as telecommunication, avionics, automotive, medical implants, domestic appliances, etc. These increasing complexities require functional constraints (computation capacities, reduced power consumption, miniaturization of the implementation area, etc.) and non-functional constraints (minimum time-to-market, reduced cost, maximum life, growth in the amount of productions, etc.). To increase the embedded systems performances, researchers and industry have focused on two areas of research. First, technological area that is based on the evolution of the integration level of integrated circuits. Second, methodological area, which is based on refinement of design methodologies. Faced with the physical limits of technical evolution, manufacturers of embedded systems had to demonstrate a different reactivity. They had to continuously improve their techniques and design approaches to increase the embedded systems performances.

In our study, we examine different design architectures of complex embedded systems. Our contribution lies in the hardware/software partitioning step starting from high level specification. Also, a new step has been added to the hardware/software partitioning which is the selection of the best configuration of the used soft-core processor.

2.1.1 Design Implementation Architectures

Traditional hardware FPGA design approaches are complex. This reduces FPGA productivity. Hardware implementation uses a low-level specification, VHDL or Verilog languages or combination of both, to implement embedded applications. Their implementation process consists of the (a) definition of application at a low-level specification (b) synthesis, (c) implementation, (d) simulation and (e) tests and verification steps. With the integration of soft-cores processors into FPGA, designers become able to implement complex systems on software architecture. As input, they employ a high-level specification, compile it and implement it into soft-cores processors.

Several researches demonstrate that software implementation of embedded systems allows flexibility (ability to modify specifications), ease of integration, reduction of design time and bad performances. However, hardware implementation of the same application greatly achieves high performance constraints in a long design time.

Now, FPGA offers many advantages. It can be used in all embedded systems fields (image processing, aerospace systems, security and industrial applications, etc.). It can be implemented on different architectures (hardware, software or both hardware/software) using different design methodologies (Joven et al. 2011) as illustrated on Fig. 1.

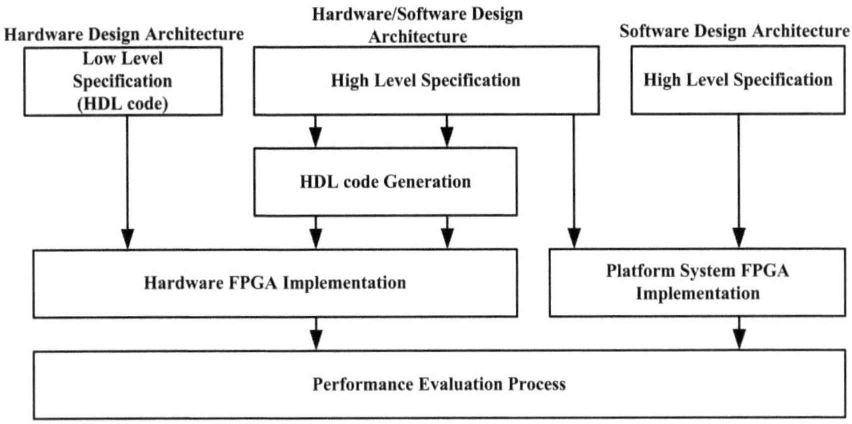

Fig. 1 FPGA architectures and methodologies design

Recently, designs approaches can be implemented using both hardware/software architectures using co-design methodology to accelerate the design process. Using this methodology, designers can incorporate co-processors, hard-cores processors and soft-cores processors. This decision of integration is taken after a hardware/software partitioning step. Hardware/Software partitioning is usually related to physical constraints (computing time, energy consumption, level of integration, area utilization) and economic constraints (cost, flexibility, design time and Time-To-Market) embedded systems constraints, as described in Table 1.

Recently, Reconfigurable devices, such as FPGAs, become highly appealing circuits for co-design methodology as they provide flexibility and ability to easily implement complex embedded applications. Using co-design methodologies, designers permit the integration of both hardware and software architectures into FPGA (Kalomiros and Lygouras 2008). Xilinx proposes its own co-design methodology using Xilinx EDK (Integrated Development Kit) environment. EDK includes both an integrated development environment (IDE) named Xilinx Platform Studio (XPS) and Software Design Kit (SDK). XPS tool allows the implementation on hardware architecture and the creation of a Microprocessor Hardware Specification (MHS) file. SDK tool permit the implementation of software architecture and the creation of the Microprocessor Software Specification (MSS) file. The MHS file defines the embedded system processor, architecture and peripherals. The MSS file defines the library customization parameters for peripherals, the processor customization parameters, the standard I/O devices, the interrupt handler routines, etc. Figure 2 depicts the co-design flow of Xilinx EDK tool.

2.1.2 Hardware/Software Partitioning Approaches

FPGAs present powerful circuits for prototyping embedded system applications, supporting both software and hardware architectures. The choice of architecture is

Table 1 Comparative studies of the software/hardware design architectures

		Software	Hardware
Physical Constraints	Execution time		*
	Energy consumption		*
	Integration		*
	Area		*
Economic Constraints	Cost	*	(Except high volume)
	Flexibility	*	
	Design time	*	
	Time-to-market	*	

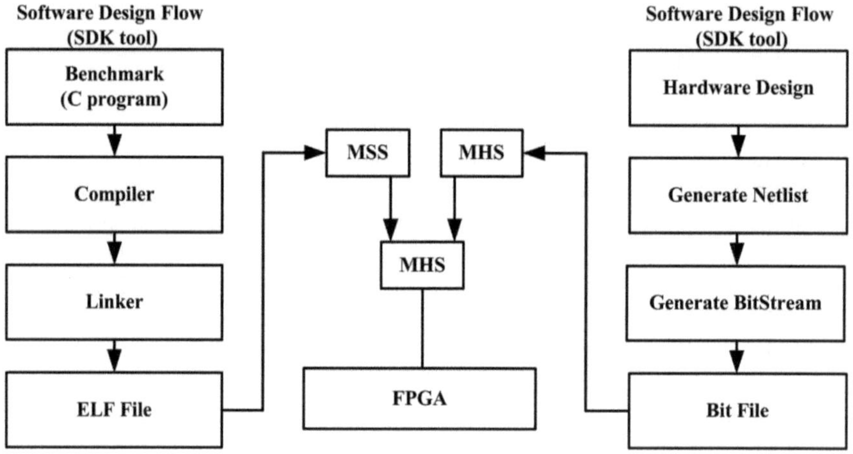

Fig. 2 Co-design flow using Xilinx EDK tool

based on the hardware/software partitioning step. The goal of that partitioning step is to determine which components of the application are suitable for hardware or software implementation. Hardware implementation is desirable to design efficient embedded systems in term of execution time and computation (co-processors). However, software implementation gives less performance in a reduced time. This partitioning is depending on embedded systems constraints such as cost, efficiency and speed. The real paradigm of co-design methodology is the great choice of hardware and software sections.

Co-design approaches promote the implementation of efficient embedded systems in a low development time by integrating hardware co-processors into software design process. During the design process, fundamental decisions have dramatically influenced the quality and the cost of the final solution. Design decisions have an impact of about 90 % of the overall cost. The most important decision is that of hardware/software partitioning.

Therefore, Partitioning is a well-known problem. During the last years, many partitioning approaches have been proposed to automate the partitioning process decision of hardware and software components (De Michell and Gupta 1997; Wiangtong et al. 2005). The feasibility of these approaches depends essentially on the system-level specification, the target architecture and the constraints parameters (hardware size, power consumption, execution time, computation, etc.). Several works were focused on the automation of the hardware/software partitioning using co-design methodologies. Many interesting approaches are presented. Some of them are described on Table 2.

As described in the table above, many partitioning hardware/software approaches exist (Madsen et al. 1997; Boßung et al. 1999; Chatha and Vemuri 2000). From the many co-design approaches, we will examine some of these. A hardware/software partitioning approach is proposed by Lysecky and Vahid (2004). This approach uses a relaxed cost function to satisfy performance in an Integer Linear Programing (ILP); it handles hardware minimization in an outer loop. Lysecky and Vahid (2004), presents a binary constraint search algorithm which determines the smaller size constraint. Vahid partitioning approach minimizes hardware, but not execution time. Kalavade and Lee (1994), proposed also a different hardware/software partitioning approach. It is based on GCLP algorithm to determine for each node iteratively the mapping to hardware or software. The used GCLP algorithm selects its appropriate objective according to critical time measure and another measure for local optimum.

Table 2 Hardware/software partitioning approaches

Approaches	Cosyma	Vulcan	Polis	CoWave	GrapeII
Specification	SDL language	HardwareC	FSM, esterel	DFL, C, etc.	DFL
Internal model	No	No	Yes	Yes	Yes
Support Y chart model?	No, Semi-automatic	Yes, with migration	No, manual	No	Yes
Support automating partitioning?	No	No	Yes (Y-chart like)	No	Yes (Y-chart like)
Supports the exploration of the design space?	Low	Low	Very high	High	Medium
Level specification of approach	No	No	Yes	No	No
Support for synthesis	No	No	Yes	No	Yes
Target architecture	Mono-processor: CPU + Co-processor	Mono-processor: CPU + ASIC with buses	Mono-processor	Multi-processors with ASICs	Multi-processors with FPGAs

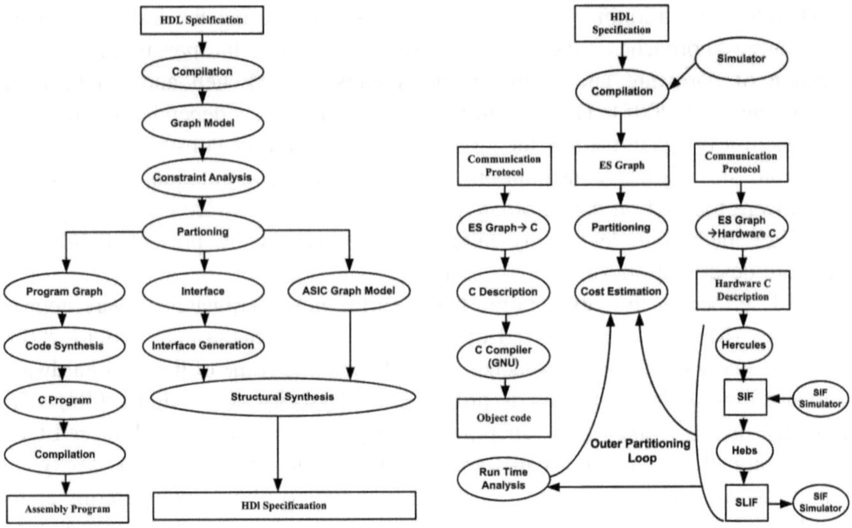

Fig. 3 Vulcan and cosyma approaches

Two representative approaches directly affecting the research of this chapter are Vulcan (Wolf 2003) and Cosyma (Co-synthesis embedded architecture) approaches (López-Vallejo and López 2003). Both Vulcan and Cosyma use partitioning approach, which iterates over hardware synthesis and software generation. Iteration, in these approaches, is necessary because there are no approaches known to accurately estimate the results of optimizing compilers and high-level synthesis tools with advanced techniques. While Vulcan is hardware oriented, starting with an all hardware implementation and moves operations to software on a given processor until time constraints are verified, COSYMA is software oriented, starting with an all software implementation on a given processor and moves operations to hardware until no time constraint is verified any more. Several studies employ these approaches to automate their co-design approach (Henkel and Ernst 1998; Gupta et al. 1992). Figure 3 illustrates these two co-design approaches.

The automation of hardware/software partitioning process allows the classification of embedded specification to determine which components can benefit from the transformation to hardware and the best configuration for getting an optimal gain of performance. Transformations of hardware nodes are provided using HLS approaches. In the next sub-section, we will introduce the HLS design approaches.

2.1.3 HLS Approaches

Using HLS approaches, complexities are managed by (a) starting the design process at a higher level of abstraction, (b) automating the hardware code generation, and (3) reusing intellectual components (IPs). Reducing the migration time from a high-level language specification to a hardware specification language presents the main objective of designers. Early works are focused on how to faster prototyping speed with automatization of the Register Transfer Level (RTL) generation process from the high level behavioural description using commercial tools (Feng et al. 2009).

During the last decades, HLS design approaches have been the main subject for research. Their principal objective is to simplify the accelerators hardware design by describing applications at high abstraction levels and generating the corresponding description of a low-level implementation. Different studies were focused to qualify the benefits of implementing HLS methodologies in terms of time-to-market, execution time and area consumption.

Thangavelu et al. (2012) evaluate the Model-Based Design approach, using XSG (Xilinx System Generator). However, Abhinvar compares the HLS approach (C-Based Design), using Catapuls, with Bluespec design approach, in order to prove that HLS is the most efficient in stage of the design development for fast prototyping complex systems (Dave et al. 2005). Indeed, it offers reduced design time and provides a generic design compared to the Bluespec design flow that generates hardware code adapted to the performance constraints and resources. The Rapid prototyping of complex systems are founded on HLS approaches such as C-Based Design approach (Dave et al. 2005), Model-Based Design approach and Architecture Based Design approach (Cherif et al. 2010) to raise their productivity (from higher levels of abstraction) and their reliability (from automatic code and test bench generation and more robust test technologies).

Model Based Design Approach

The Model-Based Design approach accentuates the use of models to increase the abstraction level of the complex systems (Lingbo et al. 2006; Wakabayashi and Okamoto 2000). This approach represents a real process of evolution in the embedded systems design. The model used, in the systems engineering, includes safety critical areas such as aerospace, automotive, etc. It is applied not only for the explanation of algorithms, but likewise, for the generation of VHDL code. The Model-based design approach level of abstraction is very high, which allows the flexibility to add, delete and modify applications in a short design time. Using this approach, designers can automate the generation, from a model to a synthesized hardware code (VHDL or Verilog), ready to be implemented on FPGA. Mode-Based Design approach emphasizes the usage of models to increase the level of abstraction to design complex systems. It allows the modelization and verification of each function separately using a low-level language or blocks. The ability to plot the progress of the application using Model-Based Design presents an advantage to detect the wrong behaviour. The design model used in the systems engineering, includes also safety critical domains like aerospace and automotive. One of the most widely used tools in these domains is Sicos-HDL, FPGA-module (LabView), Syndex-Ic and XSG.

Architecture-Based Design Approach

The Architecture Based Design approach permits an automatic generation of a synthesized hardware code (VHDL or Verilog), ready to be implemented in FPGA, from UML diagrams. The rapid prototyping approach should provide a way to accelerate the hardware language generation. It must satisfy the following features: (i) Flexibility analysis to produce different results with minimum changes such as the computing precision. (ii) Accuracy of results. The abstraction level of the Architecture Based Design approach, compared to a code written with C in the C-Based design approach and a model described in the Model-Based Design, is very high. This allows the flexibility to add, delete and modify the applications in a short design time. The efficient implementation of complex algorithms (such as a light-weight cryptographic application) in a hardware circuits (FPGA) allows a faster processing speed (parallelism) and more functionalities to support more advanced features.

C-Based Design

This approach consists in the automatic generation of hardware code like VHDL or Verilog, from a C/C++ language, ready to be implemented on FPGAs (Dave et al. 2005). Recent development of C-to-HDL tools technology has minimized the gap between software developer's experience-level, and the expertise needed to produce hardware applications. Many commercial and academic C-Based Design tools can be found in the literature: Catapult-C (Mentor Graphics), CoDeveloper™, C2H, SPARK. In this study, we chose the CoDeveloper™ tool to implement complex embedded application using hardware architecture.

2.2 System Level Specification

The choice of hardware/software partitioning, using co-design approach, presents a trade-off among various design metrics such as performance, cost, flexibility and time-to-market (López-Vallejo and López 2003; Joven et al. 2011). Several approaches of hardware/software partitioning are presented. Their classification is based on their input specifications which is it based on models or languages.

2.2.1 Model Specification

Stoy and Zebo (1994) groups indicate that initial specification can be defined as models of components such as a Finite State Machine (FSM), Discrete-Event Systems, Petri Nets, Data Flow Graphs, Synchronous/Reactive Model, and Heterogeneous Models. These models are described in the next sub-sections.

Finite State Machine (FSM)

Finite State Machine (FSM) models contain sets of states, inputs, outputs, output functions, and next-state functions. Embedded applications are described as a set of states and input values, which can activate a transition from one state to another. FSMs are usually used for modeling the control-flow dominated systems. To avoid limitations of the classical FMS, researchers have proposed several derivatives of the FSM. Some of these extensions are used in several tools such as SOLAR (Ismail et al. 1994), Hierarchical Concurrent FSM (HCFSM) (Reynari et al. 2001) and Co-design Finite State Machine (CFSM) (Cloute et al. 1999).

Discrete-Event Systems

In a Discrete-Event System, the occurrence of discrete asynchronous events triggers the transitioning from one state to another. An event is defined as an instantaneous action, and has a timestamp representation when the event took place. Events are sorted globally according to their time of arrival. A signal is defined as a set of events, and it is the main method of communication between processes (Stoy and Zebo 1994). Discrete Event modeling is often used for hardware simulation. For example, both Verilog and VHDL use Discrete Event modeling as the underlying model of Computation. Discrete Event modeling is expensive since it requires all events according to their timestamp.

Petri Nets

Petri Nets is widely used for modeling systems. Petri Nets consists of places, tokens and transitions where token are stored in places. Transition causes tokens are stored in places. Transition causes tokens to be produced and consumed. Petri Nets supports concurrency and is asynchronous; however, they lack the ability to model hierarchy. Therefore, it can be difficult to use Petri Nets to model complex systems due to its lack of hierarchy. Variation of Petri Nets has been devised to address the lack of hierarchy, such as the Hierarchal Petri Nets (HPNs) proposed by Dittrich. Hierarchical Petri Nets (HPNs) supports hierarchy in addition to maintaining the major Petri Net's features such as concurrency and asynchronously. HPNs use directed graphs as the underlying model. HPNs are suitable for modeling complex systems since they support both concurrency and hierarchy.

Data Flow Graphs

Data Flow Graphs (DFG) systems are specified using a directed graph where nodes (actors) represent inputs, outputs and operations and edges represent data paths between nodes (Reynari et al. 2001). The main usage of Data Flow is for modeling data flow dominated systems. Computations are executed only where the operands are available. Communication between processes is done via unbounded FIFO buffering Scheme (Stoy and Zebo 1994). Data Flow models support hierarchy since the nodes can represent complex functions or other Data Flow.

Several variations of Data Flow Graphs have been proposed in the literature such as Synchronous Data Flow (SDF) and Asynchronous Data Flow (ADF). In SDF, a fixed number of tokens are consumed, where in ADF the number of tokens consumed is variable.

Synchronous/Reactive Models

Synchronous modeling is based on the synchrony hypothesis. Outputs are produced instantly in reaction to inputs and there is no observable delay in the outputs. Synchronous models are used for modeling reactive real-time Systems. Stoy and Zebo (1994) mentioned two styles for modeling reactive real time systems. First multiple clocked recurrent systems (MCRS) which are suitable for data dominated by real time systems. Second, state base formalisms which are suitable for control dominated real time systems. Synchronous languages, such as Esterel, are used for capturing Synchronous/Reactive model computation.

Heterogeneous Models

Heterogeneous Models combine features of different models of computations. Two examples of heterogeneous models are presented: Programming languages and Program State Machine (PSM). Programming languages provide a heterogeneous model that can support data, activity and control modeling. Two types of programming languages are presented, imperative language such as C, and declarative languages such as LISP and PROLOG. In imperative languages, statements are executed in the same order specified in the specification. On the other hand, execution order in not specified in declarative languages since the sequence of execution is based on a set of logic rules or functions.

Program State Machine (PSM) is a merger between HCFSM and programming languages. The Spec Charts language, which was designed as an extension to VHDL, is capable of capturing the PSM model. The SpecC is another language capable of capturing the PSM model. The following Table 3 attempts to set a comparison between different models of computation.

2.2.2 Specification Using Language

The goal of a specification using language is to describe the intended functionality of non-ambiguous systems. A large number of specifications using languages are currently being used in embedded system design since there is no language that is

Table 3 Comparison of various models of computation

MOC	Origin MOC	Main application	Clock mechanism	Orientation	Time	Communication method	Hierarchy
SOLAR	FSM	Control oriented	Synch	State	No explicit time	Remote procedure call	Yes
HCSFM state charts	FSM	Control oriented/reactive real time	Synch	State	Min/Max time spent in state	Instant broadcast	Yes
CFSM	FSM	Control oriented	Async	State	Events w/t time stamp	Wire signals	Yes
Discret event	N/A	Real time	Synch	Timed	Globally sorted events w/t time stamp	Wired signals	No
HPN	Petri net	Distributed	Async	Activity	No explicit timing	N/A	Yes
SDF	DFG		Synch	Activity	No explicit timing	Unbounded FIFO	Yes
ADF	DFC		Async	Activity	No explicit timing	Bounded FIFO	Yes

MOC Model of Compilation

the best for all applications. Below is a brief overview of the widely used language specification.

- Formal Description Languages such as LOTOS (based on process algebra, and used for the specification of concurrent and distributed systems) and SDL (used for specifying distributed real time systems, and based on extended FSM).
- Real Time Languages such as Esterel (a synchronous programming language based on the synchronous hypothesis. They are used for specifying real time reactive systems. Esterel is based on FSM, with constructs to support hierarchy and concurrency) and StateCharts (the graphical specification using languages used for specifying a reactive system). StateCharts extend FSM by supporting hierarchy, accuracy and synchronization.
- Hardware Description Languages: Commonly used HDL are a VHDL (IEEE standardized hardware description language), Verilog (hardware description language, which has been standardized by the IEEE) and HardwareC (a C based language designed for hardware synthesis). It extends C by supporting structural hierarchy, concurrency, communication and synchronization.

3 FPGA Cores Processor

The emergence of soft-cores processors (implemented using logic General Purpose programmable and synthesized onto FPGA) and hard-core processors (available as embedded blocks in the silicon next to the FPGA) inside FPGA increases their efficacy. FPGAs can include various embedded processors, different communications buses, many peripherals and network interfaces. It is possible now to create a complete hardware/software system with I/O and control interfaces on a single chip (SoC). This coexistence improves the embedded system performances by reducing the communication between external processors and FPGA circuit.

Embedded systems architectures allow the coexistence between hardware and software processors working together to perform a specific application. Usually, they can contain embedded processors who are often in the form of soft-core processors (described at a higher level of abstraction, implemented and synthesized to target a given FPGA or ASIC technology) and hard-core processors. Despite the advantages of the use of hardware processors (small area and power consumption), designers of embedded systems choose the implementation using soft-core processors due to their many advantages and their different configurations. Soft-core processors offer many hardware configurations to accelerate the execution time (e.g. adding floating-point hardware as hardware components into the soft-core processor) in terms of cost, flexibility, configuration, portability and scalability.

Atmel (FPGA vendors) and Triscend firms began introducing hard-core processor on their FPGA circuits, basing on an efficient communication mechanism between hard-core and FPGA components. More recently, Altera has offered Excalibur devices hard-core using ARM9 processor, NIOS and recently NIOS II soft-cores.

Xilinx firm has proposed the Virtex II Pro device with two or more PowerPC and tens millions of programmable gates and both PicoBlaze and MicroBlaze soft-cores. OpenCore has presented OpenRISC soft-core (Bolado et al. 2004) and Gaisler Research has given LEON and LEON2 soft-cores (Denning et al. 2004). In our study, the partitioning of software/hardware components was tested on Virtex-5 FPGA circuit, which allows the integration of various MicroBlaze soft-cores processors. In this chapter, embedded soft-core processor architecture, as being examined, consists of the Xilinx MicroBlaze soft-core processor.

3.1 Xilinx MicroBlaze Soft-Core Processor Architecture

Embedded processors can be defined as software cores implemented in hardware circuits using Logic General Purpose Programmable. The most used soft-cores processors, in the designing of embedded system for Xilinx FPGA, is the Xilinx's MicroBlaze soft-core processor. MicroBlaze is a 32-bit Reduced Instruction Set Computer (RISC) architecture optimized for synthesis and implementation into Xilinx FPGAs with a separate 32-bit instruction and data buses to execute programs and access data from both on-chip and external memory at the same time. This processor includes 32-bit general-purpose registers, virtual memory management, cache software support, and FSL interfaces. It has Harvard memory architecture and uses: Two Local Memory Busses (LMB) for instruction and data memory, two-Block RAMs (BRAM) and two peripherals connected via On-chip Peripheral Bas (OPB). Three memory interfaces are supported: Local Memory Bus (LMB), the IBM Processor Local Bus (PLB), and Xilinx Cache Link (XCL): The LMB offers single-cycle access to on-chip dual-port block RAM. The PLB interfaces offer a connection to both on-chip and off-chip peripherals and memory. The CacheLink interface is proposed for use with specialized external memory controllers. The architecture of the Xilinx MicroBlaze FPGA processor, the interfaces, buses, memory, and peripherals are shown in Fig. 4.

The major advantage of choosing MicroBlaze soft-core processor, in our researches, is its higher performance and its various configurations.

3.2 Xilinx MicroBlaze Soft-Core Processor Features

The MicroBlaze Xilinx processor offers tremendous flexibility during the design process. It allows different configurations to meet the needs of their design embedded applications by adding or removing some setting parameters such as:

- Integer Multiplier Units: Add the Integer multiplication as a co-processor.
- Barrel Shifter Units: Add the Shift by bit operations as a co-processor.
- Integer Divider Units: Add the Division of Integer as a co-processor.

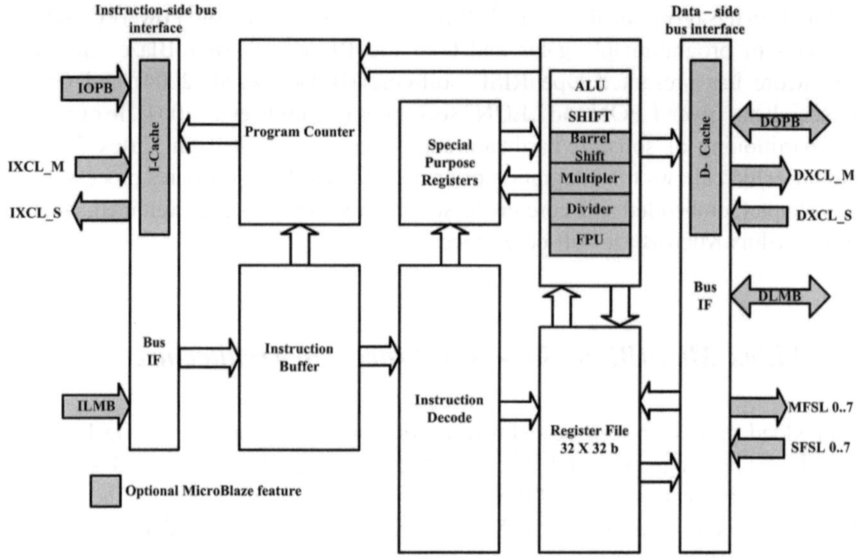

Fig. 4 Microblaze functional block diagram

- Floating-Point Units: Add Basic and Extended precision as a co-processor.
- Machine Status Register Units: Add Set and clear machine status register as a co-processor.
- Pattern Compare Unit: Add the String and pattern matching as a co-processor.

However, the designers need to select an appropriate configuration according to the application to improve the system performances. Thus, performance evaluation main function is to help embedded systems designers to answer the following questions: Does design methodology influence on the embedded system performances? Does a particular configuration affect the performance of the embedded system? How fast is the design process? What are the limits of the improvement of the design process? In the next sub-section, we will start t present our evaluation design approaches.

4 Proposed Design Approaches

The great issue of FPGA designers is that they are faced with the various problems for selecting the best architecture, the greatest hardware/software partitioning and the finest configuration of the selected soft-core processor. All these difficulties choice are constrained by execution time and area consumption. To take a decision about the final architecture design, designers need to proceed to a performance evaluation step. In our work, we propose to accelerate co-design methodology by

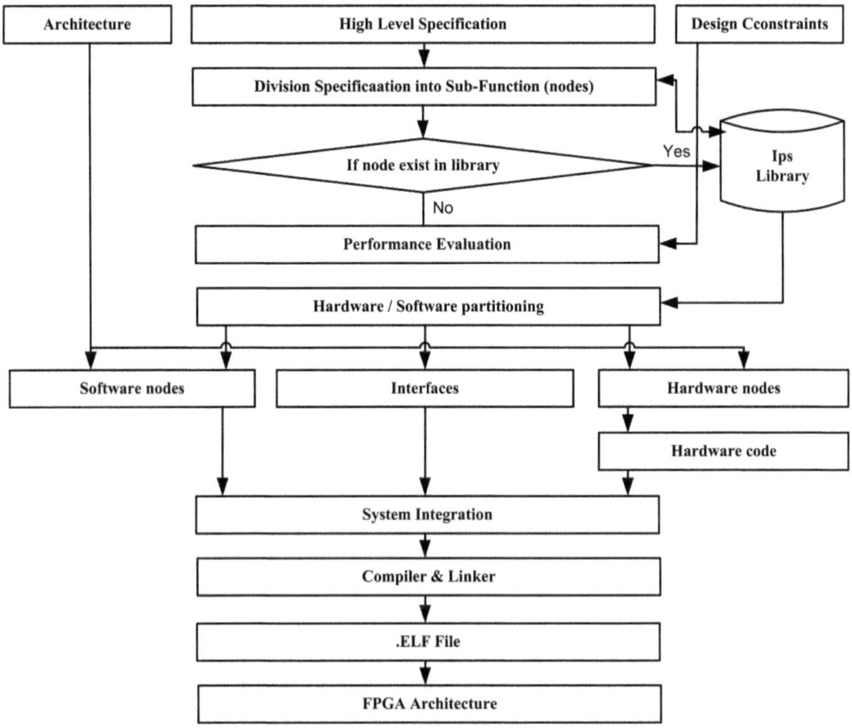

Fig. 5 Proposed co-design methodology approach

automating hardware/software partitioning step (basing of the hardware/software costs) using a high-level specification. Figure 5 illustrates our design methodology.

The low-level specification, proposed practically by all hardware/software partitioning approaches is replaced, in our approach, by a high-level specification. This high-level specification is divided into functional nodes (C functions) defined as nodes to make possible its integration on the hardware or software architecture. Beginning with a high-level specification, in the hardware/software partitioning step permits the classification of nodes on software or hardware without specifying the implementing target which allows the portability of our design process. Before partitioning, designers have to evaluate the costs of nodes (in term of execution time and area consumption) and the time taken for communication between software and hardware nodes. For software nodes, these costs are computed using profiler (e.g. compiling c code on MicroBlaze using the directive -pg, permit the generation of the profiling of each C function). However, hardware costs are measured after hardware synthesis of the high-level specification using HLS approaches. As hardware/software algorithms partitioning, we selected the Integer Linear Programing (ILP) algorithm. Figure 6 details our approach on hardware/software partitioning.

To evaluate our approach, software nodes are executed on the MicroBlaze soft-core processor with its different configurations. However, the implementation of hardware tasks is carried out, using HLS approaches. This tool allows fast prototyping of Intellectual Proprieties (IPs) that will be added to the MicroBlaze by the Fast Simplex Link (FSL) interface using Xilinx EDK tool.

5 Performance Evaluation Process

The performance evaluation of embedded systems has multiple aspects depending on the application that the system is made off. Hence, performance measurement is involved in several stages of the design process. In this chapter, we propose to evaluate the performance of the MicroBlaze FPGA soft-core processor features, in a first time, then that of our proposed design methodology, in a second time, using a lightweight cryptographic application.

5.1 Performance Evaluation Technique

Performance evaluation is the process of predicting whether the designed system satisfies the performance goal defined by the user such as area consumption and execution time (Mysore et al. 2005; Monmasson and Cristea 2007; Li and Malik 1995). Performance evaluation can be classified into two categories: Performance modeling and performance measurements as mentioned on Table 4.

5.1.1 Performance Modeling

Performance modeling approach is concerned with architecture-under-development. It can be used at an early stage of the design process where the processor is not available, or it is very expensive to prototype all possible processors architectures choices. Performance modeling may be classified into analytical-Based approach and Simulation-Based approach.

Analytical-Based Approach

The analytical modeling approach is based on probabilistic methods. Petri nets or Markov models create mathematical models of the designed embedded systems. The results of this approach are not often easy to construct. It allows predicting mainly user performance, time execution of sub-functions rapidly without compilation or execution. There has not been much study on the analytic approach for processors. Processors' structures are so complex that few analytical models can be provided for

Table 4 Performance evaluation techniques

CPU benchmarks	Synthetic benchmarks	
	Application based benchmarks	
	Algorithm based benchmarks	
Performance measurement	MP-on chip performance monitoring counters	
	Off-chip Hw monitoring	
	SW monitoring	
	Micro-coded instrumentation	
Performance modeling	Simulation	Trance driven simulation
		Execution driven simulation
		Complete system simulation
		Even driven simulation
		Software profiling
	Analytical model	Probabilistic models
		Queuing models
		Markov models
		Petri net models

them. Some research efforts are presented by Noonburg and Shen (1997) using a Markov models to model a pipelined processor, when Sorin et al. (1998) used probabilistic techniques to model a Multi-processor composed by superscalar processors.

Simulation-Based Approach

Simulation-Based approach presents the best performance modeling method in the performance evaluation of processor architectures. Model of the processor being simulated must be written in a high-level language, such as C or Java and running on some existing machine. Simulators give performance information in terms of cycles of execution, cache bit ratios, branch prediction rates, etc. Many commercial and academics simulators are presented: The SinOS simulator which presents a simple pipeline processor model and a powerful superscalar processor model. The SIMICS simulator simulates uni-processor and multi-processor models. Results of simulation approaches are not very interested in the performance evaluation of the MicroBlaze Xilinx soft-core processor because they are not exact.

5.1.2 Performance Measurement

Performance measurement approach is used for understanding systems that are already built or prototyped. Two major purposes for performance measurement approach can be used to tune systems to be built in order to understand the bottlenecks of such system. Performance measurement adjusts the application if its source code or algorithms can still be changed in order to understand the applications. This application can run on the system and tune the different design configurations. This kind of performance evaluation approach can be done using the following means:

- Microprocessor on-chip performance monitoring: can be used to understand performance of high microprocessors (Intel's Pentium III and Pentium IV, IBM Power3 and Power4 processors, AMD's Athlon, Compaq's Alpha and Sum's Ultra SPARC). Several tools are available to measure performance monitoring counters: Intel's Vtune software can be used to perform measurement when the Intel performance counters. The P6Pref utility presents a plug-in for Windows NT performance monitoring. The Compaq DIGITAL Continuous Profiling Infrastructure (DCPI) presents a very powerful tool used to profile program on the Alpha processors.
- Off-Chip hardware monitoring: Instrumentation using hardware wherewithal can be done by attaching off-chip hardware. Example Speed Tracer from AMD and Logic analyser. AMD developed hardware-trading platform to help in the design of X86 microprocessors. However, Poursepanj and Christie used a logic analyser to analyze 3D graphics workloads on AMD-K6-2 based systems.
- Software monitoring: is an important mode of performance evaluation used before the advent of on-chip performance monitoring counters. The primary advantage of software monitoring is that it is easy to execute.
- Mircocoded instrumentation: is a technique lying between trapping information on each instruction using hardware interrupts (traps) or software interrupts (traps). The tracing system modified the VAX microcode to record all instructions and data references in a reserved portion of memory.

5.1.3 CPU Benchmarks

Designers of FPGA processor have to use the CPU Benchmarks approach to get a fixed measurement of the processors 'performance, which is attempting to implement and verify the architectural and the timing behavior under a set of benchmark programs. Several open sources and commercial benchmarks are presented. Some of them are: Mibench, Paranoia, LINPACK, SPEC (Standard Performance Evaluation Corporation), and EEMBC (Embedded Microprocessor Benchmark Consortium). These Benchmarks are divided into three categories depending on the application (Korb and Noll 2010). The first category is Synthetic Benchmark (with the intention to measure one or more features of systems, processors, or compilers). The second category is application based benchmarks or "real world" benchmarks (developed to compare different processors' architectures in the same fields of applications). Finally, the third category is Algorithm Based Benchmarks (developed to compare systems architectures in special (synthetic) fields of application).

Synthetic Benchmarks

Synthetic Benchmarks are developed to measure processor specific parameters. Synthetic benchmarks are created with the intention to measure one or more features of systems, processors, or compilers. It tries to mimic instruction mixes in real-world applications. However, it is not related to how that feature will perform in a real application. Dhrystone and Whetstone benchmarks are the most-used synthetic benchmarks.

Application Based Benchmarks

Application Based Benchmarks or "real world" benchmarks are developed to compare different processor architectures in the same fields of applications. Application based or "real world" benchmarks use the code drawn from real algorithms, and they are more common in system-level benchmarking requirements.

Algorithms Based Benchmarks

Algorithm Based Benchmarks: (a compromise between the first and the second type) developed to compare systems architectures in special (synthetic) fields of application. Several studies are based on this approach to evaluate the processors' performances. Bolado et al. (2004) evaluated three soft-cores processors namely LEON2, MicroBlaze and OpenRISC to measure the execution time and the area consumption, using Dhrystone and Standford benchmarks. Berkeley Design Technology, Inc. evaluated the performance of the Texas Instruments' DSCs processors to compute the execution time using the Fast Fourier Transform (FFT) algorithms using fixed-point and floating-point data precision. Korb and Noll (2010) examined the performance of both DSPs and MCUs basing on the execution time of a number of benchmark codes included fixed-point and floating-point math operations, logic calculation, digital control, FFT, conditional jumps and recursion test algorithms. In our paper, we have chosen to adopt the performance measurement method using freely benchmark solutions. We used lightweight cryptographic secure application as a benchmark.

In the next section, we will introduce our used benchmark: The lightweight cryptographic application: Quark Hash Algorithm.

5.2 *Lightweight Cryptographic Benchmarks: Quark Hash Algorithm*

The need for Lightweight cryptographic applications have been frequently expressed by embedded systems designers, to implement a secured application such as the authentication, the password storage mechanisms, the Digital Signal Standard (DSS), the Transport Layer Security (TLS), the Internet Protocol Security (IPSec), the Random number generation algorithms; etc. Several Lightweight cryptographic algorithms are presented. Lightweight cryptographic algorithms have been designed to fit with a very compact hardware. Each algorithm can be adapted for a specific field (Korb and Noll 2010; Bogdanov et al. 2013).

- SHA family: Secure SHA Algorithms are a family of Hash Algorithms published by NIST since 1993. SHA has many derivative standards such as SHA-0, SHA-1, SHA-3
- MDA/MD5/MD6: Message-Digest Algorithm is a family of broadly used cryptographic hash function developed by Ronald Rivest that produces a 128-bit for MD4 and MD5, 256-bit for MD6.
- Quark: Family of cryptographic functions designed for resource-constrained hardware environments.
- CubeHash: A very simple cryptographic hash function designed in University of Illinois at Chicago, Department of Computer Science.
- Photon: A lightweight hash function designed for very constrained devices.
- SQUASH: Not collision resistant, suitable for RFID applications.

According to its complexity, Quark presents the most appropriate algorithm to evaluate the performance of the soft-core FPGA processor architecture. Quark can minimize area and power consumption, it offers strong security guarantees. These Hash algorithms that are efficiently implemented in low cost embedded devices are important components for securing new applications in ubiquitous computing. Quark Hash algorithm is a family of lightweight cryptographic "sponge" algorithms designed for resource-constrained hardware environments, as RFID tags. It combines a number of innovations that make it unique and optimized. In the design of Quark, designers opt for an algorithm based on bit shift. It combines a sponge construction with a capacity e equal to the digest length n, and a core permutation inspired by preceding primitives. Quark algorithm proposes three instances: u-Quark, d-Quark and s-Quark. Quark is a family of cryptographic "sponge" functions intended for resource-constrained hardware environments (Bogdanov et al. 2013). It minimizes area and power consumption, yet offers strong security guarantees.Quark function includes four functions: (1) permute function, (2) init function, (3) update function and (4) final function. These instances are parameterized by a rate r, a capacity c, an output length n and a b-bit permutation ($b = r + c$). Table 5 demonstrates the parameters of each instance of the Quark algorithms.

Table 5 Parameters of Quark hash algorithms instance

	Rate (r)	Capacity (c)	With (b)	Digest (n)
u-Quark	8	128	136	136
d-Quark	16	160	176	176
s-Quark	32	224	256	256

6 Results

As mentioned before, the main topic of this study is to evaluate and validate the effect of the Xilinx MicroBlaze features and the proposed hardware/software partitioning approach on the embedded system performances.

6.1 *Experimental Setup*

6.1.1 Hardware Experimental Setup

Performance evaluation was estimated on a first time by a basic measurement of the different MicroBlazesoft-core configurations implemented on Xilinx Virtex-5 development board (XUPV5-LX110T, xc5vlx110t, grade ff1136, speed-1), illustrated in Figure 6.

Processor performance can be measured in different metrics such as execution time, energy consumption and area utilization. The most common metric is the time required for a processor to accomplish the defined task. In some architecture using an internal CPU clock driver, execution time presents the clock driver multiplied by the total instruction cycle count. In our case, execution time is measured using a Logic Analyser to have a high-precision measurement.

Fig. 6 Our plateform

6.1.2 Software Experimental Setup

In this chapter, we propose to immediately generate a cryptographic application as a co-processor (Hardware part) that will be added to a MicroBlaze using FSL interface (Software part). EDK tool will be used to perform the integration of both Hardware and Software in our architecture design. To implement Virtex-5 embedded applications, we use CoDevelopper™ tool to generate co-processors or IPs (HLS methodology) and Xilinx Project Studio (XPS) to configure the FPGA including one MicroBlaze soft-core.

EDK Development Kit

The Xilinx EDK contains both an integrated development environment (IDE) named Xilinx Platform Studio (XPS) to create the Microprocessor Hardware Specification (MHS) file and the Software Design Kit (EDK) to create a Microprocessor Software Specification (MSS) file. The MHS file defines the embedded system processor, architecture and peripherals. The MSS file defines the library customization parameters for peripherals, the processor customization parameters, the standard I/O devices, the interrupt handler routines, etc.

CoDeveloper™

CoDeveloper™ is a commercialized by Impulse Accelerated Technologies in the CAD market. It allows designers to compile C applications directly into optimized logic ready for use with Xilinx FPGAs, in few times. ImpulseC code, the input language of CoDeveloper™, can be written and debugged in any ANSI standard C environment. The implemented algorithm can use both fixed and floating-point data point types. Impulse C is a library of functions and related data types that give a programming environment, and a programming model, for parallel applications targeting FPGA-based platforms. It has been optimized for mixed software/hardware targets, with the goal of abstracting details of inter-process communication and can allow relatively platform-independent application design. CoDeveloper™ includes the Impulse C libraries and associated software tools that help designers use standard C language for the design of highly parallel applications targeting FPGAs.

6.2 *Application of the Proposed Design Approach for Quark Benchmark*

6.2.1 Effect of MicroBlaze Soft-Core Configuration on Embedded Systems Performance

For embedded application, different MicroBlaze configurations can be provided. In real-time complex applications, both execution time and area consumption determine the efficiency and the high performance of the configured embedded soft-core processor.

The evaluation of hardware area presents one of the metric to select embedded configurations, which requires an optimal area. In a software design methodology, area consumption is independent from the implemented application. We can evaluate the performance of the soft-core processor for each configuration directly after the hardware specification step. Results prove that the average number of slices (a group of logic cell resources in FPGA) without using optimization option is very important. Table 6 depicts the area consumption recorded for some possible MicroBlaze configurations.

To evaluate the performance of the MicroBlaze soft-core processor, we have estimated the execution time in order to choose the most efficient configuration, which takes the minimum execution time onto the smaller hardware area. In our work, we compute the execution time of the configuration described in the table for three Quark hash functions (u-Quark function, d-Quark function and s-Quark function). Figure 7 illustrates the Quark hash functions execution time measurement for the 17 configurations of Xilinx MicroBlaze.

6.2.2 Automation of Partitioning Process

Designers have to specify the target architecture early in the design by defining the configuration of the software nodes to synthesize hardware nodes. Moreover, designers have, also, to determine the design constraints, performance constraints (timing) and resource constraints (area, memory). In this study, we choose to evaluate the proposed approach for a lightweight cryptographic s-Quark benchmark. We divide the C-high-level specification into four functional units (C functions) presented as nodes. We compute than nodes costs for all hardware and software possible architecture. For Software nodes, cost computation will be assured by profiling. For Hardware nodes, C functions will be transformed into Hardware specification using HLS approach, synthesized and analysed using Logical synthesis to get its costs. We propose MicroBlaze soft-core as software architecture with its different configurations (presented above). With our approach; we have to specify the costs (execution time and resources utilization) for each s-Quark node which can be implemented using soft-core (within all configurations).

Table 6 Area consumption of MicroBlaze processor synthesis

Configuration	With optimization synthesis		Without optimization synthesis	
MB + Units	LUTs	F-Fs	LUTs	F-Fs
1: Basic	1,210	1,452	1,657	1,693
2: BS	1,570	1,247	1,818	1,727
3: FPU	1,620	2,153	2,395	2,105
4: Mul	1,456	1,232	1,714	1,709
5: ID	1,581	1,326	1,801	1,805
6: MSRU	1,458	1,214	1,675	1,690
7: BS + mul + ID	1,727	1,380	1,964	1,867
8: BS + mul + FPU	2,307	1,674	2,511	2,162
9: BS + ID + FPU	2,433	1,769	2,668	2,258
10: BS + mul + MSRU	1,609	1,267	1,830	1,749
11: BS + ID + MSRU	1,734	1,365	1,966	1,846
12: BS + FPU + MSRU	2,313	1,659	2,533	2,142
13: mul + ID + FPU	2,358	1,755	2,575	2,241
14: mul + ID + MSRU	1,628	1,351	1,864	1,829
15: mul + FPU + MSRU	2,207	1,645	2,418	2,127
16: ID + MSRU + FPU	2,359	1,740	2,554	2,224
MSRU + FPU	2,202	1,625	2,417	2,107

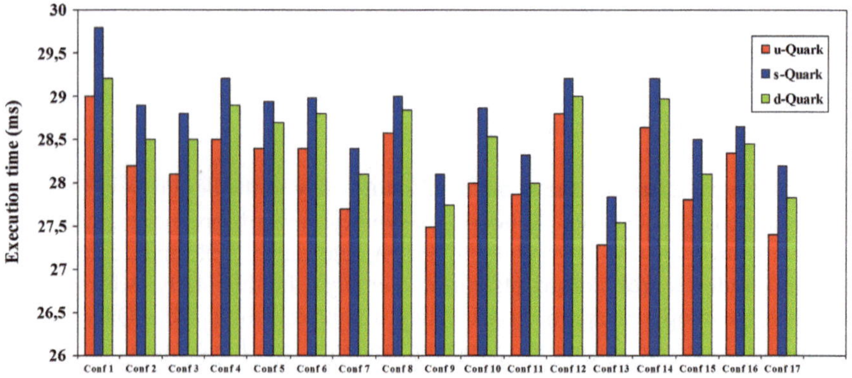

Fig. 7 Quark benchmark execution time usage for different MicroBlaze configurations

In order to select the greatest hardware/software architecture (partitioning process), we used the Integer Linear Programing (ILP) algorithm. Under the ILP algorithm, Gains of execution time and resource consumption are computed (as described in Table 7) using these two formulas:

Table 7 Temporal and Resources gain of s-Quark implementation

Task	1		2		3		4	
Gain	Gt	Gr	Gt	Gr	Gt	Gr	Gt	Gr
G1	20.94	68	0.84	10	2.91	14	6.98	145
G2	18.94	88	0.67	18	2.66	23	6.18	148
G3	19.14	109	0.72	10	2.71	21	5.78	178
G4	18.74	118	0.71	12	2.7	20	6.58	187
G5	19.94	128	0.64	14	2.63	17	5.32	198
G6	20.94	130	0.72	13	2.71	21	3.08	197
G7	19.94	132	0.82	17	2.91	19	4.18	197
G8	19.34	136	0.74	15	1.73	23	6.18	199
G9	20.14	139	0.82	17	2.81	22	4.08	201
G10	21.81	143	0.72	18	2.71	20	6.73	206
G11	18.26	146	0.71	20	2.70	22	6.18	205
G12	21.14	140	0.70	23	2.69	21	4.08	197
G13	20.74	143	0.72	14	2.71	23	6.18	207
G14	19.14	136	0.66	26	2.65	26	4.08	185
G15	18.14	148	0.64	30	2.63	27	6.73	192
G16	17.59	151	0.7	29	2.69	28	7.18	195
G17	18.14	155	0.62	26	2.7	24	6.48	206

(1) Gt = Execution time before Hw migration−Execution time after Hw migration
(2) Gr = Resources before Hw migration−Resources after Hw migration

In the designing of Quark cryptographic application, the designer has to satisfy temporal constraints while minimizing the number of the used resources. Partitioning process is based on the assignment of tasks on software and hardware units. This partitioning will be modified, with new hardware/software assignments, until the designer got the partition that meets the requirements of execution time and area consumption. The interesting parameter for partitioning is the number of nodes, which have to be partitioned. Using both hardware/software implementation, the time taken to transfer data between the soft-core and IPs (or co-processors) will be added. The cost of hardware/software communications are computed based on the width of transmitted data (8, 16 or 32 bits) and the rate of the communication buses.

As seen above, Xilinx MicroBlaze soft-core processor implements Harvard architecture. It means that it has separate bus interface for data and instruction access. The OPB interface gives a connexion to both on- and off-chip peripherals and memory. The MicroBlaze soft-core also provides 8 input and 8 output interfaces to Fast Simplex Link (FSL) buses. This FSL buses, 32 bits wide, are unidirectional non-arbitrated dedicated communication channels. In our study, we used the FSL interface due to its high performance (can reach up 300 Mb/S). EDK provides a set of Macros for reading and writing to or from FSL interface. Our

Table 8 S-Quark implementation results

	Execution time (ms)	Resources (Slices)	Design time : From architecture model to design implementation
Node 1 (Hw)	75	832	1 h
Node 2 (SW7)	32	774	5 min
Node 3 (SW5)	52	853	
Node 4 (SW1)	124	954	
Total Hw/Sw nodes	283	3413	1 h/15 min

purposed partitioning solution will determine the best partition that will reduce the number on nodes implemented on hardware and increase the number of nodes implemented on software to reduce the design time and the hardware area.

After hardware/software partitioning, we have to implement our s-Quark benchmark. Hardware nodes are implemented using HLS approach (CoDeveloper^{TM} tool). Software nodes are executed using XPS tool. The integration of the hardware nodes (co-processors or IPs) with MicroBlaze soft-core processor is achieved using EDK tool. Table 8 illustrates results of s-Quark implementation.

7 Discussions

Increasing complexities of embedded systems application sunders core the need to take design decisions at an early stage. In our study, we are based on two important decisions related to the automation of the choice of both hardware/software partitioning and soft-core processor configurations.

7.1 Soft-Core Processor Configuration

MicroBlaze is a 32-bit embedded soft-core processor with a reduced instruction set computer (RISC) architecture. It is highly configurable and specifically optimized for synthesis into Xilinx field programmable gate arrays (FPGAs). The MicroBlaze soft-core processor is available as HDL source code or structural netlist. Itcan also be integrated into ASICs. As described in Fig. 6, one of the advantages of Xilinx MicroBlazesoft-core processors is its flexibility: it uses various configurations (more than 17 configurations) required for a specific application. Another advantage is its ability to integrate customized IP cores, which can result in a dramatic acceleration in software execution time (difference between configuration 1 and configuration 17) due to applications being executed in parallel with hardware and not sequentially in software.

Quark hash functions do not use huge values. They are dominated by barrel shifter, integer arithmetic, logic decisions, and memory accesses intended to reflect the CPU activities in computing applications. It takes a huge time for memory access. As described in the Fig. 6, selecting the best configuration enables a huge gain perspective of execution time and area consumption. The performance of implemented embedded systems using basic configuration (config. 1) is very low compared to the performance using Barrel Shifter Units (BS), Integer multiplier (Mul) and Floating-Point Units (FPU) configuration (config. 8). The execution time using the basic configuration takes 29 mS (for u-Quark); 29.8 mS (for s-Quark) and 29.2 mS (for d-Quark). 8) takes 28.58 mS (for u-Quark); 29 mS (for s-Quark) and 28.84 mS (for d-Quark). Area consumption constraint has also an effect on the embedded systems performance when modifying the configuration. Using basic configuration (config. 1), with optimization, takes 1,210 LUTs and 1,452 F-Fs. However, using Barrel Shifter Units (BS), Integer multiplier (Mul) and Floating-Point Units (FPU) configuration (config. 8) takes 2,307 LUTs and 1,674 F-Fs. If the application is area-critical, the user should select the best area/execution time constraints. In real-time embedded systems, area consumption constraint is not very important compared to the execution time.

Results prove that modifying configuration have an important effect on the embedded system performances. These results are interesting to make an optimized architecture for software design, designers of embedded systems can also benefit of FPGA hardware resources to more accelerate execution time and minimize the energy consumption. Hardware/software architecture has to be used to satisfy embedded systems constraints.

The results obtained from these different configurations require approximately 20 min per configuration, so, 60 % of the time is spent by the synthesis to choose the best configuration. Automate this step using time estimation approach allows the acceleration of the design time. Also, area synthesis results can be used on the designing of other embedded application, which reduce the design time.

7.2 Hardware/Software Partitioning

Partitioning an application among software solution on a soft-core processor (MicroBlaze) and hardware co-processors (IPs) in on-chip configurable logic has been shown to improve performance in embedded systems.

The used partitioning algorithm ILP is software oriented, because it starts with only software nodes. For this reason, the initial specifications were written in a high-level language (C functions). These functions are divided into functional units named nodes (node1, node2, node3 and node4 for Quark function). The first step in hardware/software partitioning step is the computation of both nodes and communication (between hardware and software nodes) costs. The costs can be defined as the execution time and the resources using hardware implementation (Hw1) or software implementation with different configuration of Microblaze (Sw1–Sw17).

Table 9 Benefits of our design approach compared to the existing ones

		Traditional design approaches	Recent design approaches	Our design approaches'
Design time	Time required for soft-core configuration	Time required for specifying all configurations + 20 mS for synthesizing each configuration	Time required for specifying all configurations + 20 mS for synthesizing each configuration	Time required for specifying all configurations + 20 mS for synthesizing each configuration (if node is not in the library)
	Time required for hardware/software architecture selection	Manually: long time: long decision time	Long time: specification has to be written in a low-level specification	Medium: specification is written in a high-level specification
	Time required for hardware/software partitioning	Long time: time required for the manually coding for both software and hardware architecture	Medium: time required for the automatic generation of codes for a specified architecture	Medium: time required for the automatic generation of codes for choosing architecture
Flexibility		Ye	No	Yes
Portability		Yes	No	Yes
Execution time		Very high	Medium	Medium
Area consumption		High	Reduced	Reduced

Choosing the greatest partitioning is verified by ILP algorithm. As result, we select to implement the node 1 (permute C function) as hardware. Permute function (node) is dominated by barrel shifter, integer arithmetic and logic decision. Implement it as a hardware node allows the designer to minimize area and execution time at least to 1.95 % for LUTs resources and 0.86 % for execution time comparing to software implementation. In addition, the integration of hardware nodes in soft-core MicroBlaze processor did not require to inline assembler code because the FSL interface has predefined C-macros that can be used for sending and receiving data between hardware and software nodes. Results of s-Quark benchmark (illustrated on the Table 8) prove that implementing complex applications on hardware/software architecture with automatic hardware/software partitioning are better than implementing these applications on software architectures (using MicroBlaze Soft-core processors). As summary, Table 9 illustrates features of our design approach compared to the existing ones.

8 Conclusions and Perspectives

FPGA presents an interesting circuit for implementing embedded applications. The purpose of this chapter to illustrate the impact of co-design approach, on the design acceleration and architecture performance. Based on the proposed co-design approaches of hardware/software partitioning, we are contributing to specification in order to increase its level. We, also, added a step to select the finest soft-core processor configuration in order to facilitate the co-design process, improve embedded systems' performance and reduce design time.

The presented results demonstrate that the choice of the good configuration has a significant impact on the system performance. The same approach can be used to evaluate the performance of other embedded systems or other architectures. Design methodologies of embedded systems, as mentioned in this paper, can be software, hardware or both software/hardware. Using co-design methodology helps the designer to obtain a good performance in a short time-to-market based on a good hardware/software partition. In this chapter, we have also introduced the hardware/software partitioning problem from a high-level specification. Several partitioning algorithms are presented in this study: One of them is based on ILP, which is used in our empirical tests. The ILP algorithm works efficiently for graphs with several hundreds of nodes and yield optimal solutions. As perspective, we can validate our proposed approach for more complexes embedded applications using FPGA devices for other vendors such as Altera, Actel, etc. We can also study the performances and design time benefits using time estimation approach instead of real performance evaluation.

References

Arulmozhiyal, R. (2012). Design and implementation of fuzzy PID controller for BLDC motor using FPGA. In *IEEE International Conference on Power Electronics, Drives and Energy Systems (PEDES)* (pp. 1–6), December 16–19, 2012. doi:10.1109/pedes.2012.6484251.

Bogdanov, A., Knezevic, M., Leander, G., Toz, D., Varici, K., & Verbauwhede, I. (2013). Spongent: The design space of lightweight cryptographic hashing. *IEEE Transactions on Computers, 62*(10), 2041–2053.

Bolado, M., Posadas, H., Castillo, J., Huerta, P., Sanchez, P., Sanchez, C., Fouren, H., & Blasco, F. (2004). Platform based on open-source cores for industrial applications. In *Europe Conference and Exhibition on Design, Automation and Test* (pp. 1014–1019), February 16–20, 2004. doi:10.1109/date.2004.1269026.

Boßung, W., Huss, S. A., & Klaus, S. (1999). High-level embedded system specifications based on process activation conditions. *Journal of VLSI Signal Processing Systems for Signal, Image and Video Technology, 21*(3), 277–291.

Chatha, K. S., & Vemuri, R. (2000). An iterative algorithm for hardware-software partitioning, hardware design space exploration and scheduling. *Design Automation for Embedded Systems, 5*(3–4), 281–293.

Cherif, S., Quadri, I. R., Meftali, S., & Dekeyser, J. (2010). Modeling reconfigurable systems-on-chips with UML MARTE profile: An exploratory analysis. In *13th Euromicro Conference on Digital System Design: Architectures, Methods and Tools (DSD)* (pp. 706–713), September 1–3, 2010. doi:10.1109/dsd.2010.58.

Cloute, F., Contensou, J. N., Esteve, D., Pampagnin, P., Pons, P., & Favard, Y. (1999). Hardware/software co-design of an avionics communication protocol interface system: An industrial case study. In *7th International Workshop on Hardware/Software Codesign (CODES '99)* (pp. 48–52). doi:10.1109/hsc.1999.777390.

Dave, N., Ng, M. C., & Arvind. (2005). Automatic synthesis of cache-coherence protocol processors using Bluespec. In *3rd ACM and IEEE International Conference on Formal Methods and Models for Co-Design* (pp. 25–34), July 11–14, 2005. doi:10.1109/memcod.2005.1487887.

De Michell, G., & Gupta, R. K. (1997). Hardware/software co-design. *Proceedings of the IEEE, 85*(3), 349–365.

Denning, D., Irvine, J., Stark, D., & Delvin, M. (2004). Multi-user FPGA co-simulation over TCP/IP. In *15th IEEE International Workshop on Rapid System Prototyping* (pp. 151–156), June 28–30, 2004. doi:10.1109/iwrsp.2004.1311110.

Feng, W., Yuan, X., & Takach, A. (2009). Variation-aware resource sharing and binding in behavioral synthesis. In *Asia and South Pacific Design Automation Conference (ASP-DAC)* (pp. 79–84), January 19–22, 2009. doi:10.1109/aspdac.2009.4796445.

Fujita, M., & Nakamura, H. (2001). The standard SpecC language. In *Proceedings of the 14th International Symposium on Systems synthesis* (pp. 81–86).

Gruian, F., & Westmijze, M. (2008). VHDL vs. Bluespec system verilog: A case study on a java embedded architecture. In *Proceedings of the 2008 ACM Symposium on Applied Computing* (pp. 1492–1497).

Gupta, R. K., Coelho, C. N., & De Micheli, G. (1992). Synthesis and simulation of digital systems containing interacting hardware and software components. In *29th ACM/IEEE Design Automation Conference* (pp. 225–230), June 8–12, 1992. doi:10.1109/dac.1992.227832.

Henkel, J., & Ernst, R. (1998). High-level estimation techniques for usage in hardware/software co-design. In *Asia and South Pacific Design Automation Conference* (pp. 353–360), February 10–13, 1998. doi:10.1109/aspdac.1998.669500.

Ismail, T. B., Abid, M., O'brien, K., & Jerraya, A. (1994). An approach for hardware-software codesign. In *5th International Workshop on Rapid System Prototyping Shortening the Path from Specification to Prototype* (pp. 73–80), June 21–23, 1994. doi:10.1109/iwrsp.1994.315907.

Jianzhuang, W., Youping, C., Jingming, X., Bing, C., & Haiping, L. (2008). System structure for FPGA-based SOPC design using hard tasks. In *6th IEEE International Conference on Industrial Informatics, INDIN 2008* (pp. 1154–1159), July 13–16, 2008. doi:10.1109/indin.2008.4618277.

Jing-Jie, W., & Rui, H. (2011). A FPGA-based wireless security system. In *Third International Conference on Multimedia Information Networking and Security* (*MINES*) (pp. 512–515), November 4–6, 2011. doi:10.1109/mines.2011.82.

Joven, J., Strict, P., Castells-Rufas, D., Bagdia, A., De Micheli, G., & Carrabina, J. (2011). HW-SW implementation of a decoupled FPU for arm-based cortex-M1 SOCS in FPGAS. In *6th IEEE International Symposium on Industrial Embedded Systems* (*SIES*) (pp. 1–8), June 15–17, 2011. doi:10.1109/sies.2011.5953649.

Kalavade, A., & Lee, E. A. (1994). A global criticality/local phase driven algorithm for the constrained hardware/software partitioning problem. In *3rd International Workshop on Hardware/Software Codesign* (pp. 42–48), September 22–24, 1994. doi:10.1109/hsc.1994.336724.

Kalomiros, J. A., & Lygouras, J. (2008). Design and evaluation of a hardware/software FPGA-based system for fast image processing. *Microprocessors and Microsystems, 32*(2), 95–106.

Kikuchi, H., & Morioka, K. (2012). Development of wireless image sensor nodes based on FPGA for human tracking in intelligent space. In *IECON 2012—38th Annual Conference on IEEE Industrial Electronics Society* (pp. 5529–5534), October 25–28, 2012. doi:10.1109/iecon.2012.6388950.

Korb, M., & Noll, T. G. (2010). LDPC decoder area, timing, and energy models for early quantitative hardware cost estimates. In *International Symposium on System on Chip* (*SoC*) (pp. 169–172), September 29–30, 2010. doi:10.1109/issoc.2010.5625546.

Ku, D. C., & De Mitcheli, G. (1992). Relative scheduling under timing constraints: Algorithms for high-level synthesis of digital circuits. *IEEE Transactions on Computer-Aided Design of Integrated Circuits and Systems, 11*(6), 696–718.

Lach, J., Mangione-Smith, W. H., & Potkonjak, M. (1999). Robust FPGA intellectual property protection through multiple small watermarks. In *Proceedings 36th Design Automation Conference* (pp. 831–836), 1999.

Li, Y.-T. S., & Malik, S. (1995). Performance analysis of embedded software using implicit path enumeration. *ACM SIGPLAN Notices, 30*(11), 88–98.

Lingbo, Z., Fuchun, S., & Zengqi, S. (2006). Cloud model-based controller design for flexible-link manipulators. In *IEEE Conference on Robotics, Automation and Mechatronics* (pp. 1–5), December 2006. doi:10.1109/ramech.2006.252742.

López-Vallejo, M., & López, J. C. (2003). On the hardware-software partitioning problem: System modeling and partitioning techniques. *ACM Transactions on Design Automation of Electronic Systems (TODAES), 8*(3), 269–297.

Lysecky, R., & Vahid, F. (2004). A configurable logic architecture for dynamic hardware/software partitioning. In *Design, Automation and Test in Europe Conference and Exhibition* (pp. 480-485), February 16–20, 2004. doi:10.1109/date.2004.1268892.

Madsen, J., Grode, J., Knudsen, P. V., Petersen, M. E., & Haxthausen, A. (1997). LYCOS: The Lyngby co-synthesis system. *Design Automation for Embedded Systems, 2*(2), 195–235.

Mcloone, M., & Mccanny, J. V. (2003). Generic architecture and semiconductor intellectual property cores for advanced encryption standard cryptography. *IEE Proceedings on Computers and Digital Techniques, 150*(4), 239–244.

Monmasson, E., & Cirstea, M. N. (2007). FPGA design methodology for industrial control systems—A review. *IEEE Transactions on Industrial Electronics, 54*(4), 1824–1842.

Mysore, N., Akcakaya, M., Bajcsy, J., & Kobayashi, H. (2005). A new performance evaluation technique for iteratively decoded magnetic recording systems. In *Digests of the IEEE International Magnetics Conference* (*INTERMAG*) (pp. 1603–1604), April 4–8, 2005. doi:10.1109/intmag.2005.1464235.

Nasreddine, N., Boizard, J. L., Escriba, C., & Fourniols, J. Y. (2010). Wireless sensors networks emulator implemented on a FPGA. In *International Conference on Field-Programmable Technology (FPT)* (pp. 279–282), December 8–10, 2010. doi:10.1109/fpt.2010.5681484.

Noonburg, D. B., & Shen, J. P. (1997). A framework for statistical modeling of superscalar processor performance. In 3rd *International Symposium on High-Performance Computer Architecture* (pp. 298–309), February 1–5, 1997. doi:10.1109/hpca.1997.569691.

Reynari, L. M., Cucinotta, F., Serra, A., & Lavagno, L. (2001). A hardware/software co-design flow and IP library based of simulink. In *Design Automation Conference* (pp. 593–598), 2001. doi:10.1109/dac.2001.156209.

Samarawickrama, M., Rodrigo, R., & Pasqual, A. (2010). HLS approach in designing FPGA-based custom coprocessor for image preprocessing. In *5th International Conference on Information and Automation for Sustainability (ICIAFs)* (pp. 167–171), December 17–19, 2010. doi:10.1109/iciafs.2010.5715654.

Sorin, D. J., Pai, V. S., Adve, S. V., Vemon, M. K., & Wood, D. A. (1998). Analytic evaluation of shared-memory systems with ILP processors. In *The 25th Annual International Symposium on Computer Architecture* (pp. 380–391), June 27–July 1, 1998. doi:10.1109/isca.1998.694797.

Stitt, G., Lysecky, R., & Vahid, F. (2003). Dynamic hardware/software partitioning: A first approach. In *Proceedings of the 40th Annual Design Automation Conference* (pp. 250–255).

Stoy, E., & Zebo, P. (1994). A design representation for hardware/software co-synthesis. In *The 20th EUROMICRO Conference on System Architecture and Integration* (pp. 192–199), September 5–8, 1994. doi:10.1109/eurmic.1994.390391.

Talpin, J., Le Guernic, P., Shukla, S. K., Gupta, R., & Doucet, F. (2003). Polychrony for formal refinement-checking in a system-level design methodology. In *3rd International Conference on Application of Concurrency to System Design* (pp. 9–19), June 18–20, 2003. doi:10.1109/csd.2003.1207695.

Thangavelu, A., Varghese, M. V., & Vaidyan, M. V. (2012). Novel FPGA based controller design platform for DC–DC buck converter using HDL co-simulator and Xilinx system generator. In *IEEE Symposium on Industrial Electronics and Applications (ISIEA)* (pp. 270–274), September 23–26, 2012. doi:10.1109/isiea.2012.6496642.

Wakabayashi, K., & Okamoto, T. (2000). C-based SoC design flow and EDA tools: An ASIC and system vendor perspective. *IEEE Transactions on Computer-Aided Design of Integrated Circuits and Systems, 19*(12), 1507–1522.

Washington, C., & Dolman, J. (2010). Creating next generation HIL simulators with FPGA technology. In *IEEE AUTOTESTCON* (pp. 1–6), September 13–16, 2010. doi:10.1109/autest.2010.5613618.

Wiangtong, T., Cheung, P. Y., & Luk, W. (2005). Hardware/software code sign: A systematic approach targeting data-intensive applications. *IEEE Signal Processing Magazine, 22*(3), 14–22.

Wolf W. (2003). A decade of hardware/software codesign. *Computer*, *36*(4), 38–43.

Xiaoyin, S., & Dong, S. (2007). Development of a new robot controller architecture with FPGA-based IC design for improved high-speed performance. *Industrial Informatics, IEEE Transactions on, 3*(4), 312–321.

A Neural Approach to Cursive Handwritten Character Recognition Using Features Extracted from Binarization Technique

Amit Choudhary, Savita Ahlawat and Rahul Rishi

Abstract The feature extraction is one of the most crucial steps for an Optical Character Recognition (OCR) System. The efficiency and accuracy of the OCR System, in recognizing the off-line printed characters, mainly depends on the selection of feature extraction technique and the classification algorithm employed. This chapter focuses on the recognition of handwritten characters of Roman Script by using features which are obtained by using binarization technique. The goal of binarization is to minimize the unwanted information present in the image while protecting the useful information. Various preprocessing techniques such as thinning, foreground and background noise removal, cropping and size normalization etc. are also employed to preprocess the character images before their classification. A multi-layered feed forward neural network is proposed for classification of handwritten character images. The difference between the desired and actual output is calculated for each cycle and the weights are adjusted during error back-propagation. This process continues till the network converges to the allowable or acceptable error. This method involves the back propagation-learning rule based on the principle of gradient descent along the error surface in the negative direction. Very promising results are achieved when binarization features and the multilayer feed forward neural network classifier is used to recognize the off-line cursive handwritten characters.

Keywords OCR · Binarization · Feature extraction · Character recognition · Back-propagation algorithm · Neural network

A. Choudhary (✉)
Maharaja Surajmal Institute, New Delhi, India
e-mail: amit.choudhary69@gmail.com

S. Ahlawat
Maharaja Surajmal Institute of Technology, New Delhi, India
e-mail: savita.ahlawat@gmail.com

R. Rishi
UIET, Maharshi Dayanand University, Rohtak, India
e-mail: rahulrishi@rediffmail.com

Q. Zhu and A.T. Azar (eds.), *Complex System Modelling and Control Through Intelligent Soft Computations*, Studies in Fuzziness and Soft Computing 319,
DOI 10.1007/978-3-319-12883-2_26

1 Introduction

In today's life, people are constantly writing notes on paper. These notes range from the minutes of a meeting, a short reminder, a list of things to do, to a long-winded letter to a friend. In this digital age, some may wonder why people still use paper. The answer is simple: an effective substitute for paper has not been invented. Paper is still the champion when it comes to efficiency, affordability, usability, and mobility. Yet having a digital copy of these notes has many benefits, most notably the ability to search and organize the notes in many ways instantaneously. Being able to search a huge database of written notes saves time and money which are the two important things in the business world. People need a way to convert handwritten text to digital text, which can be searched and organized however the user wants. A common complaint and excuse of people is that they couldn't read their own handwriting. That makes us ask ourselves the question: If people sometimes can't read their own handwriting, with which they are quite familiar, what chance does a computer have? Fortunately, there are powerful tools that can be used that are easily implementable on a computer.

Character recognition is the ability of a computer to receive and interpret handwritten input from sources such as paper documents, photographs, touch-panels, light pen and other devices. This technology is steadily growing toward its maturity. The domain of hand written text recognition has two completely different problems of On-line and Off-line character recognition. On-line character recognition (Bharath and Madhvanath 2008) involves the automatic conversion of characters as it is written on a special digitizer or PDA, where a sensor picks up the pen-tip movements as well as pen-up/pen-down switching. That kind of data is known as digital ink and can be regarded as a dynamic representation of handwritten characters. The obtained signal is converted into letter codes which are usable within computer and text-processing applications. On the contrary, off-line character recognition involves the automatic conversion of character (as an image) into letter codes which are usable within computer and text-processing applications. The data obtained by this form is regarded as a static representation of handwritten character. The technology is successfully used by businesses which process lots of handwritten documents, like insurance companies. The quality of recognition can be substantially increased by structuring the document (by using forms). The off-line character recognition is comparatively difficult, as different people have different handwriting styles and also the characters are extracted from documents of different intensity and background (Farooq et al. 2008). Limiting the range of input can allow recognition process to improve.

Feed Forward Neural Network plays a great role in Medical Diagnostics (Azar 2013; Azar and El-Said 2013). The most important type of feed forward neural network is the Back Propagation Neural Network (BPNN). Back Propagation is a systematic method for training multi-layer artificial neural network (Sivanandam and Deepa 2008). It is a multilayer feed forward network using gradient descent based delta learning rule, commonly known as back propagation (of errors) rule. Back Propagation provides a computationally efficient method for changing the weights in a feed forward network, with differentiable activation function units, to learn a training set of input-output examples. Being a gradient descent method it minimizes the total squared error of the output computed by the net. The network is trained by supervised learning method. The aim is to train the network to achieve a balance between the ability to respond correctly to the input characters that are used for training and the ability to provide good responses to the input that are similar. The error of the output computed by network is minimized by a gradient descent method known as Back Propagation or Generalized Delta Rule.

Outline of this chapter is as follows: Sect. 2 briefs some related work already done so far by the researchers in this field. Section 3 presents the motivation behind this work and various challenges faced during the process of recognition. Section 4 describes the steps involved in the OCR experiment. Section 5 explains various preprocessing techniques employed to produce good quality image. The feature extraction technique adopted in this work is explained in Sect. 6. The process of Neural Network Training sample preparation is described in Sect. 7. Section 8 presents the recognition process and the experimental conditions in detail. Section 9 deals with implementation and functional details of the character recognition experiment. Discussion of results and interpretations are described in Sect. 10 and finally, the chapter is concluded in Sect. 11 which also presents the future path for continual work in this field.

2 Related Work

A lot of research work had been done and is still being done in character recognition for various languages. OCR is categorized into two classes, for printed characters and for handwritten characters. Compared to OCR for printed characters, very limited work can be traced for handwritten character recognition (Desai 2010). Preprocessing is the preliminary step of OCR, which transforms the data into a format that will be more easily and effectively processed. The main objective of the preprocessing stage is to normalize and remove variations that would otherwise complicate the classification and reduce the recognition rate (Alginahi 2010). The use of preprocessing techniques may enhance a document image preparing it for the next stage in a character recognition system. Thresholding, Noise Removal, Size Normalization, De-skewing and Slant Correction, Thinning and Skeletonization are the various pre-processing techniques that have been employed by various researchers in an attempt to increase the performance of the recognition process.

The Otsu method (Otsu 1979) is one of the widely used techniques used to convert a grey-level image into a binary image then calculates the optimum threshold separating those two classes so that their combined spread (intra-class variance) is minimal (Alginahi 2010).

Noise (small dots or blobs) may easily be introduced into an image during image acquisition (Verma and Blumenstein 2008). A common appearance of noise in binary images takes the form of isolated pixels, salt-and-pepper noise or speckle noise, thus; the processing of removing this type of noise is called filling, where each isolated pixel salt-and-pepper "island" is filled in by the surrounding "sea" (O'Gorman et al. 2008; Alginahi 2010).

The neural network accepted areas between the upper and lower baselines of each word as input. This area, called the core, must be of fixed height to be used in conjunction with the neural net. Therefore it was necessary to scale the words so that all cores are of an identical height (Verma and Blumenstein 2008).

De-skewing is the process of first detecting whether the handwritten word has been written on a slope, and then rotating the word if the slope's angle is too high so that the baseline of the word is horizontal (Verma and Blumenstein 2008). Some degree of skew is unavoidable either a paper is scanned manually or mechanically (Sarfraz and Rasheed 2008; Sadri and Cheriet 2009; Saba et al. 2011).

Thinning is a data reduction process that erodes an object until it is one-pixel wide, producing a skeleton of the object making it easier to recognize objects such as characters. Thinning erodes an object over and over again (without breaking it) until it is one-pixel wide. On the other hand, the medial axis transform finds the points in an object that form lines down its center (Davies 2005). The medial axis transform is similar to measuring the Euclidean distance of any pixel in an object to the edge of the object, hence, it consists of all points in an object that are minimally distant to more than one edge of the object (Russ 2007; Alginahi 2010).

The purpose of feature extraction is to achieve most relevant and discriminative features to identify a symbol uniquely (Blumenstein et al. 2007). Many feature extraction technique are proposed and investigated in the literature that may be used for numeral and character recognition. Consequently, recent techniques show very promising results for separated handwritten numerals recognition (Wang et al. 2005), however the same accuracy has not been attained for cursive character classification (Blumenstein et al. 2007). It is mainly due to ambiguity of the character without context of the entire word (Cavalin et al. 2006). Second problem is the illegibility of some characters due to nature of cursive handwriting, distorted and broken characters (Blumenstein et al. 2003).

Recently, neural network classifiers are proved to be powerful and successful for character/word recognition (Verma et al. 2004; Blumenstein et al. 2007). However, to improve the intelligence of these ANNs, huge iterations, complex computations, and learning algorithms are needed, which also lead to consume the processor time. Therefore, if the recognition accuracy is improved, the consumed learning time will increase and vice versa. Which is the main drawback of ANN based approaches (Aburas and Rehiel 2008).

3 Motivation and Various Challenges During Recognition

Now a days, English is the most commonly used language over the internet, in the entire banking system in the world, postal departments across the countries, insurance companies, space research projects, software multinational companies, business houses and research organizations. As English Language is used by a much higher percentage of the world's population, people will be benefited all over the globe if an automation system is designed for off-line handwriting recognition. The off-line handwriting recognition system can enable the automatic reading and processing of a large amount of data printed or handwritten in English script. Although such automated systems for recognizing off-line handwriting already exist, the scope of further improvement is always there. As a result, in spite of a dramatic boost in this field of research, the development of an intelligent and robust off-line handwritten words segmentation and recognition remains an open problem and continues to be an active area for research towards building an fully automated system by exploring new techniques and methodologies that would improve segmentation and recognition performance in terms of accuracy and speed.

In the process of handwritten character recognition, various challenges encountered can be described as follows:

- There can be variation in shapes and writing styles of different writers.
- Different sizes of character images written by different writers.
- Since handwriting depends on the writer and even a single writer cannot always write the same character in exactly the same way under different conditions.
- Characters may be written on a paper with colored or noisy background.
- Characters may be written by a pen having ink of different colors.

4 Overall OCR System Design

Various steps involved in the proposed handwritten character recognition system are illustrated in Fig. 1.

4.1 Character Image Acquisition

During image acquisition, the handwritten character images were acquired through a scanner or a digital camera. The input character images were saved in JPEG or BMP formats for further processing. Some of those character image samples were written on white paper with colored ink and others on a colored or a noisy background.

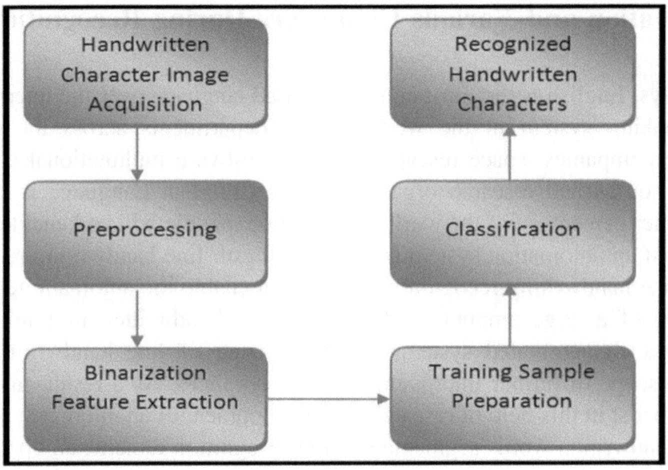

Fig. 1 Schematic diagram of the proposed character recognition system

Handwritten character images obtained by segmenting the handwritten words were also involved for the proposed feature extraction and recognition experiment. All the word images were preprocessed and the resultant thinned word images were cut vertically into the isolated characters using the proposed vertical dissection based segmentation approach.

4.2 Local Database Preparation

Character image samples written by 5 different people (age 15–50 years) were collected where each writer contributed 5 samples of the complete English alphabet (a–z). In this way 650 ($5 \times 5 \times 26 = 650$) fresh character image samples were gathered for the proposed experiment. 650 character images obtained by segmenting the handwritten words are also collected for the proposed character recognition experiment.

A database of the 1,300 handwritten character images has been prepared by combining 650 character images obtained by the segmentation of the handwritten word images and another 650 fresh handwritten character images acquired by scanning the character samples contributed by 5 different persons each contributing 5 samples (a–z). This database containing a total number of 1,300 character images was used for the proposed character feature extraction and recognition experiment. A single writer contributed 5 character samples and is shown in Fig. 2.

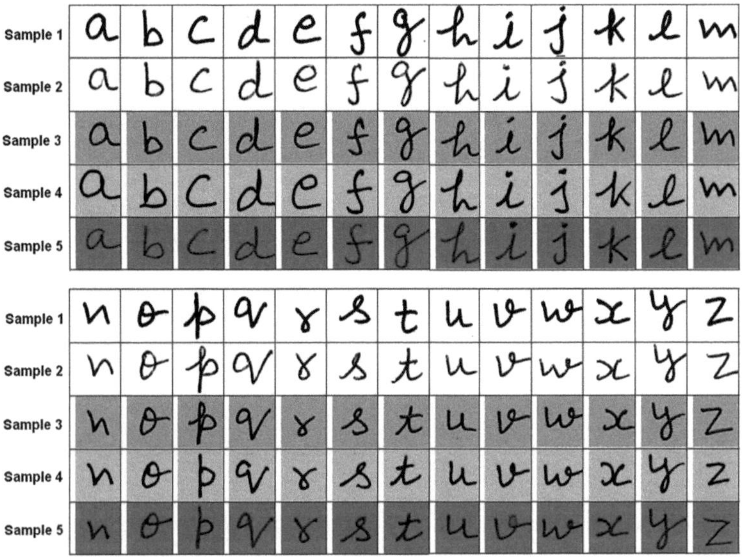

Fig. 2 Character image samples contributed by a single writer

5 Preprocessing

Preprocessing is done to remove the variability that was present in off-line handwritten characters. The preprocessing techniques that have been employed in an attempt to increase the performance of the recognition process are mentioned below.

5.1 Grayscale Conversion

The grayscale images are those images in which each pixel carries the intensity information only. Images of this type are also called black-and-white images and are composed of many shades of Gray varying from black at the weakest intensity to the white at the strongest. Grayscale images are also called monochrome images denoting the presence of only one (mono) color (crome) and are different from the bi-level binary images having only the two colors Black and White.

In this phase of preprocessing, the input image of handwritten character in BMP format from the local database as shown in Fig. 3 is converted into grayscale format by using "rgb2gray" method of MATLAB and the resultant handwritten character image is shown in Fig. 4. This preprocessing step is necessary so as to overcome the problems that may arise due to the use of pens of different colors and different intensities on various noisy and colored backgrounds.

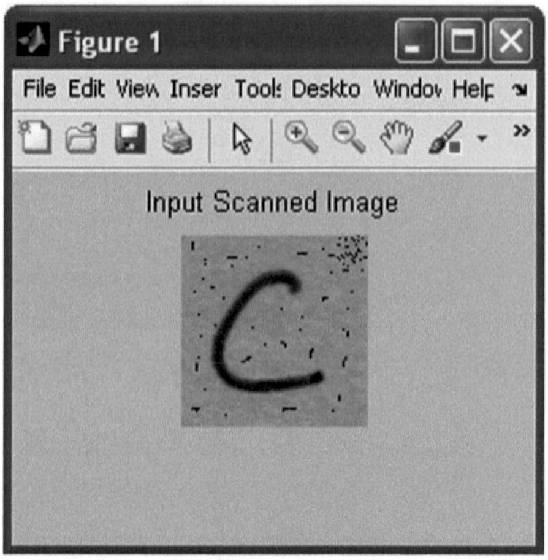

Fig. 3 Input scanned handwritten character image

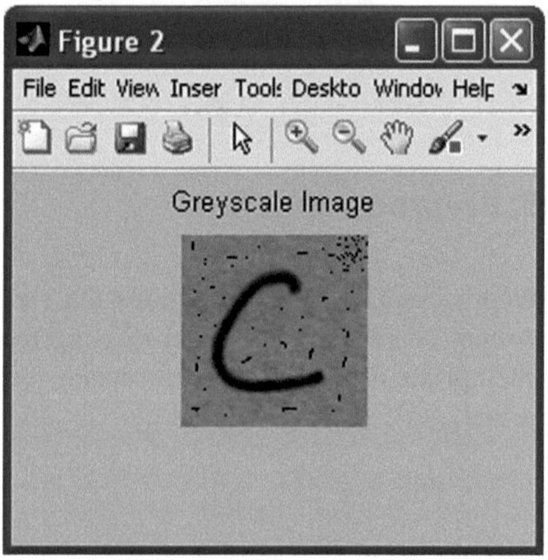

Fig. 4 Handwritten character image in grayscale format

5.2 Binarization Technique

Binarization is an important step in the image processing system and can be defined as a process in which the pixel values are separated into two groups; white as background and black as foreground. Only two colors, white and black, can be present in a binary image. The goal of binarization is to minimize the unwanted information present in the image while protecting the useful information. The performance of the document image analysis system is dependent on the outcome of the binarization algorithm so it must be carried out with maximum possible accuracy. It must preserve the maximum useful information and details present in the image, and on the other hand, it must eliminate the background noise associated with the image in an efficient way. A grayscale image after binarization can be classified into two categories: Globally Thresholded and Adaptive (Locally) Thresholded images.

5.2.1 Global Thresholding

Global thresholding methods (Otsu 1979; Kapur et al. 1985; Cheng et al. 1998; Li and Lee 1993) are most suitable in those documents in which there is uniform contrast distribution between the foreground text and the background. In other words, global Thresholding is used for binarization of those grayscale images in which the written text is of almost same intensity value and is written by a single pen on a background having almost uniform intensity but with quiet different intensity value as compared to the intensity value of the text. In such type of grayscale images, only one threshold value is used for the whole image to classify it into two categories; text and background, and the thresholding is called Fixed Global Threshold. It is used to compare the grayscale intensity of each pixel (I_x) of the image with a Global Threshold (say T = 0.5). The new intensity value of the pixel (I_y) can be calculated by the following expression as:

$$I_y = 1 \quad \text{if } I_x \geq T \text{ and } I_y = 0 \quad \text{if } I_x < T \quad \text{where } 0 < T \leq 1$$

Here, intensity value ‘0’ represents black pixels and intensity value ‘1’ represents white pixels. Hence, the pixels having intensity value greater than the global threshold will be white and the pixel having intensity value less than the global threshold will appear as black. The grayscale intensity threshold value can be assigned any value that lies between the intensity value of the foreground text and the background intensity. Generally, its value should be closer to the intensity of the foreground text as compared to the background intensity, so as to remove the maximum undesirable background noise while preserving all the important details of the foreground text.

The selection of the threshold parameter is based on the gray-scale intensity of the text in the document. More intensity leads to the more threshold value. This parameter is decided as shown in Table 1.

The first column of Table 1 represents the intensity of handwritten text present in the document. This intensity is a gray-scale value when the document is converted into grayscale format by using ‘img2gray’ method of MATLAB. The second

Table 1 Intensity/threshold comparison table

Gray-scale intensity of the text (I)	Threshold value
0.00–0.20	0.19
0.21–0.40	0.39
0.41–0.60	0.59
0.61–0.80	0.79
0.81–1.00	0.99

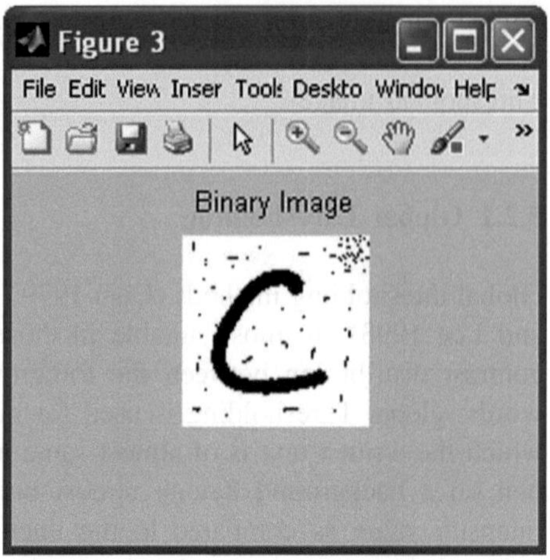

Fig. 5 Handwritten character image in binary format

column of this table represents the corresponding value of threshold intensity level for binarization process. The threshold parameter along with the grayscale image is made an input to the binarization method 'im2bw' designed in MATLAB. The output is a binary image as shown in Fig. 5.

5.2.2 Adaptive Thresholding

It has been found that the global thresholding approach gives excellent result when the text is written by a single pen with same intensity on a background of throughout uniform intensity but considerably different from the text intensity. The results start degrading when the pens of different intensities are used on various noisy backgrounds. The global thresholding method fails because of drastic variation in the text intensity and variation in contrast illumination between text and background. In such cases, the adaptive thresholding methods by Bernsen (1986), Niblack (1986), Sauvola et al. (1997) are employed and the document image is divided into small blocks and the threshold values are computed pixel by pixel or region by region. For each region or block, a local threshold value is determined for the binarization of that particular block and the binarized images of each block are combined to get the final binarized document image.

In the proposed binarization technique, the threshold value is decided based on the intensity of the text as explained in Table 1. It is assumed that the intensity of the text is less than that of background i.e. the input image has black foreground pixels and white background pixels. The colors can be inverted if the input image has text intensity more than that of background. Also, the background intensity remains almost uniform throughout the image and does not change drastically anywhere in the input image. Hence, in the proposed binarization technique, global intensity thresholding is employed and the resulting handwritten character image after background noise removal and binarization using global gray scale intensity threshold is shown in Fig. 5.

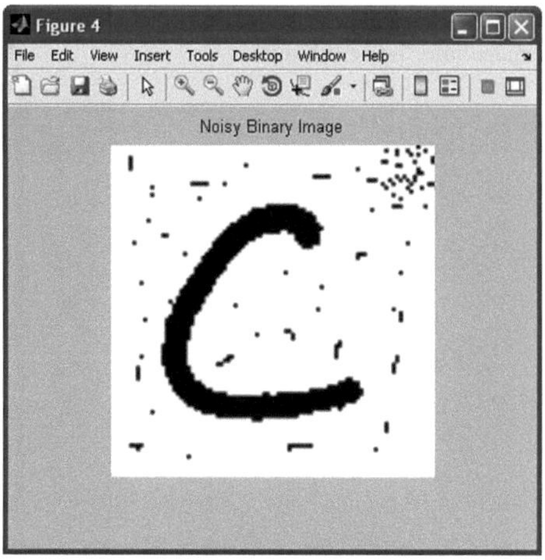

Fig. 6 Noisy handwritten character image

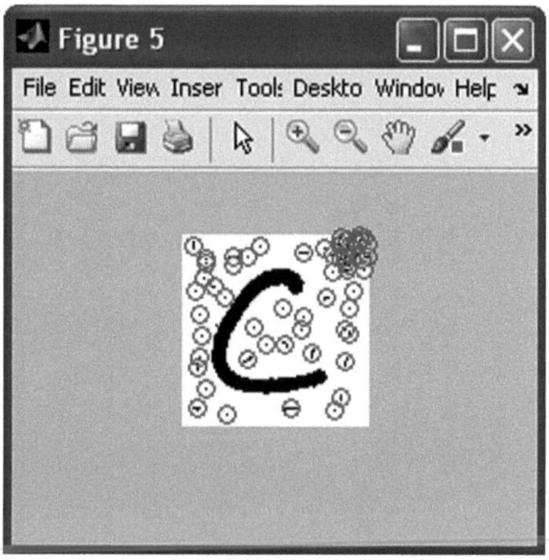

Fig. 7 Noise detection in handwritten character image

5.3 Noise Removal

Noise (small dots or foreground components) may be introduced easily into an image while scanning the handwritten character image during image acquisition. It is very necessary to eliminate the noise from the handwritten character images as shown in Fig. 6, so as to make this image fit for further processing. MATLAB's 'bwareaopen' is used to morphologically open the binary image by removing small objects that have less than a particular number (user specified) of pixels and producing another binary image. The small noise dots were removed by using 'bwareaopen' but some small portions of the characters e.g. dots '.' of characters 'i, j' etc. were also lost. The methods 'bwlabel' and 'regionprops' of MATLAB were used to highlight the pixels that were removed as shown in Fig. 7. A logical

Fig. 8 Handwritten character image after noise removal

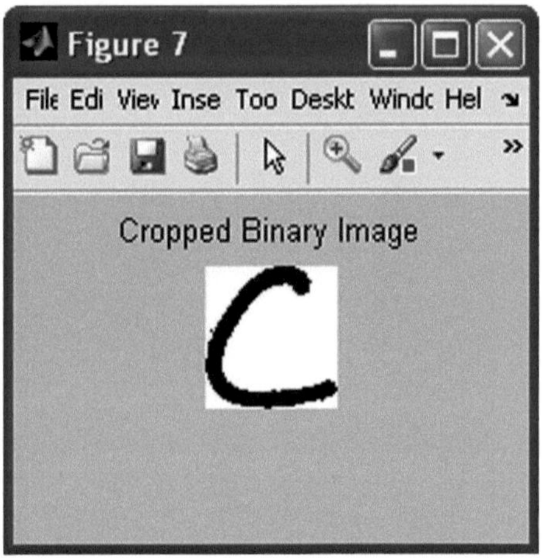

Fig. 9 Cropped handwritten character image

AND operation of the dilated characters with the pixels removed by 'bwareaopen' was performed. The portions of the characters pixels which were very near to the character image were put back. The resultant image without noise dots while retaining the portions of the characters is shown in Fig. 8.

5.4 Cropping

Image Cropping is a process in which the extra space around the handwritten character image is removed. The outcome of 'imcrop' method of MATLAB is a rectangular region with minimum area but containing the complete character image. The final image after cropping operation is shown in Fig. 9.

Fig. 10 Resized handwritten character image

5.5 Size Normalization

Since handwriting depends on the writer and even a single person cannot always write the same character in exactly the same way. The shape of the handwritten characters may be different under different conditions such as speed of writing, position of writer (sitting or standing) while writing and quality of pen and/or paper etc. The same character image (say 'c') written by different writers or by the same writer may be of different sizes. Also, there is a size difference among all the 26 characters present in the English Language. Some characters are of different height and some are of different widths. Cropped character images obtained by removing the extra space in the image around the character are also of different sizes.

There is a need to put all the handwritten character images in a uniform size i.e. the characters should be in normal form. To make all the character image samples in the normal form, all the character images are reconstructed in the size of 15 × 12 pixels by using 'imresize' method of MATLAB which employs nearest neighborhood interpolation technique. The character image after resizing is shown in Fig. 10.

6 Feature Extraction

Feature extraction is a process of studying and deriving useful information from the filtered input patterns. The derived information can be general features, which are evaluated to ease further processing. The selection of features is very important

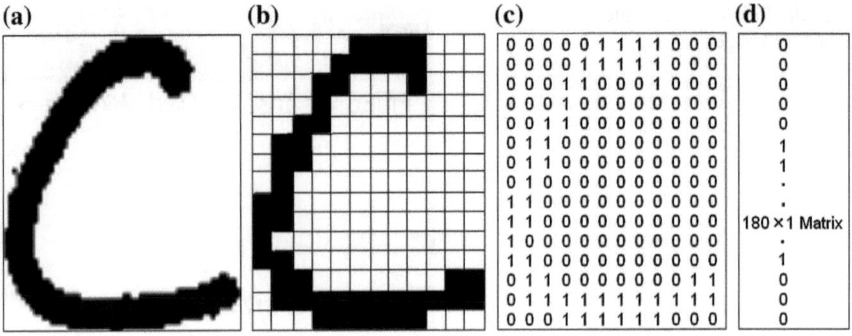

Fig. 11 **a** Binary image of character 'c' **b** resized binary image of character 'c' **c** binary matrix representation, and **d** reshaped binary matrix or feature vector of character 'c'

because there might be only one or two values, which are significant to recognize a particular character. The performance of the recognition system relies much on the quality of the features extracted as well as on the selected classifier itself.

The binary image of character 'c' as shown in Fig. 11a is resized to 15 × 12 matrixes as shown in Fig. 11b. Each cropped and size normalized character image in binary format is traced vertically column wise. A white pixel is represented by '0' and a black pixel is represented by '1' and the binary matrix representation of character 'c' is shown in Fig. 11c. This binary matrix of size 15 × 12 is then reshaped in a row first manner to a binary matrix of size 180 × 1 by using 'reshape' method of MATLAB and is shown in Fig. 11d. The column vector of size 180 × 1 as shown in Fig. 11d is a feature vector for the character image 'c' shown in Fig. 11a.

7 Sample Preparation for Neural Network Training

The feature vector of a single character is a column vector of size 180 × 1. One such feature vector of character 'c' is shown in Fig. 11d. Similarly, the feature vectors of all the 26 characters (a–z) are created in the form of binary column matrix of size 180 × 1 each. All these 26 feature vectors are combined to form a sample which is a binary matrix of size 180 × 26 as shown in Fig. 12.

In this matrix of size 180 × 26, there are 26 columns representing 26 characters of the English Language and each column represents feature vector of length 180 × 1 of a single character e.g. first column represents feature vector of character 'a', second column represents feature vector of character 'b', third column represents feature vector of character 'c' and so on.

For sample creation, 1,300 characters were gathered where each writer contributed 5 samples of the complete English alphabet (a–z). After pre-processing, these samples were considered for training such that each sample was consisting of 26 characters (a–z).

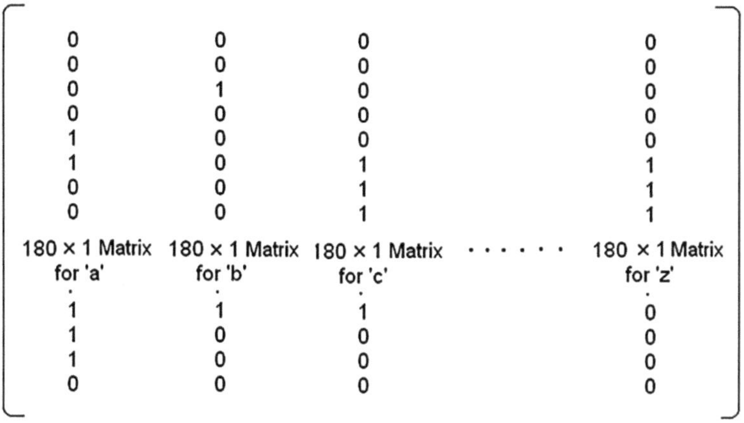

Fig. 12 Matrix representation of input sample

8 Classification and Recognition Process

This is the final step in an OCR System. The accuracy of the character recognition system depends on the preprocessing techniques adopted to clean the input character patterns. The OCR accuracy also depends on the quality of the features extracted from the input character images to be recognized. If the correct preprocessing techniques are applied efficiently and the extracted features are of good quality, the recognition accuracy of the OCR system will be high. Apart from preprocessing technique and the feature extraction technique, the OCR accuracy also depends on the type of classifier involved to do the recognition. An extensive review of the literature indicates that as far as the unconstrained handwritten character recognition is concerned, neural networks as a classifier are chosen to be the best among the others.

8.1 Methodology

To accomplish the task of character classification and mapping, the multilayer feed forward artificial neural network is considered with nonlinear differentiable function 'tansig' in all processing units of output and hidden layers. The processing units in the input layer, corresponds to the dimensionality of the input pattern, are linear. The number of output units corresponds to the number of distinct classes in the pattern classification. A method has been developed, so that network can be trained to capture the mapping implicitly in the set of input-output pattern pair collected during an experiment and simultaneously expected to modal the unknown system from which the predictions can be made for the new or untrained set of data.

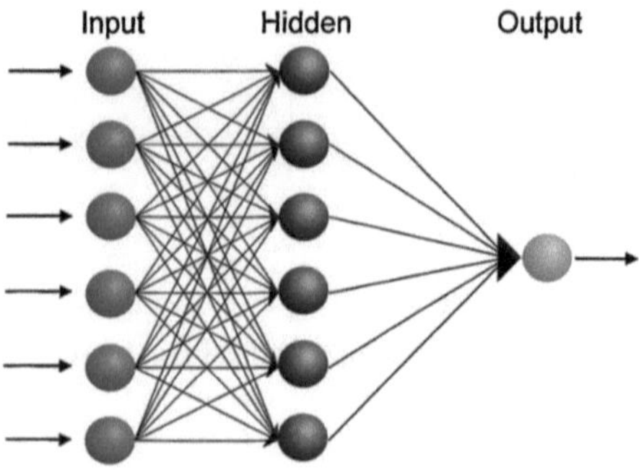

Fig. 13 Feed forward neural network with one hidden layer and a single output neuron

This method involves the back propagation-learning rule based on the principle of gradient descent along the error surface in the negative direction.

8.2 Neural Network Classifier Architecture

The difference between the desired and actual output is calculated for each cycle and the weights are adjusted during back-propagation. This process continues till the network converges to the allowable or acceptable error.

In the feed forward phase of operation, the signals are sent in forward direction and in back propagation phase of learning, the signals are sent in the reverse direction (Fig. 13).

The training algorithm of back propagation involves four stages:

(a) Initialization of weights: During this stage some random values are assigned for initialization of weights.
(b) Feed Forward: During Feed Forward stage, each input unit receives an input signal and transmits this signal to each of the hidden units. Each hidden unit then calculates the activation function and sends its signal to each output unit. The output unit calculates the activation function to form the response of the net for the given input pattern.
(c) Back Propagation of Errors: During back propagation of errors, each output unit compares its computed activation value (output) with its target value to determine the associated error for that input pattern with that unit. Based on the error, the error factor for each unit is computed and is used to distribute the error at each output unit back to all units in the previous layer. Similarly the error factor is computed for each hidden unit.
(d) Updation of the Weights and Biases: During final stage, the weights and biases are updated for the neurons at the previous levels to lower the local error.

The processing nodes of input layer used the liner activation function and the nodes of hidden and output layers used the non-liner differentiable activation function ‘tansig’ as shown in Fig. 14.

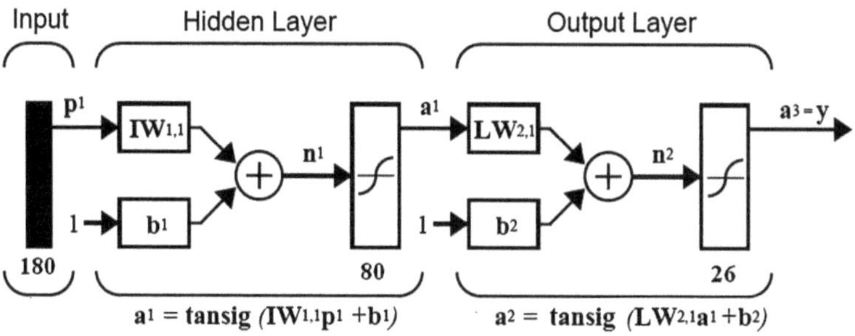

Fig. 14 Architecture of the neural network used in the recognition system

The weight update for the output neurons of the network can be determined as: The error signal at the output of neuron j at iteration n is:

$$e_j(n) = d_j(n) - y_j(n) \tag{1}$$

The instantaneous value of error for neuron j is $\frac{1}{2}e_j^2(n)$.

The instantaneous value $\varepsilon(n)$ of total error is obtained by summing $\frac{1}{2}e_j^2(n)$ of all neurons in output layer

$$\varepsilon(n) = \frac{1}{2}\sum_{j \in c} e_j^2(n) \tag{2}$$

where, 'c' includes all neurons in output layer.

Average squared error is given by

$$\varepsilon_{avg} = \frac{1}{N}\sum_{n=1}^{N} \varepsilon(n) \tag{3}$$

where, N is the total number of patterns in training set and ε_{avg} should be minimized. Back-propagation is used to update the weights. Induced local field $v_j(n)$ produced at input of activation function is given by

$$v_j(n) = \sum_{i=0}^{m} w_{ji}(n) y_i(n) \tag{4}$$

where 'm' is the number of inputs applied to neuron 'j', so the output can be written as:

$$y_j(n) = \phi_j(v_j(n)) \tag{5}$$

.

The back-propagation algorithm applies a correction $\Delta w_{ji}(n)$ to synaptic weights $w_{ji}(n)$ which is proportional to partial derivative $\frac{\partial \varepsilon(n)}{\partial w_{ji}(n)}$ which can be written as:

$$\frac{\partial \varepsilon(n)}{\partial w_{ji}(n)} = \frac{\partial \varepsilon(n)}{\partial e_j(n)} \cdot \frac{\partial e_j(n)}{\partial y_j(n)} \cdot \frac{\partial y_j(n)}{\partial v_j(n)} \cdot \frac{\partial v_j(n)}{\partial w_{ji}(n)} \tag{6}$$

Diffrentiating Eq. (2) with respect to $e_j(n)$

$$\frac{\partial \varepsilon(n)}{\partial e_j(n)} = e_j(n) \tag{7}$$

Differentiating Eq. (1) w.r.t. $y_j(n)$

$$\frac{\partial e_j(n)}{\partial y_j(n)} = -1 \tag{8}$$

Differentiating Eq. (5),

$$\frac{\partial y_j(n)}{\partial v_j(n)} = \phi_j'(v_j(n)) \tag{9}$$

Differentiating Eq. (4) w.r.t. $w_{ji}(n)$

$$\frac{\partial v_j(n)}{\partial w_{ji}(n)} = y_i(n) \tag{10}$$

Using Eqs. (7–10) in Eq. (6),

$$\frac{\partial \varepsilon(n)}{\partial w_{ji}(n)} = -e_j(n)\phi_j'(v_j(n))y_i(n) \tag{11}$$

The correction $\Delta w_{ji}(n)$ applied to $w_{ji}(n)$ is defined by

$$\begin{aligned} \Delta w_{ji}(n) &= -\eta \frac{\partial \varepsilon(n)}{\partial w_{ji}(n)} \\ &= \eta e_j(n)\phi_j'(v_j(n))y_i(n) \\ &= \eta \delta_j(n) y_i(n) \end{aligned} \tag{12}$$

Thus the weight updates for output unit can be represented as

$$\Delta w_{ji}(n) = \eta \delta_j(n) y_i(n) \tag{13}$$

where η is a constant called learning rate parameter, δ is local gradient and is a derivative of error with respect to v_j and y is the input.

8.3 Experimental Conditions

There is not any rule that gives the calculation for the ideal parameter setting for a neural network. However, the various parameters and their respective values used in the proposed training process of the handwritten character recognition experiments are shown in Table 2.

9 Implementation and Functional Details

In the current situation, the number of neurons in the input and output layers are fixed at 180 and 26 respectively. The 180 input neurons are equivalent to the input character's size as we have resized every character into a binary matrix of size 15 × 12 and then reshaped to a matrix of size 180 × 1. The number of neurons in the output layer is 26 because there are 26 English alphabets. The number of neurons in the hidden layer and the activation functions of the neurons in the hidden and output layers are to be decided.

Table 2 Experimental conditions during the recognition experiment

Parameters	Value
Input layer	
No. of input neurons	180
Transfer/activation function	Linear
Hidden layer	
No. of hidden neurons	80
Transfer/activation function	TanSig
Learning rule	Momentum
Output layer	
No. of output neurons	26
Transfer/activation function	TanSig
Learning rule	Momentum
Learning constant	0.01
Acceptable error level (MSE)	0.001
Momentum term (α)	0.90
Maximum epochs	100,000
Termination conditions	Based on minimum mean square error or maximum number of epochs allowed
Initial weights and biased term values	Randomly generated values between 0 and 1
Number of hidden layers	1

It is very difficult to determine the optimal number of hidden neurons. Too few hidden neurons will result in under-fitting and there will be high training error and statistical error due to the lack of enough adjustable parameters to map the input-output relationship. Too many hidden neurons will result in over-fitting and high variance. The network will tend to memorise the input-output relations and normally fail to generalize. Testing data or unseen data could not be mapped properly. The number of hidden neurons is directly proportional to the system resources. The bigger the number more the resources are required. The number of neurons in a hidden layer was kept 80 by trial and error method for optimal results.

Each neuron in the neural network has a transformation function. To produce an output, the neuron performs the transformation function on the weighted sum of its inputs. Various activation functions used in neural networks are compet, hardlim, logsig, poslin, purelin, radbas, satlin, softmax, tansig and tribas etc. The two transfer functions normally used in MLP are logsig and tansig.

logsig transfer function is also known as Logistic Sigmoid:

$$\log sig(x) = \frac{1}{1 + e^{-x}},$$

tansig transfer function is also known as Hyperbolic Tangent:

$$\tan sig(x) = \frac{e^{x} - e^{-x}}{e^{x} + e^{-x}}$$

The hyperbolic tangent and logistic sigmoid are related by:

$$\frac{\tan sig(x) + 1}{2} = \frac{1}{1 + e^{-2x}}$$

We get:

$$\tan sig(x) = \frac{2}{1 + e^{-2x}} - 1$$

These transfer functions are commonly used as they are easy to use mathematically and while saturating, they are close to linear near the origin. The 'tansig' activation function was used for the hidden and output layer neurons in the proposed experiment.

The training of the network employing back propagation algorithm has been done in neural network toolbox under MATLAB environment. The neural network model employing back-propagation algorithm comes under the category of supervised learning. The 'tansig' activation function has been used for the neurons of both hidden and output layers.

The adaptive learning function 'traingdx' has been used in the neural network training process. Mean Square Error (MSE) has been selected as a Cost Function in

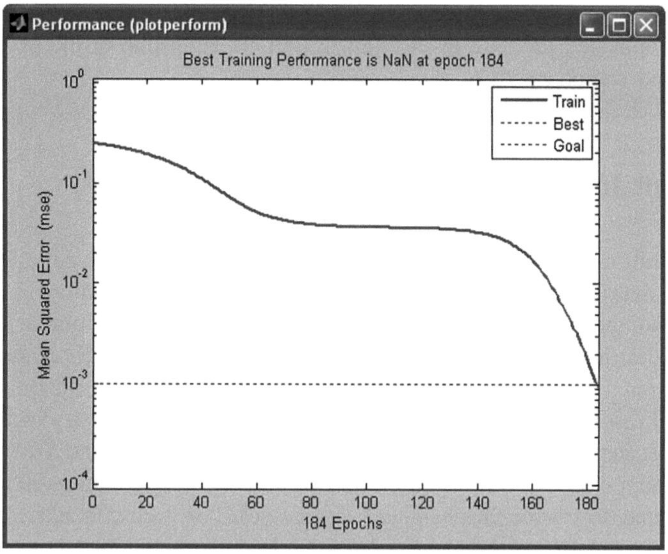

Fig. 15 The variation of MSE with the training epochs

the training process. MSE is an accepted measure of the performance index often used in backpropagation MLP networks. The lower value of MSE indicates that the network is capable of mapping the input and output accurately. The accepted error level has been set to 0.001 and the training will stop when the final value of MSE reaches at 0.001 or below this level. The performance value indicates the extent of training of the network. A low performance value indicates that the network has been trained properly.

The number of epochs required to train a network also indicates the network performance. The adjustable parameters of the network will not converge properly if the number of training epochs is insufficient and the network will not be well trained. On the other hand, the network will take unnecessary long training time if there are excessive training epochs. The number of training epochs should be sufficient enough so as to meet the aim of training. The maximum allowed epochs for the training process has been set to 100,000 (One Lac). If the network could not converge within the maximum allowed epochs count, the training will stop.

The network was trained with 50 samples of handwritten characters where each sample has 26 characters (a–z). In the proposed experiment, 1,300 ($50 \times 26 = 1,300$) handwritten characters have been involved. The process of successfully training the neural network by first training sample can be seen in Fig. 15. It is clear from the figure that the training has properly converged to the goal after 184 epochs.

If there is a saturated or horizontal straight line at the end of the MSE plot, the further training epochs are no longer beneficial and the training should be stopped by introducing the stopping criterion such as maximum number of training epochs allowed in addition to the stopping criterion of maximum acceptable error as specified in the training algorithm. This type of MSE plot is observed when there is a very complicated problem or insufficient training algorithm or the network have very limited resources.

10 Result Interpretation and Discussion

The recognition results obtained for various characters are displayed in the form of confusion matrix in Table 3. This confusion matrix shows the confusion among the recognized characters while testing the neural network's recognition accuracy.

The neural network was exposed to 50 different samples. Each character at the input will put a '1' at that neuron in the output layer in which the maximum trust is shown and rest neuron's result into '0' status. The output is a binary matrix of size 26 × 26 because each character has 26 × 1 output vector. The first 26 × 1 column stores the first character's recognition output; the following column will be for next character and so on for 26 characters (a sample). For each character the 26 × 1 vector will contain value '1' at only one place. For example, character 'a' if correctly recognized, will result in [1, 0, 0, 0 …all …0], character 'b' will result in [0, 1, 0, 0 … all …0] and so on.

In the proposed handwritten character recognition experiment, the neural network has been trained with 50 sets of each character i.e. 1,300 (50 × 26 = 1,300) character image samples from the database has been involved in the training. The confusion in recognition among the different characters is explained in Table 3. Character 'a' is presented 50 times to the neural network and is classified 43 times correctly. It is miss-classified two times as 'e' and five times as 'o'. Character 'b' is classified 49 times correctly and misidentified as character 'd' one time out of a total of fifty trials. Character 'c' is misclassified as 'e' four times and one time each as 'o' and 'u' and is classified correctly 44 times. Recognition accuracy for each character (a–z) as well as overall recognition accuracy is displayed in the last column of Table 3. The average recognition accuracy of 85.62 % is quiet good for this handwritten character recognition experiment. The three dimensional plot of confusion matrix generated in the MATLAB environment representing the performance of the classifier is shown in Fig. 16.

The recognition accuracy of 85.62 % that has been achieved here in this work is for handwritten English character recognition is very good and is better than that of many researchers but not the best among all the researchers.

The result obtained here in this work is better than the work done by Shanthi and Duraiswamy (2009), in which the 82 % recognition accuracy is achieved by using support vector machine for handwritten character recognition and image subdivision method for feature extraction. Rajashekararadhya and Ranjan (2009) proposed an off-line handwritten OCR in which the feature extraction is based on zone and image centroid. Two classifiers, nearest neighborhood and backpropagation neural network were used to achieve an accuracy which is comparable to the accuracy reported here in this work. Banashree et al. (2007) attempted classification of

Table 3 Confusion matrix representing the performance of the neural network classifier

Alphabets	a	b	c	d	e	f	g	h	i	j	k	l	m	n	o	p	q	r	s	t	u	v	w	x	y	z	Success (%)
a	43	0	0	0	0	0	0	1	0	0	0	0	2	8	5	0	0	0	0	0	0	3	0	0	0	0	86
b	0	49	0	2	0	0	0	0	0	0	3	0	0	0	0	5	0	0	0	0	0	0	0	0	0	0	98
c	0	0	44	0	4	0	0	0	0	0	0	0	0	0	0	0	0	0	0	0	0	0	0	0	0	0	88
d	0	1	0	38	0	0	0	0	0	0	0	0	1	0	0	0	0	0	0	0	0	0	0	0	0	0	76
e	2	0	4	0	44	0	0	0	0	0	0	0	0	1	0	0	0	0	0	0	0	0	0	0	0	0	88
f	0	0	0	0	0	46	0	0	0	0	0	3	0	0	0	0	0	0	0	1	0	0	0	0	0	1	92
g	0	0	0	0	0	0	39	0	0	10	0	0	0	0	0	1	4	0	0	0	0	0	0	0	13	0	78
h	0	0	0	0	0	0	0	49	0	0	5	0	4	0	0	0	0	0	0	0	0	0	0	0	0	0	98
i	0	0	0	0	0	0	0	0	37	6	0	0	0	0	0	0	0	0	0	0	0	0	0	0	0	0	74
j	0	0	0	0	0	0	1	0	5	29	0	0	0	0	0	0	0	0	0	0	0	0	0	0	0	0	58
k	0	0	0	0	0	0	0	0	0	0	42	2	0	0	0	0	0	0	0	1	0	0	0	0	1	0	84
l	0	0	0	1	0	1	0	0	3	0	0	39	0	0	0	0	0	0	0	0	0	0	0	0	0	0	78
m	0	0	0	0	0	0	0	0	0	0	0	0	41	2	0	0	0	1	1	0	0	0	0	0	0	0	82
n	0	0	0	0	0	0	0	0	0	0	0	0	0	38	0	0	0	0	1	0	0	0	0	0	0	0	76
o	5	0	1	0	1	0	0	0	0	0	0	0	0	0	43	0	0	0	1	0	0	0	0	0	0	0	86
p	0	0	0	5	0	0	0	0	0	0	0	0	0	0	0	44	0	0	0	10	0	0	0	0	0	0	88
q	0	0	0	0	0	0	2	0	0	0	0	0	0	0	0	0	46	0	0	0	0	0	0	0	1	0	92
r	0	0	0	0	0	0	0	0	0	0	0	0	0	1	0	0	0	49	0	0	0	0	0	0	0	0	98
s	0	0	0	0	0	0	0	0	0	0	0	0	0	0	0	0	0	0	46	0	0	0	0	0	0	0	92
t	0	0	0	4	0	3	0	0	5	0	0	6	0	0	0	0	0	0	0	38	0	0	0	0	0	2	76
u	0	0	1	0	1	0	0	0	0	0	0	0	0	0	2	0	0	0	0	0	48	5	0	0	0	0	96
v	0	0	0	0	0	0	0	0	0	0	0	0	0	0	0	0	0	0	0	0	2	40	0	0	0	0	80
w	0	0	0	0	0	0	0	0	0	0	0	0	0	0	0	0	0	0	0	0	0	2	50	0	0	0	100
x	0	0	0	0	0	0	0	0	0	0	0	0	2	0	0	0	0	0	1	0	0	0	0	50	0	1	100
y	0	0	0	0	0	0	8	0	0	5	0	0	0	0	0	0	0	0	0	0	0	0	0	0	35	0	70
z	0	0	0	0	0	0	0	0	0	0	0	0	0	0	0	0	0	0	0	0	0	0	0	0	0	46	92
Overall Character Recognition Accuracy =																											85.62 %

Output Class (rows); Target Class (columns)

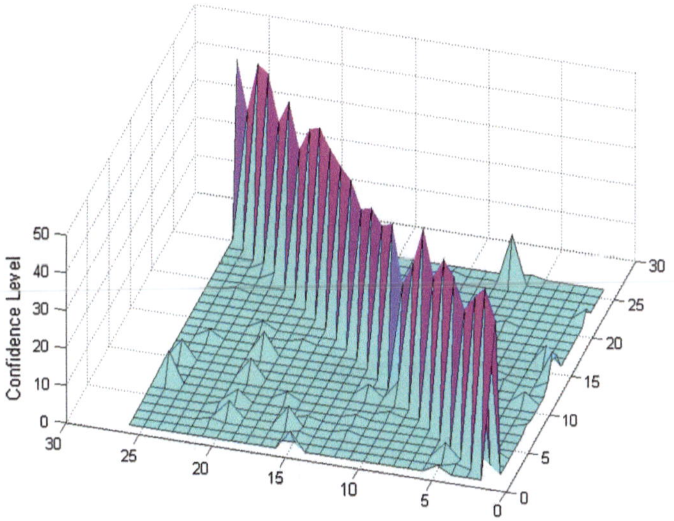

Fig. 16 Three dimensional plot of confusion matrix representing the performance of the neural network classifier

handwritten Devnagri digits using diffusion half toning algorithm. 16-segment display concept has been used here for feature extraction. They proposed a neural classifier for classification of isolated digits. Here they achieved accuracy level up

Table 4 Performance comparison of script recognition accuracy

Author	Classifier	Lexicon size (in words)	Problem domain	Recognition rate (%)
Guillevic and Suen (1998)	HMM/KNN	30	LA words (ENGLISH)	86.7
Chiang (1998)	NN	100	USPS database mail	87.4
Kim et al. (2000)	HMM/MLP	32	LA words	92.2
Oliveira et al. (2002)	MLP	12	Numerical strings	87.2
Kundu and Chen (2002)	HMM	100	Postal words	88.2
Koch et al. (2004)	MLP	1,000	Letters	67.8
Günter and Bunke (2004)	HMM + Ensembled methods		IAM	71.58
Günter and Bunke (2005)	HMM + Ensembled methods		IAM	75.61–82.28
Gatos et al. (2006a)	K-NN	3,799	IAM	81.05
Gatos et al. (2006b)	SVM		IAM	87.68
Tomoyuki et al. (2007)	Posterior probability	1,646	City names (European countries)	80.2

to 88 % which is slightly better than the accuracy achieved here in this work. Summary of recognition performances of some off-line script recognition systems in the same domain in chronological order year wise are shown in Table 4.

11 Conclusion and Future Scope

The MLP used in the proposed experiment for the handwritten character recognition employing the backpropagation algorithm performed exceptionally well with 80 neurons in the hidden layer and 'tansig' as the activation function for both hidden and output layer neurons. While preparing the training samples, each character image was resized to 15 × 12 and then reshaped into a 180 × 1 column matrix before applying it as an input to the neural network. The length of the feature vector of each character is 180. Also, there are 26 output classes (a–z) representing each character. Hence, in the MLP structure used in the proposed experiment, the number of neurons in the input and output layers has been fixed at 180 and 26 respectively.

The proposed method for the handwritten character recognition using the descent gradient approach of backpropagation algorithm showed the remarkable enhancement in the performance. The use of binarization features along with backpropagation feed forward neural network yielded the excellent recognition accuracy. Although the success rate of 85.62 % is considered excellent but a scope of improvement is always there. The performance of a recognition system mainly depends on the quality of samples used for training and the techniques employed to extract the features and the type of classifier. Preprocessing techniques, feature extraction techniques and the methodology used to select the neural network parameters can be improved to get further improved results.

Nevertheless, more work needs to be done especially on the test for more complex handwritten characters. The proposed work can be carried out to recognize English words of different character lengths after proper segmentation of the words into isolated character images. In future, better pre-processing techniques will be used. The skew and slat correction module will also be incorporated in the future work. This module was not applied here in this experiment because it was assumed that all the handwritten character samples are free from slant and skew. The character images which are rotated at a certain angle will also be included in the character recognition experiment carried out in future. Also, the various parameters of the neural network architecture such as number of hidden neurons, number of hidden layers, choice of activation function in hidden and output layers, learning rate, momentum constant, cost function, training termination criterion etc. will also be optimized by detailed rigorous investigation and experimentation.

References

Aburas, A. A., & Rehiel, S. A. (2008). New promising off-line tool for Arabic handwritten character recognition based on JPEG2000 image compression. In *Proceedings of the 3rd International Conference on Introduction and Communication Technology—from Theory to Applications (ICTTA)* (pp. 1–5), April 7–11, 2008. doi:10.1109/ICTTA.2008.4530087.

Alginahi, Y. (2010). Preprocessing techniques in character recognition. *Character Recognition.* In M. Mori (Ed.) (pp. 1–20). In Techopen Publishers. ISBN 978-953-307-105-3. doi:10.5772/9776.

Azar A. T. (2013). Fast neural network learning algorithms for medical applications. *Neural Computing and Applications, 23*(3–4). 1019–1034. doi:10.1007/s00521-012-1026-y.

Azar, A. T., & El-Said, S. A. (2013). Probabilistic neural network for breast cancer classification. *Neural Computing and Applications, 23*(6), 1737–1751. doi:10.1007/s00521-012-1134-8.

Banashree, N. P., Andhre, D., Vasanta, R., & Satyanarayana, P. S. (2007). OCR for script identification of Hindi (Devanagari) numerals using error diffusion Halftoning Algorithm with neural classifier. *International Journal of Computer, Information Science and Engineering, 1* (2), 281–285.

Bernsen, J. (1986). Dynamic thresholding of grey-level images. In *Proceedings of 8th International Conference on Pattern Recognition* (pp. 1251–1255), Paris, France.

Bharath, A. & Madhvanath, S. (2008). FreePad: A novel handwriting-based text input for pen and touch interfaces. In *Proceedings of the 13th International Conference on Intelligent User Interfaces* (pp. 297–300), New York, NY, USA. doi:10.1145/1378773.1378814.

Blumenstein, M., Liu, X. Y., & Verma, B. (2007). An investigation of the modified direction feature for cursive character recognition. *Pattern Recognition, 40*(2), 376–388. doi:10.1016/j.patcog.2006.05.017.

Blumenstein, M., Verma, B. & Basli, H. (2003). A novel feature extraction technique for the recognition of segmented handwritten characters. In *Proceedings of the 7th International Conference on Document Analysis and Recognition* (Vol. 1, pp. 137–141). Edinburgh, UK: IEEE Computer Society Press. doi:10.1109/ICDAR.2003.1227647.

Cavalin, P. R., Britto, A. S., Bortolozzi, F., Sabourin, R. & Oliveira, L. S. (2006). An implicit segmentation based method for recognition of handwritten strings of characters. In *Proceedings of ACM Symposium on Applied Computing (SAC)* (pp. 836–840), New York, NY, USA. doi:10.1145/1141277.1141468.

Cheng, H. D., Chen, J. R., & Li, J. (1998). Threshold selection based on fuzzy c-partition entropy approach. *Pattern Recognition, 31*(7), 857–870.

Chiang, J.-H. (1998). A hybrid neural network model in handwritten word recognition. *Neural Networks, 11*(2), 337–346.

Davies, E. (2005). *Machine vision—Theory algorithms practicalities* (3rd ed.). San Francisco, CA, USA: Morgan Kaufmann Publishers. ISBN 13: 978-0-12-206093-9.

Desai, A. A. (2010). Gujarati handwritten numeral optical character recognition through neural network. *Pattern Recognition, 43*(7), 2582–2589. doi:10.1016/j.patcog.2010.01.008.

Farooq, F., Bhardwaj, A., Cao, H. & Govindaraju, V. (2008). Topic based language models for OCR correction. In *Proceedings of the 2nd Workshop on Analytics for Noisy Unstructured Text Data* (pp. 107–112), New York, NY, USA. doi:10.1145/1390749.1390766

Gatos, B., Pratikakis, I. & Perantonis, S. J. (2006a). Hybrid off-line cursive handwriting word recognition. In *Proceedings of 18th International Conference on Pattern Recognition* (ICPR'06) (Vol. 2, pp. 998–1002), Hong Kong. doi:10.1109/ICPR.2006.644.

Gatos, B., Pratikakis, I., Kesidis, A. L. & Perantonis, S. J. (2006b). Efficient off-line cursive handwriting word recognition. In Proceedings of the 10th International Workshop on Frontiers in Handwriting Recognition. October 2006, La Baule.

Guillevic, D., Suen, C. Y. (1998). HMM-KNN word recognition engine for bank check processing. In *Proceedings of International Conference on Pattern Recognition*, Brisbane. Washington DC, USA: IEEE Computer Society, pp. 1526–1529. DOI: 10.1109/ICPR.1998.711998.

Günter, S., & Bunke, H. (2004). Feature selection algorithms for the generation of multiple classifier systems and their application to handwritten word recognition. *Pattern Recognition Letters, 25*(11), 1323–1336.

Günter, S., & Bunke, H. (2005). Off-line cursive handwriting recognition using multiple classifier systems—On the influence of vocabulary, ensemble, and training set size. *Optics and Lasers in Engineering, 43*(3–5), 437–454.

Kapur, J. N., Sahoo, P. K., & Wong, A. K. C. (1985). A New method for gray-level picture threshold using the entropy of the histogram. *Computer Vision, Graphics, and Image Processing, 29*, 273–285.

Kim, J. H., Kim, K. K., & Suen, C. Y. (2000). An HMM-MLP hybrid model for cursive script recognition. *Pattern Analysis and Application, 3*, 314–324.

Koch, G, Paquet, T., Heutte, L. (2004). Combination of contextual information for handwritten word recognition. In *Proceedings of* 9th *International Workshop on Frontiers in Handwriting Recognition*, Kokubunji (pp. 468–473). doi:10.1109/IWFHR.2004.27.

Kundu, Y. H., & Chen, M. (2002). Alternatives to variable duration HMM in handwriting recognition. *IEEE Transactions on Pattern Analysis and Machine Intelligence, 20*(11), 1275–1280. doi:10.1109/34.730561.

Li, C. H., & Lee, C. K. (1993). Minimum cross entropy thresholding. *Pattern Recognition, 26*(4), 617–625.

Niblack, W. (1986). *An introduction to digital image processing*. Englewood Cliffs: Prentice Hall.

O'Gorman, L., Sammon, M., & Seul, M. (2008). *Practical algorithms for image analysis*. New York, NY, USA: Cambridge University Press. ISBN 978-0 = 521-88411-2.

Oliveira, L. S., Sabourin, R., Bortolozzi, F., & Suen, C. Y. (2002). Automatic recognition of handwritten numerical strings: A recognition and verification strategy. *IEEE Transactions on Pattern Analysis and Machine Intelligence, 24*(11), 1438–1454.

Otsu, N. (1979). A threshold selection method from gray level histogram. *IEEE Transaction on System, Man, Cybernetics, 9*(1), 62–66.

Rajashekararadhya, S. V., & Ranjan, P. V. (2009). Efficient zone based feature extraction algorithm for handwritten numeral recognition of four popular south Indian scripts. *Journal of Theoretical and Applied Information Technology, 4*(12), 1171–1180.

Russ, J. (2007). *The image processing handbook* (5th ed.). Boca Raton, FL, USA: CRC Press. ISBN 0849372542.

Saba, T., Sulong, G., & Rehman, A. (2011). Document image analysis: Issues, comparison of methods and remaining problems. *Artificial Intelligence Review, 35*(2), 101–118.

Sadri, J., Cheriet, M. (2009). A new approach for skew correction of documents based on particle swarm optimization. In *Proceedings of 10th International Conference on Document Analysis and Recognition ICDAR '09* (pp. 1066–1070). IEEE Computer Society, Washington, DC, USA .doi:10.1109/ICDAR.2009.268.

Sarfraz, M., Rasheed, Z. (2008). Skew estimation and correction of text using bounding box. In *Proceedings of 5th International Conference on Computer Graphics, Imaging and Visualization (CGIV '08)* (pp. 259–264). Washington, DC, USA: IEEE Computer Society. doi: 10. 1109/CGIV.2008.10.

Sauvola, J., Seppänen. T., Haapakoski, S. & Pietikänen, M. (1997). Adaptive document binarization. In *Fourth International Conference Document Analysis and Recognition (ICDAR)* (pp. 147–152), Ulm, Germany.

Shanthi, N., & Duraiswamy, K. (2009). A novel SVM-based handwritten Tamil character recognition. *Pattern Analysis and Application, 13*, 173–180. doi:10.1007/s10044-009-0147-0.

Sivanandam, S. N., & Deepa, S. N. (2008). *Principals of soft computing* (pp. 71–83). New Delhi, India: Wiley-India. ISBN 978812652741.

Tomoyuki, H., Takuma, A. & Bunpei, I. (2007). An analytic word recognition algorithm using a posteriori probability. In *Proceedings of the 9th International Conference on Document Analysis and Recognition* (Vol. 2, pp. 669–673), September 23–26, 2007, Tokyo. doi:10.1109/ ICDAR.2007.4376999.

Verma, B., Blumenstein, M. (2008). Pattern recognition technologies and applications: Recent advances pp. 1–16. Hershey, New York: Information Science Reference (An Imprint of IGI Global Publications).

Verma, B., Blumenstein, M., & Ghosh, M. (2004). A novel approach for structural feature extraction: Contour vs direction. *Pattern Recognition Letter, 25*(9), 975–988.

Wang, X., Ding, X., & Liu, C. (2005). Gabor filters-based feature extraction for character recognition. *Pattern Recognition, 38*(3), 369–379.

System Identification Technique and Neural Networks for Material Lifetime Assessment Application

Mas Irfan P. Hidayat

Abstract Modeling of a material lifetime to assess the material useful lifetime during its service in design has been always challenging task. In the present study, a framework of system identification technique based upon nonlinear autoregressive exogenous inputs (NARX) was introduced and presented for material lifetime assessment using neural networks (NN). Using the framework, the task of material lifetime assessment was accomplished in a fashion of one-step ahead prediction with respect to stress level. In addition, by sliding over one-step to one-step of the stress level, the task of prediction dynamically covered all loading spectrum. As a result, material lifetime assessment can be fashioned for a wide spectrum of loading in an efficient manner based upon limited material lifetime data as the basis of the NARX regressor. The multilayer perceptron (MLP)-NARX and radial basis functions NN (RBFNN)-NARX models were developed to predict fatigue lives of composite materials under multiaxial and multivariable loadings. Several multidirectional laminates of polymeric based composites were examined in this study.

1 Introduction

It is always crucial to understand the fatigue degradation of materials in many applications to ensuring the long term reliability of a component or structure. It is well known that in design of structures, fatigue failure is the most important aspect because it is closely related to performance, reliability, and durability of the structures (Reifsnider 1991). In addition, when a new class of materials has been introduced in structural applications, characterization of the new class of materials under simulated loading conditions is also indispensable. Modeling of material lifetime to assess the useful lifetime of a material during its service in design

M.I.P. Hidayat (✉)
Department of Materials and Metallurgical Engineering-FTI, Institut Teknologi Sepuluh Nopember Surabaya, Kampus ITS Keputih Sukolilo, 60111 Surabaya, Indonesia
e-mail: irfan@mat-eng.its.ac.id

Q. Zhu and A.T. Azar (eds.), *Complex System Modelling and Control Through Intelligent Soft Computations*, Studies in Fuzziness and Soft Computing 319,
DOI 10.1007/978-3-319-12883-2_27

therefore has been always challenging task. Moreover, the ability to make accurate predictions of fatigue durability is critical to the related design optimization process (Post et al. 2008).

Recently, composite materials in particular fiber reinforced polymer (FRP) composites have become more popular materials instead of metals in many structural applications due to their excellent properties, such as high strength to weight ratio, tailored properties along preferred direction and high corrosion resistance. For instances, the use of composite materials has been common in many components of automotive, aircraft, ship hull as well as wind turbine blade structures. Fatigue characterization of the composite materials therefore is also important. In particular, it is also desirable to understand and assess the fatigue behaviour of the materials for an expected or anticipated spectrum or variable amplitude fatigue loading.

Nonetheless, modeling of composites fatigue life under complex and spectrum loading conditions comes with a greater challenge to researchers in this field. Different with metals, more considerations must be taken into account in the modeling of composites lifetime, such as wide variety of component materials or fiber and matrix types, laminate design or lay-ups, anticipated failure modes, fatigue states governed by stress ratios-R, on-axis/off-axis orientation as well as manufacturing methods. As a result, such a modeling task becomes complicated and developing a universal understanding of the performance of composite materials under spectrum fatigue loadings is also very difficult because many factors should be included and anticipated in the model (Reifsnider 1991; Harris 2003; Passipoularidis et al. 2011). On the other hand, in most cases, authors only had limited experimental fatigue data in hands. It thus makes the model development is also frequently impeded by a large amount of fatigue testing data needed, which is very costly and time consuming to collect.

Numerous empirical and phenomenological models have been introduced over the past 40 years of fatigue studies for composite materials (Post et al. 2008; Philippidis and Passipoularidis 2007; Passipoularidis and Philippidis 2009). A well known common approach is that the fatigue behaviour of composite materials is to be modeled or predicted using readily collected constant amplitude fatigue data for a material or system of materials of interest, as traditionally fatigue characterization of a material is performed under constant amplitude sinusoidal loading. Nonetheless, as stated by Post et al. (2008), it is often that any empirically determined model parameters are to be fitted to the variable amplitude fatigue data modeled. Thus the relative accuracy remains uncertain between the empirical and phenomenological approaches reported in the literature. As a result, many models are developed with a specific material and loading configuration and their generalization to other cases remains uncertain.

Driven by the requirement for speeding up time frame from research stage to market place and also cutting down the associated cost, in recent years there has been increasingly interest in pursuing and utilizing alternative approaches based upon soft computing framework, in particular neural networks (NN), to develop efficient and robust predictive model for fatigue life assessment of composite materials. It is interesting to note that the characteristic and capability of soft computing techniques are lying on emulating relationships in sets of input data to subsequently predict the outcome of another new set of input data, for examples, another composite system or stress environment.

NN have been previously employed for elevated temperature creep-fatigue life prediction (Venkatesh and Rack 1999), fracture toughness and tensile strength of microalloy steel evaluation (Haque and Sudhakar 2002), prediction of fatigue crack growth rate in welded tubular joints (Fathi and Aghakouchak 2007), while genetic algorithm (GA) has been employed as parameterization tool for fatigue crack growth of Al-5052 (Bukkapatnam and Sadananda 2005) as well as optimization tool for fuzzy logic and NN models in life prediction of boiler tubes (Majidian and Saidi 2007). Moreover, recently NN has been also employed to build a probability distribution function for fatigue life prediction of steel under step-stress conditions (Pujol and Pinto 2011).

In recent years, soft computing techniques have found their applications in the field of fatigue life assessment of composite materials in particular under variable amplitude loading conditions (Aymerich and Serra 1998; Lee and Almond 2003). The use of soft computing techniques in fatigue life assessment of composite materials has a wide range of applications from unidirectional (Al-Assaf and El-Kadi 2001; El-Kadi and Al-Assaf 2002) to multidirectional laminate (Freire Junior et al. 2005; Vassilopoulos et al. 2007, 2008; Freire Junior et al. 2007, 2009).

In the present chapter, a framework of system identification technique based upon nonlinear autoregressive exogenous inputs (NARX) for material lifetime assessment using neural networks (NN) will be presented. Using the proposed framework, material lifetime assessment can be fashioned for a wide spectrum of loading in an efficient manner based upon limited material lifetime data as the basis of the NARX regressor. The key aspect of the new approach is that sliding over one-step to one-step of the stress level so that the task of prediction dynamically covered all loading spectrum.

The remaining of this chapter is organized as follows. Comprehensive reviews in the modeling of fatigue life of composite materials along with the motivation and objective for the present study are presented in Sect. 1. In Sect. 2, concept of constant life diagrams (CLD) as rational of the use of the NARX structure in the present application is briefly described. NN architectures developed and employed in this study are presented in Sect. 3. Section 4 describes composite materials examined and numerical procedures employed for the developed NN structures. Results and discussion are presented in Sect. 5, followed by conclusions in Sect. 6.

1.1 Soft Computing Techniques for Fatigue Life Assessment of Composite Materials

Al-Assaf and El-Kadi (2001) and El-Kadi and Al-Assaf (2002) assessed the fatigue life of unidirectional glass fiber/epoxy laminae using different neural network paradigms, namely feed forward (FF), modular (MN), radial basis function (RBF) and principal component analysis (PCA) networks, and compared the prediction results to the experimental data. Specimens with five fiber angle orientations of 0°, 19°, 45°, 71° and 90° were tested under three stress ratio-*R* conditions of −1, 0 and 0.5. Ninety two experiment data made up the application data for the networks. They found that NN can be trained to model the nonlinear behaviour of composite laminate subjected to cyclic loading and the prediction results were comparable to other current fatigue-life prediction methods.

Freire Junior et al. (2005) followed different approach, by which NN was utilized to build constant life diagrams (CLD) of fatigue. The researchers built CLD of a plastic reinforced with fiberglass (DD16 material) with $[90/0/\pm45/0]_S$ lay-up. Four training data sets (each set consists of 3*R*, 4*R*, 5*R* and 6*R* values, respectively) were set up from twelve stress ratio-*R* values. It was found that the use of NN to build CLD was very promising where the NN model trained using only three *S-N* curves could generalize and construct other remaining *S-N* curves of the CLD building. For better generalization, however, six *S-N* curves should be utilized in NN training.

Vassilopoulos et al. (2007) criticized that the determination of six *S-N* curves was a costly task for the NN prediction purpose. Instead, these authors used a small portion, namely 40–50 %, of the experimental data. It was shown that it is possible to build CLD using the small portion data and NN was proven to be a sufficient tool for modelling fatigue life of GFRP multidirectional laminates.

Further, Vassilopoulos et al. (2008) have employed genetic programming for modeling the fatigue life of several fiber–reinforced composite material systems. It was shown that if the genetic programming tool is adequately trained, it can produce theoretical predictions that compare favorably with corresponding predictions by conventional methods for the interpretation of fatigue data. It was also pointed out that the modeling accuracy of this computational technique was very high. In addition, the proposed modeling technique presented certain advantages compared to conventional methods. The new technique was a stochastic process that led straight to a multi-slope *S-N* curve following the trend of the experimental data, without the need for any assumptions.

Bezazi et al. (2007) have investigated fatigue life prediction of sandwich composite materials under flexural tests using a Bayesian trained artificial neural network. The authors noticed the good generalization of NN trained with Bayesian technique in comparison to that with maximum likelihood approach in predicting fatigue behaviour of the sandwich structure. Nonetheless, only one lay-up configuration was considered in the work.

Freire Junior et al. (2007, 2009), in their next attempts, showed that the use of modular networks (MN) gives more satisfactory results than feed-forward (FF) neural network. However, it was still necessary to increase the training sets for better results.

Hidayat and Melor (2009) have noticed the potential use of limited number of fatigue data in the NN modelling of fatigue life of composite materials with Bayesian regularization. The authors have investigated E-glass/epoxy ($[\pm45/0_4/\pm45]$) and DD16 or E-glass/polyester ($[90/0/\pm45/0]_S$) composites under fatigue loadings with various stress ratio values. It was found that although only two stress ratios were used in the training set, the NN model developed was able to generalize well and gave reasonably accurate fatigue life prediction under a wide range of stress ratio values. The reliability and accuracy of the NN prediction were quantified by a small MSE value.

Klemenc and Fajdiga (2012) have employed a class of evolutionary algorithms to estimate *S-N* curves and their scatter using a differential ant-stigmergy algorithm (DASA). In Klemenc and Fajdiga (2013), the authors have extended the use of evolutionary algorithms of GA and DASA for estimating *E-N* curves and their scatter.

From the recent investigations, soft computing techniques have been proven to be a sufficient tool for modelling fatigue life of composite materials, ranging from unidirectional to multidirectional laminate, with the potential for fatigue life assessment under variable amplitude loadings. In addition, the use of soft computing techniques has been also a new route in the task of fatigue life assessment where the main aim is to develop soft computing models that produce reliable prediction using a limited body of fatigue data, which in turn can support design decisions very soon and reliably.

However, it is still necessary to optimize NN prediction of composite fatigue life under variable amplitude loading by utilizing less fatigue data but at the same time ensuring reasonably accurate prediction to take full advantage of the NN potential for much more efficient fatigue life assessment. Therefore, further confirmation and examination for various composite materials or different NN architectures and frameworks are required. Moreover, it will be also valuable to find the sufficient training data set for various types of composites in relation to the full utilization of fatigue data available. It is worth to point out here that fatigue behavior is still so complicated that the problem requires more effort before NN can be used with more confidence (Zhang and Friedrich 2003).

1.2 Motivation and Objectives

In the previous works, investigations were only focused on the application of various soft computing models, in particular NN, in the task of material lifetime assessment. As a matter of fact, however, no further attempt has been made to link and ground the use of NN models to one important concept of fatigue life analysis and design, namely constant life diagrams or CLD. It will be shown that the utilization of CLD concept could lead to the use of non-linear auto-regressive exogenous inputs (NARX) structure as a natural of choice of NN configuration for the task of material lifetime assessment.

Moreover, in term of fatigue life assessment of composite materials under multivariable amplitude loading, previous investigations were only focused on the utilization of NN for multivariable amplitude loadings with respect to different stress ratio values-*R*. No further attempt, however, so far has been devoted to the utilization of NN for fatigue life assessment of composite materials under both multiaxial and multivariable loading conditions, in which factors of stress ratios and on-axis/off-axis orientations are taken into account and treated in simultaneously.

To introduce and implement new perspective and efficient approach based upon artificial intelligence and system identification technique in the field of composite materials lifetime assessment are the main motivation and objective of the present study. It is preferred to handle the fatigue life assessment in such a way that fatigue lives of different stress ratio and on-axis/off-axis orientation values are predicted based upon fatigue data from, if possible, just limited number of stress ratio(s) and on-axis/off-axis orientation(s) as the basis of training data.

In this study, the multilayer perceptron (MLP)-NARX and radial basis functions NN (RBFNN)-NARX models were developed and further applied for composite materials lifetime assessment application. Rational of the use of the NARX structure in the application was emphasized and linked to the concept of CLD as described in Sect. 2. Fatigue life assessment was then performed and realized as one-step ahead prediction with respect to each stress level corresponding to stress ratio values arranged in such a way that transition took place from a fatigue region to another one in the CLD. As a result, material lifetime assessment can be fashioned for a wide spectrum of loading in an efficient manner. Such an analysis constitutes to a variable amplitude or spectrum fatigue loading (Vassilopoulos et al. 2010).

2 Constant Life Diagrams (CLD)

Constant life diagrams (CLD) are graphical representations of the safe regime of constant amplitude loading for a given specified life, e.g. the endurance limit or infinite life (Sendeckyj 2001; Vassilopoulos et al. 2010). CLD also serves as a convenient way in fatigue life assessment analysis under spectrum loading. CLD is also another way to represent the *S-N* curve, with which design engineers are very familiar. Stress ratio *R*, which is a ratio between minimum and maximum alternating stresses, now in CLD also indicates what fatigue region the stress ratio value belongs to. Figure 1 represents the CLD schematic.

The points along each radial line are the points of *S-N* curve for a specific stress ratio. Moreover, as one can see in Fig. 1, fatigue region moves from tensile-tensile to compressive-compressive sector in CCW direction forming a spectrum of loading conditions and all points with the same fatigue life *N* are connected with lines in a plane of amplitude stress (S_a)–mean stress (S_m) axes. The transition regions are marked by stress ratio values of $R = 1$ (ultimate static strength), $R = 0$ (minimum alternating stress equals zero), $R = -1$ (maximum alternating stress equals the absolute value of minimum alternating stress) and $R = \pm \sim$ (the absolute value of minimum alternating stress is much higher than the value of maximum alternating stress, which can be either positive or negative).

Dynamic nature of both the CLD and the NARX model to result in the first application of the NARX model in spectrum fatigue analysis will be explored in the developed NN architectures and procedures in the subsequent sections.

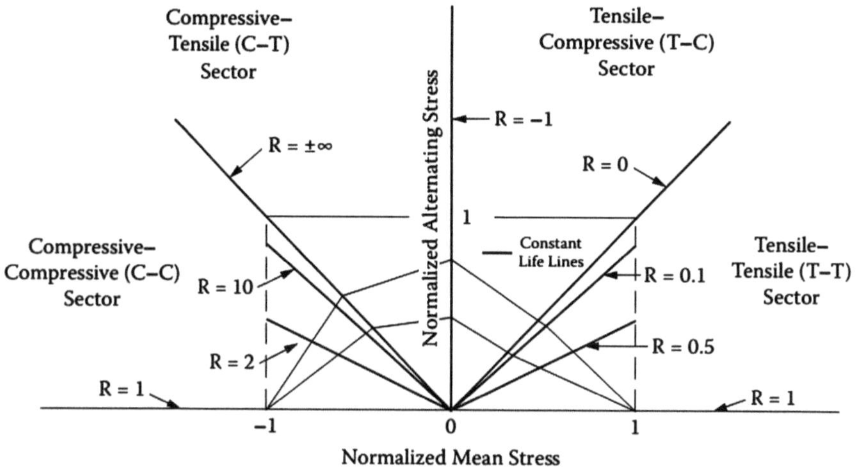

Fig. 1 The schematic of CLD for fatigue life assessment analysis

3 NN Architectures

3.1 MLP

Figure 2 shows an MLP with one hidden layer and single output, which is the most popular NN architecture commonly employed in the simulation with NN.

The notations presented in Fig. 2 are: p input sets, L number of elements in input vector, s number of hidden nodes, n the summed up of weighted inputs, a the output of activation function in the corresponding layer, $w^1_{j,i}$ and b^1_j input weight and bias ($i = 1$ to L, $j = 1$ to s), $w^2_{1,j}$ and b_o layer weight and output bias, and y the MLP output. Superscripts 1 and 2 represent the first layer of hidden and the second layer of output, respectively.

Learning in NN is achieved by adjusting the corresponding weights in response to external environment of input sets. The weights adjustment is accomplished by a set of learning rule by which an objective function is minimized. In what follows, problem formulation of NN learning will be concisely presented, particularly from the supervised learning context of the MLP. Nonetheless, the formulation can be also extended for RBFNN.

Let (P,T) be a pair of random variables with values in $P = \Re^m$ and $T = \Re$, respectively. The regression of T on P is a function of P, $f : P \to T$, giving the mean value of T conditioned on P, $E(T|P)$.

Let random samples $O_1^Q = \{(P_1, T_1), \ldots, (P_Q, T_Q)\}$ of size Q can be drawn from the distribution of (P,T) as an observation set. For $Q \geq 1$, $\hat{f}_Q$ will denote an estimator of f based on the random samples, that is a map $\hat{f}_Q : O_1^Q \to \hat{f}_Q(O_1^Q, .)$, where for fixed $O_1^Q, p \to \hat{f}_Q(O_1^Q, p)$, is an estimate of the regression function $f(p)$.

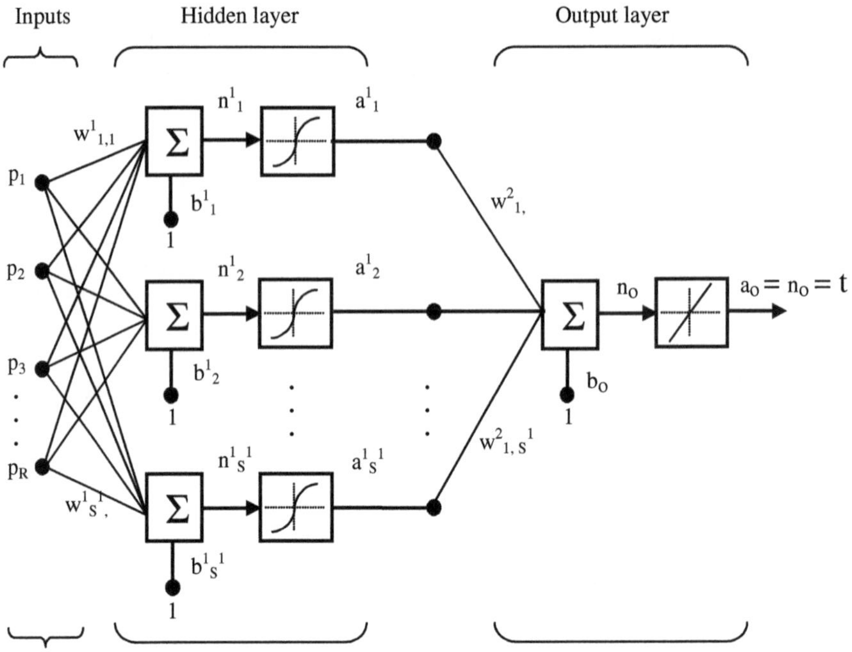

Fig. 2 MLP with one hidden layer and single output

Furthermore, for convenience P and T will be referred to as the sets of input and variable output, respectively. Given the observation set O, learning in NN for realization of the estimate $\hat{f}$ means adjusting to vector of parameters weight $\mathbf{w}$ and biases $\mathbf{b}$ using a set of learning rule or learning algorithm in such a way that $\hat{f}$ minimizes the objective function or empirical error defined as:

$$E(\mathbf{w}) = \sum_{q=1}^{Q} \left[t_q - \hat{f}(\mathbf{p}_q; \mathbf{w})\right]^2 \tag{1}$$

and generalizes well or outputs properly when a novel input vector $\mathbf{p}_{test}$ never seen before is fed into the network.

The estimate $\hat{f}$ realized by the MLP shown in Fig. 2 given the training set O can be written as:

$$\hat{f}(\mathbf{p}; \mathbf{w}) = \sum_{i=1}^{s} \mathbf{w}_{1,i}^2 \tau\left(\mathbf{w}_{i,j}^1 \mathbf{p} + \mathbf{b}_i\right) + \mathbf{b}_o \tag{2}$$

where $\tau(.)$ is a sigmoidal function used in the nodes of hidden layer.

In addition, by keeping in mind that learning in NN is principally updating the network weights based on the given set of examples so that the network will give proper response to new examples, below are two limiting factors of the NN learning. First, only a finite number of observation points (example pairs) are available. This means that the available examples sometimes must be fully utilized for the NN learning purpose to provide proper learning of the underlying process. Hence, the practicability and feasibility of using limited examples for NN learning to yield accurate prediction output are assured. The second is that the realization of target at the points of observation $p_q, q = 1, \ldots, Q$, is observed with an additive noise e_q:

$$e_q = T_q - f(P_q) \tag{3}$$

The observations are then noisy and the target noises e_q introduce a random component in the estimation error.

3.2 RBFNN

Figure 3 describes a schematic diagram of the RBFNN with the distance function of Euclidean distance denoted by $\|\mathbf{x} - c_i\|$, which will be further explained in this section. In Fig. 3, the input sets are denoted by x, the target outputs are denoted by y and the number of hidden nodes is represented by s.

As depicted in Fig. 3, it is clear that in RBFNN, the inputs reach the hidden layer nodes unchanged. In addition, the output estimate $\hat{f}$ realized by the RBFNN given the training examples can be expressed as:

$$\hat{f} = \sum_{i=1}^{s} \phi_i(\|\mathbf{x} - c_i\|).\mathbf{w}_{j,i} \tag{4}$$

where: $\mathbf{x}$ is the vector of input sets, c_i is the ith center node in the hidden layer and $\mathbf{w}_{j,i}$ is the vector of weights from the output nodes to the center nodes, ϕ_i are the radial basis functions of the center nodes, and $\|\mathbf{x} - c_i\|$ is the distance between the point representing the input $\mathbf{x}$ and the center of the ith hidden node.

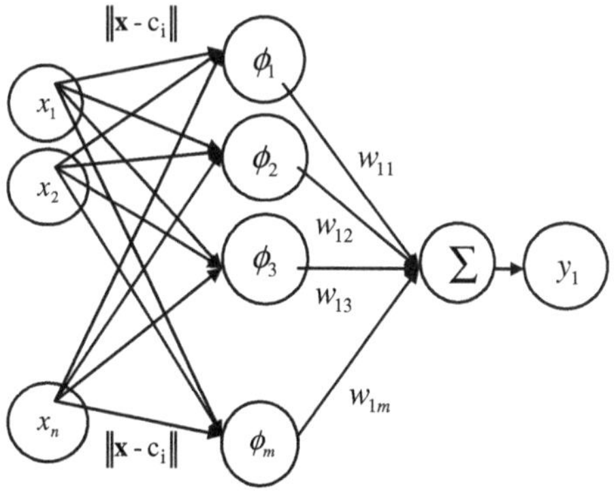

Fig. 3 Schematic diagram of RBFNN

The most widely used radial basis function ϕ was Gaussian function:

$$\phi(\|\mathbf{x} - c_i\|) = \exp\left(-\frac{(\chi - \gamma)^2}{\psi^2}\right) \tag{5}$$

γ and ψ are the parameters that control the “position” and “width” of the RBF centers, respectively. It is clear that there are four sets of parameters to be determined in the training of the RBFNN. The parameters are governing the network mapping properties, namely the number of centers (c) in the hidden layer, the position of RBF centers (γ), the width of RBFs (ψ), and the RBFNN weights (**w**). Furthermore, training of RBFNN involves both supervised and unsupervised learning methods, with respect to, respectively, the weights of RBFNN and the parameters of c, γ and ψ. The output layer is trained by a supervised learning method, where the synaptic weights are updated as usual with respect to the objective function chosen. On the other hand, training of the hidden layer involves the determination of the first three parameters mentioned. The parameters are dependent only on the inputs and are independent of the outputs, thus making this part of the learning process an unsupervised one. The readers are directed to Haykin (2009).

RBFNN has been successfully applied for different purposes such as, among others, (Catelani and Fort 2000; Mollah and Pratihar 2008). Its application in the field of fatigue life assessment of composite materials, to the author best knowledge, is nonetheless still limited (Al-Assaf and El-Kadi 2001; El-Kadi and Al-Assaf 2002). In addition, several advantages on using RBFNN are: simplicity of the architecture, reduction in training time and the capability to deal with unseen data (Haykin 2009).

3.3 Neural Networks with NARX Structure

NN with NARX structure has the signal vector applied to the NN input layer consisting of a data window made up by present and past values of *exogenous* (independent) inputs and by delayed values of the outputs. The NN model belongs to a class of recurrent neural networks (RNN) with one feed-back loop from the NN output layer to the input layer. Moreover, the presence of the feed-back loop has enabled such a configuration to acquire state representations. It also provides a unified representation for a wide class of discrete-time nonlinear systems (Chen et al. 1990; Narendra and Parthasarathy 1990).

Mathematically, a NARX model can be represented as:

$$\begin{aligned} y(n+1) &= f[\mathbf{y}(n); \mathbf{u}(n)] \\ y(n+1) &= f\left[y(n), \ldots, y(n-d_y+1); u(n), u(n-1), \ldots, u(n-d_u+1)\right] \end{aligned} \tag{6}$$

where $u(n)$ and $y(n)$, respectively, state the input and output of the model at discrete time n; $u(n), y(n) \in \Re$.

Moreover, d_y and d_u are the output-memory and input-memory orders. d_y represents the number of lagged output values, which is often referred to as the order of the model, d_u represents the number of lagged input values (d_u, $d_y \geq 1$ and $d_u \leq d_y$). The vectors $\mathbf{y}(n)$ and $\mathbf{u}(n)$, therefore, form the output and input regressors, respectively.

The NARX model is commonly trained using two basic modes, namely:

1. Parallel (P) Mode

Using this mode, the output regressor utilized the estimated outputs which are fed back to the regressor.

$$\hat{y}(n+1) = \hat{f}\left[\hat{y}(n), \ldots, \hat{y}(n-d_y+1); u(n), u(n-1), \ldots, u(n-d_u+1)\right] \quad (7)$$

2. Series-Parallel (SP) Mode

Using this mode, the output regressor utilized the actual output values.

$$\hat{y}(n+1) = \hat{f}\left[y(n), \ldots, y(n-d_y+1); u(n), u(n-1), \ldots, u(n-d_u+1)\right] \quad (8)$$

It is worth to note that standard feed-forward architecture trained with back-propagation (BP) technique can be used directly in the NARX mode of SP. In addition, various learning algorithms are also widely applicable. A form of regularization may also be employed because the additive measurement errors, ϵ_n, which are zero-mean Gaussian variables with $\mathrm{Var}[\epsilon_n] = \sigma^2$, can be also present.

Figure 4 illustrates the NARX with input and output tapped delay lines (TDL), in parallel and series-parallel architectures (Neural Network Toolbox User's Guide 1992).

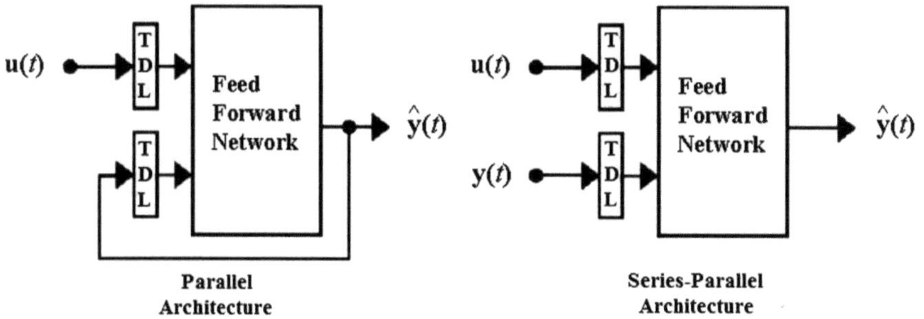

Fig. 4 Parallel and series-parallel architectures of NARX network

4 Materials and Methods

4.1 *Materials for Fatigue Life Assessment of Multivariable Amplitude Loadings*

The investigated materials were two multidirectional laminates of E-glass/polyester, typical materials used in wind turbine blade applications. The first material had the corresponding lay-up of $[90/0/\pm45/0]_S$ and is called as DD16 (Mandell and Samborsky 2010). The second material had the corresponding lay-up of $[0/(\pm45)_2/0]_T$ and were cut by diamond saw wheel at on-axis (0°) and off-axis (15°, 30°, 45°, 60°, 75° and 90°) orientations (Vassilopoulos and Philippidis 2002). The corresponding database containing fatigue data of various *R* values (R = 0.1, 0.5, 0.7, 0.8, 0.9, −0.5, −1, −2 and 10) makes it suitable for the study purpose.

From the fatigue data, stress ratio *R*, maximum stress (S_{max}) and minimum stress (S_{min}) values were used as input set and the output was the corresponding fatigue cycles (log *N*) for the input set. For each particular *R* value, mean fatigue life values were used. Also, all the data were normalized so that they fall in the range of −1 to 1 using the following formula:

$$\mathbf{x}_n = \frac{2\mathbf{x} - \mathbf{x}_{max} - \mathbf{x}_{min}}{\mathbf{x}_{max} - \mathbf{x}_{min}} \tag{9}$$

where: $\mathbf{x}_n$ is the normalized value of the input variables (R, σ_{max} and σ_{min}) or the output variable (log *N*), $\mathbf{x}$ is the un-normalized data and $\mathbf{x}_{min}$, $\mathbf{x}_{max}$ are the minimum and maximum values of the variables, respectively.

Table 1 summarizes the materials examined together with the orientation and the training and testing sets employed. Note that stress ratio values *R* were arranged in CCW direction according to the CLD, moving across from tensile-tensile sector to compressive-compressive sector. In addition, fatigue data of R = 0.1 and 10 were chosen as training set because the best relative positions of the *R* values in the CLD, as shown in Fig. 1. Moreover, the fatigue testing on the *R* values are also commonly conducted. Note also that number of stress levels in each *R* value employed was 5.

Table 1 Materials examined with the orientations and the training and testing sets employed for fatigue life assessment of multivariable amplitude loadings

Material	Angle orientation	Fatigue data as training set	Fatigue data as testing set
E-glass/polyester $[90/0/\pm45/0]_S$ (Mandell and Samborsky 2010)	On-axis	R = 0.1 and 10	R = 0.9, 0.8, 0.7, 0.5, −0.5, −1 and −2
E-glass/polyester $[0/(\pm45)_2/0]_T$ (Vassilopoulos and Philippidis 2002)	On-axis	R = 0.1 and 10	R = 0.5 and −1
	45°	R = 0.1 and 10	R = 0.5 and −1

4.2 Materials for Fatigue Life Assessment of Multivariable and Multiaxial Loadings

The investigated materials were multidirectional laminates of E-glass/polyester (Vassilopoulos and Philippidis 2002) and E-glass fabrics/epoxy (Mandell and Samborsky 2010), typical materials used in wind turbine blade applications. The corresponding lay-ups were $[0/(\pm 45)_2/0]_T$ and $[\pm 45/0_4/\pm 45/]$, respectively. The materials were cut by diamond saw wheel at on-axis (0°) and off-axis orientations. For E-glass/polyester material, the corresponding off-axis orientations were 15°, 30°, 45°, 60°, 75° and 90° (Vassilopoulos and Philippidis 2002), while for E-glass fabrics/epoxy material, the only off-axis orientation was 90° (Mandell and Samborsky 2010).

In addition, the corresponding database containing fatigue data of various stress ratio values and the corresponding on-axis/off-axis orientations of $\boldsymbol{R} = \mathbf{0.1}$: $\theta = 0°$, 15°, 45°, 75° and 90°; $\boldsymbol{R} = \mathbf{0.5}$: $\theta = 0°$ and 45°; $\boldsymbol{R} = -\mathbf{1}$: $\theta = 0°$, 30°, 45°, 60° and 90°; and $\boldsymbol{R} = \mathbf{10}$: $\theta = 0°$, 30°, 45°, 60° and 90° for E-glass/polyester, and of $\boldsymbol{R} = \mathbf{0.1}$: $\theta = 0°$ and 90°; $\boldsymbol{R} = \mathbf{0.5}$: $\theta = 0°$ and 90°; $\boldsymbol{R} = -\mathbf{0.5}$: $\theta = 0°$ and 90°; $\boldsymbol{R} = -\mathbf{1}$: $\theta = 0°$ and 90°; $\boldsymbol{R} = -\mathbf{2}$: $\theta = 0°$ and 90°; and $\boldsymbol{R} = \mathbf{10}$: $\theta = 0°$ and 90° for E-glass fabrics/epoxy. The database comprised, respectively, 85 and 96 fatigue data, making the database suitable for the study purpose. Note that number of stress levels in each stress ratio value employed were 5 and 8 for E-glass/polyester and E-glass fabrics/epoxy, respectively.

From the fatigue data, stress ratio (R), on-axis/off-axis orientation (θ) and maximum stress (S_{max}) values were used as input set and the output was the corresponding fatigue cycles (log N) for the input set. For each particular R value, mean fatigue life values were used. As in the previous section, all the data were also normalized into the range of −1 to 1 using Eq. (9).

Table 2 summarizes the materials examined together with the training and testing sets employed. Note that for the assessment task, stress ratio values-R were arranged in CCW direction according to the CLD, moving across from tensile-tensile sector to compressive-compressive sector. In addition, fatigue data as training set of $R = 0.1$ and 10 were chosen because the best relative positions of the R values in the CLD (Hidayat and Melor 2009; Hidayat et al. 2011; Hidayat and Berata 2011). The corresponding θ value chosen for both the stress ratios was 0°. With the training and testing data, the NN model will develop multivariable and multiaxial fatigue life assessment analysis.

4.3 Methods

In the present study, the training algorithm of Levenberg-Marquardt was chosen and utilized to result in fast and efficient NN model (Nocedal and Wright 2006). The use of Levenberg-Marquardt algorithm for fast and efficient NN modeling has

Table 2 Materials examined with the orientations and the training and testing sets employed for fatigue life assessment of multivariable and multiaxial loadings

Material	Fatigue data as training set: R and θ values	Fatigue data as testing set: R and θ values
E-glass/polyester $[0/(\pm45)_2/0]_T$ (Vassilopoulos and Philippidis 2002)	$R = 0.1$: $\theta = 0°$ $R = 10$: $\theta = 0°$	$R = 0.5$: $\theta = 0°$
		$R = -1$: $\theta = 0°$
		$R = 0.1$: $\theta = 15°$
		$R = -1$: $\theta = 30°$
		$R = 10$: $\theta = 30°$
		$R = 0.1$: $\theta = 45°$
		$R = 0.5$: $\theta = 45°$
		$R = -1$: $\theta = 45°$
		$R = 10$: $\theta = 45°$
		$R = -1$: $\theta = 60°$
		$R = 10$: $\theta = 60°$
		$R = 0.1$: $\theta = 75°$
		$R = 0.1$: $\theta = 90°$
		$R = -1$: $\theta = 90°$
		$R = 10$: $\theta = 90°$
E-glass fabrics/epoxy $[\pm45/0_4/\pm45/]$ (Mandell and Samborsky 2010)	$R = 0.1$: $\theta = 0°$ $R = 10$: $\theta = 0°$	$R = 0.5$: $\theta = 0°$
		$R = -0.5$: $\theta = 0°$
		$R = -1$: $\theta = 0°$
		$R = -2$: $\theta = 0°$
		$R = 0.1$: $\theta = 90°$
		$R = 0.5$: $\theta = 90°$
		$R = -0.5$: $\theta = 90°$
		$R = -1$: $\theta = 90°$
		$R = -2$: $\theta = 90°$
		$R = 10$: $\theta = 90°$

been also shown in other field of application, for instance in Azar (2013) for medical applications. Moreover, Bayesian regularization was incorporated (Foresee and Hagan 1997) to accommodate the noise which may be present in the target data as well as to deal with limited training data that may lead to ill-posed problem.

4.3.1 Adaptation of Bayesian Framework Within the Levenberg-Marquardt Algorithm

Bayesian regularization was incorporated in the Levenberg-Marquardt algorithm through the modified objective function of NN, $E(\mathbf{w})$, as follows:

$$E(\mathbf{w}) = \beta \sum_{q=1}^{Q} \left[t_q - \hat{f}(\mathbf{p}_q; \mathbf{w})\right]^2 + \alpha \sum_{i=1}^{W} w_i^2 \tag{10}$$

where: α is a weight decay parameter, β is an inverse noise variance parameter, t_q is the target data, the estimate $\hat{f}$ realized by the NN, $\mathbf{p}$ is the vector of input sets, $\mathbf{w}$ is the vector of weights (and biases), Q is the number of training examples and W is the total number of weights.

Equation (10) can be further rewritten as:

$$E(\mathbf{w}) = \beta E_D + \alpha E_w \tag{11}$$

where:

$$E_D = \sum_{q=1}^{Q} \left[t_q - \hat{f}(\mathbf{p}_q; \mathbf{w})\right]^2 \tag{12}$$

$$E_w = \sum_{i=1}^{W} w_i^2 \tag{13}$$

Using the modified cost function, the gradient $\mathbf{g}$ and Hessian $\mathbf{H}$, respectively, are:

$$\mathbf{g} = 2\beta \mathbf{J}^{\mathrm{T}} \mathbf{r} + 2\alpha \mathbf{w} \tag{14}$$

$$\mathbf{H} = 2\beta \mathbf{J}^{\mathrm{T}} \mathbf{J} + 2\alpha \mathbf{I} \tag{15}$$

Thus, the increment of weights $\Delta\mathbf{w}$ becomes:

$$\Delta\mathbf{w} = -\left[\beta \mathbf{J}^T \mathbf{J} + (\lambda + \alpha)\mathbf{I}\right]^{-1} \mathbf{g} \tag{16}$$

Furthermore, for the purpose of updating α and β, the Hessian formulation was utilized through the following equations:

$$\gamma = I - 2\alpha\, trace\left(\mathbf{H}^{-1}\right) \tag{17}$$

$$\alpha = \frac{\gamma}{2E_w} \tag{18}$$

$$\beta = \frac{Q - \gamma}{2E_D} \tag{19}$$

where: γ is the effective number of parameters, that is a measure of how many parameters or weights are effectively used in the NN learning with respect to the cost function reduction, I is the total number of initial weights during initialization, Q is the number of training examples and λ is lambda parameter (the parameter of LM).

Based on the above formulas, the Levenberg-Marquardt algorithm implementing Bayesian regularization can be stated as follows:

Step 1. The weights $\mathbf{w}$ and parameters λ, α and β were initialized. For example: $\lambda = 0.005$, $\alpha = 0$ and $\beta = 1$. The algorithm is not too sensitive to the initial choice of the parameters. In addition, the choice of $\alpha = 0$ and $\beta = 1$ means that the NN is starting from the original cost function.

Step 2. One step of the Levenberg-Marquardt algorithm to minimize the objective function was taken as per Eq. (16).

Step 3. If $E(\mathbf{w} + \Delta\mathbf{w}) < E(\mathbf{w})$, then $\mathbf{w}_{new} = \mathbf{w} + \Delta\mathbf{w}$ was accepted as a new iteration.

Step 4. The effective number of parameter γ was computed using Eq. (17) and the Hessian formulation of Eq. (15) was utilized.

Step 5. The parameters α and β were updated using Eq. (18) for α and Eq. (19) for β.

Step 6. Steps 2–5 were repeated until the stopping criterion was satisfied or convergence was achieved.

It is clear that the NARX-SP architecture is being currently employed and by sliding over one-step to one-step of stress level, the prediction will be dynamically covering all the spectrum loadings of the testing sets according to the CLD. As a result, material lifetime assessment can be fashioned for a wide spectrum of loading in an efficient manner based upon solely the training data as the basis of the NARX regressor, thus developed variable amplitude or spectrum fatigue analysis.

It is important to note that the number of hidden nodes employed was 10 and the parameter c of the RBFNN also took the same number. For the RBFNN, the rest of the corresponding parameters were determined accordingly using K-means technique (Haykin 2009).

NN parameters used in the present NARX modeling using the Levenberg-Marquardt algorithm with Bayesian regularization are described in Table 3.

Table 3 NN parameters used in the present NARX modeling using the Levenberg-Marquardt algorithm with Bayesian regularization

NN parameters	Value
Initial lambda, λ_{init}	0.005
Initial weight decay, α_{init}	0
Initial inverse noise, β_{init}	−1
Maximum number of iterations	200
Minimum gradient, $\mathbf{g}_{min}$	1×10^{-10}
Maximum lambda, λ_{max}	1×10^{10}
Performance goal	0
Number of hidden nodes	10

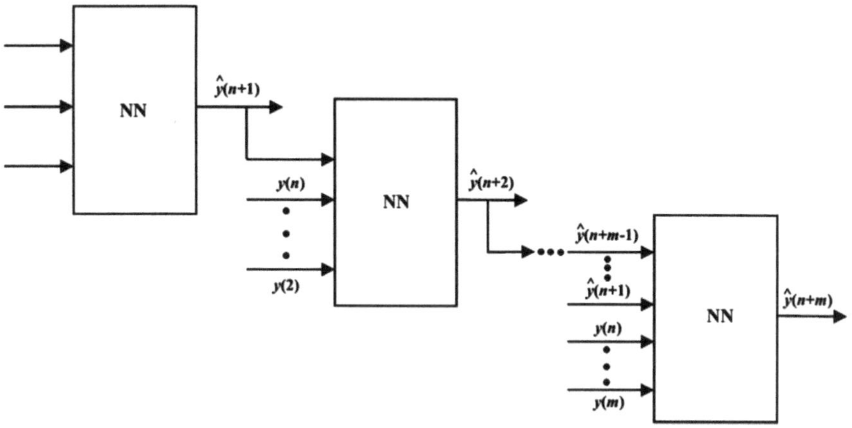

Fig. 5 Spectrum fatigue life prediction made up by one-step ahead prediction using NN with NARX-series parallel structure

4.3.2 Spectrum Fatigue Life Prediction

Fatigue life assessment of the materials was performed and realized as one-step ahead prediction with respect to each stress level S corresponding to stress ratio values R arranged in such a way that transition took place from a fatigue region to another one in the CLD. Figure 5 describes the lifetime assessment process using NN with NARX model in the study.

Using the methods described previously, all simulation results of fatigue life assessment of the composite materials are presented and discussed in following sections.

5 Simulation Results and Discussion

5.1 Fatigue Life Assessment of Multivariable Amplitude Loadings with MLP-NARX Model

Here, E-glass/polyester of $[90/0/\pm45/0]_S$, E-glass/polyester of $[0/(\pm45)_2/0]_T$ whose on-axis orientation, E-glass/polyester of $[0/(\pm45)_2/0]_T$ whose 45° orientation are denoted as Material I, Material II-on-axis and Material II-45°, respectively.

Figures 6, 7, 8, 9, 10, 11 and 12 show the *S-N* curves obtained by the NN-NARX model and the experimental data for the tested stress ratios $R = 0.9$, 0.8, 0.7, 0.5, −0.5, −1 and −2 of Material I, respectively. Note that the NN fatigue life prediction results of $R = -0.5$ in Fig. 10 represented "the worst prediction", while those of $R = 0.7$ in Fig. 8 represented the best one.

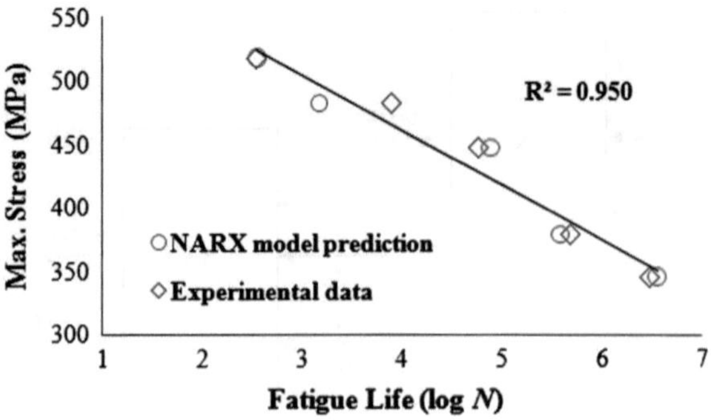

Fig. 6 *S-N* curves obtained by the NN-NARX model and the experimental data for $R = 0.9$ of Material I

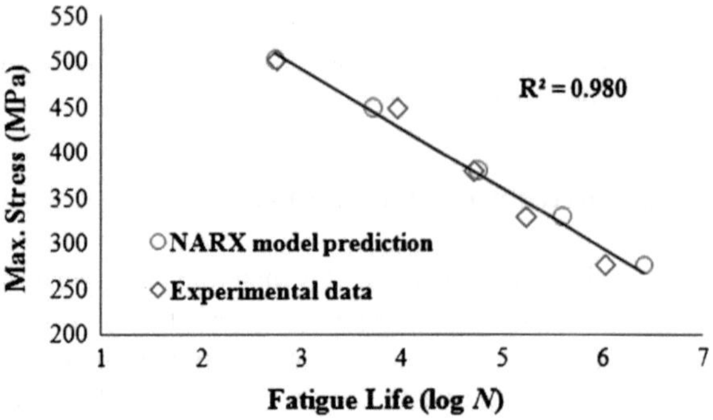

Fig. 7 *S-N* curves obtained by the NN-NARX model and the experimental data for $R = 0.8$ of Material I

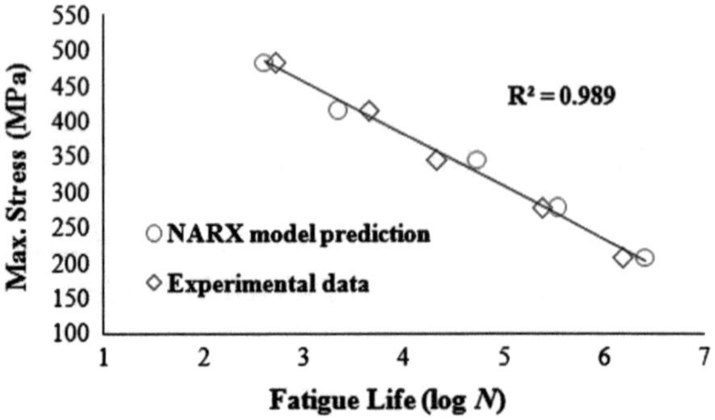

Fig. 8 *S-N* curves obtained by the NN-NARX model and the experimental data for $R = 0.7$ of Material I

It can be seen that the NN-NARX model prediction results were consistent with the experimental data showing the NN applicability and capability to model the problem considered. The NN model also showed its ability to dynamically predict the fatigue lives sliding over each stress level in a fashion of spectrum loading made up by several R values. Looking at the fatigue life prediction results for Material I produced by the NN architecture, it may be worth to also note that the results

Fig. 9 *S-N* curves obtained by the NN-NARX model and the experimental data for $R = 0.5$ of Material I

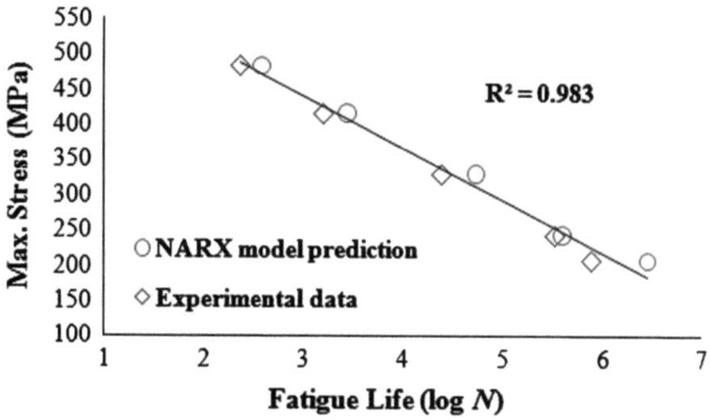

Fig. 10 *S-N* curves obtained by the NN-NARX model and the experimental data for $R = -0.5$ of Material I

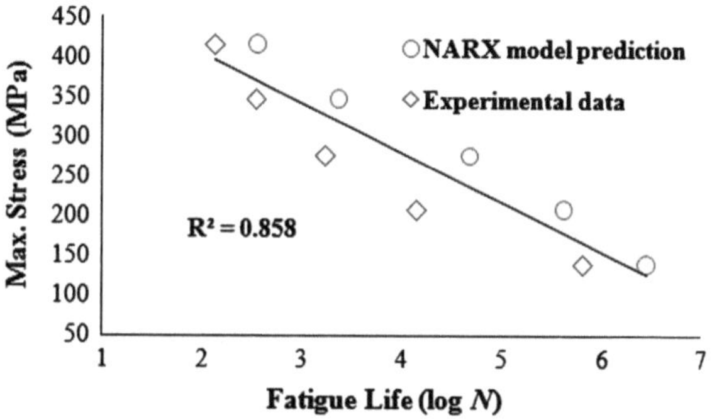

Fig. 11 *S-N* curves obtained by the NN-NARX model and the experimental data for $R = -1$ of Material I

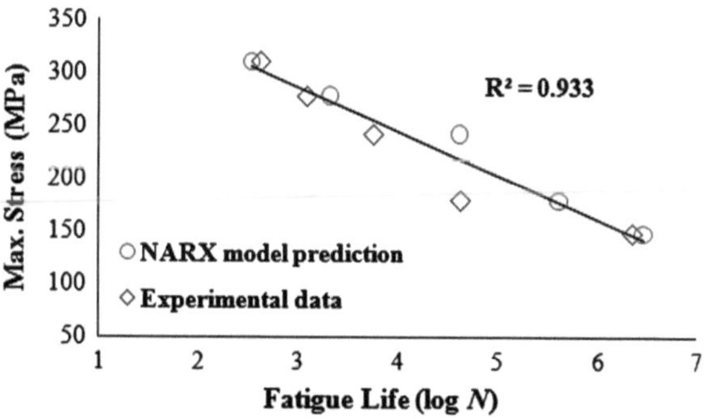

obtained were comparable with those obtained by the previous researchers investigating also the same material using NN. See, for instance, in Freire Junior et al. (2005, 2007) in particular for the NN fatigue life prediction results of $R = -0.5$ and 0.7, which were further casted in the corresponding *S-N* curves in the papers.

Furthermore, Figs. 13 and 14 depict respectively the *S-N* curves obtained by the NN-NARX model and the experimental data for stress ratios $R = 0.5$ and -1 of Material II-on axis and of Material II-45°. It can be also seen that in general the

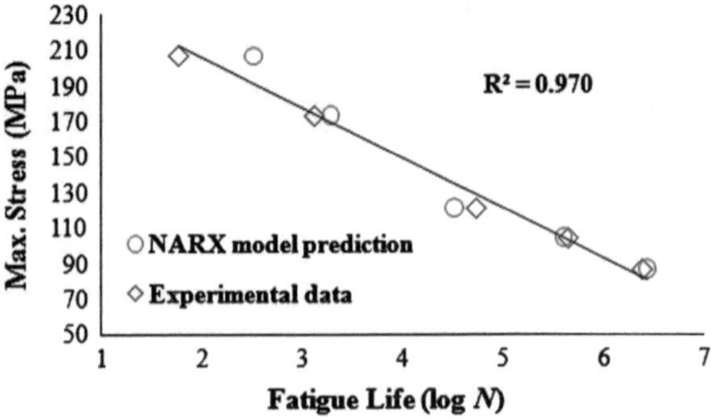

Fig. 12 *S-N* curves obtained by the NN-NARX model and the experimental data for $R = -2$ of Material I

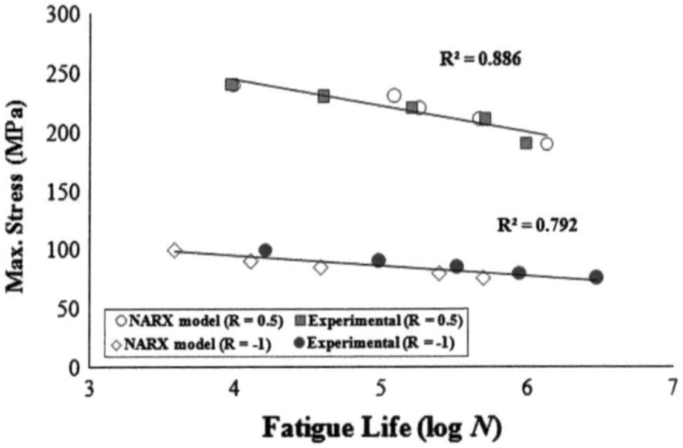

Fig. 13 *S-N* curves obtained by the NN-NARX model and the experimental data for, respectively, $R = 0.5$ and -1 of Material II-on-axis

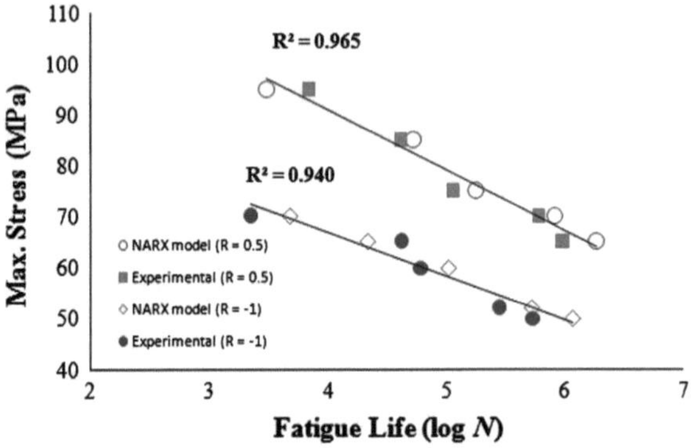

Fig. 14 *S-N* curves obtained by the NARX model and the experimental data for, respectively, $R = 0.5$ and -1 of Material II-45°

NN-NARX model prediction results can fairly follow the experimental data trend. For Material II-on-axis, the NN fatigue life prediction results of $R = 0.5$ were good at some stress levels, while those of $R = -1$ were all underestimated compared to those of the experimental data. For the later, it can be said nonetheless that although the corresponding NN fatigue life prediction results were far enough with those of the experimental data, the conservative prediction results were still preserved. For Material II-45°, the NN fatigue life prediction results of $R = 0.5$ and -1 were good for all stress levels involved. As can be seen, the NN fatigue life prediction results can closely follow the experimental data trend. The coefficient of determination (R^2) of the fatigue life prediction can be also observed. For Materials I and II, best values of the coefficient of determination was 0.989 and 0.9653, respectively. For all the materials examined, the values of R^2 was ranging from 0.7923 to 0.989.

5.2 *Fatigue Life Assessment of Multivariable Amplitude Loadings with RBFNN-NARX Model*

In this section, fatigue life assessment of multivariable amplitude loading with the RBFNN-NARX model is presented.

Figure 15 presents the NN fatigue life prediction of Material I at stress ratios *R* of the testing set using the RBFNN-NARX model. It can be seen that the RBFNN-NARX model prediction results were consistent with the experimental data showing also the applicability of the RBFNN-NARX model for this problem. The RBFNN-NARX model also showed its ability to dynamically predict the fatigue lives sliding over each stress level in a fashion of spectrum loading made up by several *R* values.

Figures 16 and 17 further depict respectively fatigue life prediction of Material II-on-axis and Material II-45°. It can be also seen that in general the RBFNN-NARX model prediction results can fairly follow the experimental data trend.

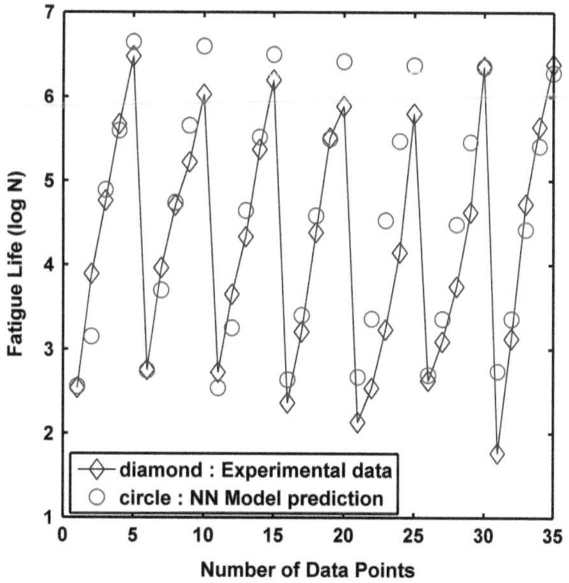

Fig. 15 Fatigue lives predicted by the RBFNN-NARX structure for the tested sets: R = 0.9, 0.8, 0.7, 0.5, −0.5, −1 and −2 of Material I (from *left* to *right*)

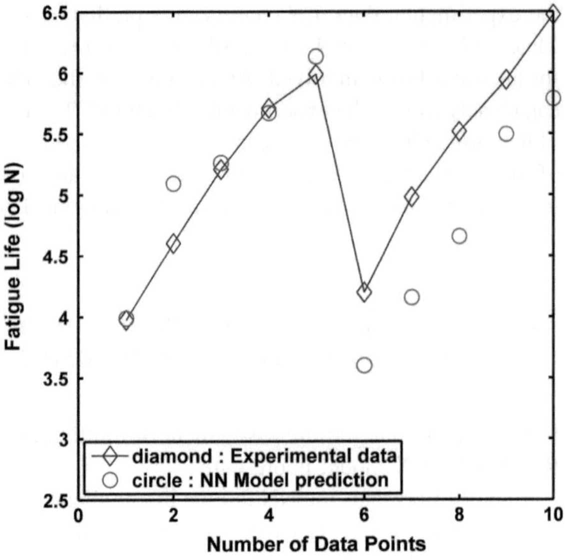

Fig. 16 Fatigue lives predicted by the RBFNN-NARX model for the tested sets: $R = 0.5$ and -1 of Material II-on-axis (from *left* to *right*)

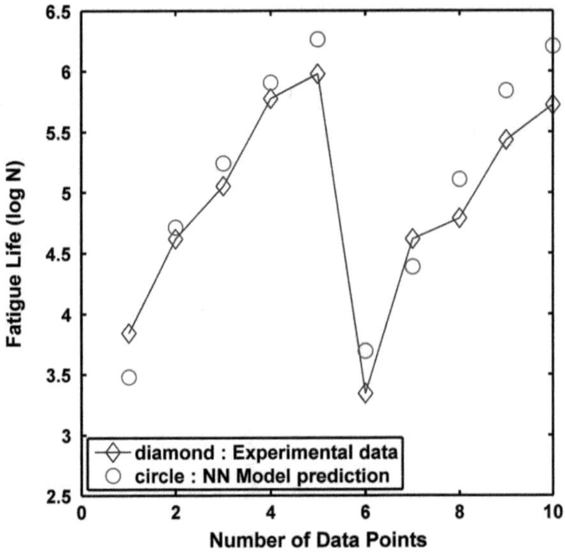

Fig. 17 Fatigue lives predicted by the RBFNN-NARX model for the tested sets: $R = 0.5$ and -1 of Material II-45° (from *left* to *right*)

The accuracy of the RBFNN model prediction compared to the experimental data was measured in mean squared error (MSE). The MSE prediction values for the Material I, Material II-on-axis and Material II-45° were 0.258, 0.285 and 0.088, respectively. It is important to note here that the produced mean squared error (MSE) values of fatigue life prediction results of the RBFNN-NARX model competed favorably, even better, with those of the MLP-NARX model. The comparison was shown in Table 4.

Table 4 Comparison of MSE values of fatigue life prediction results of the MLP-NARX and RBFNN-NARX models

Material and angle orientation		MSE of the MLP-NARX model	MSE of the RBFNN-NARX model
E-glass/polyester $[90/0/\pm45/0]_S$	On-axis	0.27	0.258
E-glass/polyester $[0/(\pm45)_2/0]_T$	On-axis	0.32	0.285
	45°	0.07	0.088

The good accuracy of the RBFNN models may be attributed to the use of radial basis functions having parameters that control the positions of the RBFs among the data or sample points and also their widths of influence (spread) to the sample points, namely the parameters of γ and ψ, contributing further to the RBFNN models resolution capability, at the expense of evaluating and determining more parameters suitably.

5.3 *Fatigue Life Assessment of Multivariable and Multiaxial Loadings with MLP-NARX Model*

It is noted here that for this fatigue life assessment task, there are 15 and 10 testing sets to be predicted for E-glass/polyester and E-glass fabrics/epoxy, respectively. Note also the arrangement of fatigue data as training set and fatigue data as testing set, in particular those of R and θ values, as shown in Table 2. The NN simulation results and the related discussion will be referred to what Table 2 described.

In addition, it is important to note again that only one information value of axial orientation-θ was utilized in the training set employed, while two values of stress ratio-R were employed.

Figures 18 and 19 present respectively multivariable and multiaxial fatigue life predictions of E-glass/polyester and E-glass fabrics/epoxy materials at the testing sets examined. It can be seen that the present NN model indeed showed its ability to dynamically predict the fatigue lives from the testing sets examined by sliding over each stress level in a fashion of spectrum loading and multiaxial orientation, made up by several R and θ values.

The accuracy of the NN-NARX model prediction was stated by the produced mean squared errors (MSE) values of 0.123 and 0.27 for E-glass/polyester and E-glass fabrics/epoxy, respectively. It is worth to note that the training sets employed were a very small number of fatigue data.

It is noted here that there are some noticeable discrepancies observed between fatigue lives predicted by the NN-NARX model and those of experimental data. For E-glass/polyester, they belong to fatigue lives of $R = -1$: $\theta = 0$, 60 and 90°, respectively. For E-glass fabrics/epoxy, the noticeable discrepancies belong to

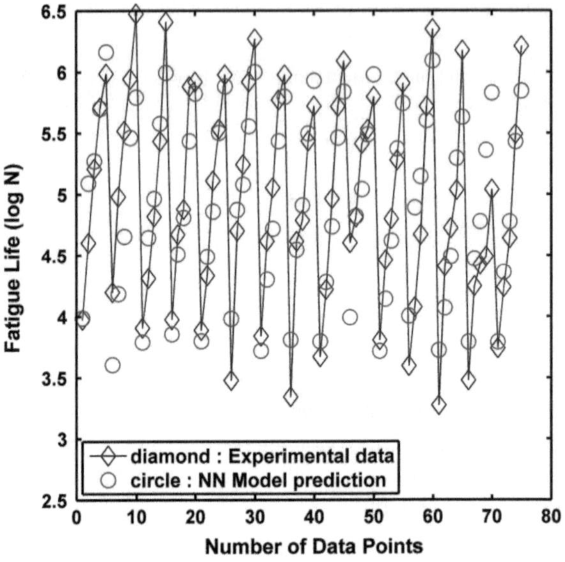

Fig. 18 Multivariable and multiaxial fatigue life prediction of the NN-NARX model for R and θ values of the testing sets: $R0.5\theta0°$, $R\text{-}1\theta0°$, $R0.1\theta15°$, $R\text{-}1\theta30°$, $R10\theta30°$, $R0.1\theta45°$, $R0.5\theta45°$, $R\text{-}1\theta45°$, $R10\theta45°$, $R\text{-}1\theta60°$, $R10\theta60°$, $R0.1\theta75°$, $R0.1\theta90°$, $R\text{-}1\theta90°$ and $R10\theta90°$ for E-glass/polyester

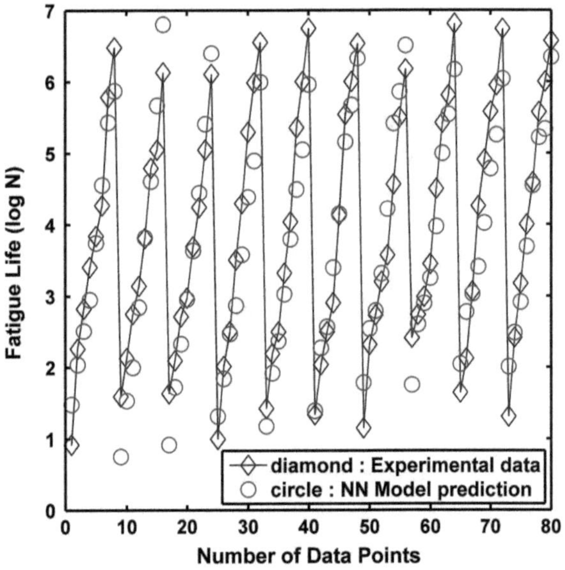

Fig. 19 Multivariable and multiaxial fatigue life prediction of the NN-NARX model for R and θ values of the testing sets: $R0.5\theta0°$, $R\text{-}0.5\theta0°$, $R\text{-}1\theta0°$, $R\text{-}2\theta0°$, $R0.1\theta90°$, $R0.5\theta90°$, $R\text{-}0.5\theta90°$, $R\text{-}1\theta90°$, $R\text{-}2\theta90°$ and $R10\theta90°$ for E-glass fabrics/epoxy

$R = -2$: $\theta = 90°$. The discrepancies will be examined further in the related *S-N* curves.

Figures 20 and 21 show the *S-N* curves of E-glass/polyester obtained by the NN-NARX model and the experimental data for stress ratios $R = -1$: $\theta = 90°$ and $R = 10$: $\theta = 30°$, respectively. Furthermore, Figs. 22 and 23 show the *S-N* curves of E-glass fabrics/epoxy obtained by the NN-NARX model and the experimental data

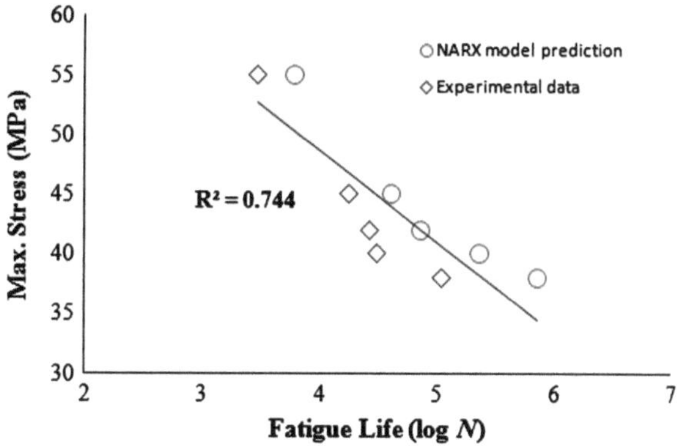

Fig. 20 *S-N* curves obtained by the NN-NARX model and the experimental data for $R = -1$: $\theta = 90°$ of E-glass/polyester

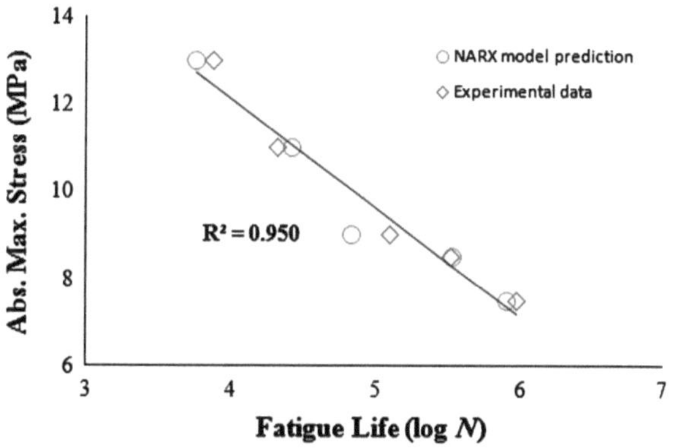

Fig. 21 *S-N* curves obtained by the NN-NARX model and the experimental data for $R = 10$: $\theta = 30°$ of E-glass/polyester

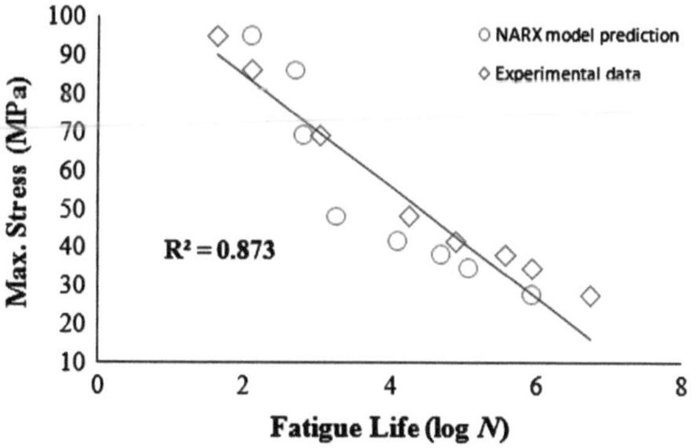

Fig. 22 *S-N* curves obtained by the NN-NARX model and the experimental data for $R = -2$: $\theta = 90°$ of E-glass fabrics/epoxy

for stress ratios $R = -2$: $\theta = 90°$ and $R = 0.5$: $\theta = 90°$, respectively. Note that the NN simulation results were selected and presented because the results respectively corresponded to the lowest and the highest of R^2 values which represent the "goodness" of the NN-NARX model in modeling fatigue lives for the problem considered.

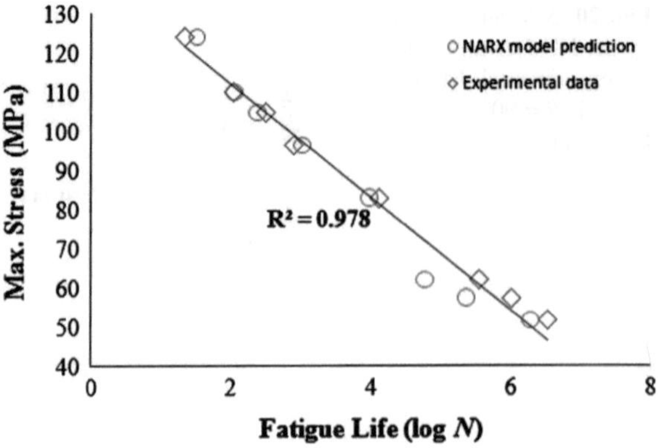

Fig. 23 *S-N* curves obtained by the NN-NARX model and the experimental data for $R = 0.5$: $\theta = 90°$ of E-glass fabrics/epoxy

In the form of *S-N* curves, it can be seen that the discrepancies observed in fatigue lives predicted by the NN-NARX model and those of experimental data were not excessively large. The results may justified that the coefficient of determination (R^2) of the fatigue life prediction can be considered high for the materials examined. The values of R^2 was ranging from 0.7445 to 0.9504 and from 0.8737 to 0.9788 for E-glass/polyester and E-glass fabrics/epoxy, respectively. In addition, it is interesting to note that the highest R^2 values for the materials were coming from the fatigue data related to the non-on-axis orientations i.e. $\theta = 30°$ and $\theta = 90°$ for E-glass/polyester and E-glass fabrics/epoxy, respectively. Furthermore, it is also obvious that the fatigue life assessment of multivariable and multiaxial loadings is fashioned in an efficient manner. This emphasized again the applicability and the feasibility of the present NN-NARX model and the procedure developed in the present study for multivariable and multiaxial fatigue life assessment of the composite materials.

For remarks on the presented fatigue life prediction results, the following discussions are highlighted. Firstly, because fatigue life assessment is realized as one-step ahead prediction with respect to each stress level-*S* corresponding to the related *R* and θ values, the accuracy of the NN prediction results hence will depend on the accuracy of the NN prediction result on each stress level-*S* examined. The obtained NN fatigue life prediction for a stress level will affect that of the next stress level. With respect to this matter, two aspects may be considered that the change in the selection of training fatigue data will also change the NN fatigue life prediction results. Also, the sequence of fatigue data of the testing sets examined can affect the NN fatigue life prediction results obtained. To be consistent, in the present study, stress ratio-*R* values were arranged according to their positions and transitions in the CLD, while on-axis/off-axis orientation-θ values were arranged based on the magnitude value from longitudinal to transverse direction, as shown in Table 2.

Secondly, only one value of θ was employed in the training set here i.e. $\theta = 0°$. It is intended here to employ the value to examine the feasibility of the present approach. Looking at the NN prediction results obtained, it appears that the NN-NARX model was able to perform the fatigue life prediction of the material fairly well using the starting point of limited training fatigue data. Moreover, different values of θ, say $\theta = 0$ and 90°, can be of course selected and employed in the training set, thus the training set could consist of, for examples, fatigue data from $R = 0.1$: $\theta = 0°$ and $R = 10$: $\theta = 90°$. Using the selection, the training data will be based upon two different values of R and θ. It becomes clear that multivariable and multiaxial aspects of fatigue life assessment are emphasized.

Thirdly, the discrepancies observed between fatigue lives predicted by the NN-NARX model and those of experimental data may be reduced by using different selection of training fatigue data. It is hoped that using different training data, NN would give better prediction results of fatigue lives which is indicated by improvement in the corresponding MSE values. Related to this, the improvement of the MSE prediction values may also be produced with respect to the variation of the hidden nodes number in a sensitivity analysis, which is however still not further considered in the present study and left as the subject for further study.

In the following section, informative bounds of NN prediction to further describe the discrepancies observed between fatigue lives predicted by the NN-NARX model and those of experimental data will be presented and discussed.

5.4 Informative Bounds of NN Prediction for Fatigue Life Assessment of Multivariable and Multiaxial Loadings with MLP-NARX Model

To better describe the produced discrepancies in fatigue lives, the informative bounds of NN prediction would be also important to be examined. With such information, the noticeable discrepancies in fatigue lives can be better described and the obtained NN fatigue life prediction will be strongly supported by comprehensive information of fatigue lives and therefore further support any subsequent product design decisions. The informative bounds of NN prediction may be regarded as the error bars for the NN prediction results. In the present study, the informative bounds of NN prediction have been computed and developed for the Levenberg-Marquardt algorithm with Bayesian regularization following the work by Nabney (2002) and MacKay (2004). The readers are also directed to (MacKay 2004) for further reference of Bayesian techniques.

Noting the objective function of NN incorporating Bayesian regularization in Eq. (10) with the parameters of weight decay α and of inverse noise variance β, the variance for Gaussian distribution is stated as (Nabney 2002):

$$\sigma^2 = \frac{1}{\beta} + \mathbf{g}^T \mathbf{A}^{-1} \mathbf{g} \tag{20}$$

where: $\mathbf{g}$ and $\mathbf{A}$ are respectively the gradient matrix and the Hessian of the error function. It is important to note here that the variance has contributions from both the output noise model ($1/\beta$) and the posterior distribution in the weights.

Here, the informative bounds of NN prediction in the fatigue life assessment of E-glass/polyester ($[90/0/\pm45/0]_S$) and E-glass/polyester ($[0/(\pm45)_2/0]_T$) as listed respectively in Tables 1 and 2 are described as the fatigue life prediction tasks represent multivariable and multiaxial fatigue life assessment.

Figures 24 and 25 depict the informative bounds of NN fatigue life prediction for these materials. It can be seen from Figs. 24 and 25 that the NN-NARX model employing the Bayesian regularization can conveniently provide the informative upper and lower bounds of prediction as the error bars for the mean lifetimes. It is interesting to note that the error bars for fatigue lives of the first testing set ($R = 0.9$ in Fig. 24 and $R0.5\theta0°$ in Fig. 25) are narrowest in comparison to the others.

The presented results clearly show the feasibility of the present spectrum fatigue life analysis as the testing stress ratio sets are moving from the tensile-tensile to the compressive-compressive sector in the CLD region, thus forming a spectrum of loading conditions. In addition, as the lifetime prediction will slide over the succession of testing stress ratio sets, the error bars will also vary accordingly based upon the one-step ahead prediction with respect to each stress level S corresponding to stress ratio values R.

It should be pointed out here that the bounds of NN prediction are also useful in describing the scatter of fatigue lives due to variability in aspects of material (material from different batches or manufacturers), specimen (production and surface quality), fatigue load (types of load, frequency or equipment), environment (controlled temperature and humidity), personal or laboratory technicians skill (Schijve 2009) as well as other arbitrary set of influential factors (Bucar et al. 2007).

Figures 26, 27 and 28 depict the fatigue lives scatter for E-glass/polyester ($[90/0/\pm45/0]_S$) from experiments (Mandell and Samborsky 2010) along with the upper

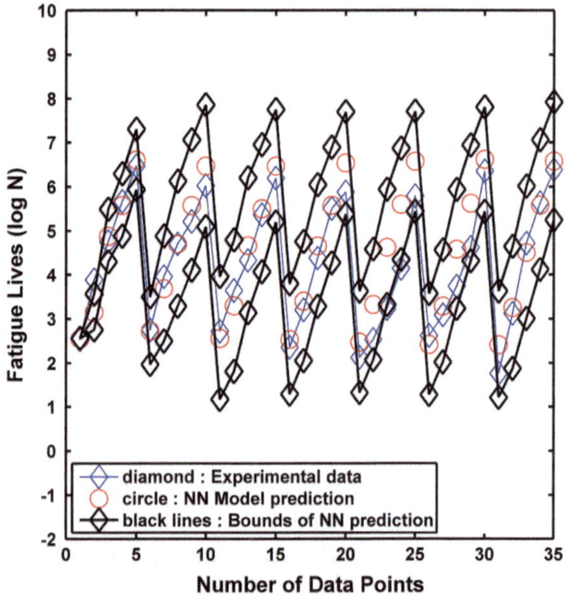

Fig. 24 Bounds of MLP-NARX prediction for multivariable fatigue life assessment of E-glass/polyester ($[90/0/\pm45/0]_S$) for the testing sets: $R = 0.9, 0.8, 0.7, 0.5, -0.5, -1$ and -2 (see Table 1)

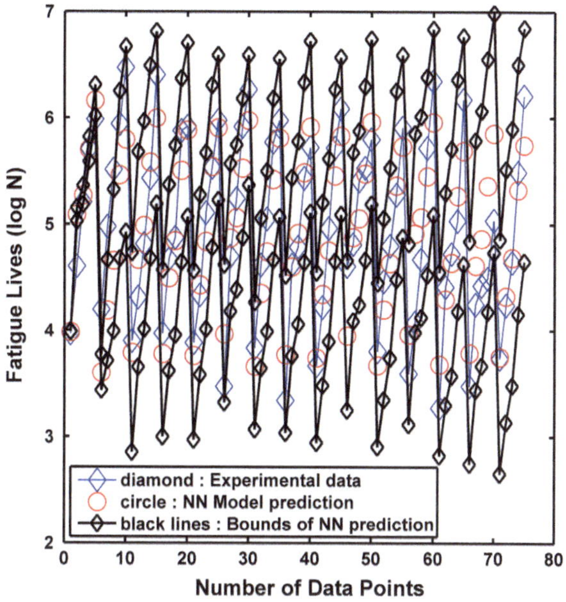

Fig. 25 Bounds of MLP-NARX prediction for multivariable and multiaxial fatigue life assessment of E-glass/polyester ($[0/(\pm45)_2/0]_T$). See the corresponding Fig. 18

and lower bounds of NN-NARX model prediction for the stress ratios of $R = 0.5$, -0.5 and -2, respectively. Note that the stress ratio values of the material were chosen as representative examples as their corresponding fatigue lives scatters are wide in comparison to those of other stress ratio values.

One can observe from Figs. 26, 27 and 28 that the upper and lower bounds of NN-NARX prediction can adequately describe the fatigue lives scatters produced from fatigue testing, while at the same time the obtained mean lifetimes from the NN-NARX model can give good prediction of the experimental data. Clearly, the results show the applicability of the present NN modeling for multivariable and multiaxial fatigue life assessment thus building spectrum fatigue analysis. Moreover, it is worth noting the clear concept of CLD which has been employed and linked to the selection of NARX structure in the modeling approach.

In addition, considering that the present NARX modeling is made up of one-step ahead prediction, it is also worth noting that the presented results give confidence level to the user in using NN with NARX structure for describing scatter of fatigue lives of composite materials. Moreover, the simulation task can be also extended for other probabilities of failures (10 and 90 % probabilities of failures). As has been stated by Bucar et al. (2007), once an NN configuration has been designed and optimized, the network can be used suitably for describing the scatter of *S-N* curves for arbitrary set of influential factors. Hence, it would be interesting to investigate the fatigue lives scatter of composite materials associated to various probabilities of failure in comparison to those of isotropic materials such as metals (Klemenc and Fajdiga 2012, 2013) using the present NN-NARX modeling framework. Such a modeling task will be the further subject of the NN-NARX modeling study.

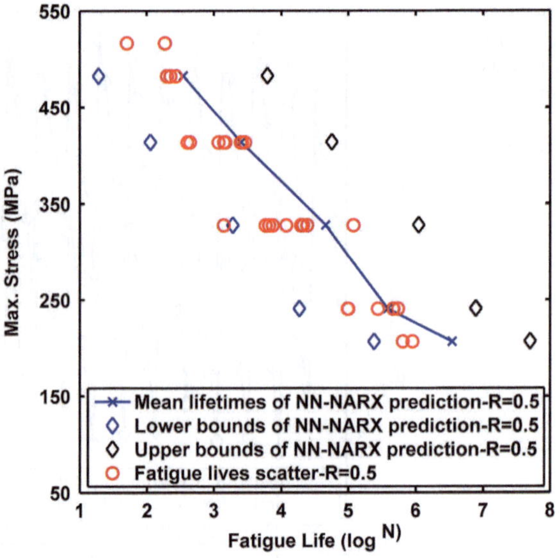

Fig. 26 Fatigue lives scatter for E-glass/polyester ($[90/0/\pm45/0]_S$) from experiments (Mandell and Samborsky 2010) along with the *upper* and *lower* bounds of NN-NARX model prediction for the stress ratio $R = 0.5$

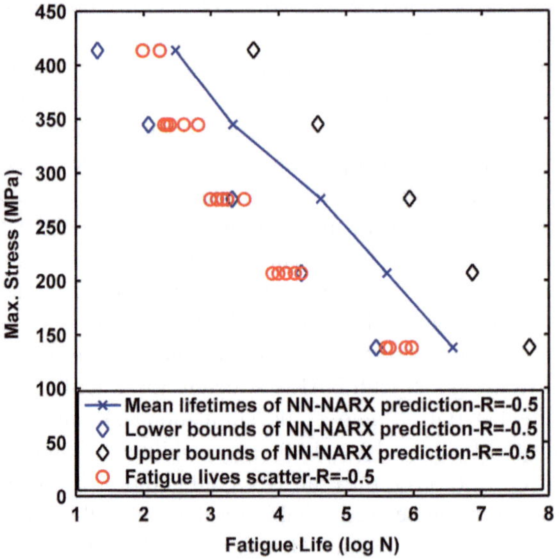

Fig. 27 Fatigue lives scatter for E-glass/polyester ($[90/0/\pm45/0]_S$) from experiments (Mandell and Samborsky 2010) along with the *upper* and *lower* bounds of NN-NARX model prediction for the stress ratio $R = -0.5$

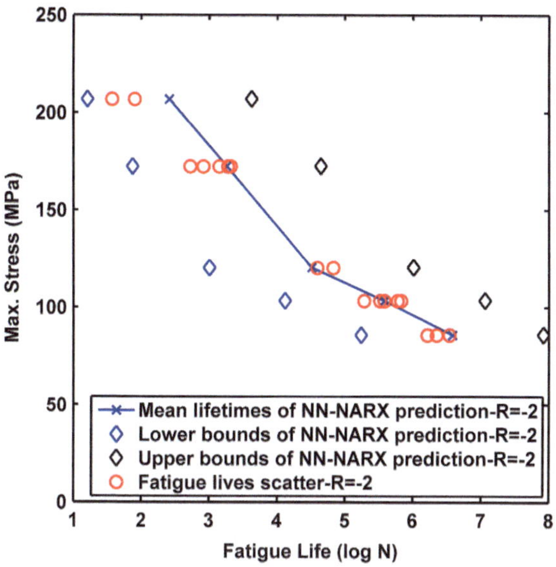

Fig. 28 Fatigue lives scatter for E-glass/polyester ($[90/0/\pm45/0]_S$) from experiments (Mandell and Samborsky 2010) along with the *upper* and *lower* bounds of NN-NARX model prediction for the stress ratio $R = -2$

For final remarks, NARX model has been employed in many fields of applications, among others, (Hung and Kao 2002; Basso et al. 2005). Now, its application has been extended in the first time to spectrum fatigue life analysis and assessment of composite materials.

6 Conclusions

Composite materials lifetime assessment using NN with NARX structure has been presented in the present study. Dynamic nature of both the CLD and the NARX model was emphasized and linked, resulting in the first application of the NARX model in spectrum fatigue analysis. Fatigue life assessment was realized as one-step ahead prediction with respect to each stress level corresponding to stress ratio values arranged in such a way that transition took place from a fatigue region to another one in the CLD region. As a result, composite materials lifetime assessment can be fashioned for a wide spectrum of loading and orientation in an efficient manner. The simulation results hence showed the applicability and the efficiency of the NN-NARX model when applied for different materials and loading situations as well as axis orientations. A framework of artificial intelligence and system identification techniques therefore has been introduced and presented as a new approach and perspective in the field of fatigue life assessment of composite materials.

In addition to the aspects of training data such as variation of training data set, sequence of fatigue data and the anticipated bounds of prediction, one may also give consideration to the initialization of network weights, where classes of evolutionary algorithms such as genetic algorithms (GA) may be employed in the NN model to obtain an optimum initialization of network weights. This can result in further optimization for the current NN models, which is also interesting for the subject of further study.

Acknowledgment The present author would like to thank the Montana State University and A.P. Vassilopoulos and T.P. Philippidis (doi:10.1016/S0142-1123(02)00003-8) for the fatigue database published through the internet. The author also would like to thank to editors and reviewers for their useful suggestions and comments that further improve the presentation of this research work.

References

Al-Assaf, Y., & El-Kadi, H. (2001). Fatigue life prediction of unidirectional glass fiber/epoxy composite laminae using neural networks. *Composites Structures, 53*(6), 65–71.

Aymerich, F., & Serra, M. (1998). Prediction of fatigue strength of composite laminates by means of neural networks. In The third seminar on experimental techniques and design in composite materials (pp. 231–240), October 30–31, 1996, Cagliari, Italy. doi:10.4028/www.scientific.net/KEM.144.231.

Azar, A. T. (2013). Fast neural network learning algorithms for medical applications. *Neural Computing and Applications, 23*(3–4), 1019–1034.

Basso, M., Giarre, L., Groppi, S., & Zappa, G. (2005). NARX models of an industrial power plant gas turbine. *IEEE Transactions on Control System Technology, 13*(4), 599–604.

Bezazi, A., Pierce, S. G., Worden, K., & Harkati, E. H. (2007). Fatigue life prediction of sandwich composite materials under flexural tests using a Bayesian trained artificial neural network. *International Journal of Fatigue, 29*(4), 738–747.

Bucar, T., Nagode, M., & Fajdiga, M. (2007). An improved neural computing method for describing the scatter of S-N curves. *International Journal of Fatigue, 29*(12), 2125–2137.

Bukkapatnam, S. T. S., & Sadananda, K. (2005). A genetic algorithm for unified approach-based predictive modeling of fatigue crack growth. *International Journal of Fatigue, 27*(10–12), 1354–1359.

Catelani, M., & Fort, A. (2000). Fault diagnosis of electronic analog circuits using a radial basis function network classifier. *Measurement, 28*(3), 147–158.

Chen, S., Billings, S. A., & Grant, P. M. (1990). Non-linear system identification using neural networks. *International Journal of Control, 51*(6), 1191–1214.

El-Kadi, H., & Al-Assaf, Y. (2002). Prediction of the fatigue life of unidirectional glass fiber/epoxy composite laminae using different neural network paradigms. *Composites Structures, 55*(1), 239–246.

Fathi, A., & Aghakouchak, A. A. (2007). Prediction of fatigue crack growth rate in welded tubular joints using neural network. *International Journal of Fatigue, 29*(2), 261–275.

Foresee, F. D., & Hagan, M. T. (1997). Gauss-Newton approximation to Bayesian learning. In The 1997 IEEE international conference on neural networks (ICNN) (pp. 1930–1935), June 9–12, 1997, Houston, TX, USA. doi:10.1109/ICNN.1997.614194.

Freire Junior, R. C. S., Neto, A. D. D., & de Aquino, E. M. F. (2005). Building of constant life diagrams of fatigue using artificial neural networks. *International Journal of Fatigue, 27*(7), 746–751.

Freire Junior, R. C. S., Neto, A. D. D., & de Aquino, E. M. F. (2007). Use of modular networks in the building of constant life diagrams. *International Journal of Fatigue, 29*(3), 389–396.

Freire Junior, R. C. S., Neto, A. D. D., & de Aquino, E. M. F. (2009). Comparative study between ANN models and conventional equations in the analysis of fatigue failure of GFRP. *International Journal of Fatigue, 31*(5), 831–839.

Haque, M. E., & Sudhakar, K. V. (2002). ANN back-propagation prediction model for fracture toughness in microalloy steel. *International Journal of Fatigue, 24*(9), 1003–1010.

Harris, B. (Ed.). (2003). *Fatigue in composites*. Cambridge: Woodhead Publishing Ltd.

Haykin, S. (2009). *Neural networks and learning machines* (3rd ed.). USA: Pearson Prentice Hall.

Hidayat, M. I. P., & Melor, P. S. (2009). Optimizing neural network prediction of composite fatigue life under variable amplitude loading using Bayesian regularization chap. 9. In S. M. Sapuan & I. M. Mujtaba (Eds.), *Composite materials technology: Neural network applications*. USA: CRC Press, Taylor and Francis LLC.

Hidayat, M. I. P., Yusoff, P. S. M. M., & Berata, W. (2011). Neural networks with NARX structure for material lifetime assessment application. In The 2011 IEEE symposium on computers and informatics (ISCI) (pp. 273–278), March 20–23, 2011, Kuala Lumpur, Malaysia. doi:10.1109/ISCI.2011.5958926.

Hidayat, M. I. P., & Berata, W. (2011). Neural networks with radial basis function and NARX structure for material lifetime assessment application. In The 12th international conference on quality in research (QiR 12) (pp. 143–150), July 4–7, 2011, Bali, Indonesia. doi:10.4028/www.scientific.net/AMR.277.143.

Hung, S. L., & Kao, C. Y. (2002). Structural damage detection using the optimal weights of the approximating artificial neural networks. *Earthquake Engineering and Structural Dynamics, 31*(2), 217–234.

Klemenc, J., & Fajdiga, M. (2012). Estimating S-N curves and their scatter using a differential ant-stigmergy algorithm. *International Journal of Fatigue, 43*(1), 90–97.

Klemenc, J., & Fajdiga, M. (2013). Joint estimation of E-N curves and their scatter using evolutionary algorithms. *International Journal of Fatigue, 56*(1), 42–53.

Lee, J. A., & Almond, D. P. (2003). A neural-network approach to fatigue life prediction chap. 21. In B. Harris (Ed.), *Fatigue in composites*. Cambridge: Woodhead Publishing Ltd.

MacKay, D. J. C. (2004). *Information theory, inference and learning algorithms*. England: Cambridge University Press.

Majidian, A., & Saidi, M. H. (2007). Comparison of fuzzy logic and neural network in life prediction of boiler tubes. *International Journal of Fatigue, 29*(3), 489–498.

Mandell, J. F., & Samborsky, D. D. (2010). DOE/MSU composite material fatigue database: Test, methods, material and analysis. SAND97-3002. Albuquerque, NM: Sandia National Laboratories.

Mollah, A. A., & Pratihar, D. K. (2008). Modeling of TIG welding and abrasive flow machining processes using radial basis function networks. *International Journal of Advanced Manufacturing Technology, 37*(9–10), 937–952.

Nabney, I. T. (2002). *NETLAB algorithms for pattern recognition*. London: Springer.

Narendra, K., & Parthasarathy, K. (1990). Identification and control of dynamic systems using neural networks. *IEEE Transactions on Neural Networks, 1*(1), 4–27.

Neural Network Toolbox™ User's Guide © COPYRIGHT 1992–2010. USA: The MathWorks, Inc.

Nocedal, J., & Wright, S. J. (2006). *Numerical optimization* (2nd ed.). New York: Springer.

Philippidis, T. P., & Passipoularidis, V. A. (2007). Residual strength after fatigue in composites: Theory vs. experiment. *International Journal of Fatigue, 29*(12), 2104–2116.

Passipoularidis, V. A., & Philippidis, T. P. (2009). A study of factors affecting life prediction of composites under spectrum loading. *International Journal of Fatigue, 31*(3), 408–417.

Passipoularidis, V. A., Philippidis, T. P., & Brondsted, P. (2011). Fatigue life prediction in composites using progressive damage modelling under block and spectrum loading. *International Journal of Fatigue, 33*(2), 132–144.

Post, N. L., Case, S. W., & Lesko, J. J. (2008). Modeling the variable amplitude fatigue of composite materials: A review and evaluation of the state of the art for spectrum loading. *International Journal of Fatigue, 30*(12), 2064–2086.

Pujol, J. C. F., & Pinto, J. M. A. (2011). A neural network approach to fatigue life prediction. *International Journal of Fatigue, 33*(3), 313–322.

Reifsnider, K. L. (Ed.). (1991). *Fatigue of composite materials*. Amsterdam: Elsevier.

Schijve, J. (Ed.). (2009). *Fatigue of structures and materials* (pp. 373–394). Netherlands: Springer.

Sendeckyj, G. P. (2001). Constant life diagrams—A historical review. *International Journal of Fatigue, 23*(4), 347–353.

Vassilopoulos, A. P., & Philippidis, T. P. (2002). Complex stress state effect on fatigue life of GRP laminates. Part I, experimental. *International Journal of Fatigue, 24*(8), 813–823.

Vassilopoulos, A. P., Georgopoulos, E. F., & Dionysopoulos, V. (2007). Artificial neural networks in spectrum fatigue life prediction of composite materials. *International Journal of Fatigue, 29* (3), 20–29.

Vassilopoulos, A. P., Georgopoulos, E. F., & Keller, T. (2008). Comparison of genetic programming with conventional methods for fatigue life modeling of FRP composite materials. *International Journal of Fatigue, 30*(9), 1634–1645.

Vassilopoulos, A. P., Manshadi, B. D., & Keller, T. (2010). Influence of the constant life diagram formulation on the fatigue life prediction of composite materials. *International Journal of Fatigue, 32*(4), 659–669.

Venkatesh, V., & Rack, H. J. (1999). A neural network approach to elevated temperature creep–fatigue life prediction. *International Journal of Fatigue, 21*(3), 225–234.

Zhang, Z., & Friedrich, K. (2003). Artificial neural networks applied to polymer composites: A review. *Composites Science and Technology, 63*(14), 2029–2044.

Measuring Software Reliability: A Trend Using Machine Learning Techniques

Nishikant Kumar and Soumya Banerjee

Abstract It has become inevitable for every software developer to understand, to follow that how and why software fails, and to express reliability in quantitative terms. This has led to a proliferation of software reliability models to estimate and predict reliability. The basic approach is to model past failure data to predict future behavior. Most of the models have three major components: assumptions, factors and a mathematical function, usually high order exponential or logarithmic used to relate factors to reliability. Software reliability models are used to forecast the curve of failure rate by statistical evidence available during testing phase. They also can indicate about the extra time required to carry out the test procedure in order to meet the specifications and deliver desired functionality with minimum number of defects. Therefore there are challenges whether, autonomous or machine learning techniques like other predictive methods could be able to forecast the reliability measures for a specific software application. This chapter contemplates reliability issue through a generic Machine Learning paradigm while referring the most common aspects of Support Vector Machine scenario. Couples of customized simulation and experimental results have been presented to support the proposed reliability measures and strategies.

Keywords Reliability measure · Software defects · Machine learning · Support vector machine · Statistical validations

N. Kumar · S. Banerjee (✉)
Department of Computer Science, Birla Institute of Technology, Mesra, Deoghar Campus, Jasidih, Deogfhar 814142, Jharkhand, India
e-mail: dr.soumya@ieee.org

N. Kumar
e-mail: Nishi27in@gmail.com

Q. Zhu and A.T. Azar (eds.), *Complex System Modelling and Control Through Intelligent Soft Computations*, Studies in Fuzziness and Soft Computing 319,
DOI 10.1007/978-3-319-12883-2_28

1 Introduction

The trend of growth of the size and complexity of software systems shows an exponential growth and will continue in the future. With the increase in requirements and demands for, and dependencies on computers, the probability of failures in software also increase which may cause serious, even fatal consequences to various systems, especially time-critical systems. As a result, decent quality of information and support systems has become a major concern for our society. Reliability is the most important aspect of quality, and software reliability is defined as the probability of failure-free software operation for a specified period of time in a specified environment (Musa 1973; Musa and Okumoto 1973; Huang and Lyu 2011). Hence, software reliability engineering (SRE) has become an interesting topic of research and is developing very rapidly. Reliability assessment methods and improvement techniques are of significance to software managers, users as well as practitioners.

In this chapter, an effort has been initiated to measure reliability of software application while incorporating machine learning techniques. The motivation of such work is primarily bi-focal: Firstly, conventional metric measures and statistical estimation could be questioned in terms of efficacy and moreover machine learning could be a suitable choice; hence search based software engineering could yield an emerging vertical as reliability measure. Primarily it will comprise of broad activities in terms measuring the reliability of the software product or application (Gupta et al. 2011).

- Estimation: Determination of current software reliability by applying statistical inference techniques to failure obtained during system test or system operation. It is measure regarding achieved reliability from past until current point.
- Prediction: Determination of future software reliability based on available software metrics and measures. There are two possible cases:
  - Failure data is available: This is applicable to testing and operation stage wherein estimation techniques can be used to parameterize and verify software reliability models to perform future reliability prediction.
  - Failure data not available: This is applicable to design and coding stage where metrics obtained from software development process and characteristics of developed product can be used to determine reliability of software upon testing or delivery.

Considering the growth model of reliability analysis (Gholizadeh et al. 2012) the recent reliability artifacts concentrate several verticals of reliability including the fuzzy Bayesian system reliability assessment also has been initiated. The process is based on prior two-parameter exponential distribution for estimating fuzzy map of reliability under squared error for any given soft-ware domain under test. Although software reliability growth model (SRGM) basically predicts the fault detection coverage in software testing phases. The general problem that is encountered is to minimize the number of remaining faults for a given fixed amount of testing effort and reliability objective. To encompass the procedure viable Poisson process namely Goel-Okumoto and Delayed S Shaped model are presented to estimate the faults (Roy 2014). Significant recent development in software reliability modelling and its applications in the form of quantitative techniques for software quality/reliability measurement and assessment are discussed in recent monographs. It is observed that, a quality engineering analysis of human factors affecting software reliability during the design-review phase, positioned as upper stream of software development. On contrast software reliability growth models based on stochastic differential equations and discrete calculus during the testing phase, the lower one could be addressed. Multivariate analysis method also has been adopted as an effective tool for investigating quality-oriented software management analysis (Yamada 2014). Even there are commercial breakthroughs where the cause of reliability of software product has been examined. Reliability of software depends not only on intrinsic factors such as its code properties, but also on extrinsic factors that is, the properties of the operational environment (Bird et al. 2014). We also introspect that recovery under crash could also enhance the reliability of system to some extent. This leads to the development of prediction model for high possible crash prone methods, which may impress reliability of software system as well as components. Predicts crash-prone methods by extracting byte code features, which includes operation codes, or op codes, in order to represent the execution sequence of instructions from the byte code of the method bodies. The intuition is that certain byte code sequences are more likely to lead to crashes than others. After this induction, a model has been developed to train a classifier while learning patterns from sequential op codes, in order to classify methods as crash-prone or non-crash-prone (i.e., less possibility to crash) (Kim et al. 2013). Even relevant learning paradigms are well traversed in recent literature to quantify fault and defect prediction of software system, that could be an estimate and assurance of reliability (Suresh et al. 2014; Kapila and Singh 2013). As reliability is generally measured by the number of faults found in the developed software hence, software fault prediction is a challenging task for researchers before the software is released. Hence, accurate fault prediction is one of the major goals so as to release software having the least possible faults. In this context, introduction of CK metrics (Chidamber and Kemerer (CK) metrics), is to assess the status for predicting faults for an open-source software product. Statistical methods such as linear regression and other relevant neural network techniques could be effectively model the reliability measurement simulation. Practically, in depth study exhibits that, The CK metric suite consists of six metrics, namely, weighted method per class (WMC), depth of inheritance tree (DIT), number of children (NOC), coupling between objects (CBO), response for class (RFC), and lack of cohesion (LCOM) (Chidamber and Kemerer 1994). The trend of implementation (Ren and Qin 2014) for measuring software defects and reliability has been strongly supported through machine learning which is significantly closer to statistical community encompassed so far. The reason to deploy machine learning process in software engineering is to combat the class imbalanced problem. It suggests that the majority of defects in a software system are located in a small percentage of the program modules. In order address this class imbalanced attribute, the data level or algorithm based measure both are reported in corresponding literature (Seiert et al. 2009). The measures suggested are emphasizing on dimension reduction issues. Improved dimension reduction methods are suggested for the class imbalanced problem by means of Partial Least Squares (PLS) (Barker and Rayens 2003), Linear Discriminant Analysis (LDA) (Xue and Titterington 2008) and Principle Component Analysis (PCA) (Jiang 2009; Ma et al. 2012).

These postulates of software engineering motivate to deploy couple of machine learning e.g. Support Vector Machine which could provide immediate advantage of statistical estimation of reliability. As the proposed method has been gradually shifted from statistical simulation to machine learning techniques, therefore certain relevant observations have been recorded. At the beginning, the relative error versus time graph obtained from SVR shows that a as more failure data from previous weeks are collected during the software development and testing process, learning is more efficient as a result of which parameters are estimated to values that make the regression function better fit the given dataset. This fosters more accurate prediction about the number of failures in succeeding week. The highlight of the chapter is to motivate the inclusion of many other relevant machine learning techniques. The remaining part of the chapter has been organized as follows: Sect. 2 lists out the different metrics those are significant while presenting the analysis, followed by the Sect. 3 with all relevant works on reliability measure. Section 4 concentrates on the data set concerning the importance for empirical and theoretical validation of software reliability and Sect. 5 elaborates the inclusion of machine learning and experimental analysis. Finally Sect. 6 gives the conclusion and discusses further scope of research.

2 Matrices to Measure Software Reliability: List of Definitions

Conventionally, the concept of reliability in terms of failure data needs to be properly measured by various means during software development and operational phases. However, software failures are always design failures. Often the system continues to be available in spite of the fact that a failure has occurred. Various metrics used for measuring software reliability are described in Table 1.

3 Relevant Recent Works: Software Reliability Measures

Various models of software reliability have been proposed, discussed, modified, and formalized mainly since 1970s, although some of them have also suffered from a lot of critiques. Software reliability analysis which was at the beginning (in sixties) based on proof of correctness had passed to a period of stochastic modeling

Table 1 Reliability metrics (Aggarwal et al. 2009b)

Metric on reliability test	Description
Probability of failure on demand (POFOD)	It is a measure of likelihood that the system will fail when a service request is made and is relevant for safety-critical and non-stop systems. POFOD = 0.001 means 1 out of 1,000 service requests result in failure
Rate of fault occurrence (ROCOF)	It is the frequency of occurrence of unexpected behaviour and is relevant for operating systems and transaction processing systems. ROCOF = 0.02 means 2 failures are likely in each 100 operational time units
Mean time to failure (MTTF)	It is a measure of the time between observed failures and is relevant for systems with long transactions e.g. CAD systems. MTTF = 500 means that the time between failures is 500 time units
Availability (AVAIL)	It is a measure of how likely the system is available for use taking into account repair/restart time. It is relevant for continuously running systems e.g. telephone switching systems. Availability = 0.998 means software is available for 998 out of 1,000 time units
Mean time to repair (MTTR)	Expected time until a system will be repaired after a failure is observed. When MTTF and MTTR of a system is measured, its availability (probability that a system is available when needed) can be determined as follows: $A_{vail} = \frac{MTTF}{MTTF+MTTR}$
Failure functions	Cumulative failure, intensity failure, failure Rate function

of the failure process and statistical analysis of failure data during the seventies. The major works during the recent years (2012–2013) could be taken as road map and has been given below: Wider varieties of approaches using discretized nonhomogeneous Poisson process (NHPP) model have been found in measuring the direct possibility of reliability measure. The conceptual bootstrapping method is known as one of the useful Monte Carlo methods for obtaining probability distributions for estimators by a re-sampling method (Inoue 2013). Continuous efforts have been solicited with Particle Swarm Optimization and Support Vector Machine (PSO-SVM) model and the characteristics of software reliability prediction have been significantly improved over statistical measures (Xiaonan et al. 2013). The aim of another approach is to introduce a Fuzzy random field environment (FRFE) reliability model that covers both the testing and operating phases in the development cycle. The proposed model is based on Weibull distribution (Garmabaki et al. 2013). Pietrantuono et al. (2010) developed an architecture based approach for software reliability and testing time allocation (Pietrantuono et al. 2010). Similarly, software test prioritization, fixing of bug and defect analysis has been well envisaged by software engineering industrial community and several important quality methods are evolved from widow operating system point of view (Murphy-Hill et al. 2013; Czerwonka et al. 2011; Wohlin 2013). Certain updated work provides an overview of Software Reliability measurement and improvement policies then examines different improvement policies for software reliability, however, they also declared that there is no single model that is universal to all the situations (Toor and Bahl 2013). With another approach of statistical estimation problems which consider the probability distributions of estimators of system failure rate, optimal checkpoint interval and its associated minimum expected system cost (Tokumoto et al. 2012). Interested readers are also suggested to check back different appropriate contribution of support vector machine approach already envisaged in software reliability verification (Tian and Noore 2005; Pai and Hong 2006; Yang and Li 2007; Lo 2010). After introspecting the statistical and machine learning approach, present work provides a sample error normalization strategy as to quantify minimum rate of error under each application. This also assures better reliability estimates for the specific software components.

4 Data Descriptions

There are two basic types of input data used in most Software Reliability Growth Models (SRGMs): time-domain data and interval-domain data. The time-domain data is obtained by recording the separate times when the failure occurred like Mean Time To Failure (MTTF), Mean Time Between Failures (MTBF), etc. The interval-domain data is obtained by counting the cumulative number of failures that occurred over a given period. This project makes use of interval-domain data set which gives the billing solution (mediation system) for a telecom service provider company of Hyderabad, India, where the software has failed to calculate correct billing in beta release discovered during weekly progress evaluation in the year 2009–2010. The dataset contains numbers of cumulative software faults per week given for 28 weeks and a total of 234 failures were observed in the whole time period (Table 2).

Table 2 Sample random data set

Week	Failures	Week	Failures
1	3	15	7
2	3	16	0
3	38	17	2
4	19	18	3
5	12	19	2
6	13	20	5
7	26	21	2
8	32	22	3
9	8	23	4
10	8	24	1
11	11	25	2
12	14	26	1
13	7	27	0
14	7	28	1

4.1 Experimental Setting in Statistical Method

Prior to migrate towards the Machine Learning approach, a brief estimation of statistical parameter of reliability measure could be accumulated. Hence, both NHPP and Newton Raphson model is envisaged.

4.1.1 Parameter Estimation

The software failure process is modeled by NHPP model with mean value function given as: $\mu(t) = a(1 - e^{-bt})$ a and b other parameters to be estimated from the failure data (the parameter is interpreted as the number of initial faults in the software and the parameter b is the fault detection rate which is related to the reliability growth rate in the testing process.

Once the model has been constructed, its parameters a and b can be estimated using maximum-likelihood estimation technique. The parameter b can be determined numerically using a technique known as Newton-Raphson. Newton-Raphson method is one of the most popularly used numerical techniques for solving complex non-linear equations which has good convergence. It is a method for finding successively better linear approximations to the roots (or zeroes) of a real-valued function. The idea of the method is as follows: one starts with an initial guess which is reasonably close to the true root, then the function is approximated by its tangent line (which can be computed using the tools of calculus), and one computes the x-intercept of this tangent line (which is easily done with elementary algebra). This x-intercept will typically be a better approximation to the function's root than the original guess, and the method can be iterated. Given a function f over the real values of x and its derivative f, the method begins with a first guess x0 for a root of the function f. Then the better approximation of the root is calculated as follows:

$$x_1 = x_0 - \frac{f(x_o)}{f'(x_o)} \tag{1}$$

The process is repeated as:

$$x_{n+1} = x_n - \frac{f(n)}{f'(x_n)} \tag{2}$$

until a sufficiently accurate value is reached.

4.2 Machine Learning: Prologue on Support Vector Regression

Support Vector Machines (SVM), originally used for classification purposes, can also be applied to regression problems by introducing an alternative loss function as already stated in the previous section. Support Vector Regression (SVR) maps the input data x into a higher dimensional feature space F by nonlinear mapping and then a linear regression problem is obtained and solved in this feature space. The goal is to find a function f(x) that has a maximum deviation ϵ from the actually obtained output y for all the training data, that is, all errors less than are accepted but not more than that. Given a set of N training data $(x_i, y_i)|x_i \in R^n, y_i \in$ R, i $= 1, 2, \ldots,$ N, where x_i denotes the input vector of dimension n, y_i is the corresponding target value, and n is the total number of data patterns. The linear regression function is:

$$f(x) = \langle w, x \rangle + b, \quad w \in R^n \text{ and } b \in R. \tag{3}$$

Here, w is the weight vector and b is the bias term. To estimate the value of w and b for the selection of the best hype plane, we need to minimize the following regularized risk function:

$$R = \frac{1}{2}||w||^2 + c\sum_{i=1}^{n} L_\epsilon(y_i, f(x_i)) \tag{4}$$

where, the first term is the regularized term which represents the ability of prediction for regression, and the second term is the empirical error or risk, wherein the constant C > 0 determines the trade-off between the training errors and the model complexity. The ϵ-loss insensitive function, present in the second term of the risk function, is defined as:

$$L_\epsilon(y, f(x)) = \begin{cases} 0, & |y - f(x)| \leq a \\ y - f(x) - \epsilon, & |y - f(x)| > a \end{cases}$$

This function gives the loss incurred by predicting f(x) instead of y. Now, we introduce two slack variables i and i* in the above regression estimation problem to transform it into an equivalent constrained optimization problem. The loss function and the slack variables allow the presence of noisy data; here noisy data refers to those data points which lie outside the ϵ-tube. If the observed point is above the Îµ tube, i is the positive difference between the observed value and Îµ, and if the observed point is below the ϵ-tube, ϵ_i^* is the negative difference between the observed value and ϵ. Hence, the constrained optimization problem formed amounts to minimizing the following equation:

$$R = \frac{1}{2}||w||^2 + c\sum_{i=1}^{n}(\varepsilon_i + \varepsilon_i^*) \quad (6)$$

subject to the following requirements:

$$y_i - f(x_i) \le \epsilon + \varepsilon_i$$
$$f(x_i) - y_i \le \epsilon + \varepsilon_i^*$$
$$\varepsilon_i \ge 0,\ \varepsilon_{i^*} \ge 0, \quad \mathrm{i} = 1, 2, \ldots, \mathrm{N}$$

4.3 Experimental Setting for Generic Machine Learning Strategy of Testing Reliability

Algorithm 4.1 The Basic ML Algorithm line up

Input : Parameterized policy of test domain with initial test parameters $\theta = \theta_o$ and we evaluate the derivative of the deviation of scope of test and actual parameter as $\nabla\theta log\pi$
Set parameters for different time steps, error and deviation
for $t = 0, 1, 2, 3, \ldots$ **do**
input the sample with domain scopeπ and set already visited bug points to a new time step a_t.
observe the deviation of sample with error
Update basis function as transpose of new test domain after deviation as $[\phi]^T$ and thus $\nabla\theta log\pi$ becomes $[\nabla\theta log\pi]^T$
Compute natural gradient and update all time stamps of sample data
Modify policy of test parameters if any.
end for

4.3.1 Data Preprocessing

The real world databases are highly susceptible to noisy and missing data. So various preprocessing techniques can be used to improve the quality of data and thereby improve the prediction results. Data cleaning can be used to fill in the missing values. Data transformation can be used to improve the accuracy, speed and efficiency of the algorithms used. Here the data is normalized using the Z-score normalization where the values of an attribute, A, are normalized based on the mean $(\bar{v})$ and standard deviation (σ_a) of the attribute. The normalized value v of v can be obtained as:

$$v' = (v - \bar{v})/\sigma_a \quad (8)$$

4.3.2 Input and Output

The given data set is divided into two sets: training set and testing set. The training set is used to build the model and the testing set is used to evaluate the model. The training data set is fed into the SVR model and the parameters that lead to the best accuracies are selected. We use the recent k data elements seen so far to assess the software reliability. The training model can reflect the mapping of input and output of this process by learning a set of training data pairs (x_i, x_{i+1}) where the observed data is within the sliding window of size k. The input $(x_{i-k}, x_{i-k+1}, \ldots, x_{i-1})$ is fed into the SVR model and the corresponding target value is $(x_{i-k+1}, x_{i-k+2}, \ldots, x_i)$. After the training process, the SVR model has learnt the inherent correspondence of the software failure process between these two vectors. Therefore, on giving an input value x_i, the predicted value of x_{i+1} can be obtained. When each new data element is arrived, the training and prediction processes of SVR model are performed alternately. For example, if the (i + 1)th data element is arrived, the model could be trained again with new input vector $(x_{i-k+1}, x_{i-k+2}, \ldots, x_i)$ and target vector $(x_{i-k+2}, x_{i-k+3}, \ldots, x_{i+1})$ and then the trained model can be used to predict the value of x_{i+2}. In this approach, all available failure data are not used, instead only the data elements in the sliding window are used. This is because the early failure behavior may have less impact on the later failure process (Figs. 1 and 2).

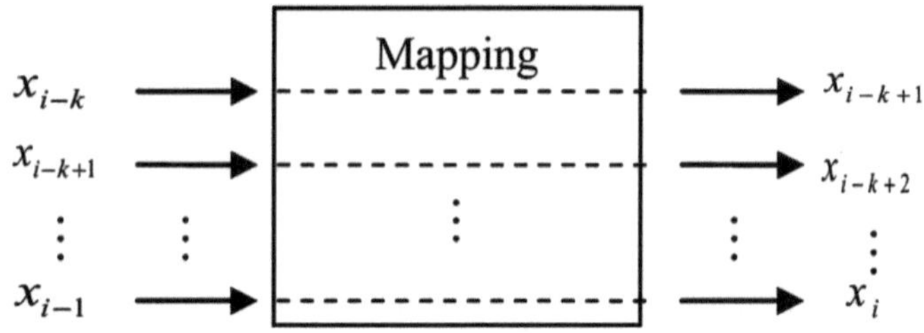

Fig. 1 SVR training process

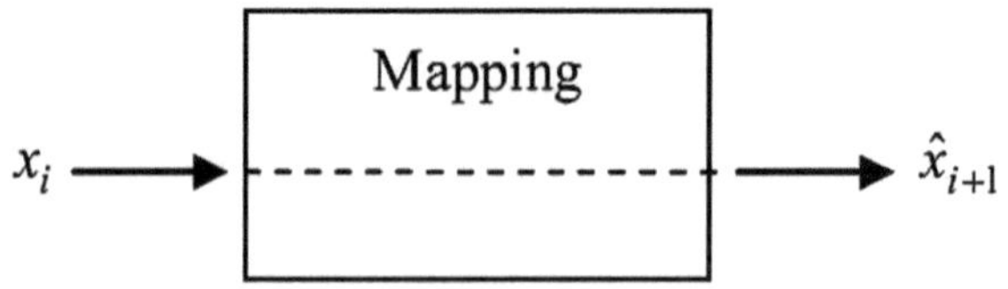

Fig. 2 SVR prediction process

4.3.3 Parameter Calculation

As there are no general rules to identify the free parameters σ^2, C, and ϵ, the optimum values are determined by the approach as follows. First, the values of any two parameters are fixed in any constants, and then adjusting the value of the third parameter until minimum forecasting error occurs. Second, set the value of the parameter determined in previous step; the value of another parameter is also fixed, and similarly, the value of the third parameter is adjusted until minimum forecasting error appears. Third, repeat both the previous steps until these three parameters have been identified.

4.3.4 Reliability Measurement

Two commonly used measures of software reliability are goodness-of-fit and next-step predictability. The goodness-of-fit is determined by fitting a curve corresponding to the proposed approach to all the data points in the training data set. The deviation between the observed and the fitted values of the number of cumulative failures per week is then evaluated. The next step predictability is determined by feeding the unknown data set to the training model. The input values $x_{i-k}, x_{i-k+1}, \ldots, x_i$ are used to predict the value of x_{i+1} where k denotes the number of input nodes considered. Then the predicted and actual values of the cumulative number of failures per week are compared. Relative error (RE) is used to represent the results of the above mentioned measures. The relative error (RE) is defined as

$$RE = |\frac{x_{i'} - x_i}{x_i}| \tag{9}$$

where x_i' denotes the predicted value of the number of cumulative failures per week, and x_i denotes the actual value of the number of cumulative failures per week.

5 Results and Analysis

After accomplishing the algorithm is previous section, we obtain various test simulation results by using MATLAB, [1]Sparse Modelling Software; with R package (refer Appendix).

[1] http://spams-devel.gforge.inria.fr/.

5.1 Results of NHPP Modeling Using MLE

Goel-Okumoto model, which is a non-homogeneous Poisson process model, was implemented on the failure count dataset available from telecom service provider company because it is considered to be most suitable for handling billing and other inventory related data. The parameters a and b were estimated using Maximum Likelihood Estimator (MLE) method. The non-linear equation for parameter b was solved using a widely used numerical root-solving technique namely Newton-Raphson. Initial approximation for b given in this method was a small value like 0.1 or 0.2. The resulting value of b was then used to compute parameter a. Parameters were estimated for each week based on failure data of previous weeks and subsequently reliability was calculated from the values of the parameters. A graph was plotted between the reliability function and time to verify the concave nature of the GO model. Graphs were also plotted for a versus time and b versus time which showed the characteristic curves for total number of faults at end of each week resulting from the testing process and failure occurrence rate respectively. We observe:

- Maximum Likelihood Estimation technique results in instable and non-existent values for parameter a and b for certain weeks due to which reliability function values are incorrect for those weeks.
- Erroneous portions in the concave graph of reliability versus time.
- Erroneous portions in graphs for a versus time and b versus *time.*[2]

Simulation plots presented here take 11th week as their starting week. The starting value is user defined for more flexibility.

Figure 3 shows how the graphs for weeks 11 and 12 have incorrect reliability values due to NaN values of parameter b. This b value when substituted into the equation for r(t|s) made the power of exponential term 0 finally making value of reliability zero. For rest of the weeks, parameter values being stable resulted in correct concave graph which is shown without errors in graph of Fig. 4.

Parameter a represents the final number of faults that can be detected during testing. It increases as certain numbers of failures are detected in each succeeding week and gets added up to the total count. This is shown correctly for weeks 13–28 as depicted in graph of Fig. 5. However, MLE results in incorrect estimation of values of a for week 11 and 12 which produce erroneous graph of Fig. 6. Graph in Fig. 6 excludes the values of a for weeks 11 and 12 to give correct characteristics.

Figure 7 demonstrates graph for parameter b which denotes the failure occurrence rate, versus time. The graph has neglected the NaN values for week 11 and 12. Comparing it with the software reliability revised bathtub curve, it can be found that it resembles the useful life portion of the graph. As upgrades are made, failure

[2] Graphs presented here take 11th week as their starting week. The starting value can be user defined for more flexibility.

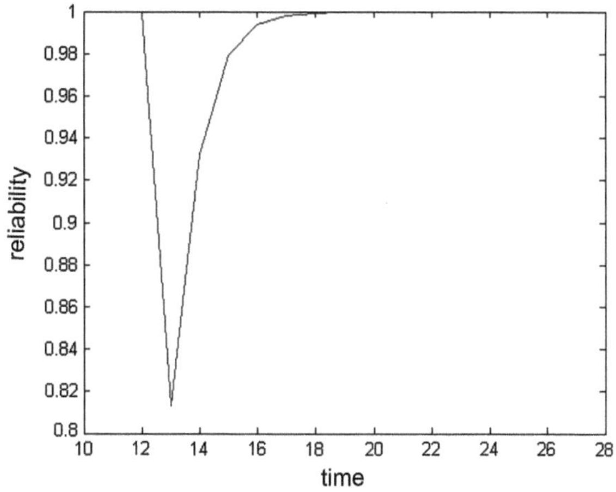

Fig. 3 Reliability versus time showing error

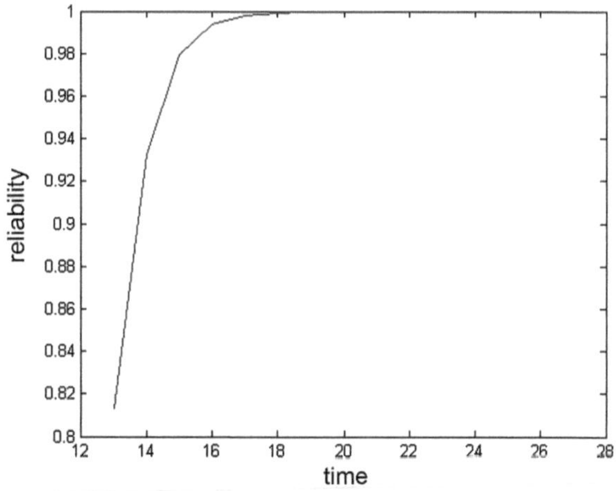

Fig. 4 Reliability versus time without error

occurrence rate drastically increases, following which the failure rate decreases gradually as failures are detected and fixed. Each upward slope of the graph represents an upgrade.

5.2 Results of Support Vector Algorithm

As already mentioned in Sect. 3, an interval-domain data set representing the cumulative number of failure per week is used here for the purpose of software reliability analysis. Support Vector Regression is the machine learning technique used for the same which, given the number of failures that occur in the current week, predicts the expected number of failures for the next week. The training data

Fig. 5 a versus t with error

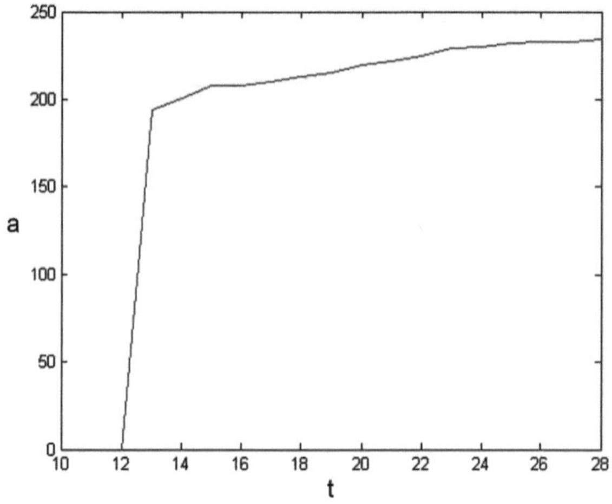

Fig. 6 a versus t without error

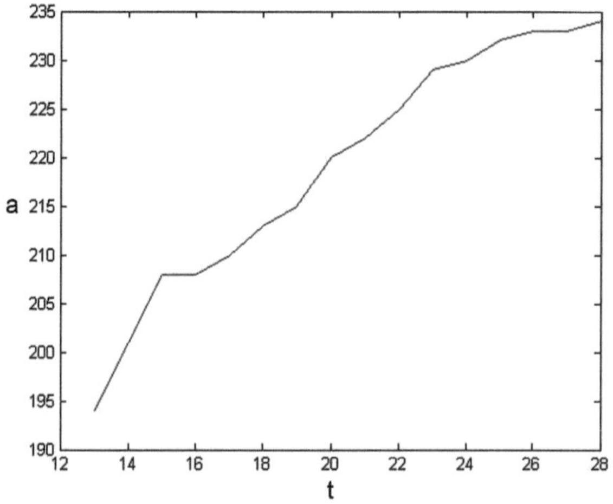

set is taken as three quarters of the given data set and the remaining one quarter is used for the testing purpose. In the training process, the whole data set is normalized using Z-score normalization and the standard deviation of the normalized values is taken as the standard deviation, σ, for the Gaussian kernel function which is used for the mapping of the input data elements into a higher-dimensional space, and the value obtained is 1.0545. The value of C is then tuned by comparing the predicted and observed values of the failure data of each week in the training data set such that minimum loss occurs. The same function is used to choose appropriate values for the Lagrange multipliers, i, i^*, used to solve the optimization problem, which are then used for the rest of the training process. Using these values, the value of the bias term, b, is tuned in a similar way as that of C, that is, by comparing the predicted and observed values of the data elements in the training data set so

Fig. 7 b versus t

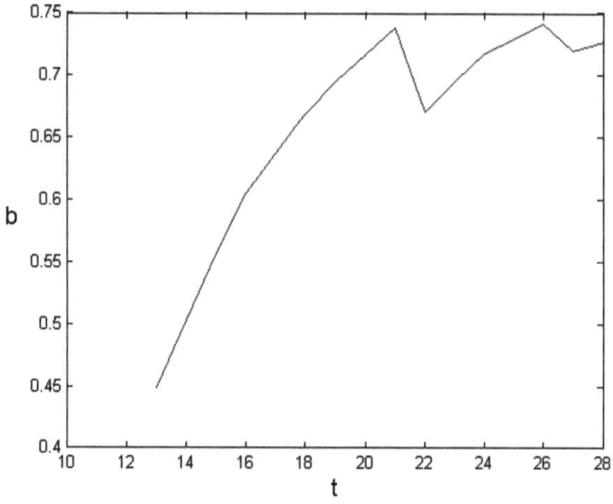

that minimum error is incurred. The estimated and tuned values of all the parameters are then used in the testing process to predict the failure for the next week given the failure in the current week, and the relative error for the predicted output is computed.

The graph in Fig. 7 shows the variation of error of prediction with time and is generated for the testing portion of the data set (from 22nd to 28th week) using the parameter values obtained in the training process. The decreasing amplitude of this graph implies that with the availability of more data, the heuristics become stronger and the prediction becomes more accurate giving lesser error. Figure 8 gives the graph of relative error versus time for 11th–28th week, and this also shows decreasing amplitude, validating the fact as mentioned above. However, steep fluctuations are visible in the graph owing to the fact that random values are chosen for Lagrange multipliers (Fig. 9).

5.3 Discussion

Software reliability analysis has been arranged using two methods; one is a statistical method while the other is based on machine learning. Above results clearly reveal that support vector regression overcomes the limitation of NHPP modeling. The parameters in SVR can be appropriately tuned through learning over the data set itself therefore giving no space to non-existent or unstable parameters. This is evident from the fact that 11th and 12th week GO model parameters could not be estimated by MLE but using SVR, learning parameters could be appropriately estimated for these weeks. The plot produced below is concentrating on the sample error normalization strategy as to quantify minimum rate of error under each

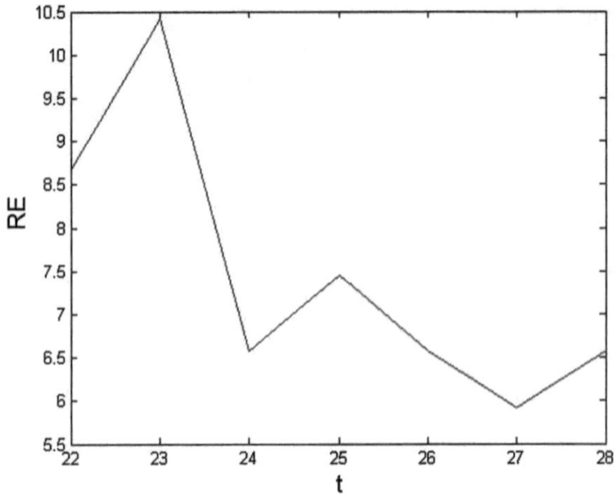

Fig. 8 RE versus t (1st phase version)

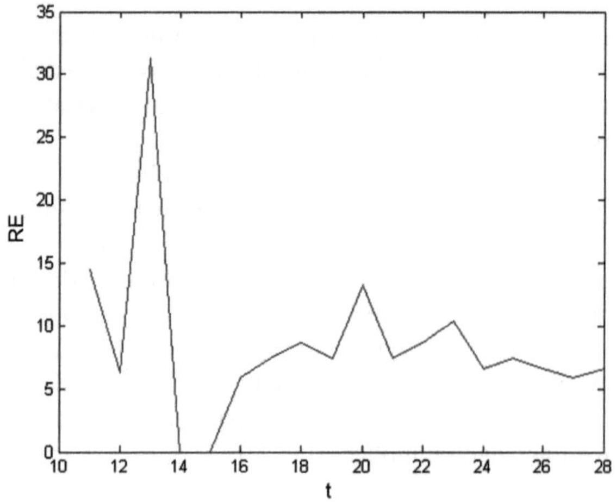

Fig. 9 RE versus t (2nd phase version with time estimation as variable)

application. Sample normalization has been denoted with different color representation. It has been observed that there is few overlapping status over test iterations of 12 weeks. The generic plot for the code is given Appendix (Fig. 10).

5.3.1 Comparative Approaches of ML: SVM Classification Plot Versus Regression Based ML

Encountering the contemporary methodologies of ML, for reliability measures of software components, it has been emphasized that the classification of faults or defects can be expressed as general multivariate logistic regression formula, which is as follows (Aggarwal et al. 2009a):

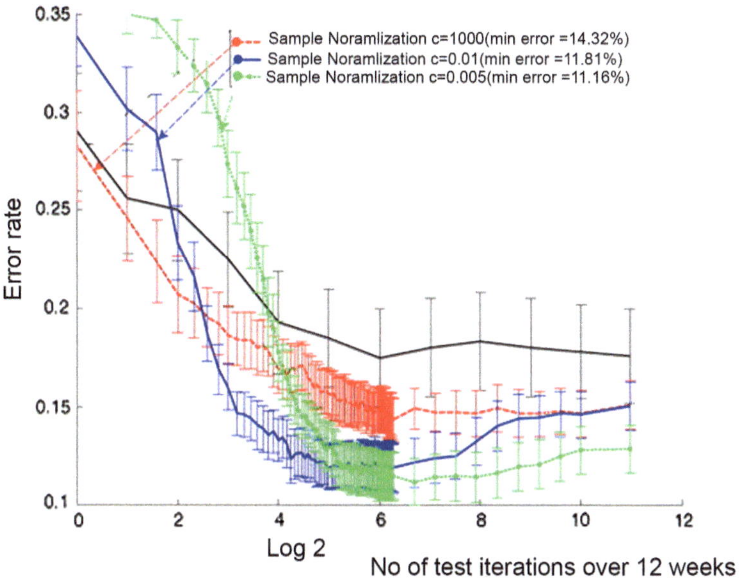

Fig. 10 Measuring SVM plot with parametric values

$$Prob(X_1, X_2, \ldots, X_n) = \frac{e^{g(x)}}{1 + e^{g(x)}} \tag{10}$$

where $g(x) = B_0 + B_1 \times X_1 + B_2 \times X_2 + \cdots + B_n \times X_n$ is the probability of a class being faulty, Xi, $(1 \leq i \leq n)$ are independent variables. Considering this validation, different parametric choices are made:

1. Sensitivity (correctness of the predicted model) Specificity (It is defined as the ratio of number of classes (including faulty and non-faulty) that are predicted correctly to the total number of classes.)
2. Precision
3. Area under the curve
4. Cut Off points

These parametric choices open up extensive following elaborated statistical values derived from the mentioned bill data defects: From the actual state of the art of billing software under study from 11th to 12th week. This system consists of 452 classes. Out of 452 classes, there are 291 faulty classes containing 513 numbers of faults. It can be seen from that 71.53 % of classes contain 1 fault, 15.3 % of classes contain 2 faults and so on. Hence the validation under statistical measure could be

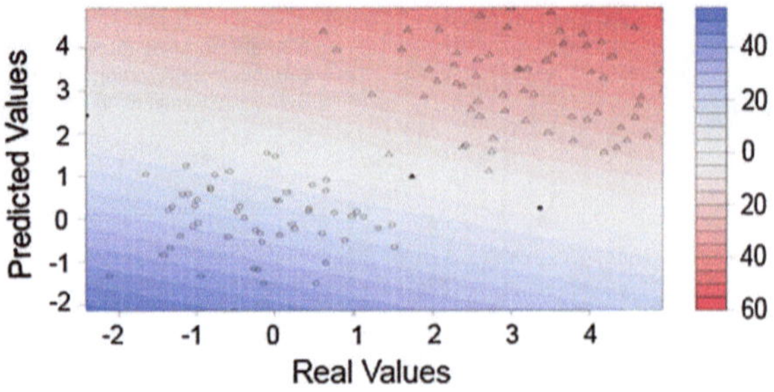

Fig. 11 SVM classification plot with real values

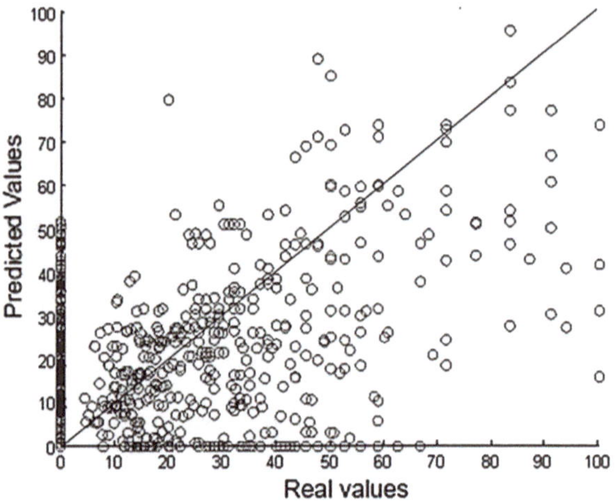

Fig. 12 Regression plot based decision

initial step to foster further extension under different software *metrics*[3] (refer Fig. 12). Further, for both SVM real values and regression based real values, the regression based ML slightly out performs in terms of precision and therefore can be further chosen as standard model. For this present study data set was restricted and thus given python code can only be extended towards few metrics mentioned instead testing all the other statistical measures suggested (Figs. 11 and 13).

The entire simulation and validation is performed in restricted environment under open source *NetWorkX*[4] environment supported by Python Language data structures.

---

[3] LOC: Lines of Code (not a C&K metric), DAM: Data Access Metric (QMOOD metric suite) MOA: Measure of Aggregation (QMOOD metric suite), MFA: Measure of Functional Abstraction (QMOOD metric suite), CAM: Cohesion Among Methods of Class (QMOOD metric suite), IC: Inheritance Coupling (quality oriented extension for the C&K metric suite), CBM: Coupling Between Methods (quality oriented extension for the C&K metric suite), AMC: Average Method Complexity (quality oriented extension to C&K metric suite), CC: McCabe's Cyclomatic Complexity.

[4] https://networkx.github.io/.

Metric	Mean	Std. Error of Mean	Median	Std. Deviation	Minimum	Maximum	Percentiles		
							25	50	75
WMC	13.501	0.698	10	14.677	0	134	5	10	16
DIT	1.869	0.040	2	0.850	1	6	1	2	2
NOC	0.738	0.331	0	6.963	0	134	0	0	0
CBO	10.120	0.932	6	19.585	0	214	4.75	6	9
RFC	30.351	1.763	21	37.067	0	390	13	21	36.25
LCOM	100.464	21.017	22	441.849	0	7059	1	22	53.25
CA	5.233	0.838	2	17.620	0	212	1	2	4
CE	5.224	0.431	4	9.059	0	133	2	4	6
NPM	11.600	0.606	9	12.747	0	101	4	9	14
LCOM3	0.999	0.025	0.85	0.534	0	2	0.749	0.85	1.129
LOC	292.595	30.046	124.5	631.675	0	9886	59.75	124.5	321.25
DAM	0.459	0.019	0.5	0.404	0	1	0	0.5	0.889
MOA	0.814	0.121	0	2.551	0	34	0	0	1
MFA	0.358	0.015	0.361	0.318	0	1	0	0.361	0.572
CAM	0.376	0.010	0.311	0.208	0	1	0.253	0.311	0.467
IC	0.577	0.026	1	0.555	0	3	0	1	1
CBM	1.952	0.116	1	2.439	0	20	0	1	4
AMC	19.362	1.880	12.192	39.516	0	616.375	6.375	12.192	20.544
MAX_CC	3.704	0.367	2	7.713	0	126	1	2	3
AVG_CC	1.188	0.052	0.976	1.090	0	17.125	0.814	0.975	1.289

Fig. 13 Emperical data set emperical

6 Conclusion

Machine learning has become phenomenal over statistical estimation techniques by transforming various industrial and scientific fields for over the last decade. Equipped with the capability to handle large, complex and categorical data; cope with non-existent or inconsistent data values; and reason on the basis of previous data, machine learning has offered tremendously improved results. This project shows how support vector regression, a machine learning based method when applied to software reliability domain gives better estimation and prediction in comparison to its statistical counterpart, Goel-Okumoto modeling. However, performance of SVR is limited by the heuristic basis used in parameter estimation. The range of values used for tuning the learning parameters must be carefully chosen for desired accuracy. Fully automating this aspect is still an associated challenge which needs to be handled in future. Another area of future study concerns the computation of Lagrange Multipliers for support vectors. As their values can lie anywhere over the range [0, C], it becomes difficult to determine their exact values. Hence, instead of choosing random values, appropriate heuristic function must be designed for this purpose to make the method more deterministic.

Appendix: Sample Snap Code for Machine Learning Classifier Used by Software Fault Data Set of Billing System

```
## A Test Classifier
x < – rnorm(100, mean = 5)
probplot(x)
## the same with horizontal tickmarks at the y-axis
opar < – par(“las”)
par(las = 1)
probplot(x)
## this should show the lack of fit at the tails
probplot(x, “qunif”)
## for increasing degrees of freedom the t-distribution converges to rbridge
## normal
probplot(x, qt, df = 1) probplot(x, qt, df = 3) probplot(x, qt, df = 10)
probplot(x, qt, df = 100) ## manually add the line through the quartiles
p < – probplot(x, line = FALSE)
lines(p, col = “green”, lty = 2, lwd = 2)
## Make the line at prob = 0.5 red
lines(p, h = 0.5, col = “red”)
### The following use the estimated distribution given by the green
### line:
## What is the probability that x is smaller than 7?
lines(p, v = 7, bend = TRUE, col = “blue”)
## Median and 90lines(p, h = 0.5, col = “red”, lwd = 3, bend = TRUE)
lines(p, h = c(0.05, 0.95), col = “red”, lwd = 2, lty = 3, bend = TRUE)
par(opar)
attach(Sample data Set)
## classification mode
# default with factor response:
model < – svm(Species ~., data = iris)
# alternatively the traditional interface:
x < – subset(iris, select = −Species)
y < – Species
model < – svm(x, y)
print(model)
summary(model)
# test with train data
pred < – predict(model, x)
# (same as:)
pred < – fitted(model)
```

```
# Check accuracy:
table(pred, y)
# compute decision values and probabilities:
pred < – predict(model, x, decision.values = TRUE)
attr(pred, "decision.values")[1:4,]
# visualize (classes by color, SV by crosses):
plot(cmdscale(dist(iris,[−5])),
col = as.integer(iris,[5]),
pch = c("o", " + ")[1:150 tune 53
## try regression mode on two dimensions
# create data
x < – seq (0.1, 5, by = 0.05)
y < – log(x) + rnorm(x, sd = 0.2)
# estimate model and predict input values
m < – svm(x, y)
new < –predict(m, x)
# visualize
plot(x, y)
points(x, log(x), col = 2)
points(x, new, col = 4)
## density-estimation
# create 2-dim. normal with rho = 0:
X < –data.frame(a = rnorm(1000), b = rnorm(1000))
attach(X)
# traditional way:
m < – svm(X, gamma = 0.1)
# formula interface:
m < – svm( ~ ., data = X, gamma = 0.1)
# or:
m < – svm( ~ a + b, gamma = 0.1)
# test:
newdata < – data.frame(a = c(0, 4), b = c(0, 4))
predict (m, newdata)
# visualize:
plot(X, col = 1:1000 points(newdata, pch = " + ", col = 2, cex = 5)
# weights: (example not particularly sensible)
i2 < – iris
levels(i2$Species)[3] < – "versicolor"
summary(i2$Species)
wts < – 100 /table(i2$Species)
wts
m < – svm(Species ~ ., data = i2, class.weights = wts)
```

References

Aggarwal, K., Singh, Y., Kaur, A., & Malhotra, R. (2009a). Empirical analysis for investigating the effect of object-oriented metrics on fault proneness: A replicated case study. *Software Process: Improvement and Practice, 16*(1), 39–62.

Aggarwal, K., Singh, Y., Kaur, A., & Malhotra, R. (2009b). Empirical analysis for investigating the effect of object-oriented metrics on fault proneness: A replicated case study. *Software Process: Improvement and Practice, 16*(1), 39–62.

Barker, M., & Rayens, W. (2003). Partial least squares for discrimination. *Journal of Chemometrics, 17*(3), 166–173.

Bird, C., Ranganath, V.-P., Zimmermann, T., Nagappan, N., & Zeller, A. (2014s). Extrinsic influence factors in software reliability: A study of 200,000 windows machines. In *Proceedings of the 36th International Conference on Software Engineering (ICSE 2014)*. Hyderabad, India: ACM.

Chidamber, S. R., & Kemerer, C. F. (1994). Metrics suite for object oriented design. *IEEE Transactions on Software Engineering, 20*(6), 476–493.

Czerwonka, J., Das, R., Nagappan, N., Tarvo, A., & Teterev, A. (2011). Crane: Failure prediction, change analysis and test prioritization in practice—experiences from windows. In *Proceedings of the 2011 Fourth IEEE International Conference on Software Testing, Verification and Validation* (pp 357–366).

Garmabaki, A. H. S., Ahmadi, A., Kapur, P. K., & Kumar, U. (2013). Predicting software reliability in a fuzzy field environment. *International Journal of Reliability, Quality and Safety Engineering, 20*(3), 1–13.

Gholizadeh, R., Shirazi, A. M., & Gildeh, B. S. (2012). Fuzzy bayesian system reliability assessment based on prior two-parameter exponential distribution under different loss functions. *Journal of Software Testing, Verification And Reliability, 22*(3), 203–217.

Gupta, A., Choudhary, D., & Saxena, S. (2011). Software reliability estimation using yamada delayed s shaped model under imperfect debugging and time lag. *International Journal of Computer Applications, 23*(7), 0975–8887.

Inoue, S., & Yamada, S. (2013). A bootstrapping approach for software reliability measurement based on a discretized NHPP model. *Journal of Software Engineering and Applications, 6*, 1–7. doi:10.4236/jsea.2013.64A001.

Huang, C.-Y., & Lyu, M. R. (2011). Estimation and analysis of some generalized multiple change-point software reliability models. *IEEE Transactions on Reliability, 60*(2), 498–515.

Jiang, X. (2009). Asymmetric principle component and discriminant analyses for pattern recognition. *IEEE Transactions on Pattern Analysis and Machine Intelligence, 31*(5), 931–937.

Kapila, H., & Singh, S. (2013). Analysis of ck metrics to predict software fault-proneness using bayesian inference. *International Journal of Computer Applications, 74*(2), 1–4.

Kim, S., Zimmermann, T., Bettenburg, N., Premraj, R., & Shivaji., S. (2013). Predicting method crashes with bytecode operations. In *6th India Software Engineering Conference, ISEC'13,* (pp. 3–12), February 21–23, 2013, New Delhi, India: ACM Publication.

Lo, J.-H. (2010). Predicting software reliability with support vector machines. In *The 2010 IEEE Second International Conference on Computer Research and Development* (pp. 765–769).

Ma, Y., Luo, G., & Chen, H. (2012). Kernel based asymmetric learning for software defect prediction. *IEICE Transactions on information and systems, E95-D*(1), 267–270.

Murphy-Hill, E., Zimmermann, T., Bird, C., & Nagappan, N. (2013). The design of bug fixes. In *Proceedings of the 35th International Conference on Software Engineering* (*ICSE 2013*). IEEE.

Musa, J. D. (1973). A theory of software reliability and its application. *IEEE Transactions on Software Engineering, 1*(3), 312–327. doi:10.1109/TSE.1975.6312856.

Musa, J. D. & Okumoto, K. (1973). A logarithmic poisson execution time model for software reliability measurement. *Software System Design Methods*, *22*, 275–298. Bell Laboratories, Whippany, N. J. 07981.

Pai, P. F., & Hong, W. C. (2006). Software reliability forecasting by support vector machines with simulated annealing algorithms. *Journal of Systems and Software, 79*(6), 747–755.

Pietrantuono, R., Russo, S., & Trivedi, K. S. (2010). Software reliability and testing time allocation: An architecture-based approach. *IEEE Transactions on Software Engineering, 36* (3), 323–337.

Ren, J. & Qin, K. (2014). On software defect prediction using machine learning. *Journal of Applied Mathematics, 2014,* 1–8. Article ID 785435.

Roy, B. (2014). A quantitative analysis of NHPP based software reliability growth models. *International Journal of Innovative Research in Computer and Communication Engineering, 2* (1), 2338–2432.

Seiert, C., Khoshgoftaar, T. M., & Hulse, J. V. (2009). Improving software-quality predictions with data sampling and boosting. *IEEE Transactions on Systems, Man, and Cybernetics Part A: Systems and Humans, 39*(6), 1283–1294.

Suresh, Y., Kumar, L., & Rath, S. K. (2014). Statistical and machine learning methods for software fault prediction using ck metric suite: A comparative analysis, *2014*, 1–16. doi:10.1155/2014/251083. Article ID 251083 (Hindawi Publishing Corporation ISRN Software Engineering).

Tian, L., & Noore, A. (2005). Dynamic software reliability prediction: An approach based on support vector machines. *International Journal of Reliability, Quality and Safety Engineering, 12*(4), 309–321.

Tokumoto, S., Dohi, T., & Yun, W. Y. (2012). Toward development of risk-based checkpointing scheme via parametric bootstrapping. In *Proceedings of the 2012 Workshop on Recent Advances in Software Dependability* (pp. 50–55).

Toor, S. & Bahl, K. (2013). Software reliability measurement and improvement policies. In *IJCA Proceedings on International Conference on Advances in Management and Technology 2013 iCAMT* (pp. 41–44). New York, USA: Foundation of Computer Science.

Wohlin, C. (2013). Empirical software engineering research with industry: Top 10 challenges. In *Proceedings of the First International Workshop on Conducting Empirical Studies in Industry (CESI 2013)—An ICSE 2013 Workshop, San Francisco, USA* (pp. 43–46).

Xiaonan, Z., Junfeng, Y., Siliang, D., & Shudong, H. (2013). A new method on software reliability prediction. *Mathematical Problems in Engineering, 2013*, 1–8, doi:10.1155/2013/385372. Article ID 385372.

Xue, J., & Titterington, D. (2008). Do unbalanced data have a negative effect on LDA? *Pattern Recognition, 41*(5), 1558–1571.

Yamada, S. (2014). *Software Reliability Modeling Fundamentals and Applications*. Berlin: Springer.

Yang, B. & Li, X. (2007). A study on software reliability prediction based on support vector machines. In *IEEE International Conference on Industrial Engineering and Engineering Management* (pp. 1176–1180).

Hybrid Metaheuristic Approach for Scheduling of Aperiodic OS Tasks

Hamza Gharsellaoui and Samir Ben Ahmed

Abstract This book chapter deals with the purpose of one hybrid approach for solving the real-time embedded systems scheduling composed of aperiodic OS tasks which are used to control physical processes that range in complexity from automobile ignition systems to controllers for flight systems and nuclear power plants. In these systems, the correctness of system functions depends upon not only the results of computation but also on the times at which results are produced. This book chapter presents real-time scheduling techniques for reducing the response time of aperiodic tasks scheduled with real-time periodic tasks on uniprocessor systems where two problems are addressed: (i) the scheduling of aperiodic tasks when they arrive in order to obtain a feasible system, and (ii) the scheduling of periodic and aperiodic tasks to minimize their response time. Indeed, in order to improve the responsiveness to both types of problems, our approach proposed in this book chapter presents an efficient hybrid methaheuristic based on the combination of the Polling Server (PS) and the Background Server (BS). The effectiveness and the performance of the designed approach are evaluated through simulation studies. A tool named RT-Reconfiguration is developed in our research laboratory at INSAT Institute to support this new proposed approach.

H. Gharsellaoui
Higher School of Computer Science and Technology (ESTI),
Carthage University, Carthage, Tunisia
e-mail: Gharsellaoui.hamza@gmail.com; hamzacampus@yahoo.fr

H. Gharsellaoui · S. Ben Ahmed
National Institute of Applied Sciences and Technology (INSAT),
Carthage University, Carthage, Tunisia

H. Gharsellaoui
Al-Jouf College of Technology, TVTC, Al-Jouf, Kingdom of Saudi Arabia

S. Ben Ahmed (✉)
Faculty of Mathematical, Physical and Natural Sciences of Tunis, FST,
University of Tunis El Manar, Tunis, Tunisia
e-mail: Samir.benahmed@fst.rnu.tn

Q. Zhu and A.T. Azar (eds.), *Complex System Modelling and Control Through Intelligent Soft Computations*, Studies in Fuzziness and Soft Computing 319,
DOI 10.1007/978-3-319-12883-2_29

1 Introduction

The use of computers for control and monitoring of industrial processes has expanded greatly in recent years, and will probably expand even more dramatically in the near future Layland and Liu (1973). Indeed, Real-time systems are used to control physical processes that range in complexity from automobile ignition systems to controllers for flight systems and nuclear power plants. In these systems, the correctness of system functions depends upon not only the results of computation but also on the times at which results are produced. Also, a process control computer performs one or more control and monitoring functions. The IEEE Real-Time Systems Symposium has in the last two decades been the main forum for publishing the key results in aperiodic real-time scheduling theory that are reviewed in this book chapter work. A number of other initiatives were funded also including a series of influential studies commissioned by the European Space Agency. For example, the works appear in Abdelzaher et al. (2004a, b); Abeni and Buttazzo (1998, 2001, 2004); Aggarwal and Chraibi (1995). Various scheduling policies for real-time multiprocessors systems have been constantly evolving. Scheduling of real-time applications/tasks on multiprocessor systems often has to be combined with task to processor mapping. With the current design trends moving towards multicores and multiprocessor systems for high performance and embedded system applications, the need to develop design techniques to maximize utilization of processor time, and at the same time minimize power consumption, have gained importance. General design techniques to achieve the above goals have been fueled by the development of both hardware and software solutions. Hardware solutions to minimize power and energy consumption include: dynamic voltage and frequency scaling (DVFS) processors, dynamic power management modules Advanced Configuration and Power Interface (ACPI), thermal management modules, intelligent energy management (IEM), and heterogeneous multicore or multiprocessor systems. Software solutions for maximizing processor utilization and energy minimizations include: parallelization of instructions, threads, and tasks; effective thread/task scheduling; and mapping algorithms for multicore and multiprocessor systems Naveen and Venkatesan (2013). In Layland and Liu (1973), the authors used the example of the antenna pointing to track a spacecraft in its orbit. In this use case, each function to be performed has associated with it a set of one or more tasks where some of these tasks are executed in response to events in the equipment controlled by or monitored by the computer. The remainder are executed in response to events in other tasks and none of the tasks may be executed before the event which requests it occurs and each of them must be completed before some fixed time has elapsed following the request for it. This kind of tasks is called a real-time tasks. A real-time task is generally placed into one of four categories based upon its arrival pattern and its deadline. If meeting a given task's deadline is critical to the system's operation, then the task's deadline is considered to be hard. If it is desirable to meet a task's deadline but occasionally missing the deadline can be tolerated, then the deadline is considered to be soft. Tasks with regular arrival times are called periodic tasks. A common use of periodic tasks is to process sensor data and update the current state of the real-time system on a regular basis. Periodic tasks, typically used in control and signal-processing applications, have hard deadlines. Tasks with irregular arrival times are aperiodic tasks. Aperiodic tasks are used to handle the processing requirements of random events such as operator requests. An aperiodic task typically has a soft deadline. Aperiodic tasks that have hard deadlines are called sporadic tasks. We assume that each task has a known worst-case execution time. In summary, we have hard and soft deadline periodic tasks. A periodic task has a regular inter-arrival time equal to its period and a deadline that coincides with the end of its current period. Periodic tasks usually have hard deadlines, but in some applications the deadlines can be soft.

Soft deadline aperiodic tasks. An aperiodic task is a stream of jobs arriving at irregular intervals. Soft deadline aperiodic tasks typically require a fast average response time.

Sporadic tasks. A sporadic task is an aperiodic task with a hard deadline and a minimum inter-arrival time. Note that without a minimum inter-arrival time restriction, it is impossible to guarantee that a sporadic task's deadline would always be met. To meet the timing constraints of the system, a scheduler must coordinate the use of all system resources using a set of well-understood real-time scheduling algorithms that meet the following objectives: Guarantee that tasks with hard timing constraints will always meet their deadlines. Attain a high degree of schedulable utilization for hard deadline tasks (periodic and sporadic tasks). Schedulable utilization is the degree of resource utilization at or below which all hard deadlines can be guaranteed. The schedulable utilization attainable by an algorithm is a measure of the algorithm's utility: the higher the schedulable utilization, the more applicable the algorithm is for a range of real-time systems. Provide fast average response times for tasks with soft deadlines (aperiodic tasks). Ensure scheduling stability under transient overload.

In some applications, such as radar tracking, an overload situation can develop in which the computation requirements of the system exceed the schedulable resource utilization. A scheduler is said to be stable if during overload it can guarantee the deadlines of critical tasks even though it is impossible to meet all task deadlines. The quality of a scheduling algorithm for real-time systems is judged by how well the algorithm meets these objectives. This book chapter work develops advanced hybrid approach to schedule aperiodic tasks. For soft deadline aperiodic tasks, the goal is to provide fast average response times. For hard deadlines aperiodic tasks (sporadic tasks), the goal is to guarantee that their deadlines will always be met. The new hybrid approach presented here meet both of these goals and are still able to guarantee the deadlines of hard deadline periodic tasks.

Each periodic task τ_i is characterized according to Layland and Liu (1973), by an initial offset S_i (a release time), a worst-case execution time C_i, a relative deadline D_i and a period T_i.

Each aperiodic task τ_i is characterized by a worst-case execution time C_i and a relative deadline D_i. A task is synchronous if its release time is equal to 0. Otherwise, it's asynchronous.

The issue regarding the duration of each scheduling time slot addresses problems such as the frequency at which the scheduled tasks are delivered to working processors and the frequency with which newly-arrived tasks are sought for consideration in the scheduling process. This is an important issue in dynamic scheduling of aperiodic, real-time tasks and has mostly been ignored. The existing approaches concentrate on finding a feasible solution for the entire batch of tasks in the current scheduling period without regard for arriving tasks, for keeping other processors idle, and/or for missing the deadlines of scheduled tasks in the current period, due to long scheduling times Hamidzadeh and Atif (1996). When scheduling real-time tasks dynamically on the processors of a multiprocessor architecture, it is very important to address several issues about the scheduling algorithm, such as the time slots at which the algorithm is invoked, the duration of each invocation time slot, the distribution of the scheduling task itself among different processors, and the complexity of the algorithm. These major issues in real-time task scheduling have rarely been addressed. They are important factors because each one by itself and in relation with the other factors, creates several tradeoffs that can directly affect the quality of the answers produced by the scheduler and the degree of predictability and guarantee that the scheduling algorithm can provide in meeting the task time constraints Hamidzadeh and Atif (1996). Many real-time applications involve combined scheduling of hard and soft real-time tasks. Hard real-time tasks have critical deadlines that are to be met in all working scenarios to avoid catastrophic consequences. In contrast, soft real-time tasks (e.g., multimedia tasks) are those whose deadlines are less critical such that missing the deadlines occasionally has minimal effect on the performance of the system. In military applications, such as attack helicopters, multimedia information is being used to provide tracking and monitoring capabilities, that can be used directly to engage a threat and avoid crashing unexpectedly. It is possible to allocate separate resources for each of hard and soft tasks. However, sharing the available resources among both types of tasks would have enormous economical and functional impact. Therefore, it is necessary to support combined scheduling of hard and soft real-time tasks in such systems, in which multiprocessors are increasingly being used to handle the compute intensive applications. Real-time tasks, beside being hard or soft, can be periodic or aperiodic. Therefore, in combined scheduling we may encounter the following task combinations: (1) periodic hard tasks with aperiodic soft tasks, (2) periodic hard tasks with periodic soft tasks, (3) aperiodic hard tasks with aperiodic soft tasks, or (4) aperiodic hard tasks with periodic soft tasks. We assume in this work that all the tasks are independent, periodic and aperiodic. A tool named RT-Reconfiguration is developed in our research laboratory at INSAT university to support this new proposed approach. The organization of this original book chapter work is as follows. The next section presents a related work and formalizes some known concepts in the real-time scheduling theory. Section 3 presents the system model of a real-time task. In Sect. 4, we define a new theoretical approach about the aperiodic task scheduling contribution and our proposed approach is implemented, simulated and analyzed also. Finally, Sect. 5 presents a conclusion of this work and a future trends.

2 Related Work

Scheduling algorithms have been classically categorized into online or offline, priority or non-priority, preemptive or non preemptive, and hard or soft deadline scheduling algorithms. Two of the classical scheduling algorithms for uniprocessor are fixed priority rate-monotonic scheduling (RMS), and dynamic priority EDF algorithms.

A real-time system often has both periodic and aperiodic tasks. Lehoczky et al. (1987) developed the Deferrable Server algorithm which is compatible with the rate monotonic scheduling algorithm and provides a greatly improved average response time for soft deadline aperiodic tasks over polling or background service algorithms while still guaranteeing the deadlines of periodic tasks.

In Marouf et al. (2012), the authors proposed scheduling algorithms in the case of harmonic and non-harmonic strict periodic tasks. In Marouf and Sorel (2010, 2011), a schedulability analysis for such tasks were proposed where software fault-tolerance has been considered through the primary/alternate task models. When a primary task cannot meet its deadline, an alternate task is run. The alternate task can be the same task (the task is re-executed).

The scheduling problem for aperiodic tasks is very different from the scheduling problem for periodic tasks. Scheduling algorithms for aperiodic tasks must be able to guarantee the deadlines for hard deadline aperiodic tasks and provide good average response times for soft deadline aperiodic tasks even though the occurrences of the aperiodic requests are nondeterministic. The impact also of soft deadline aperiodic requests on tasks with hard deadlines can be reduced by using an aperiodic server task. Aperiodic requests are queued for the server when they arrive, and executed as soon as the scheduling algorithm permits the server to execute them.

According to Ghazalie and Baker (1994), a simple form of aperiodic server is a background server where the server is scheduled at lower priority than every hard deadline task. In this way, soft deadline requests never cause a hard deadline to be missed. The main drawback of this model is that both average and worst case response times for the server may be unacceptably long.

Another simple form of aperiodic server is a polling server. With polling, the server is treated as a hard deadline periodic task with a fixed execution time budget, whose deadline is equal to its period in which the server executes and serves all the requests that have been enqueued up to that time. In the case when there are more requests than can be served in the budgeted time, they are carried over to the budgeted time, they are carried over to the next period. In one hand, the period must be short enough in this case in order to achieve the desired average response time and to guarantee any associated hard deadline. On the other hand, the queueing (accumulation) of requests allows the period of the server to be longer than the interarrival time of the requests and as a consequence, the adverse effect on schedulability of other tasks is reduced.

In Ghazalie and Baker (1994), the authors consider that the advantage of the polling over background processing is that hard deadlines can be guaranteed, since the server period is treated as a hard deadline. Also, by using multiple servers at different priority levels, one can accommodate a set of tasks with a range of hard and soft deadline requirements. In contrast, the main disadvantage of polling server is that, under reasonable assumptions about the distribution of aperiodic requests, the average response time is at least half the server period plus the average execution time. Thus, the only way to improve response time for soft-deadline tasks is to reduce the server period before using the scheduling algorithm.

According to Layland and Liu (1973), a scheduling algorithm is a set of rules that determine the task to be executed at a particular moment. The scheduling algorithms to be studied in this book chapter are preemptive and priority driven ones. Indeed, whenever there is a request for a task that is of higher priority than the one currently being executed, the running tasks is immediately interrupted and the newly requested task is started. Thus, a scheduling algorithm is said to be *static* if priorities are assigned to tasks once and for all. It is also called a *fixed priority* scheduling algorithm. In contrast, a scheduling algorithm is said to be a *dynamic* if priorities of tasks might change from request to request.

A scheduling algorithm is said to be *a mixed or hybrid* scheduling algorithm if the priorities of some tasks are fixed yet the priorities of the remaining tasks vary from request to request.

In this book chapter, we adapt this approach of hybrid scheduling as described in the following Sect. 4.

For a detailed analysis of aperiodic servers see Guillem (2001) and Burns and Guillem (1999). The aperiodic scheduling algorithm must also accomplish these goals without compromising the hard deadlines of the periodic tasks. For the aperiodic scheduling, authors presented Slack stealing Thuel and Lehoczky (1994) and aperiodic servers, such as the sporadic server Sprunt et al. (1989) and the deferrable server Strosnider et al. (1995), allow aperiodic tasks to be handled within a periodic task framework. Lipari and Buttazzo (1999) proposed a hybrid method that combined periodic and aperiodic tasks which shared many resources. Our approach try by allowing periodic tasks to be handled with an aperiodic ones by an hybrid approach in the same framework. To the author's knowledge, no result is available in the state of the art for scheduling both periodic and aperiodic tasks, except that we propose in our original work where an approach to deal with complex timing constraints and with minimizing the response time is proposed.

3 System Model

A task model is required as the basis for discussing scheduling. A real-time task is a basic executable entity, which can be scheduled; it can be either periodic or aperiodic, with soft or hard timing constraint. A task is best defined with its main timing parameters. For our work, we shall assume that time parameters have the domain the set of positive real numbers (non-negative integers). We present the following well-known concepts in the theory of aperiodic real-time scheduling Layland and Liu (1973):

- An aperiodic task τ_i (C_i; D_i) is an infinite collection of jobs that have their request times constrained by a Worst Case Execution Time (WCET) C_i and a relative deadline D_i.
- Deadline: The time when a task must be finished executing.
- Worst Case Execution Time (WCET): The longest possible execution time for a task on a particular type of system.
- Response time: The time it takes a task to finish execution. Measured from release time to execution completes, including preemptions.
- Preemptive scheduling: an executing task may be interrupted at any instant in time and have its execution resumed later.
- Release/ready time: The time a task is ready to run and just waits for the scheduler to activate it.
- A busy period is defined as a time interval $[a, b)$ such that there is no idle time in $[a, b)$ (the processor is fully busy) and such that both a and b are idle times.
- $U = \sum_{i=1}^{n} \frac{C_i}{T_i}$ is the processor utilization factor. In the case of synchronous and asynchronous, independent and periodic tasks. $U = \sum_{i=1}^{n} \frac{C_i}{\min(T_i, D_i)} \leq 1$ is a sufficient condition but not necessary for the EDF-based scheduling of real-time tasks.
- A hard real-time task is never allowed to miss a deadline because that can lead to complete failure of the system. A hard real-time task can be safety-critical and this means that if a deadline is missed it can lead to catastrophically consequences which can harm persons or the environment.
- A soft real-time task is a task when a deadline is allowed to be missed, while there is no complete failure of the system, it can lead to decreased performance.
- Mapping is the process of assigning each task in an application to a processor, such that the processor executes the task and satisfies the task deadline, as well as other constraints (power, throughput, completion time), if any.
- **Polling Server** is a periodic task whose purpose is to provide relatively high priority service to aperiodic task requests with a period T_S, a computation time C_S (capacity) and scheduled in the same way as periodic tasks. It is ready to run at the start of its period and services pending arriving aperiodic tasks over the interval from the beginning of its period until C_S time units later. The polling server task is subject to preemption by higher priority tasks, until either it exhausts its execution time or there is no execution time of aperiodic tasks left to be executed. Otherwise, the polling server loses any of the unused execution time and is unavailable to service aperiodic tasks until the start of its next period. The polling server task is scheduled as if it was a periodic task with period T_S. Aperiodic tasks that arrive or remain when the polling server is unavailable can be serviced at background priority.

- **Deferrable Server** a Deferrable Server (DS) is a periodic task with period T_{DS} and capacity C_{DS}. The DS is used to provide high priority service to aperiodic tasks. It is ready at the start of its period and services aperiodic task arrivals, subject to preemption by higher priority tasks, until it exhausts its execution time C_{DS} or the end of its period is reached Strosnider et al. (1995). Unlike the polling server which loses any unused execution time when there is no aperiodic work remaining, the DS execution time C_{DS}, is available for servicing aperiodic arrivals throughout its entire period. It loses any unused execution time at the end of its period when its full capacity C_{DS} is restored. The DS task is scheduled as if it was a periodic task with period T_{DS}. Aperiodic tasks that arrive when the DS execution time, C_{DS}, has been exhausted can be serviced at background priority. In general, the DS task is assigned a priority according to the rate monotonic algorithm based on its period, T_{DS}, relative to the other periodic tasks. While the DS task can execute at any priority level, assigning the DS task the highest priority allows one to guarantee that the deadlines of aperiodic alerts are met as well as enhancing the responsiveness of the soft deadline aperiodic tasks. At intermediate priority levels, the DS is less capable of providing responsive aperiodic service. Moreover, DS capacity can be lost because of higher priority preemptions even when aperiodic tasks are ready for processing Strosnider et al. (1995).
- **Background Server** a Background Server (BS) schedules aperiodic tasks in background (when no periodic task is running) and schedule of periodic tasks is not changed. BS Treats aperiodic tasks as lowest-priority tasks and have the following advantages:

  - Simple,
  - Aperiodic tasks has no impact on the schedulability of periodic tasks

  In contrast, the disadvantages of the background scheduling are the following:

  - Aperiodic tasks have very long response times when the utilization of periodic tasks is high,
  - Acceptable only if the system is not busy or the aperiodic tasks can tolerate long delays

- **Sporadic Server** Sporadic Server (SS) is a real-time scheduling algorithm used to control the execution of processes/threads on a system. This scheduler allows one to set the maximum amount of time a process may receive in a specified time window. The basic idea is that the server is given a time budget at the server's priority that it consumes when executing on the processor, and which is replenished according to some rules:

  - bounds the CPU time consumed by a process
  - guarantees CPU time given to a process
  - conforms process's execution to simpler model

4 Aperiodic Task Scheduling Contribution

This book chapter deals with the problem of uniprocessor scheduling of both periodic and sporadic/aperiodic tasks on uniprocessor. We introduce in this work an EDF-based scheduling algorithm to optimize the response times of reconfigurable tasks while ensuring first that all the periodic tasks meet their deadlines and second that all the sporadic and aperiodic tasks can respect their constraints. A necessary and sufficient schedulability test is presented, and an efficient O(n + m) guarantee algorithm is proposed. Indeed, to obtain this goal, this system should be changed and automatically adapted to its environment on the occurrence of random disturbances such as hardware-software faults. A random disturbance is defined in this work as any random internal or external event allowing additions, removals or updates of tasks at run time to adapt the system's behavior. Therefore, the system's implementation is dynamically changed and should meet all considered deadlines of the current combination of tasks. Nevertheless, when an automatic reconfiguration scenario is applied, the deadlines of new and old tasks can be violated. We mean by reconfiguration scenario in our work, the removal, update or addition of new aperiodic tasks when they arrive at run-time without prior knowledge, in order to save the whole system on the occurrence of hardware-software faults in a safe state (feasible system), or also to improve its performance when random disturbances happen at run-time. This scheduling algorithm is used at run-time to provide dynamic solutions when deadlines are violated after a particular reconfiguration scenario. We propose an agent-based architecture where an intelligent software agent is used to evaluate the response times, to calculate the processor utilization factor and also to verify the satisfaction of real-time deadlines.

4.1 First Method: An EDF Based Scheduling Approach

Nowadays, due to the growing class of portable systems, such as personal computing and communication devices, embedded and real-time systems contain complex software which is increasing by the time. This complexity is growing because many available software development models don't take into account the specific needs of embedded and systems development. The software engineering principles for embedded system should address specific constraints such as hard timing constraints, limited memory and power use, predefined hardware platform technology, and hardware costs. The new generations of embedded control systems are addressing new criteria such as flexibility and agility Gharsellaoui et al. (2012). For these reasons, there is a need to develop tools, methodologies in embedded software engineering and reconfigurable embedded control systems as an independent discipline. By response for this requirement of developing a reconfigurable systems, many interesting academic and industrial studies have been made in recent years. Feasibility Conditions (FC) for the dimensioning of a real-time system enables a designer to grant that timeliness constraints associated to an application run by the system are always met for all possible configurations. The goal of the FC is to ensure a deterministic respect of the timeliness constraints Gharsellaoui et al. (2012).

4.1.1 Reconfiguration of Sporadic Tasks

A reconfiguration can be decided either off-line or on-line. In the first case, the goal is to check if several hardware platforms or several hardware configurations can be used to run a specific application while preserving the timeliness constraints of the tasks. In the second case, a reconfiguration might result from a system mode change to adapt the system to the context of its execution or to handle hardware or software faults. Sensitivity analysis aims at studying the ability to introduce more flexibility in the specifications. In this work, we study many sensitivity analysis (more than one task parameter can evolve).

This book chapter focuses on the dynamic reconfigurations of assumed mixture of off-line and on-line workloads that should meet deadlines defined according to user requirements. We propose an intelligent agent-based architecture in which a software agent is deployed to dynamically adapt the system to its environment by applying reconfiguration scenarios. The agent dynamically provides technical solutions for users when the system becomes unfeasible (e.g. deadlines are violated), by sending sporadic tasks to idle times, by modifying the deadlines of tasks, the worst case execution times (WCETs), the activation time, by tolerating some non critical tasks m among n according to the (m, n) firm model (Hamdaoui and Ramanathan 1995) and a reasonable cost, or in the worst case by removing some non hard (soft) tasks according to predefined heuristic. We implement the agent to support these services which are applied to a running example with real-life design examples in order to demonstrate the effectiveness and the excellent performance of the new proposed algorithm.

On the other hand, the scheduling of tasks is an essential requirement in most real-time embedded systems, but invariably leads to unwanted CPU overheads. This book chapter work presents also real-time scheduling techniques for reducing the response times of uniprocessor aperiodic tasks to be scheduled with real-time periodic tasks. Two problems are addressed in this part: (i) the scheduling of aperiodic tasks when they arrive in order to obtain a feasible system, and (ii) the scheduling of periodic and aperiodic tasks to minimize their response times. In order to improve the responsiveness to both types of problems, an efficient approach is proposed by using the Poisson distribution which is a discrete distribution. It is often used as a model for the number of events in a specific time period. Instead, it uses the fixed interval of time or space in which the number of successes is recorded. The space of feasible deadlines (D-space) is then assumed to be equal to one time unit in our proposed approach. The effectiveness and the performance of the designed approach is evaluated through simulation studies.

We assume that for all the real-time tasks, we have M automatic reconfiguration scenarios (ψ_1, ψ_2, …, ψ_M). We mean in this thesis by an automatic reconfiguration scenario ψ_h (h $\in$ 1…M), any operation that adds, removes or also updates tasks at run-time which corresponds to this automatic reconfiguration scenario ψ_h. Automatic updates of tasks mean modifications of their temporal parameters e.g. Periods and/or deadlines, or modifications of their Worst Case Execution Times.

In our model, a sporadic task is represented by $\sigma_i^{\psi_h}$ for each reconfiguration scenario ψ_h and it's completely characterized by specifying its worst case execution time $C_i^{\psi_h}$, its relative deadline $D_i^{\psi_h}$, the deadline tolerance value $m_i^{\psi_h}$, and its value $I_i^{\psi_h}$ for each reconfiguration scenario ψ_h. In the following, we assume that the task class can be derived from the task value. For instance, tasks with value $I_i^{\psi_h}$ = H can be considered as HARD and others with value $I_i^{\psi_h}$ = S are considered as SOFT.

For the proofs which provide a necessary and sufficient condition for the schedulability of task sets at each task activation, we require a notation that identifies the *k*th task in the current ordered list at time t. For this, we use the notation $\sigma_{i,1}^{\psi_h}, \sigma_{i,2}^{\psi_h}, \ldots, \sigma_{i,m}^{\psi_h}$ where $\sigma_{i,k}^{\psi_h}$ is the task in the *k*th order for each reconfiguration scenario ψ_h.

In our model, we assume that the minimum interarrival time of each sporadic task is equal to its relative deadline $D_i^{\psi_h}$, thus a sporadic task $\sigma_i^{\psi_h}$ can be completely characterized by specifying its worst case execution time $C_i^{\psi_h}$ and its relative deadline $D_i^{\psi_h}$. Hence, a sporadic task set will be denoted as follows: $\xi^{\psi_h} = \{\sigma_i^{\psi_h}(C_i^{\psi_h}, D_i^{\psi_h})\}$, $i = 1$ to m, h = 1 to M.

In summary and in our work, a sporadic task set will be denoted for each reconfiguration scenario ψ_h as follows:

$$\xi^{\psi_h} = \{\sigma_i^{\psi_h}(C_i^{\psi_h}, D_i^{\psi_h}, m_i^{\psi_h}, I_i^{\psi_h})\}, \quad \text{i} = 1 \text{ to m, h} = 1 \text{ to M}.$$

Within this framework, different solutions are used for handling sporadic tasks in an optimal fashion. In particular, the tasks are scheduled based on their deadline, guaranteed based on C_i, D_i, m_i, I_i, and rejected based on I_i value.

The other known method is to use response time analysis, which consists of computing the worst-case response time (WCRT) of all tasks in a system and ensuring that each tasks WCRT is less than its relative deadline. To avoid these problems, and to have a feasible system in this thesis work, our proposed tool RT-Reconfiguration can be used. For this reason, we present the following relationships among the parameters defined above for each reconfiguration scenario ψ_h:

$$d_i^{\psi_h} = a_i^{\psi_h} + D_i^{\psi_h} \tag{1}$$

$$L_i^{\psi_h} = d_i^{\psi_h} - a_i^{\psi_h} - C_i^{\psi_h} \tag{2}$$

$$R_i^{\psi_h} = d_i^{\psi_h} - f_i^{\psi_h} \tag{3}$$

$$f_1^{\psi_h} = t + c_1^{\psi_h};\quad f_i^{\psi_h} = f_{i-1}^{\psi_h} + c_i^{\psi_h} \quad \forall i > 1 \tag{4}$$

Moreover, our approach combines many nice scheduling features, further enhancing its optimality. The main contribution of this work is the development and the performance evaluation of an efficient version of the EDF algorithm.

Guarantee Algorithm

Buttazzo and Stankovic (1993) present a dynamic on-line guarantee test in terms of residual time. Based on their proposed algorithm, we will extend this algorithm by including tolerance indicator and task rejection policy for each reconfiguration scenario ψ_h. The basic properties stated by the following lemma and theorem are used to derive an efficient $O(n+m)^2$ algorithm for analyzing the schedulability of the sporadic task set whenever a new task arrives in the systems after each reconfiguration scenario ψ_h.

Lemma *Given a set* $\xi^{(\psi_h)} = \{\sigma_1^{(\psi_h)}, \sigma_2^{(\psi_h)}, \ldots, \sigma_n^{(\psi_h)}\}$ *of active sporadic tasks ordered by increasing deadline in a linked list, the residual time* $R_i^{(\psi_h)}$ *of each task* $\sigma_i^{(\psi_h)}$ *at time t can be computed by the following recursive formula:*

$$R_1^{(\psi_h)} = d_1^{(\psi_h)} - t - c_1^{(\psi_h)} \tag{5}$$

$$R_i^{(\psi_h)} = R_{i-1}^{(\psi_h)} + (d_i^{(\psi_h)} - d_{i-1}^{(\psi_h)}) - c_i^{(\psi_h)}. \tag{6}$$

Now we introduce a new framework for handling real-time sporadic tasks under overload conditions, and we propose an efficient version of the Earliest Deadline First algorithm (EDF).

For sporadic tasks, the utilization factor could be computed by considering the minimum interarrival time as a sort of period. However, this would lead to an overestimation of the workload, since it would refer to the (very pessimistic) case in which all sporadic tasks have the maximum arrival rate.

One main purpose of our EDF-based algorithm is to operate well even in overload conditions. However, it is difficult to develop a good measure of load in a real-time system because each task has a unique start time and deadline. The idea is to iteratively identify the cpu utilization required by all the tasks up to the *i*th task.

In many real applications, such as robotics, the deadline timing semantics is more flexible than scheduling theory generally permits. Basically, our approach minimizes the pessimism found in a basic guarantee algorithm.

Another real application issue is that once some task has to miss a deadline, it should be the least valuable task. Again, many algorithms do not address this fact.

Contribution: Algorithm for Feasibility Testing of Sporadic Task Systems

In the current work, we use as a scheduling policy, the EDF algorithm. After a reconfiguration scenario ψ_h was applied at run time, the intelligent agent proposes useful solutions for users by sending sporadic tasks to idle times which are considered as non-productive times of the processor, by modifying the deadlines of tasks, the worst case execution time (WCETs), the activation time, by tolerating some non critical k among m tasks to miss their deadlines or by removing some tasks according to a predefined heuristic.

We suppose that each system ξ can be automatically and repeatedly reconfigured. ξ is initially considered as $\xi^{(0)}$ and after the h_{th} reconfiguration ξ turns into $\xi^{(\psi_h)}$, where $h \in \aleph_+^{(*)}$. We define VP_1 as the first virtual processor to virtually execute old periodic tasks and VP_2 as the second virtual processor to virtually execute new sporadic tasks, implementing the system after the hth reconfiguration scenario ψ_h. In $\xi^{(\psi_h)}$, all old tasks from $\xi^{(\psi_{h-1})}$ are executed by the newly updated $VP_1^{(\psi_h)}$ and the added sporadic tasks are executed by $VP_2^{(\psi_h)}$. The proposed intelligent agent is trying to minimize the response time $R^{(\psi_h)}$ of ξ^{ψ_h} after each reconfiguration scenario ψ_h.

Formalization

We assume in this work a system ξ composed of a mixture of n periodic and m sporadic tasks. The initial processor utilization factor U before any addition scenario of new sporadic tasks to ξ is $U = \sum_{i=1}^{n} \frac{C_i}{T_i}$. An assumed system $\xi^{(\psi_{h-1})} = \{\tau_1, \tau_2, \ldots, \tau_n\}$ turns after a reconfiguration scenario ψ_h to $\xi^{(\psi_h)} = \{\tau_1, \tau_2, \ldots, \tau_n\ \sigma_{n+1}, \sigma_{n+2}, \ldots, \sigma_m\}$ by considering that m-n new sporadic tasks are added to $\xi^{(\psi_{h-1})}$. After each addition scenario ψ_h, the tasks are logically divided into two subsets. One contains the so called new sporadic tasks which are added to the system, and the rest of tasks taken from $\xi^{(\psi_{h-1})}$ are considered as old tasks to form the second subset.

Problem

After any addition scenario ψ_h, the response time can be increased and/or some old/new tasks can miss their deadlines.

When a reconfiguration scenario is automatically applied at run-time, the proposed intelligent agent logically decomposes the physical processor of $\xi^{(\psi_h)}$ into two virtual processors $VP_1^{(\psi_h)}$ and $VP_2^{(\psi_h)}$ with different utilization factors $UVP_1^{(\psi_h)}$ and $UVP_2^{(\psi_h)}$ to adapt the system to its environment with a minimum response times. $UVP_1^{(\psi_h)}$ corresponds to the processor utilization of the system before any addition scenario, and $UVP_2^{(\psi_h)}$ can be assigned to any value lower than 1 according to user requirements.

Therefore, based on the research work in Buttazzo and Stankovic (1993) which provides a window-constrained-based method to determine how much a task can increase its computation time without missing its deadline under EDF scheduling [for more informations about the window-constrained-based method, you can see Buttazzo and Stankovic (1993)]. We propose in our thesis work, a window constrained schedule which is used to separate old and new tasks. Old and new tasks are located in different windows to schedule the system with a minimum response times. Idle periods in $VP_1^{(\psi_h)}$, which appear alternatively with busy periods, are considered as logical windows for the execution of the second virtual processor $VP_2^{(\psi_h)}$. In this case, a first logical window corresponding to $VP_1^{(\psi_h)}$ is reserved for old tasks (periodic and sporadic) that should be reconfigured to meet their deadlines and to reduce their response time, and a second window corresponding to $VP_2^{(\psi_h)}$ is reserved for new sporadic tasks with an optimal response time after any sporadic tasks addition. The physical processor of $\xi^{(\psi_h)}$ will be running with this solution in two phases for the minimization of response times.

We assume in the following that new sporadic tasks are dynamically added to a system and request the processor at a time t which is not smaller than P_i (=D_i). After any reconfiguration scenario ψ_h and in order to keep only two virtual processors in the system $\xi^{(\psi_h)}$, the proposed intelligent agent automatically merges $VP_1^{(\psi_{h-1})}$ and $VP_2^{(\psi_{h-1})}$ into $VP_1^{(\psi_h)}$ and creates also a new VP_2 for the reconfiguration scenario ψ_h named $VP_2^{(\psi_h)}$, to adapt old and new tasks, respectively. The $VP_2^{(\psi_h)}$ is assumed to be a located logical pool in idle periods of $VP_1^{(\psi_h)}$ and used to execute new added tasks. The old tasks are assumed to be executed by the first virtual processor $VP_1^{(\psi_h)}$.

Running Example

To illustrate the key point of the proposed reconfiguration approach, we consider the Volvo task system shown in Tables 1 and 2, as a motivational example noted ξ composed of 2 characterized periodic tasks (τ_A and τ_B) and 3 sporadic tasks (σ_D, σ_E, and σ_H) as a first task set (1) and 3 added sporadic tasks as a second task set to be added later (2).

For example, the processor utilization factor of ξ in Table 1 is U $= \sum_{i=1}^{5} \frac{C_i}{T_i\, or\, D_i}$, ($D_i$, is in the case of sporadic tasks) $= \frac{C_A}{T_A} + \frac{C_B}{D_B} + \frac{C_D}{D_D} + \frac{C_E}{D_E} + \frac{C_H}{D_H} = 0.2 + 0.4 + 0.12 + 0.08 + 0.004 = 0.804 \leq 1$, so the system ξ is feasible.

Table 1 The first volvo case study

Task	T_i	C_i	D_i
A	10	2	10
B	20	2	5
D	50	6	50
E	100	8	100
H	2,000	8	2,000

Table 2 The added tasks to volvo case study

Task	T_i	C_i	D_i
C	50	1	2
F	2,000	7	100
G	2,000	8	100

Five independent computations τ_A, τ_B, σ_D, σ_E, and σ_H, to be executed on an embedded processor core. τ_A and τ_B are periodic tasks, σ_D, σ_E, and σ_H are sporadic ones. Each task can be executed immediately after its arrival and must be finished by its deadline. These tasks are feasible because the processor utilization factor U = 0.804 ≤ 1. and should meet all required deadlines defined in user requirements and we have $Feasibility(Current_{\xi}(t)) \equiv True.$

We suppose now, that a reconfiguration scenario ψ_1 is applied at time t1 to add 3 new tasks *C*; *F*; *G*; as described in Table 2 to the initial Volvo case study system composed of the five characterized tasks (τ_A, τ_B, σ_D, σ_E, and σ_H). The new processor utilization becomes U = 1.454 > 1. Therefore the system is unfeasible and $Feasibility(Current_{\xi^{\psi_1}}(t)) \equiv False.$

In our running example, $\xi = \{\tau_A, \tau_B, \sigma_D, \sigma_E, \text{ and } \sigma_H\}$ is initially considered as $\xi^{(0)}$ and after the 1th reconfiguration scenario $((\psi_1))$ which corresponds to the addition of the new sporadic tasks $(\{\sigma_C, \sigma_F, \text{and } \sigma_G\})$, $\xi = \xi^{(0)}$ turns into $\xi^{(\psi_1)}$. We define VP_1 and VP_2 two virtual processors to virtually execute old (periodic and sporadic) and new sporadic tasks implementing the system after the 1th reconfiguration scenario ψ_1. e.g. VP_1 will execute τ_A, τ_B, σ_D, σ_E, and σ_H and VP_2 will execute σ_C, σ_F, and σ_G. In other words, in $\xi^{(\psi_1)}$, all old tasks from $\xi^{(\psi_0)}$ (τ_A, τ_B, σ_D, σ_E, and σ_H) are executed by the newly updated $VP_1^{(\psi_1)}$ with the processor utilization factor $UVP_1^{(\psi_1)}$ and the added sporadic tasks (σ_C, σ_F, and σ_G) are executed by $VP_2^{(\psi_1)}$ with the processor utilization factor $UVP_2^{(\psi_1)}$. The proposed intelligent agent is trying to minimize the response time $R^{(\psi_1)}$ of ξ^{ψ_1} after the reconfiguration scenario ψ_1.

Proposed Solutions

After each addition scenario ψ_h, the proposed intelligent agent proposes to modify the virtual processors, to modify the deadlines of old and new tasks, the WCETs and the activation time of some tasks or to remove some soft tasks as the following steps:

- *Step 1*: the agent automatically merges $VP_1^{(\psi_{h-1})}$ and $VP_2^{(\psi_{h-1})}$ into $VP_1^{(\psi_h)}$ and calculates $VP_1^{(\psi_h)}$ and $R^{(\psi_h)}$,
- *Step 2*: it calculates new deadlines for old and new sporadic tasks in order to obtain the feasibility of the system and to reduce the response time,
- *Step 3*: it calculates new WCETs,
- *Step 4*: it calculates new activation times $a_i^{(\psi_h)}$,
- *Step 5*: it defines the tasks can miss their deadlines (k among (n + m)),
- *Step 6*: it defines which tasks can be removed from the system $\xi^{(\psi_h)}$ in order to obtain the system's feasibility.
- **Solution 1:** Moving some arrival tasks to be scheduled in idle times. (idle times are caused when some tasks complete before its worst case execution time) (7)
- **Solution 2:** maximize the $\mathrm{d}_\mathrm{i}^{(\psi_h)}$ (8)

By applying Eq. (3) that notices:

$$R_i^{(\psi_h)} = d_i^{(\psi_h)} - f_i^{(\psi_h)}, \text{ we have:}$$

and by applying Eq. (4) that notices:

$$f_1^{\psi_h} = \mathrm{t} + c_1^{\psi_h}, \text{ we can deduce that } f_i^{(\psi_h)} = \mathrm{t} + c_i^{\psi_h} \text{and we have:}$$
$$R_i^{(\psi_h)} = d_i^{(\psi_h)} - \mathrm{t} - C_i^{(\psi_h)}.$$

Or, to obtain a feasible system after a reconfiguration scenario (ψ_h), the following formula must be enforced:

$$R_i^{(\psi_h)} \geq 0.$$

By this result we can write:

$$d_{inew}^{(\psi_h)} - \mathrm{t} - C_i^{(\psi_h)} \geq 0, \text{ (where } d_{inew} \text{ is the new deadline value) or}$$
$$d_{inew}^{(\psi_h)} = d_i^{(\psi_h)} + \theta_i^{(\psi_h)}.$$
$$\text{So, } d_i^{(\psi_h)} + \theta_i^{(\psi_h)} - \mathrm{t} - C_i^{(\psi_h)} \geq 0$$
$$\Rightarrow \theta_i^{(\psi_h)} \geq \mathrm{t} + C_i^{(\psi_h)} - d_i^{(\psi_h)}.$$

- **Solution 3**: minimize the $c_i^{(\psi_h)}$ (9)

  By applying Eq. (3) that notices:

$$R_i^{(\psi_h)} = d_i^{(\psi_h)} - f_i^{(\psi_h)}, \text{ we have:}$$
$$R_i^{(\psi_h)} = d_i^{(\psi_h)} - \mathrm{t} - C_i^{(\psi_h)}.$$

Or, to obtain a feasible system after a reconfiguration scenario (ψ_h), the following formula must be enforced:

$$R_i^{(\psi_h)} \geq 0.$$

By this result we can write:

$$d_i^{(\psi_h)} - \mathrm{t} - C_{inew}^{(\psi_h)} \geq 0, \text{ (where } C_{inew} \text{ is the new worst case execution time value)}$$
$$\text{Or } C_{inew}^{(\psi_h)} = C_i^{(\psi_h)} + \beta_i^{(\psi_h)}.$$
$$\text{So, } d_i^{(\psi_h)} - \mathrm{t} - C_i^{(\psi_h)} - \beta_i^{(\psi_h)} \geq 0$$
$$\Rightarrow d_i^{(\psi_h)} - \mathrm{t} - C_i^{(\psi_h)} \geq \beta_i^{(\psi_h)}$$
$$\Rightarrow \beta_i^{(\psi_h)} \leq d_i^{(\psi_h)} - \mathrm{t} - C_i^{(\psi_h)}$$

- **Solution 4**: Enforcing the release time to come back:

$$a_i^{(\psi_h)} \rightarrow a_{inew}^{(\psi_h)} \text{(where } a_{inew} \text{ is the new activation time value)}$$
$$\rightarrow (a_{inew}^{(\psi_h)} = a_i^{(\psi_h)} + \Delta\mathrm{t}) \quad (10)$$

By applying Eq. (1) that notices:

$$d_i^{(\psi_h)} = a_i^{(\psi_h)} + D_i^{(\psi_h)},$$

we have:

$$R_i^{(\psi_h)} = a_i^{(\psi_h)} + D_i^{(\psi_h)} - \mathrm{t} - C_i^{(\psi_h)}.$$

Or, to obtain a feasible system after a reconfiguration scenario (ψ_h), the following formula must be enforced:

$$R_i^{(\psi_h)} \geq 0$$
$$\Rightarrow a_i^{(\psi_h)} + D_i^{(\psi_h)} - \mathrm{t} - C_i^{(\psi_h)} \geq 0.$$

By this result we can write:

$$a_{inew}^{(\psi_h)} + D_i^{(\psi_h)} - \mathrm{t} - C_i^{(\psi_h)} \geq 0,$$
$$\text{where } a_{inew}^{(\psi_h)} = a_i^{(\psi_h)} + \Delta\mathrm{t}.$$

So, we obtain:

$$a_i^{(\psi_h)} + \Delta\mathrm{t} + D_i^{(\psi_h)} - \mathrm{t} - C_i^{(\psi_h)} \geq 0$$
$$\Rightarrow \Delta\mathrm{t} \geq \mathrm{t} + C_i^{(\psi_h)} - a_i^{(\psi_h)} - D_i^{(\psi_h)}.$$

- **Solution 5**: Tolerate some non critical tasks k among (m + n) based on (m, n) firm model proposed by Hamdaoui and Ramanathan (1995)

$$\begin{aligned} &\xi^{(\psi_h)} = \{\tau_i^{(\psi_h)}(C_i^{(\psi_h)}, D_i^{(\psi_h)}, m_i^{(\psi_h)}, I_i^{(\psi_h)})\},\ \mathrm{i} = 1 \text{ to m},\ \mathrm{h} = 1 \text{ to M}. \\ &m_i^{(\psi_h)} = 1,\ \text{it tolerates missing deadline}, \\ &m_i^{(\psi_h)} = 0,\ \text{it doesn't tolerate missing deadline}, \\ &I_i^{(\psi_h)} = \mathrm{H},\ \text{Hard task}, \\ &I_i^{(\psi_h)} = \mathrm{S},\ \text{Soft task}, \end{aligned} \tag{11}$$

- **Solution 6**: Removal of some non critical tasks (to be rejected)

$$\begin{aligned} &\xi^{(\psi_h)} = \{\tau_i^{(\psi_h)}(C_i^{(\psi_h)}, D_i^{(\psi_h)}, m_i^{(\psi_h)}, I_i^{(\psi_h)})\},\ \mathrm{i} = 1 \text{ to m},\ \mathrm{h} = 1 \text{ to M}. \\ &m_i^{(\psi_h)} = 1,\ \text{it tolerates missing deadline}, \\ &m_i^{(\psi_h)} = 0,\ \text{it doesn't tolerate missing deadline}, \\ &I_i^{(\psi_h)} = \mathrm{H},\ \text{Hard task}, \\ &I_i^{(\psi_h)} = \mathrm{S},\ \text{Soft task}, \end{aligned} \tag{12}$$

For every solution the corresponding response time is:

$Resp_{k,1}^{(\psi_h)}$ the response time calculated by the first solution,
$Resp_{k,2}^{(\psi_h)}$ the response time calculated by the second solution,
$Resp_{k,3}^{(\psi_h)}$ the response time calculated by the third solution,
$Resp_{k,4}^{(\psi_h)}$ the response time calculated by the fourth solution,
$Resp_{k,5}^{(\psi_h)}$ the response time calculated by the fifth solution,
$Resp_{k,6}^{(\psi_h)}$ the response time calculated by the sixth solution,

We define now, $Resp_k^{(\psi_h)}$ optimal noted $Resp_k^{(\psi_h,opt)}$ according to the previous six solutions calculated by the intelligent agent (Solution 1, Solution 2, Solution 3, Solution 4, Solution 5 and Solution 6) by the following expression:

$$Resp_k^{(\psi_h,opt)} = \min(Resp_{k,1}^{(\psi_h)},\ Resp_{k,2}^{(\psi_h)},\ Resp_{k,3}^{(\psi_h)},\ Resp_{k,4}^{(\psi_h)},\ Resp_{k,5}^{(\psi_h)},\ \text{and}\ Resp_{k,6}^{(\psi_h)})\ \text{(the minimum of the six values)}.$$

So, the calculation of $Resp_k^{(\psi_h,opt)}$ allows us to obtain and to calculate the minimizations of response times values and to get the optimum of these values.

Running Example

In our running example, the agent proposes after the arrival of new sporadic tasks σ_C, σ_F and σ_G to be added to $\xi^{(0)}$ that evolves into $\xi^{(\psi_1)} = \{\tau_A, \tau_B, \sigma_C, \sigma_D, \sigma_E, \sigma_F, \sigma_G$ and $\sigma_H\}$ six solutions in order to re-obtain the feasibility of the system.

The General EDF-based Scheduling Strategy

When dealing with the deadline tolerance factor m_i, each task has to be computed with respect to the deadline tolerance factor m_i.

Algorithm GUARANTEE($\xi^{(\psi_h)}$; $\sigma_a^{(\psi_h)}$**)**
begin
t = get current time();
$R_0^{(\psi_h)} = 0$;
$d_0^{(\psi_h)} = t$;
Insert $\sigma_a^{(\psi_h)}$ in the ordered task list;
$\xi^{(\psi_h)`} = \xi^{(\psi_h)} \cup \sigma_a^{(\psi_h)}$;
k = position of $\sigma_a^{(\psi_h)}$ in the task set $\xi^{(\psi_h)`}$;
for each task $\sigma_i^{(\psi_h)`}$ such that i ≥ k do
{
$R_i^{(\psi_h)} = R_{i-1}^{(\psi_h)} + (d_i^{(\psi_h)} - d_{i-1}^{(\psi_h)}) - c_i^{(\psi_h)}$;
if ($R_i^{(\psi_h)}$ ≥ **0) then**
{
return ("Guaranteed");}
}
else

return
{ ("You can try by using solution 1, or,
You can try by using solution 2, or,
You can try by using solution 3, or,
You can try by using solution 4, or,
You can try by using solution 5, or,
You can try by using solution 6 !");
}
Compute($Resp_{k,1}^{(\psi_h)}$);
Compute($Resp_{k,2}^{(\psi_h)}$);
Compute($Resp_{k,3}^{(\psi_h)}$);
Compute($Resp_{k,4}^{(\psi_h)}$);
Compute($Resp_{k,5}^{(\psi_h)}$);
Compute($Resp_{k,6}^{(\psi_h)}$);
Generate($Resp_{k}^{(\psi_h,opt)}$)
end

Complexity

This algorithm assumes that sporadic tasks span no more than one hyperperiod of the periodic tasks $hp = [0,\ 2^{*}\mathrm{LCM} + \max_k(a_{k,1}^{(\psi_h)})]$, where *LCM* is the well-known Least Common Multiple of all task periods and $(a_{k,1}^{(\psi_h)})$ is the earliest activation time of each task $\tau_k^{(\psi_h)}$. The extension of the proposed algorithm should be straightforward, when this assumption does not hold and its running time is O(n + m) Gharsellaoui et al. (2012). The EDF-based schedulability in the case of a mixture of periodic (synchronous and asynchronous tasks) and sporadic tasks, i.e. each task has an offset S_i, such that the jobs are released at $k^*T_i + S_i$ ($k \in \aleph$) is strongly coNP-hard Buttazzo and Stankovic (1993). This complexity was decreased in our approach to efficient $O(n + m)^2$ guarantee algorithm. This optimal algorithm results in the dynamic scheduling solutions. These solutions are presented by a proposed intelligent agent-based architecture where a software agent is used to evaluate the response time, to calculate the processor utilization factor and also to verify the satisfaction of real-time deadlines. On the other hand, the busy period, which is computed for every analyzed task set and has a pseudo-polynomial complexity for T_S, is decreased also by the optimization of the response time. The most important results are presented in our work. So, we can deduce that using our proposed approach under such conditions may be advantageous.

Experimental Results: Discussion and Evaluation

In this subsection, in order to quantify the benefits of the proposed approach over the other works from the state of the art and in order to check also the suggested configurations of tasks allowing the system's feasibility and the response time minimization, we simulate the agent's behavior on several test sets in order to rate the performance of the optimal scheduling algorithm in our proposed approach. The most important observation was obtained by the comparison of our proposed approach against the others from the literature about the current values. We tested the feasibility of the *Volvo* task system by another algorithms, so that we can compare the results directly. We carried out several test runs and examined them under different aspects. The total utilization of the static schedule is 75 %, the classic one is 145.4 % and the other proposed by our method is 63.1 %.

Therefore, we can confirm that this method is nowadays very advantageous given the fast response time and the performance of the RT-Reconfiguration tool.

By applying the six solutions of this tool RT-Reconfiguration, we conclude that our approach can allow more reactive and also more efficient feasible systems. This advantage can be important in many cases where critical control tasks should be intensively executed in small periods of time. This work also, concentrates on the context of systems containing a set of tasks which is not feasible; the reconfiguration was applied in order not only to obtain the system's feasibility but also to get the performance of the system by reducing the response time of the processes to be tolerated in interactive environment and by avoiding unnecessarily frequent context switch leading to more overheads resulting in less throughput. This advantage was increased and proved clearly by the volvo case study. This advantage was illustrated by examples even for the case of aperiodic tasks also.

Moreover, with the revolution of semiconductors technology and the development of efficient reconfiguration tools, the use of our method and the RT-Reconfiguration tool will become increasingly important, and very advantageous for rapid and efficient response time of the aperiodic/sporadic reconfigurable OS tasks, especially when the user has no other choice than to choose the previous proposed solutions and to decide the proper values of each reconfigured task's parameters in order to obtain the system's feasibility and to minimize the response time of the studied systems.

4.1.2 Reconfiguration of Aperiodic Tasks

The scheduling problem for aperiodic tasks in real-time embedded systems is very different from that for periodic tasks. Scheduling algorithms for aperiodic tasks must be able to guarantee the deadlines for hard aperiodic tasks (also called a sporadic tasks) and provide good average response times for soft aperiodic tasks. Our aperiodic scheduling approach must also accomplish these goals without compromising the hard deadlines of the periodic tasks in the case when we have a mixture of periodic, sporadic and aperiodic tasks. We assume also in our original work, that an invocation of each aperiodic task will follow the Poisson arrival model. Our new proposed contribution is described in the following paragraph.

Contribution

In our proposed approach we are interested in automatic reconfigurations of real-time embedded control systems that should meet deadlines defined by user requirements. Automatic reconfigurations are applied by Intelligent Agents Gharsellaoui et al. (2012). These systems are implemented by sets of tasks that we assume independent, periodic, sporadic and aperiodic.

The goal of our original approach applied to real-time system reconfiguration and scheduling is to construct systems that are guaranteed to meet all hard deadlines and that minimize the response time for all soft deadlines. We define an agent-based architecture that checks the system's evolution and defines useful solutions for users when deadlines are not satisfied after any reconfiguration scenario. We assume also that the invocation of each aperiodic task will follow the Poisson arrival model. The application domain of the Poisson law was limited for a long time to that of the rare events as the suicides of children, the arrivals of boats in a bearing or the accidents due to the kicks of horse in the armies. But since a few decades its scope considerably widened. At present, we use it in telecommunications a lot (to count the number of communications in an interval of given time), the statistical quality control, the description of certain phenomena bound to the radioactive destruction (the destruction of the radioactive pits(cores) following, besides, an exponential law of noted parameter so average), the biology, the meteorology, etc. Two cases of suggestions are possible to be provided by our intelligent agent: modification of worst case execution times of tasks or modification of their deadlines. The users should choose one of these solutions to re-obtain the system's feasibility and to minimize the response time of the soft aperiodic tasks. We developed a tool RT-Reconfiguration and tested it in order to support the agent's services. We will well formalize this approach in the following subsection.

Formalization

We now formally describe our proposed concept on which our work is based. We define an agent-based architecture for reconfigurable real-time embedded systems that should classically meet different deadlines defined in user requirements. The agent controls the system's evolution and provides solutions for users when deadlines are violated after any reconfiguration scenario. In our contribution for the aperiodic real-time tasks, we will restrict to only one reconfiguration scenario (M = 1).

Let *Sys* be a set of n_1 real-time tasks composed of n_1^i tasks τ^i of type i, n_1^j tasks τ^j of type j and n_1^k tasks τ^k of type k; i.e., $Sys = \{\tau_1^i, \tau_2^i, \ldots, \tau_{n_1}^i; \tau_1^j, \tau_2^j, \ldots,$

$\tau^j_{n_1}$; τ^k_1, τ^k_2, …, $\tau^k_{n_1}$} such that $n^i_1 + n^j_1 + n^k_1 = n_1$. These real-time tasks support different functionalities. Let Ω_{Sys} be the set of all tasks that can possibly implement the system, and let us denote by $Current_{Sys}(t)$ the set of tasks implementing the system *Sys* at *t* time units. These tasks should meet all deadlines defined in user requirements. In this case, we note that $Feasibility(Current_{Sys}(t)) \equiv True$.

Problem

Now, we suppose the arrival of n_2 new aperiodic tasks at run-time at time t_1 for each reconfiguration scenario ψ_h. By considering a feasible System *Sys* before the application of the reconfiguration scenario ψ_h, each one of the tasks of ξ_{old} is feasible, e.g. the execution of each instance is finished before the corresponding deadline. When the reconfiguration scenario is applied at time t_1, two cases exist,

- If tasks of $Current_{Sys}(t_1) = \xi_{new} \cup \xi_{old}$ are feasible, then no reaction should be done by the agent,
- Otherwise, the agent should provide two solutions for users (find the new parameters α, β and γ of tasks of ξ_{new} and ξ_{old} as the first solution and find the parameter λ of the Poisson distribution as a second solution) in order to re-obtain the system's feasibility.

The Poisson law with parameter λ, or law of the rare events, corresponds to the following model:

Over a period T, an event arrives on average λ time. We call X the random variable determining the number of times when the event occurs for the period T.

X takes whole values: 0, 1, 2…

This random variable follows a law of probability defined by:

$$P(k) = P(X = k) = \frac{\lambda^k}{k!}e^{-\lambda}$$

where,

- λ is a strictly positive real number
- 1/λ is a rate of the occurrences of aperiodic tasks per hyperperiod.

Characterization

In our proposed and original work, the occurrence of aperiodic tasks (events) follows a Poisson law with a constant parameter λ, and a variable $1/\lambda$ called its occurrence rate per time unit. In our model, we assume that the time unit is equal to an hyperperiod $hp = [0, 2^*\mathrm{LCM} + max_k(a_{i,1})]$, where *LCM* is the well-known Least Common Multiple and $(a_{i,\ 1})$ is the earliest release time (activation time) of each task τ_i. For this reason, we note $1/\lambda_{hp}$ the occurrence of aperiodic tasks rate per hyperperiod hp on a uniprocessor system. Moreover, aperiodic tasks occurrences are statistically independent events.

Proposed Solutions

The agent should react to propose useful solutions for users in order to make reconfigurable systems meet deadline requirements for real-time systems. The value of λ determines the distribution of the tasks. Hence, in order to reduce the response time of aperiodic tasks, and to obtain a feasible system, we assume that the hyperperiod hp is equal to one time unit and we will tolerate the arrival of some distinct types of tasks. In our current example, we will use three distinct types i, j, k tasks as a simple example, e.g. n_2^i tasks τ^i of type i, n_2^j tasks τ^j of type j, and n_2^k tasks τ^k of type k in order to obtain the feasibility and to reduce the response time of a system under study at run-time.

First solution

In order to obtain a feasible system by the first proposed solution and as supposed to work for the hyperperiod = 1 time unit, then $T_i = T_j = T_k = 1$ and the following formula should be satisfied:

$$\sum_{i=1}^{n^i} \frac{C_i}{1} + \sum_{j=1}^{n^j} \frac{C_j}{1} + \sum_{k=1}^{n^k} \frac{C_k}{1} \leq 1$$

where i, j and k are the marks of new arrival aperiodic tasks of type i, j and k.

So, we have

$$\begin{aligned} &\sum_{i=1}^{n^i} WCET_i + \sum_{j=1}^{n^j} WCET_j + \sum_{k=1}^{n^k} WCET_k \leq 1 \\ &\Rightarrow \sum_{i=1}^{n^i} C_i + \sum_{j=1}^{n^j} C_j + \sum_{k=1}^{n^k} C_k \leq 1 \\ &\Rightarrow \alpha C_i + \beta C_j + \gamma C_k \leq 1. \end{aligned} \tag{13}$$

The agent proceeds in this case, as a first solution, to find the new parameters α, β and γ of tasks of ξ_{new} and ξ_{old} in order to reconfigure the system at run-time. The question now, is how to do to calculate the values of α, β and γ to have the adequate solution and consequently the whole system becomes feasible?

Running Example

To more illustrate this point and according to the expression (13), we have the following expression after changing the variables $WCET_i$, $WCET_j$ and $WCET_k$ by 0.3, 0.2 and 0.5:

$$\Rightarrow 0.3\alpha + 0.2\beta + 0.5\gamma \leq 1$$

in this case and in order to keep and guarantee this disparity, thus mathematically the following conditions must be satisfied:

α has to be ≤3, that means $\alpha \in \{0, 1, 2, 3\}$
β has to be ≤5, that means $\beta \in \{0, 1, 2, 3, 4, 5\}$
γ has to be ≤2, that means $\gamma \in \{0, 1, 2\}$
In this case, we shall have 72 triplet (4 × 6 × 3), that means 72 possibilities for α, β and γ combinations.

Example (0, 0, 0); (0, 0, 1); … (2, 2, 0); (2, 1, 0); … (3, 0, 0); (1, 1, 1). But, not all these mentioned combinations satisfy the basic condition for real-time task scheduling. There are some combinations, which cannot satisfy the basic utilization condition. So, the intelligent proposed agent will propose to the user the appropriate combinations which can fit to the basic condition.

Second solution

The agent proceeds as a second solution to find the parameter λ of the Poisson distribution to model the arrival of aperiodic tasks. Indeed, according to the law of mathematical probability, to have certain event it is necessary that the following formula should be satisfied:

$$P(X_i = n_1^i) + P(X_j = n_2^j) + P(X_k = n_3^k) = 1 \tag{14}$$

By applying the Poisson law during one time unit with a mean of λ. We will find the value of this parameter λ in order to reach the system's feasibility after a reconfiguration scenario was applied.

According to the law of mathematical probability and in order to have certain event, the following expression must be verified:

$$
\begin{aligned}
& P(X_i = n_1^i) + P(X_j = n_2^j) + P(X_k = n_3^k) = 1 \\
& \Rightarrow \frac{\lambda^{n_1^i}}{n_1^i!} e^{-\lambda} + \frac{\lambda^{n_2^j}}{n_2^j!} e^{-\lambda} + \frac{\lambda^{n_3^k}}{n_3^k!} e^{-\lambda} = 1 \\
& \Rightarrow e^{-\lambda} \left(\frac{\lambda^{n_1^i}}{n_1^i!} + \frac{\lambda^{n_2^j}}{n_2^j!} + \frac{\lambda^{n_3^k}}{n_3^k!} \right) = 1 \\
& \Rightarrow \underbrace{\left(\frac{\lambda^{n_1^i}}{n_1^i!} + \frac{\lambda^{n_2^j}}{n_2^j!} + \frac{\lambda^{n_3^k}}{n_3^k!} \right)}_{P(\lambda)} = e^{\lambda} \\
& \Rightarrow P(\lambda) = e^{\lambda}.
\end{aligned}
$$

Thus to have a feasible system, it is necessary to find the value of λ, i.e., that is to solve the equation: $P(\lambda) = e^{\lambda}$. In this case, we restraint to the mathematical problem said problem of the fixed point (f(x) = x), or the problem of the contracting function f: E $\rightarrow$ E which contracts on a point, the unique point fixes of f.

The purpose of this work is to establish this theorem, said about fixed point.

Here, we have:

$$
\begin{aligned}
& P(\lambda) = e^{\lambda} \\
& \Rightarrow Log(P(\lambda)) = Log(e^{\lambda}) = \lambda \\
& \Rightarrow f(\lambda) = \lambda, \text{ with f} = \text{Log}(\text{P}(\lambda)).
\end{aligned}
$$

Diagrammatically the unique value of λ is the intersection of both curves y = P (λ) and z $= e^{\lambda}$; it is the focused solution.

Running Example

To more illustrate this point, we assume that $n_1^i = 1$, $n_2^j = 2$ and $n_3^k = 3 \Rightarrow P(\lambda) = \lambda + \frac{\lambda^2}{2} + \frac{\lambda^3}{6}$.

So,

$$
P(\lambda) = e^{\lambda} \Rightarrow \lambda + \frac{\lambda^2}{2} + \frac{\lambda^3}{6} = e^{\lambda}.
$$

In this case and mathematically, λ = 0 is the unique solution and even by graphic resolution we obtain the intersection of both curves $y = P(\lambda) = \lambda + \frac{\lambda^2}{2} + \frac{\lambda^3}{6}$ and $z = e^{\lambda}$, is the unique point $\lambda = 0$. Thus to have a feasible system with one task of type i, two tasks of type j and three tasks of type k, i fix the value of lambda to 0 in my program. In this case the program would return the optimal solution.

Algorithm and Complexity

In the following paragraph we will describe our proposed algorithm for this contribution for the aperiodic tasks reconfiguration.

Algorithm begin
t = get current time();
$U = 0$;
For each partition
Compute(α);
Compute(β);
Compute(γ);
Display_parameters(α, β, γ);
save (α, β, γ);
if (feasible) then
{
return ("Guaranteed");
}
else
return
("You can try by using solution 1, or,
You can try by using solution 2,
Compute($Resp_{k,1}$);
Compute($Resp_{k,2}$);
Generate($Resp_k^{opt}$);
end

Complexity

The reconfigurable schedulability in the case of aperiodic tasks, i.e., each task has an offset S_i, such that is strongly coNP-hard. This complexity was decreased in our approach from Np-hard to efficient $O(n + m)^2$ guarantee algorithm when we have M reconfigurations scenarios ψ_h. This efficient algorithm results in the dynamic scheduling solutions. These solutions are presented by a proposed intelligent agent-based architecture where a software agent is used to evaluate the response time, to calculate the processor utilization factor and also to verify the satisfaction of real-time deadlines.

On the other hand, the busy period, which is computed for every analyzed task set and has a pseudo-polynomial complexity for $U \leq 1$, is decreased also by the optimization of the response times in our work.

	Period / Activation	Calcul	Priority	Utilisation / Response
e1	7	3		17
e2	11	4		33
T1	20	6	3	0.3
T2	10	4	2	0.4
Polling Server	8	2	1	0.25

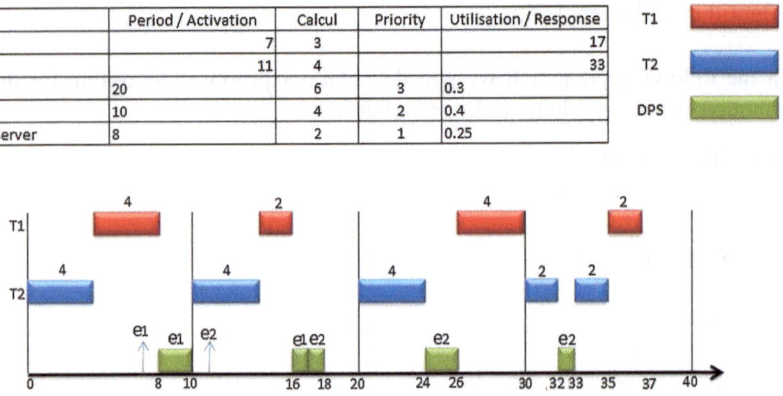

Fig. 1 The simulation with only polling server

4.2 Second Method: Hybrid Scheduling Approach

The scheduling problem of aperiodic tasks is very different from that of periodic tasks. Indeed, scheduling algorithms for aperiodic tasks must be able to guarantee the deadlines for hard deadline aperiodic tasks and provide good average response times for soft deadline aperiodic tasks even though the occurrence of the aperiodic requests are non deterministic. The aperiodic scheduling algorithm must also accomplish these goals without compromising the hard deadlines of the periodic tasks.

4.2.1 Contribution

One hybrid approach composed of the combination of two common approaches for servicing aperiodic requests are background processing and polling tasks. Background servicing of aperiodic requests occurs whenever the processor is idle (i.e., not executing any periodic tasks and no periodic tasks pending). If the load of the periodic task set is high, then utilization left for background service is low, and background service opportunities are relatively infrequent.

Polling consists of creating a periodic task for servicing aperiodic requests. At regular intervals, the polling task is started and services any pending aperiodic requests. However, if no aperiodic requests are pending, the polling task suspends itself until its next period and the time originally allocated for aperiodic service is not preserved for aperiodic execution but is instead used by periodic tasks. Note that if an aperiodic request occurs just after the polling task has suspended, then the aperiodic request must wait until the beginning of the next polling task period or until background processing resumes before being serviced. Even though polling tasks and background processing can provide time for servicing aperiodic requests,

	Period / Activation	Calcul	Priority	Utilisation / Response
e1	7	3		17
e2	11	4		25
T1	20	6	3	0.3
T2	10	4	2	0.4
Polling Server	8	2	1	0.25
Background Server	*	*	10	*

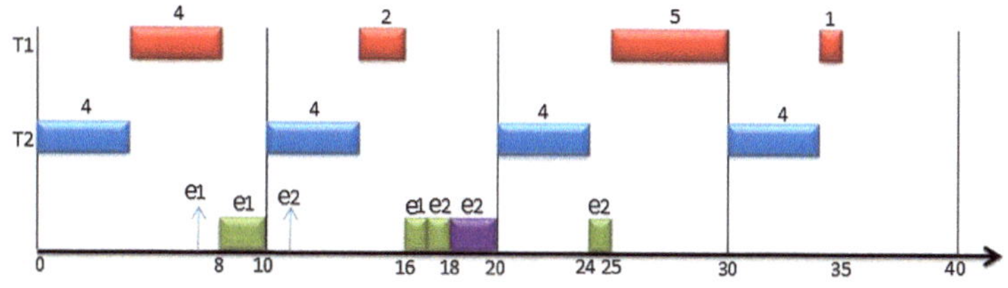

Fig. 2 The simulation with polling server and background server

they have the drawback that the average wait and response times for these algorithms can be long, especially for background processing. Figure 1 illustrates the operation of background and polling aperiodic service using the periodic task set presented in the table of the same picture (Fig. 1).

4.2.2 Motivating Example

Let us suppose a real-time embedded system *Sys*1 to be initially implemented by 2 characterized tasks as shown in Fig. 1. These tasks are feasible because the processor utilization factor U $= 0.7 \leq 1$. These tasks should meet all required deadlines defined in user requirements and we have $\mathit{Feasibility}(\mathit{Current}_{Sys1}(t)) \equiv \mathit{True}$.

	Period / Activation	Calcul	Priority	Utilisation / Response
e1	7	3		10
e2	11	4		25
T1	20	6	3	0.3
T2	10	4	2	0.4
Deferrable server	8	2	1	0.25
Background Server	*	*	10	*

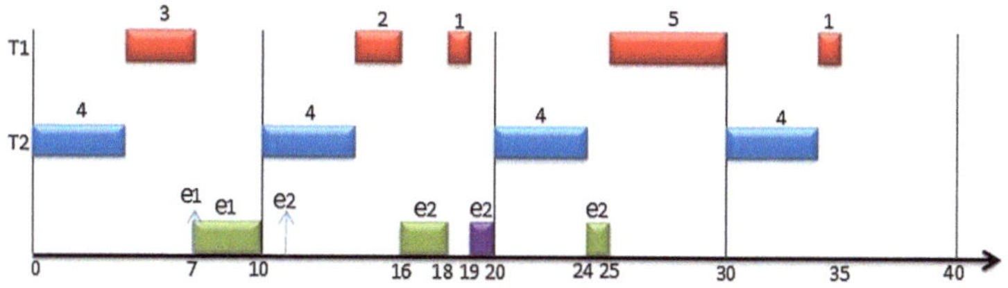

Fig. 3 The simulation with deferrable server and background server

	Period / Activation	Calcul	Priority	Utilisation / Response
e1	7	3		16
e2	11	4		24
T1	20	6	3	0.3
T2	10	4	2	0.4
Sporadic server	8	2	1	0.25
Background Server	*	*	10	*

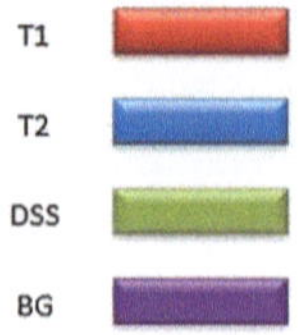

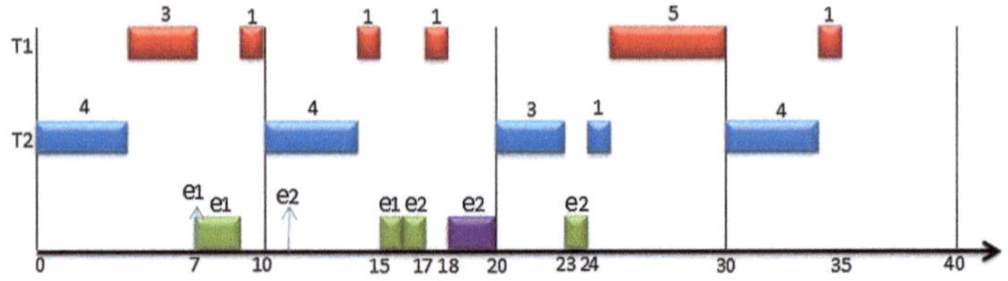

Fig. 4 The simulation with sporadic server and background server

We suppose that a reconfiguration scenario is applied at t1 and t2 time units with the arrival of 2 new aperiodic tasks e_1 at t1 = 7 and e_2 at t2 = 11 time units. Therefore the system is feasible by applying the polling server to schedule the system but the response time is equal to 17 and 33 for both e_1 and e_2 respectively. Now by applying our new hybrid approach, the response time of the second arrival aperiodic task is decreased from 33 to 25 time units as we observe in Fig. 2. By applying the new hybrid approach with Deferrable Server and Background server, the response time of the first arrival aperiodic task is decreased from 17 to 10 time units and the response time of the second arrival aperiodic task is decreased from 33 to 25 time units as we observe in Fig. 3. Finally, by applying the new hybrid approach with Sporadic Server and Background server, the response time of the first arrival aperiodic task is decreased from 17 to 16 time units and the response time of the second arrival aperiodic task is decreased from 33 to 24 time units as we observe in Fig. 4.

4.2.3 Formalization

By considering real-time operating system (OS) tasks scheduling, let $n = n_1 + n_2$ be the number of a mixed workload of periodic and aperiodic tasks in $Current_\Gamma$ (t). The reconfiguration of the system $Current_\Gamma$ (t) means the modification of its implementation that will be as follows at t time units:

$$Current_\Gamma(t) = \xi_{new} \cup \xi_{old}$$

where ξ_{old} is a subset of n_1 old periodic tasks which are periodic and not affected by the reconfiguration scenario (e.g. they implement the system before the time t), and ξ_{new} is a subset of n_2 new aperiodic tasks in the system. We assume that an updated task is considered as a new one at t time units. By considering a feasible System Sys before the application of the reconfiguration scenario, each task of ξ_{old} is feasible, e.g. the execution of each instance is finished before the corresponding deadline.

4.2.4 Experimental Analysis and Discussion

In this section, in order to check the suggested configurations of tasks allowing the system's feasibility and the response time minimization, we simulate the agent's behavior on several test sets in order to rate the performance of the polling server and the background server in our hybrid scenario.

Simulation

We have conducted several test sets in order to rate the performance of the polling server and the background server in our hybrid scenario. We have set up a real-time reconfiguration tool named RT-Reconfiguration that allows us to randomly generate task sets, schedule them according to the proposed hybrid method, and displays the schedules for visual control. Our test rows have been on each 1,000 randomly generated task sets, while the number of tasks is significantly higher. We have scheduled task sets with the polling server and the proposed hybrid method.

Discussion

In each of these examples, many aperiodic requests occur at any moment of the time. The response time performance of only polling service or only background service for the aperiodic requests is poor. Since background service occurs when the resource is idle, with the polling server, the response time performance for the aperiodic requests is better than both single background service and single polling service for all requests. For these examples, a polling server is created with an execution time of 1 time unit and a period of 5 time units. Also note that since any aperiodic request only needs half of the polling server's capacity, the remaining half is discarded because no other aperiodic tasks are pending. Thus, these examples demonstrate how polling and background can provide an improvement in aperiodic response time performance over background service or polling one and are always able to provide immediate service for aperiodic requests. The proposed hybrid scheduling algorithm may thus be appropriate for many applications.

The other most important thing is the complexity which is decreased in our approach to O(nlog(n)) because the proposed algorithm is recursive, and the Earliest Deadline First algorithm also, would be maintaining all tasks that are ready for execution in a queue. Any freshly arriving task would be inserted at the end of queue. Each task insertion will be achieved in O(1) or constant time, but task selection (to run next) and its deletion would require O(n) time, where n is the number of tasks in the queue. When a task arrives, a record for it can be inserted into the heap in O(log(n)) time where n is the total number of tasks in the priority queue. Therefore, the time complexity of our hybrid algorithm is equal to that of a typical sorting algorithm which is O(n log(n)). So O(nlog(n)) time is required. Finally, for both the polling server and the background server in our hybrid scenario approach performs best and yield improved average response times for aperiodic requests and the most important results are presented in our work. So, we can deduce that using our proposed approach under such conditions may be advantageous.

5 Conclusion and Future Works

In this book chapter, we propose a new theory for the minimization of the response time of aperiodic real-time tasks with the polling server and the background server that can be applied to uniprocessor systems and proved it correct. We showed that this theory is capable to reconfigure the whole system. Previous work in this area has been described, several and best solution has been suggested. This hybrid solution is primarily intended to reduce the processor demand and the response time of each task set independent of the number of tasks in a uniprocessor system. A tool is developed and tested to support all these services.

At the end, we present a few inadequacies and propose directions of research to extend our study to the case of distributed systems and, we plan also to apply this contribution to other complex reconfigurable systems that we have chosen to not cover in this book chapter. We hope that this work will serve as a good starting point and a useful reference for researchers working on the development of real-time scheduling approches of aperiodic tasks in embedded systems.

References

Abdelzaher, T. S. V., Sharma, V., & Lu, C. (2004a). A utilization bound for aperiodic tasks and priority driven scheduling. *IEEE Transactions on Computers, 53*(3), 334–350.

Abdelzaher, T., Thaker, G., & Lardieri, P. (2004b). A feasible region for meeting aperiodic end-to-end deadlines in source pipelines. In *IEEE International Conference on Distributed Computing Systems* (pp. 436–445).

Abeni, L., & Buttazzo, G. (1998). Integrating multimedia applications in hard real-time systems. In *Proceedings of the 19th IEEE Real-Time Systems Symposium* (pp. 4–13).

Abeni, L., & Buttazzo, G. (2001). Hierarchical QoS management for time sensitive applications. In *proceedings of the IEEE Real-Time Technology and Applications Symposium* (pp. 63–72).

Abeni, L., & Buttazzo, G. (2004). Resource reservations in dynamic real-time systems. *Real-Time Systems, 27*(2), 123–165.

Aggarwal, S., & Chraibi, C. (1995). Scheduling of hyperperiodic tasks in a multiprocessor environment. In *Proceedings of the 2nd ISSAT Conference on Reliability and Quality in Design.*

Burns, A., & Guillem, B. (1999). New results on fixed priority aperiodic servers. In *20th IEEE Real-Time Systems Symposium, RTSS* (pp. 68–78).

Buttazzo, G., & Stankovic, J. (1993). RED: A robust earliest deadline scheduling. 3rd International Workshop on Responsive Computing.

Gharsellaoui, H., Khalgui, M., & Ben Ahmed, S. (2012). New optimal preemptively scheduling for real-time reconfigurable sporadic tasks based on earliest deadline first algorithm. *Journal of International Advanced Pervasive and Ubiquitous Computing, IJAPUC, 4*(2), 65–81.

Ghazalie, T., & Baker, T. (1994). *Aperiodic servers in a deadline scheduling environment*. Tallahassee, FL: Department of Computer Science, Florida State University. (32306).

Guillem, B. (2001). Weakly hard real-time systems. *IEEE Transactions on Computers, 50*(4), 308–321.

Hamdaoui, M., & Ramanathan, P. (1995). A dynamic priority assignment technique for streams with (m, k)-firm deadlines. *IEEE Transactions on Computers, 44*(1), 1443–1451.

Hamidzadeh, B., & Atif, Y. (1996). Dynamic scheduling of real-time aperiodic tasks on multiprocessor architectures. In *Proceedings of the 29th Annual Hawaii International Conference on System Sciences* (pp. 469–478).

Layland, J., & Liu, C. (1973). Scheduling algorithms for multi-programming in a hard-real-time environment. *Journal of the ACM, 20*(1), 46–61.

Lehoczky, J. P., Sha L., & Strosnider, J. K. (1987). Enhanced aperiodic responsiveness in hard-real-time environments. In *Proceedings of IEEE Real-Time Systems Symposium* (pp. 261–270).

Lipari, G., & Buttazzo, G. (1999). Schedulability analysis of periodic and aperiodic tasks with resource constraints. *International Journal of Systems Architecture, 46*(4), 327–338.

Marouf, M., George, L., & Sorel, Y. (2012). Schedulability analysis for a combination of non-preemptive strict periodic tasks and preemptive sporadic tasks. In *ETFA'12—17th IEEE International Conference on Emerging Technologies and Factory Automation* (pp. 1–8).

Marouf, M., & Sorel, B. Y. (2010). Schedulability conditions for non-preemptive hard real-time tasks with strict period. In *18th International Conference on Real-Time and Network Systems, RTNS10*.

Marouf, M., & Sorel, B. Y. (2011). Scheduling non-preemptive hard real-time tasks with strict periods. In *16th IEEE International Conference on Emerging Technologies and Factory Automation ETFA2011* (pp. 1–8).

Naveen, A., & Venkatesan, M. (2013). Energy aware scheduling of aperiodic real-time tasks on multiprocessor systems. *Journal of Computing Science and Engineering JCSE, 7*(1), 30–43.

Sprunt, B., Sha, L., & Lehoczky, J. (1989). Aperiodic task scheduling for hard-real-time systems. *Real-Time Systems, 1*(1), 27–60.

Strosnider, K., Lehoczky, J. P., & Sha, L. (1995). The deferrable server algorithm for enhanced aperiodic responsiveness in hard real-time environments. *IEEE Transactions on Computers, 44* (1), 73–91.

Thuel, S. R., & Lehoczky, J. (1994). Algorithms for scheduling hard aperiodic tasks in fixed-priority systems using slack stealing. In *Real-Time Systems Symposium* (pp. 22–33).

MIX
Papier aus verantwortungsvollen Quellen
Paper from responsible sources
FSC® C105338

FSC
www.fsc.org

If you have any concerns about our products,
you can contact us on
ProductSafety@springernature.com

In case Publisher is established outside the EU,
the EU authorized representative is:
Springer Nature Customer Service Center GmbH
Europaplatz 3, 69115 Heidelberg, Germany

Printed by Libri Plureos GmbH
in Hamburg, Germany